BUSINESS/SCIENCE/TECHNOLOGY DIVISION
CHICAGO PUBLIC LIBRARY
400 SOUTH STATE STREET
CHICAGO, ILLINOIS 60605

Chicago Public Library

Y0-DLD-663

Atlas of human central nervous system de

DISCARD

Form 178 rev. 11-00

# *The* HUMAN BRAIN *during the* THIRD TRIMESTER

# ATLAS OF
# HUMAN CENTRAL NERVOUS SYSTEM DEVELOPMENT
# SERIES

## Shirley A. Bayer *and* Joseph Altman

### VOLUME 1

*The Spinal Cord from Gestational Week 4 to the 4th Postnatal Month*

### VOLUME 2

*The Human Brain during the Third Trimester*

### VOLUME 3

*The Human Brain during the Second Trimester*

### VOLUME 4

*The Human Brain during the Late First Trimester*

### VOLUME 5

*The Human Brain during the Early First Trimester*

# The HUMAN BRAIN *during the* THIRD TRIMESTER

**Shirley A. Bayer** *and* **Joseph Altman**

## CRC PRESS

Boca Raton   London   New York   Washington, D.C.

### Library of Congress Cataloging-in-Publication Data

Catalog record is available from the Library of Congress under LCCN: 2003065455

This book contains information obtained from authentic and highly regarded sources. Reprinted material is quoted with permission, and sources are indicated. A wide variety of references are listed. Reasonable efforts have been made to publish reliable data and information, but the author and the publisher cannot assume responsibility for the validity of all materials or for the consequences of their use.

Neither this book nor any part may be reproduced or transmitted in any form or by any means, electronic or mechanical, including photocopying, microfilming, and recording, or by any information storage or retrieval system, without prior permission in writing from the publisher.

The consent of CRC Press LLC does not extend to copying for general distribution, for promotion, for creating new works, or for resale. Specific permission must be obtained in writing from CRC Press LLC for such copying.

Direct all inquiries to CRC Press LLC, 2000 N.W. Corporate Blvd., Boca Raton, Florida 33431.

**Trademark Notice:** Product or corporate names may be trademarks or registered trademarks, and are used only for identification and explanation, without intent to infringe.

### Visit the CRC Press Web site at www.crcpress.com

© 2004 by CRC Press LLC

No claim to original U.S. Government works
International Standard Book Number 0-8493-1421-6
Library of Congress Card Number
Printed in the United States of America  1  2  3  4  5  6  7  8  9  0
Printed on acid-free paper

## DEDICATION

We dedicate this volume to the memory of Dr. Paul Ivan Yakovlev (1894-1983) who spent 42 years (1930-1972) of his productive life in assembling and processing more than 1000 normal and pathological human brain specimens that now constitute the precious Yakovlev Collection.

## ACKNOWLEDGMENTS

We thank Dr. William DeMyer, pediatric neurologist at Riley Hospital for Children, Indiana University Medical Center, Indianapolis, Indiana, for access to his personal library on human central nervous system development. We also thank the staff of the National Museum of Health and Medicine at the Armed Forces Institute of Pathology, Walter Reed Hospital, Washington, D.C.: Dr. Adrianne Noe, Director; Archibald J. Fobbs, Curator of the Yakovlev Collection; Elizabeth C. Lockett, and William Discher. We are most grateful to Dr. James M. Petras at the Walter Reed Institute of Research who made his dark room facilities available so that we could develop all the photomicrographs on location rather than in our laboratory in Indiana. Finally, we thank Barbara Norwitz, Erika Dery, Samar Haddad, Shayna Murry, and Jonathan Pennell at CRC Press for their personal attention to us and for expert help during production of the manuscript.

# CONTENTS

**PART I. INTRODUCTION** --------------------------------------------------- 1
    A. Organization of the Atlas ------------------------------------- 1
    B. Specimens Used ------------------------------------------------ 1
    C. Photography and Computer Processing --------------------------- 1
    D. Identification of Brain Structures ----------------------------- 2
    E. References ----------------------------------------------------- 3

**PART II. GW37 CORONAL** --------------------------------------------- 4
    Low Magnification Plate 1 (Level 1: Section 161) ------------------ 6
    Low Magnification Plate 2 (Level 2: Section 281) ------------------ 7
    Low Magnification Plate 3 (Level 3: Section 361) ------------------ 8
    Low Magnification Plate 4 (Level 4: Section 511) ------------------ 9
    Low Magnification Plate 5 (Level 5: Section 611) ------------------ 10
    Low Magnification Plate 6 (Level 6: Section 691) ------------------ 11
    Low Magnification Plate 7 (Level 7: Section 721) ------------------ 12
    Low Magnification Plate 8 (Level 8: Section 761) ------------------ 13
    Low Magnification Plate 9 (Level 9: Section 831) ------------------ 14
    Low Magnification Plate 10 (Level 10: Section 881) ---------------- 15
    Low Magnification Plate 11 (Level 11: Section 981) ---------------- 16
    Low Magnification Plate 12 (Level 12: Section 1021) --------------- 17
    Low Magnification Plate 13 (Level 13: Section 1081) --------------- 18
    Low Magnification Plate 14 (Level 14: Section 1141) --------------- 19
    Low Magnification Plate 15 (Level 15: Section 1181) --------------- 20
    Low Magnification Plate 16 (Level 16: Section 1221) --------------- 21
    Low Magnification Plate 17 (Level 17: Section 1311) --------------- 22
    Low Magnification Plate 18 (Level 18: Section 1371) --------------- 23
    Low Magnification Plate 19 (Level 19: Section 1501) --------------- 24
    Low Magnification Plate 20 (Levels 20, 21: Sections 1611, 1711) --- 25
    High Magnification Cortex Plate 21: Frontal Cortex ---------------- 26
    High Magnification Cortex Plate 22: Frontal Cortex ---------------- 27
    High Magnification Cortex Plate 23: Precentral Gyrus -------------- 28
    High Magnification Cortex Plate 24: Precentral Gyrus -------------- 29
    High Magnification Cortex Plate 25: Postcentral Gyrus ------------- 30
    High Magnification Cortex Plate 26: Postcentral Gyrus ------------- 31
    High Magnification Cortex Plate 27: Parietal Cortex --------------- 32
    High Magnification Cortex Plate 28: Parietal Cortex --------------- 33
    High Magnification Cortex Plate 29:
    Striate/Peristriate Cortical Transition Area ---------------------- 34
    High Magnification Cortex Plate 30:
    Striate/Peristriate Cortical Transition Area ---------------------- 35
    High Magnification Plates 31A, 31B: (Level 5: Section 611) -------- 36, 37
    High Magnification Plates 32A, 32B: (Level 6: Section 691) -------- 38, 39
    High Magnification Plates 33A, 33B: (Level 7: Section 721) -------- 40, 41
    High Magnification Plates 34A, 34B: (Level 8: Section 761) -------- 42, 43
    High Magnification Plates 35A, 35B: (Level 9: Section 831) -------- 44, 45
    High Magnification Plates 36A, 36B: (Level 10: Section 881) ------- 46, 47
    High Magnification Plates 37A, 37B: (Level 11: Section 981) ------- 48, 49
    High Magnification Plates 38A, 38B: (Level 12: Section 1021) ------ 50, 51
    High Magnification Plates 39A, 39B: (Level 13: Section 1081) ------ 52, 53
    High Magnification Plates 40A, 40B: (Level 14: Section 1141) ------ 54, 55
    High Magnification Plates 41A, 41B: (Level 15: Section 1181) ------ 56, 57

    High Magnification Plates 42A, 42B: (Level 16: Section 1221) ------ 58, 59
    High Magnification Plates 43A, 43B: (Level 17: Section 1311) ------ 60, 61
    High Magnification Plates 44A, 44B: (Level 18: Section 1371) ------ 62, 63
    High Magnification Plates 45A, 45B: (Level 19: Section 1501) ------ 64, 65
    High Magnification Plates 46A, 46B:
    (Levels 20, 21: Sections 1611, 1711) ------------------------------ 66, 67

**PART III. GW37 SAGITTAL** ------------------------------------------ 68
    Low Magnification Plate 47 (Level 1: Section 961) ----------------- 70
    Low Magnification Plate 48 (Level 2: Section 941) ----------------- 71
    Low Magnification Plate 49 (Level 3: Section 841) ----------------- 72
    Low Magnification Plate 50 (Level 4: Section 781) ----------------- 73
    Low Magnification Plate 51 (Level 5: Section 721) ----------------- 74
    Low Magnification Plate 52 (Level 6: Section 661) ----------------- 75
    Low Magnification Plate 53 (Level 7: Section 621) ----------------- 76
    Low Magnification Plate 54 (Level 8: Section 501) ----------------- 77
    High Magnification Plates 55A, 55B: (Level 1: Section 961) -------- 78, 79
    High Magnification Plates 56A, 56B: (Level 2: Section 941) -------- 80, 81
    High Magnification Plates 57A, 57B: (Level 3: Section 841) -------- 82, 83
    High Magnification Plates 58A, 58B: (Level 4: Section 781) -------- 84, 85
    High Magnification Plates 59A, 59B: (Level 5: Section 721) -------- 86, 87
    High Magnification Plates 60A, 60B: (Level 6: Section 661) -------- 88, 89
    High Magnification Plates 61A, 61B: (Level 7: Section 621) -------- 90, 91
    High Magnification Plates 62A, 62B: (Level 8: Section 501) -------- 92, 93
    High Magnification Cerebellum Plate 63: Nodulus ------------------- 94
    High Magnification Cerebellum Plate 64: Paramedian Lobule --------- 95

**PART IV. GW37 HORIZONTAL** ----------------------------------------- 96
    Low Magnification Plate 65 (Level 1: Section 341) ----------------- 98
    Low Magnification Plate 66 (Level 2: Section 441) ----------------- 99
    Low Magnification Plate 67 (Level 3: Section 531) ----------------- 100
    Low Magnification Plate 68 (Level 4: Section 561) ----------------- 101
    Low Magnification Plate 69 (Level 5: Section 631) ----------------- 102
    Low Magnification Plate 70 (Level 6: Section 731) ----------------- 103
    Low Magnification Plate 71 (Level 7: Section 761) ----------------- 104
    Low Magnification Plate 72 (Level 8: Section 801) ----------------- 105
    Low Magnification Plate 73 (Level 9: Section 851) ----------------- 106
    Low Magnification Plate 74 (Level 10: Section 921) ---------------- 107
    Low Magnification Plate 75 (Level 11: Section 981) ---------------- 108
    Low Magnification Plate 76 (Level 12: Section 1051) --------------- 109
    Low Magnification Plate 77 (Level 13: Section 1111) --------------- 110
    Low Magnification Plate 78 (Level 14: Section 1161) --------------- 111
    Low Magnification Plate 79 (Levels 15, 16: Sections 1251, 1311) --- 112
    Low Magnification Plate 80
    (Levels 17, 18, 19: Sections 1331, 1381, 1441) -------------------- 113
    High Magnification Plates 81A, 81B: (Level 3: Section 531) -------- 114, 115
    High Magnification Plates 82A, 82B: (Level 4: Section 561) -------- 116, 117
    High Magnification Plates 83A, 83B: (Level 5: Section 631) -------- 118, 119
    High Magnification Plates 84A, 84B: (Level 6: Section 731) -------- 120, 121
    High Magnification Plates 85A, 85B: (Level 7: Section 761) -------- 122, 123
    High Magnification Plates 86A, 86B: (Level 8: Section 801) -------- 124, 125

# CONTENTS

High Magnification Plates 87A, 87B: (Level 9: Section 851) ----------- 126, 127
High Magnification Plates 88A, 88B: (Level 10: Section 921) ----------- 128, 129
High Magnification Plates 89A, 89B: (Level 11: Section 981) ----------- 130, 131
High Magnification Plates 90A, 90B: (Level 12: Section 1051) ---------- 132, 133
High Magnification Plates 91A, 91B: (Level 13: Section 1111) ---------- 134, 135
High Magnification Plates 92A, 92B: (Level 14: Section 1161) ---------- 136, 137
High Magnification Plates 93A, 93B: (Level 15: Section 1251) ---------- 138, 139
High Magnification Plates 94A, 94B:
(Levels 16, 17: Sections 1311, 1331) ------------------------------------- 140, 141
High Magnification Plates 95A, 95B:
(Levels 18, 19: Sections 1381, 1441) ------------------------------------- 142, 143

**PART V.  GW32 SAGITTAL** ------------------------------------------------------- **144**
    Low Magnification Plate 96 (Level 1: Section 801) ----------------------- 146
    Low Magnification Plate 97 (Level 2: Section 741) ----------------------- 147
    Low Magnification Plate 98 (Level 3: Section 681) ----------------------- 148
    Low Magnification Plate 99 (Level 4: Section 581) ----------------------- 149
    Low Magnification Plate 100 (Level 5: Section 481) ---------------------- 150
    Low Magnification Plate 101 (Level 6: Section 421) ---------------------- 151
    High Magnification Plates 102A, 102B: (Level 1: Section 801) --------- 152, 153
    High Magnification Plates 103A, 103B: (Level 2: Section 741) --------- 154, 155
    High Magnification Plates 104A, 104B: (Level 3: Section 681) --------- 156, 157
    High Magnification Plates 105A, 105B: (Level 4: Section 581) --------- 158, 159
    High Magnification Plates 106A, 106B: (Level 5: Section 481) --------- 160, 161
    High Magnification Plates 107A, 107B: (Level 6: Section 421) --------- 162, 163
    High Magnification Cerebellum Plate 108: Culmen, Lingula ----------------- 164
    High Magnification Cerebellum Plate 109: Folium -------------------------- 165
    High Magnification Cerebellum Plate 110: Uvula, Nodulus ------------------ 166
    High Magnification Cerebellum Plate 111: Germinal Trigone ---------------- 167

**PART VI.  GW30 HORIZONTAL** --------------------------------------------------- **168**
    Low Magnification Plate 112 (Level 1: Section 261) ---------------------- 170
    Low Magnification Plate 113 (Level 2: Section 400) ---------------------- 171
    Low Magnification Plate 114 (Level 3: Section 441) ---------------------- 172
    Low Magnification Plate 115 (Level 4: Section 521) ---------------------- 173
    Low Magnification Plate 116 (Level 5: Section 621) ---------------------- 174
    Low Magnification Plate 117 (Level 6: Section 671) ---------------------- 175
    Low Magnification Plate 118 (Level 7: Section 701) ---------------------- 176
    Low Magnification Plate 119 (Level 8: Section 721) ---------------------- 177
    Low Magnification Plate 120 (Level 9: Section 761) ---------------------- 178
    Low Magnification Plate 121 (Level 10: Section 801) --------------------- 179
    Low Magnification Plate 122 (Level 11: Section 841) --------------------- 180
    Low Magnification Plate 123 (Level 12: Section 861) --------------------- 181
    Low Magnification Plate 124 (Level 13: Section 881) --------------------- 182
    Low Magnification Plate 125 (Level 14: Section 941) --------------------- 183
    Low Magnification Plate 126 (Level 15: Section 1001) -------------------- 184
    Low Magnification Plate 127 (Level 16: Section 1041) -------------------- 185
    Low Magnification Plate 128 (Levels 17, 18: Sections 1081, 1121) -------- 186
    Low Magnification Plate 129 (Levels 19, 20: Sections 1161, 1201) -------- 187
    Low Magnification Plate 130 (Levels 21, 22: Sections 1241, 1281) -------- 188
    Low Magnification Plate 131 (Levels 23, 24: Sections 1351, 1391) -------- 189

High Magnification Plates 132A, 132B: (Level 3: Section 441) -------- 190, 191
High Magnification Plates 133A, 133B: (Level 4: Section 521) -------- 192, 193
High Magnification Plates 134A, 134B: (Level 5: Section 621) -------- 194, 195
High Magnification Plates 135A, 135B: (Level 6: Section 671) -------- 196, 197
High Magnification Plates 136A, 136B: (Level 7: Section 701) -------- 198, 199
High Magnification Plates 137A, 137B: (Level 8: Section 721) -------- 200, 201
High Magnification Plates 138A, 138B: (Level 9: Section 761) -------- 202, 203
High Magnification Plates 139A, 139B: (Level 10: Section 801) ------- 204, 205
High Magnification Plates 140A, 140B: (Level 11: Section 841) ------- 206, 207
High Magnification Plates 141A, 141B: (Level 12: Section 861) ------- 208, 209
High Magnification Plates 142A, 142B: (Level 13: Section 881) ------- 210, 211
High Magnification Plates 143A, 143B: (Level 14: Section 941) ------- 212, 213
High Magnification Plates 144A, 144B: (Level 15: Section 1001) ------ 214, 215
High Magnification Plates 145A, 145B: (Level 16: Section 1041) ------ 216, 217
High Magnification Plates 146A, 146B: (Level 17: Section 1081) ------ 218, 219
High Magnification Plates 147A, 147B: (Level 18: Section 1121) ------ 220, 221
High Magnification Plates 148A, 148B: (Level 19: Section 1161) ------ 222, 223
High Magnification Plates 149A, 149B:
(Levels 20, 21: Sections 1201, 1241) ------------------------------------ 224, 225
High Magnification Plates 150A, 150B:
(Levels 22, 23, 24: Sections 1281, 1351, 1391) -------------------------- 226, 227

**PART VII.  GW29 CORONAL** ----------------------------------------------------- **228**
    Low Magnification Plate 151 (Level 1: Section 321) ---------------------- 230
    Low Magnification Plate 152 (Level 2: Section 501) ---------------------- 231
    Low Magnification Plate 153 (Level 3: Section 621) ---------------------- 232
    Low Magnification Plate 154 (Level 4: Section 701) ---------------------- 233
    Low Magnification Plate 155 (Level 5: Section 741) ---------------------- 234
    Low Magnification Plate 156 (Level 6: Section 781) ---------------------- 235
    Low Magnification Plate 157 (Level 7: Section 821) ---------------------- 236
    Low Magnification Plate 158 (Level 8: Section 861) ---------------------- 237
    Low Magnification Plate 159 (Level 9: Section 901) ---------------------- 238
    Low Magnification Plate 160 (Level 10: Section 981) --------------------- 239
    Low Magnification Plate 161 (Level 11: Section 1021) -------------------- 240
    Low Magnification Plate 162 (Level 12: Section 1061) -------------------- 241
    Low Magnification Plate 163 (Level 13: Section 1101) -------------------- 242
    Low Magnification Plate 164 (Level 14: Section 1161) -------------------- 243
    Low Magnification Plate 165 (Level 15: Section 1201) -------------------- 244
    Low Magnification Plate 166 (Level 16: Section 1261) -------------------- 245
    Low Magnification Plate 167 (Level 17: Section 1301) -------------------- 246
    Low Magnification Plate 168 (Level 18: Section 1361) -------------------- 247
    Low Magnification Plate 169 (Level 19: Section 1521) -------------------- 248
    Low Magnification Plate 170 (Level 20: Section 1621) -------------------- 249
    High Magnification Cortex Plate 171: Frontal Cortex --------------------- 250
    High Magnification Cortex Plate 172: Precentral Gyrus ------------------- 251
    High Magnification Cortex Plate 173: Parietal Cortex -------------------- 252
    High Magnification Cortex Plate 174: Striate Cortex --------------------- 253
    High Magnification Plates 175A, 175B: (Level 4: Section 701) --------- 254, 255
    High Magnification Plates 176A, 176B: (Level 5: Section 741) --------- 256, 257
    High Magnification Plates 177A, 177B: (Level 6: Section 781) --------- 258, 259
    High Magnification Plates 178A, 178B: (Level 7: Section 821) --------- 260, 261

# CONTENTS

High Magnification Plates 179A, 179B: (Level 8: Section 861) -------- 262, 263
High Magnification Plates 180A, 180B: (Level 9: Section 901) -------- 264, 265
High Magnification Plates 181A, 181B: (Level 10: Section 981) -------- 266, 267
High Magnification Plates 182A, 182B: (Level 11: Section 1021) ------ 268, 269
High Magnification Plates 183A, 183B: (Level 12: Section 1061) ------ 270, 271
High Magnification Plates 184A, 184B: (Level 13: Section 1101) ------ 272, 273
High Magnification Plates 185A, 185B: (Level 14: Section 1161) ------ 274, 275
High Magnification Plates 186A, 186B: (Level 15: Section 1201) ------ 276, 277
High Magnification Plates 187A, 187B: (Level 16: Section 1261) ------ 278, 279
High Magnification Plates 188A, 188B: (Level 17: Section 1301) ------ 280, 281
High Magnification Plates 189A, 189B: (Level 18: Section 1361) ------ 282, 283

**PART VIII.  GW26 SAGITTAL** ------------------------------------------------------- **284**
Low Magnification Plate 190 (Level 1: Section 501) ------------------------- 286
Low Magnification Plate 191 (Level 2: Section 481) ------------------------- 287
Low Magnification Plate 192 (Level 3: Section 461) ------------------------- 288
Low Magnification Plate 193 (Level 4: Section 421) ------------------------- 289
Low Magnification Plate 194 (Level 5: Section 381) ------------------------- 290
Low Magnification Plate 195 (Level 6: Section 361) ------------------------- 291
Low Magnification Plate 196 (Level 7: Section 341) ------------------------- 292
Low Magnification Plate 197 (Level 8: Section 321) ------------------------- 293
Low Magnification Plate 198 (Level 9: Section 261) ------------------------- 294
Low Magnification Plate 199 (Level 10: Section 221) ------------------------- 295
High Magnification Plates 200A, 200B: (Level 1: Section 501) -------- 296, 297
High Magnification Plates 201A, 201B: (Level 2: Section 481) -------- 298, 299
High Magnification Plates 202A, 202B: (Level 3: Section 461) -------- 300, 301
High Magnification Plates 203A, 203B: (Level 4: Section 421) -------- 302, 303
High Magnification Plates 204A, 204B: (Level 5: Section 381) -------- 304, 305
High Magnification Plates 205A, 205B: (Level 6: Section 361) -------- 306, 307
High Magnification Plates 206A, 206B: (Level 7: Section 341) -------- 308, 309
High Magnification Plates 207A, 207B: (Level 8: Section 321) -------- 310, 311
High Magnification Plates 208A, 208B: (Level 9: Section 261) -------- 312, 313
High Magnification Plates 209A, 209B: (Level 10: Section 221) -------- 314, 315
High Magnification Cerebellum Plate 210: Lingula ---------------------------- 316
High Magnification Cerebellum Plate 211: Culmen ---------------------------- 317
High Magnification Cerebellum Plate 212: Declive ---------------------------- 318
High Magnification Cerebellum Plate 213: Uvula ---------------------------- 319

High Magnification Cerebellum Plate 214: Nodulus -------------------------- 320
High Magnification Cerebellum Plate 215: Germinal Trigone ---------------- 321

**PART IX.  GW26 HORIZONTAL** ------------------------------------------------------ **322**
Low Magnification Plate 216 (Level 1: Section 240) ------------------------- 324
Low Magnification Plate 217 (Level 2: Section 340) ------------------------- 325
Low Magnification Plate 218 (Level 3: Section 400) ------------------------- 326
Low Magnification Plate 219 (Level 4: Section 460) ------------------------- 327
Low Magnification Plate 220 (Level 5: Section 520) ------------------------- 328
Low Magnification Plate 221 (Level 6: Section 560) ------------------------- 329
Low Magnification Plate 222 (Level 7: Section 620) ------------------------- 330
Low Magnification Plate 223 (Level 8: Section 660) ------------------------- 331
Low Magnification Plate 224 (Level 9: Section 700) ------------------------- 332
Low Magnification Plate 225 (Level 10: Section 740) ----------------------- 333
Low Magnification Plate 226 (Level 11: Section 780) ----------------------- 334
Low Magnification Plate 227 (Level 12: Section 820) ----------------------- 335
Low Magnification Plate 228
(Levels 13, 14: Sections 860, 880) ----------------------------------------- 336
Low Magnification Plate 229
(Levels 15, 16, 17: Sections 900, 940, 960) ------------------------------ 337
High Magnification Plates 230A, 230B: (Level 3: Section 400) -------- 338, 339
High Magnification Plates 231A, 231B: (Level 4: Section 460) -------- 340, 341
High Magnification Plates 232A, 232B: (Level 5: Section 520) -------- 342, 343
High Magnification Plates 233A, 233B: (Level 6: Section 560) -------- 344, 345
High Magnification Plates 234A, 234B: (Level 7: Section 620) -------- 346, 347
High Magnification Plates 235A, 235B: (Level 8: Section 660) -------- 348, 349
High Magnification Plates 236A, 236B: (Level 9: Section 700) -------- 350, 351
High Magnification Plates 237A, 237B: (Level 10: Section 740) -------- 352, 353
High Magnification Plates 238A, 238B: (Level 11: Section 780) -------- 354, 355
High Magnification Plates 239A, 239B: (Level 12: Section 820) -------- 356, 357
High Magnification Plates 240A, 240B: (Level 13: Section 860) -------- 358, 359
High Magnification Plates 241A, 241B: (Level 14: Section 880) -------- 360, 361
High Magnification Plates 242A, 242B: (Level 15: Section 900) -------- 362, 363
High Magnification Plates 243A, 243B: (Level 16: Section 940) -------- 364, 365
High Magnification Plates 244A, 244B: (Level 17: Section 960) -------- 366, 367

**GLOSSARY** ------------------------------------------------------------------------------------- **368**

# PART I: INTRODUCTION

### A. Organization of the Atlas

This Atlas focuses on the development of the human brain during the third trimester, and is Volume 2 in the *Atlas of Human Central Nervous System Development* series. Volume 1 (Bayer and Altman, 2002) provides a record of the development of the spinal cord. The brain specimens presented here are nearing anatomical maturity (at the light microscopic level) and nearly all the structures present in the adult brain are recognizable. Our rationale for starting our brain studies with the most mature specimens is to begin with the known (this volume), then gradually proceed to the unknown – the uncharted territory of human brain development during the midfetal and early embryonic periods (Volumes 3, 4 and 5). The journey will provide a record of when and how known brain structures differentiate, when they first become recognizable, and when they are still absent in less mature specimens. By the third trimester, as seen in the late-fetal specimens illustrated in this volume, only remnants of the embryonic nervous system – the primary neuroepithelium, the secondary germinal matrices, the streams of migrating neurons, and the transitional brain areas – remain. These embryonic structures gradually become more prominent in the younger fetuses to be dealt with in the forthcoming volumes, and are the sole components of the brain in the youngest embryos.

This volume contains grayscale photographs of Nissl-stained sections of brains cut in three cardinal planes (coronal, sagittal, and horizontal) of normal specimens from gestational week (GW) 37 in the late third trimester to GW26 in the early third trimester. In this, and in the forthcoming volumes, older specimens precede younger specimens. Virtually every named brain structure is shown in at least one plane, and most structures can be discerned in two or three planes from specimens of the same nominal age or closely matched ages. Hence, this volume provides a comprehensive atlas of the maturing brain as well as of its late-stage development.

In this volume, each specimen is presented in a separate part of the Atlas: GW37 in the coronal plane in **Part II**; GW37 in the sagittal plane in **Part III**; GW37 in the horizontal plane in **Part IV**; GW32 in the sagittal plane in **Part V**; GW30 in the horizontal plane in **Part VI**; GW29 in the coronal plane in **Part VII**; GW26 in the sagittal plane in **Part VIII**; and GW26 in the horizontal plane in **Part IX**. Selected coronal plates are presented in order from rostral to caudal; the dorsal part of each section is always toward the top of the page, the ventral part at the bottom, and the midline is in the vertical center of each section. Sagittal plates are ordered from medial to lateral; the anterior part of each section is always facing to the left, posterior to the right. Horizontal plates are ordered from dorsal to ventral; the anterior part of each section is always facing to the left, posterior to the right, and the midline is in the horizontal center of each section. Each Part contains *low magnification plates* and *high magnification companion plates*. The *low magnification plates* appear on single pages that show unlabeled full contrast photos of entire sections on the left side and low contrast copies on the right with superimposed outlines and unabbreviated labels. The main purpose of the low magnification plates is to identify the large structures of the brain, such as the various lobes and gyri of the cerebral cortex, and large subdivisions of the brain core, such as the basal ganglia, thalamus, hypothalamus, midbrain, pons, cerebellum, and medulla. The *high magnification companion plates* are designated as **A** and **B** on facing pages and feature enlarged views of the brain core. The **A** part of each plate on the left page shows the full contrast photograph without labels; the **B** part shows low contrast copies of the same photograph on the right page with superimposed outlines of structures and unabbreviated labels. The main purpose of these plates is to identify smaller structures within the brain core. In addition, every coronally-sectioned specimen contains high magnification plates of the cortical plate in different areas of the cerebral cortex, and every sagittally-sectioned specimen contains high magnification plates of different lobules of the cerebellar cortex in the midline vermis. Because our emphasis is on development, transient structures that appear only in immature brains are labeled in *italics*, either directly in some of the high magnification plates or in **bold numbers** that refer to *italicized* labels in a Table. During dissection, embedding, cutting, and staining, some of the sections illustrated were torn; that damage is usually surrounded by *dashed lines*. Finally, an alphabetized **Glossary** gives brief definitions of most labels used in the Plates with expanded definitions of all transient developmental structures.

### B. Specimens Used

All specimens in this volume (designated by the prefix **Y**) are from the Yakovlev Collection housed in the National Museum of Health and Medicine, Armed Forces Institute of Pathology, Washington, D.C. (Forthcoming volumes will also contain specimens from the Carnegie and Minot Collections.) Over a 40-year span, Paul I. Yakovlev (1894-1983) collected brains and some spinal cords from about 1500 normal and abnormal human specimens, ranging in age from the early second trimester of fetal development through old age (Haleem, 1990). The specimens were fixed in formalin, embedded in celloidin, and cut at 35 µm. The sections were arranged in two sets of consecutively numbered slides. One set was stained for cell bodies (Nissl) using cresyl violet; the other set was stained for myelin using the Loyez modification of Weigert's hematoxylin. In selecting the specimens, we rejected those with evidence of pathology, and used only those that were well-preserved during preparation and have not deteriorated since then. We ended up with six Nissl-stained coronal specimens, six horizontal specimens, and sixteen sagittal specimens in the age range between GW37 and GW26. From that group, eight specimens were chosen for presentation in this Atlas as most representative of their respective age groups: two coronally sectioned brains, three sagittally sectioned brains, and three horizontally sectioned brains.

### C. Photography and Computer Processing

The Nissl-stained sections were photographed at low magnification, using a macro lens (Vivitar, 55 mm 1:2.8 auto macro) attached to a Nikkormat 35-mm camera that was mounted on a stand over a fluorescent light board. The magnification varied for each specimen, according to brain size: the section with the largest area that could be accommodated within the field of view set the magnification for all sections of a particular specimen. For better resolution of select regions and structures, most sections were also photographed

# INTRODUCTION

at higher magnifications with a Nikkormat 35-mm camera attached to a 12-cm Leitz/Wetzlar lens with bellows. All photographs were taken with a green filter to increase the contrast of the black and white film (Kodak technical pan #TP442). The film was developed at 20°C for 6 to 7 minutes in Kodak HC110 developer (dilution F), followed by Kodak stop bath for 30 seconds, Kodak fixer for 5 minutes, Kodak hypo clearing agent for 1 minute, running water rinse for 10 minutes, and a brief rinse in Kodak photo-flo before drying.

The negatives were scanned at 2700 dots-per-inch (dpi) with a Nikon Coolscan-1000 35-mm film scanner which was interfaced to a PowerPC G3 Macintosh computer, running Adobe Photoshop (version 5.02) with a plug-in Nikon driver. To capture the subtle shades of gray, the negatives were scanned as color positives, inverted, and converted to grayscale. Using the enhancement features built into Adobe Photoshop and the additional features of Extensis Intellihance, adjustments were made to increase contrast and sharpness. When the image resolution was set to 300 dpi, a full-size photographic file printed at approximately 12 × 10 inches. For the low magnification plates, the images were reduced to fit two images side-by-side on a single plate. For the high magnification plates, the images are shown at full size on separate pages to see structural details in the brain core. Adobe Illustrator was used to superimpose labels and outline structural details on low contrast copies of the Adobe Photoshop files.

## D. Identification of Brain Structures

The late-fetal brain specimens illustrated in this volume contain virtually all the structures found in the mature brain. Hence, these could be identified with some assurance by relying on classical textbooks of neuroanatomy (Ranson and Clark, 1959; Ariëns Kappers et al., 1967; Crosby et al., 1962; Truex and Carpenter, 1969; Brodal, 1981). We also consulted treatises dealing with specific regions of the mature and maturing human brain: the cerebral cortex (Mai et al., 1997; Warner, 2001); the basal telencephalon (Ulfig, 1989; Martin et al., 1991); the amygdala (Sims and Williams, 1990; Amaral et al., 1992; Setzer and Ulfig, 1999), the thalamus (Walker, 1938; Forutan et al., 2001); the hypothalamus (Nauta and Haymaker, 1969; Koutcherov et al., 2002); the visual system (Polyak, 1957); the auditory brainstem nuclei (Moore, 1987; Moore et al., 1999); the nuclei of the pons and the medulla (Paxinos and Huang, 1995); and the cerebellum (Angevine et al., 1961). Since fiber tracts were not stained in our material, many identifications (particularly in the brainstem) posed some difficulties; any label followed by a question mark indicates that we are not completely sure of the identity.

The late-fetal specimens examined in this volume contain only remnants of embryonic brain structures that are prominent at earlier stages of development. In contrast to the mature human brain, there are no comprehensive textbooks available on the morphogenesis of the developing human brain through its entire course. There are several book chapters and reviews available, but these deal with select brain regions and a few stages of human brain development (Moore, 1987; Moore et al., 1999; Setzer and Ulfig, 1999; Forutan et al., 2001; Koutcherov et al., 2002). Therefore we relied heavily on our experimental work in the developing rat brain to identify embryonic structures. For those studies, we labeled proliferating cells in the germinal matrices with $^3$H-thymidine injections at daily intervals through the entire prenatal period, and every other day up to weaning. By varying survival times after $^3$H-thymidine exposure, we used autoradiography to establish timetables of neurogenesis, traced the speed and route of neuronal and glial migration, and documented settling patterns in the maturing brain. The results of these studies were published over a period of 30 years in various journals (*Journal of Comparative Neurology*, *Experimental Neurology*, *International Journal of Developmental Neuroscience*, *Brain Research*, and others), and were summarized in monographs (Altman and Bayer, 1982, 1984, 1986), review articles (Altman, 1992; Altman and Bayer, 2003a; Bayer and Altman, 1995a, 1995b; 2004; Bayer et al., 1995), and books (Bayer and Altman, 1991; Altman and Bayer, 1997, 2001). In an early study (Bayer et al., 1993), we presented evidence for some parallels in the early development of the rat brain and human brain. That motivated us to undertake this extensive and in-depth survey of human brain development and interpret it in the light of our experimental rat data. Based mostly on an analysis of younger fetuses and embryos, we identify several *embryonic* brain structures that either disappear or leave only vestiges behind in the mature brain. These embryonic structures in third trimester fetuses include: (i) vestiges of the primary germinal matrix, the *neuroepithelium*, at a few sites; (ii) the secondary germinal matrix, the *subventricular zone*, that replaces the neuroepithelium at other sites; (iii) other secondary germinal matrices that form at some distance from the ventricles, such as the external germinal layer of the cerebellar cortex or the subgranular zone of the hippocampal dentate gyrus; (iv) *migratory streams*, such as the rostral and lateral migratory streams of the cerebral cortex; (v) areas that contain neurons that sojourn in transitory fields before settling in their final locations, such as the *stratified transitional fields* in the neocortex; (vi) *glioepithelia* along several fiber tracts; and (vii) dispersed sites of glial proliferation that precede the myelination of fiber tracts, known as *myelination gliosis*.

At the outset of its development, the proliferative neuroepithelium is the sole constituent of the brain; it is composed of pluripotent stem cells that are the ultimate source of all the neurons and glia of the central nervous system. Following a precise spatio-temporal pattern, the neuroepithelial cells generate postmitotic large and midsize neurons that move out to form the brain's gross circuitry. While the cells of the neuroepithelium look alike along the entire neuraxis, regions and patches can be distinguished that differ in the time course of their growth (thickening and expansion) and decline (thinning and shrinking), and in the dynamics of their cell proliferation. At first approximation, these neuroepithelial patches or mosaics are identified by their position and are distinguished in such general terms as the amygdaloid neuroepithelium, hippocampal neuroepithelium, or occipital neuroepithelium. The virtual disappearance of the neuroepithelium by the third trimester indicates that the production of the brain's gross-circuitry neurons has ended earlier. At many sites, the neuroepithelium also gives rise to secondary germinal matrices with more limited potencies. Several secondary germinal matrices persist in the third-trimester brain. Some interneurons may still be generated in the subventricular zone of the cerebral cortex. The cortical subventricular zone is also generating glial cells that will disperse throughout the cortex. There are several secondary germinal matrices at some distance from the ventricles that produce neurons during the third trimester and even for some time after birth. The external germinal layer of the cerebellar cortex is generating granule, basket and stellate cells. The subgranular zone of the hippocampus is generating granule cells of the dentate gyrus. Also prominent are several migratory streams during the third trimester. One of these, the rostral migratory stream contains, among other cell types, granule cells that settle in the olfactory bulb and it is also a source of glia. Migrating cells in some brain regions stop in sojourn zones for varying lengths of time before settling in their final locations. In the human neocortex these sojourning cells form alternating layers with afferent, efferent and commissural (callosal) fibers that we call *stratified transitional fields*. Banding patterns in these fields differ considerably in the frontal, paracentral, occipital and temporal lobes (Altman and Bayer, 2003b) and their vestiges are still present throughout the third trimester. These fields have been postulated to be sites where connections of the cerebral cortex are established before neurons settle in the cortical plate; that process is still in progress during the third trimester. The mature stratification of the cortical plate, the primordium of the cortical gray matter, starts to be evident during the third trimester and finishes after birth. Glioepithelia are fate-restriced precursors of oligodendrocytes responsible for the myelination of axons in fiber tracts. Several of these are still prominent during the third trimester beneath the corpus callosum, lining the fornix, and adjacent to some fiber tracts. At other sites glia dispersed within fiber tracts appear to proliferate locally before the onset of myelination. This *myelination gliosis* is evident at the beginning of the third trimester in the cuneate and gracile fasciculi of the medulla and spinal cord (Bayer and Altman, 2002).

A final note. In the first volume of this series dealing with spinal cord development from the early embryonic period through infancy (Bayer and Altman, 2002), we provided quantitative summaries of several ontogenetic trends, such as the area of the neuroepithelium, the area of the ventricular lumen, and the expansion of the white matter and gray matter over the entire span of development. Because this volume deals only with the latest phase of fetal brain development, comprehensive quantitative summaries will be featured in Volume 5 of this series when all the brain studies are completed.

# INTRODUCTION

## E. References

Altman, J. (1992) The early stages of nervous system development: Neurogenesis and neuronal migration. In: A. Björklund, T. Hökfelt and M. Tohyama (eds.) *Handbook of Chemical Neuroanatomy.* Volume 10. *Ontogeny of Transmitters and Peptides in the CNS*, pp. 1-31. Amsterdam: Elsevier.

Altman, J. and S. A. Bayer (1982) Development of the Cranial Nerve Ganglia and Related Nuclei in the Rat. (*Advances in Anatomy, Embryology and Cell Biology,* Vol. 74). Berlin: Springer-Verlag.

Altman, J. and S. A. Bayer (1984) The Development of the Rat Spinal Cord. (*Advances in Anatomy, Emryology and Cell Biology,* Vol. 85). Berlin: Springer-Verlag.

Altman, J. and S. A. Bayer (1986) The Development of the Rat Hypothalamus. (*Advances in Anatomy, Embryology and Cell Biology,* Vol.100). Berlin: Springer-Verlag.

Altman, J. and S. A. Bayer (1997) *Development of the Cerebellar System in Relation to its Evolution, Structure, and Functions.* Boca Raton, FL: CRC Press.

Altman, J. and S. A. Bayer (2001) *Development of the Human Spinal Cord: An Interpretation Based on Experimental Studies in Animals.* New York, NY: Oxford University Press.

Altman, J. and S. A. Bayer (2003a) Neuroembryology. In: G. Adelman and B. H. Smith (eds.) *Encyclopedia of Neuroscience.* (Third edition) Amsterdam: Elsevier. (In press)

Altman, J. and S. A. Bayer (2003b) Regional differences in the stratified transitional field and the honeycomb matrix of the developing human cerebral cortex. *Journal of Neurocytology,* 31:613-632.

Amaral, D. G., J. L Price, A. Pitkänen, and S. T. Carmichael (1992) Anatomical organization of the primate amygdaloid complex. In: J. P. Aggleton (ed.) *The Amygdala: Neurobiological Aspects of Emotion, Memory, and Mental Dysfunction.* New York, NY: Wiley-Liss, pp. 1-66.

Angevine, J. B., E. L. Mancall, and P. I. Yakovlev (1961) *The Human Cerebellum: An Atlas of Gross Topography in Serial Sections.* Boston, MA: Little, Brown.

Ariëns Kappers, C. U., G. C. Huber, and E. C. Crosby (1936) *The Comparative Anatomy of the Nervous System of Vertebrates, Including Man.* Volumes 1-3. New York, NY: Hafner Publishing Company. [Reprint, 1967].

Bayer, S. A. and J. Altman (1991) *Neocortical Development.* New York, NY: Raven Press.

Bayer, S. A and J. Altman (1995a) Neurogenesis and neuronal migration. In: G. Paxinos (ed.) *The Rat Nervous System* (Second edition), pp. 1041-1078. San Diego, CA: Academic Press.

Bayer, S. A and J. Altman (1995b) Principles of neurogenesis, neuronal migration, and neural circuit formation. In: G. Paxinos (ed.) *The Rat Nervous System* (Second edition), pp. 1079-1098. San Diego, CA: Academic Press.

Bayer, S. A. and J. Altman (2002) *Atlas of Human Central Nervous System Development.* Volume 1: *The Spinal Cord from Gestational Week 4 to the 4th Postnatal Month.* Boca Raton, FL: CRC Press.

Bayer, S. A. and J. Altman (2004) Development of the telencephalon: Neural stem cells, neurogenesis and neuronal migration. In: G. Paxinos (ed.) *The Rat Nervous System* (Third edition). San Diego, CA: Academic Press. pp. 27-73.

Bayer, S. A., J. Altman, R. J. Russo and X. Zhang (1993) Timetables of neurogenesis in the human brain based on experimentally determined patterns in the rat. *Neurotoxicology,* 14:83-144.

Bayer, S. A., J. Altman, R. J. Russo and X. Zhang (1995) Embryology. In: S. Duckett (ed.) *Pediatric Neuropathology,* pp. 54-107. Baltimore, MD: Williams & Wilkins.

Brodal, A. (1981) *Neurological Anatomy in Relation to Clinical Medicine.* (Third edition). New York, NY: Oxford University Press.

Crosby, E. C., T. Humphrey, and E. W. Lauer (1962) *Correlative Anatomy of the Nervous System.* New York, NY: The Macmillan Company.

Forutan, F., J. K. Mai, K. W. S. Ashwell, S. Lensing-Höhn, D. Nohr, T. Voss, J. Bohl, and C. Andressen (2001) Organisation and maturation of the human thalamus as revealed by CD15. *Journal of Comparative Neurology,* 437:476-495.

Haleem, M. (1990) *Diagnostic Categories of the Yakovlev Collection of Normal and Pathological Anatomy and Development of the Brain.* Washington, D.C.: Armed Forces Institute of Pathology.

Koutcherov, Y., J. K. Mai, K. W. S. Ashwell, and G. F. Paxinos (2002) Organisation of human hypothalamus in fetal development. *Journal of Comparative Neurology,* 446:301-324.

Mai, J. K., J. Assheuer, and G. Paxinos (1997) *Atlas of the Human Brain.* San Diego, CA: Academic Press.

Martin, L. J., R. E. Powers, T. L. Dellovade, and D. L. Price (1991) The bed nucleus-amygdala continuum in human and monkey. *Journal of Comparative Neurology,* 309:445-485.

Moore, J. K. (1987) The human auditory brainstem: a comparative view. *Hearing Research*, 29:1-32.

Moore, J. K., D. D. Simmons, and Y-L. Guan (1999) The human olivocochlear system: organization and development. *Audiology and Neuro-otology*, 4:311-325.

Nauta, W. J. H., and W. Haymaker (1969) Hypothalamic nuclei and fiber connections. In: *The Hypothalamus*, W. Haymaker, E. Anderson, and W. J. H. Nauta (eds.). Springfield, IL: Charles C. Thomas. pp. 136-209.

Paxinos, G., and X-F. Huang (1995) *Atlas of the Human Brainstem.* San Diego, CA: Academic Press.

Polyak, S. (1957) *The Vertebrate Visual System*, H. Klüver (ed.). Chicago, IL: The University of Chicago Press.

Ranson, S. W. and S. L. Clark (1959) *The Anatomy of the Nervous System: Its Development and Function* (Tenth edition). Philadelphia, PA: W. B. Saunders Company.

Setzer, M., and N. Ulfig (1999) Differential expression of calbindin and calretinin in the human fetal amygdala. *Microscopy Research and Technique*, 46:1-17.

Sims, K. S., and R. S. Williams (1990) The human amygdaloid complex: A cytologic and histochemical atlas using Nissl, myelin, acetylcholinesterase and nicotinamide adenine dinucleotide phosphate diaphorase staining. *Neuroscience,* 36:449-472.

Truex, R. C., and M. B. Carpenter (1969) *Human Neuroanatomy* (6th edition). Baltimore, MD: Williams & Wilkins Company.

Ulfig, N. (1989) Configuration of the magnocellular nuclei in the basal forebrain of the human adult. *Acta Anatomica,* 134:100-105.

Walker, A. E. (1938) *The Primate Thalamus.* Chicago, IL: The University of Chicago Press.

Warner, J. J., (2001) *Atlas of Neuroanatomy with Systems Organization and Case Correlation.* Boston, MA: Butterworth-Heinemann.

# PART II: GW37 CORONAL

This specimen is case number W-217-65 (Perinatal RPSL) in the Yakovlev Collection. A male infant, weighing 3,057 grams, was born alive on August 10, 1965, and survived for 44 hours. Death occurred because a hyaline membrane obstructed the airway to the lungs. Upon autopsy, a subarachnoid hemmorhage in the left temporal/parietal area of the cerebral cortex was noted. The brain itself, cut in the coronal (frontal) plane in 35-μm thick sections, is classified as a Normative Control in the Yakovlev Collection (Haleem, 1990). Although there is no photograph of this brain before it was embedded and cut, the photograph of the lateral view of another GW37 brain that Larroche published in 1967 (**Figure 1**) is similar to the features of the brain in Y217-65. There are definite divisions between major gyri in all lobes of the cerebral hemispheres (labeled in **Figure 1**).

The approximate cutting plane of this brain is indicated in **Figure 2** (facing page) with lines superimposed on the GW37 brain from the Larroche (1967) series. The dorsal part of each section is posterior to the ventral part. This brain is nearly even in mediolateral orientation. For example, the temporal poles on right vs. left sides of the brain appear at the same time (**Level 5, Section 611**). The sections chosen for illustration are more closely spaced to show small structures in the diencephalon, midbrain, pons, and medulla. Illustrated sections are spaced farther apart when they contain only large brain structures, such as the cerebral cortex, basal ganglia, and cerebellum. Photographs of 21 different Nissl-stained sections (**Levels 1-21**) are shown at low magnification in **Plates 1-20**. Different areas of the cerebral cortex are shown at very high magnification in **Plates 21-30**. The core of the brain and the cerebellum are shown at high magnification in companion **Plates 31AB-46AB** for **Levels 5-21**.

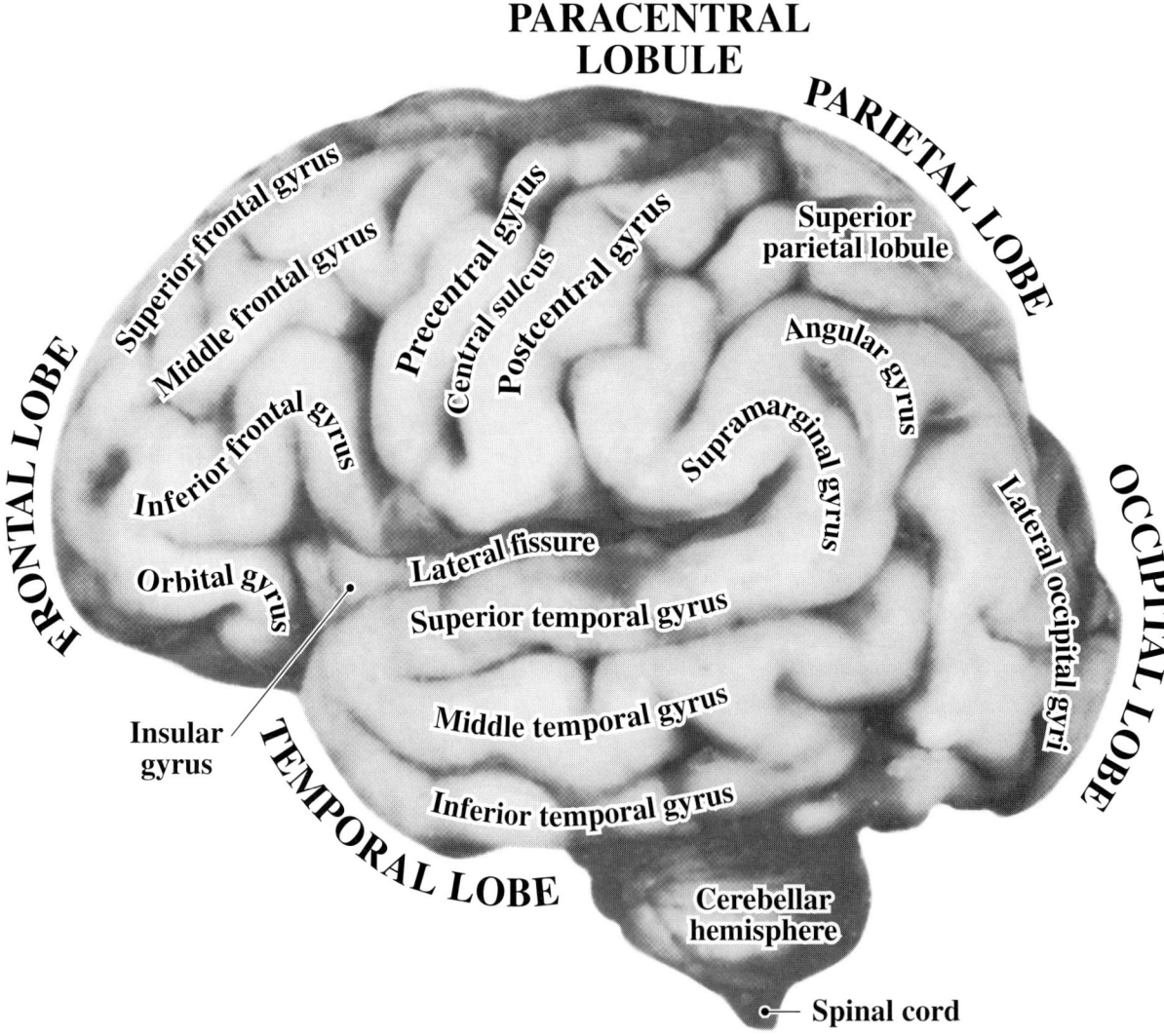

**Figure 1.** Lateral view of a GW37 brain with major structures in the cerebral hemispheres labeled. (From the photographic series of J. C. Larroche (1967) Maturation morphologique du système nerveux central: ses rapports avec le développement pondéral du foetus et son age gestationnel. In: *Regional Development of the Brain in Early Life*, A. Minkowski (ed.), London: Blackwell, page 248.)

## GW37 CORONAL SECTION PLANES

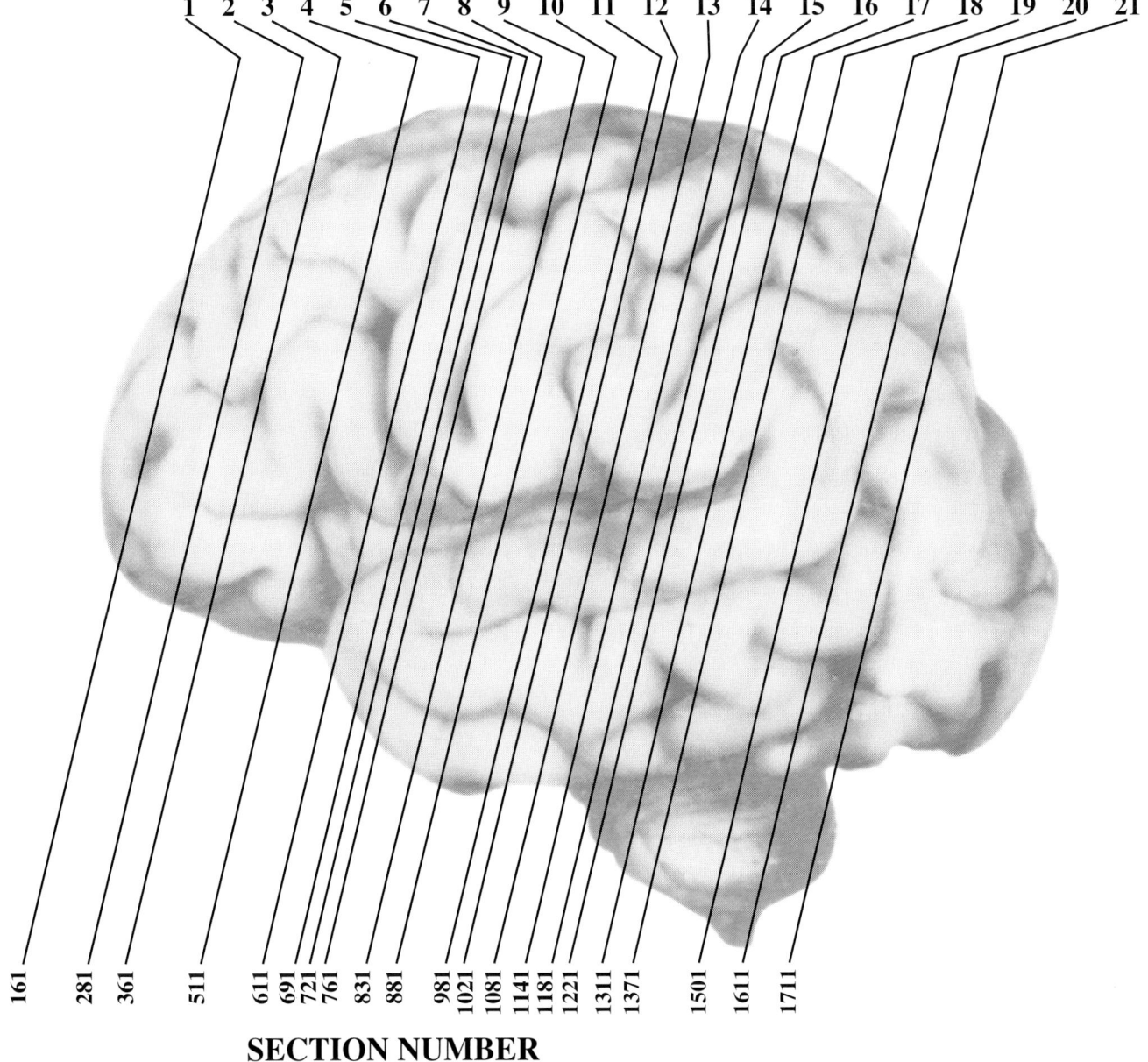

**Figure 2.** Lateral view of the same GW37 brain shown in **Figure 1** with the approximate locations and cutting angle of the sections of Y217-65. (From the photographic series of J. C. Larroche (1967) Maturation morphologique du système nerveux central: ses rapports avec le développement pondéral du foetus et son age gestationnel. In: *Regional Development of the Brain in Early Life*, A. Minkowski (ed.), London: Blackwell, page 248.)

Y217-65 contains several immature structures. In the cortical regions of the telencephalon, remnants of the germinal matrices are present in all lobes of the cerebral cortex where the ***neuroepithelium/subventricular zone*** may still be generating neocortical interneurons. Remnants of migrating and sojourning neurons and/or glia are visible in all lobes of the cerebral cortex as ***stratified transitional fields***, thin in the occipital lobe, and thicker in the frontal, parietal and temporal lobes. Indeed, thickness variations can be used as identifying criteria for designating a particular cortical germinal matrix as occipital or otherwise. Many neurons, glia, and their mitotic precursor cells are still migrating through the olfactory peduncle toward the olfactory bulb (***rostral migratory stream***) from a presumed source area in the germinal matrix at the junction between the cerebral cortex, striatum, and nucleus accumbens. Within the lateral parts of the cerebral cortex, streams of neurons and glia are still in the ***lateral migratory stream***. That stream percolates through the claustrum, endopiriform nucleus, external capsule, and uncinate fasciculus, and the cells appear to be heading toward the insular cortex, primary olfactory cortex, temporal cortex, and basolateral parts of the amygdaloid complex. In the basal ganglia, there is a prominent ***neuroepithelium/ subventricular zone*** overlying the striatum and nucleus accumbens where neurons are probably still being generated. The striatal portion can be subdivided into anterolateral, anteromedial, and posterior parts. Another region of active neurogenesis in the telencephalon is the ***subgranular zone*** in the hilus of the dentate gyrus that is the source of granule cells. Other structures in the telencephalon, such as the septum, fornix, and Ammon's horn part of the hippocampus, have only a thin, darkly staining layer at the ventricle, and these are presumed to be generating glia, cells of the choroid plexus, and the ependymal lining of the ventricle.

Most of the structures in the diencephalon appear to be settled and are maturing, and the third ventricle is lined by a thin ***glioepithelium/ ependyma***. In the midbrain and anterior pons, there is a slightly thicker and more convoluted ***glioepithelium/ependyma*** lining the posterior cerebral aqueduct and anterior fourth ventricle. The posterior pons and entire medulla have a thin ***glioepithelium/ependyma*** lining the rest of the fourth ventricle. The ***external germinal layer*** is prominent over the entire surface of the cerebellar cortex and is still producing basket, stellate, and granule cells. The ***germinal trigone*** is still visible at the base of the nodulus and along the floccular peduncle; choroid plexus cells and glia may still be originating here.

# PLATE 1

**GW37 Coronal**
**CR 350 mm**
**Y217-65**
**Level 1: Section 161**

*Remnants of the germinal matrix, migratory streams, and transitional fields*

**1** *Frontal stratified transitional field*
**2** *Subpial granular layer (cortical)*

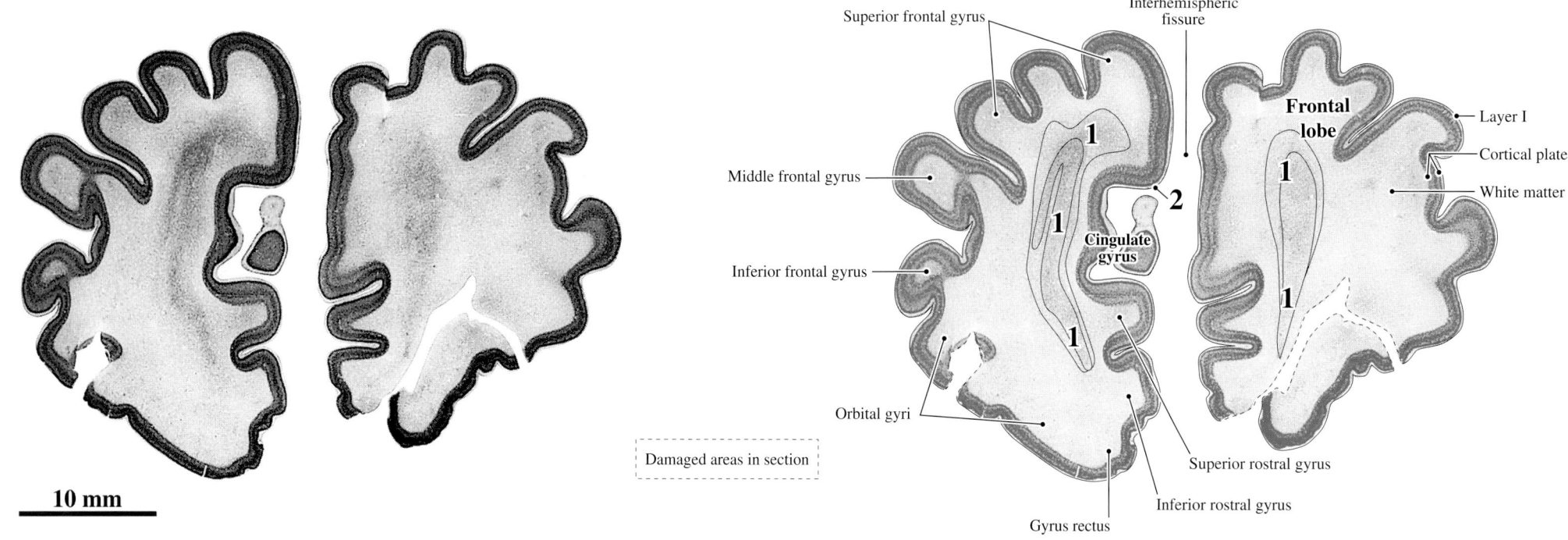

# PLATE 2

**GW37 Coronal**
**CR 350 mm**
**Y217-65**
**Level 2: Section 281**

*Remnants of the germinal matrix, migratory streams, and transitional fields*

1. *Frontal neuroepithelium and subventricular zone*
2. *Frontal stratified transitional field*
3. *Subpial granular layer (cortical)*

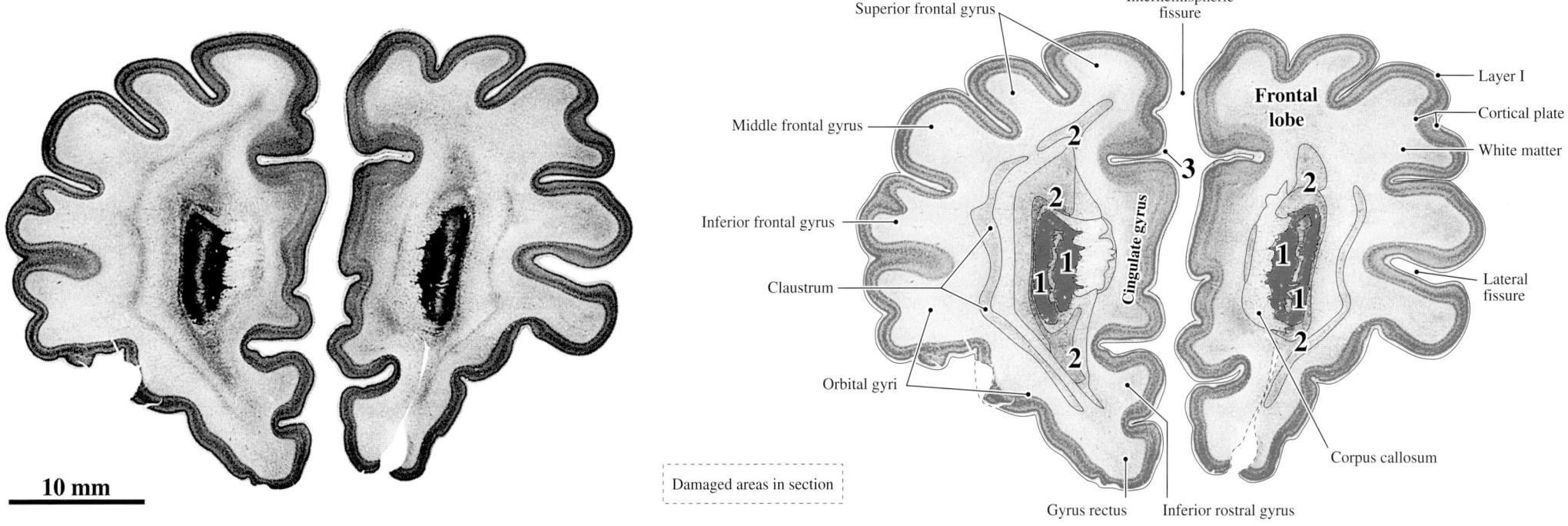

# PLATE 3

GW37 Coronal
CR 350 mm
Y217-65
Level 3: Section 361

**Remnants of the germinal matrix, migratory streams, and transitional fields**

| | | | |
|---|---|---|---|
| 1 | *Rostral migratory stream (source area)* | 4 | *Callosal glioepithelium* |
| 2 | *Frontal neuroepithelium and subventricular zone* | 5 | *Anterolateral striatal neuroepithelium and subventricular zone* |
| 3 | *Frontal stratified transitional field* | 6 | *Subpial granular layer (cortical)* |

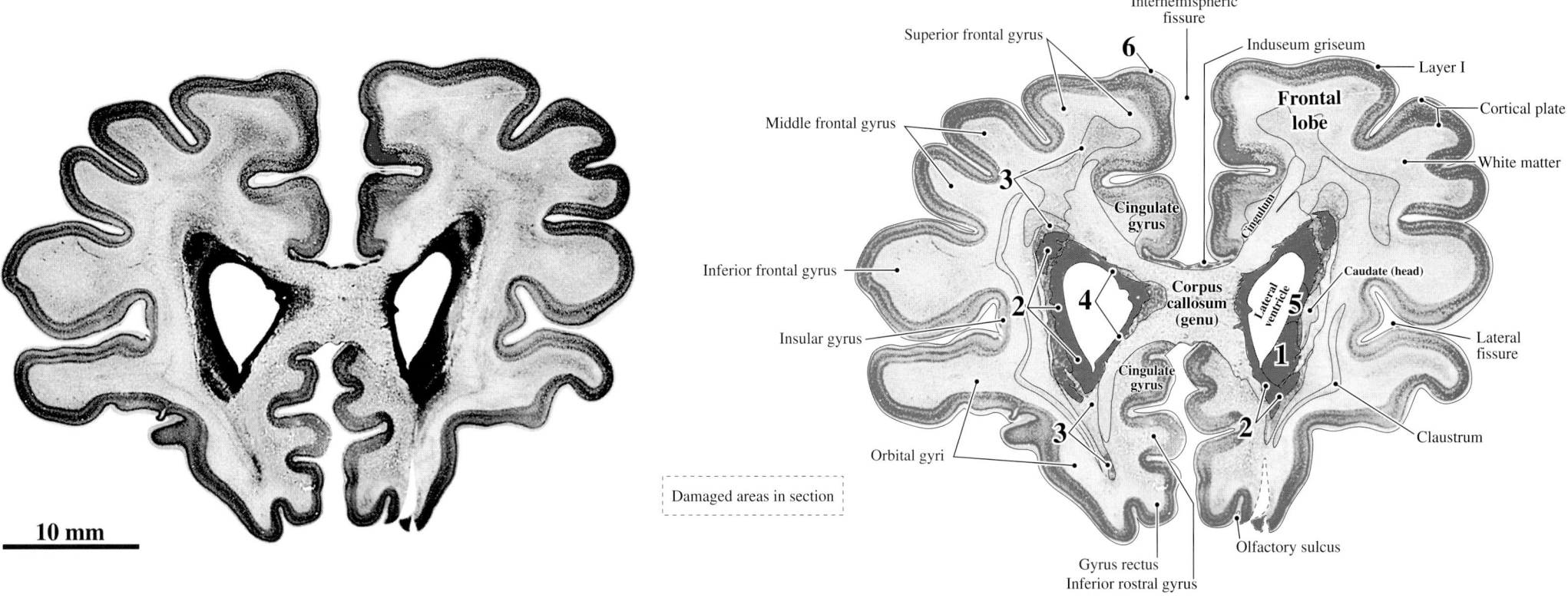

# PLATE 4

**GW37 Coronal**
**CR 350 mm**
**Y217-65**
**Level 4: Section 511**

### Remnants of the germinal matrix, migratory streams, and transitional fields

| | | | |
|---|---|---|---|
| 1 | Rostral migratory stream | 6 | Callosal sling |
| 2 | Frontal neuroepithelium and subventricular zone | 7 | Fornical glioepithelium |
| 3 | Frontal stratified transitional field | 8 | Anterolateral striatal neuroepithelium and subventricular zone |
| 4 | Frontal neuroepithelium and subventricular zone (intermingled with the source of the rostral migratory stream) | 9 | Accumbent neuroepithelium and subventricular zone (intermingled with the source of the rostral migratory stream) |
| 5 | Callosal glioepithelium | 10 | Subpial granular layer (cortical) |

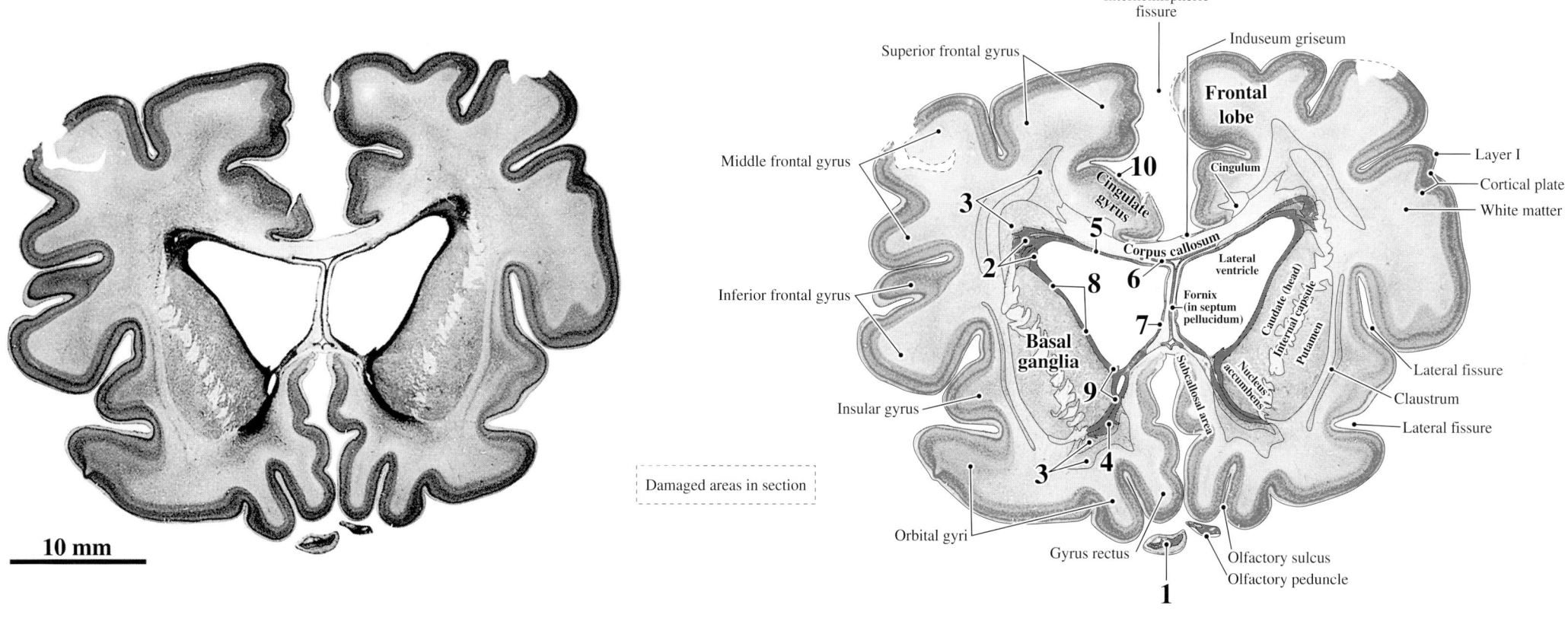

# PLATE 5

GW37 Coronal
CR 350 mm
Y217-65
Level 5: Section 611

See detail of brain core
in Plates 31A and B.

### Remnants of the germinal matrix, migratory streams, and transitional fields

| | | | |
|---|---|---|---|
| 1 | Rostral migratory stream | 7 | Anterolateral striatal neuroepithelium and subventricular zone |
| 2 | Frontal neuroepithelium and subventricular zone | 8 | Anteromedial striatal neuroepithelium and subventricular zone |
| 3 | Frontal stratified transitional field | 9 | Accumbent neuroepithelium and subventricular zone (intermingled with the source of the rostral migratory stream) |
| 4 | Callosal glioepithelium | | |
| 5 | Callosal sling | 10 | Septal glioepithelium/ependyma |
| 6 | Fornical glioepithelium | 11 | Subpial granular layer (cortical) |

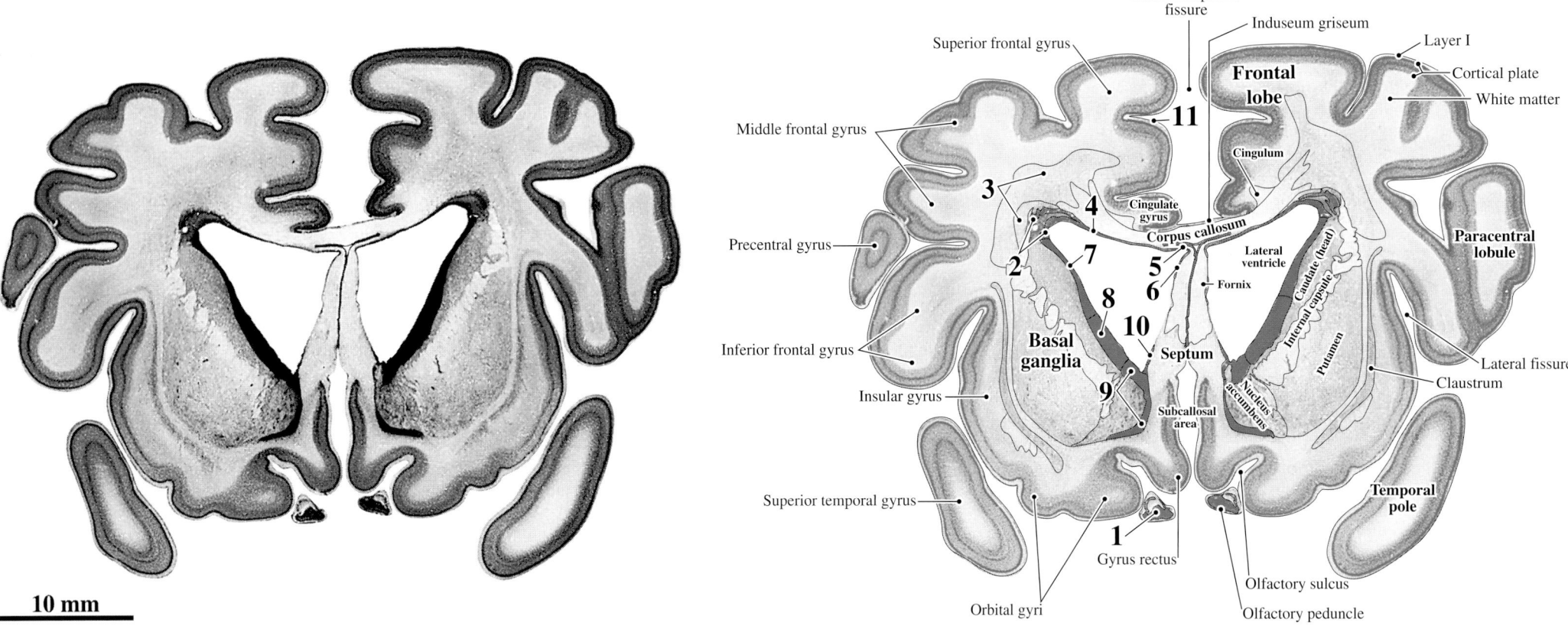

# PLATE 6

**GW37 Coronal**
**CR 350 mm**
**Y217-65**
**Level 6: Section 691**

### Remnants of the germinal matrix, migratory streams, and transitional fields

| | |
|---|---|
| 1 *Rostral migratory stream* | 7 *Lateral migratory stream (cortical)* |
| 2 *Frontal neuroepithelium and subventricular zone* | 8 *Anterolateral striatal neuroepithelium and subventricular zone* |
| 3 *Frontal stratified transitional field* | 9 *Anteromedial striatal neuroepithelium and subventricular zone* |
| 4 *Callosal glioepithelium* | 10 *Strionuclear glioepithelium* |
| 5 *Callosal sling* | 11 *Diencephalic (preoptic) glioepithelium/ependyma* |
| 6 *Fornical glioepithelium* | 12 *Subpial granular layer (cortical)* |

**See detail of brain core in Plates 32A and B.**

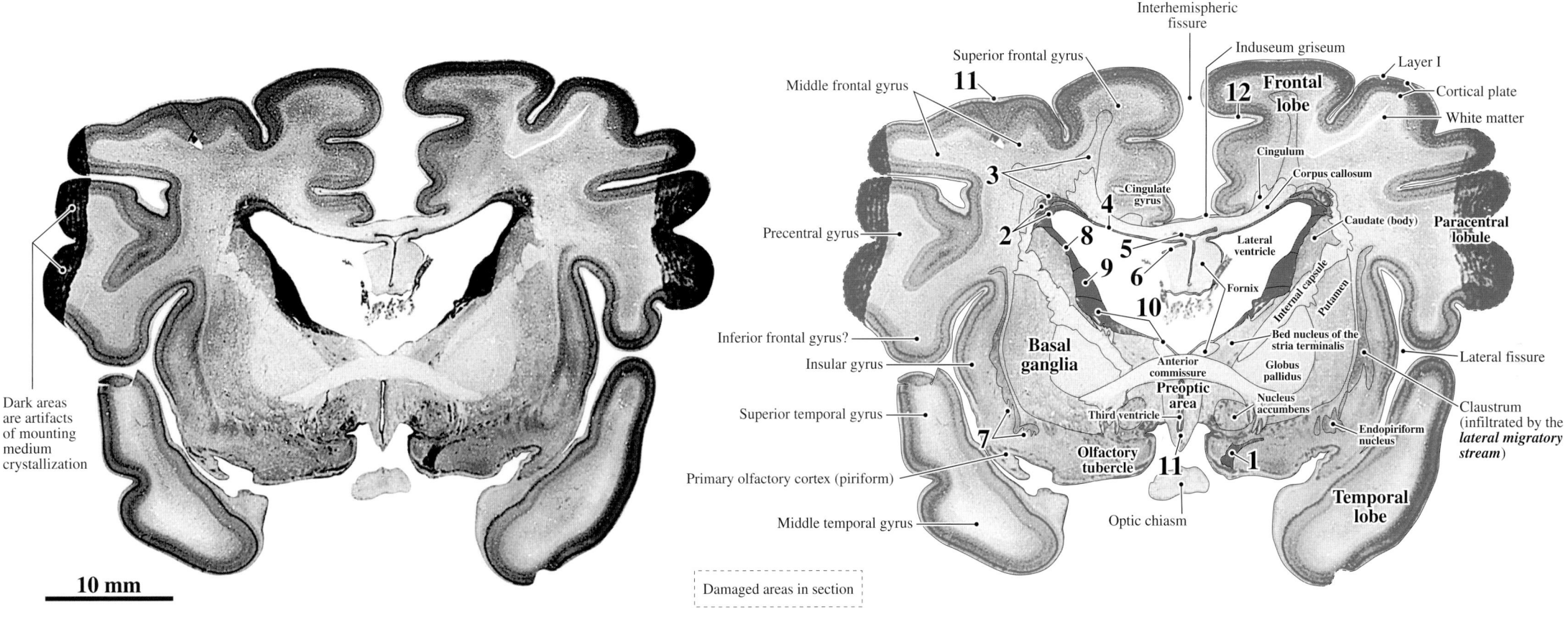

## PLATE 7

GW37 Coronal
CR 350 mm
Y217-65
Level 7: Section 721

See detail of brain core
in Plates 33A and B.

*Remnants of the germinal matrix, migratory streams, and transitional fields*

| | | | |
|---|---|---|---|
| 1 | Frontal neuroepithelium and subventricular zone | 7 | Anterolateral striatal neuroepithelium and subventricular zone |
| 2 | Frontal stratified transitional field | 8 | Anteromedial striatal neuroepithelium and subventricular zone |
| 3 | Callosal glioepithelium | 9 | Strionuclear glioepithelium |
| 4 | Callosal sling | 10 | Diencephalic (thalamic) glioepithelium/ependyma |
| 5 | Fornical glioepithelium | 11 | Diencephalic (preoptic) glioepithelium/ependyma |
| 6 | Lateral migratory stream (cortical) | 12 | Subpial granular layer (cortical) |

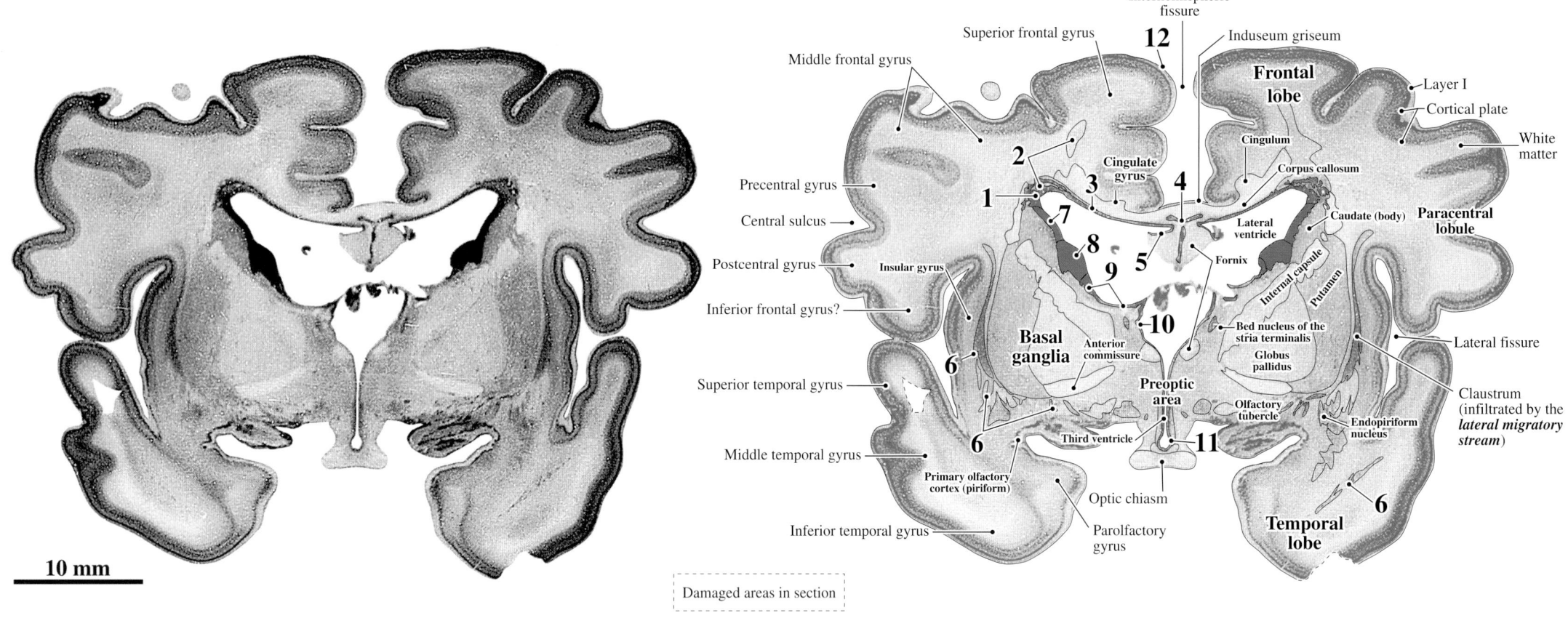

# PLATE 8

**GW37 Coronal**
**CR 350 mm**
**Y217-65**
**Level 8: Section 761**

*Remnants of the germinal matrix, migratory streams, and transitional fields*

| | | | |
|---|---|---|---|
| 1 | Frontal neuroepithelium and subventricular zone | 7 | Anterolateral striatal neuroepithelium and subventricular zone |
| 2 | Frontal stratified transitional field | 8 | Anteromedial striatal neuroepithelium and subventricular zone |
| 3 | Callosal glioepithelium | 9 | Strionuclear glioepithelium |
| 4 | Callosal sling | 10 | Diencephalic (thalamic) glioepithelium/ependyma |
| 5 | Fornical glioepithelium | 11 | Diencephalic (hypothalamic) glioepithelium/ependyma |
| 6 | Lateral migratory stream (cortical) | 12 | Subpial granular layer (cortical) |

See detail of brain core in Plates 34A and B.

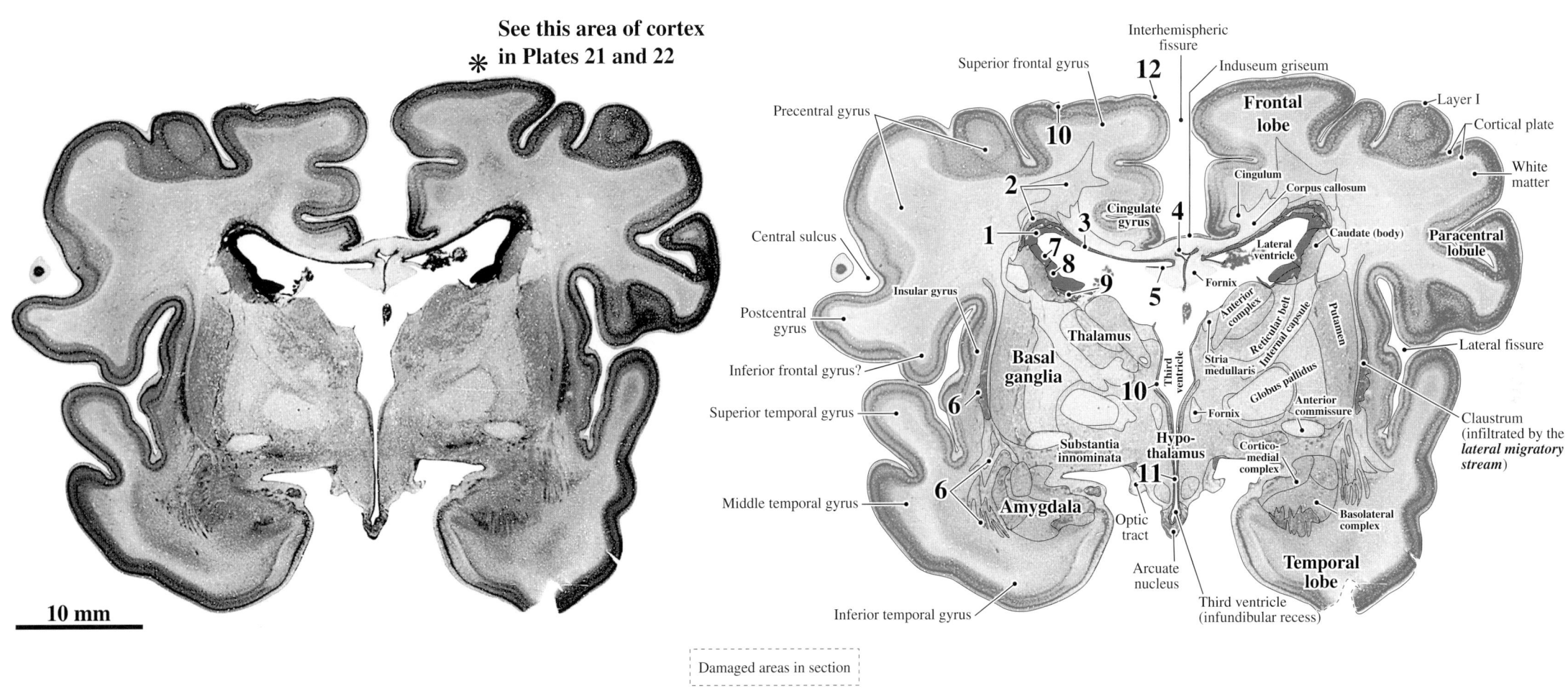

# PLATE 9

**GW37 Coronal**
**CR 350 mm**
**Y217-65**
**Level 9: Section 831**

**See detail of brain core in Plates 35A and B.**

### Remnants of the germinal matrix, migratory streams, and transitional fields

| | |
|---|---|
| 1 *Frontal neuroepithelium and subventricular zone* | 7 *Lateral migratory stream (cortical)* |
| 2 *Frontal stratified transitional field* | 8 *Amygdaloid glioepithelium/ependyma* |
| 3 *Callosal glioepithelium* | 9 *Posterior striatal neuroepithelium and subventricular zone* |
| 4 *Callosal sling* | 10 *Strionuclear glioepithelium* |
| 5 *Fornical glioepithelium* | 11 *Diencephalic (thalamic) glioepithelium/ependyma* |
| 6 *Parahippocampal neuroepithelium, subventricular zone, and stratified transitional field* | 12 *Diencephalic (hypothalamic) glioepithelium/ependyma* |
| | 13 *Subpial granular layer* |

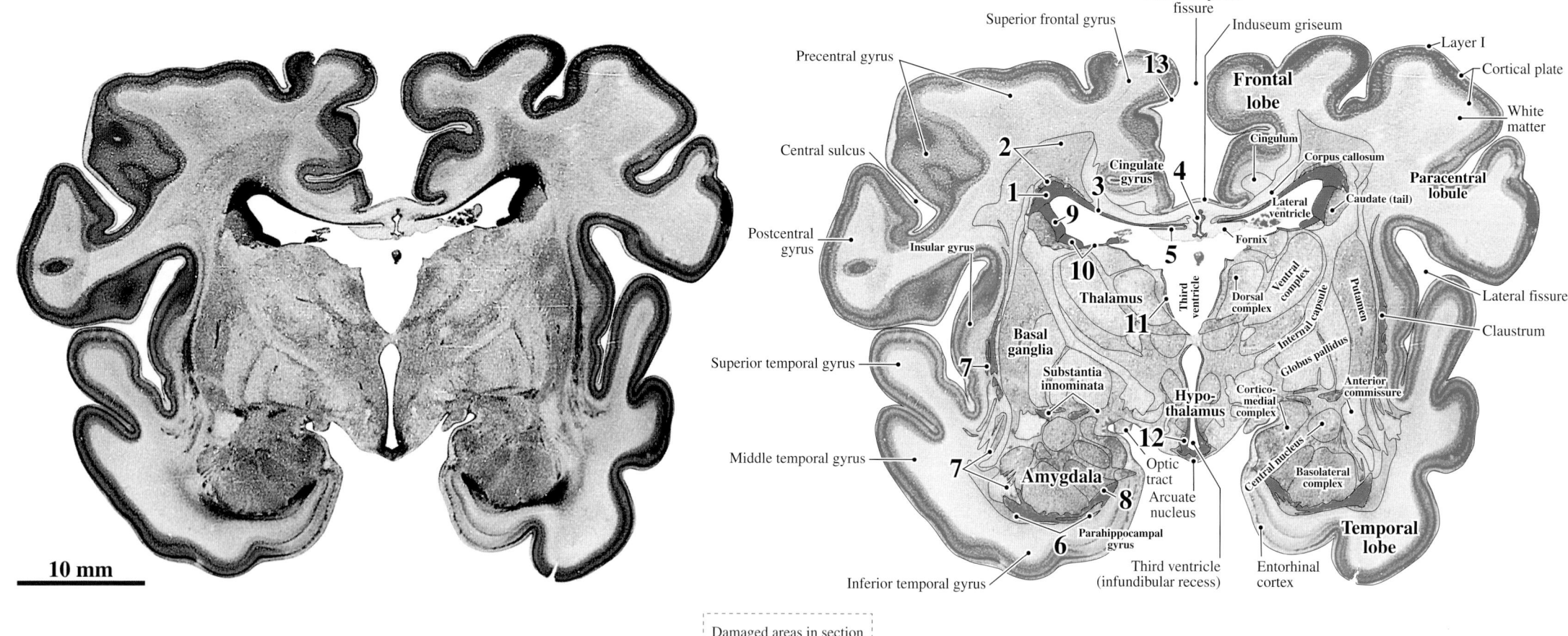

# PLATE 10

**GW37 Coronal**
**CR 350 mm**
**Y217-65**
**Level 10: Section 881**

See detail of brain core in Plates 36A and B.

### Remnants of the germinal matrix, migratory streams, and transitional fields

1. Frontal/paracentral neuroepithelium and subventricular zone
2. Frontal/paracentral stratified transitional field
3. Callosal glioepithelium
4. Callosal sling
5. Fornical glioepithelium
6. Parahippocampal neuroepithelium, subventricular zone, and stratified transitional field
7. Temporal neuroepithelium and subventricular zone
8. Temporal stratified transitional field
9. Alvear glioepithelium
10. Lateral migratory stream (cortical)
11. Amygdaloid glioepithelium/ependyma
12. Posterior striatal neuroepithelium and subventricular zone
13. Strionuclear glioepithelium
14. Diencephalic (thalamic) glioepithelium/ependyma
15. Diencephalic (hypothalamic) glioepithelium/ependyma
16. Subpial granular layer

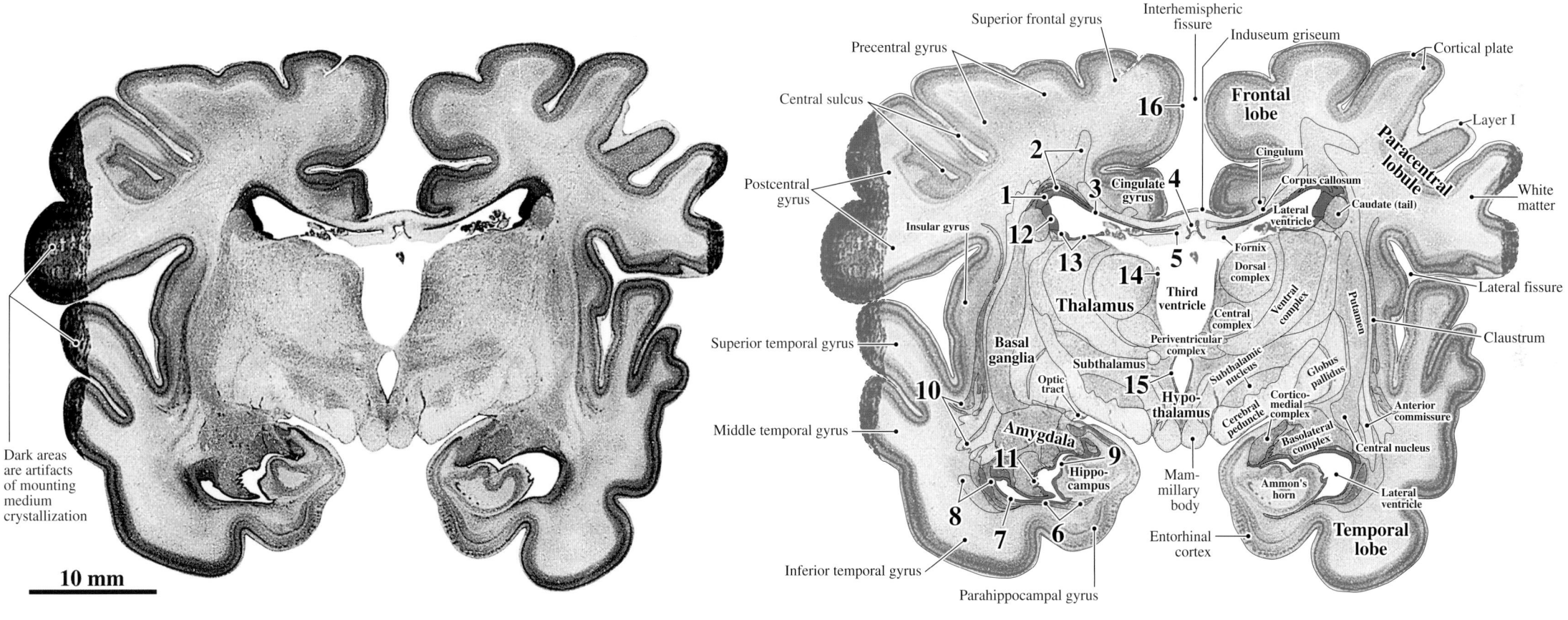

# PLATE 11

**GW37 Coronal
CR 350 mm
Y217-65
Level 11: Section 981**

**See detail of brain core in Plates 37A and B.**

***Remnants of the germinal matrix, migratory streams, and transitional fields***

| | | | |
|---|---|---|---|
| 1 | *Frontal/paracentral neuroepithelium and subventricular zone* | 9 | *Alvear glioepithelium* |
| 2 | *Frontal/paracentral stratified transitional field* | 10 | *Subgranular zone (dentate)* |
| 3 | *Callosal glioepithelium* | 11 | *Lateral migratory stream (cortical)* |
| 4 | *Callosal sling* | 12 | *Posterior striatal neuroepithelium and subventricular zone* |
| 5 | *Fornical glioepithelium* | 13 | *Strionuclear glioepithelium* |
| 6 | *Parahippocampal neuroepithelium, subventricular zone, and stratified transitional field* | 14 | *Diencephalic (thalamic) glioepithelium/ependyma* |
| 7 | *Temporal neuroepithelium and subventricular zone* | 15 | *Subpial granular layer (cortical)* |
| 8 | *Temporal stratified transitional field* | | |

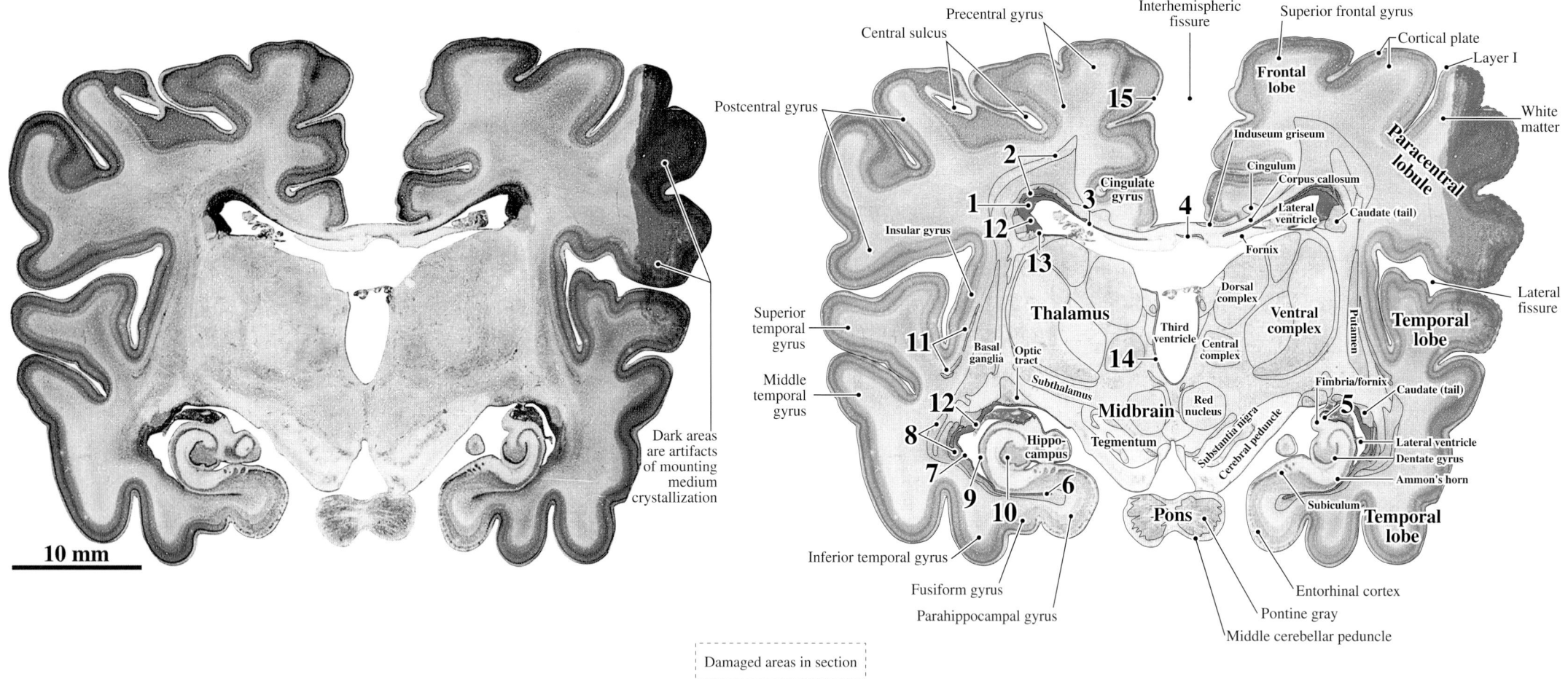

# PLATE 12

**GW37 Coronal**
**CR 350 mm**
**Y217-65**
**Level 12: Section 1021**

*Remnants of the germinal matrix, migratory streams, and transitional fields*

| | | | |
|---|---|---|---|
| 1 | Paracentral neuroepithelium and subventricular zone | 9 | Alvear glioepithelium |
| 2 | Paracentral stratified transitional field | 10 | Subgranular zone (dentate) |
| 3 | Callosal glioepithelium | 11 | Lateral migratory stream (cortical) |
| 4 | Callosal sling | 12 | Posterior striatal neuroepithelium and subventricular zone |
| 5 | Fornical glioepithelium | 13 | Strionuclear glioepithelium |
| 6 | Parahippocampal neuroepithelium, subventricular zone, and stratified transitional field | 14 | Mesencephalic glioepithelium/ependyma |
| 7 | Temporal neuroepithelium and subventricular zone | 15 | Subpial granular layer (cortical) |
| 8 | Temporal stratified transitional field | | |

**See detail of brain core in Plates 38A and B.**

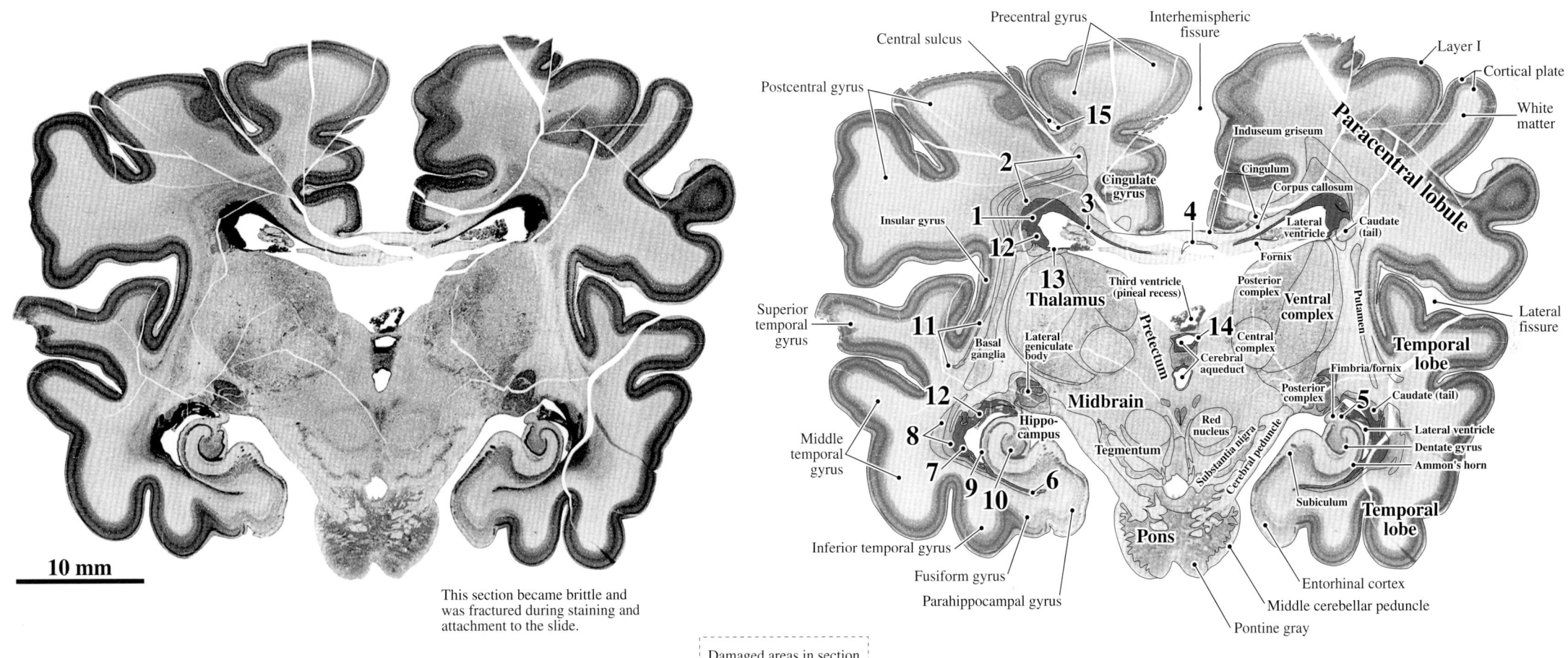

This section became brittle and was fractured during staining and attachment to the slide.

Damaged areas in section

# PLATE 13

**GW37 Coronal**
**CR 350 mm**
**Y217-65**
**Level 13: Section 1081**

See detail of brain core
in Plates 39A and B.

*Remnants of the germinal matrix, migratory streams, and transitional fields*

| | |
|---|---|
| 1 *Paracentral neuroepithelium and subventricular zone* | 8 *Temporal stratified transitional field* |
| 2 *Paracentral stratified transitional field* | 9 *Alvear glioepithelium* |
| 3 *Callosal glioepithelium* | 10 *Subgranular zone (dentate)* |
| 4 *Callosal sling* | 11 *Posterior striatal neuroepithelium and subventricular zone* |
| 5 *Fornical glioepithelium* | 12 *Strionuclear glioepithelium* |
| 6 *Parahippocampal neuroepithelium, subventricular zone, and stratified transitional field* | 13 *Mesencephalic glioepithelium/ependyma* |
| 7 *Temporal neuroepithelium and subventricular zone* | 14 *Subpial granular layer (cortical)* |

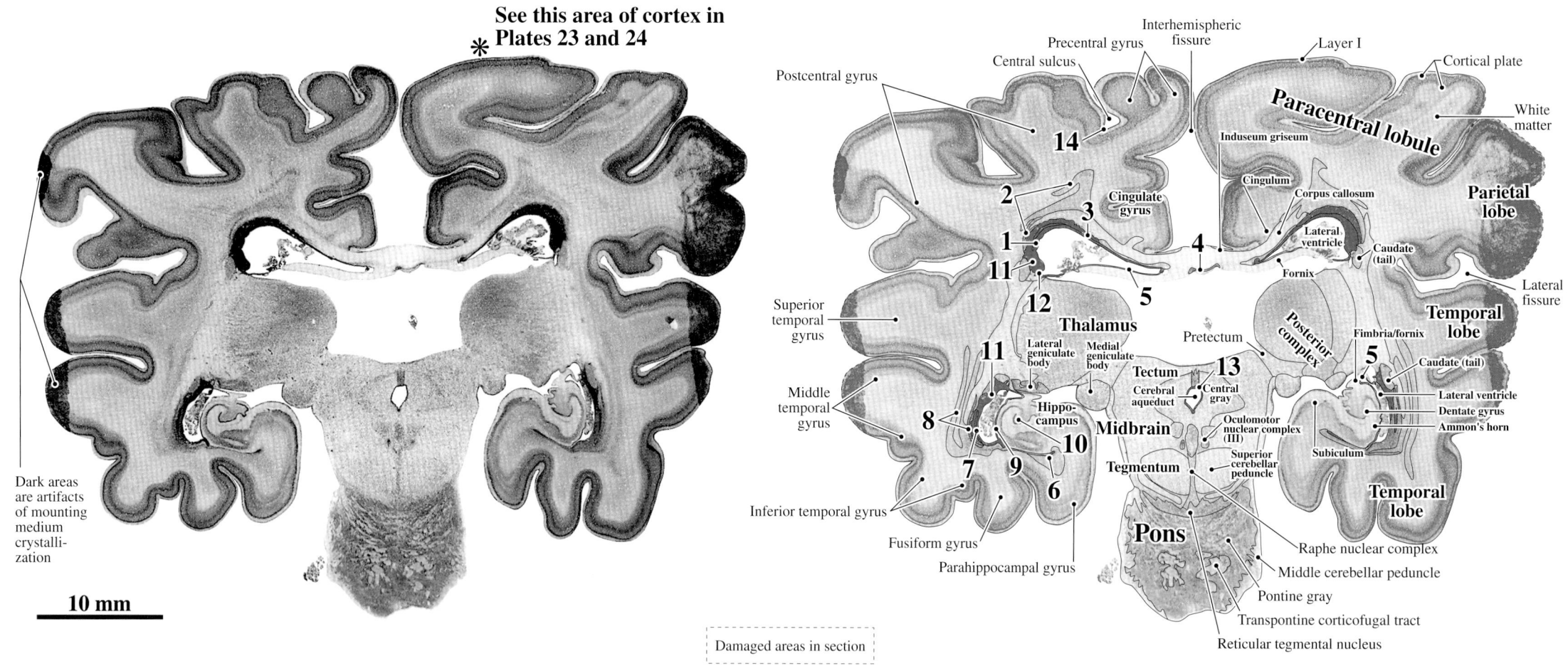

# PLATE 14

**GW37 Coronal
CR 350 mm
Y217-65
Level 14: Section 1141**

### Remnants of the germinal matrix, migratory streams, and transitional fields

1. Paracentral/parietal neuroepithelium and subventricular zone
2. Paracentral/parietal stratified transitional field
3. Callosal glioepithelium
4. Fornical glioepithelium
5. Parahippocampal neuroepithelium, subventricular zone, and stratified transitional field
6. Temporal neuroepithelium and subventricular zone
7. Temporal stratified transitional field
8. Alvear glioepithelium
9. Subgranular zone (dentate)
10. Posterior striatal neuroepithelium and subventricular zone
11. Strionuclear glioepithelium
12. Mesencephalic glioepithelium/ependyma
13. Subpial granular layer (cortical)

**See detail of brain core in Plates 40A and B.**

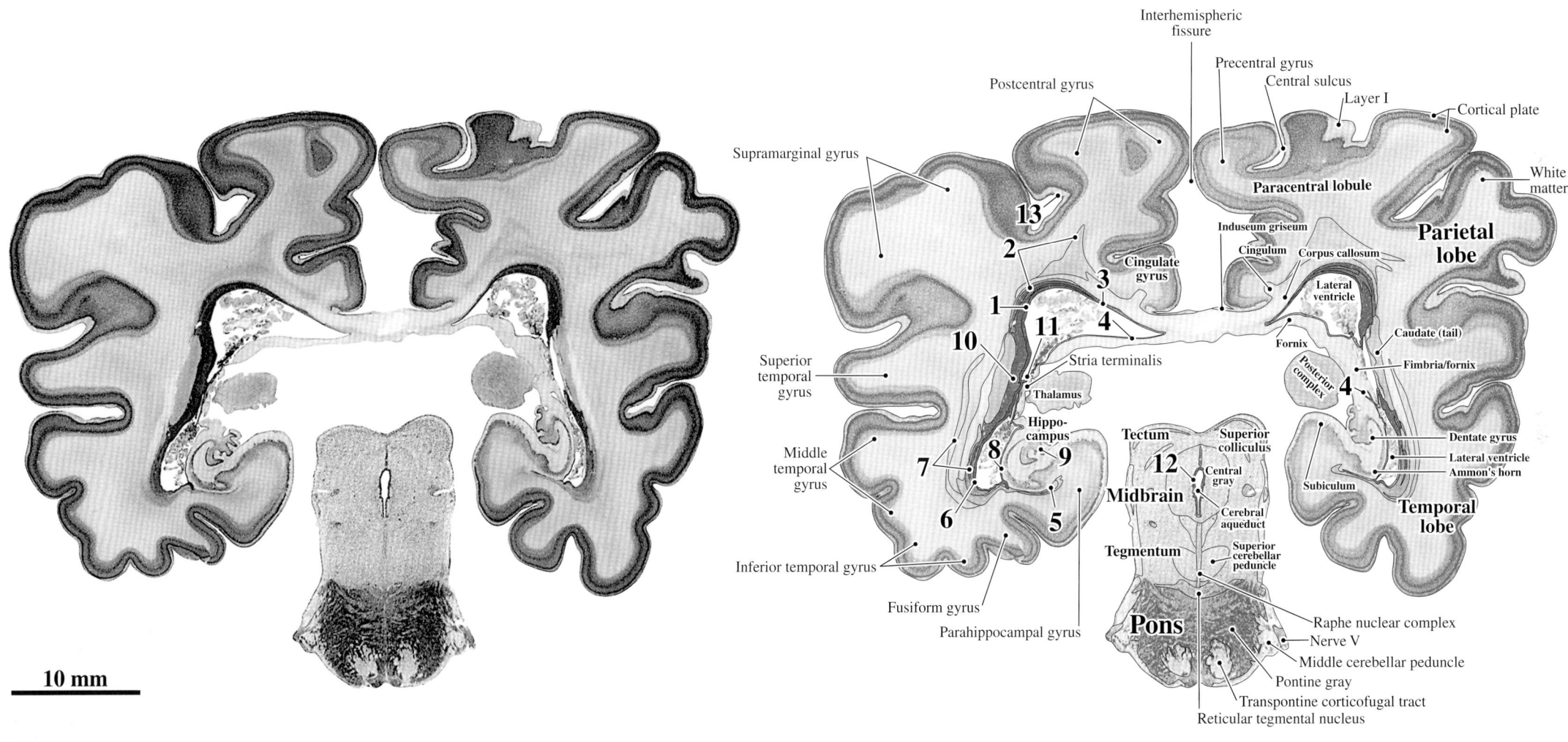

# PLATE 15

GW37 Coronal
CR 350 mm
Y217-65
Level 15: Section 1181

See detail of brain core and cerebellum in Plates 41A and B.

### Remnants of the germinal matrix, migratory streams, and transitional fields

| | |
|---|---|
| 1 *Paracentral/parietal neuroepithelium and subventricular zone* | 7 *Temporal stratified transitional field* |
| 2 *Paracentral/parietal stratified transitional field* | 8 *Alvear glioepithelium* |
| 3 *Callosal glioepithelium* | 9 *Subgranular zone (dentate)* |
| 4 *Fornical glioepithelium* | 10 *Mesencephalic glioepithelium/ependyma* |
| 5 *Parahippocampal neuroepithelium, subventricular zone, and stratified transitional field* | 11 *External germinal layer (cerebellum)* |
| 6 *Temporal neuroepithelium and subventricular zone* | 12 *Subpial granular layer (cortical)* |

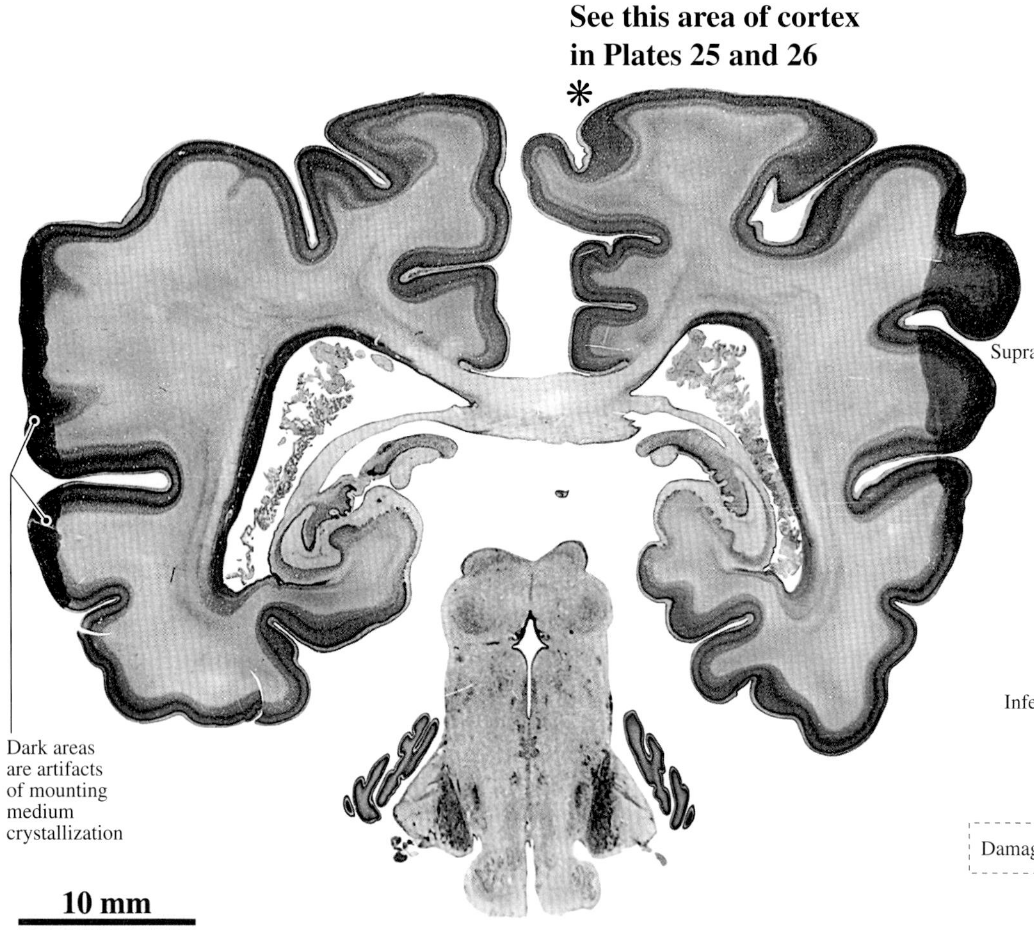

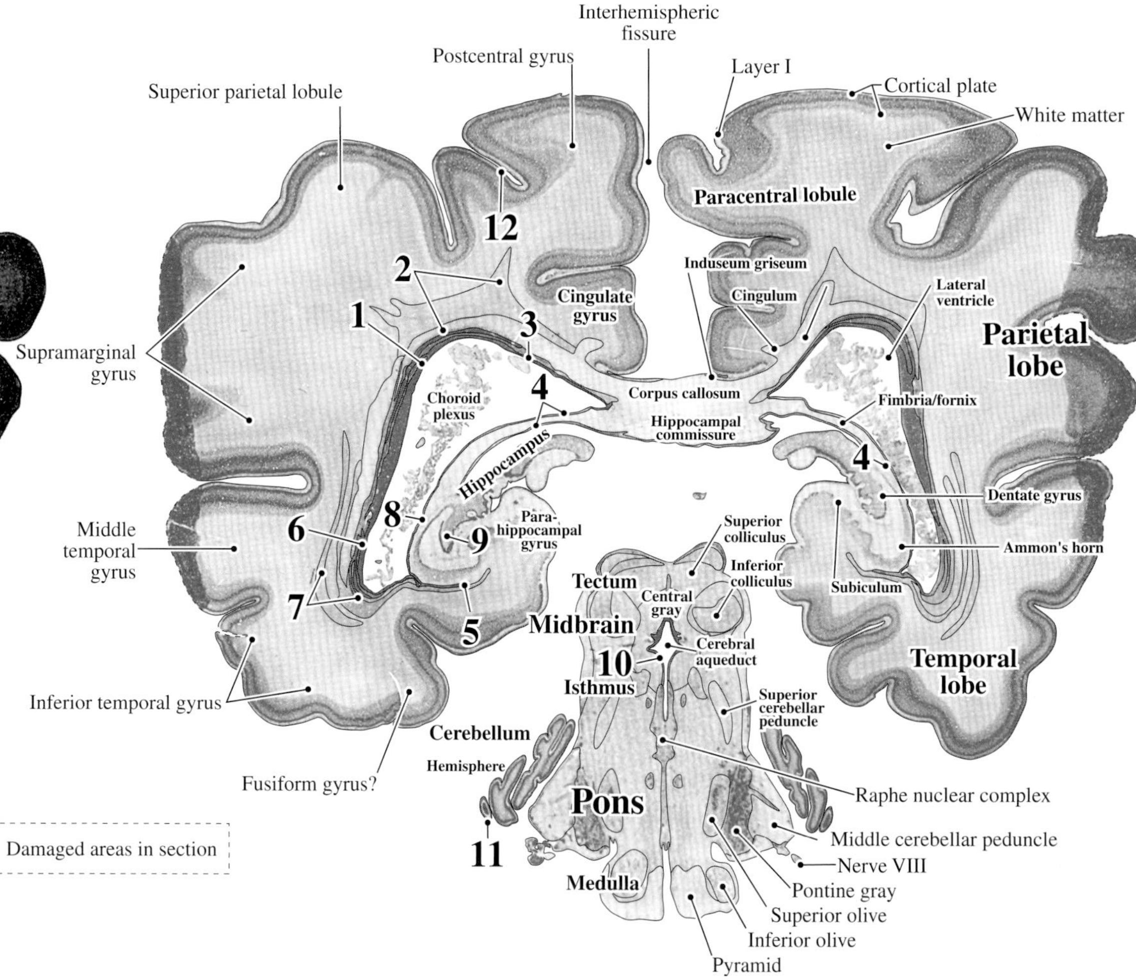

## PLATE 16

**GW37 Coronal**
**CR 350 mm**
**Y217-65**
**Level 16: Section 1221**

### Remnants of the germinal matrix, migratory streams, and transitional fields

1. Paracentral/parietal neuroepithelium and subventricular zone
2. Paracentral/parietal stratified transitional field
3. Callosal glioepithelium
4. Fornical glioepithelium
5. Parahippocampal neuroepithelium, subventricular zone, and stratified transitional field
6. Temporal neuroepithelium and subventricular zone
7. Temporal stratified transitional field
8. Alvear glioepithelium
9. Mesencephalic glioepithelium/ependyma
10. External germinal layer (cerebellum)
11. Subpial granular layer (cortical)

See detail of brain core and cerebellum in Plates 42A and B.

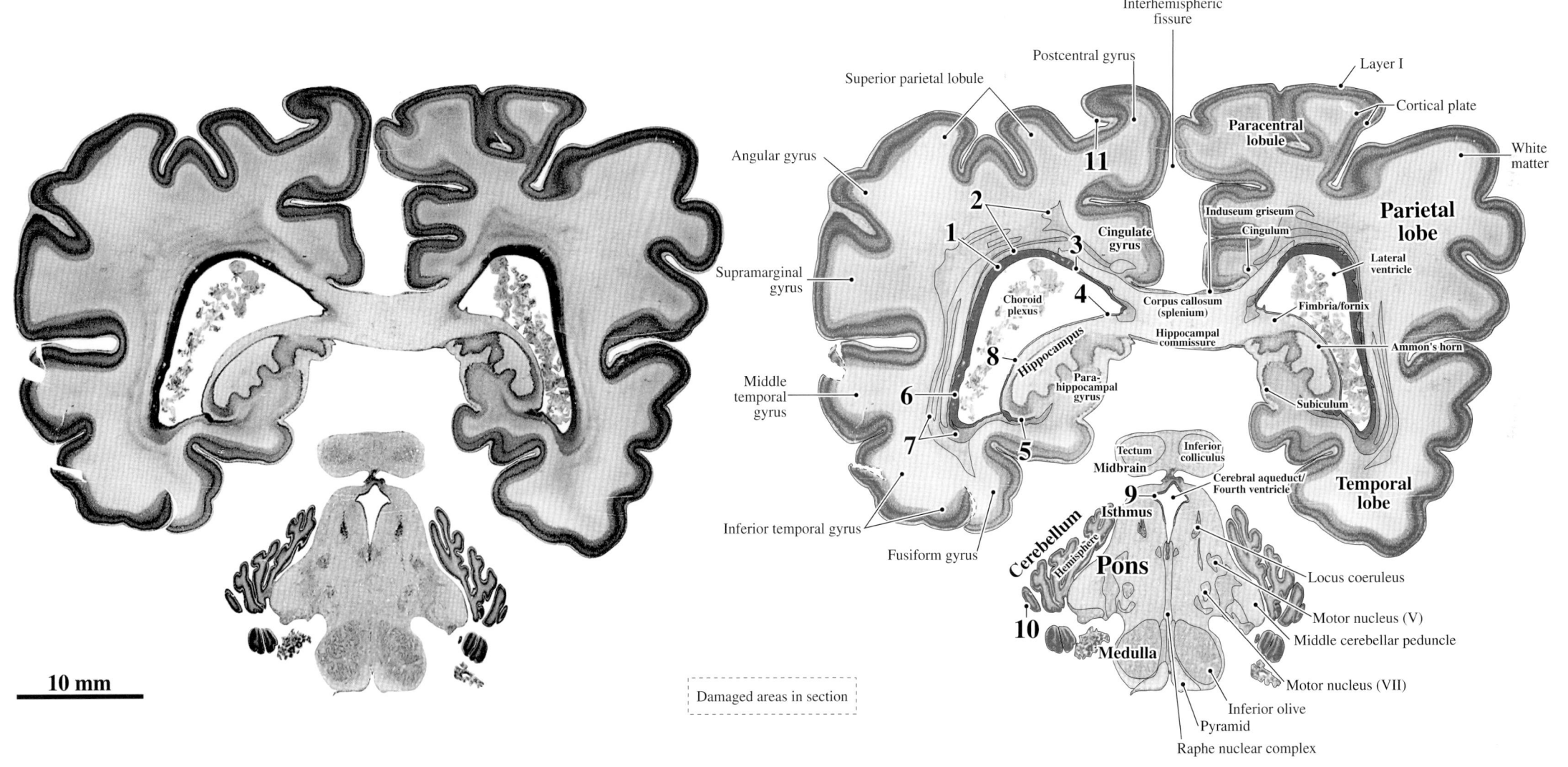

# PLATE 17

**GW37 Coronal**
**CR 350 mm**
**Y217-65**
**Level 17: Section 1311**

**See detail of brain core and cerebellum in Plates 43A and B.**

*Remnants of the germinal matrix, migratory streams, and transitional fields*

| | |
|---|---|
| 1 *Parietal neuroepithelium and subventricular zone* | 7 *Temporal neuroepithelium and subventricular zone* |
| 2 *Parietal stratified transitional field* | 8 *Temporal stratified transitional field* |
| 3 *Callosal glioepithelium* | 9 *Medullary glioepithelium/ependyma* |
| 4 *Cingulate neuroepithelium, subventricular zone, and stratified transitional field* | 10 *Pontine glioepithelium/ependyma* |
| 5 *Occipital neuroepithelium and subventricular zone* | 11 *External germinal layer (cerebellum)* |
| 6 *Occipital stratified transitional field* | 12 *Subpial granular layer (cortical)* |

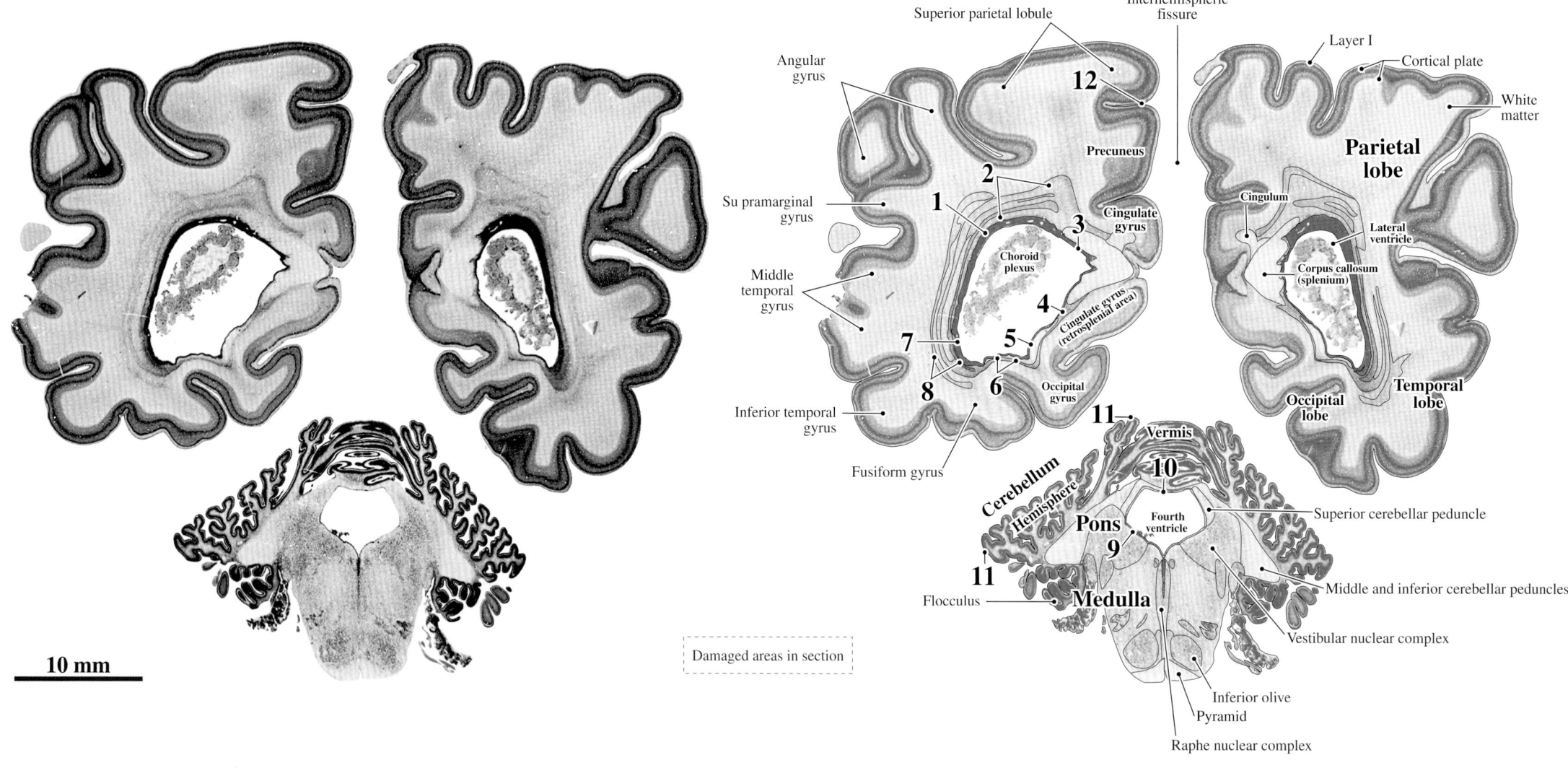

# PLATE 18

**GW37 Coronal**
**CR 350 mm**
**Y217-65**
**Level 18: Section 1371**

*Remnants of the germinal matrix, migratory streams, and transitional fields*

| | | | |
|---|---|---|---|
| 1 | Parietal neuroepithelium and subventricular zone | 7 | Temporal neuroepithelium and subventricular zone |
| 2 | Parietal stratified transitional field | 8 | Temporal stratified transitional field |
| 3 | Callosal glioepithelium (intermingled with cingulate neuroepithelium and subventricular zone) | 9 | External germinal layer (cerebellum) |
| 4 | Cingulate stratified transitional field | 10 | Germinal trigone (cerebellum) |
| 5 | Occipital neuroepithelium and subventricular zone | 11 | Medullary glioepithelium/ependyma |
| 6 | Occipital stratified transitional field | 12 | Subpial granular layer (cortical) |

See detail of brain core and cerebellum in Plates 44A and B.

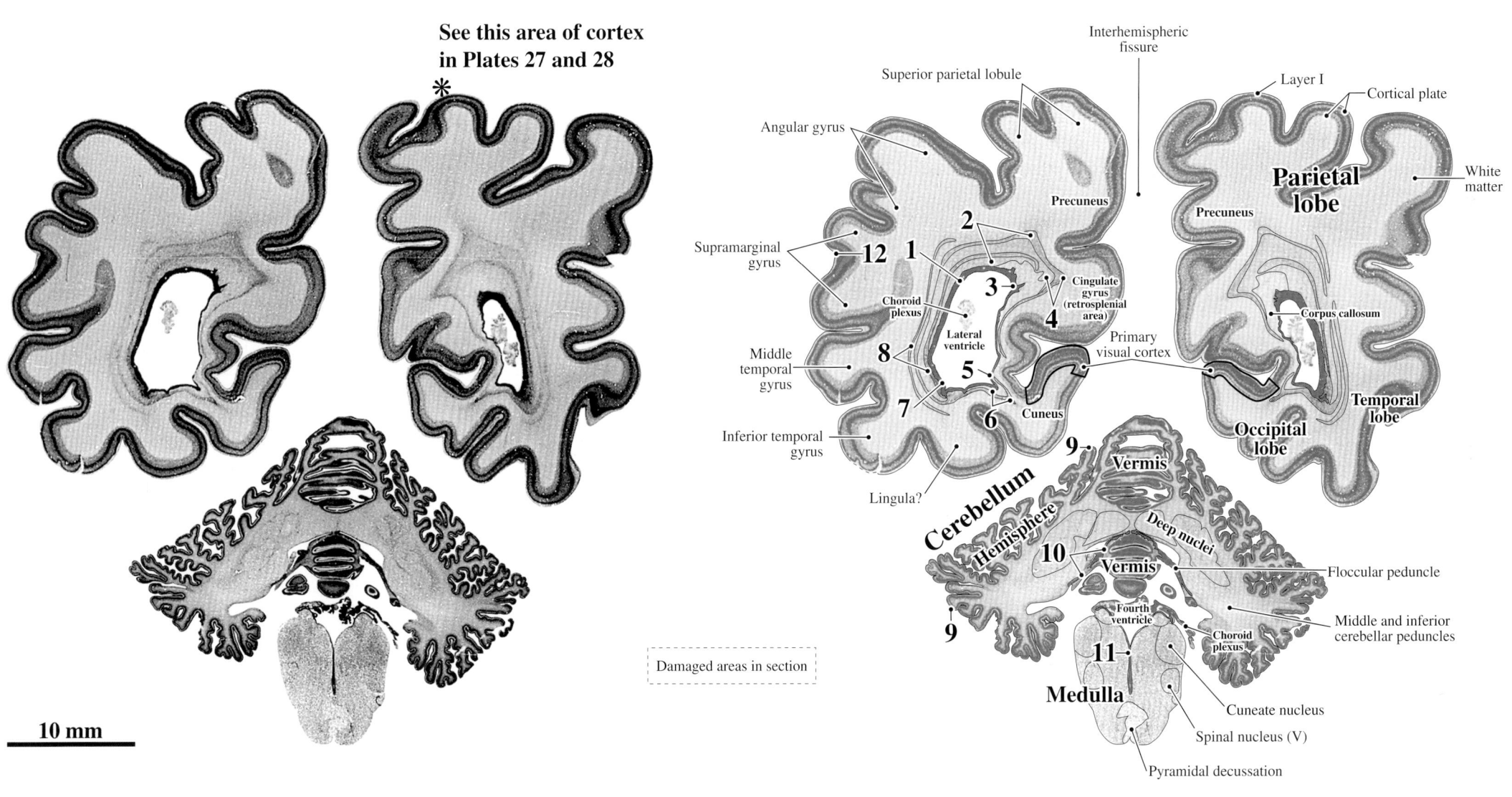

# PLATE 19

GW37 Coronal
CR 350 mm
Y217-65
Level 19: Section 1501

*Remnants of the germinal matrix, migratory streams, and transitional fields*

| 1 *Parietal neuroepithelium and subventricular zone* | 5 *Temporal neuroepithelium and subventricular zone* |
|---|---|
| 2 *Parietal stratified transitional field* | 6 *Temporal stratified transitional field* |
| 3 *Occipital neuroepithelium and subventricular zone* | 7 *External germinal layer (cerebellum)* |
| 4 *Occipital stratified transitional field* | 8 *Subpial granular layer (cortical)* |

**See detail of cerebellum in Plates 45A and B.**

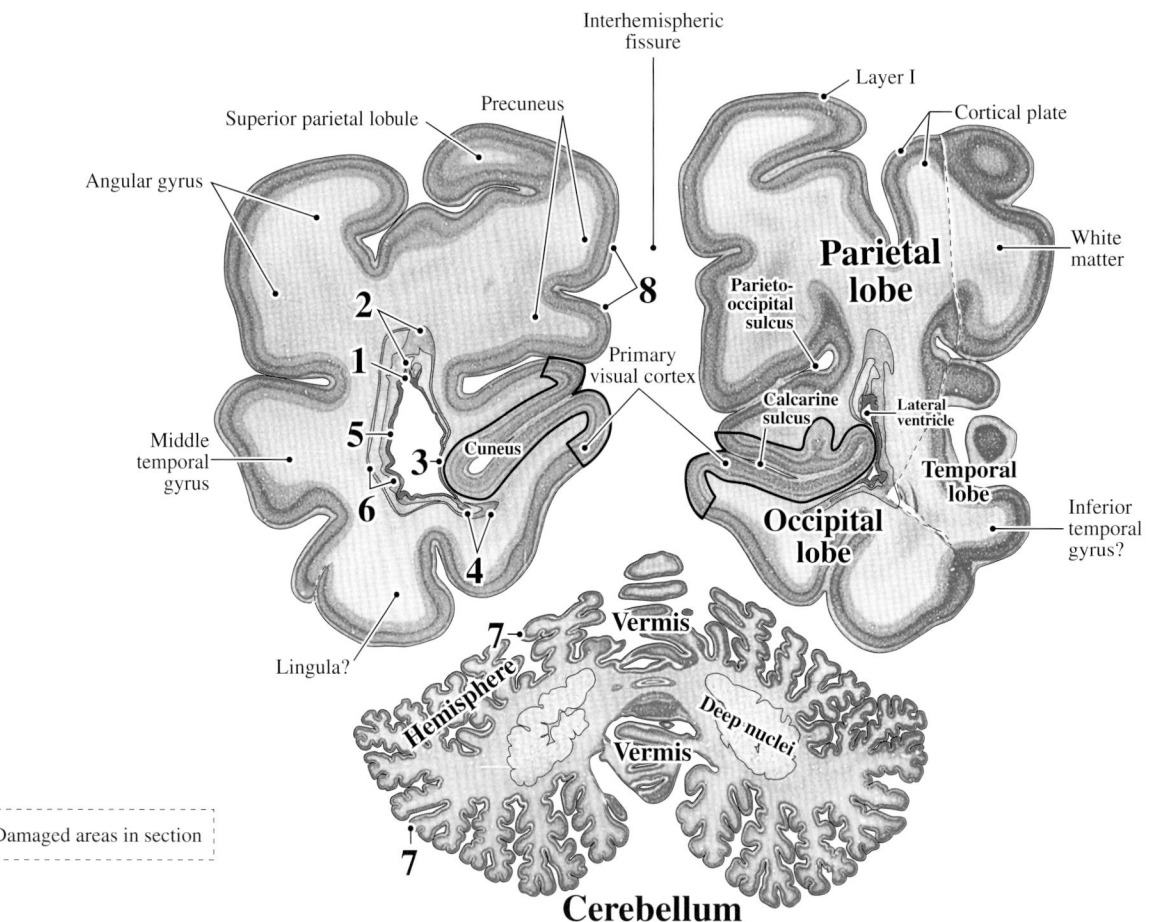

# PLATE 20

GW37 Coronal, CR 350 mm, Y217-65

### Remnants of the germinal matrix, migratory streams, and transitional fields

1 *Parietal neuroepithelium and subventricular zone*
2 *Parietal stratified transitional field*
3 *Occipital neuroepithelium and subventricular zone*
4 *Occipital stratified transitional field*
5 *Temporal neuroepithelium and subventricular zone*
6 *Temporal stratified transitional field*
7 *External germinal layer (cerebellum)*
8 *Subpial granular layer (cortical)*

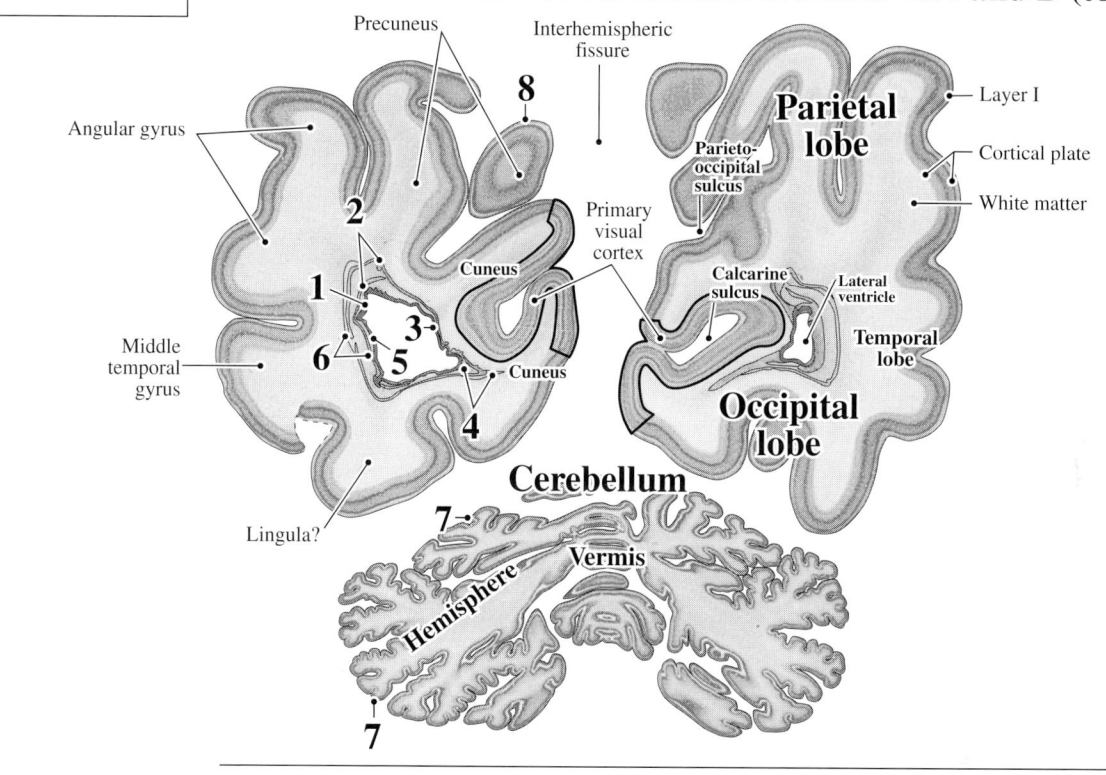

Level 20: Section 1611
See detail of cerebellum in Plates 46A and B (top).

Level 21: Section 1711
See detail of cerebellum in Plates 46A and B (bottom).

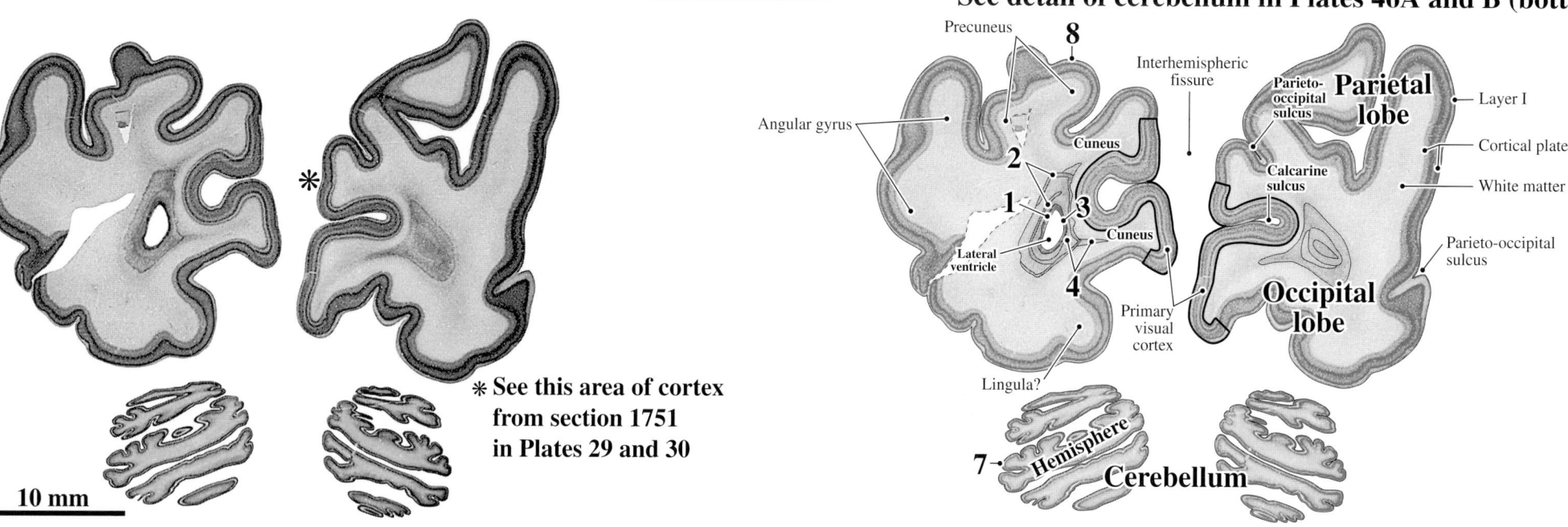

∗ See this area of cortex from section 1751 in Plates 29 and 30

**PLATE 21**

GW37 Coronal
CR 350 mm
Y217-65
Level 8: Section 761
FRONTAL CORTEX

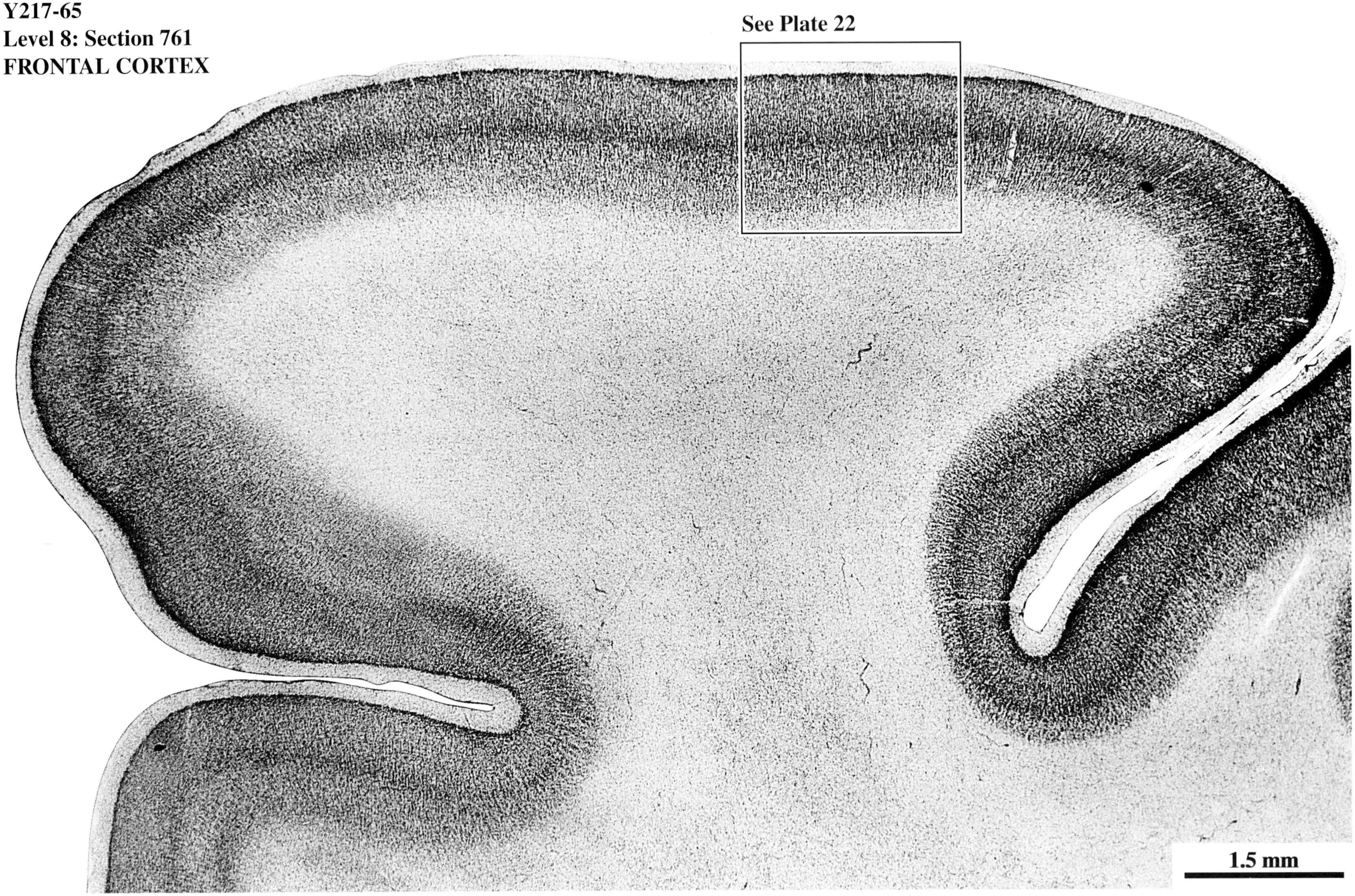

See Plate 22

1.5 mm

See the entire section in Plate 8. For detail, see Plate 22.

**PLATE 22**

**FRONTAL CORTEX**

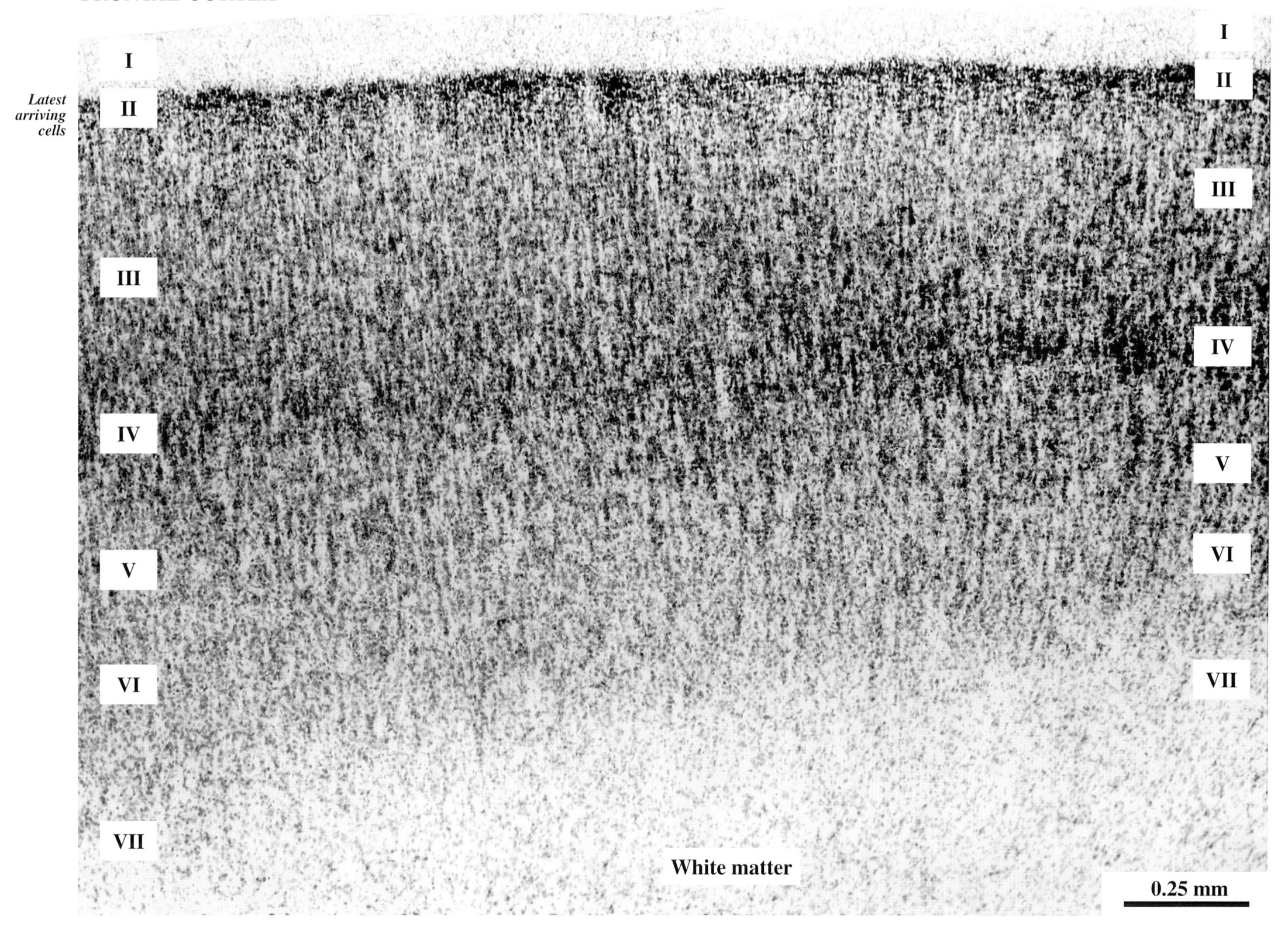

*Latest arriving cells*

I
II
III
IV
V
VI
VII

I
II
III
IV
V
VI
VII

White matter

0.25 mm

## PLATE 23

**GW37 Coronal**
**CR 350 mm**
**Y217-65**
**Level 13: Section 1081**
**CORTEX OF THE PRECENTRAL GYRUS**
**(Primary motor cortex)**

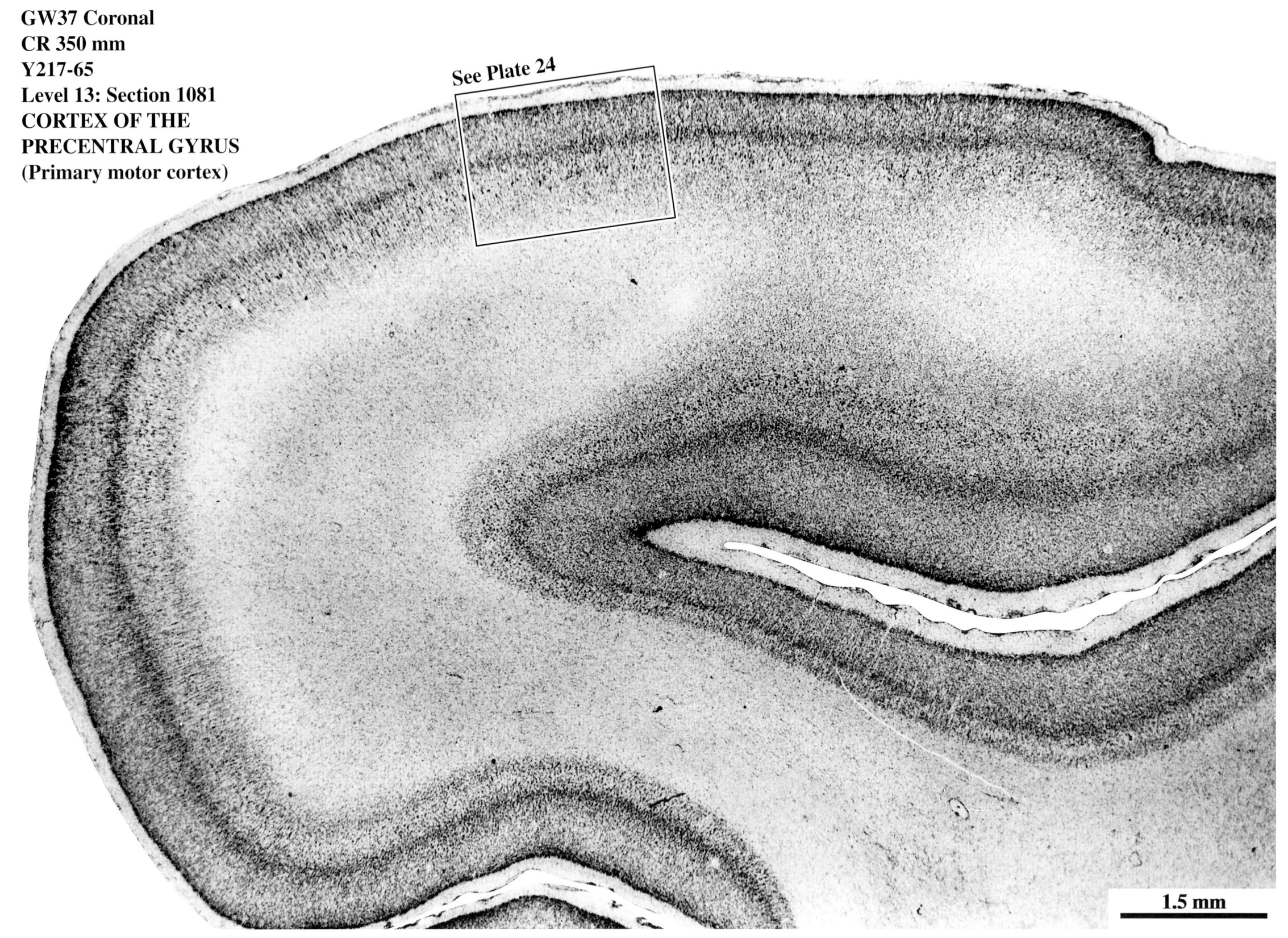

See Plate 24

1.5 mm

See the entire section in Plate 14. For detail, see Plate 24.

**PLATE 24**

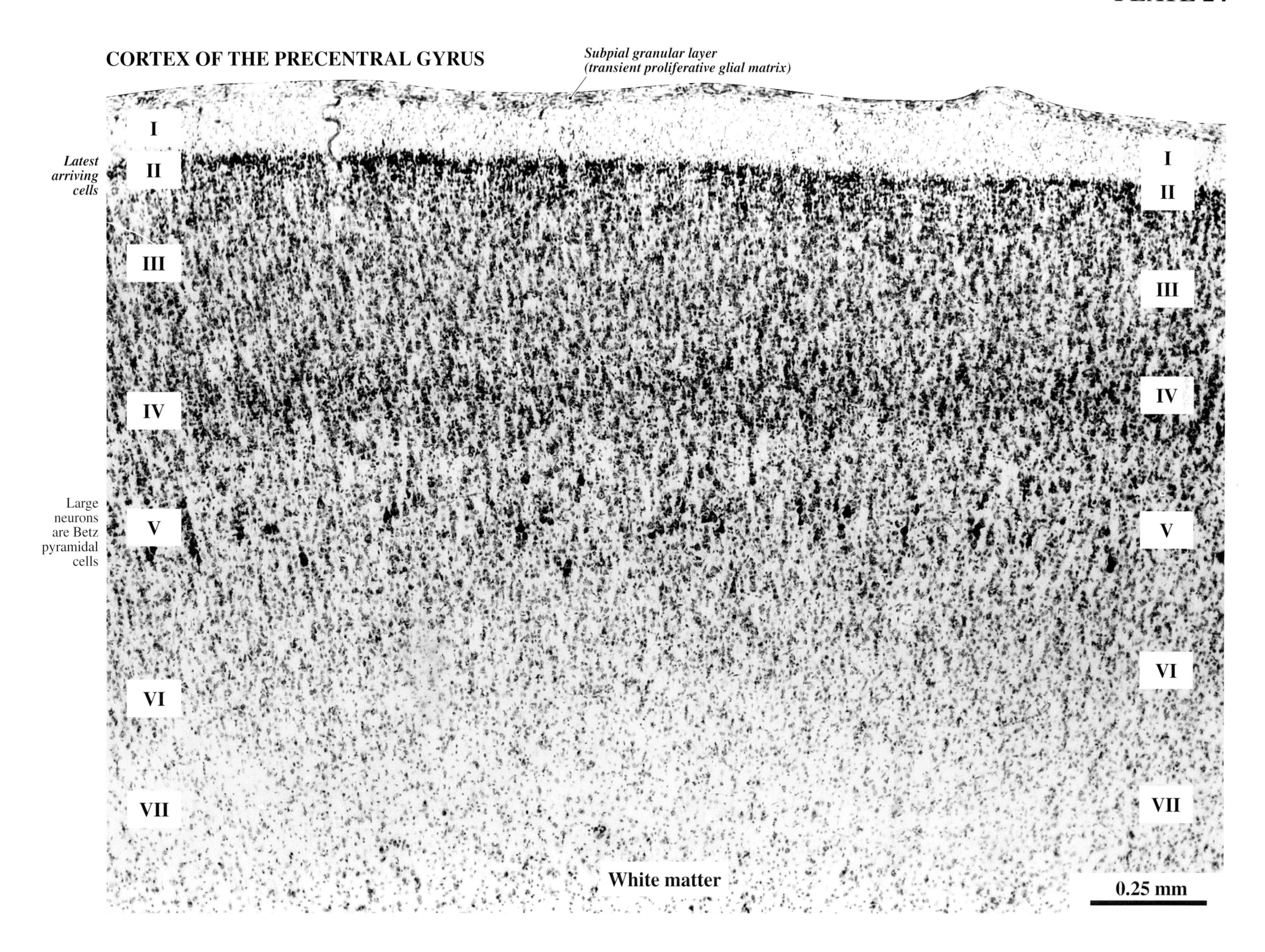

**CORTEX OF THE PRECENTRAL GYRUS**

*Subpial granular layer (transient proliferative glial matrix)*

*Latest arriving cells*

*Large neurons are Betz pyramidal cells*

I, II, III, IV, V, VI, VII

White matter

0.25 mm

## PLATE 25

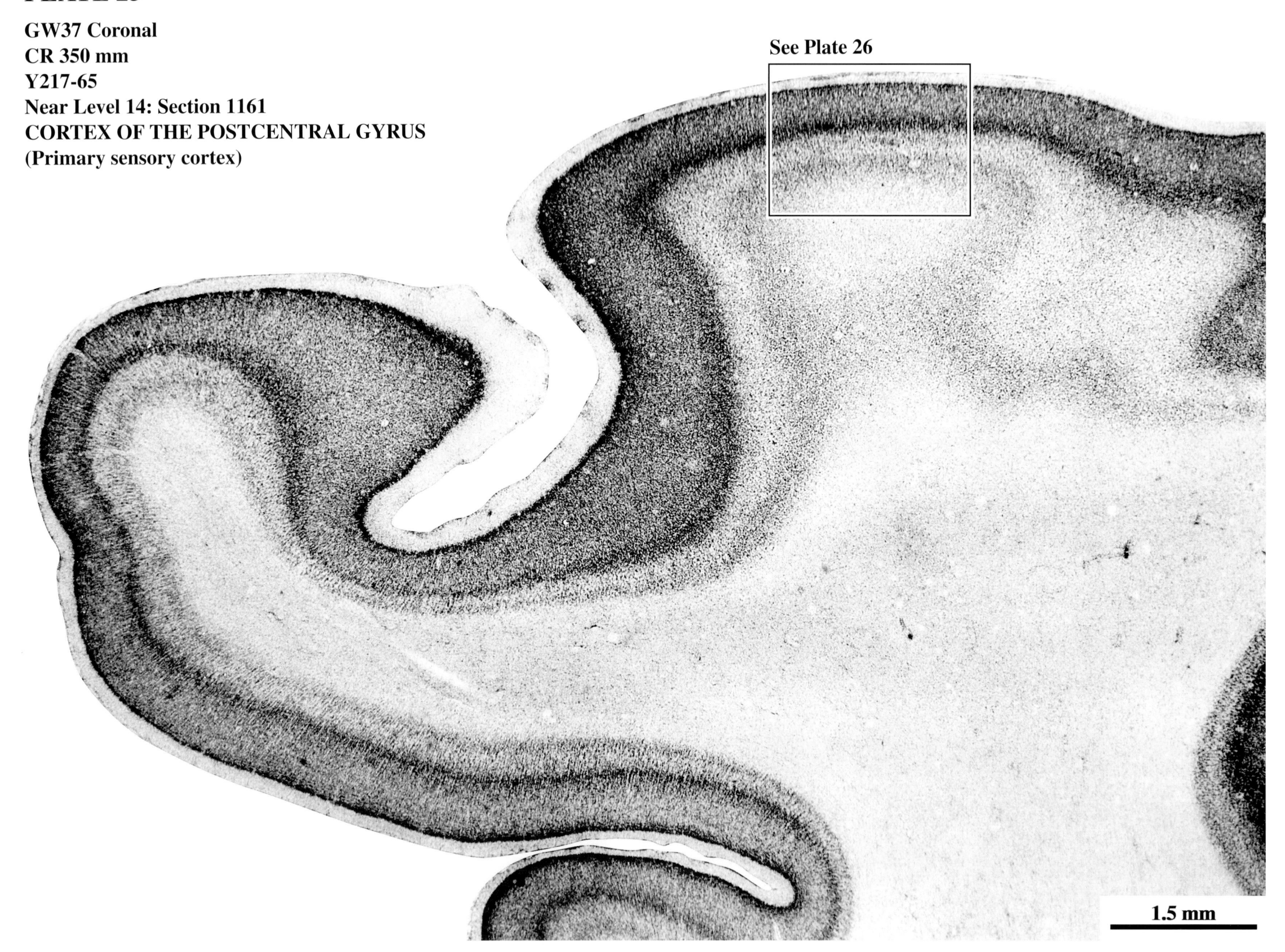

**GW37 Coronal**
**CR 350 mm**
**Y217-65**
**Near Level 14: Section 1161**
**CORTEX OF THE POSTCENTRAL GYRUS**
(Primary sensory cortex)

See Plate 26

1.5 mm

PLATE 26

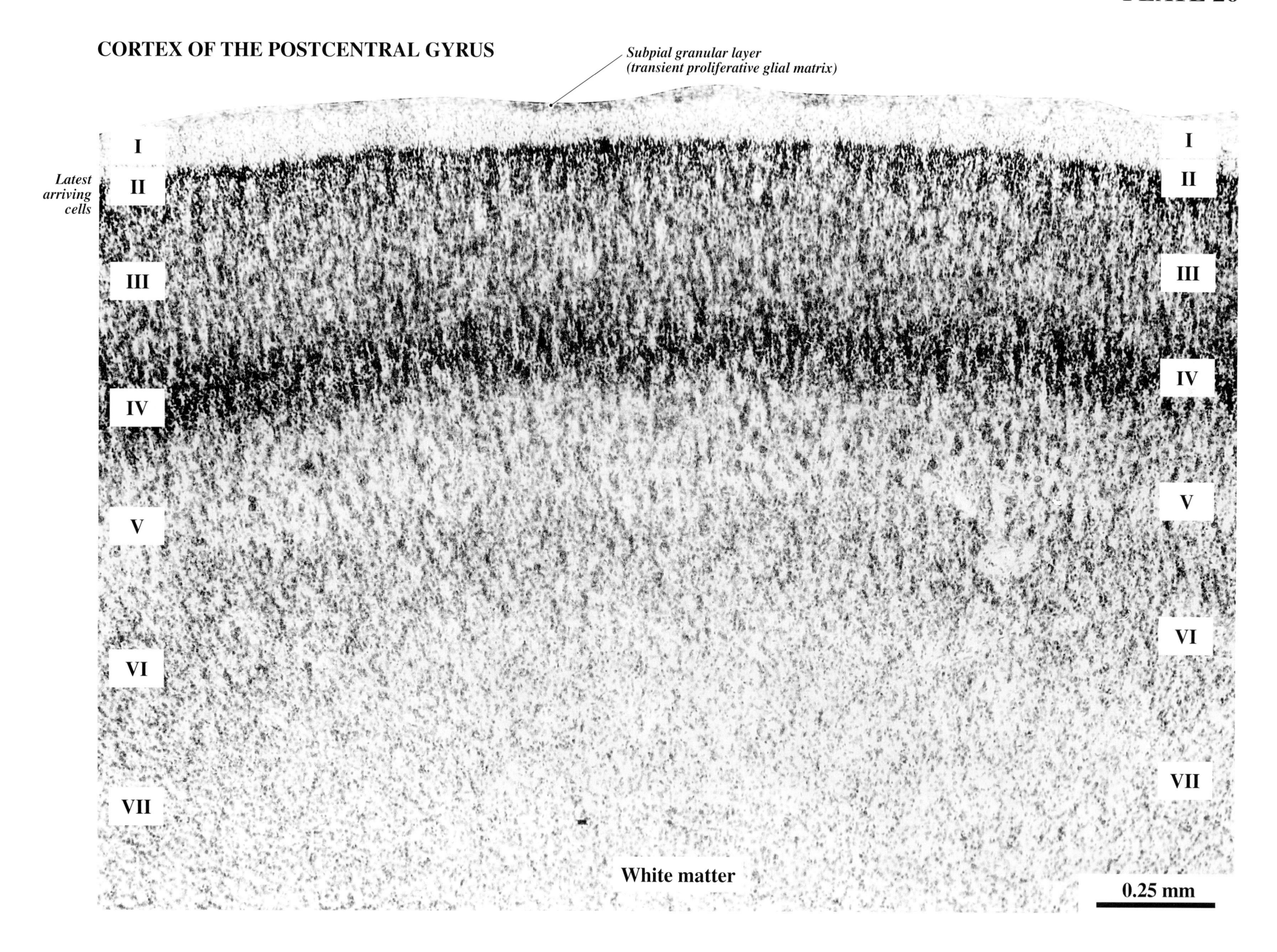

**CORTEX OF THE POSTCENTRAL GYRUS**   *Subpial granular layer (transient proliferative glial matrix)*

I
II
*Latest arriving cells*
III
IV
V
VI
VII

I
II
III
IV
V
VI
VII

White matter

0.25 mm

## PLATE 27

GW37 Coronal
CR 350 mm
Y217-65
Level 18: Section 1371
**PARIETAL CORTEX**

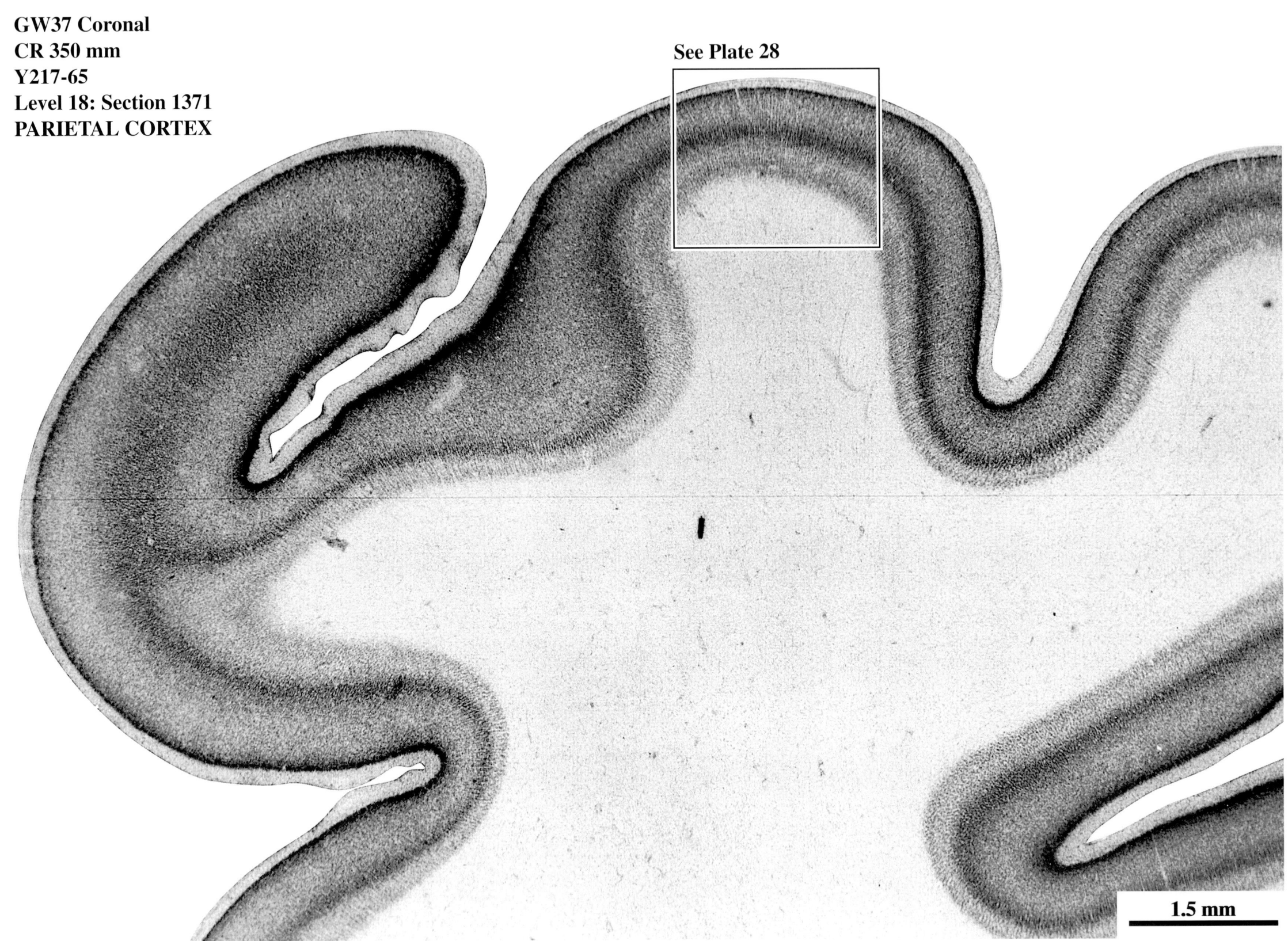

See Plate 28

1.5 mm

See the entire section in Plate 18. For detail, see Plate 28.

PLATE 28

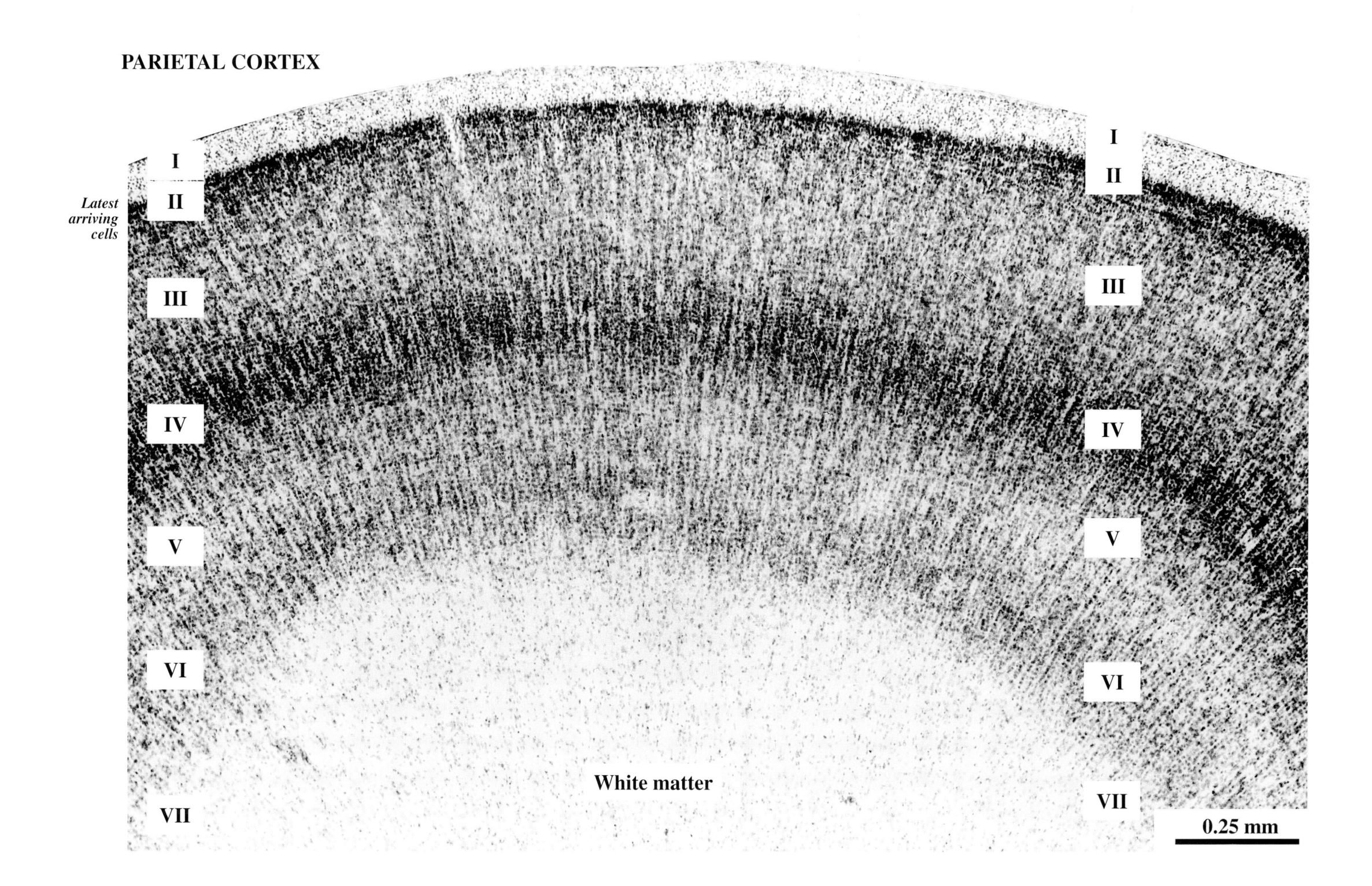

**PARIETAL CORTEX**

I
*Latest arriving cells* II
III
IV
V
VI
VII

I
II
III
IV
V
VI
VII

White matter

0.25 mm

## PLATE 29

GW37 Coronal
CR 350 mm
Y217-65
Near Level 21: Section 1751
**STRIATE/PERISTRIATE
CORTICAL TRANSITION AREA**

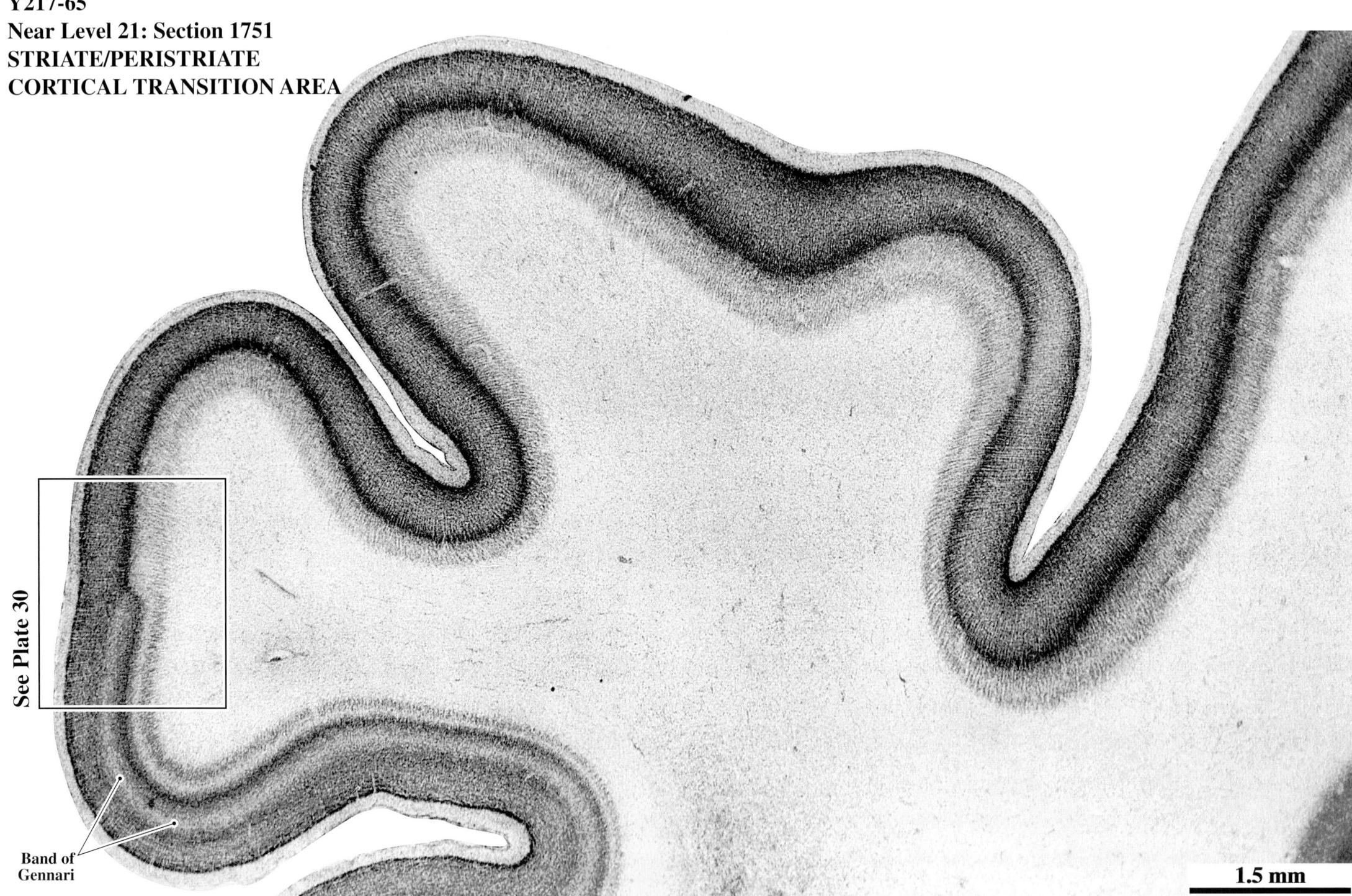

See Plate 30

Band of Gennari

1.5 mm

PLATE 30

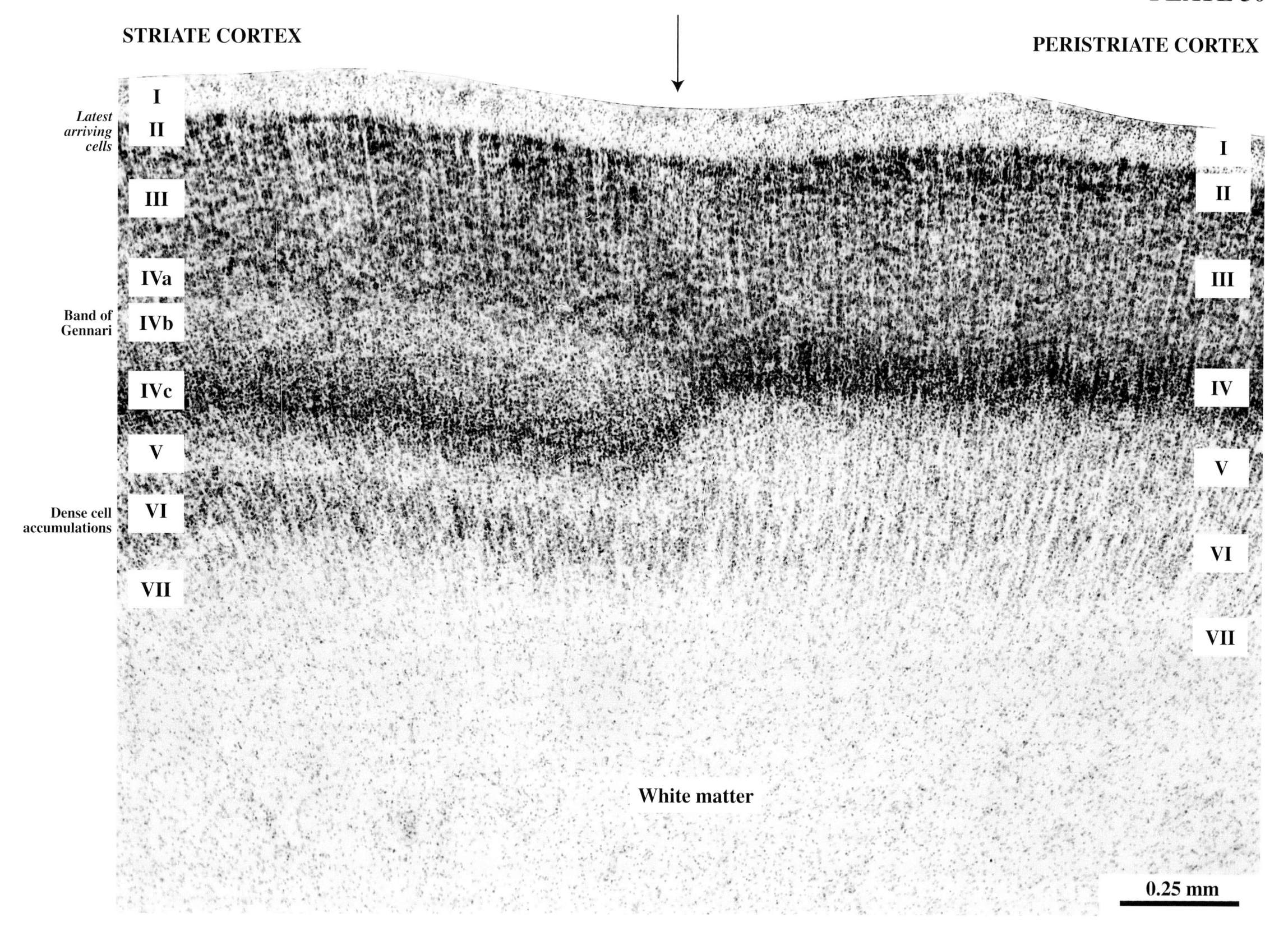

STRIATE CORTEX

PERISTRIATE CORTEX

I
II
*Latest arriving cells*

III

IVa
**Band of Gennari**
IVb

IVc

V

**Dense cell accumulations**
VI

VII

I
II
III
IV
V
VI
VII

White matter

0.25 mm

## PLATE 31A

**GW37 Coronal
CR 350 mm
Y217-65
Level 5:
Section 611**

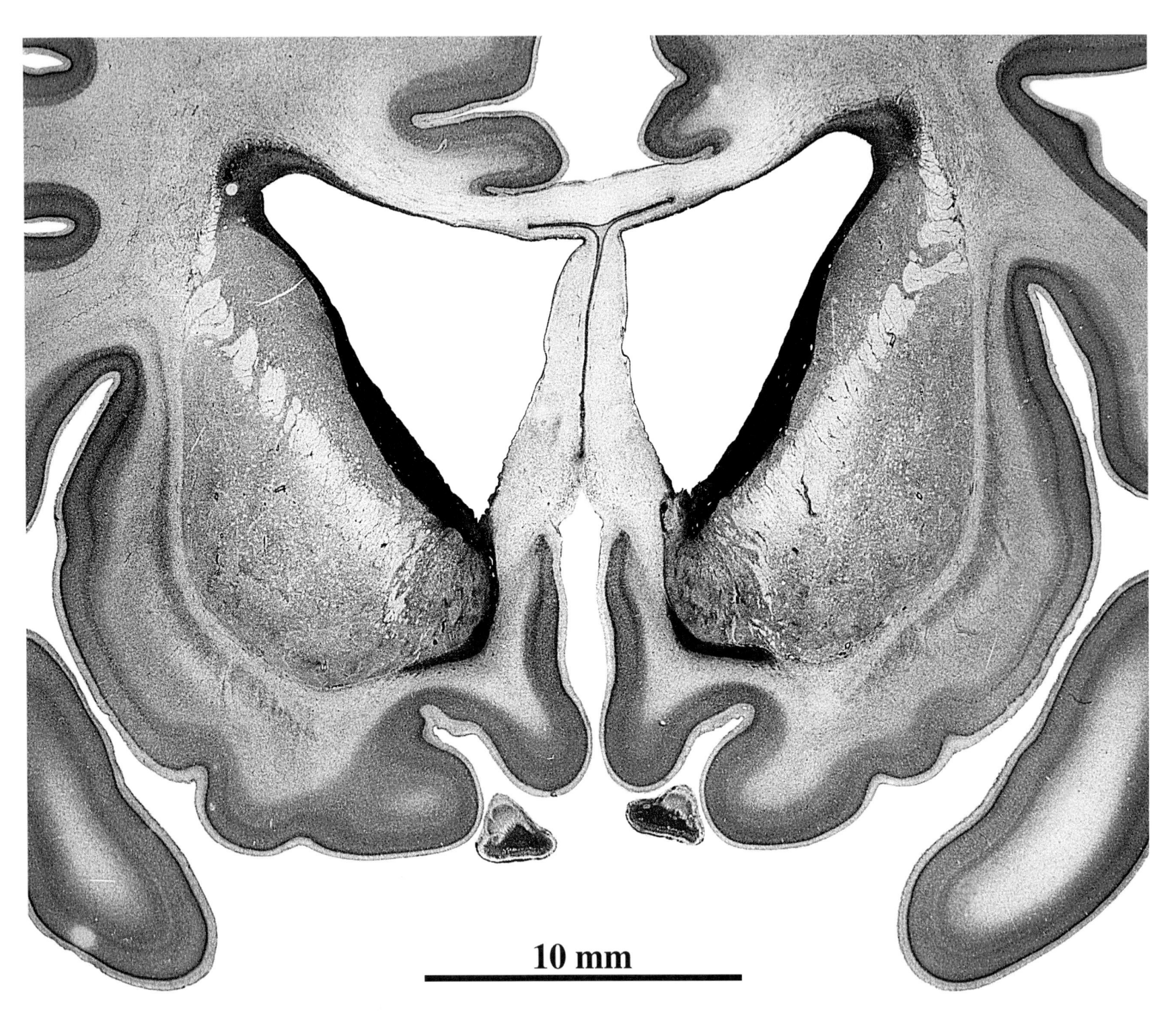

10 mm

See the entire section in Plate 5.

**PLATE 31B**

**Germinal and transitional structures in *italics***

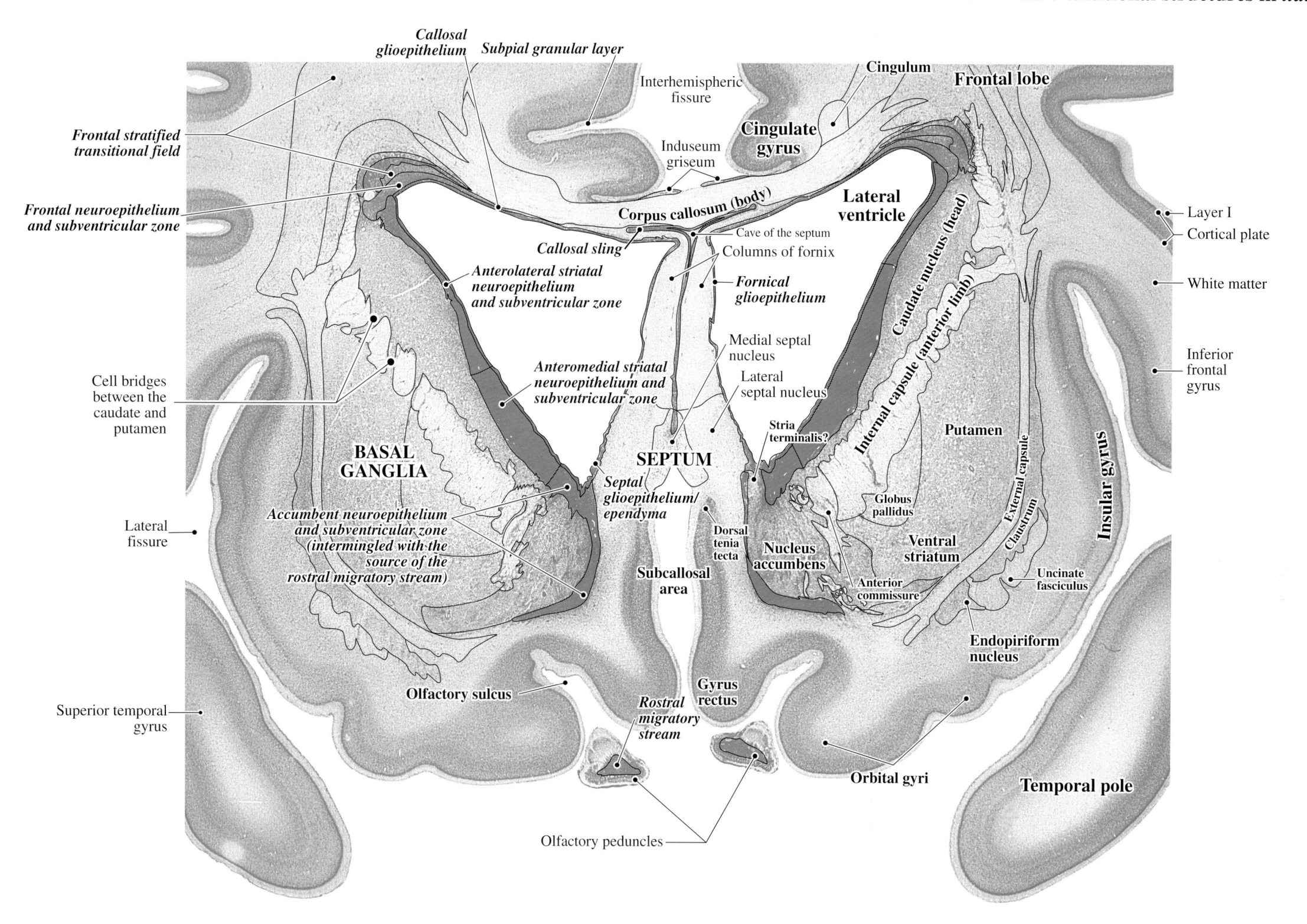

*Callosal glioepithelium* · *Subpial granular layer* · Interhemispheric fissure · Cingulum · **Frontal lobe**

*Frontal stratified transitional field*

*Frontal neuroepithelium and subventricular zone*

Induseum griseum

**Cingulate gyrus**

Corpus callosum (body)

**Lateral ventricle**

Cave of the septum

*Callosal sling* · Columns of fornix

*Anterolateral striatal neuroepithelium and subventricular zone*

*Fornical glioepithelium*

Caudate nucleus (head)

Layer I
Cortical plate
White matter

Cell bridges between the caudate and putamen

*Anteromedial striatal neuroepithelium and subventricular zone*

Medial septal nucleus
Lateral septal nucleus

Internal capsule (anterior limb)

Inferior frontal gyrus

**BASAL GANGLIA**

**SEPTUM**

Stria terminalis?

Putamen

*Septal glioepithelium/ ependyma*

*Accumbent neuroepithelium and subventricular zone (intermingled with the source of the rostral migratory stream)*

Dorsal tenia tecta

**Nucleus accumbens**

Globus pallidus

**Ventral striatum**

External capsule
Claustrum

**Insular gyrus**

Uncinate fasciculus

Lateral fissure

Anterior commissure

**Endopiriform nucleus**

Olfactory sulcus

Subcallosal area

Gyrus rectus

Superior temporal gyrus

*Rostral migratory stream*

Orbital gyri

**Temporal pole**

Olfactory peduncles

**PLATE 32A**

GW37 Coronal
CR 350 mm, Y217-65
Level 6: Section 691

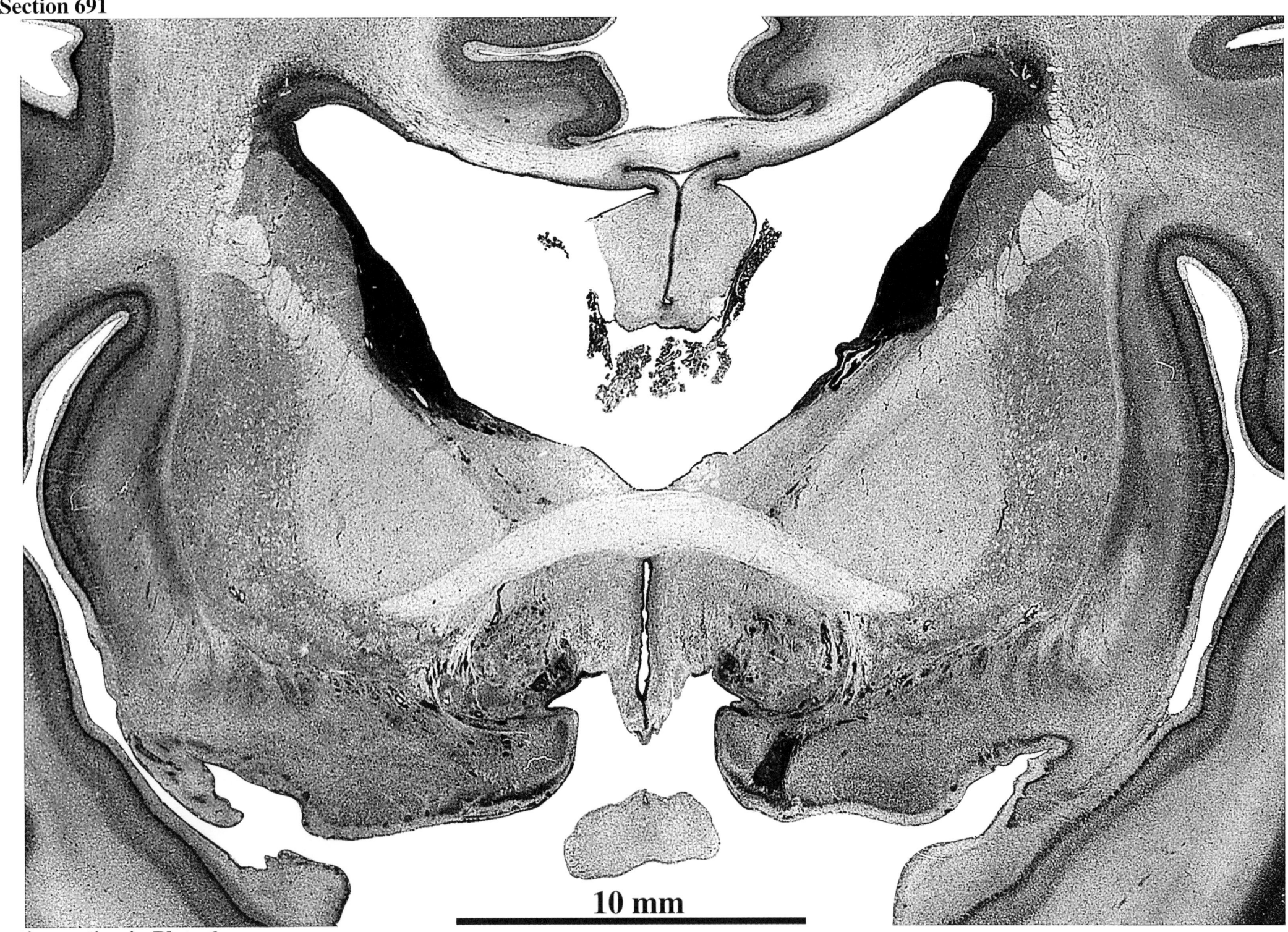

See the entire section in Plate 6.

10 mm

# PLATE 32B

**Germinal and transitional structures in *italics***

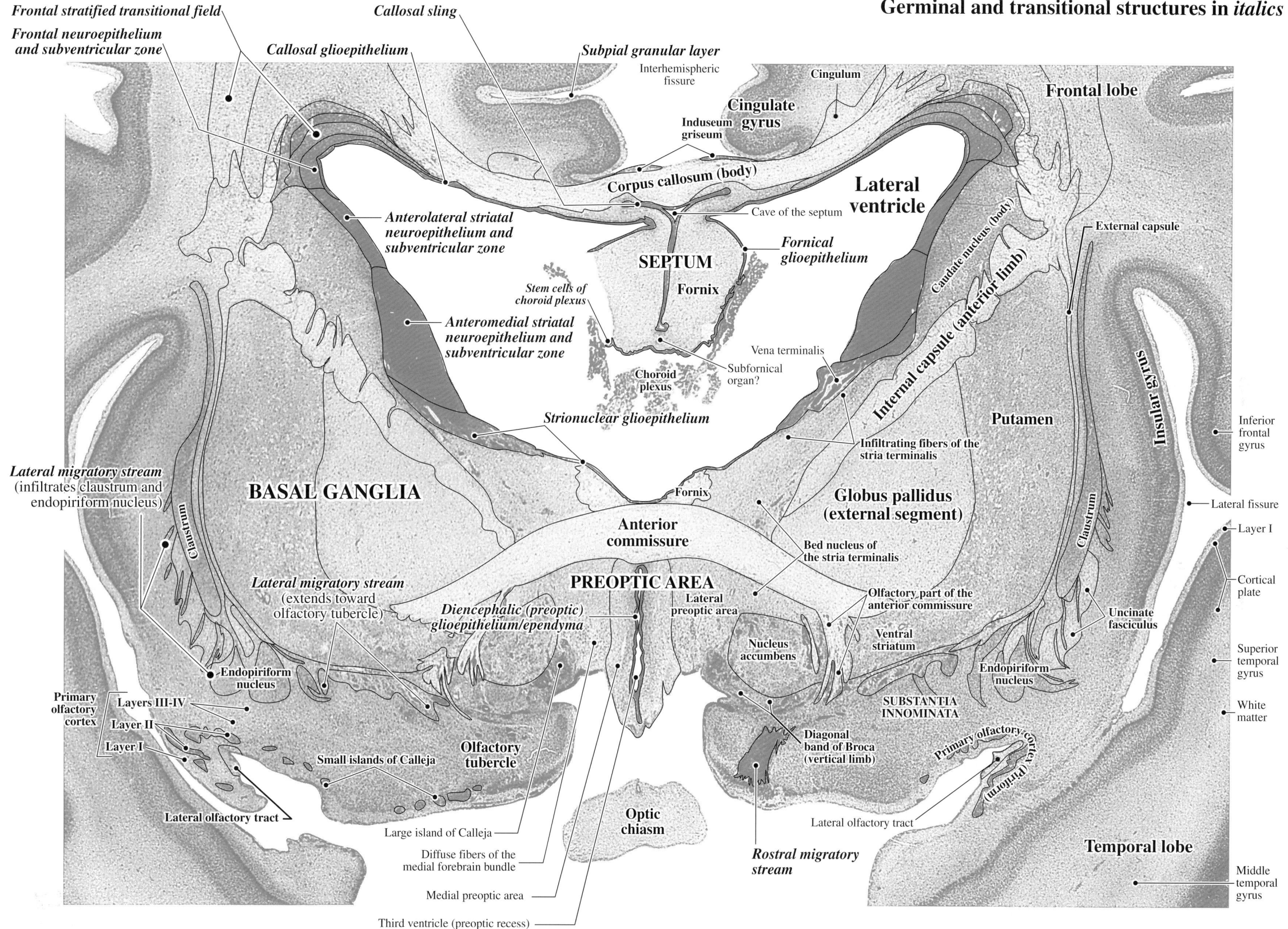

**PLATE 33A**

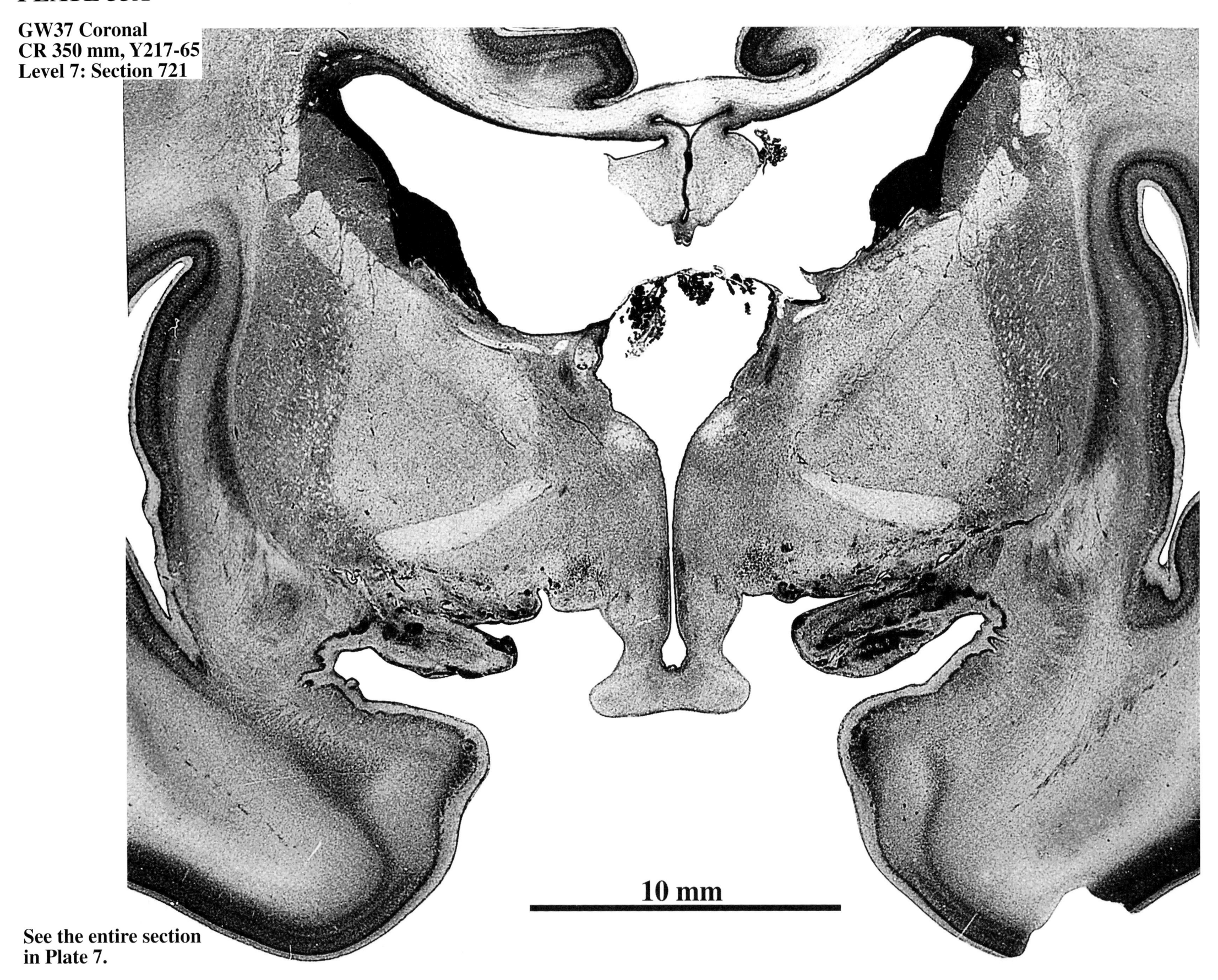

GW37 Coronal
CR 350 mm, Y217-65
Level 7: Section 721

10 mm

See the entire section in Plate 7.

**PLATE 33B**

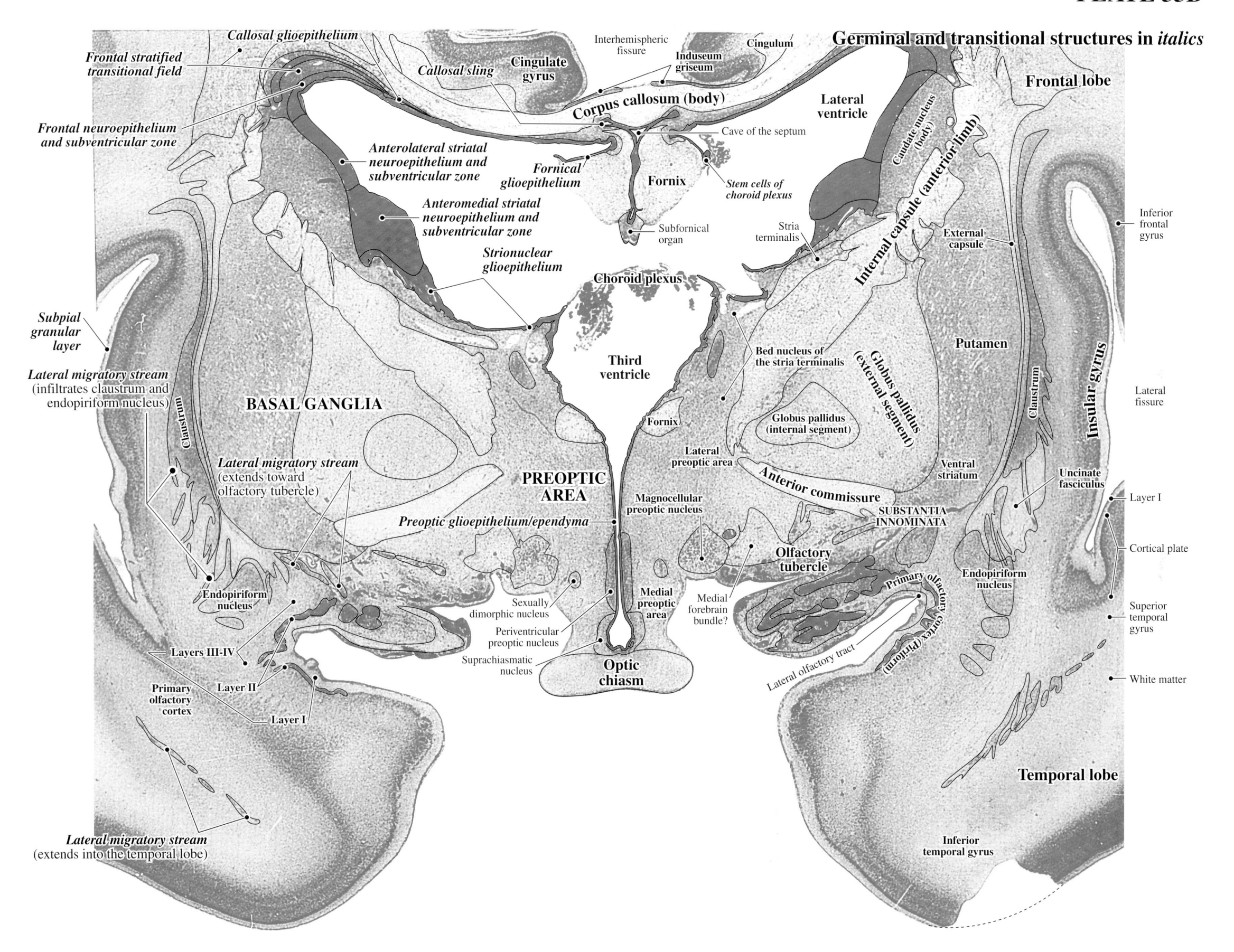

## PLATE 34A

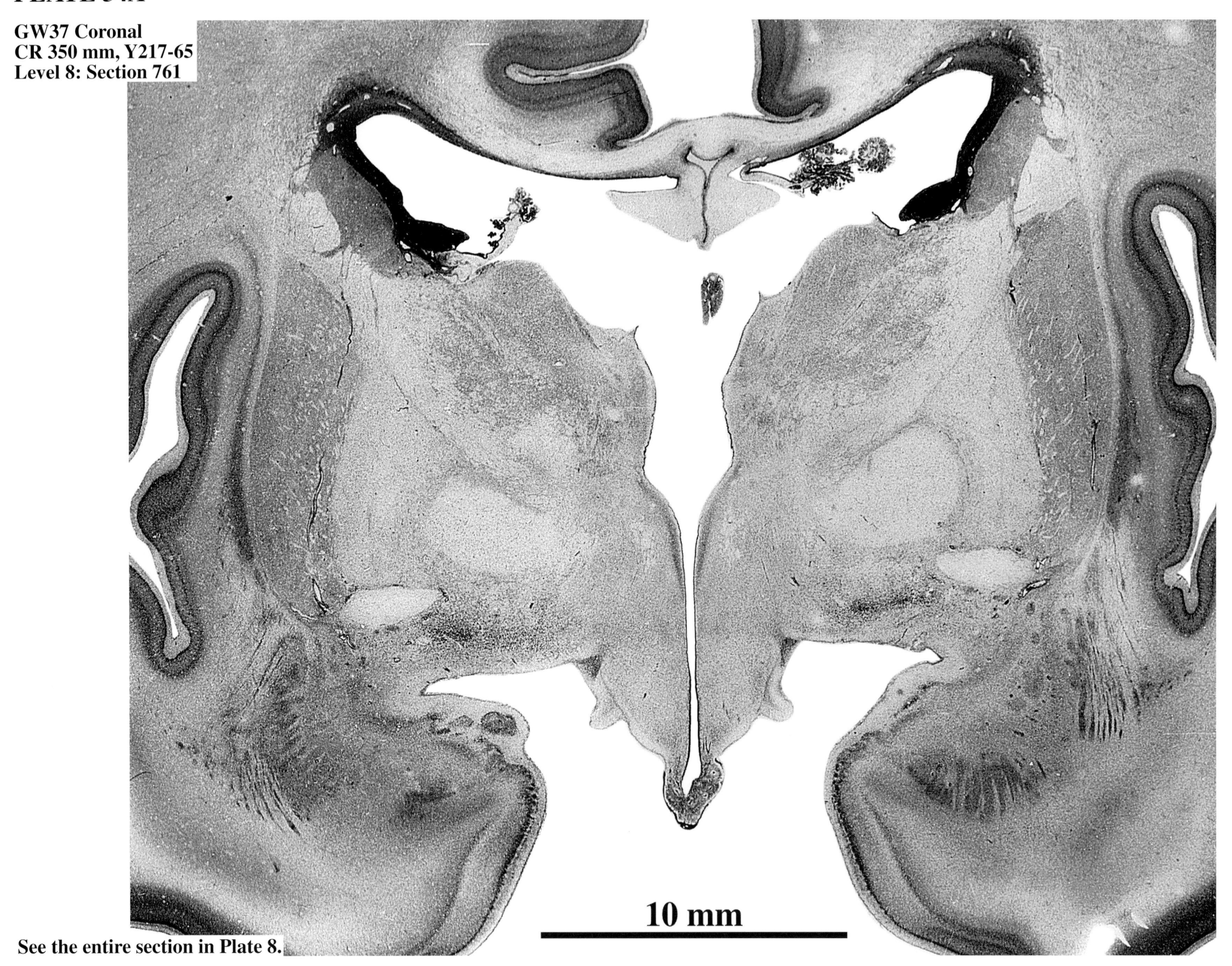

**GW37 Coronal**
**CR 350 mm, Y217-65**
**Level 8: Section 761**

10 mm

See the entire section in Plate 8.

PLATE 34B

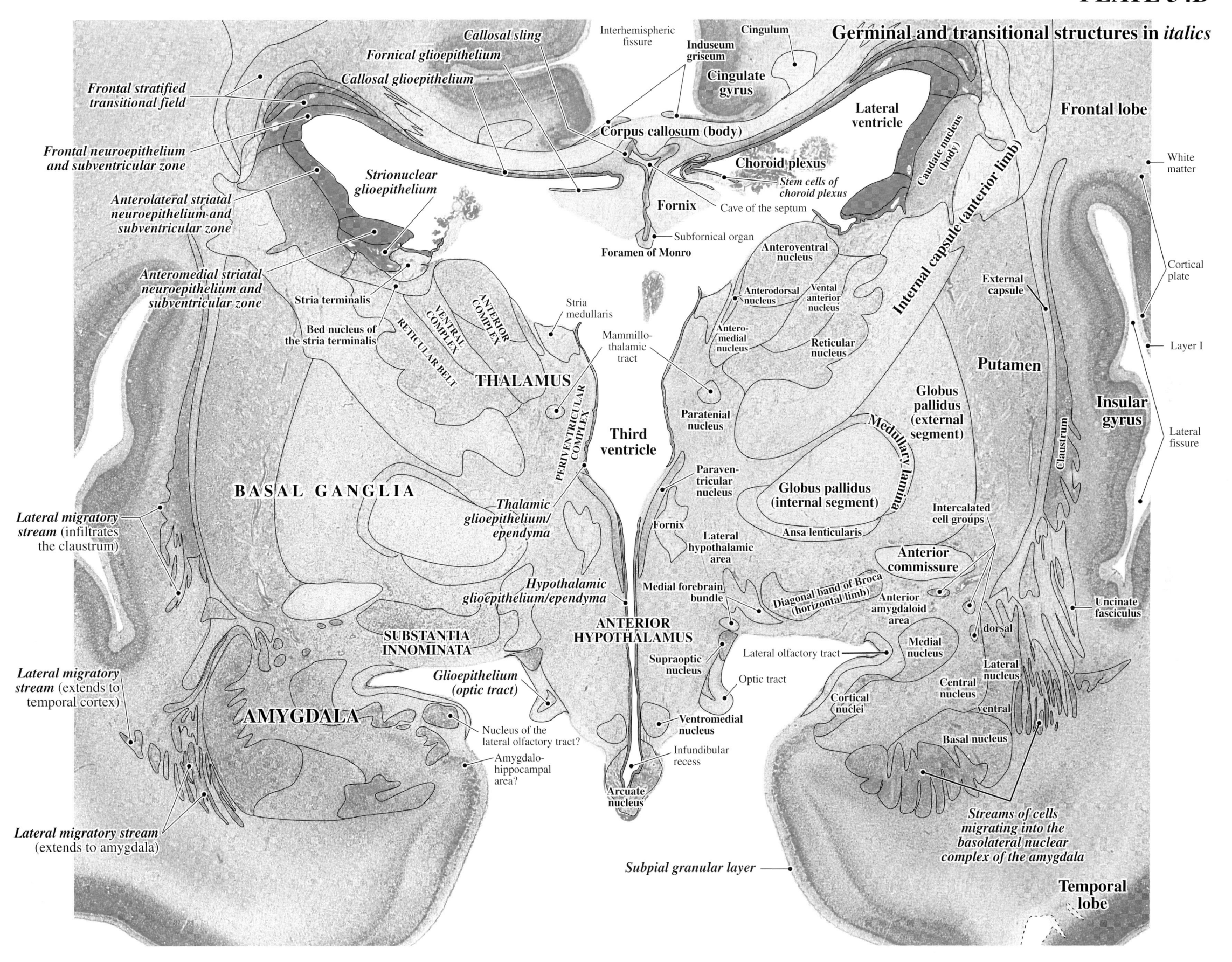

**PLATE 35A**

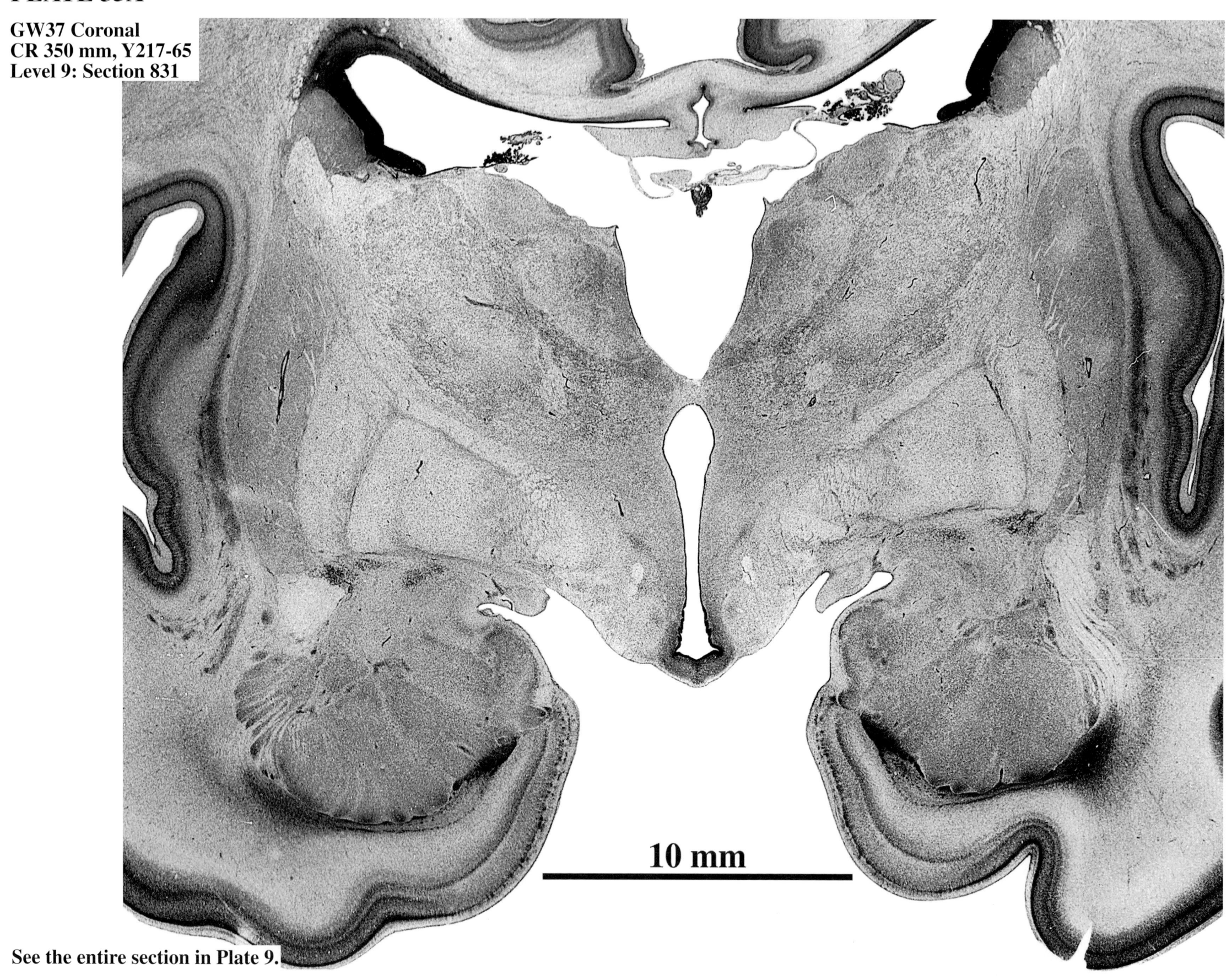

GW37 Coronal
CR 350 mm, Y217-65
Level 9: Section 831

10 mm

See the entire section in Plate 9.

# PLATE 35B

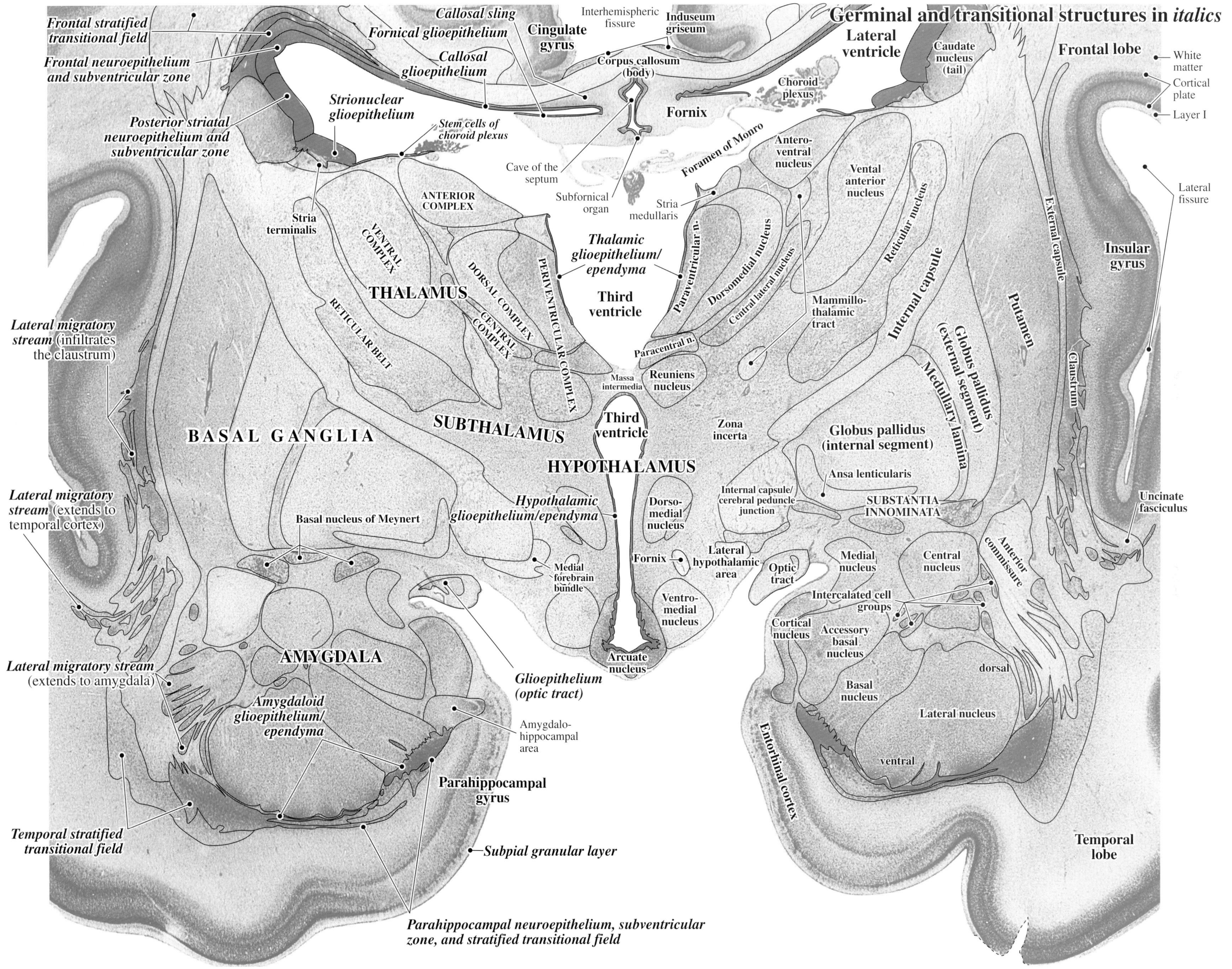
Germinal and transitional structures in *italics*
Frontal stratified transitional field
Frontal neuroepithelium and subventricular zone
Posterior striatal neuroepithelium and subventricular zone
Callosal sling
Fornical glioepithelium
Callosal glioepithelium
Strionuclear glioepithelium
Stem cells of choroid plexus
Cingulate gyrus
Interhemispheric fissure
Induseum griseum
Corpus callosum (body)
Lateral ventricle
Caudate nucleus (tail)
Frontal lobe
White matter
Cortical plate
Layer I
Choroid plexus
Fornix
Foramen of Monro
Cave of the septum
Subfornical organ
Stria medullaris
Stria terminalis
ANTERIOR COMPLEX
VENTRAL COMPLEX
DORSAL COMPLEX
CENTRAL COMPLEX
PERIVENTRICULAR COMPLEX
RETICULAR BELT
THALAMUS
Thalamic glioepithelium/ ependyma
Third ventricle
Antero-ventral nucleus
Vental anterior nucleus
Reticular nucleus
Paraventricular n.
Dorsomedial nucleus
Central lateral nucleus
Mammillo-thalamic tract
Internal capsule
Globus pallidus (external segment)
Medullary lamina
Putamen
External capsule
Claustrum
Insular gyrus
Lateral fissure
Paracentral n.
Massa intermedia
Reuniens nucleus
Lateral migratory stream (infiltrates the claustrum)
BASAL GANGLIA
SUBTHALAMUS
Third ventricle
Zona incerta
Globus pallidus (internal segment)
HYPOTHALAMUS
Ansa lenticularis
Lateral migratory stream (extends to temporal cortex)
Hypothalamic glioepithelium/ependyma
Dorso-medial nucleus
Internal capsule/ cerebral peduncle junction
SUBSTANTIA INNOMINATA
Uncinate fasciculus
Basal nucleus of Meynert
Medial forebrain bundle
Fornix
Lateral hypothalamic area
Optic tract
Medial nucleus
Central nucleus
Anterior commissure
Ventro-medial nucleus
Intercalated cell groups
Cortical nucleus
Accessory basal nucleus
AMYGDALA
Arcuate nucleus
Lateral migratory stream (extends to amygdala)
Glioepithelium (optic tract)
dorsal
Basal nucleus
Amygdaloid glioepithelium/ ependyma
Amygdalo-hippocampal area
Lateral nucleus
Entorhinal cortex
ventral
Parahippocampal gyrus
Temporal stratified transitional field
Subpial granular layer
Temporal lobe
Parahippocampal neuroepithelium, subventricular zone, and stratified transitional field

**PLATE 36A**

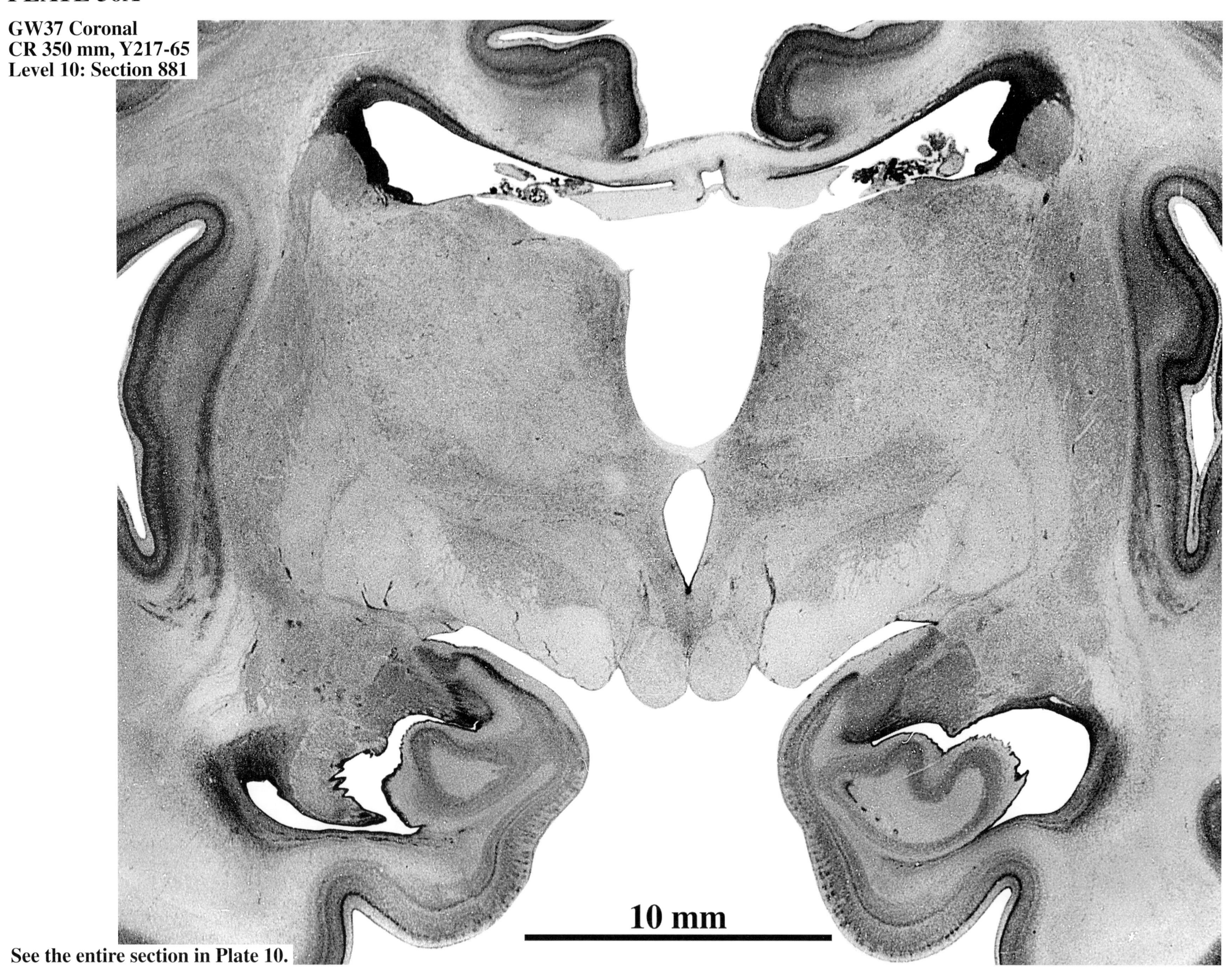

GW37 Coronal
CR 350 mm, Y217-65
Level 10: Section 881

10 mm

See the entire section in Plate 10.

PLATE 36B

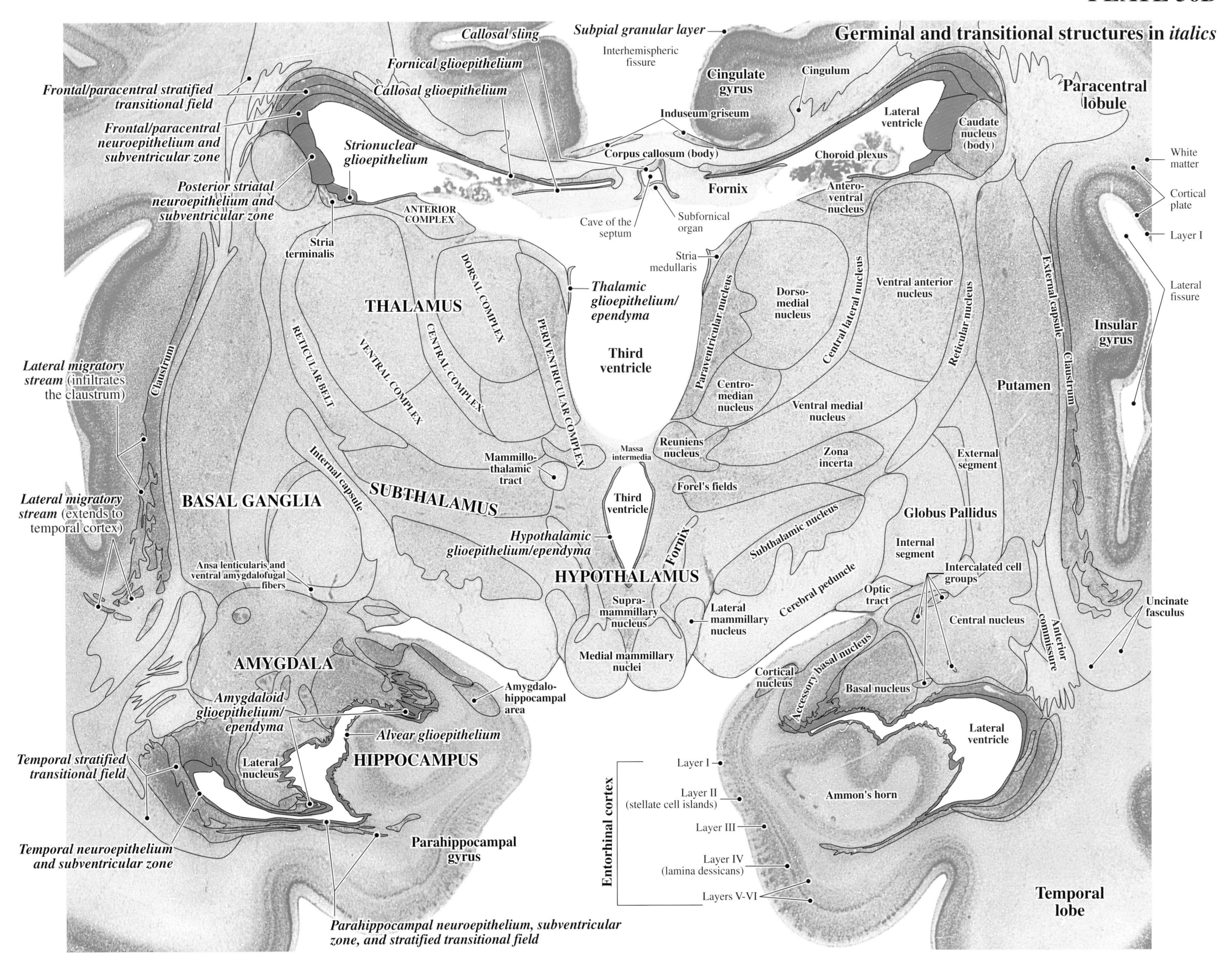

**PLATE 37A**

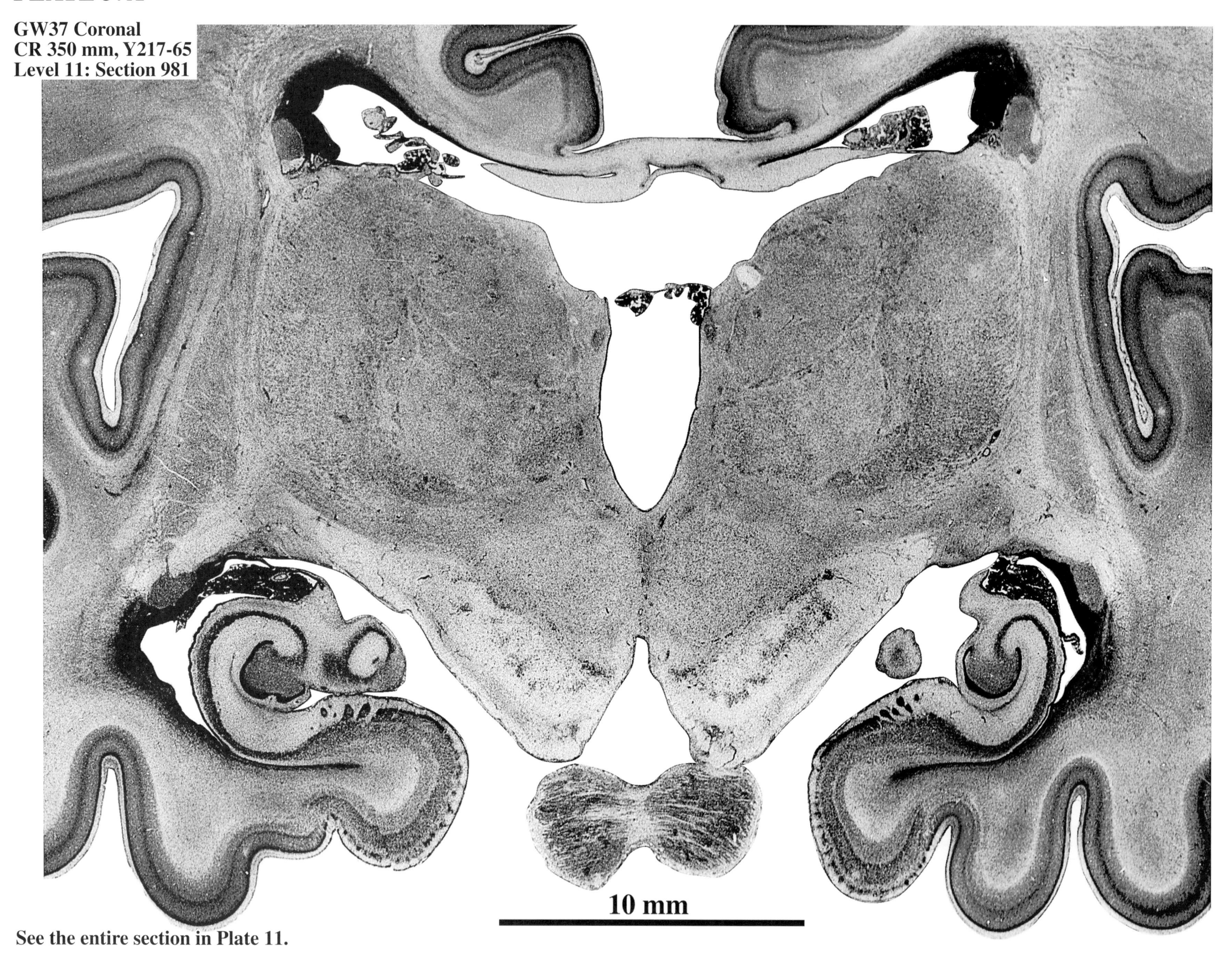

GW37 Coronal
CR 350 mm, Y217-65
Level 11: Section 981

See the entire section in Plate 11.

10 mm

### PLATE 37B

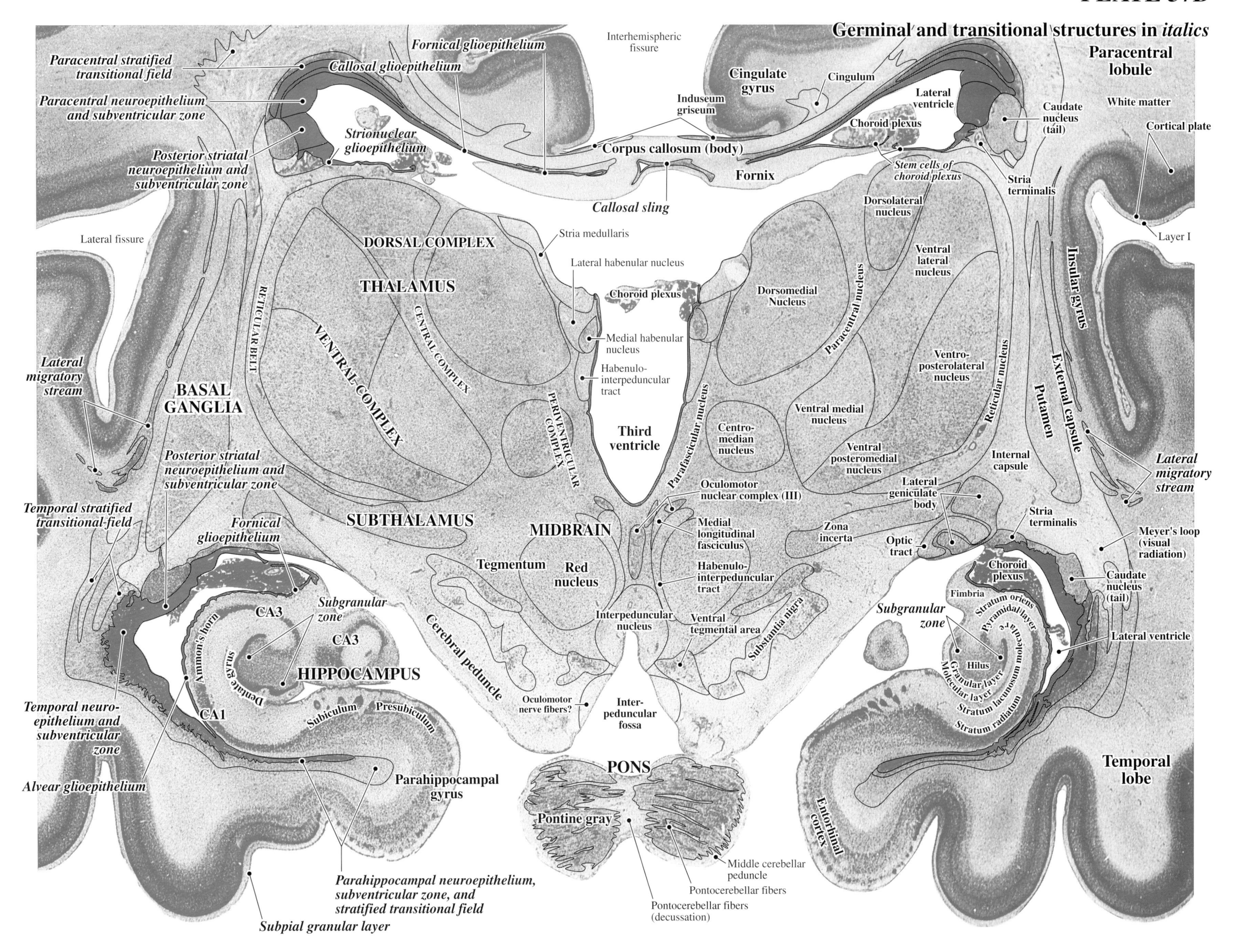

## PLATE 38A

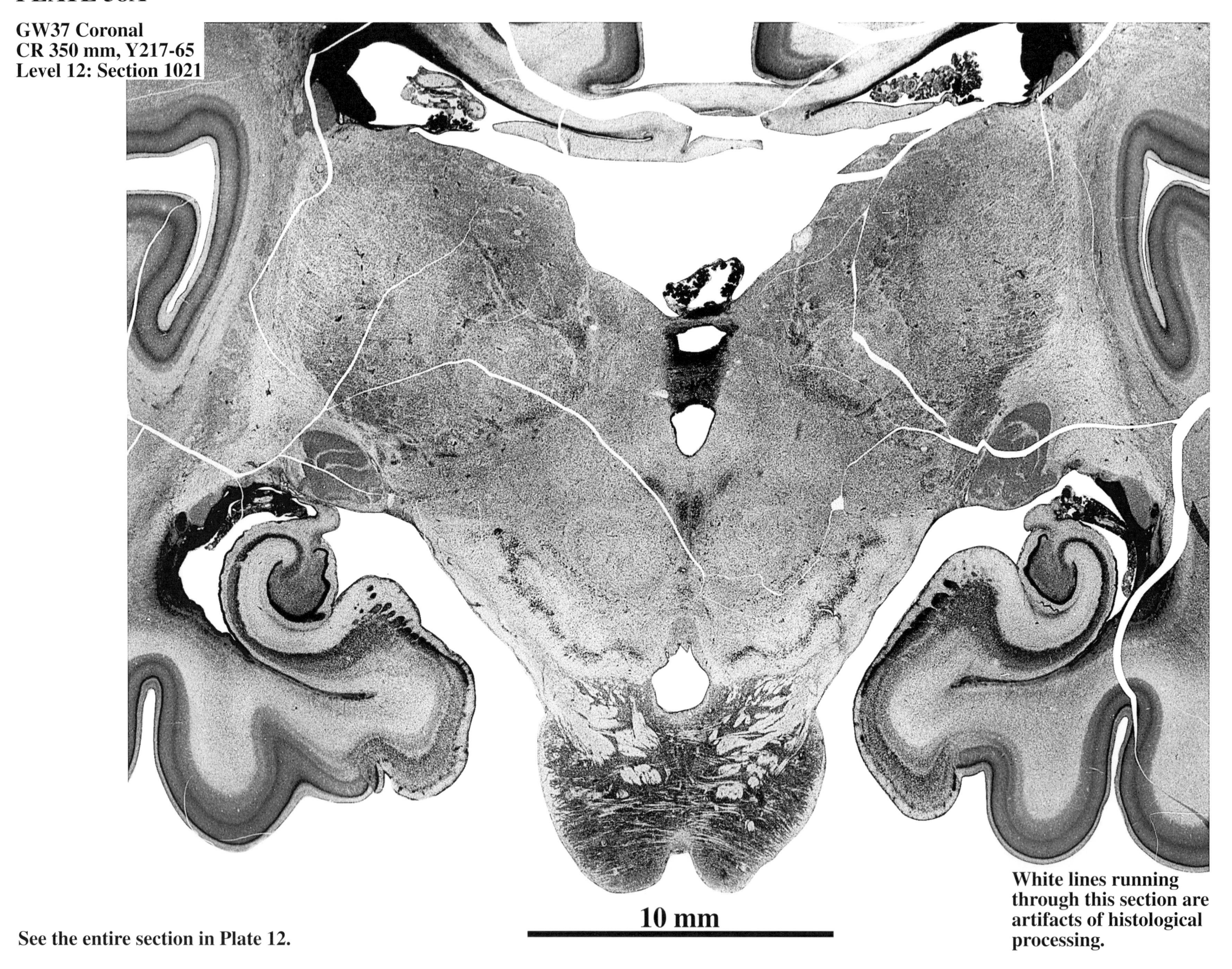

GW37 Coronal
CR 350 mm, Y217-65
Level 12: Section 1021

See the entire section in Plate 12.

10 mm

White lines running through this section are artifacts of histological processing.

**PLATE 38B**

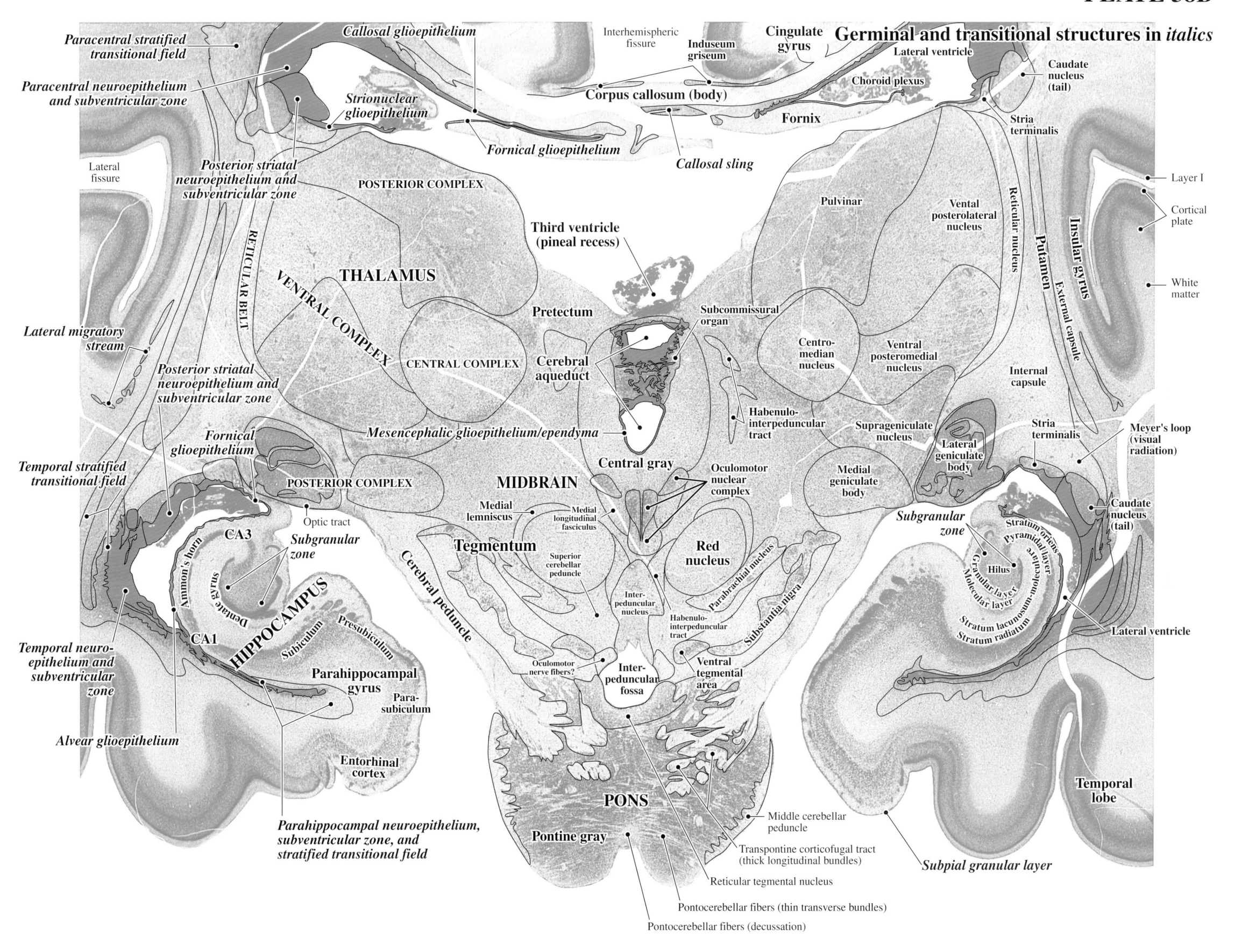

**PLATE 39A**

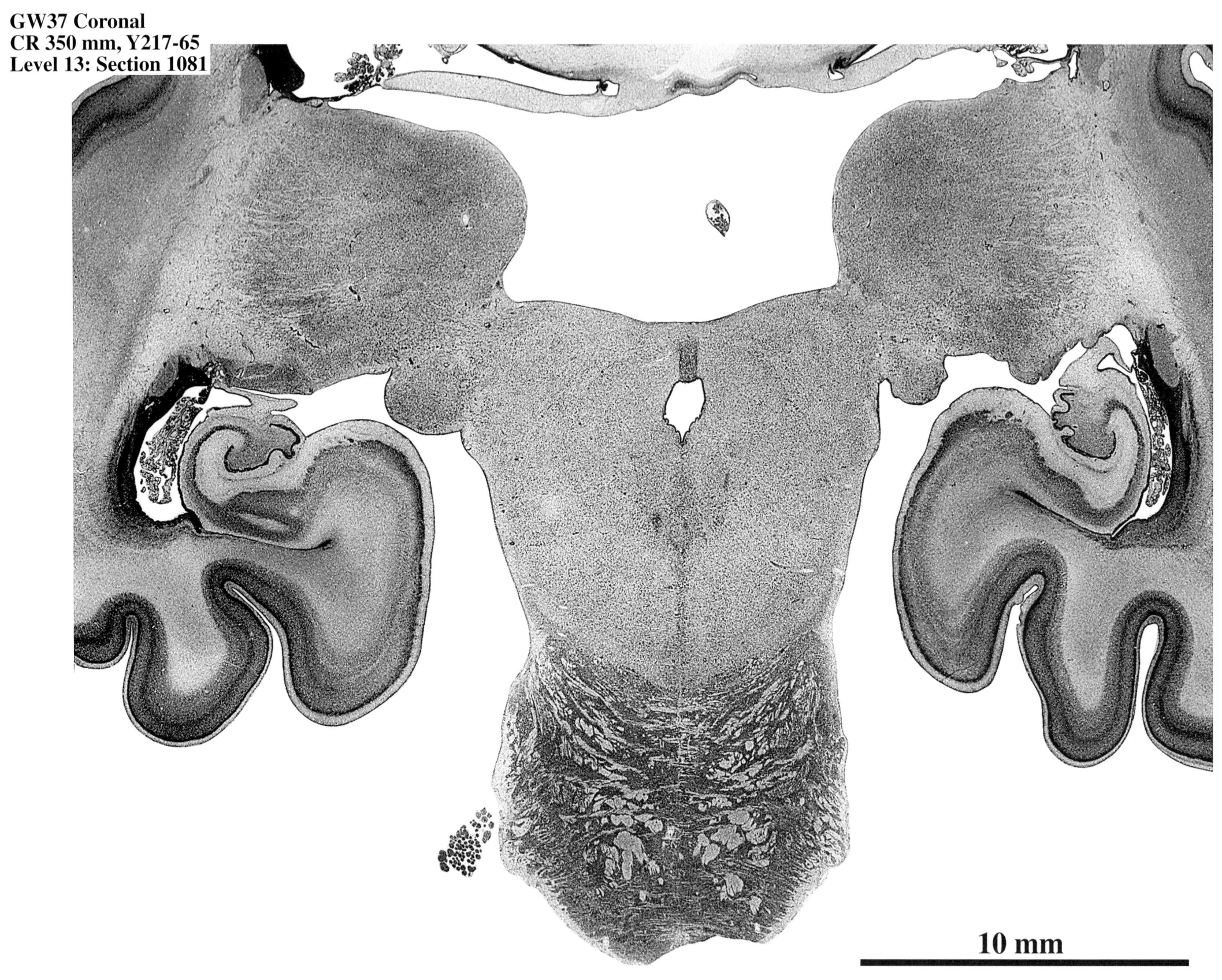

GW37 Coronal
CR 350 mm, Y217-65
Level 13: Section 1081

10 mm

See the entire section in Plate 13.

## PLATE 39B
**Germinal and transitional structures in *italics***

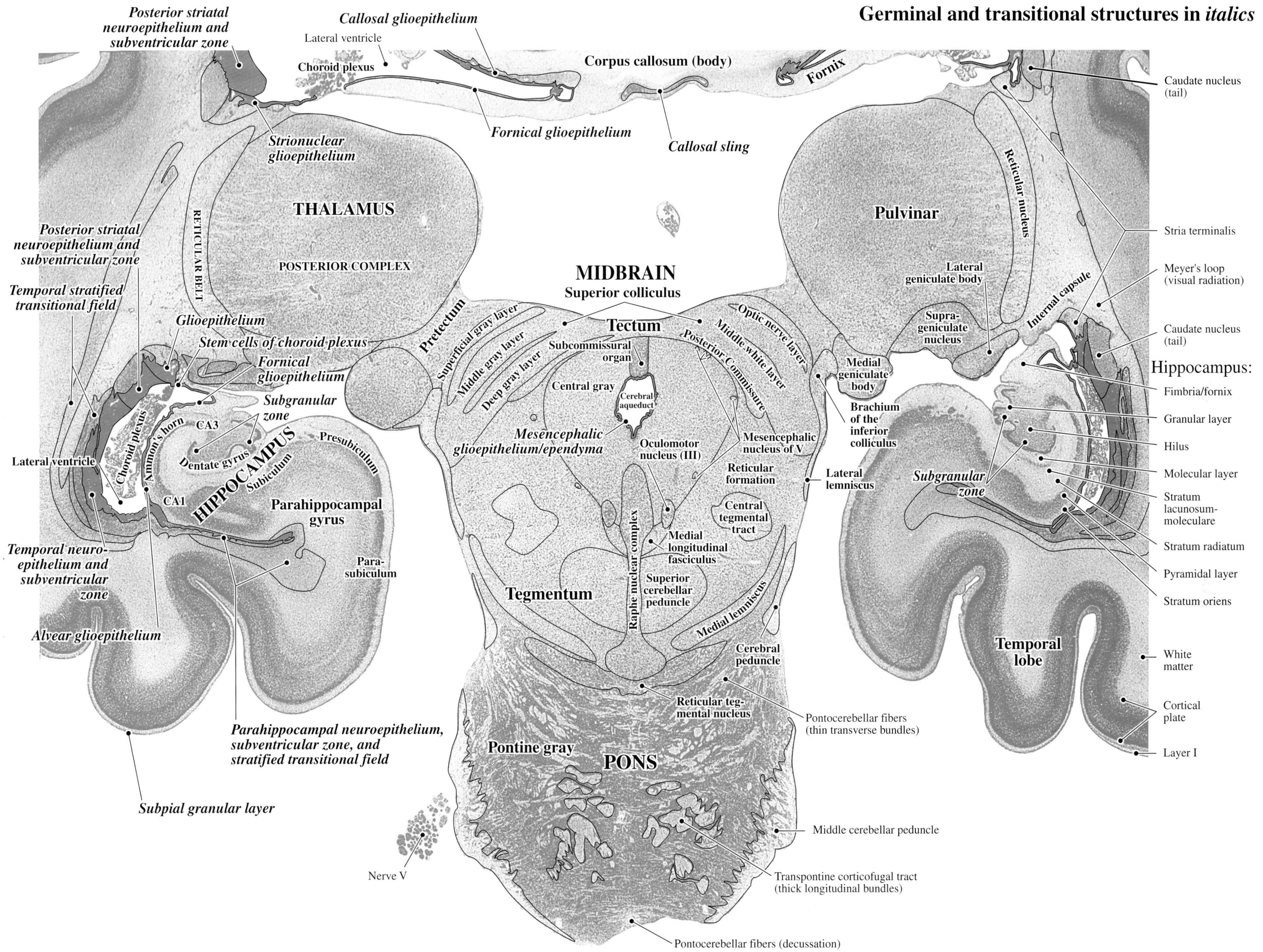

## PLATE 40A

**GW37 Coronal**
CR 350 mm, Y217-65, Level 14: Section 1141

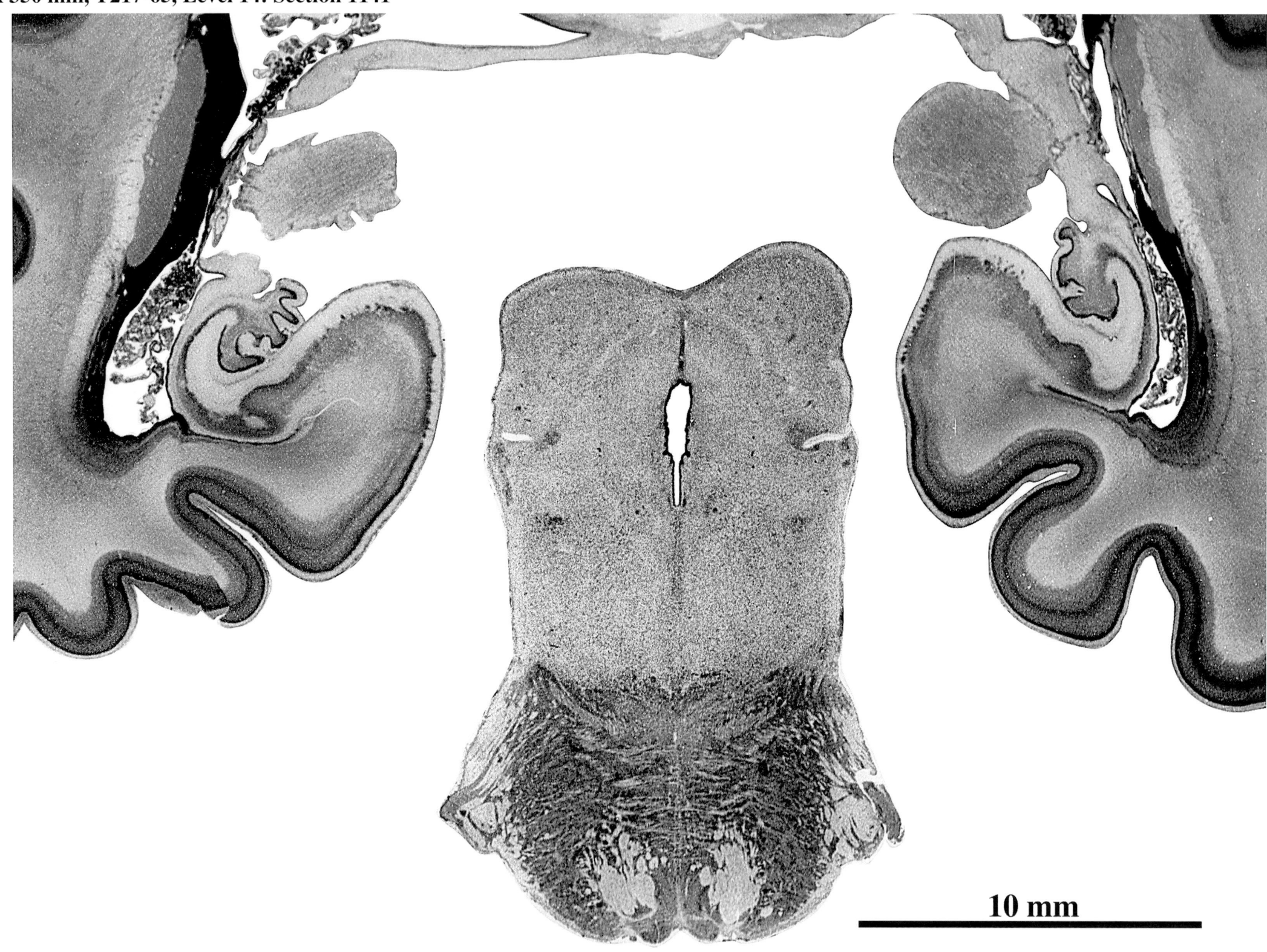

10 mm

See the entire section in Plate 14.

**PLATE 40B**

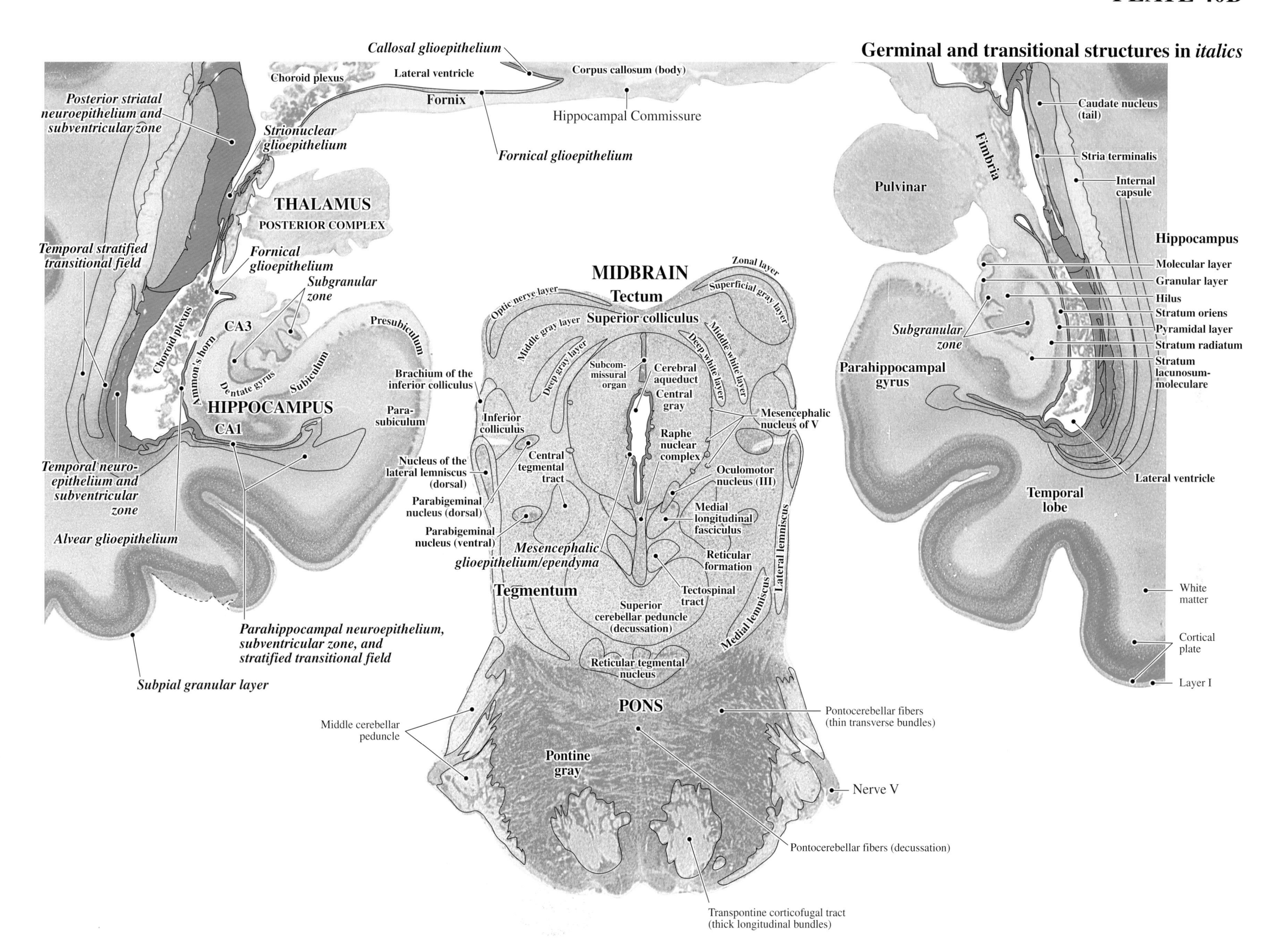

**Germinal and transitional structures in *italics***

## PLATE 41A

**GW37 Coronal**
CR 350 mm, Y217-65, Level 15: Section 1181

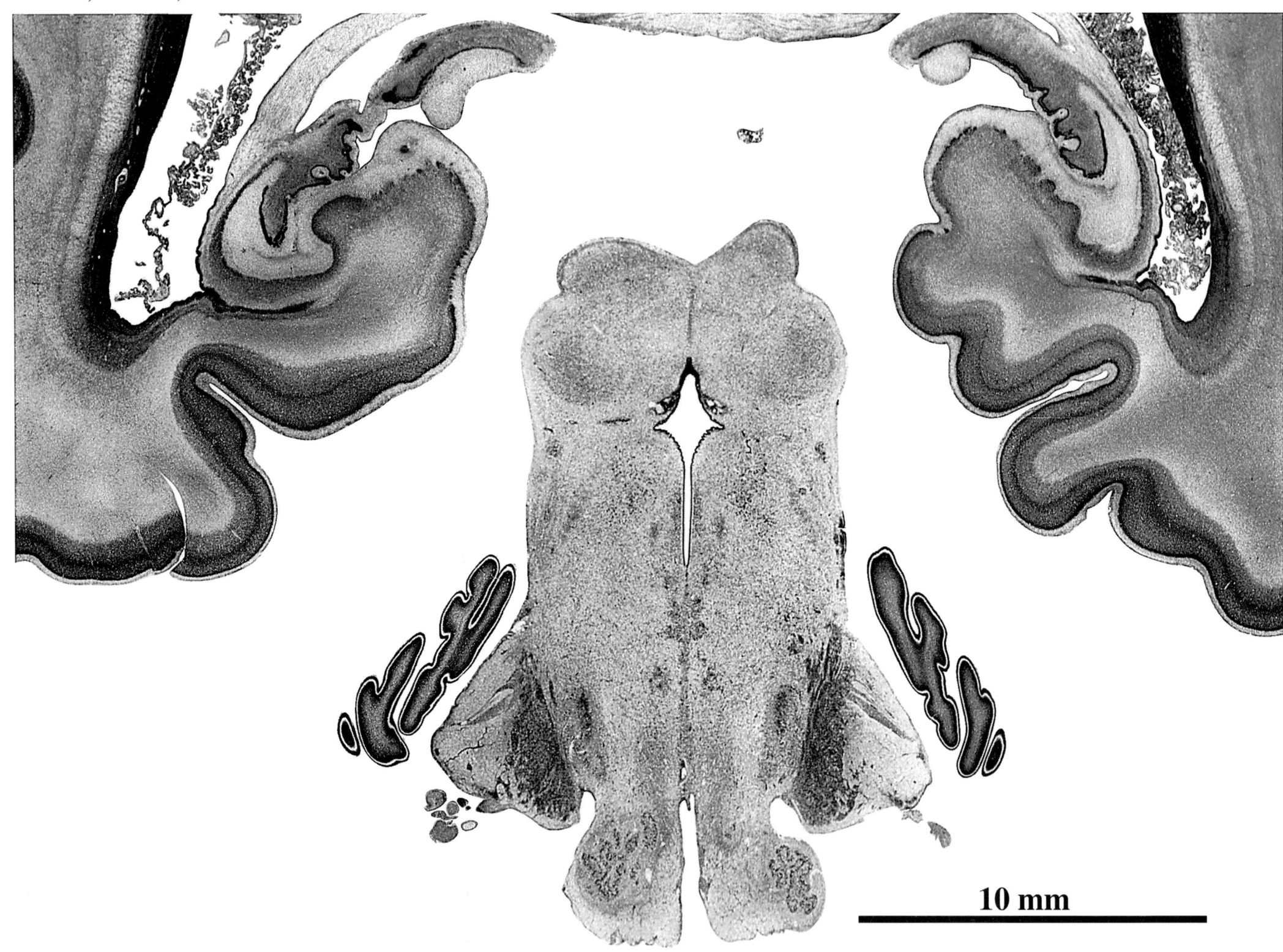

10 mm

See the entire section in Plate 15.

**PLATE 41B**

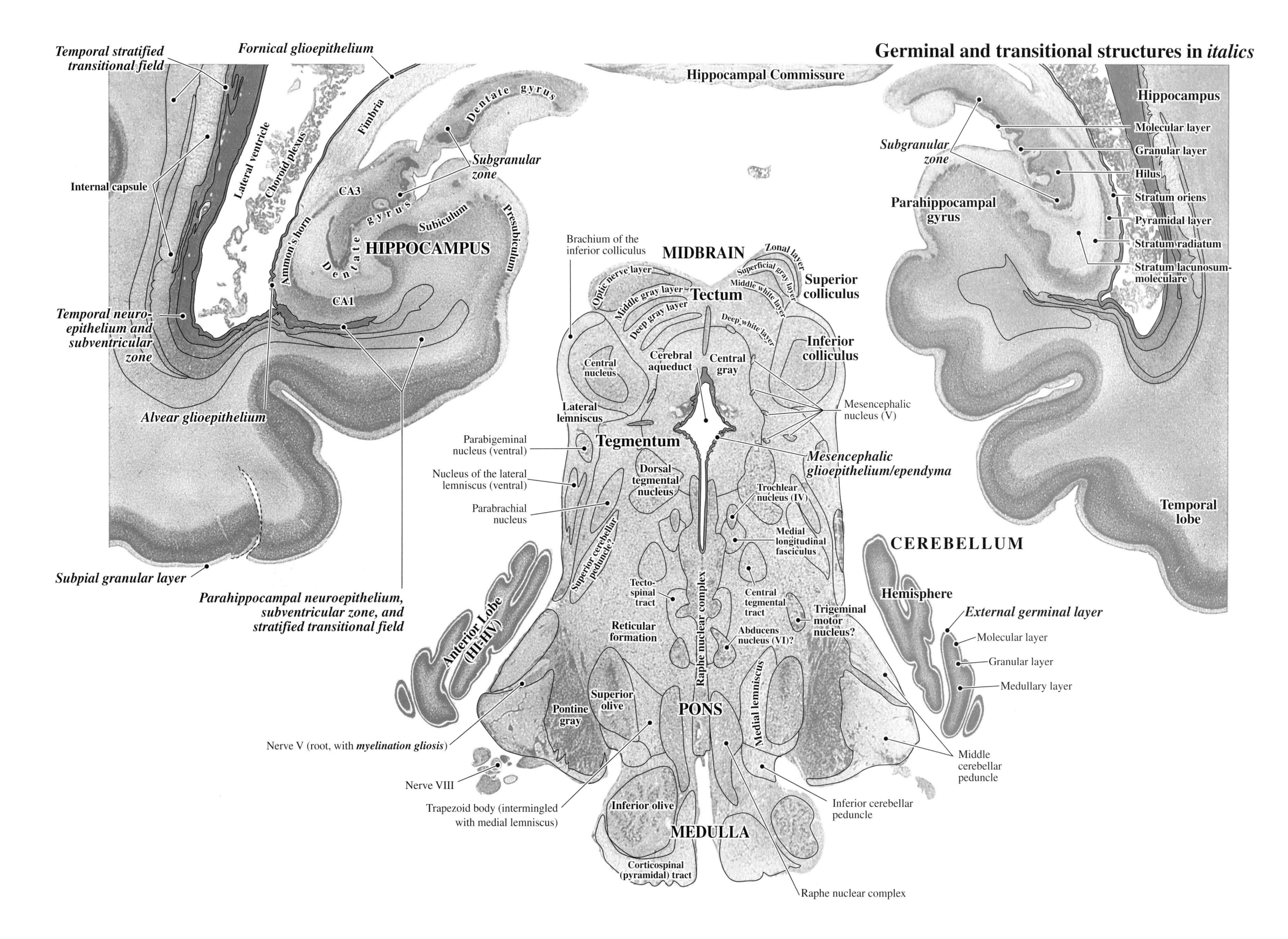

Germinal and transitional structures in *italics*

**PLATE 42A**

GW37 Coronal
CR 350 mm, Y217-65
Level 16: Section 1221

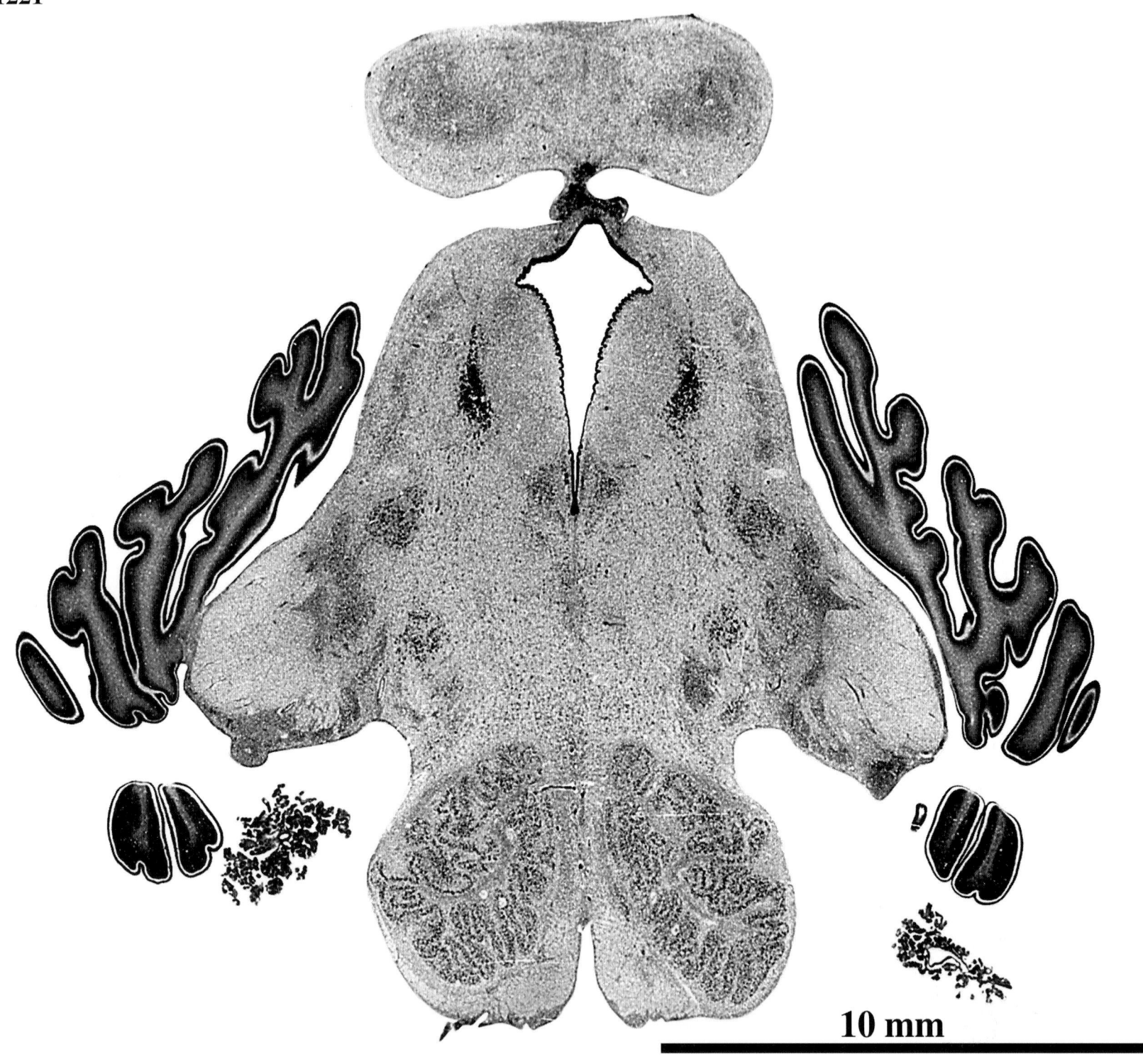

10 mm

See the entire section in Plate 16.

**PLATE 42B**

Germinal and transitional structures in *italics*

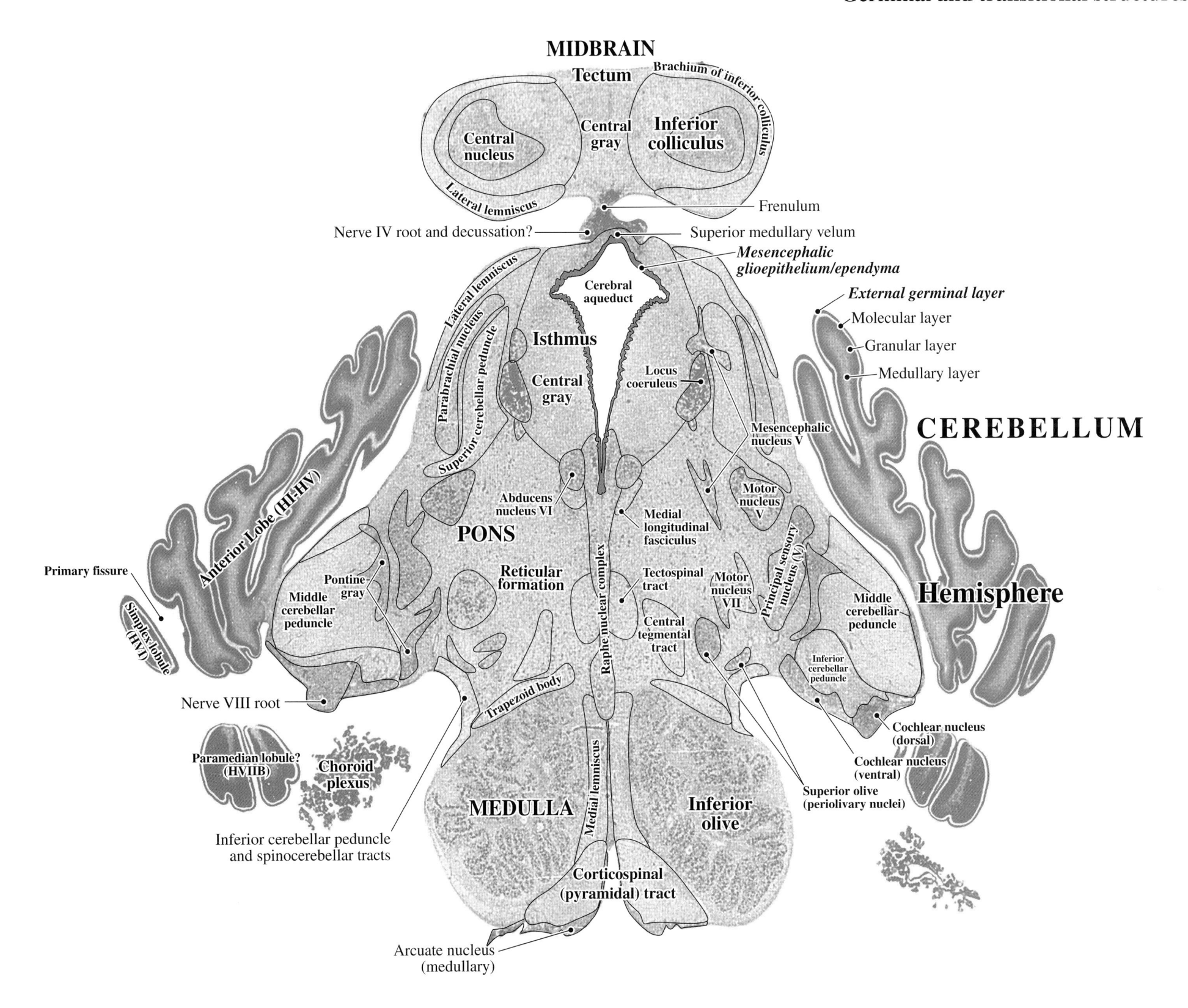

**PLATE 43A**

GW37 Coronal
CR 350 mm
Y217-65
Level 17: Section 1311

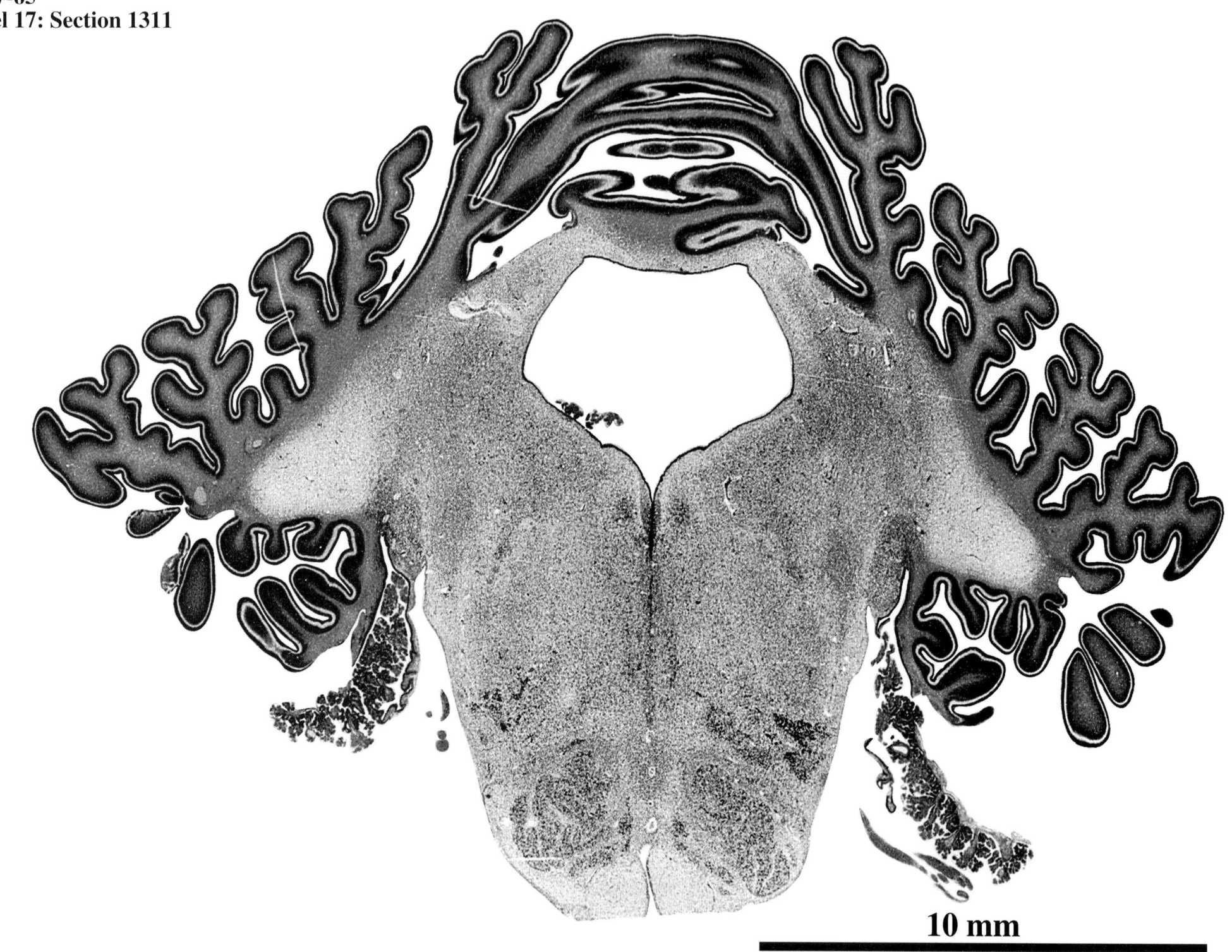

10 mm

See the entire section in Plate 17.

## PLATE 43B

**Germinal and transitional structures in *italics***

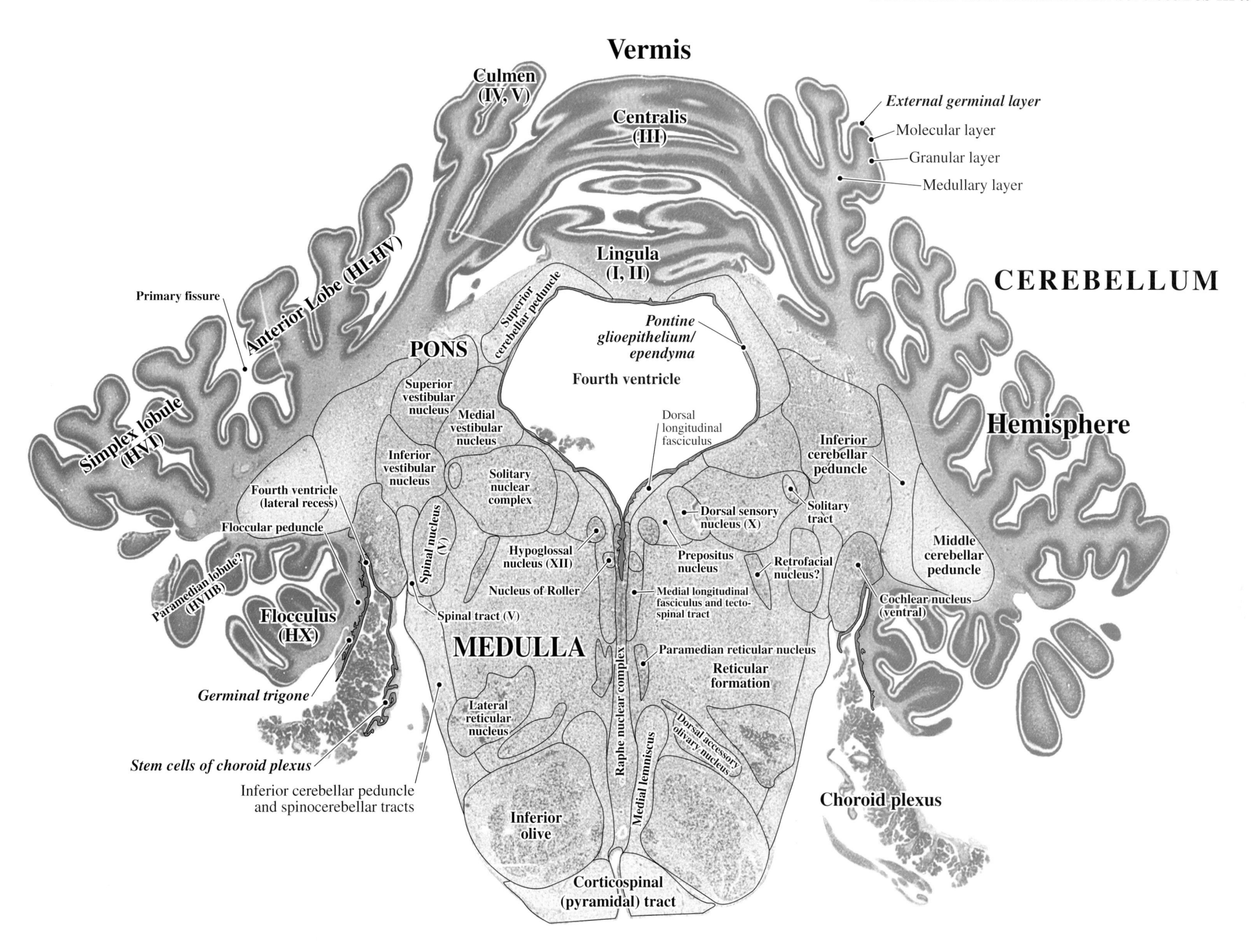

**PLATE 44A**

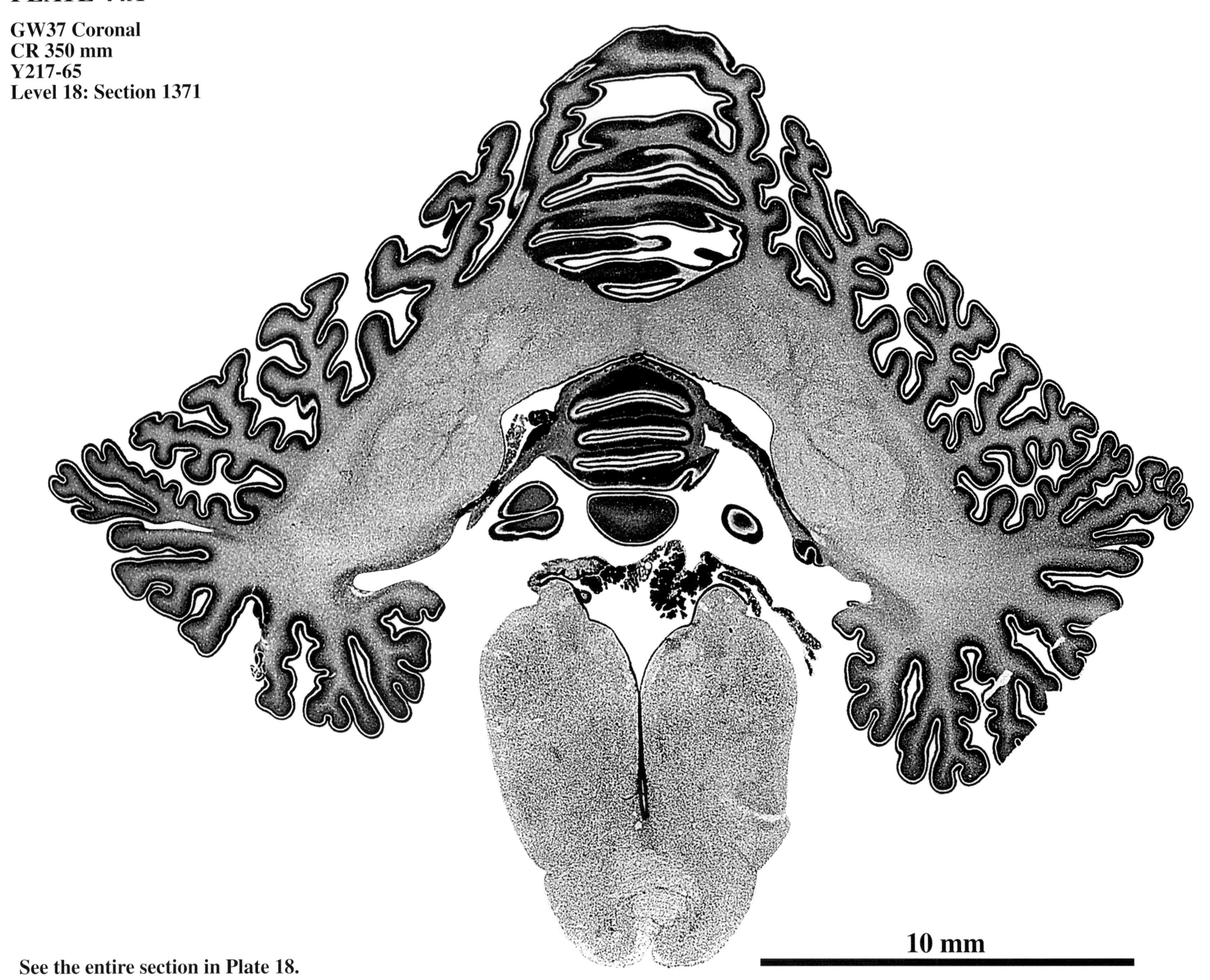

GW37 Coronal
CR 350 mm
Y217-65
Level 18: Section 1371

See the entire section in Plate 18.

10 mm

## PLATE 44B

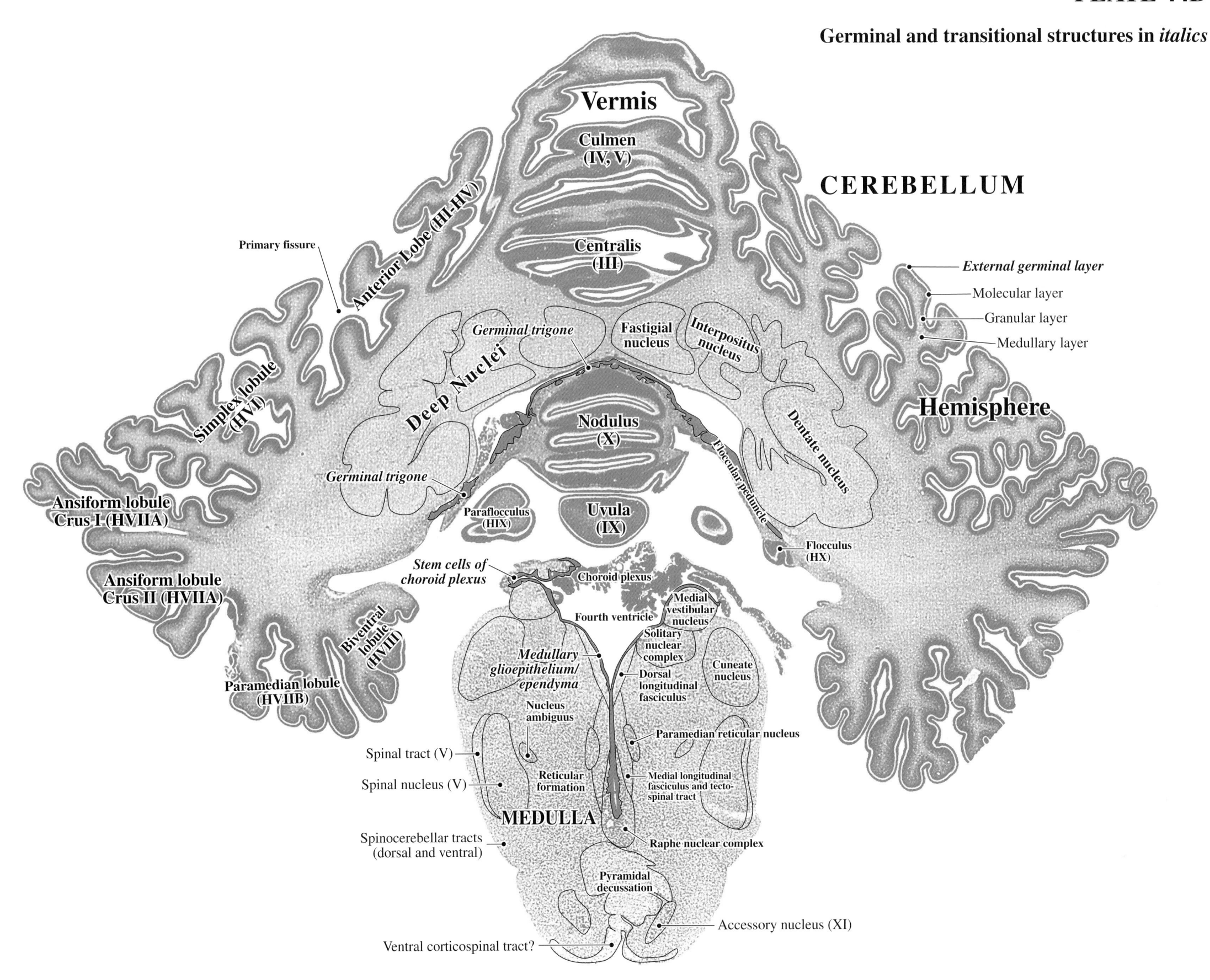

**Germinal and transitional structures in *italics***

# CEREBELLUM

**Vermis** — Culmen (IV, V), Centralis (III)

Anterior Lobe (HI-HV), Primary fissure

*External germinal layer*, Molecular layer, Granular layer, Medullary layer

**Hemisphere**

Simplex lobule (HVI)

*Deep Nuclei* — *Germinal trigone*, Fastigial nucleus, Interpositus nucleus, Dentate nucleus

Nodulus (X), *Germinal trigone*, Parafloccular peduncle

Ansiform lobule Crus I (HVIIA)

Parafloculus (HIX), Uvula (IX), Flocculus (HX)

Ansiform lobule Crus II (HVIIA)

Biventral lobule (HVIII)

Paramedian lobule (HVIIB)

*Stem cells of choroid plexus*, Choroid plexus, Fourth ventricle, Medial vestibular nucleus, Solitary nuclear complex, Cuneate nucleus

*Medullary glioepithelium/ependyma*, Dorsal longitudinal fasciculus

Nucleus ambiguus, Paramedian reticular nucleus

Spinal tract (V), Spinal nucleus (V), Reticular formation, Medial longitudinal fasciculus and tecto-spinal tract

## MEDULLA

Spinocerebellar tracts (dorsal and ventral), Raphe nuclear complex, Pyramidal decussation, Accessory nucleus (XI), Ventral corticospinal tract?

**PLATE 45A**

GW37 Coronal
CR 350 mm
Y217-65
Level 19: Section 1501

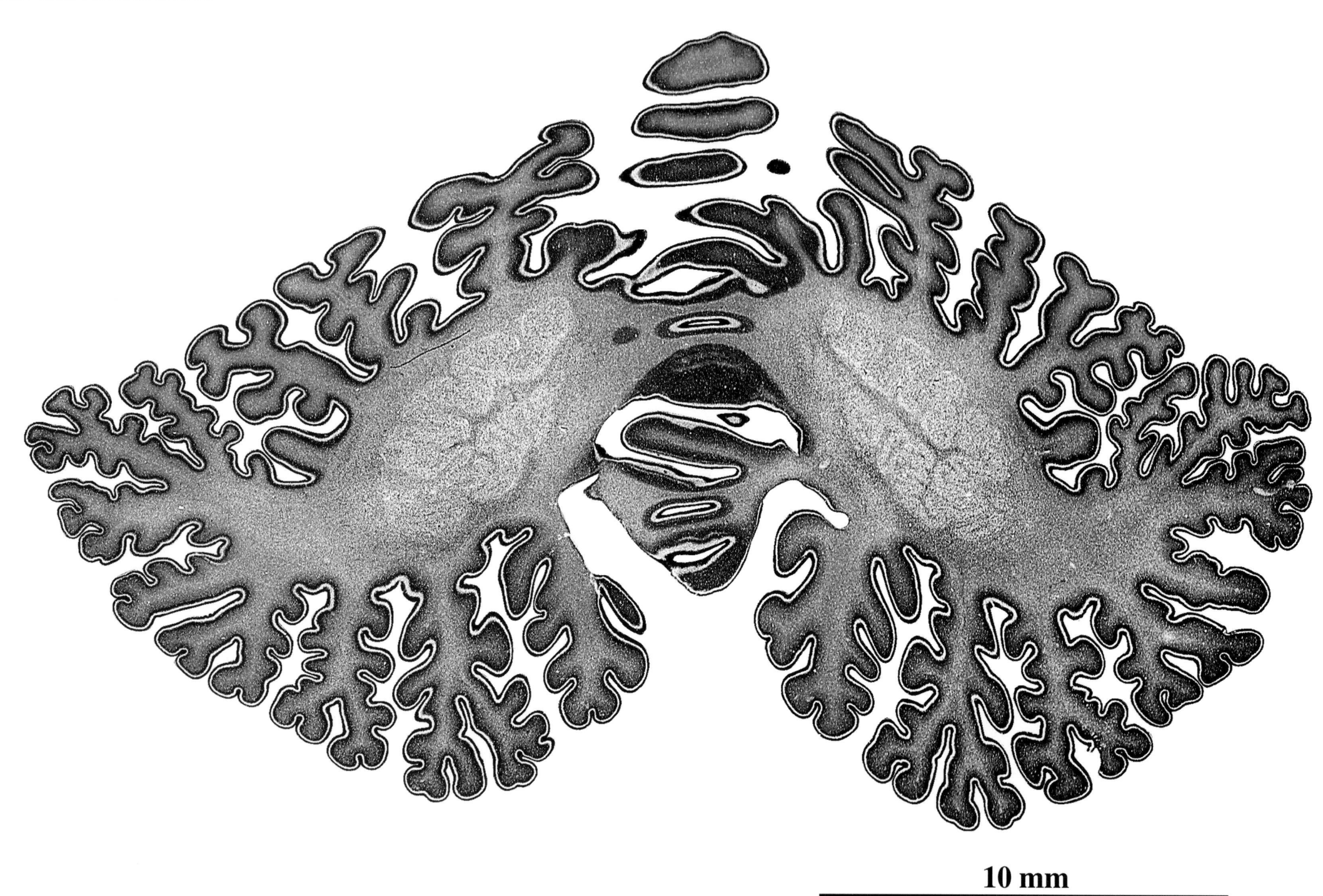

10 mm

See the low magnification view in Plate 19.

**PLATE 45B**

**Germinal and transitional structures in *italics***

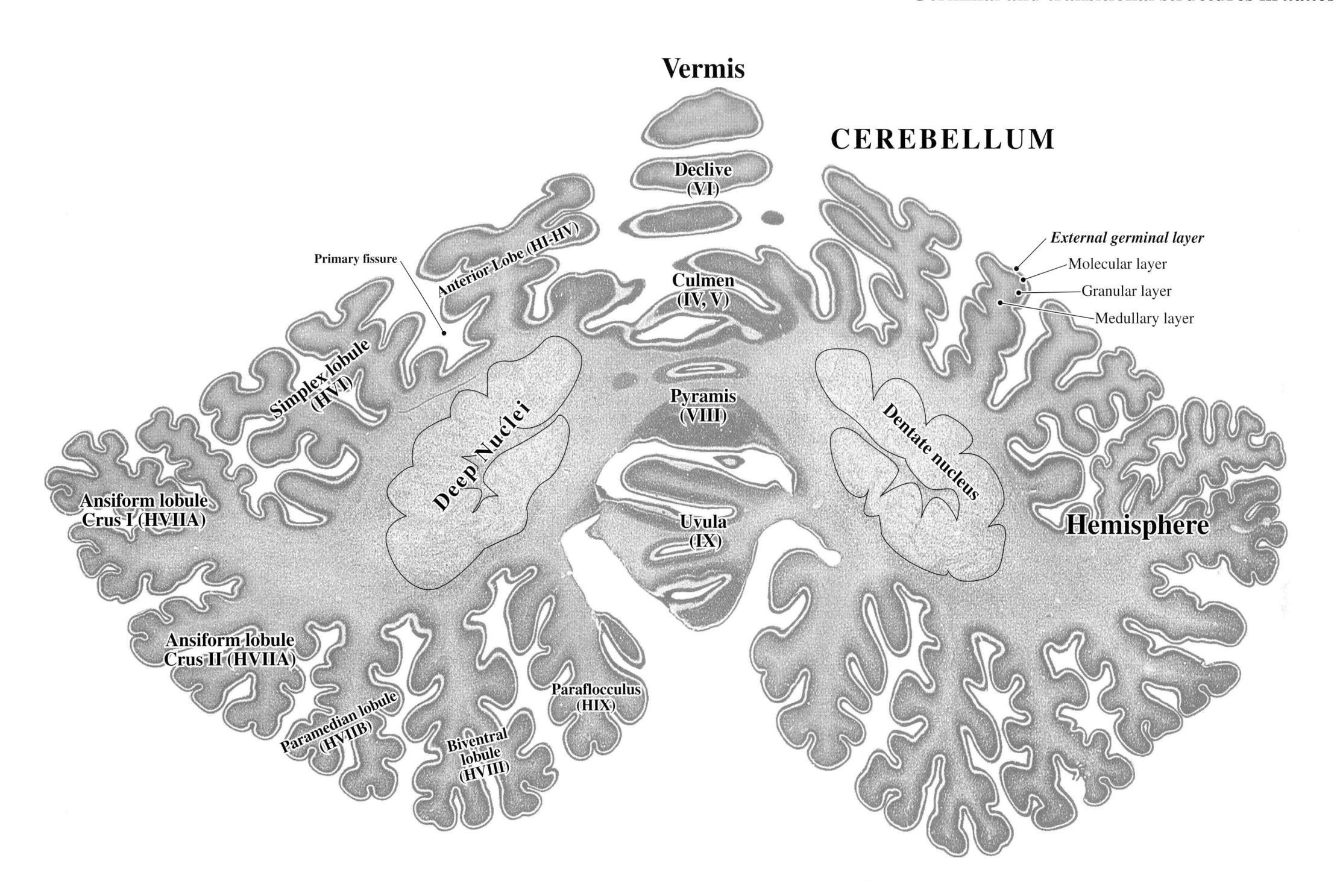

**CEREBELLUM**

**Vermis**

- Declive (VI)
- Culmen (IV, V)
- Pyramis (VIII)
- Uvula (IX)

**Hemisphere**

- Anterior Lobe (HI-HV)
- Primary fissure
- Simplex lobule (HVI)
- Ansiform lobule Crus I (HVIIA)
- Ansiform lobule Crus II (HVIIA)
- Paramedian lobule (HVIIB)
- Biventral lobule (HVIII)
- Paraflocculus (HIX)

- Deep Nuclei
- Dentate nucleus

- *External germinal layer*
- Molecular layer
- Granular layer
- Medullary layer

## PLATE 46A

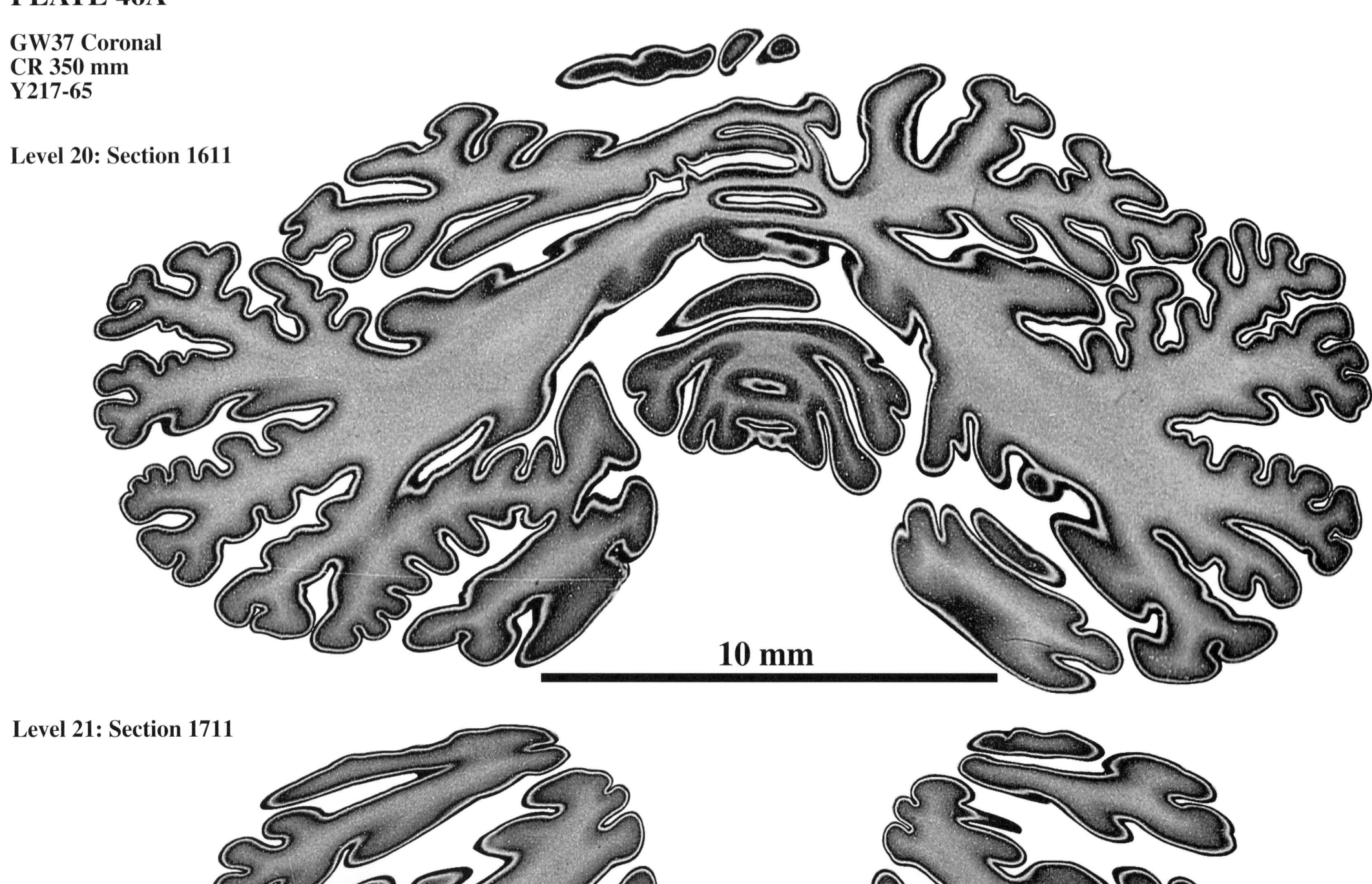

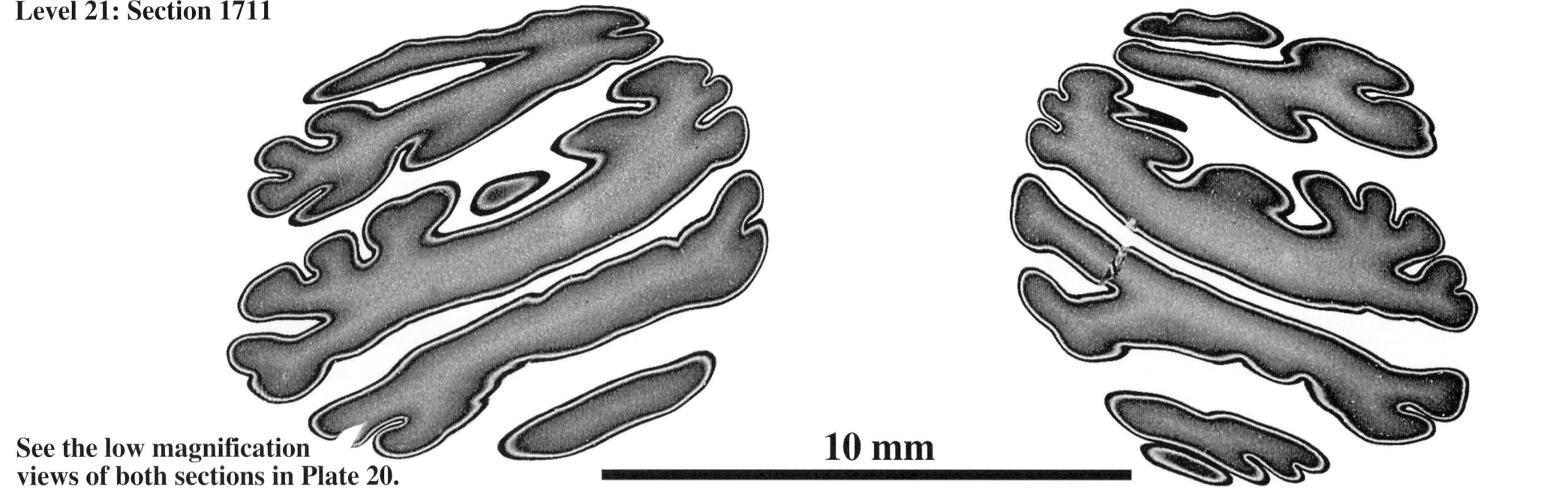

**GW37 Coronal**
**CR 350 mm**
**Y217-65**

**Level 20: Section 1611**

10 mm

**Level 21: Section 1711**

See the low magnification
views of both sections in Plate 20.

10 mm

## PLATE 46B

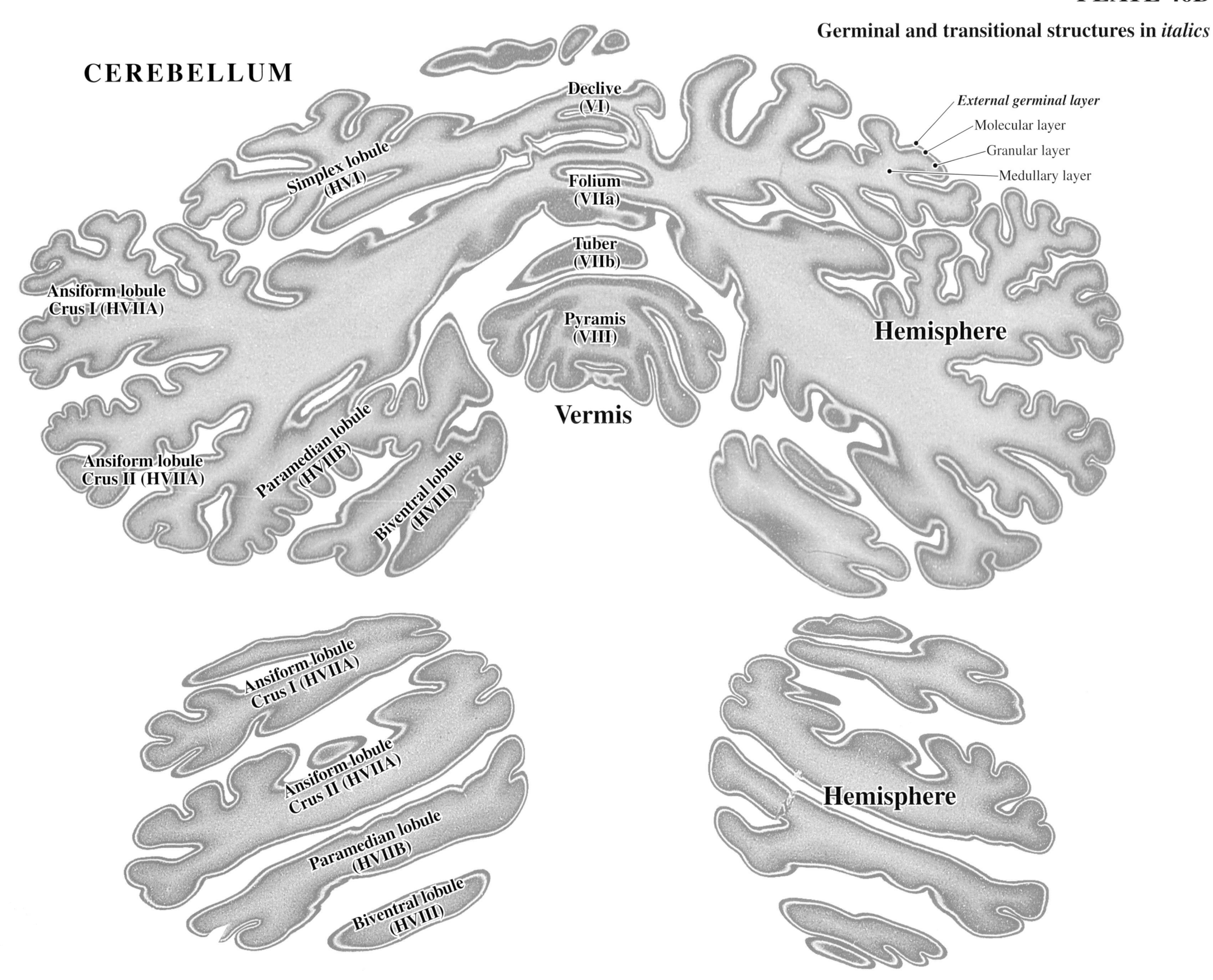

**Germinal and transitional structures in *italics***

**CEREBELLUM**

- *External germinal layer*
- Molecular layer
- Granular layer
- Medullary layer

Declive (VI)
Folium (VIIa)
Tuber (VIIb)
Pyramis (VIII)

Simplex lobule (HVI)
Ansiform lobule Crus I (HVIIA)
Ansiform lobule Crus II (HVIIA)
Paramedian lobule (HVIIB)
Biventral lobule (HVIII)

**Vermis**

**Hemisphere**

Ansiform lobule Crus I (HVIIA)
Ansiform lobule Crus II (HVIIA)
Paramedian lobule (HVIIB)
Biventral lobule (HVIII)

**Hemisphere**

# PART III: GW37 SAGITTAL

This specimen is case number B-180-61 (Perinatal RPSL) in the Yakovlev Collection. A female infant survived for one hour after birth. Death occurred because a hyaline membrane obstructed the airway to the lungs. The brain was cut in the sagittal plane in 35-μm thick sections and is classified as a Normative Control in the Yakovlev Collection (Haleem, 1990). There is no photograph of this brain before it was embedded and cut, so the photograph of another GW37 brain that Larroche published in 1966 (**Figure 3**) is used to show medial surface features.

Photographs of 8 different Nissl-stained sections (**Levels 1-8**) are shown at low magnification in **Plates 47-54**. The core of the brain and the cerebellum are shown at high magnification in companion **Plates 55AB-62AB** for **Levels 1-8**. Very high magnification views of different regions of the cerebellar cortex are shown in **Plates 63 and 64**. Because the section numbers decrease from Level 1 (most medial) to Level 8 (most lateral), they are from the left side of the brain; the right side has higher section numbers proceeding medial to lateral. The cutting plane of this brain is nearly parallel to the midline. However, the posterior part of each section is angled more toward the right than the anterior part. For example, in **Level 1** (**Plate 47**) part of the paraflocculus from the right side of the brain is beneath the cerebellar vermis. The paraflocculus is very small in **Level 2** (**Plate 48**) and the left paraflocculus appears in **Level 3** (**Plate 49**). The sections chosen for illustration are spaced closer together near the midline to show small structures in the diencephalon, midbrain, pons, and medulla.

Y180-61 contains the same group of immature structures that are in the other GW37 brains. In the cortical regions of the telencephalon, remnants of the germinal matrices are present in all lobes of the cerebral cortex where the ***neuroepithelium/subventricular zone*** may still be generating neocortical interneurons. Remnants of migrating and sojourning neurons and/or glia are visible in all lobes of the cerebral cortex as ***stratified transitional fields***, thin in the occipital lobe, and thicker in the frontal, parietal and temporal lobes. Indeed, thickness variations can be used as identifying criteria for designating a particular cortical germinal matrix as occipital or otherwise. Many neurons, glia, and their mitotic precursor cells are still migrating through the olfactory peduncle toward the olfactory bulb (***rostral migratory stream***) from a presumed source area in the germinal matrix at the junction between the cerebral cortex, striatum, and nucleus accumbens. Within the lateral parts of the cerebral cortex, streams of neurons and glia are still in the ***lateral migratory stream*** that percolates through the claustrum, endopiriform nucleus, external capsule, and uncinate fasciculus. These cells appear to be heading toward the insular cortex, primary olfactory cortex, temporal cortex, and basolateral parts of the amygdaloid complex. In the basal ganglia, there is a prominent ***neuroepithelium/subventricular zone*** overlying the striatum and nucleus accumbens where neurons are probably still being generated. Another region of active neurogenesis in the telencephalon is the ***subgranular zone*** in the hilus of the dentate gyrus that is the source of granule cells. Other structures in the telencephalon, such as the septum, fornix, and Ammon's horn part of the hippocampus, have only a thin, darkly staining layer at the ventricle, and these are presumed to be generating glia, cells of the choroid plexus, and the ependymal lining of the ventricle.

Most of the structures in the diencephalon appear to be settled and are maturing, and the third ventricle is lined by a thin ***glioepithelium/ependyma***. In the midbrain and anterior pons, there is a slightly thicker and more convoluted ***glioepithelium/ependyma*** lining the posterior cerebral aqueduct and anterior fourth ventricle. The posterior pons and entire medulla have a thin ***glioepithelium/ependyma*** lining the rest of the fourth ventricle. The ***external germinal layer*** is prominent over the entire surface of the cerebellar cortex and is still producing basket, stellate, and granule cells. The ***germinal trigone*** is still visible at the base of the nodulus and along the floccular peduncle; choroid plexus cells and glia may still be originating here.

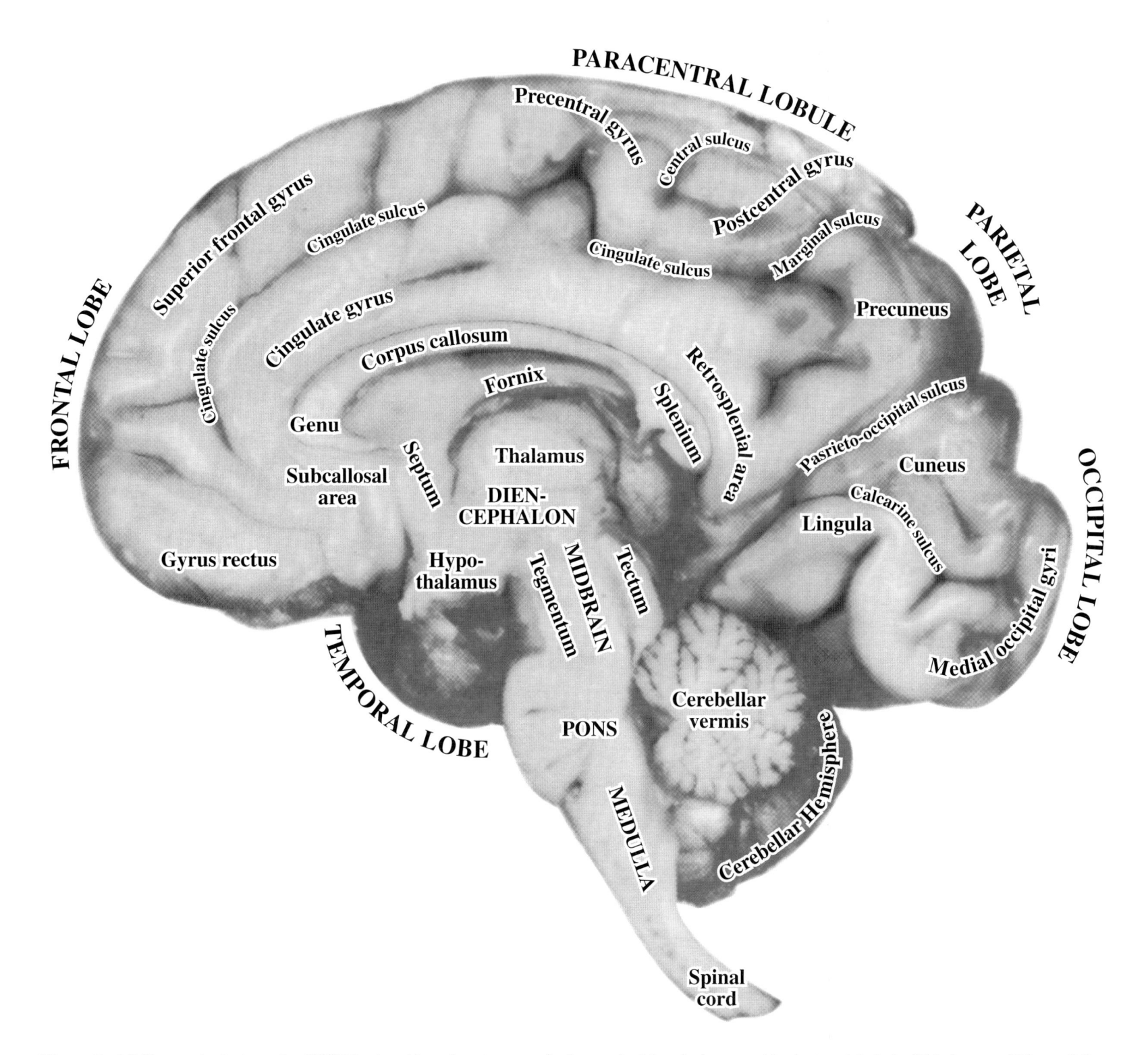

**Figure 3.** Midline sagittal view of a GW37 brain with major structures in the cerebral hemispheres and brainstem labeled. (This is part of Figure 2-9 on page 27 in B. A. Curtis, S. Jacobson, and E. M. Marcus (1972) *An Introduction to the Neurosciences*, Philadelphia: W. B. Saunders. The photograph was originally published by J. C. Larroche (1966) The development of the central nervous system during intrauterine life. In: *Human Development*, F. Falkner (ed.), Philadelphia: W. B. Saunders, page 259.)

# PLATE 47

GW37 Sagittal
CR 310 mm
Y180-61
Level 1: Section 961

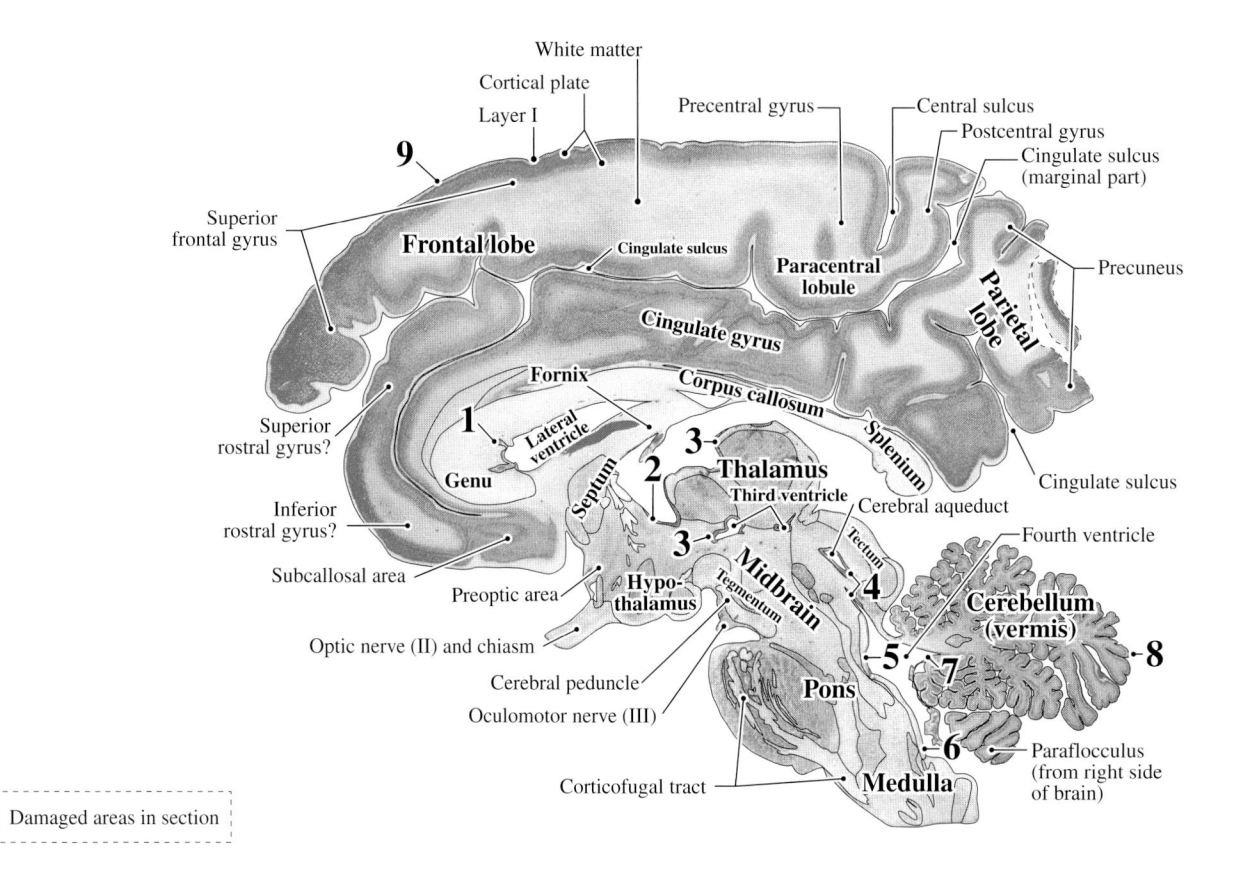

***Remnants of the germinal matrix, migratory streams, and transitional fields***

| | | | |
|---|---|---|---|
| 1 | *Callosal sling* | 6 | *Medullary glioepithelium/ependyma* |
| 2 | *Strionuclear glioepithelium* | 7 | *Germinal trigone (cerebellum)* |
| 3 | *Diencephalic glioepithelium/ependyma* | 8 | *External germinal layer (cerebellum)* |
| 4 | *Mesencephalic glioepithelium/ependyma* | 9 | *Subpial granular layer (cortical)* |
| 5 | *Pontine glioepithelium/ependyma* | | |

**See detail of brain core and cerebellum in Plates 55A and B.**

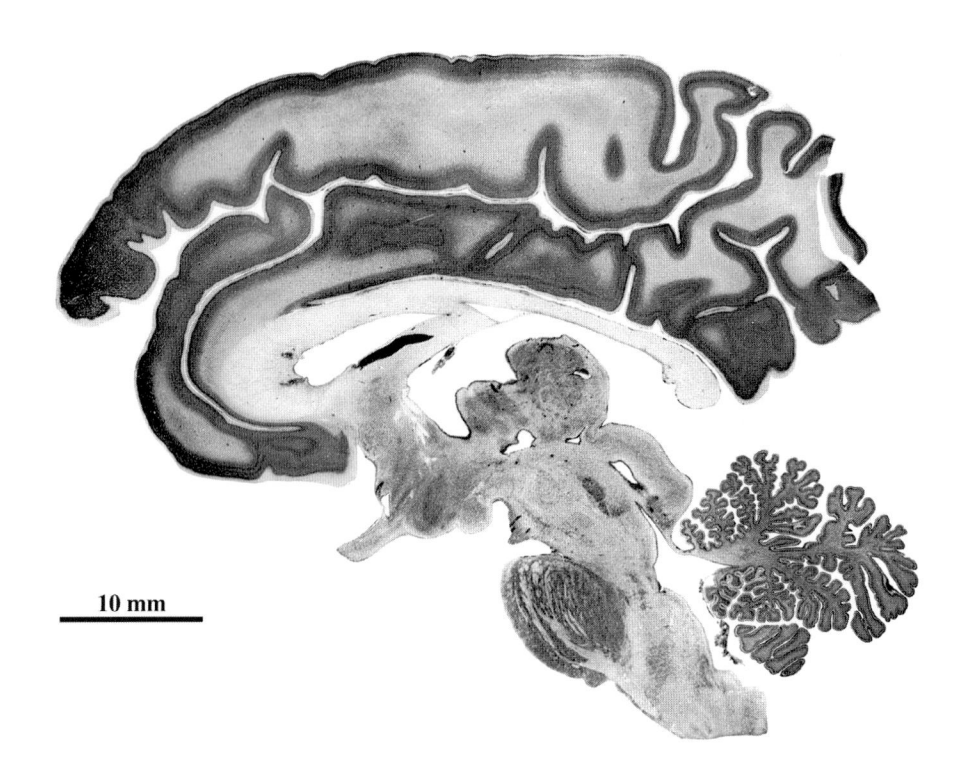

# PLATE 48

**GW37 Sagittal**
**CR 310 mm**
**Y180-61**
**Level 2: Section 941**

### Remnants of the germinal matrix, migratory streams, and transitional fields

| | | | |
|---|---|---|---|
| 1 | Callosal sling | 5 | Medullary glioepithelium/ependyma |
| 2 | Strionuclear glioepithelium | 6 | Germinal trigone |
| 3 | Mesencephalic glioepithelium/ependyma | 7 | External germinal layer (cerebellum) |
| 4 | Pontine glioepithelium/ependyma | 8 | Subpial granular layer (cortical) |

**See detail of brain core and cerebellum in Plates 56A and B.**

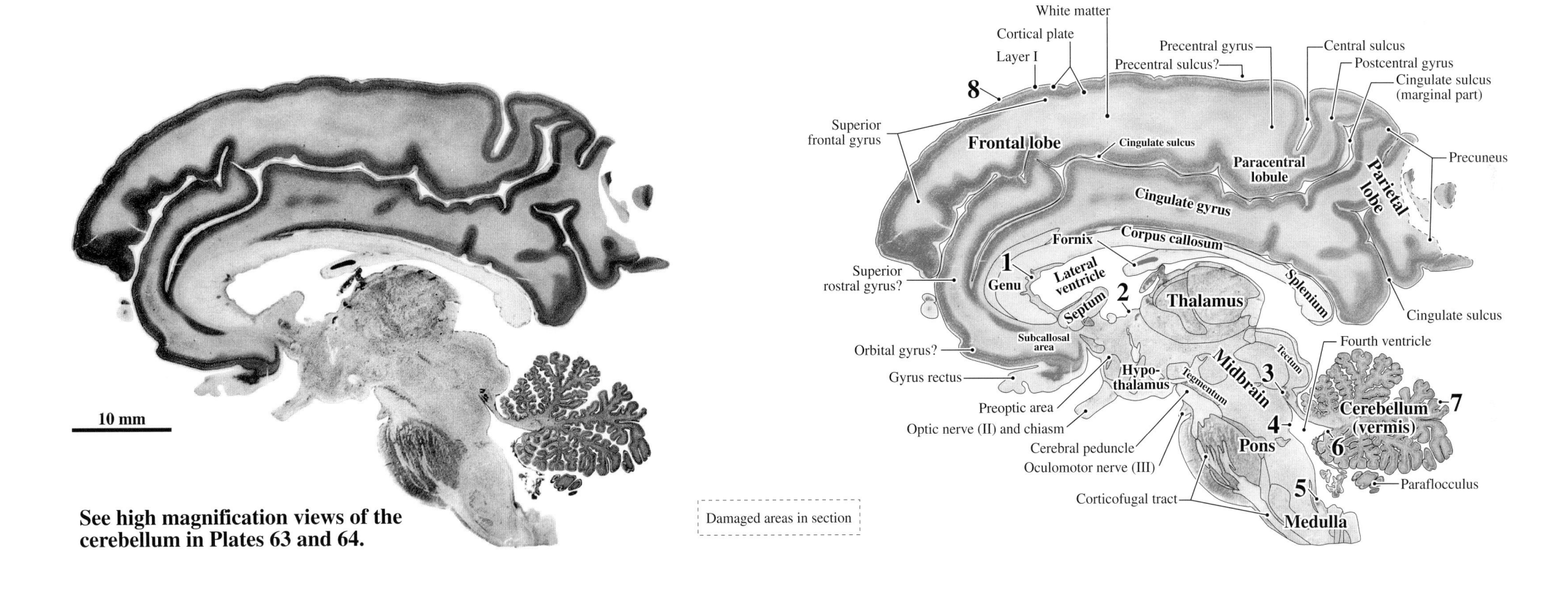

**See high magnification views of the cerebellum in Plates 63 and 64.**

Damaged areas in section

# PLATE 49

GW37 Sagittal
CR 310 mm
Y180-61
Level 3: Section 841

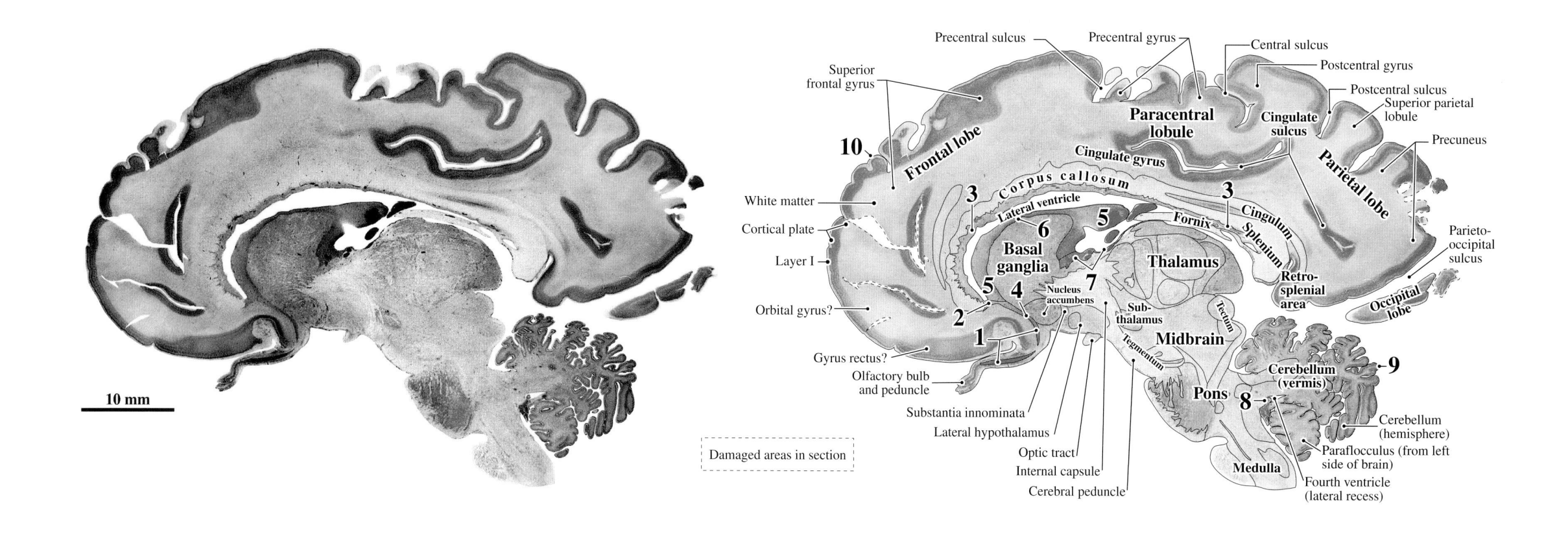

*Remnants of the germinal matrix, migratory streams, and transitional fields*

| | | | |
|---|---|---|---|
| 1 | *Rostral migratory stream* | 6 | *Anterolateral striatal neuroepithelium and subventricular zone* |
| 2 | *Rostral migratory stream (source area)* | 7 | *Strionuclear glioepithelium* |
| 3 | *Callosal glioepithelium* | 8 | *Pontine and medullary glioepithelium/ependyma* |
| 4 | *Accumbent neuroepithelium and subventricular zone (infiltrated by the rostral migratory stream)* | 9 | *External germinal layer (cerebellum)* |
| 5 | *Anteromedial striatal neuroepithelium and subventricular zone* | 10 | *Subpial granular layer (cortical)* |

**See detail of brain core and cerebellum in Plates 57A and B.**

# PLATE 50

**GW37 Sagittal**
**CR 310 mm**
**Y180-61**
**Level 4: Section 781**

## Remnants of the germinal matrix, migratory streams, and transitional fields

| | | | |
|---|---|---|---|
| 1 | Rostral migratory stream | 6 | Anterolateral striatal neuroepithelium and subventricular zone |
| 2 | Frontal stratified transitional field (intermingled with the source of the rostral migratory stream) | 7 | Anteromedial striatal neuroepithelium and subventricular zone |
| 3 | Frontal stratified transitional field | 8 | Strionuclear glioepithelium |
| 4 | Callosal glioepithelium | 9 | External germinal layer (cerebellum) |
| 5 | Fornical glioepithelium | 10 | Subpial granular layer (cortical) |

See detail of brain core and cerebellum in Plates 58A and B.

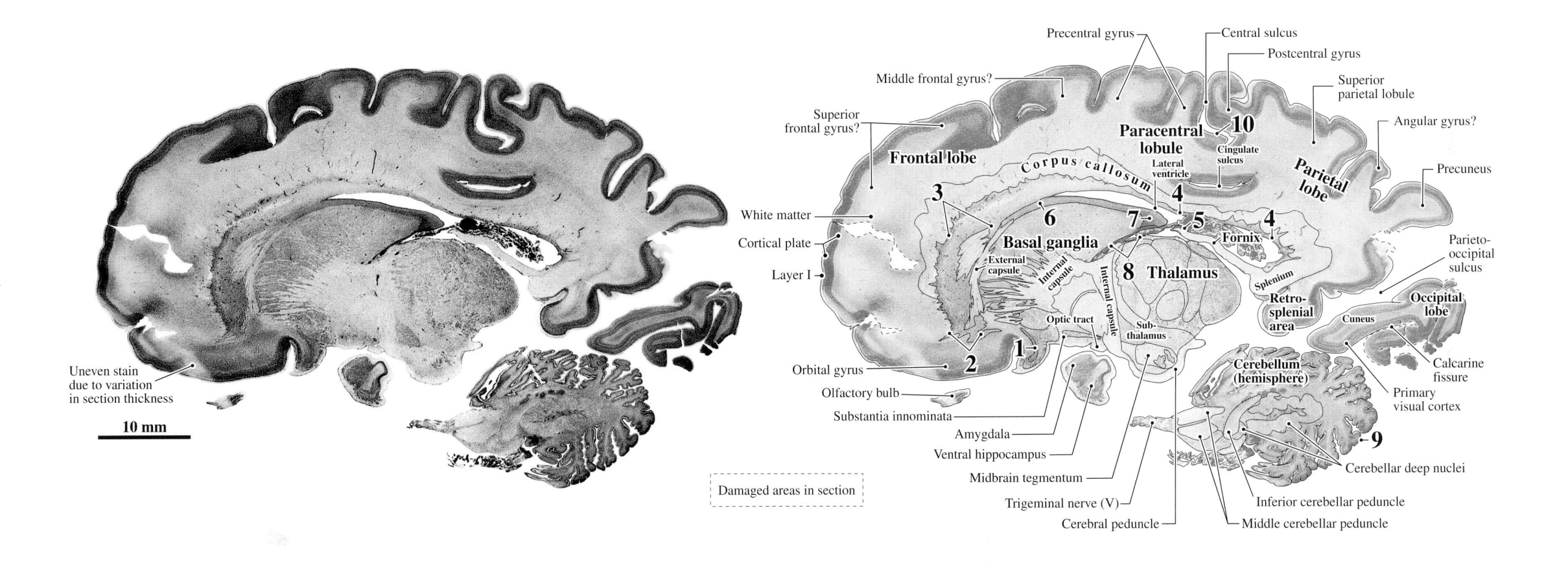

## PLATE 51

**GW37 Sagittal**
**CR 310 mm**
**Y180-61**
**Level 5: Section 721**

**See detail of brain core and cerebellum in Plates 59A and B.**

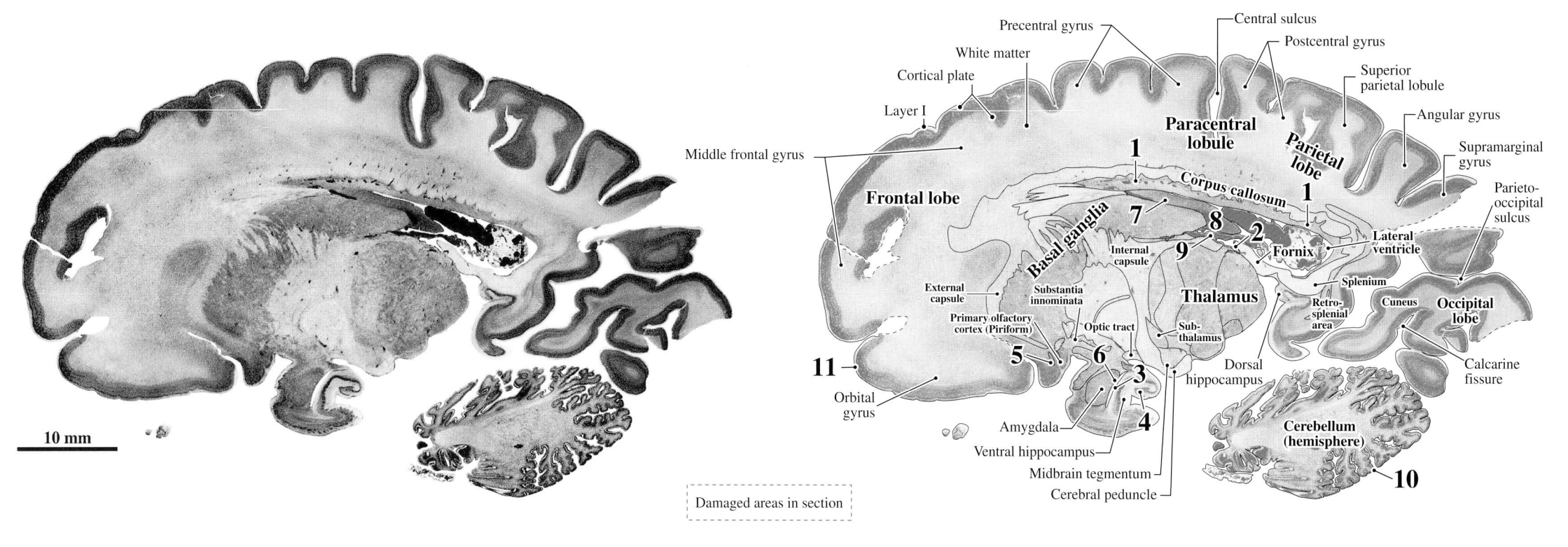

*Remnants of the germinal matrix, migratory streams, and transitional fields*

| 1 | Callosal glioepithelium | 6 | Amygdaloid glioepithelium/ependyma |
| --- | --- | --- | --- |
| 2 | Fornical glioepithelium | 7 | Anterolateral striatal neuroepithelium and subventricular zone |
| 3 | Alvear glioepithelium | 8 | Posterior striatal neuroepithelium and subventricular zone |
| 4 | Subgranular zone (dentate) | 9 | Strionuclear glioepithelium |
| 5 | Lateral migratory stream (cortical) | 10 | External germinal layer (cerebellum) |
|  |  | 11 | Subpial granular layer (cortical) |

## PLATE 52

**GW37 Sagittal
CR 310 mm
Y180-61
Level 6: Section 661**

***Remnants of the germinal matrix, migratory streams, and transitional fields***

| | | | |
|---|---|---|---|
| 1 | Callosal glioepithelium | 6 | Amygdaloid glioepithelium/ependyma |
| 2 | Fornical glioepithelium | 7 | Posterior striatal neuroepithelium and subventricular zone |
| 3 | Alvear glioepithelium | 8 | Strionuclear glioepithelium |
| 4 | Subgranular zone (dentate) | 9 | External germinal layer (cerebellum) |
| 5 | Lateral migratory stream (cortical) | 10 | Subpial granular layer (cortical) |

See detail of brain core and cerebellum in Plates 60A and B.

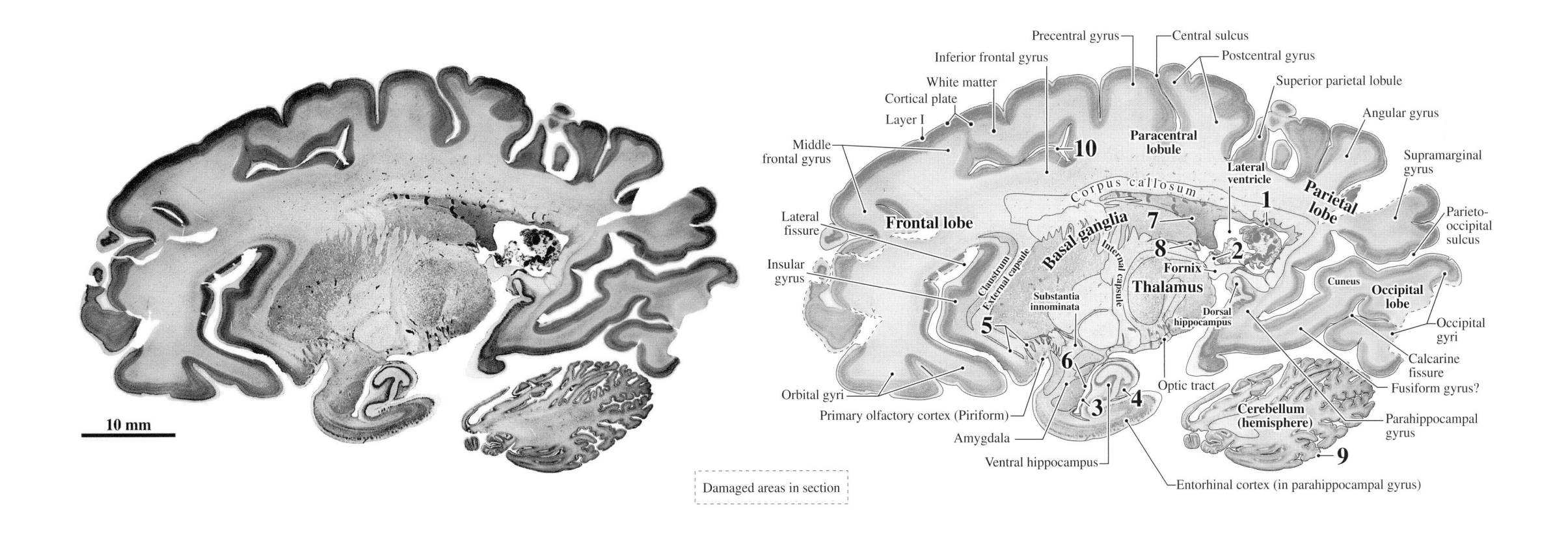

# PLATE 53

GW37 Sagittal
CR 310 mm
Y180-61
Level 7: Section 621

See detail of brain core and cerebellum in Plates 61A and B.

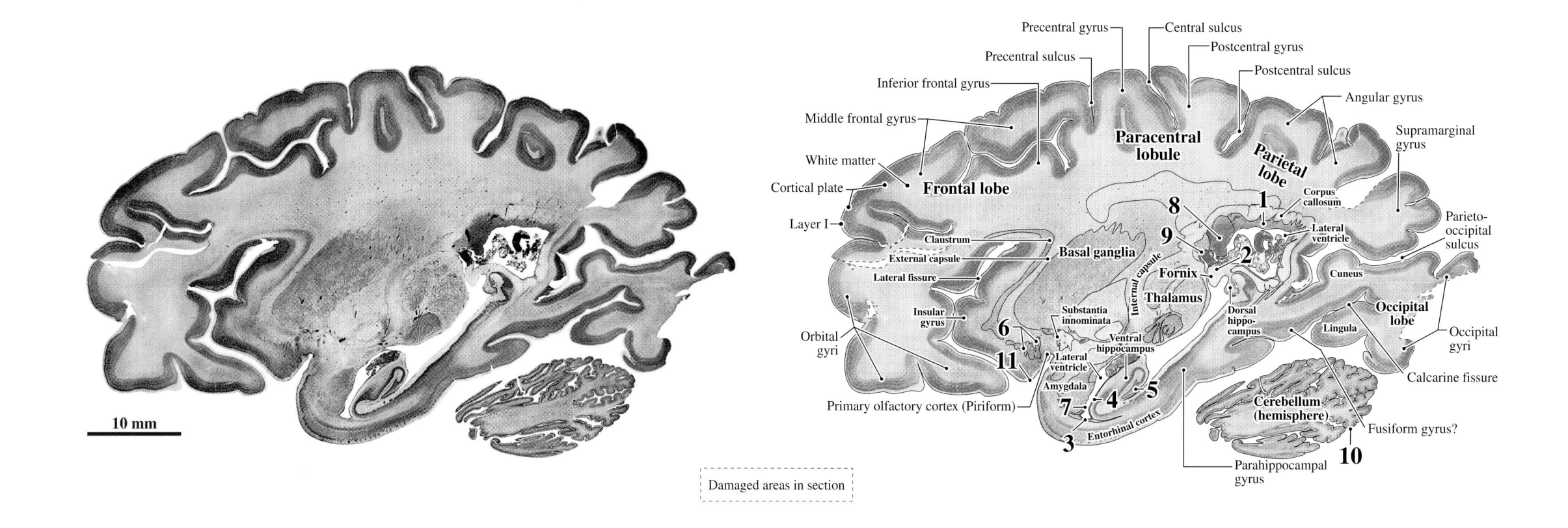

### Remnants of the germinal matrix, migratory streams, and transitional fields

1 Callosal glioepithelium
2 Fornical glioepithelium
3 Parahippocampal neuroepithelium, subventricular zone, and stratified transitional field
4 Alvear glioepithelium
5 Subgranular zone (dentate)
6 Lateral migratory stream (cortical)
7 Amygdaloid glioepithelium/ependyma
8 Posterior striatal neuroepithelium and subventricular zone
9 Strionuclear glioepithelium
10 External germinal layer (cerebellum)
11 Subpial granular layer (cortical)

# PLATE 54

**GW37 Sagittal
CR 310 mm
Y180-61
Level 8: Section 501**

## Remnants of the germinal matrix, migratory streams, and transitional fields

1 Parietal neuroepithelium, subventricular zone, and stratified transitional field
2 Occipital neuroepithelium, subventricular zone, and stratified transitional field
3 Parahippocampal neuroepithelium, subventricular zone, and stratified transitional field
4 Alvear glioepithelium
5 Subgranular zone (dentate)
6 Lateral migratory stream (cortical)
7 Amygdaloid glioepithelium/ependyma
8 Posterior striatal neuroepithelium and subventricular zone
9 Strionuclear glioepithelium
10 External germinal layer (cerebellum)
11 Subpial granular layer (cortical)

See detail of brain core and cerebellum in Plates 62A and B.

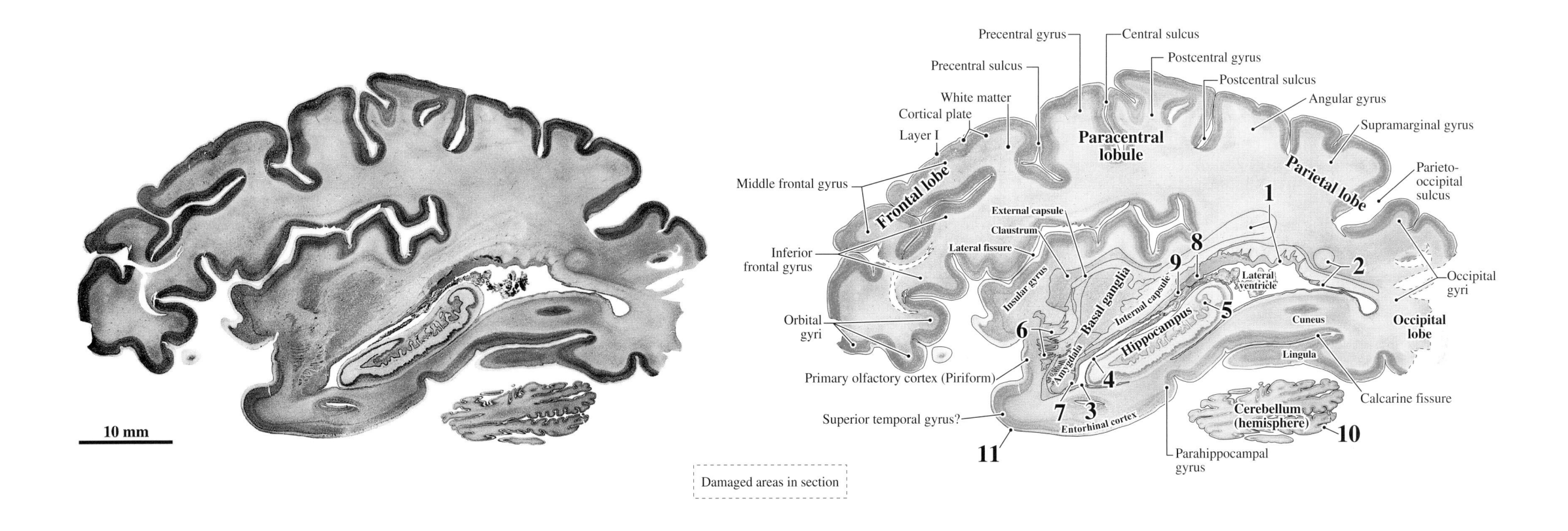

**PLATE 55A**
GW37 Sagittal
CR 310 mm
Y180-61
Level 1: Section 961

2.5 mm

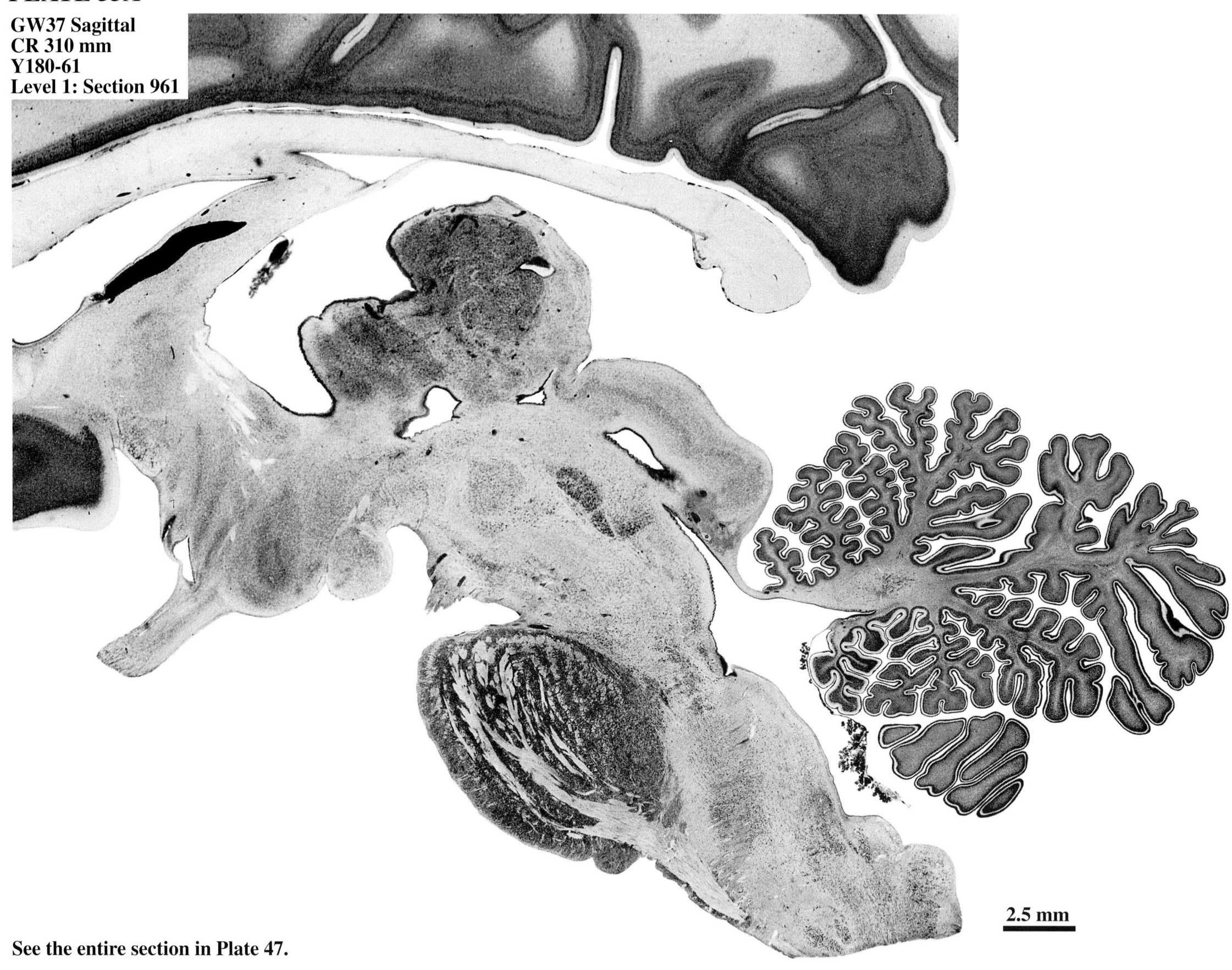

See the entire section in Plate 47.

PLATE 55B

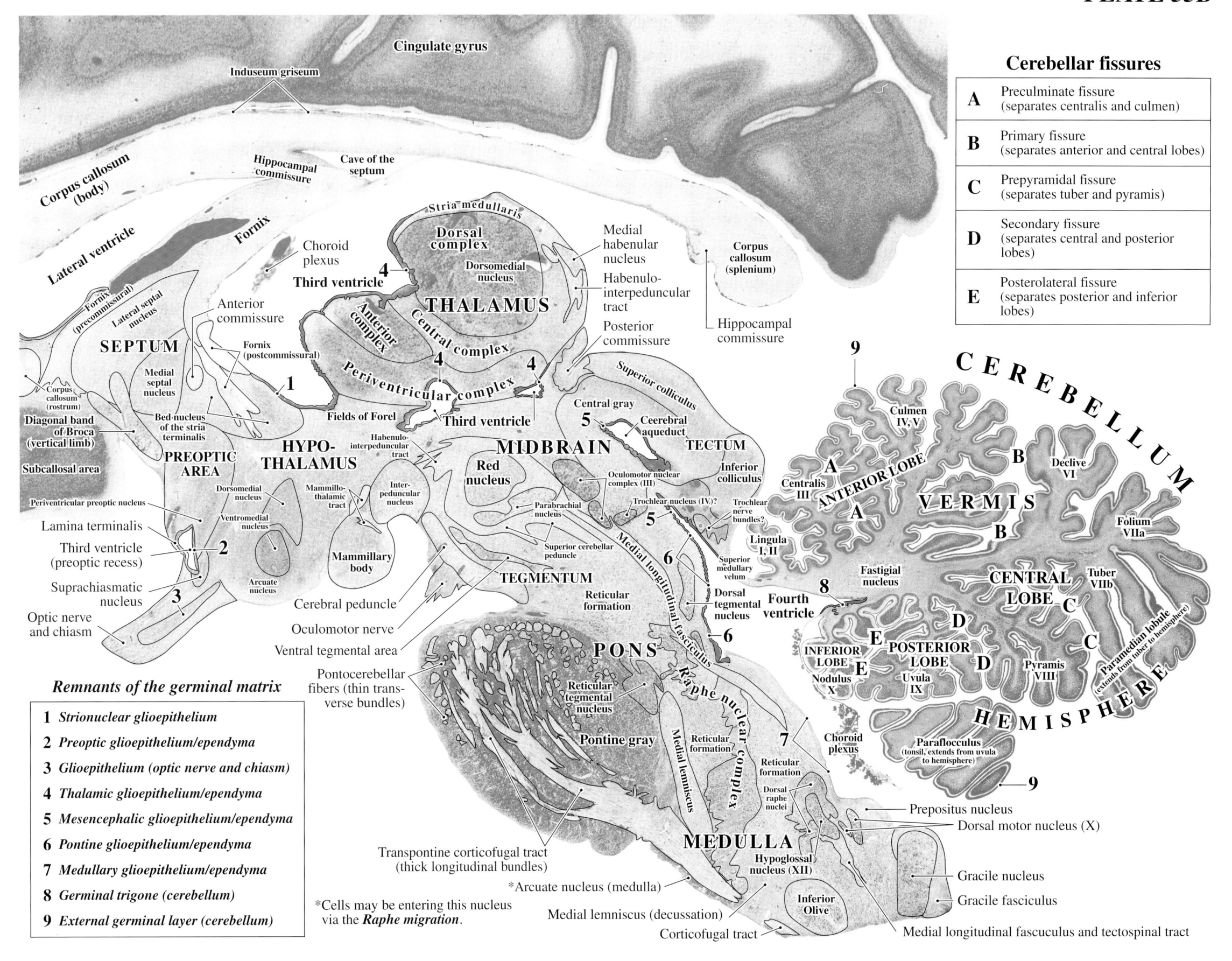

### Cerebellar fissures

| | | |
|---|---|---|
| A | Preculminate fissure | (separates centralis and culmen) |
| B | Primary fissure | (separates anterior and central lobes) |
| C | Prepyramidal fissure | (separates tuber and pyramis) |
| D | Secondary fissure | (separates central and posterior lobes) |
| E | Posterolateral fissure | (separates posterior and inferior lobes) |

### Remnants of the germinal matrix

1. *Strionuclear glioepithelium*
2. *Preoptic glioepithelium/ependyma*
3. *Glioepithelium (optic nerve and chiasm)*
4. *Thalamic glioepithelium/ependyma*
5. *Mesencephalic glioepithelium/ependyma*
6. *Pontine glioepithelium/ependyma*
7. *Medullary glioepithelium/ependyma*
8. *Germinal trigone (cerebellum)*
9. *External germinal layer (cerebellum)*

*Cells may be entering this nucleus via the **Raphe migration**.

**PLATE 56A**

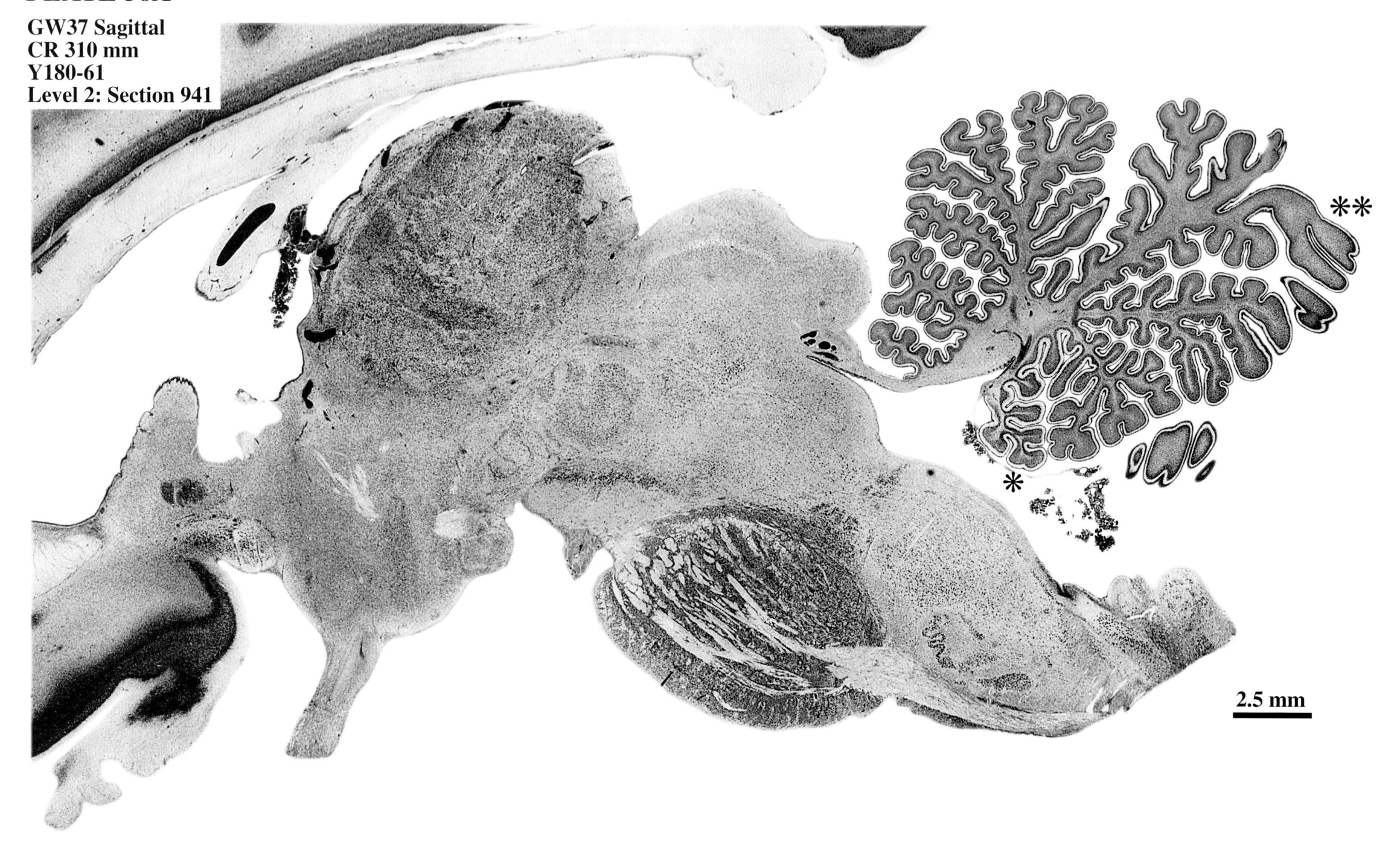

GW37 Sagittal
CR 310 mm
Y180-61
Level 2: Section 941

2.5 mm

See the entire section in Plate 48.

✻ Area of high magnification in Plate 63

✻✻ Area of high magnification in Plate 64

# PLATE 56B

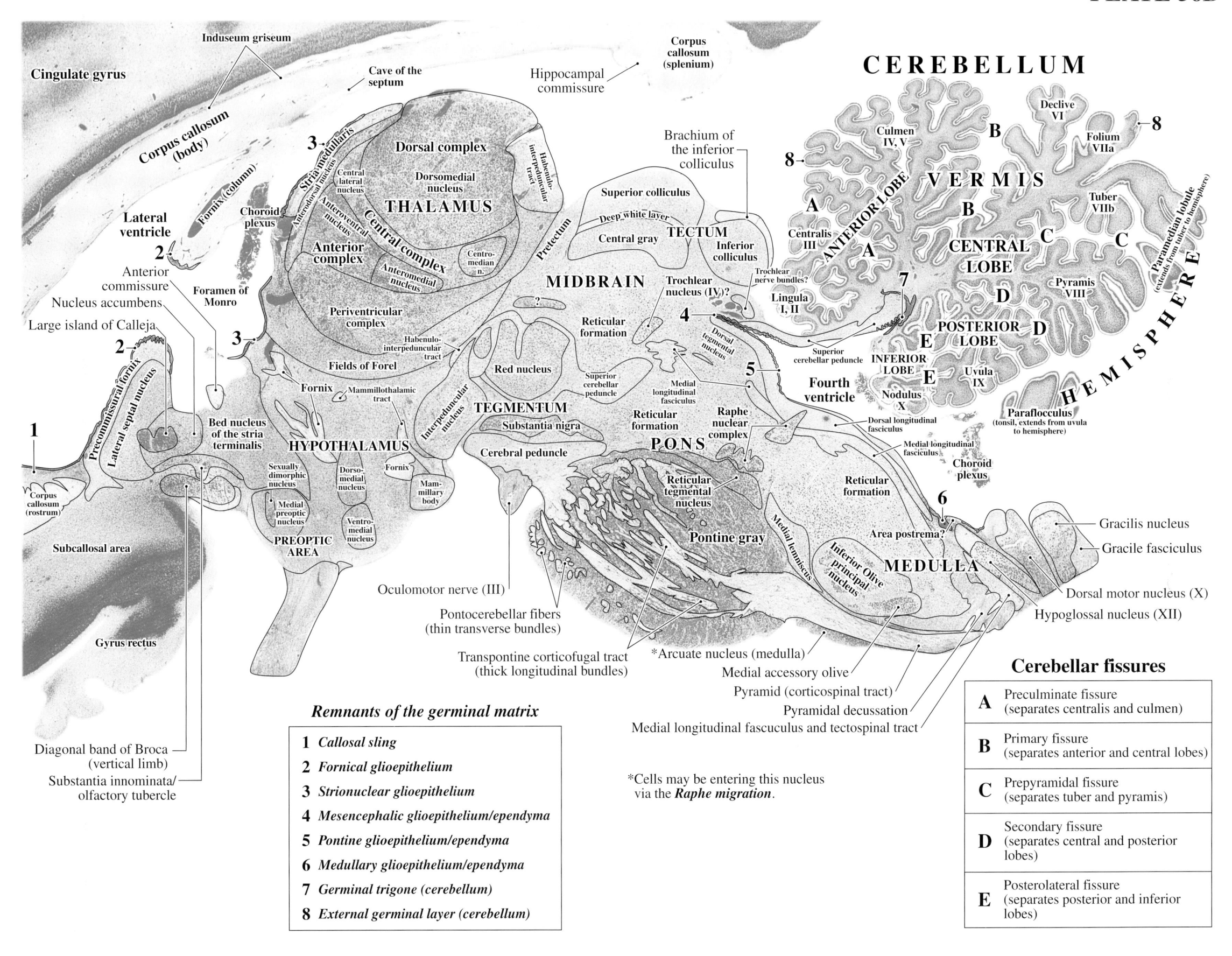

## Cerebellar fissures

| | | |
|---|---|---|
| **A** | Preculminate fissure | (separates centralis and culmen) |
| **B** | Primary fissure | (separates anterior and central lobes) |
| **C** | Prepyramidal fissure | (separates tuber and pyramis) |
| **D** | Secondary fissure | (separates central and posterior lobes) |
| **E** | Posterolateral fissure | (separates posterior and inferior lobes) |

### Remnants of the germinal matrix

1. *Callosal sling*
2. *Fornical glioepithelium*
3. *Strionuclear glioepithelium*
4. *Mesencephalic glioepithelium/ependyma*
5. *Pontine glioepithelium/ependyma*
6. *Medullary glioepithelium/ependyma*
7. *Germinal trigone (cerebellum)*
8. *External germinal layer (cerebellum)*

*Cells may be entering this nucleus via the **Raphe migration**.

## PLATE 57A

GW37 Sagittal
CR 310 mm
Y180-61
Level 3: Section 841

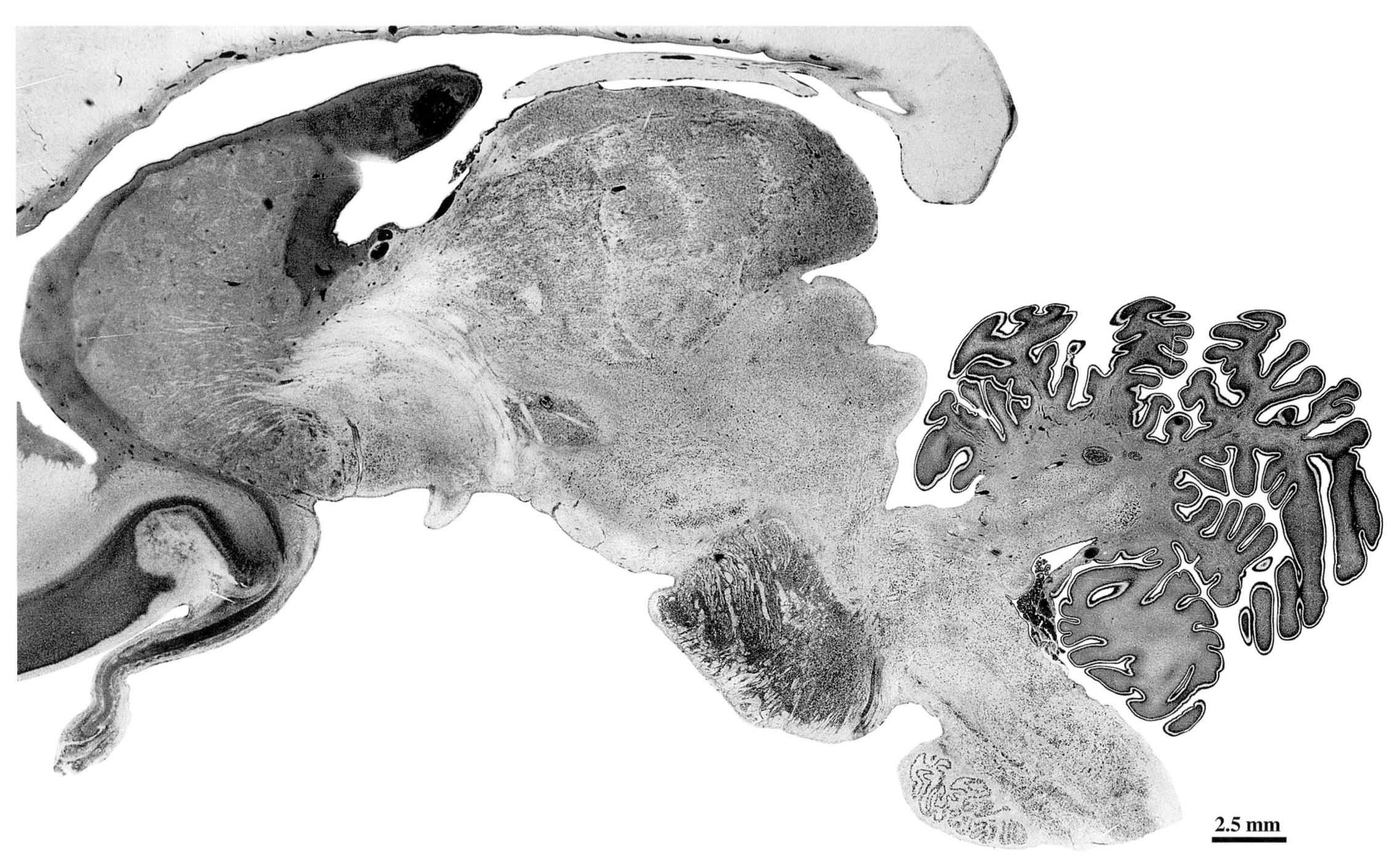

2.5 mm

See the entire section in Plate 49.

PLATE 57B

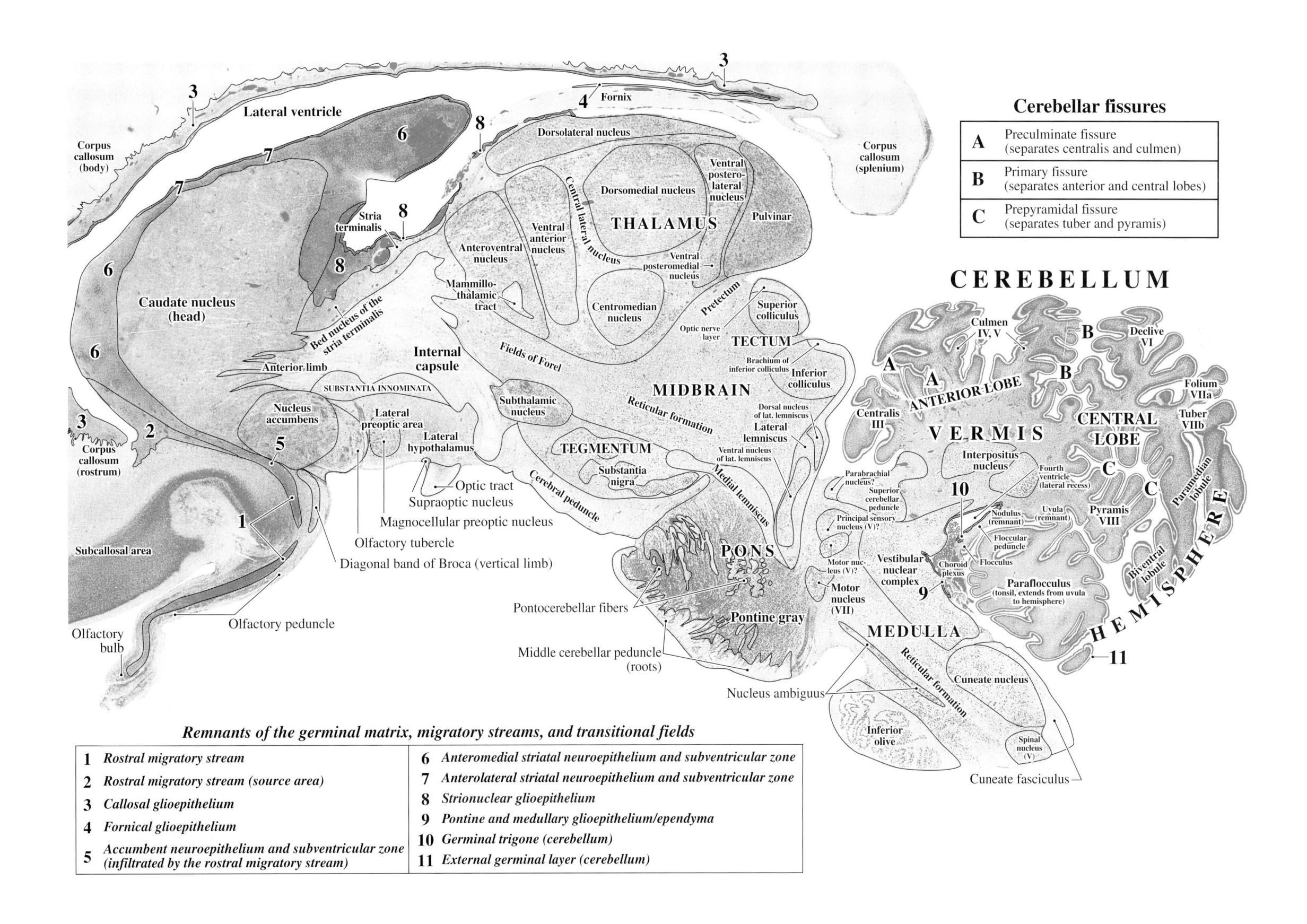

**Cerebellar fissures**

| | | |
|---|---|---|
| **A** | Preculminate fissure (separates centralis and culmen) | |
| **B** | Primary fissure (separates anterior and central lobes) | |
| **C** | Prepyramidal fissure (separates tuber and pyramis) | |

*Remnants of the germinal matrix, migratory streams, and transitional fields*

| | | | | |
|---|---|---|---|---|
| 1 | *Rostral migratory stream* | 6 | *Anteromedial striatal neuroepithelium and subventricular zone* | |
| 2 | *Rostral migratory stream (source area)* | 7 | *Anterolateral striatal neuroepithelium and subventricular zone* | |
| 3 | *Callosal glioepithelium* | 8 | *Strionuclear glioepithelium* | |
| 4 | *Fornical glioepithelium* | 9 | *Pontine and medullary glioepithelium/ependyma* | |
| 5 | *Accumbent neuroepithelium and subventricular zone (infiltrated by the rostral migratory stream)* | 10 | *Germinal trigone (cerebellum)* | |
| | | 11 | *External germinal layer (cerebellum)* | |

## PLATE 58A

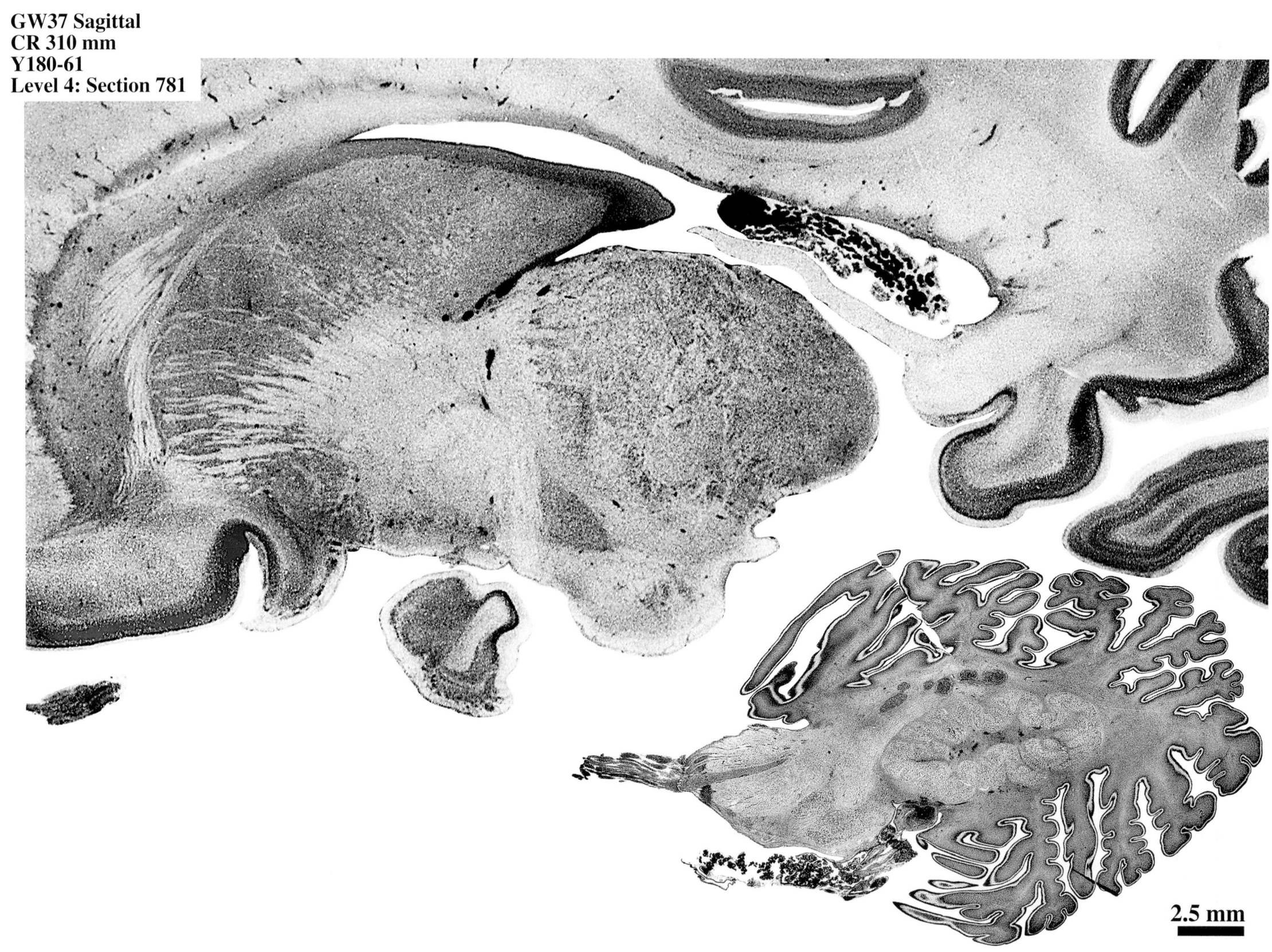

GW37 Sagittal
CR 310 mm
Y180-61
Level 4: Section 781

2.5 mm

See the entire section in Plate 50.

PLATE 58B

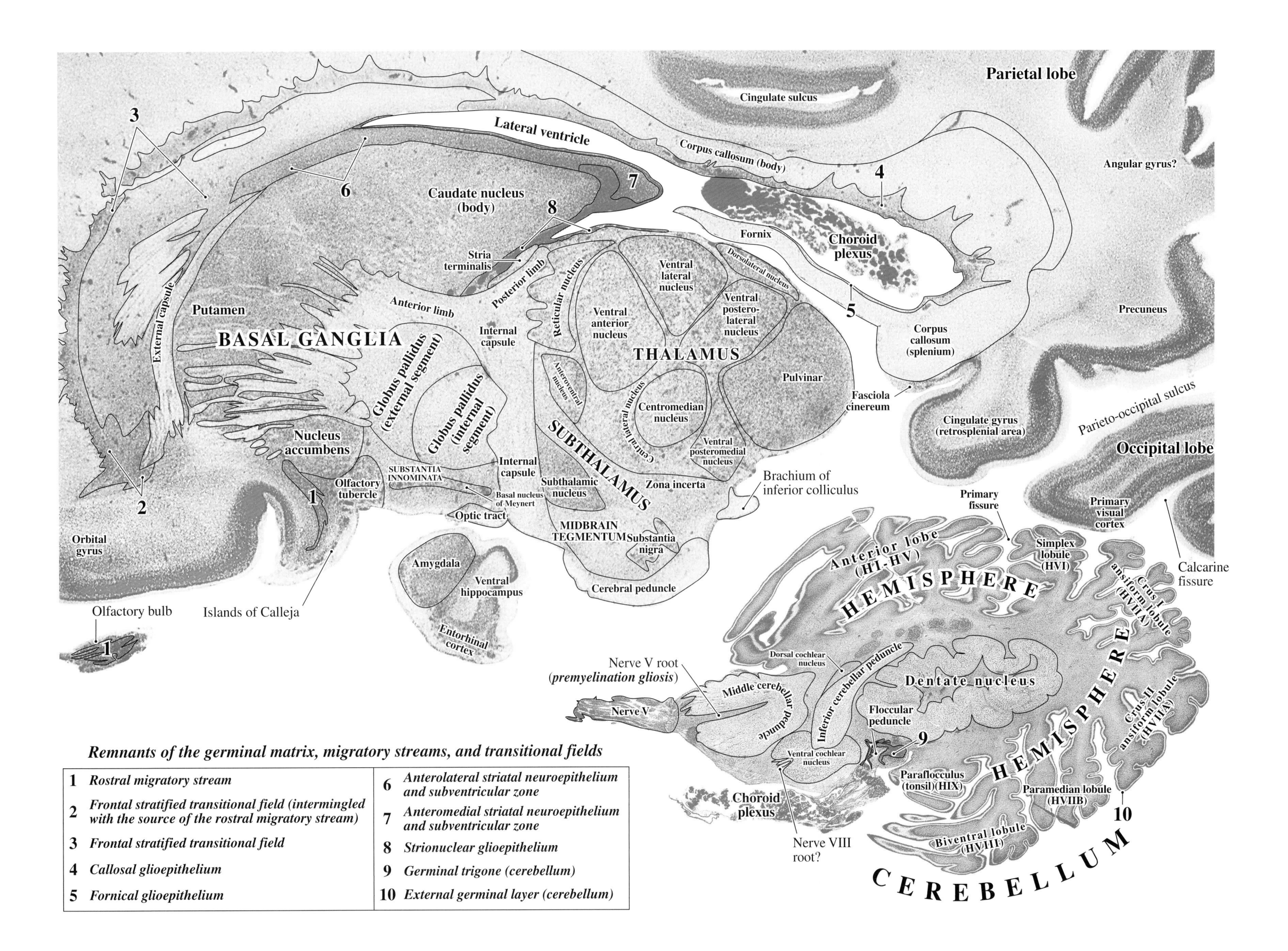

**Remnants of the germinal matrix, migratory streams, and transitional fields**

| | | | |
|---|---|---|---|
| 1 | *Rostral migratory stream* | 6 | *Anterolateral striatal neuroepithelium and subventricular zone* |
| 2 | *Frontal stratified transitional field (intermingled with the source of the rostral migratory stream)* | 7 | *Anteromedial striatal neuroepithelium and subventricular zone* |
| 3 | *Frontal stratified transitional field* | 8 | *Strionuclear glioepithelium* |
| 4 | *Callosal glioepithelium* | 9 | *Germinal trigone (cerebellum)* |
| 5 | *Fornical glioepithelium* | 10 | *External germinal layer (cerebellum)* |

**PLATE 59A**

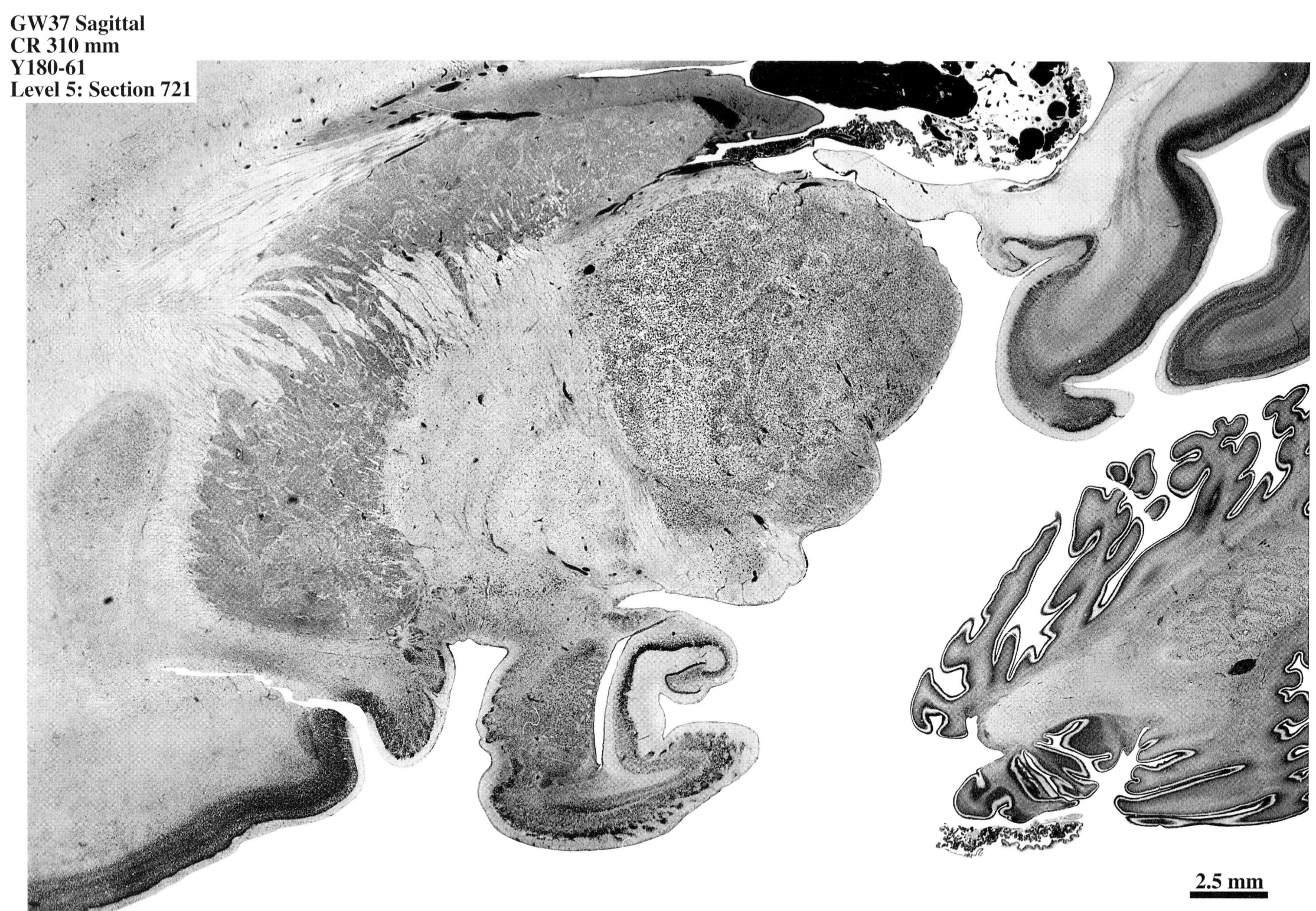

GW37 Sagittal
CR 310 mm
Y180-61
Level 5: Section 721

2.5 mm

See the entire section in Plate 51.

**PLATE 59B**

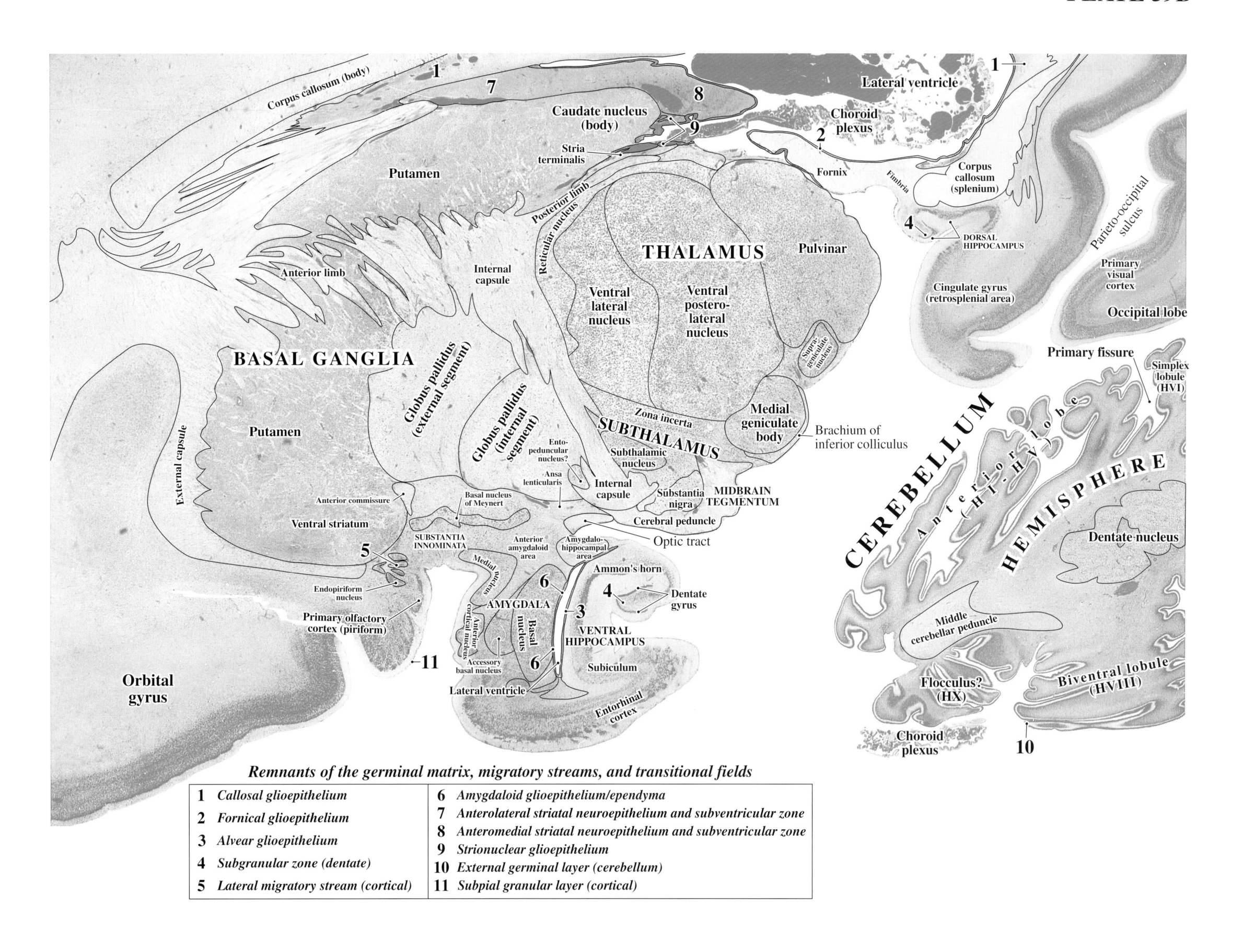

*Remnants of the germinal matrix, migratory streams, and transitional fields*

| | |
|---|---|
| 1 Callosal glioepithelium | 6 Amygdaloid glioepithelium/ependyma |
| 2 Fornical glioepithelium | 7 Anterolateral striatal neuroepithelium and subventricular zone |
| 3 Alvear glioepithelium | 8 Anteromedial striatal neuroepithelium and subventricular zone |
| 4 Subgranular zone (dentate) | 9 Strionuclear glioepithelium |
| 5 Lateral migratory stream (cortical) | 10 External germinal layer (cerebellum) |
| | 11 Subpial granular layer (cortical) |

## PLATE 60A

**GW37 Sagittal
CR 310 mm, Y180-61
Level 6: Section 661**

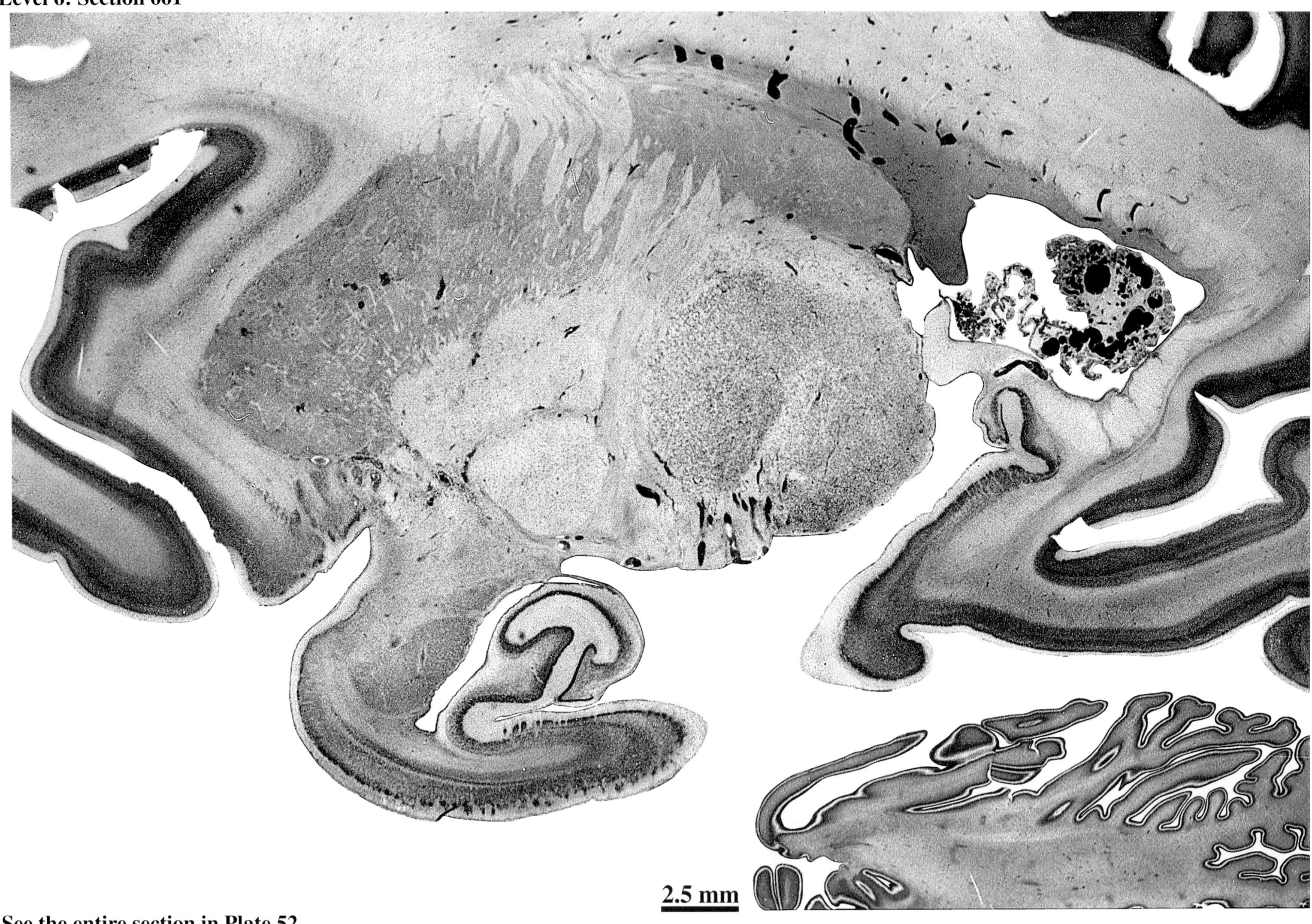

2.5 mm

See the entire section in Plate 52.

**PLATE 60B**

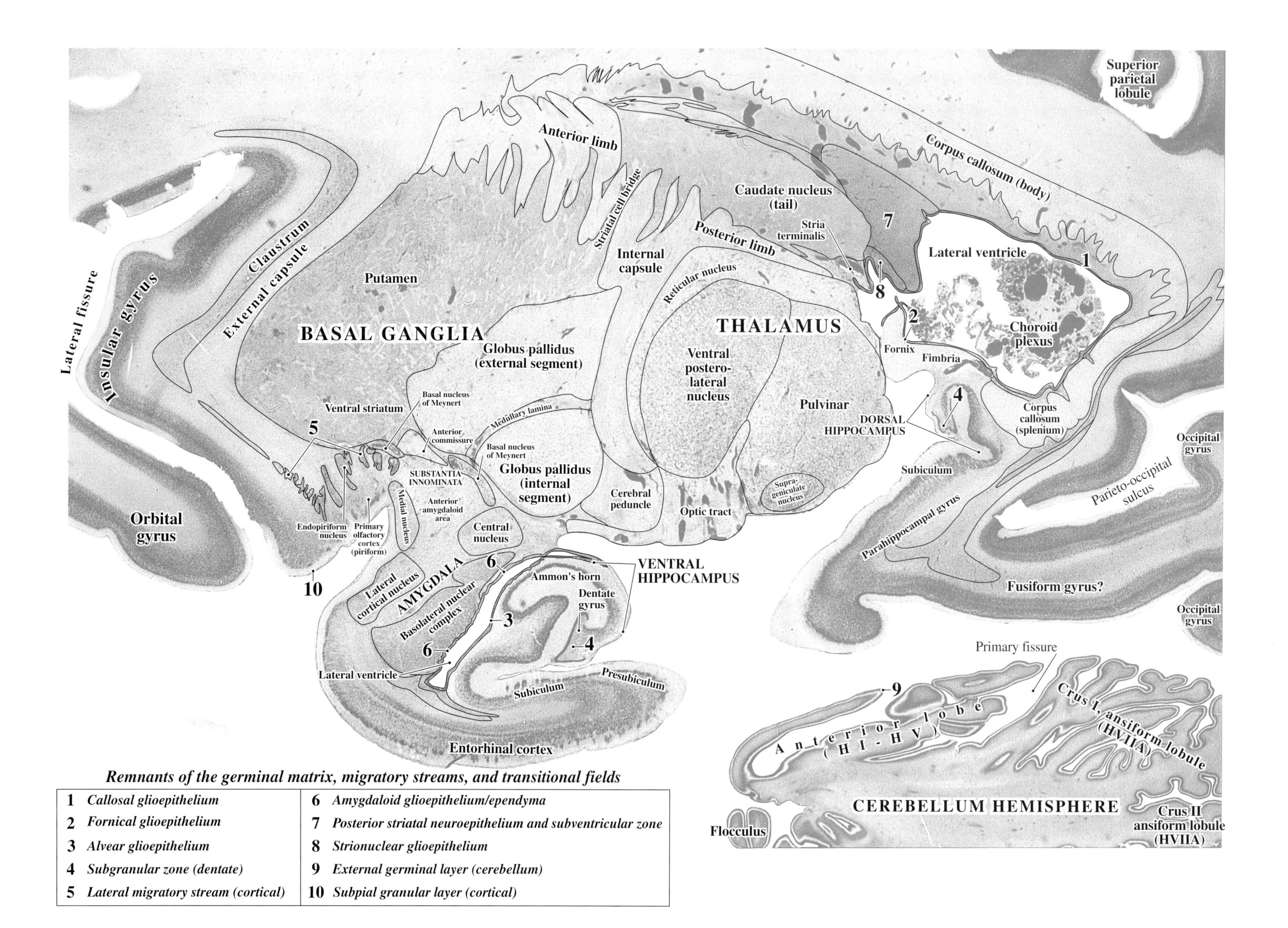

**Remnants of the germinal matrix, migratory streams, and transitional fields**

| | |
|---|---|
| 1 *Callosal glioepithelium* | 6 *Amygdaloid glioepithelium/ependyma* |
| 2 *Fornical glioepithelium* | 7 *Posterior striatal neuroepithelium and subventricular zone* |
| 3 *Alvear glioepithelium* | 8 *Strionuclear glioepithelium* |
| 4 *Subgranular zone (dentate)* | 9 *External germinal layer (cerebellum)* |
| 5 *Lateral migratory stream (cortical)* | 10 *Subpial granular layer (cortical)* |

## PLATE 61A

**GW37 Sagittal
CR 310 mm, Y180-61
Level 7: Section 621**

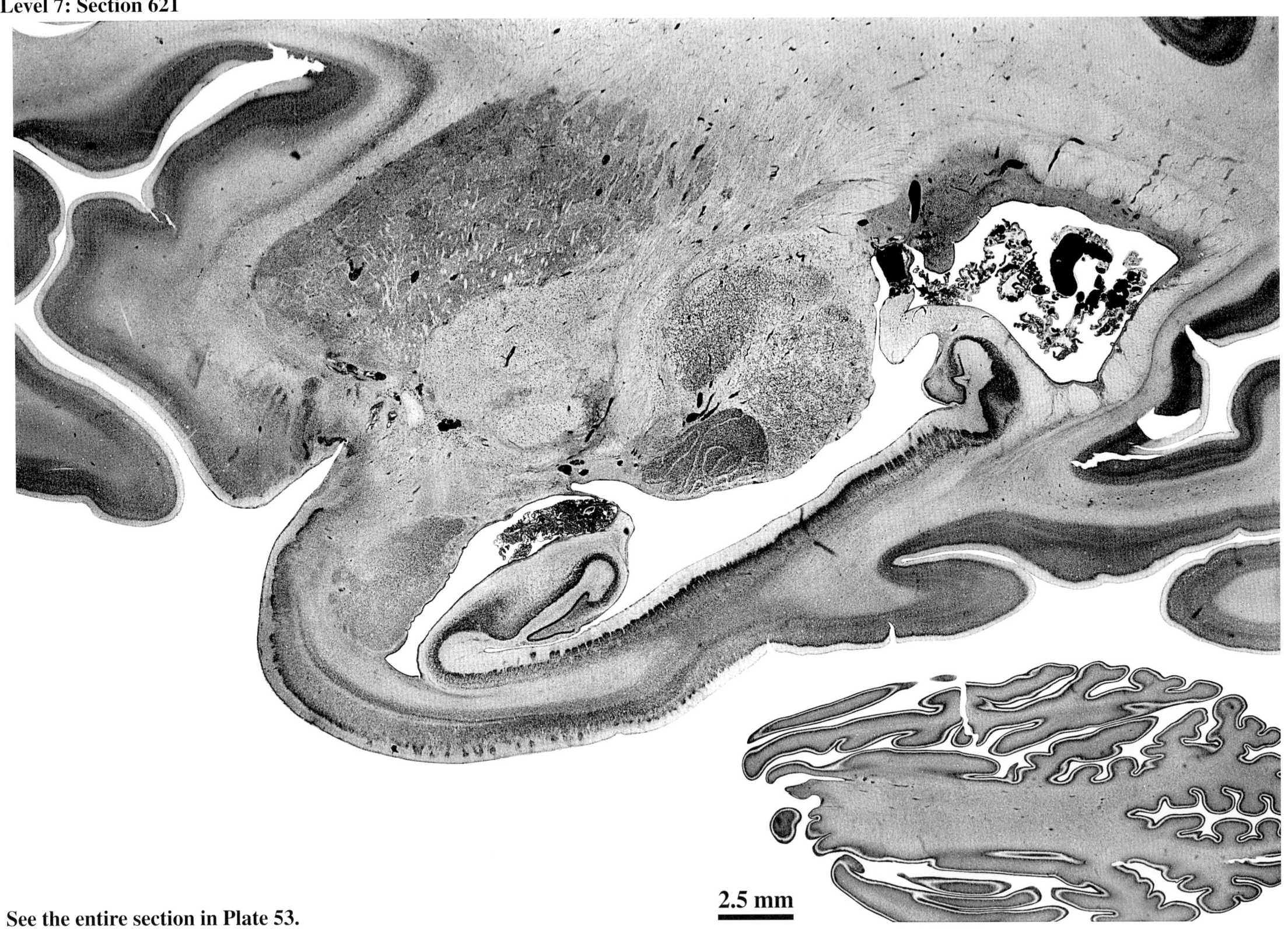

2.5 mm

See the entire section in Plate 53.

# PLATE 61B

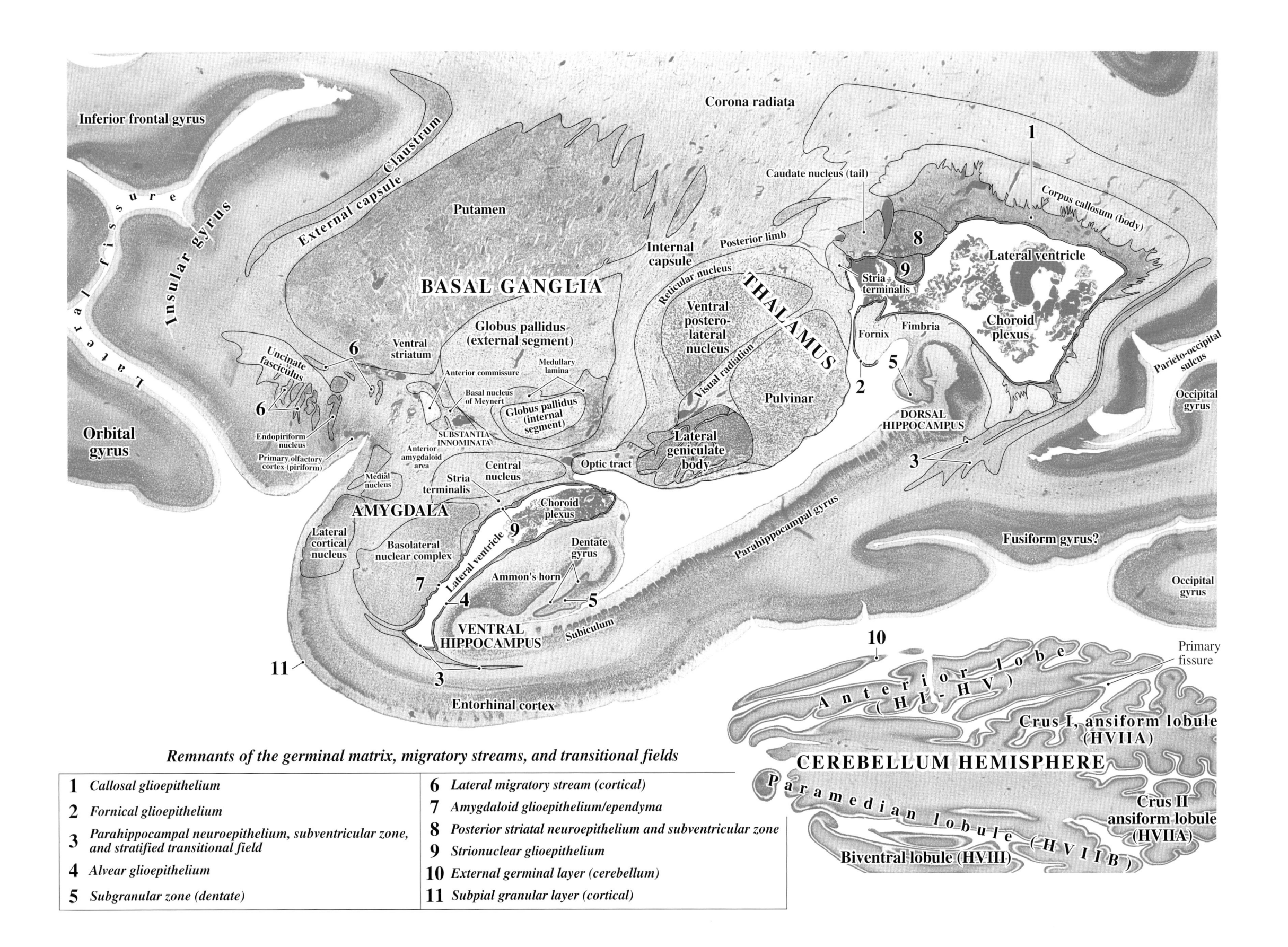

*Remnants of the germinal matrix, migratory streams, and transitional fields*

| | |
|---|---|
| 1 *Callosal glioepithelium* | 6 *Lateral migratory stream (cortical)* |
| 2 *Fornical glioepithelium* | 7 *Amygdaloid glioepithelium/ependyma* |
| 3 *Parahippocampal neuroepithelium, subventricular zone, and stratified transitional field* | 8 *Posterior striatal neuroepithelium and subventricular zone* |
| | 9 *Strionuclear glioepithelium* |
| 4 *Alvear glioepithelium* | 10 *External germinal layer (cerebellum)* |
| 5 *Subgranular zone (dentate)* | 11 *Subpial granular layer (cortical)* |

## PLATE 62A

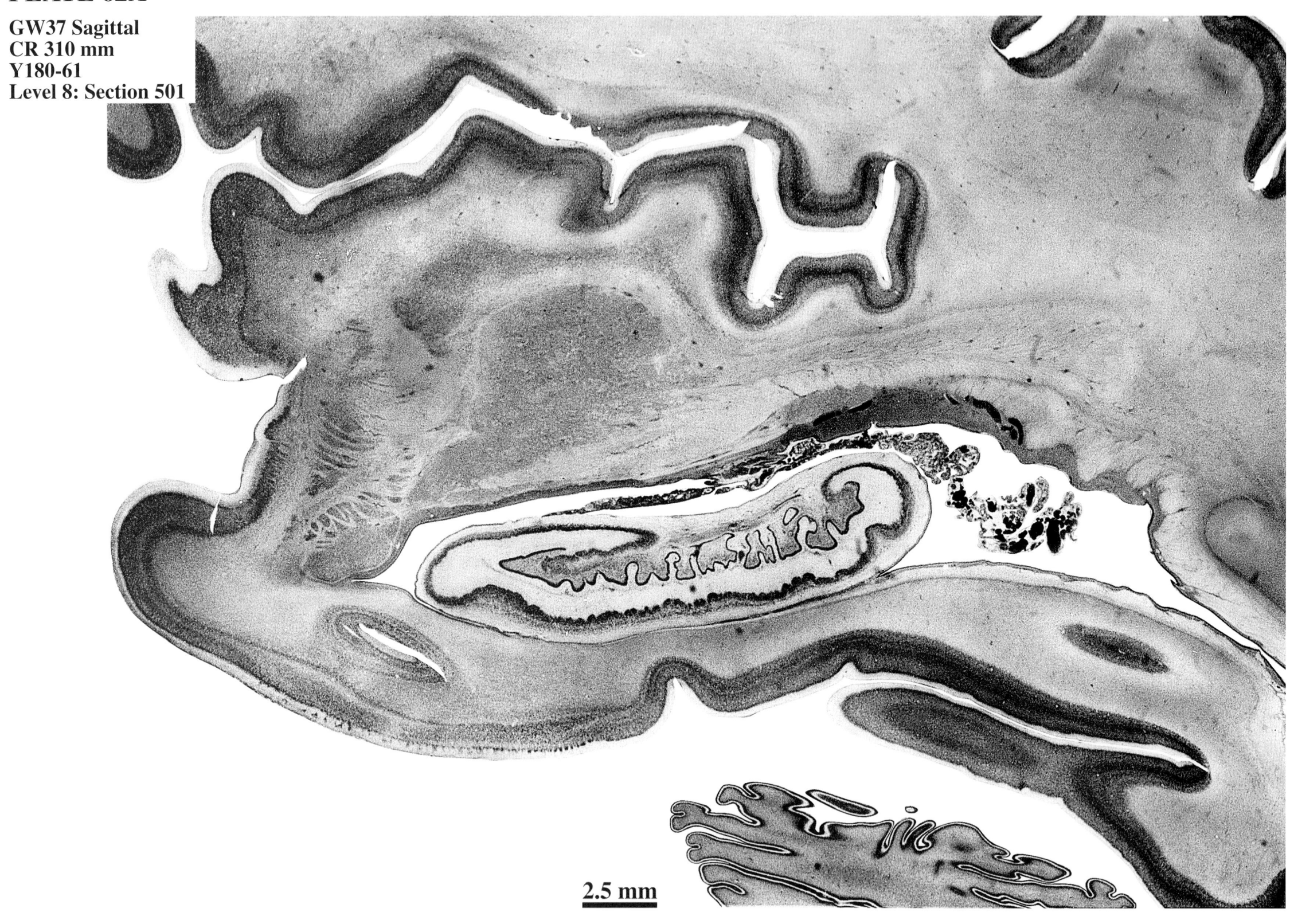

GW37 Sagittal
CR 310 mm
Y180-61
Level 8: Section 501

2.5 mm

See the entire section in Plate 54.

## PLATE 62B

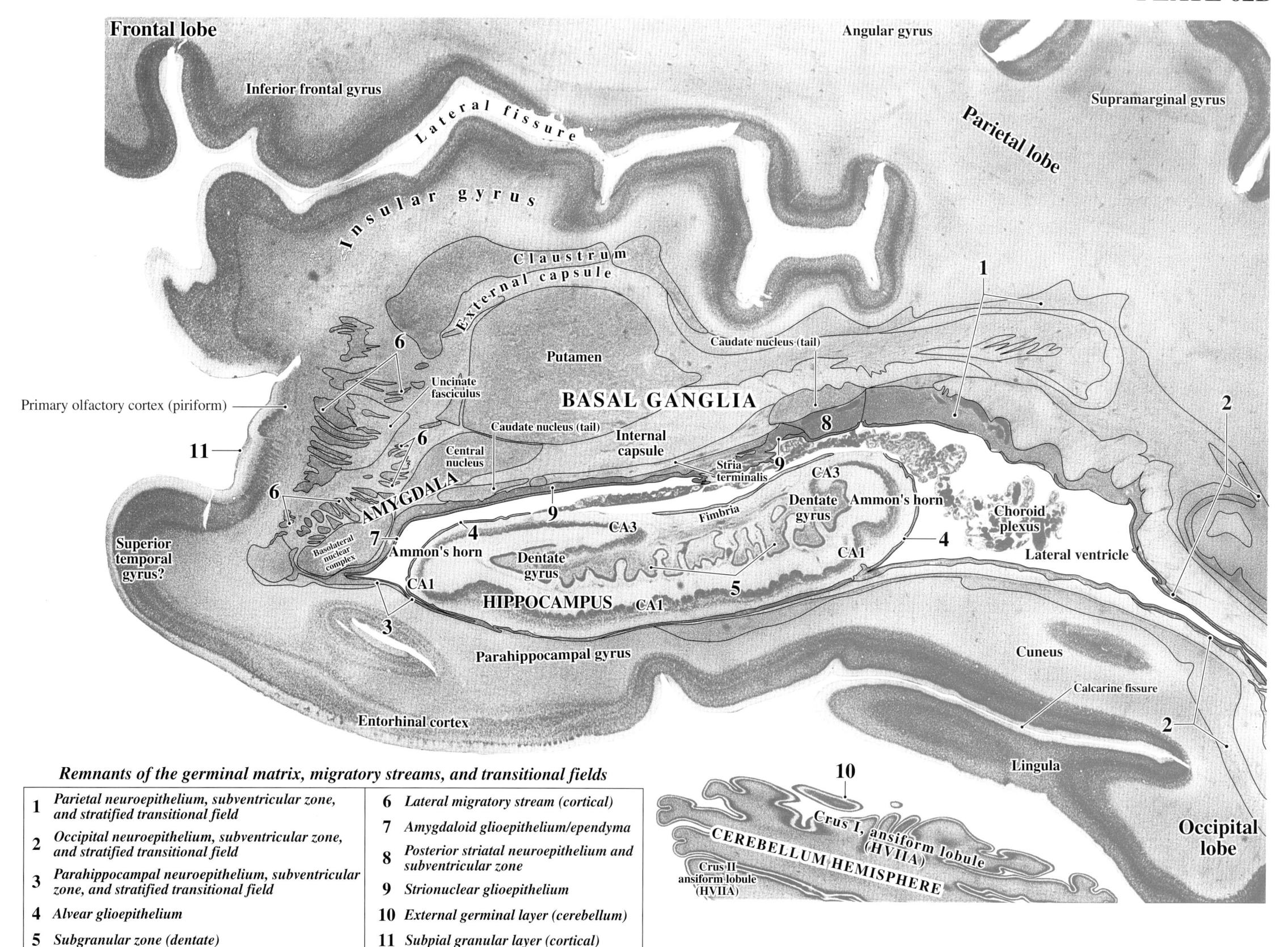

**Remnants of the germinal matrix, migratory streams, and transitional fields**

| | | | |
|---|---|---|---|
| 1 | Parietal neuroepithelium, subventricular zone, and stratified transitional field | 6 | Lateral migratory stream (cortical) |
| 2 | Occipital neuroepithelium, subventricular zone, and stratified transitional field | 7 | Amygdaloid glioepithelium/ependyma |
| 3 | Parahippocampal neuroepithelium, subventricular zone, and stratified transitional field | 8 | Posterior striatal neuroepithelium and subventricular zone |
| 4 | Alvear glioepithelium | 9 | Strionuclear glioepithelium |
| 5 | Subgranular zone (dentate) | 10 | External germinal layer (cerebellum) |
| | | 11 | Subpial granular layer (cortical) |

# PLATE 63

**GW37, CR 310 mm, Y180-61 Level 2, Section 941**

**CEREBELLUM, VERMIS, INFERIOR LOBE, LOBULE X: NODULUS**
(A relatively mature region of the cerebellar cortex)

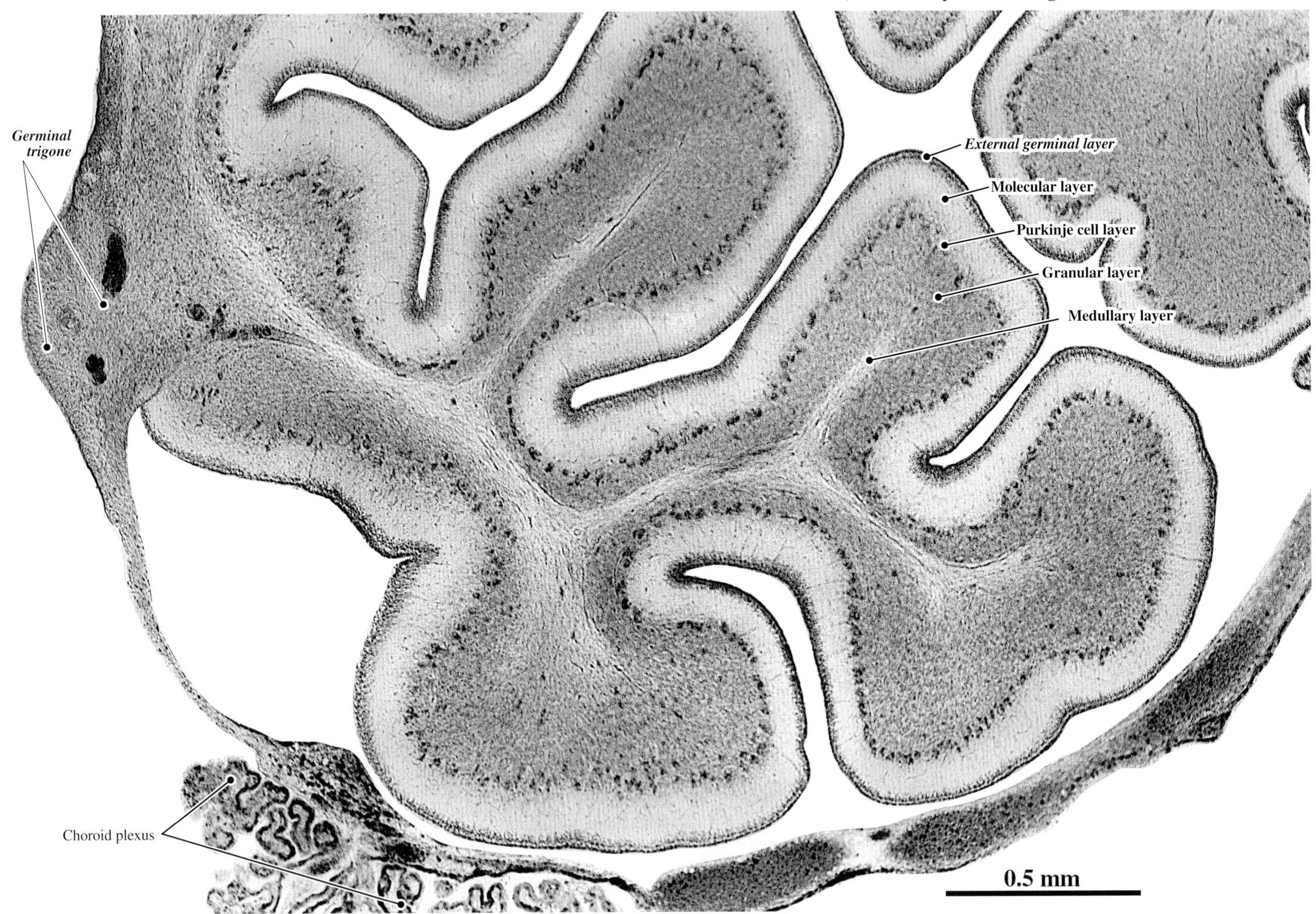

*Germinal trigone*

*External germinal layer*

Molecular layer

Purkinje cell layer

Granular layer

Medullary layer

Choroid plexus

0.5 mm

See the entire section in Plate 48.
See larger view of the entire cerebellum in Plates 56A and 56B.

## PLATE 64

**GW37, CR 310 mm, Y180-61**
**Level 2, Section 941**
**CEREBELLUM, HEMISPHERE, PARAMEDIAN LOBULE**
(A relatively immature region of the cerebellar cortex)

**GW37, CR 310 mm, Y180-61**
**Level 4, Section 781**
**CEREBELLUM, HEMISPHERE, CORTEX**
(very high magnification)

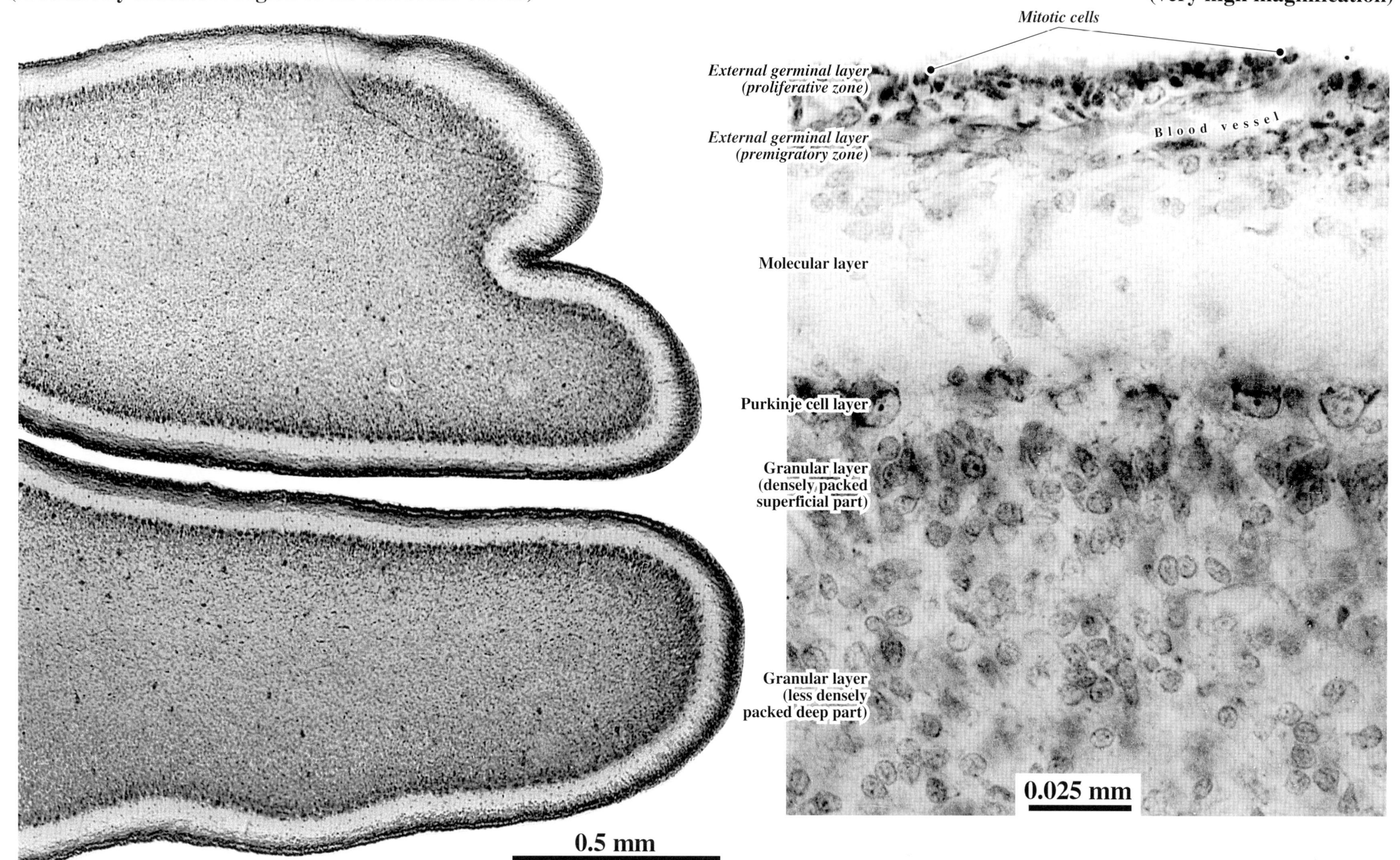

*Mitotic cells*

*External germinal layer (proliferative zone)*

*External germinal layer (premigratory zone)*

Blood vessel

Molecular layer

Purkinje cell layer

Granular layer (densely packed superficial part)

Granular layer (less densely packed deep part)

0.025 mm

0.5 mm

See the entire section in Plate 48.
See larger view of the entire cerebellum in Plates 56A and 56B.

See the entire section in Plate 50.
See larger view of the entire cerebellum in Plates 58A and 58B.

# PART IV: GW37 HORIZONTAL

This specimen is case number B-301-62 (Perinatal RPSL) in the Yakovlev Collection. A male infant was stillborn after a premature separation of the placenta. This brain is classified as a Normative Control in the Yakovlev Collection (Haleem, 1990). It was cut in the horizontal plane in 35-μm thick sections. Since there is no available photograph of this brain before it was embedded and cut, a photograph of the lateral view of another GW37 brain that Larroche published in 1967 (**Figure 4**) is used to show gross anatomical features.

The approximate cutting plane of this brain is indicated in **Figure 5** (facing page) with lines superimposed on the GW37 brain from the Larroche (1967) series. The anterior part of each section (on the left in all photographs) is slightly dorsal to the posterior part. In terms of the medial-lateral plane of sectioning, the part of each section near the bottom of the page shows structures cut slightly dorsal to those on the other side of the midline. As in all other specimens, the sections chosen for illustration are spaced closer together to show small structures in the diencephalon, midbrain, pons, and medulla. Illustrated sections are spaced farther apart when they contain only large brain structures, such as the cerebral cortex, basal ganglia, and cerebellum. Photographs of 19 different Nissl-stained sections (**Levels 1-19**) are shown at low magnification in **Plates 65-80**. The core of the brain and the cerebellum are shown at high magnification in companion **Plates 81AB-95AB** for **Levels 3-19**.

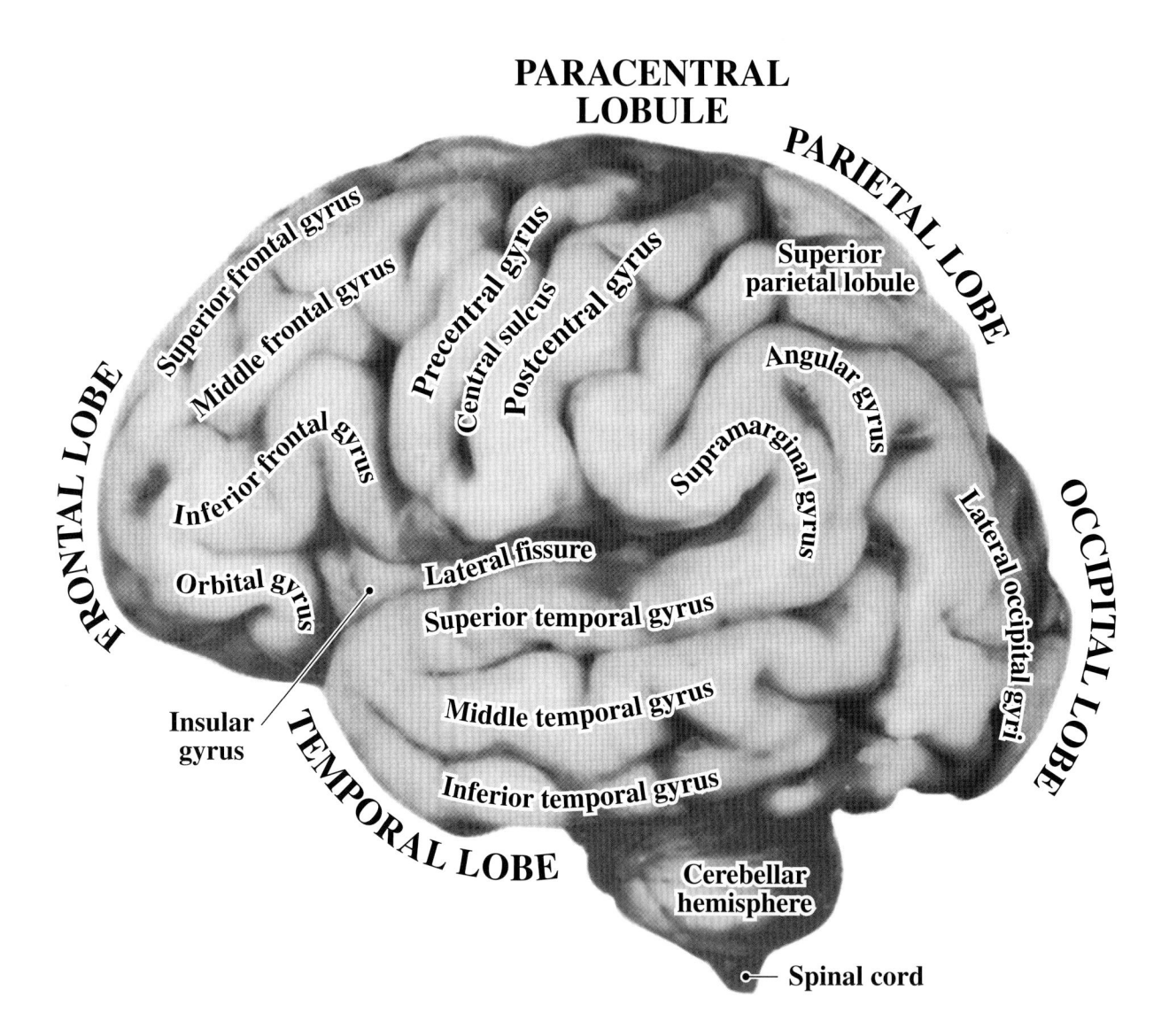

**Figure 4.** Lateral view of a GW37 brain with major structures in the cerebral hemispheres labeled (the same as **Figure 1** repeated here for convenience). (From the photographic series of J. C. Larroche (1967) Maturation morphologique du système nerveux central: ses rapports avec le développement pondéral du foetus et son age gestationnel. In: *Regional Development of the Brain in Early Life*, A. Minkowski (ed.), London: Blackwell, page 248.)

## GW37 HORIZONTAL SECTION PLANES

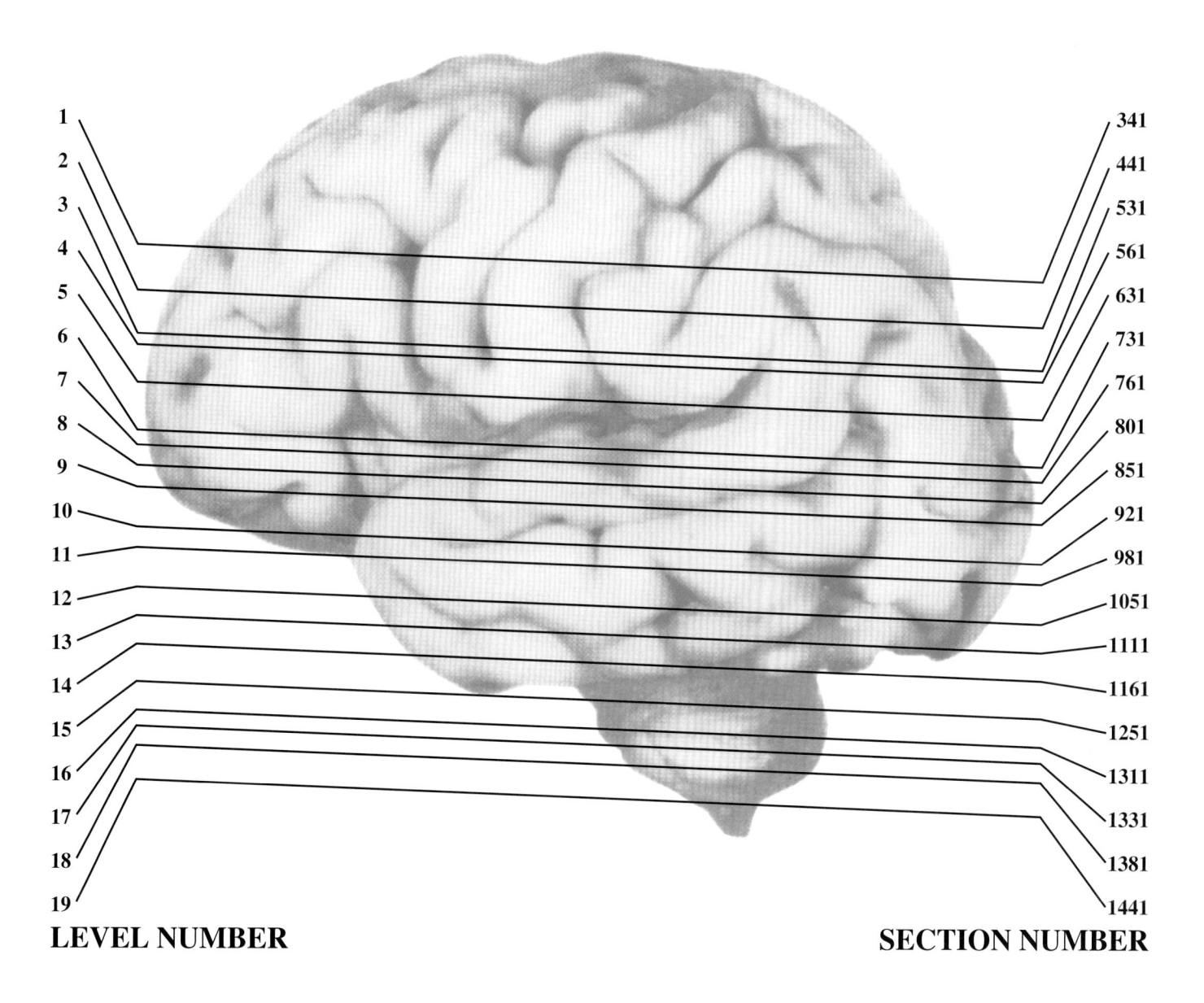

| Level Number | Section Number |
|---|---|
| 1 | 341 |
| 2 | 441 |
| 3 | 531 |
| 4 | 561 |
| 5 | 631 |
| 6 | 731 |
| 7 | 761 |
| 8 | 801 |
| 9 | 851 |
| 10 | 921 |
| 11 | 981 |
| 12 | 1051 |
| 13 | 1111 |
| 14 | 1161 |
| 15 | 1251 |
| 16 | 1311 |
| 17 | 1331 |
| 18 | 1381 |
| 19 | 1441 |

**Figure 5.** Lateral view of the same GW37 brain shown in **Figure 4** with the approximate locations and cutting angle of the sections of Y301-62. (From the photographic series of J. C. Larroche (1967) Maturation morphologique du système nerveux central: ses rapports avec le développement pondéral du foetus et son age gestationnel. In: *Regional Development of the Brain in Early Life*, A. Minkowski (ed.), London: Blackwell, page 248.)

Y301-62 contains several immature structures that are found in the other two GW37 specimens. In the cortical regions of the telencephalon, remnants of the germinal matrices are present in all lobes of the cerebral cortex where the ***neuroepithelium/subventricular zone*** may still be generating neocortical interneurons. Remnants of migrating and sojourning neurons and/or glia are visible in all lobes of the cerebral cortex as ***stratified transitional fields***, thin in the occipital lobe, and thicker in the frontal, parietal and temporal lobes. Because of its plane of sectioning, this specimen shows the most complete view of the occipital lobes at GW37. Many neurons, glia, and their mitotic precursor cells are still migrating through the olfactory peduncle toward the olfactory bulb (***rostral migratory stream***) from a presumed source area in the germinal matrix at the junction between the cerebral cortex, striatum, and nucleus accumbens. Within the lateral parts of the cerebral cortex, streams of neurons and glia are still in the ***lateral migratory stream***. That stream percolates through the claustrum, endopiriform nucleus, external capsule, and uncinate fasciculus, and cells appear to be heading toward the insular cortex, primary olfactory cortex, temporal cortex, and basolateral parts of the amygdaloid complex. In the basal ganglia, there is a prominent ***neuroepithelium/subventricular zone*** overlying the striatum and nucleus accumbens where neurons are probably still being generated. Another region of active neurogenesis in the telencephalon is the ***subgranular zone*** in the hilus of the dentate gyrus that is the source of granule cells. Other structures in the telencephalon, such as the septum, fornix, and Ammon's horn have only a thin, darkly staining layer at the ventricle, and these are presumed to be generating glia, cells of the choroid plexus, and the ependymal lining of the ventricle.

Most of the structures in the diencephalon appear to be settled and are maturing, and the third ventricle is lined by a thin ***glioepithelium/ependyma***. In the midbrain and anterior pons, there is a slightly thicker and more convoluted ***glioepithelium/ependyma*** lining the posterior cerebral aqueduct and anterior fourth ventricle. The posterior pons and entire medulla have a thin ***glioepithelium/ependyma*** lining the rest of the fourth ventricle. The ***external germinal layer*** is prominent over the entire surface of the cerebellar cortex and is still producing basket, stellate, and granule cells. The ***germinal trigone*** is still visible at the base of the nodulus and along the floccular peduncle; choroid plexus cells and glia may still be originating here.

# PLATE 65

**GW37 Horizontal**
**CR 320 mm**
**Y301-62**
**Level 1: Section 341**

*Remnants of the germinal matrix, migratory streams, and transitional fields*

| | |
|---|---|
| **1** | *Frontal stratified transitional field* |
| **2** | *Paracentral stratified transitional field* |
| **3** | *Cingulate stratified transitional field* |
| **4** | *Subpial granular layer* |

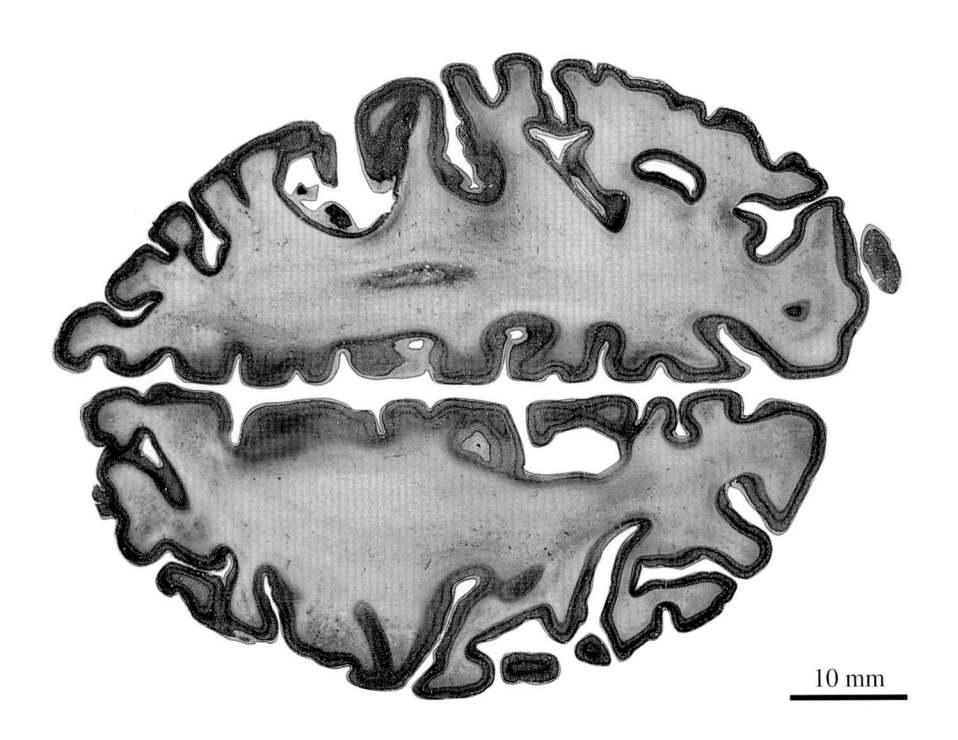

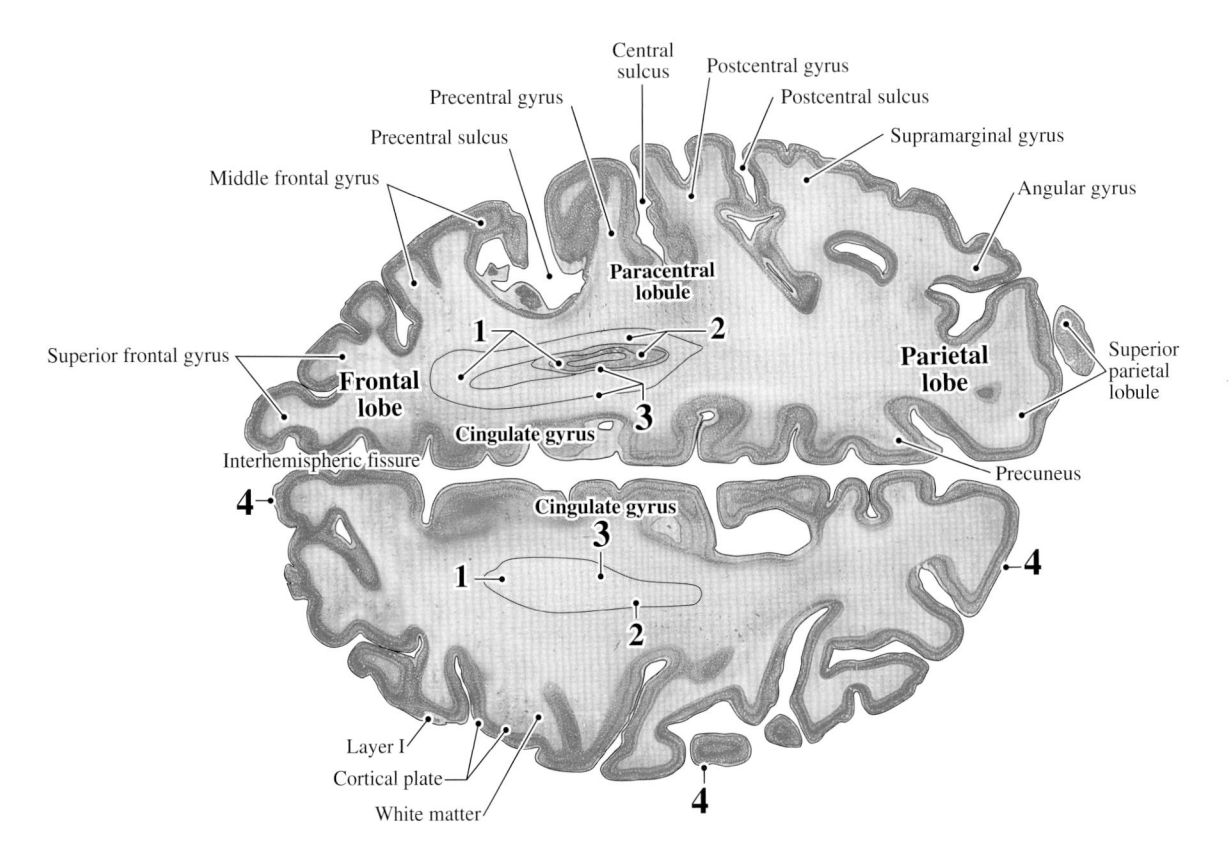

## PLATE 66

**GW37 Horizontal**
**CR 320 mm**
**Y301-62**
**Level 2: Section 441**

### Remnants of the germinal matrix, migratory streams, and transitional fields

| | |
|---|---|
| 1 Frontal neuroepithelium and subventricular zone | 6 Paracentral neuroepithelium and subventricular zone |
| 2 Frontal stratified transitional field | 7 Paracentral stratified transitional field |
| 3 Callosal glioepithelium | 8 Anterolateral striatal neuroepithelium and subventricular zone |
| 4 Parietal neuroepithelium and subventricular zone | 9 Posterior striatal neuroepithelium and subventricular zone |
| 5 Parietal stratified transitional field | 10 Strionuclear glioepithelium |
| | 11 Subpial granular layer |

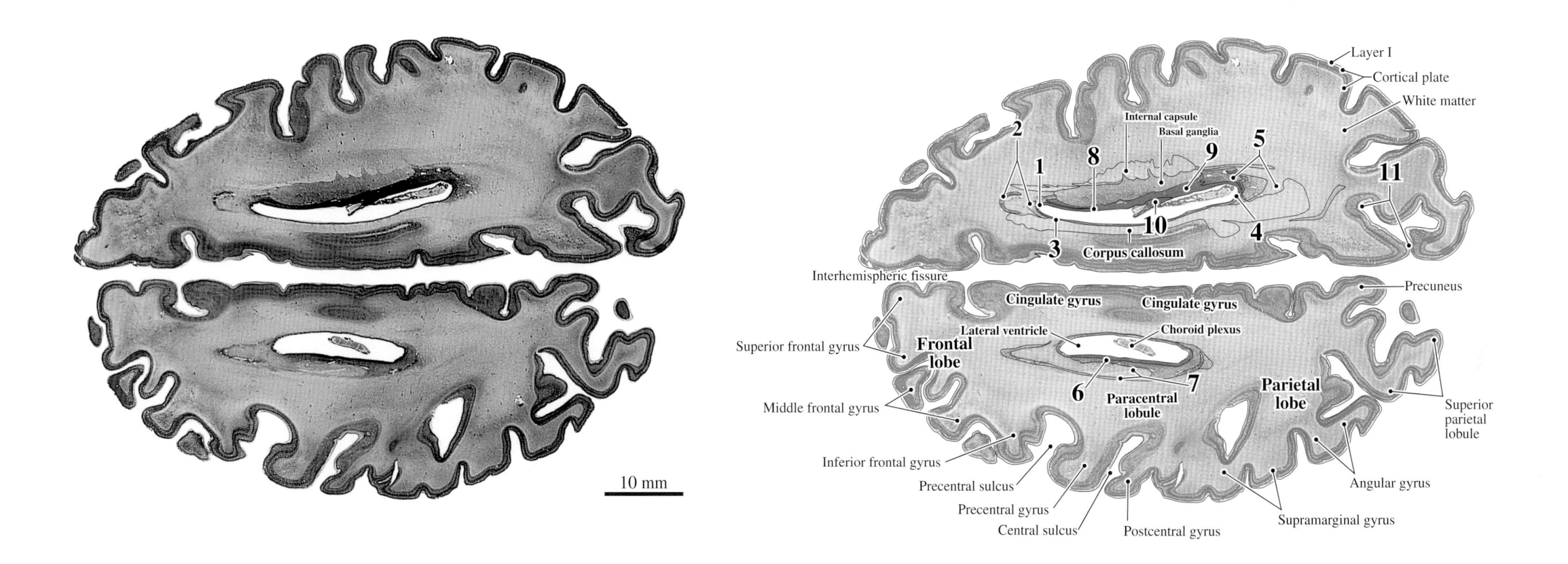

# PLATE 67

**GW37 Horizontal**
**CR 320 mm**
**Y301-62**
**Level 3: Section 531**

**See detail of brain core in Plates 81A and B.**

*Remnants of the germinal matrix, migratory streams, and transitional fields*

| | | | |
|---|---|---|---|
| 1 | *Rostral migratory stream (source area)* | 7 | *Parietal neuroepithelium and subventricular zone* |
| 2 | *Frontal neuroepithelium and subventricular zone* | 8 | *Parietal stratified transitional field* |
| 3 | *Frontal stratified transitional field* | 9 | *Posterior striatal neuroepithelium and subventricular zone* |
| 4 | *Callosal glioepithelium* | 10 | *Anterolateral striatal neuroepithelium and subventricular zone* |
| 5 | *Callosal sling* | 11 | *Anteromedial striatal neuroepithelium and subventricular zone* |
| 6 | *Fornical glioepithelium* | 12 | *Strionuclear glioepithelium* |

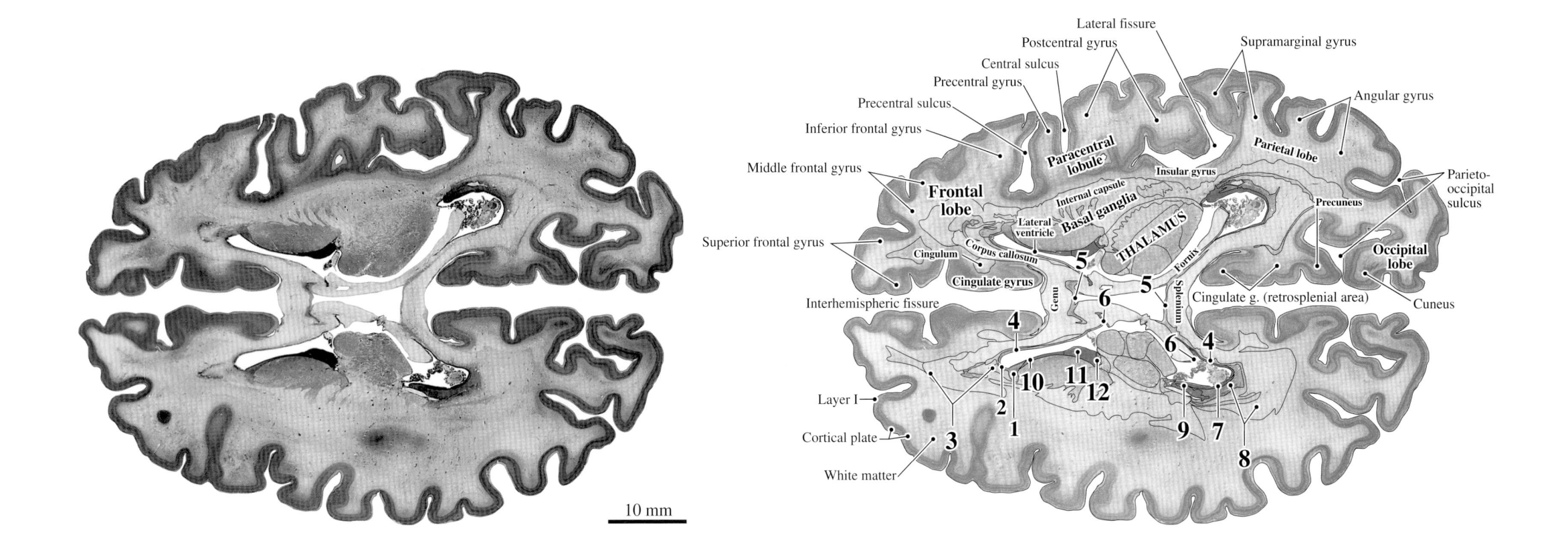

# PLATE 68

**GW37 Horizontal**
**CR 320 mm**
**Y301-62**
**Level 4: Section 561**

*Remnants of the germinal matrix, migratory streams, and transitional fields*

| | | | |
|---|---|---|---|
| 1 | Rostral migratory stream (source area) | 7 | Parietal/temporal neuroepithelium and subventricular zone |
| 2 | Frontal neuroepithelium and subventricular zone | 8 | Parietal/temporal stratified transitional field |
| 3 | Frontal stratified transitional field | 9 | Posterior striatal neuroepithelium and subventricular zone |
| 4 | Callosal glioepithelium | 10 | Anterolateral striatal neuroepithelium and subventricular zone |
| 5 | Callosal sling | 11 | Anteromedial striatal neuroepithelium and subventricular zone |
| 6 | Fornical glioepithelium | 12 | Strionuclear glioepithelium |

See detail of brain core in Plates 82A and B.

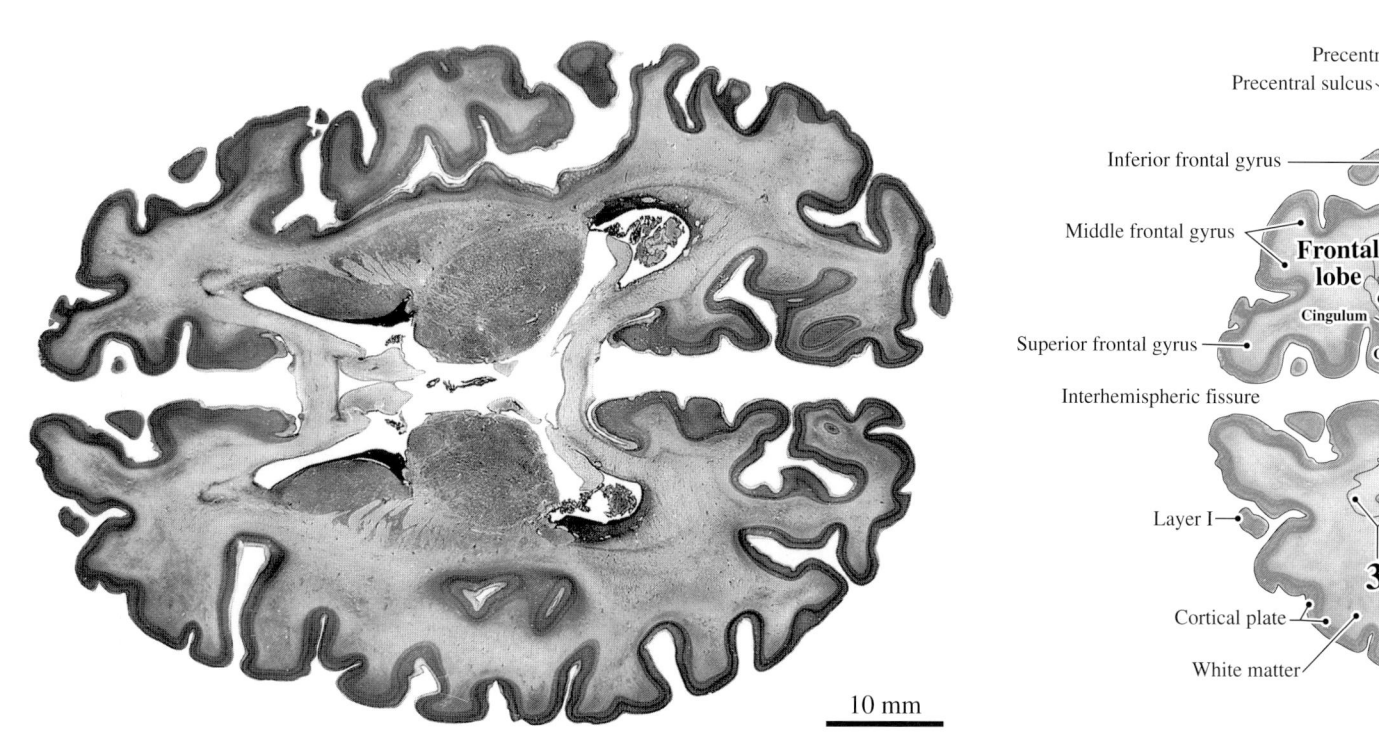

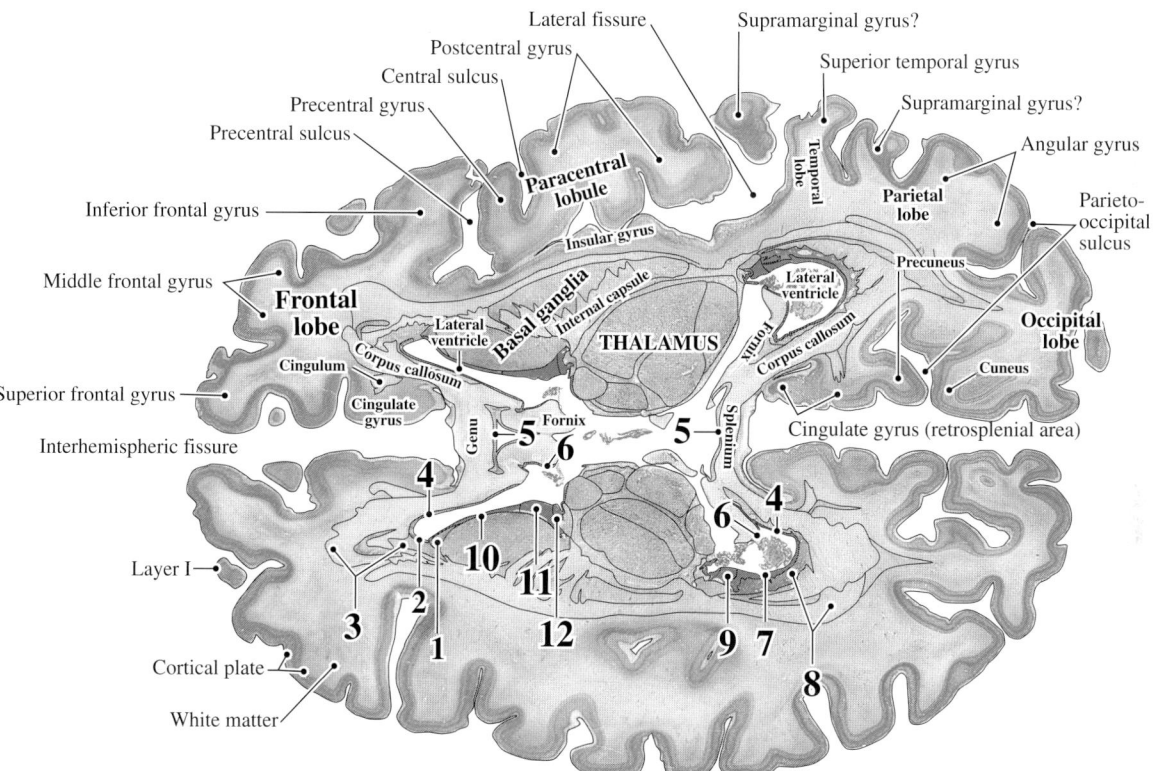

# PLATE 69

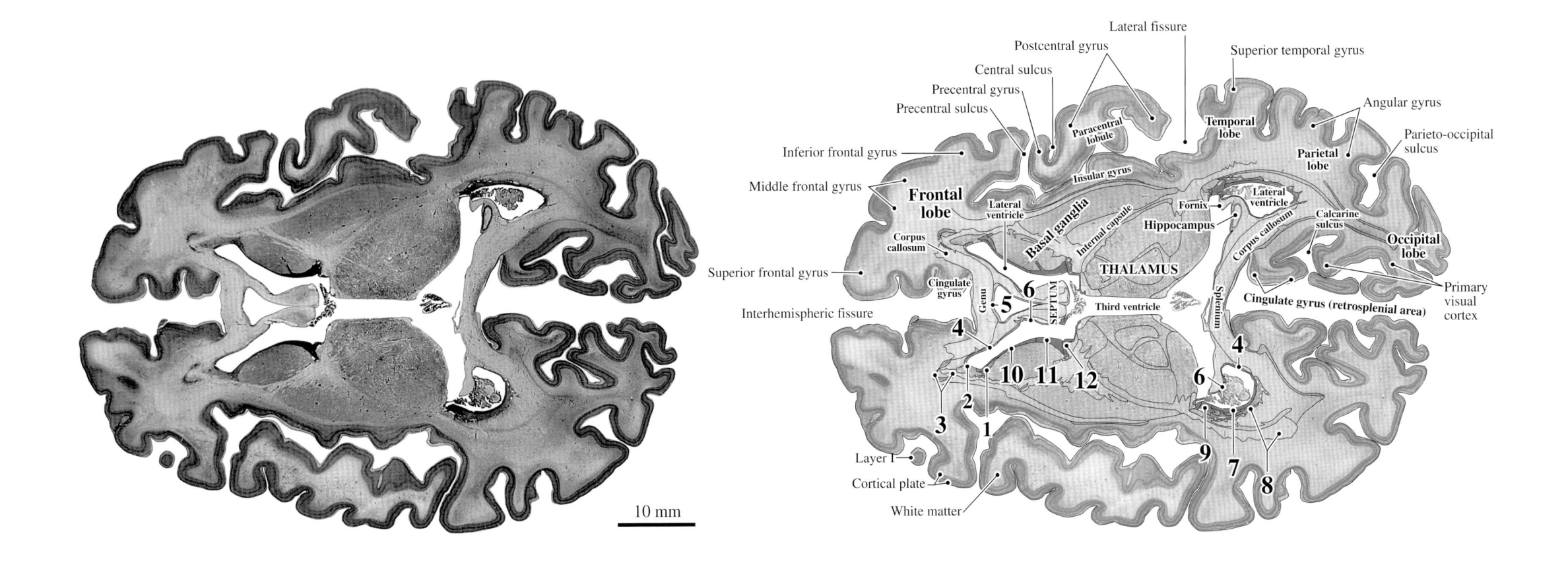

**GW37 Horizontal**
**CR 320 mm**
**Y301-62**
**Level 5: Section 631**

*Remnants of the germinal matrix, migratory streams, and transitional fields*

| | | | |
|---|---|---|---|
| 1 | *Rostral migratory stream (source area)* | 7 | *Parietal/temporal neuroepithelium and subventricular zone* |
| 2 | *Frontal neuroepithelium and subventricular zone* | 8 | *Parietal/temporal stratified transitional field* |
| 3 | *Frontal stratified transitional field* | 9 | *Posterior striatal neuroepithelium and subventricular zone* |
| 4 | *Callosal glioepithelium* | 10 | *Anterolateral striatal neuroepithelium and subventricular zone* |
| 5 | *Callosal sling* | 11 | *Anteromedial striatal neuroepithelium and subventricular zone* |
| 6 | *Fornical glioepithelium* | 12 | *Strionuclear glioepithelium* |

**See detail of brain core in Plates 83A and B.**

## PLATE 70

**GW37 Horizontal**
**CR 320 mm**
**Y301-62**
**Level 6: Section 731**

*Remnants of the germinal matrix, migratory streams, and transitional fields*

1. Rostral migratory stream (source area)
2. Frontal neuroepithelium and subventricular zone
3. Frontal stratified transitional field
4. Callosal glioepithelium
5. Fornical glioepithelium
6. Occipital neuroepithelium and subventricular zone
7. Occipital stratified transitional field
8. Parietal/temporal neuroepithelium and subventricular zone
9. Parietal/temporal stratified transitional field
10. Alvear glioepithelium
11. Posterior striatal neuroepithelium and subventricular zone
12. Anterolateral striatal neuroepithelium and subventricular zone
13. Anteromedial striatal neuroepithelium and subventricular zone
14. Strionuclear glioepithelium
15. Septal glioepithelium/ependyma

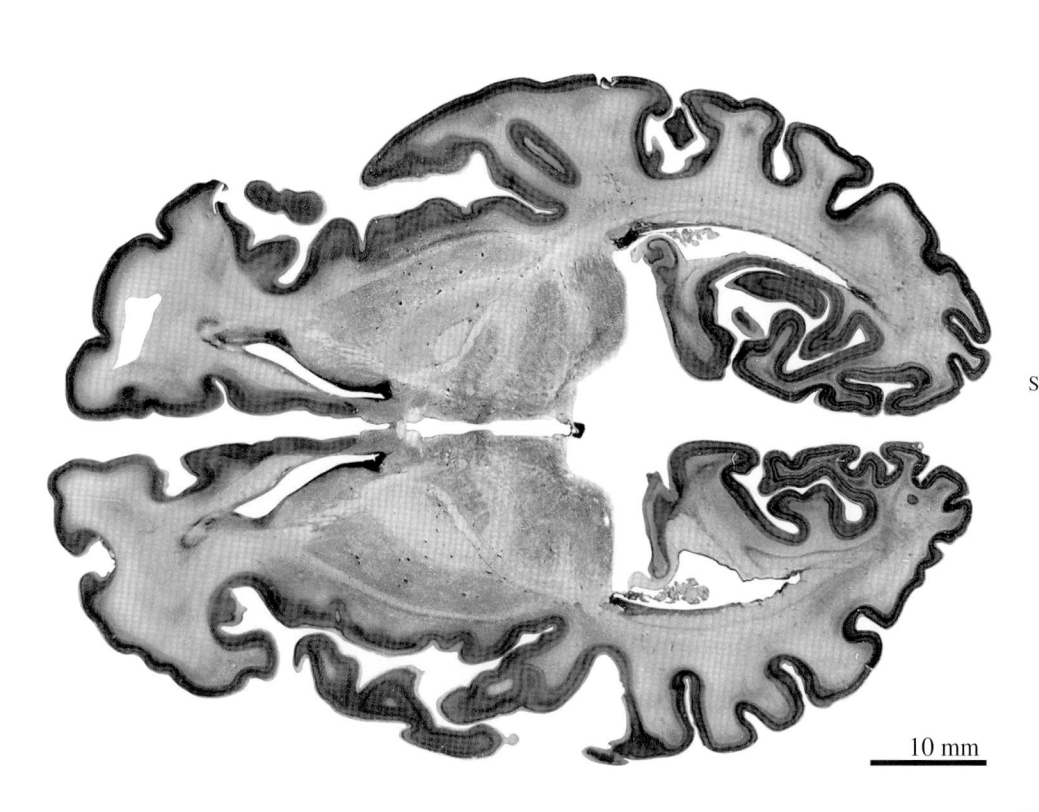

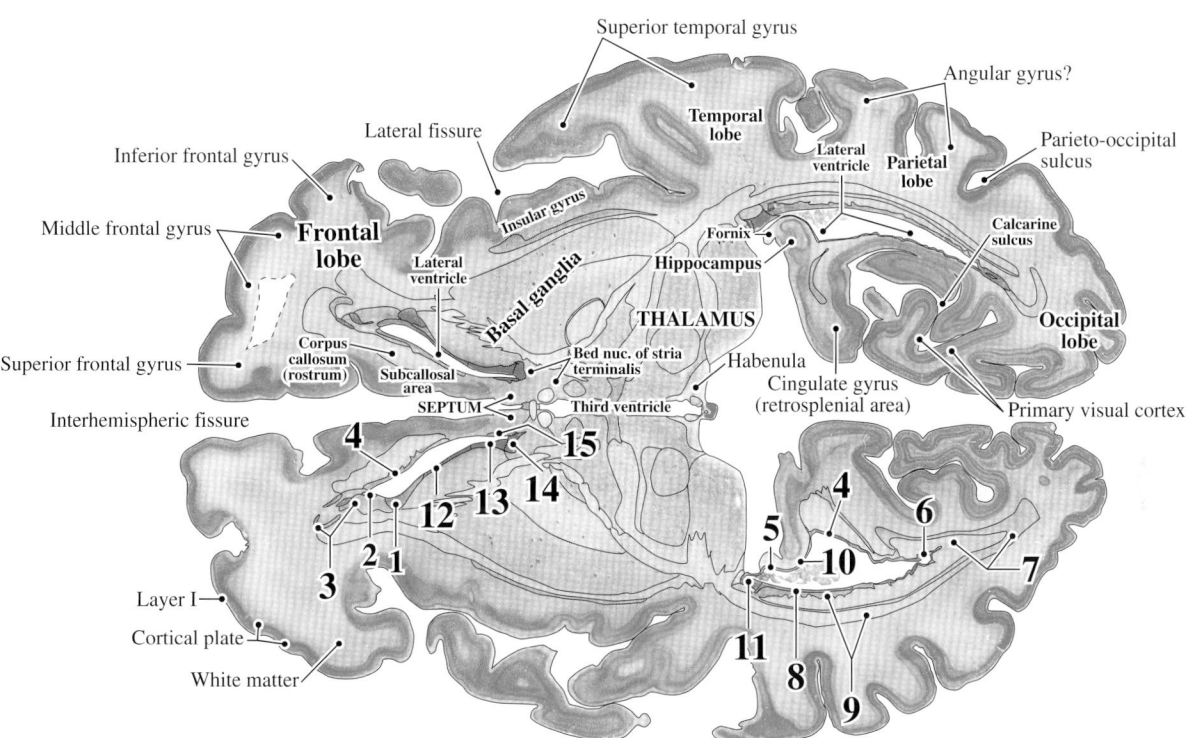

See detail of brain core in Plates 84A and B.

# PLATE 71

**GW37 Horizontal**
**CR 320 mm**
**Y301-62**
**Level 7: Section 761**

**See detail of brain core in Plates 85A and B.**

*Remnants of the germinal matrix, migratory streams, and transitional fields*

| | | | |
|---|---|---|---|
| 1 | *Rostral migratory stream (source area?)* | 8 | *Temporal neuroepithelium and subventricular zone* |
| 2 | *Frontal neuroepithelium and subventricular zone* | 9 | *Temporal stratified transitional field* |
| 3 | *Frontal stratified transitional field* | 10 | *Alvear glioepithelium* |
| 4 | *Callosal glioepithelium* | 11 | *Subgranular zone (dentate)* |
| 5 | *Fornical glioepithelium* | 12 | *Posterior striatal neuroepithelium and subventricular zone* |
| 6 | *Occipital neuroepithelium and subventricular zone* | 13 | *Anterolateral striatal neuroepithelium and subventricular zone* |
| 7 | *Occipital stratified transitional field* | 14 | *Anteromedial striatal neuroepithelium and subventricular zone* |
| | | 15 | *Strionuclear glioepithelium* |

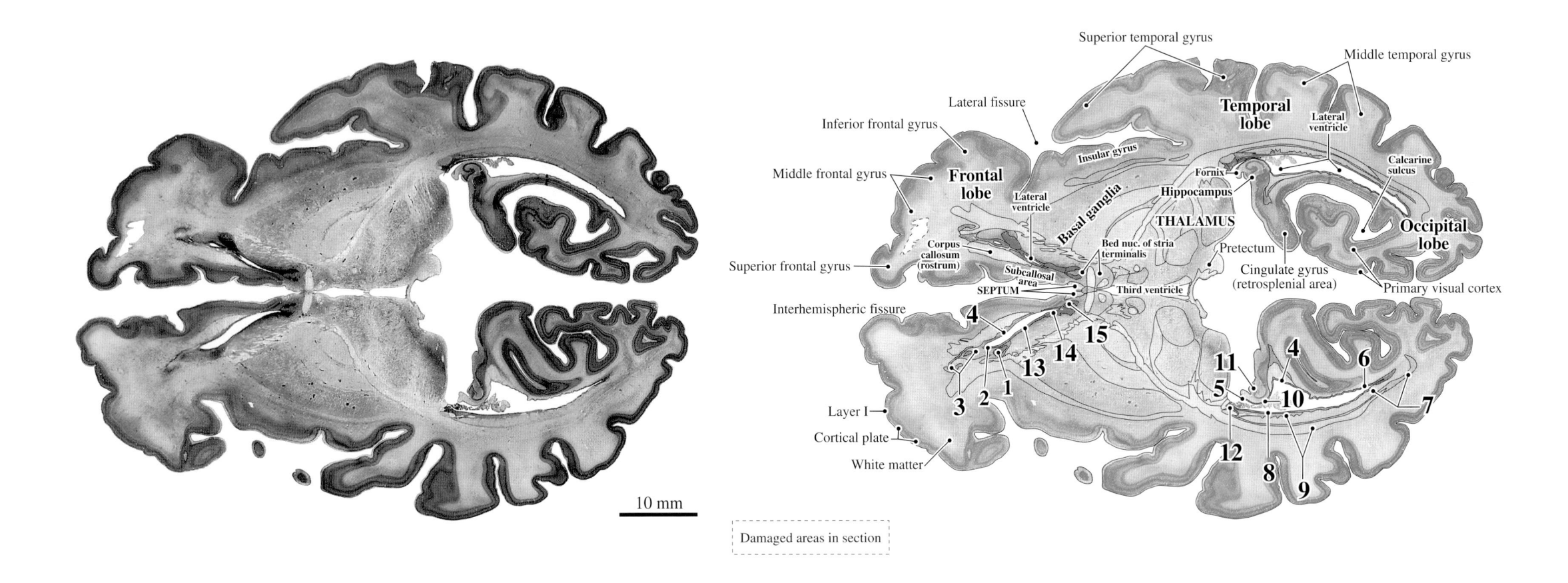

# PLATE 72

**GW37 Horizontal**
**CR 320 mm**
**Y301-62**
**Level 8: Section 801**

## Remnants of the germinal matrix, migratory streams, and transitional fields

1. Rostral migratory stream (source area)
2. Frontal neuroepithelium and subventricular zone
3. Frontal stratified transitional field
4. Callosal glioepithelium
5. Fornical glioepithelium
6. Parahippocampal neuroepithelium, subventricular zone and stratified transitional field
7. Occipital neuroepithelium and subventricular zone
8. Occipital stratified transitional field
9. Temporal neuroepithelium and subventricular zone
10. Temporal stratified transitional field
11. Alvear glioepithelium
12. Subgranular zone (dentate)
13. Lateral migratory stream (cortical)
14. Posterior striatal neuroepithelium and subventricular zone
15. Accumbent neuroepithelium and subventricular zone (infiltrated by the rostral migratory stream)

**See detail of brain core in Plates 86A and B.**

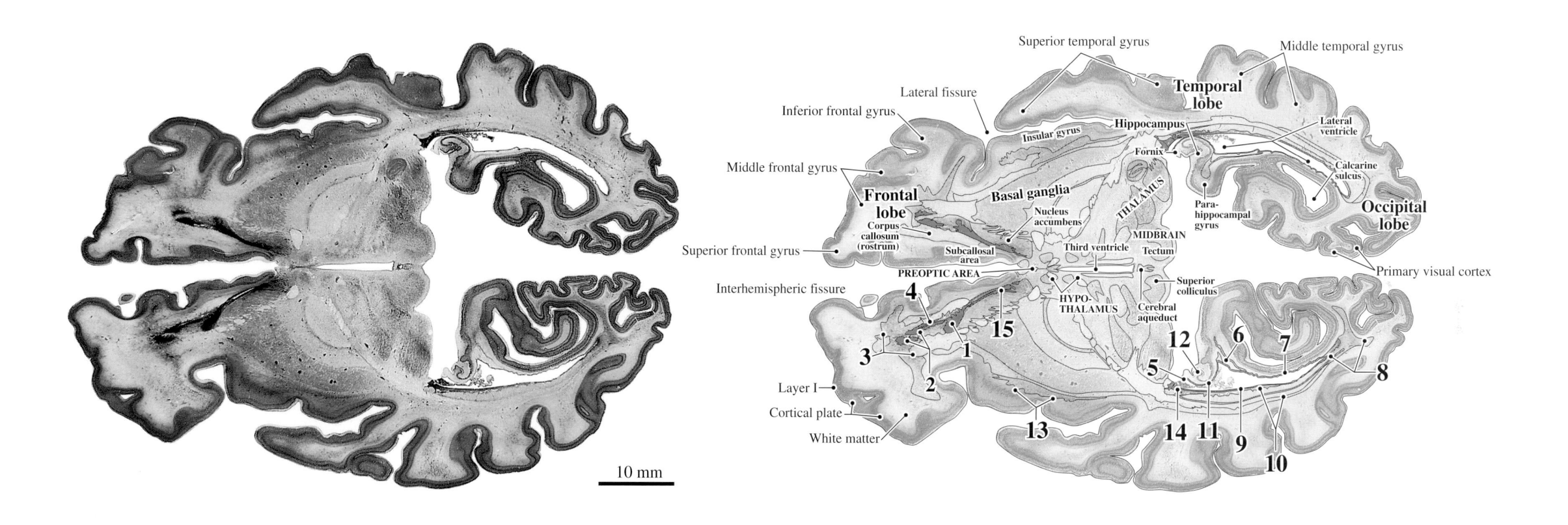

10 mm

## PLATE 73

GW37 Horizontal
CR 320 mm
Y301-62
Level 9: Section 851

See detail of brain core and cerebellum in Plates 87A and B.

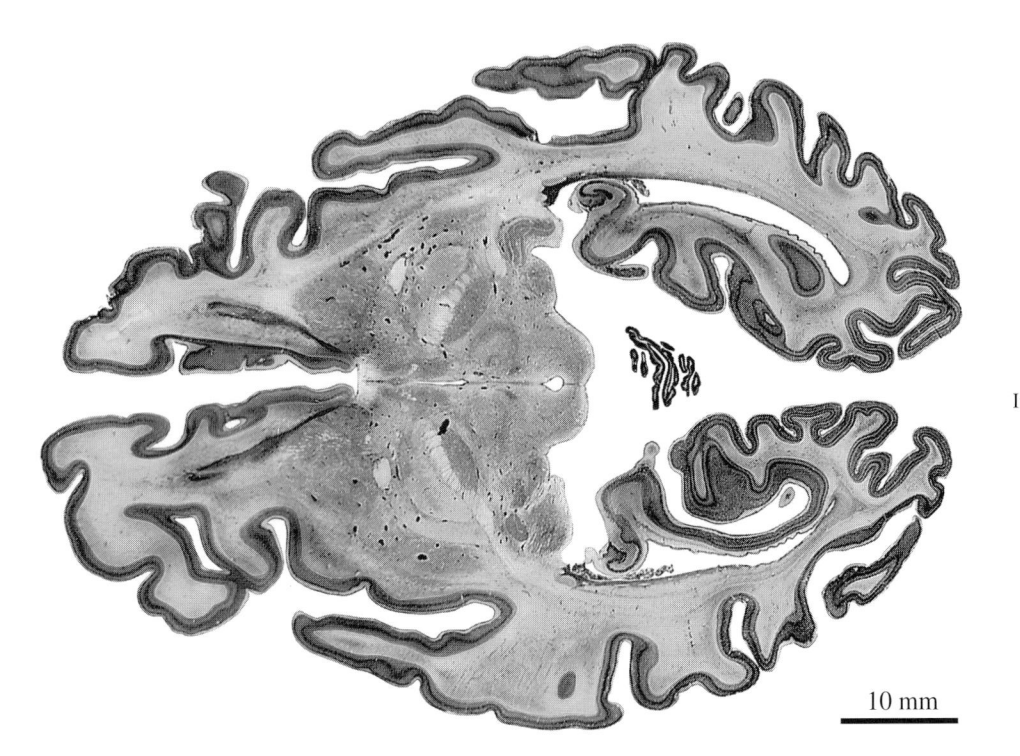

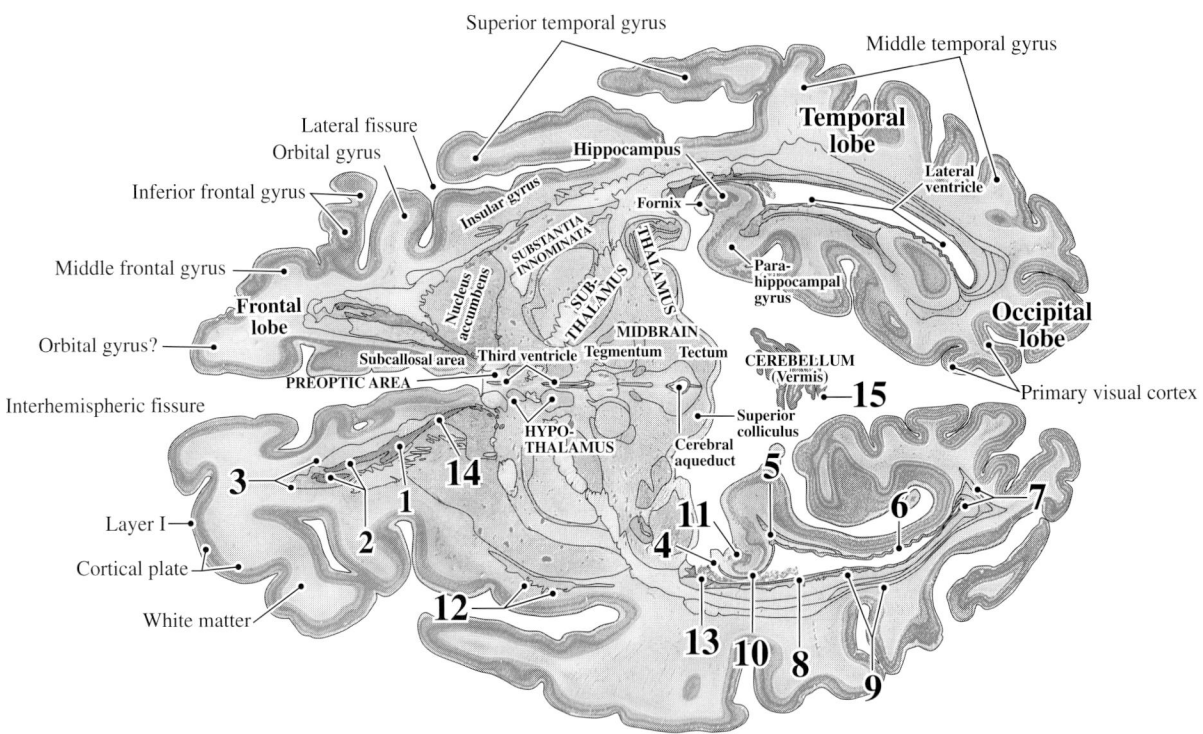

*Remnants of the germinal matrix, migratory streams, and transitional fields*

| | | | |
|---|---|---|---|
| 1 | *Rostral migratory stream (source area)* | 8 | *Temporal neuroepithelium and subventricular zone* |
| 2 | *Frontal neuroepithelium and subventricular zone* | 9 | *Temporal stratified transitional field* |
| 3 | *Frontal stratified transitional field* | 10 | *Alvear glioepithelium* |
| 4 | *Fornical glioepithelium* | 11 | *Subgranular zone (dentate)* |
| 5 | *Parahippocampal neuroepithelium, subventricular zone and stratified transitional field* | 12 | *Lateral migratory stream (cortical)* |
| | | 13 | *Posterior striatal neuroepithelium and subventricular zone* |
| 6 | *Occipital neuroepithelium and subventricular zone* | 14 | *Accumbent neuroepithelium and subventricular zone (infiltrated by the rostral migratory stream)* |
| 7 | *Occipital stratified transitional field* | 15 | *External germinal layer (cerebellum)* |

## PLATE 74

**GW37 Horizontal**
**CR 320 mm**
**Y301-62**
**Level 10: Section 921**

### Remnants of the germinal matrix, migratory streams, and transitional fields

| | | | |
|---|---|---|---|
| 1 | *Rostral migratory stream* | 7 | *Temporal stratified transitional field* |
| 2 | *Frontal stratified transitional field* | 8 | *Alvear glioepithelium* |
| 3 | *Fornical glioepithelium* | 9 | *Subgranular zone (dentate)* |
| 4 | *Parahippocampal neuroepithelium, subventricular zone and stratified transitional field* | 10 | *Lateral migratory stream (cortical)* |
| 5 | *Occipital neuroepithelium, subventricular zone and stratified transitional field* | 11 | *Posterior striatal neuroepithelium and subventricular zone* |
| 6 | *Temporal neuroepithelium and subventricular zone* | 12 | *External germinal layer (cerebellum)* |

See detail of brain core and cerebellum in Plates 88A and B.

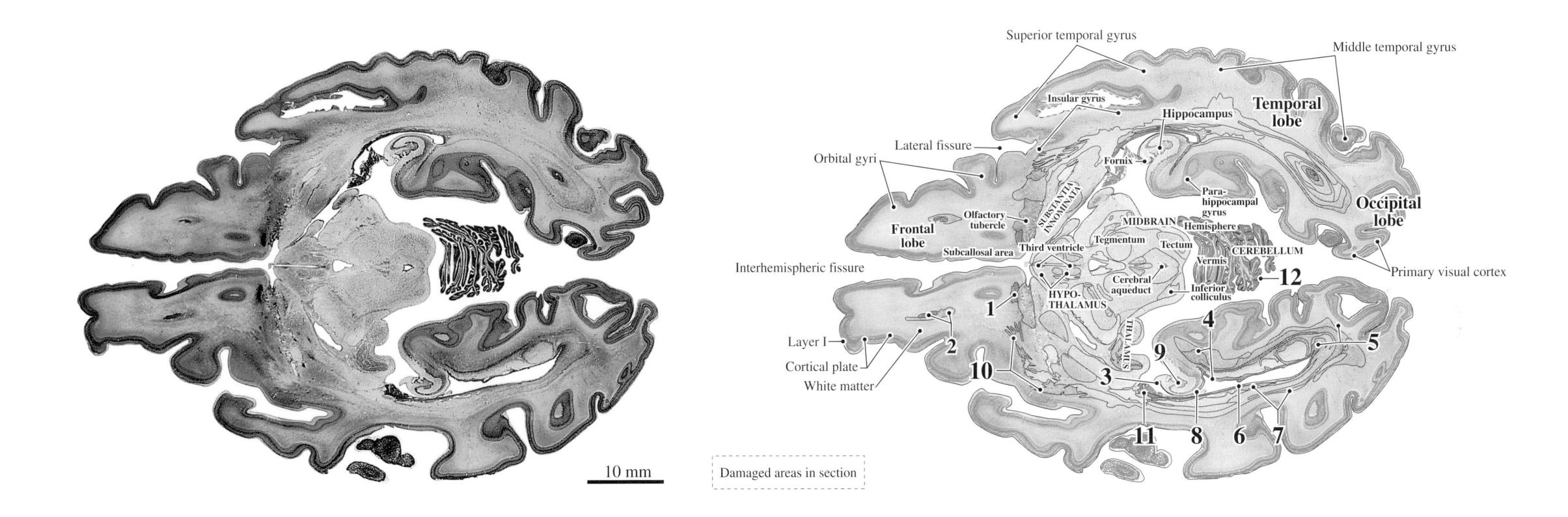

## PLATE 75

**GW37 Horizontal
CR 320 mm
Y301-62
Level 11: Section 981**

See detail of brain core and cerebellum in Plates 89A and B.

### Remnants of the germinal matrix, migratory streams, and transitional fields

| | | | |
|---|---|---|---|
| 1 | *Rostral migratory stream* | 6 | *Alvear glioepithelium* |
| 2 | *Fornical glioepithelium* | 7 | *Subgranular zone (dentate)* |
| 3 | *Parahippocampal neuroepithelium, subventricular zone and stratified transitional field* | 8 | *Lateral migratory stream (cortical)* |
| 4 | *Temporal neuroepithelium and subventricular zone* | 9 | *Amygdaloid glioepithelium* |
| 5 | *Temporal stratified transitional field* | 10 | *External germinal layer (cerebellum)* |

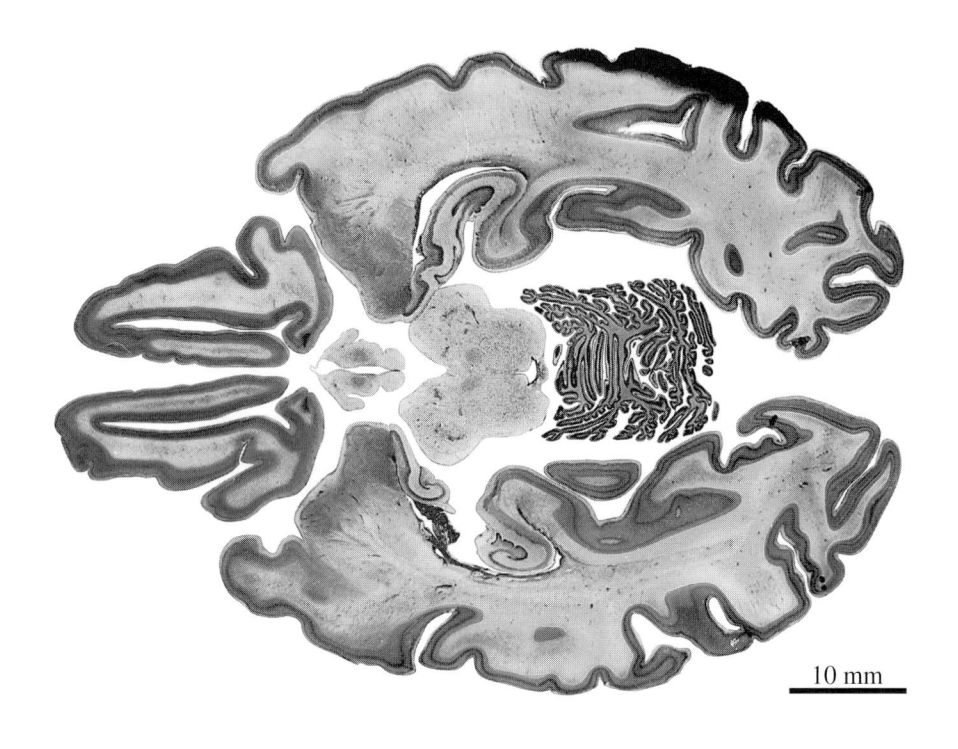

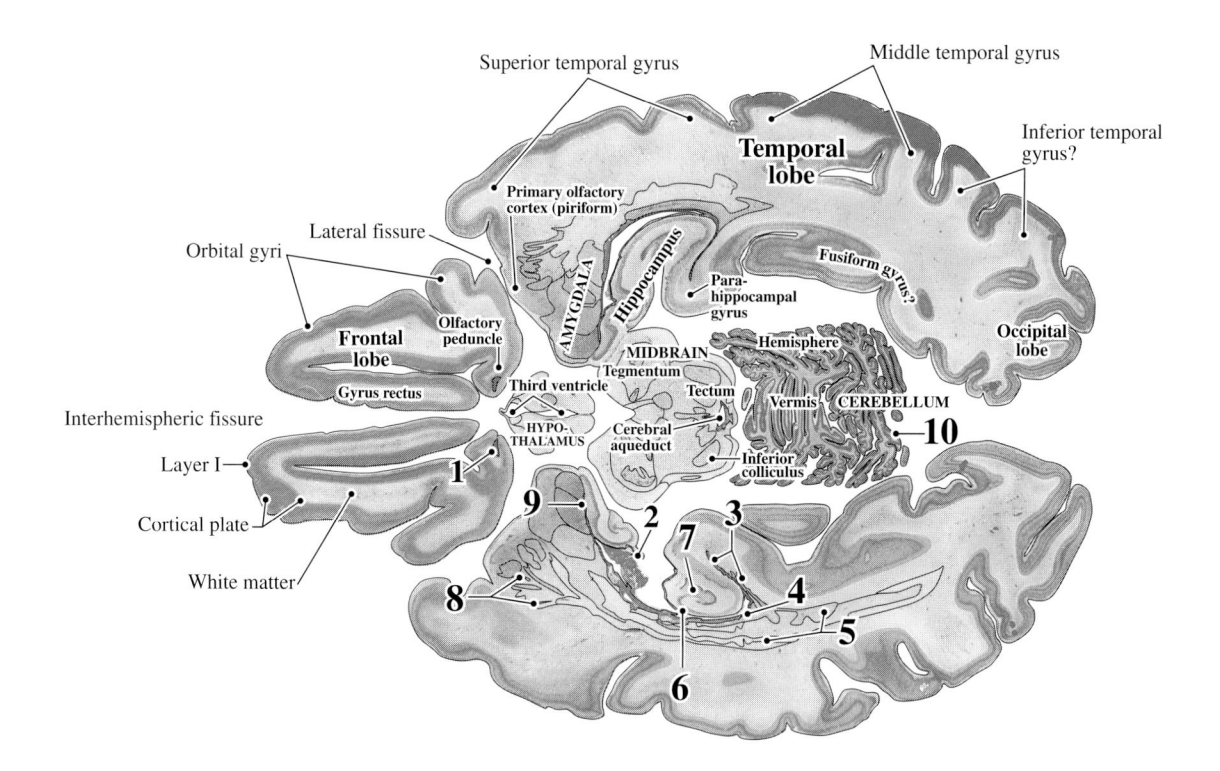

# PLATE 76

**GW37 Horizontal**
**CR 320 mm**
**Y301-62**
**Level 12: Section 1051**

## Remnants of the germinal matrix, migratory streams, and transitional fields

| | | | |
|---|---|---|---|
| 1 | Rostral migratory stream | 5 | Alvear glioepithelium |
| 2 | Parahippocampal neuroepithelium, subventricular zone and stratified transitional field | 6 | Subgranular zone (dentate) |
| | | 7 | Lateral migratory stream (cortical) |
| 3 | Temporal neuroepithelium and subventricular zone | 8 | Amygdaloid glioepithelium/ependyma |
| 4 | Temporal stratified transitional field | 9 | External germinal layer (cerebellum) |

**See detail of brain core and cerebellum in Plates 90A and B.**

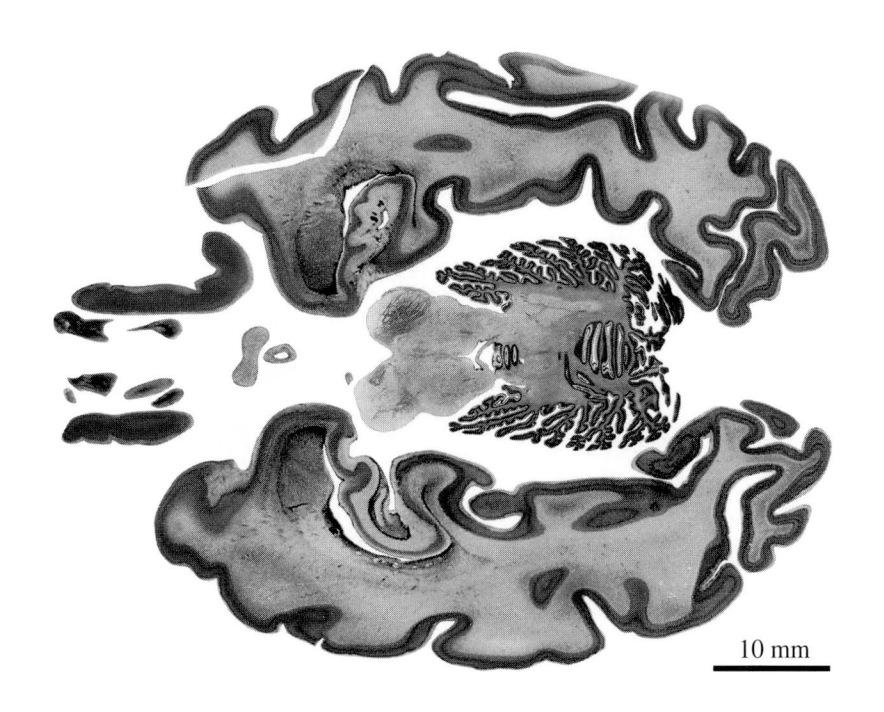

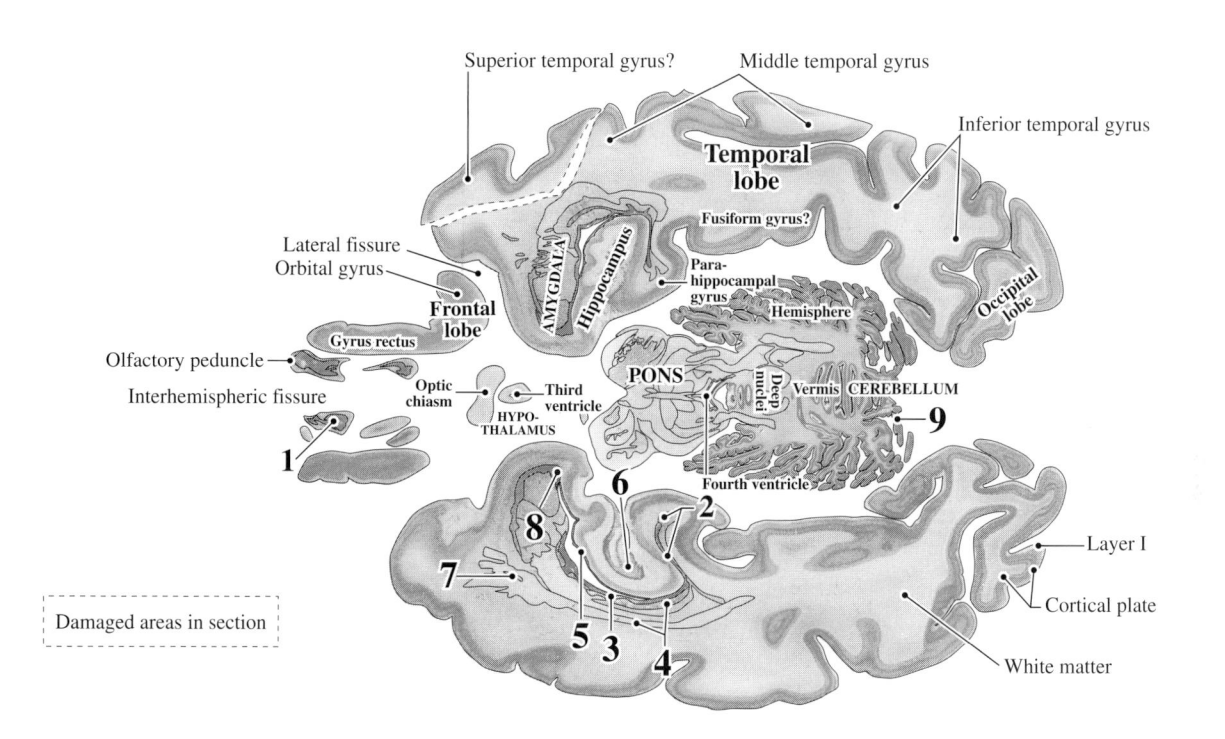

# PLATE 77

**GW37 Horizontal**
**CR 320 mm**
**Y301-62**
**Level 13: Section 1111**

*Remnants of the germinal matrix, migratory streams, and transitional fields*

| | | | |
|---|---|---|---|
| 1 | Parahippocampal neuroepithelium, subventricular zone and stratified transitional field | 4 | Alvear glioepithelium |
| 2 | Temporal neuroepithelium and subventricular zone | 5 | Lateral migratory stream (cortical) |
| | | 6 | Amygdaloid glioepithelium/ependyma |
| 3 | Temporal stratified transitional field | 7 | External germinal layer (cerebellum) |

**See detail of brain core and cerebellum in Plates 91A and B.**

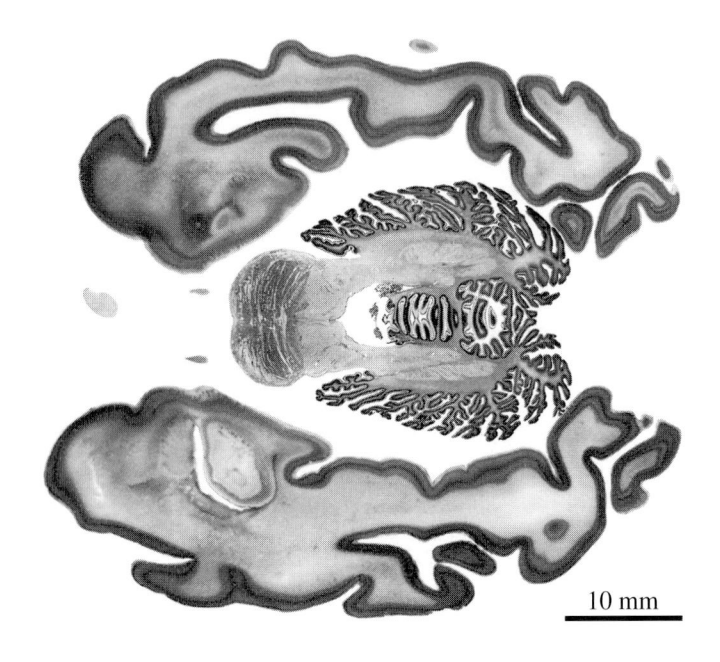

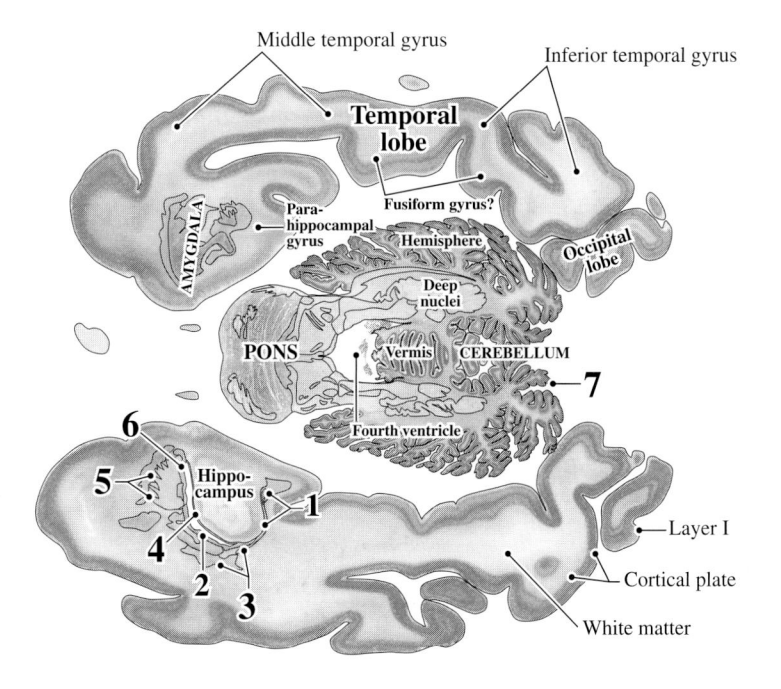

# PLATE 78

**GW37 Horizontal**
**CR 320 mm**
**Y301-62**
**Level 14: Section 1161**

*Remnants of the germinal matrix, migratory streams, and transitional fields*

1. *Parahippocampal/temporal neuroepithelium, subventricular zone and stratified transitional field*
2. *External germinal layer (cerebellum)*

See detail of brain core and cerebellum in Plates 92A and B.

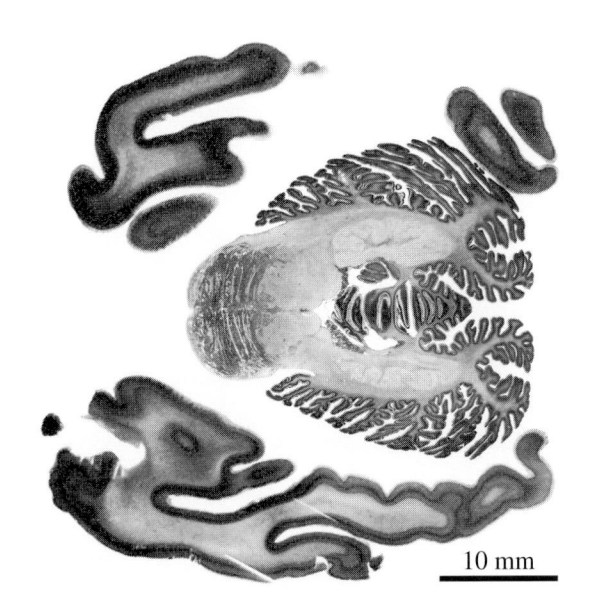

10 mm

Damaged areas in section

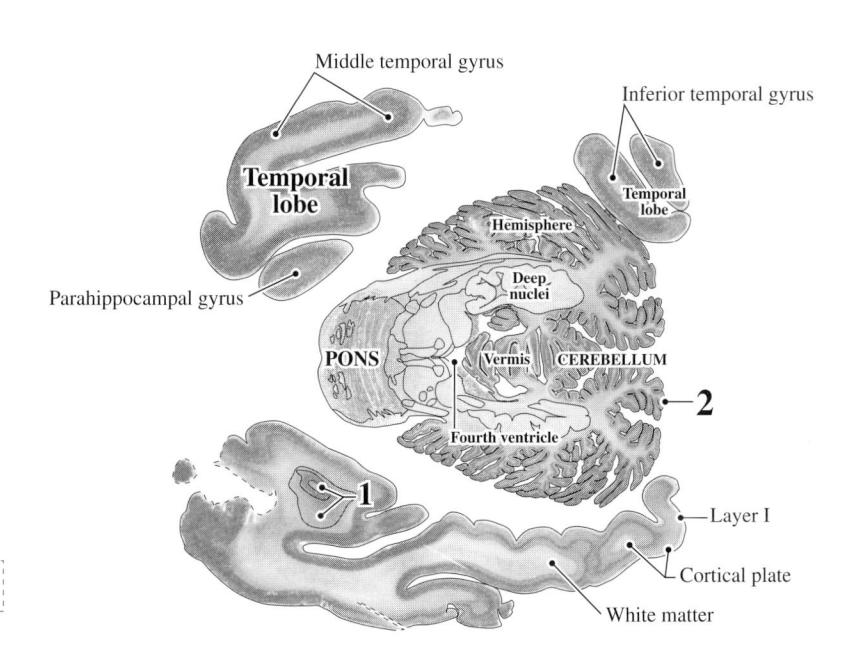

Middle temporal gyrus
Inferior temporal gyrus
Temporal lobe
Hemisphere
Deep nuclei
Temporal lobe
Parahippocampal gyrus
PONS
Vermis
CEREBELLUM
Fourth ventricle
Layer I
Cortical plate
White matter

## PLATE 79

**GW37 Horizontal, CR 320 mm, Y301-62**

*Remnants of the germinal matrix, migratory streams, and transitional fields*

**1** *External germinal layer (cerebellum)*

---

**Level 15: Section 1251**

See detail of brain core and cerebellum in Plates 93A and B.

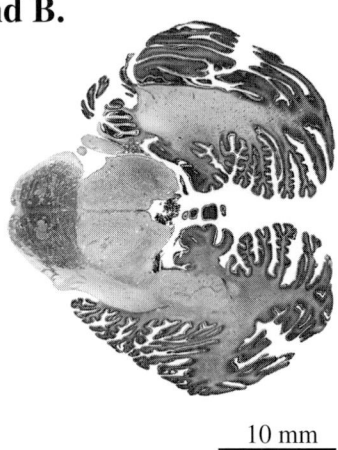

10 mm

Hemisphere
CEREBELLUM
Flocculus
Hemisphere
Transpontine corticofugal tract
MEDULLA
Paraflocculus
Pontine gray
Vermis
PONS
Fourth ventricle
Nerve V
Deep nuclei
Middle cerebellar peduncle
— 1

---

**Level 16: Section 1311**

See higher magnification view in Plates 94A and B (left panel).

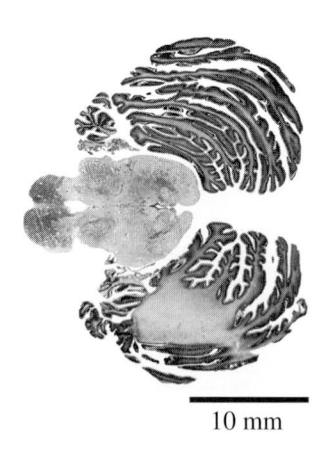

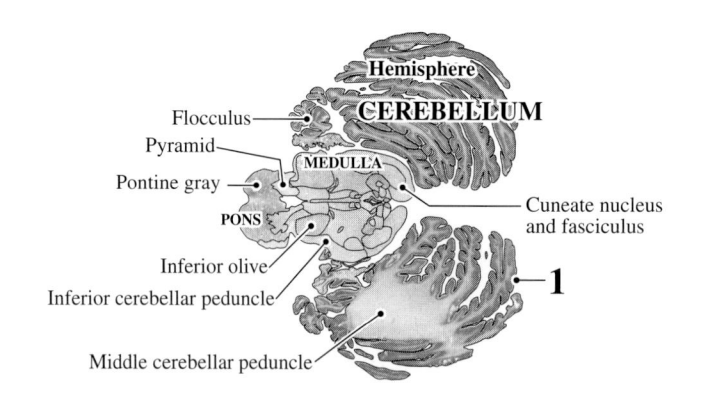

10 mm

Hemisphere
Flocculus
CEREBELLUM
Pyramid
MEDULLA
Pontine gray
Cuneate nucleus and fasciculus
PONS
Inferior olive
Inferior cerebellar peduncle
— 1
Middle cerebellar peduncle

## PLATE 80

*Remnants of the germinal matrix, migratory streams, and transitional fields*

GW37 Horizontal, CR 320 mm, Y301-62

| 1 | *External germinal layer (cerebellum)* |

**Level 17: Section 1331**

See higher magnification view in Plates 94A and B (right panel).

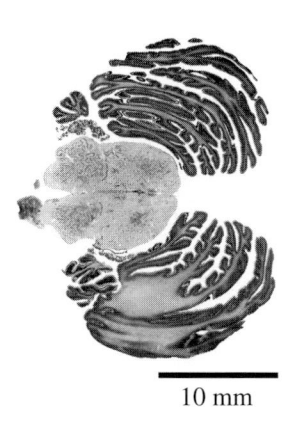

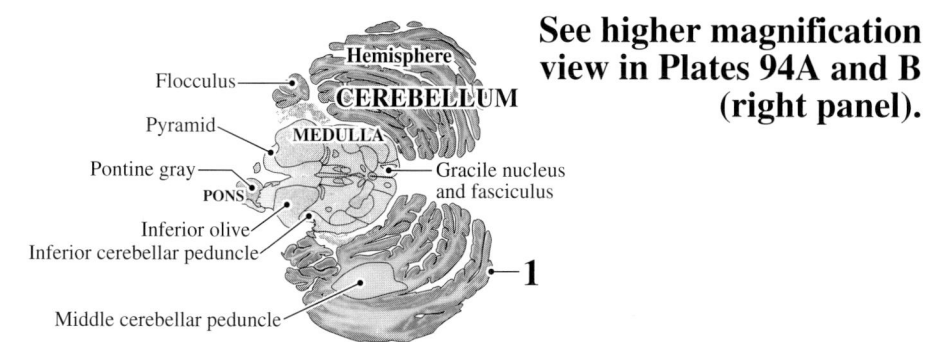

Labels: Flocculus, Hemisphere, CEREBELLUM, Pyramid, MEDULLA, Pontine gray, PONS, Gracile nucleus and fasciculus, Inferior olive, Inferior cerebellar peduncle, Middle cerebellar peduncle

10 mm

**Level 18: Section 1381**

See higher magnification view in Plates 95A and B (left panel).

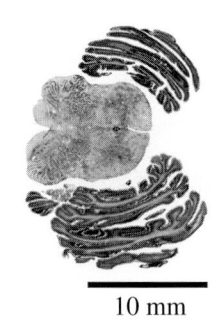

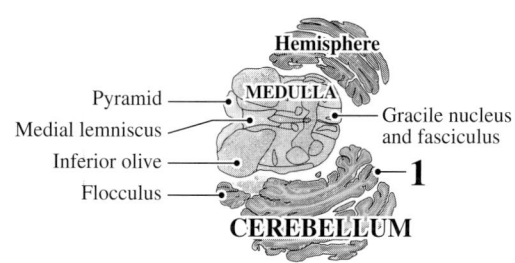

Labels: Hemisphere, Pyramid, MEDULLA, Medial lemniscus, Gracile nucleus and fasciculus, Inferior olive, Flocculus, CEREBELLUM

10 mm

**Level 19: Section 1441**

See higher magnification view in Plates 95A and B (right panel).

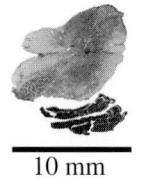

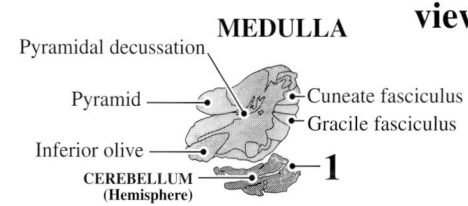

Labels: Pyramidal decussation, MEDULLA, Pyramid, Cuneate fasciculus, Gracile fasciculus, Inferior olive, CEREBELLUM (Hemisphere)

10 mm

## PLATE 81A

**GW37 Horizontal**
CR 320 mm, Y301-62, Level 3: Section 531

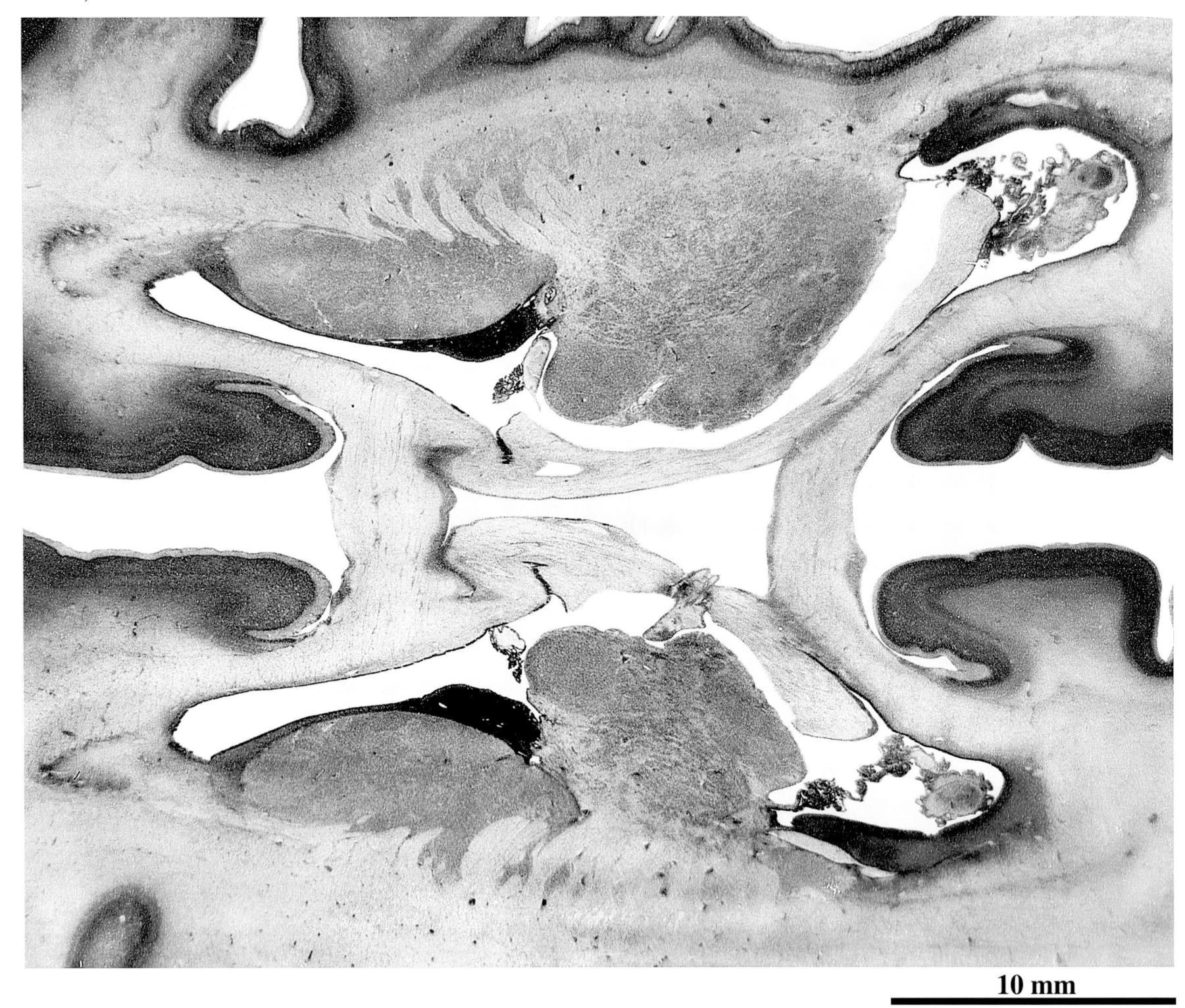

10 mm

See the entire section in Plate 67.

115

**PLATE 81B**

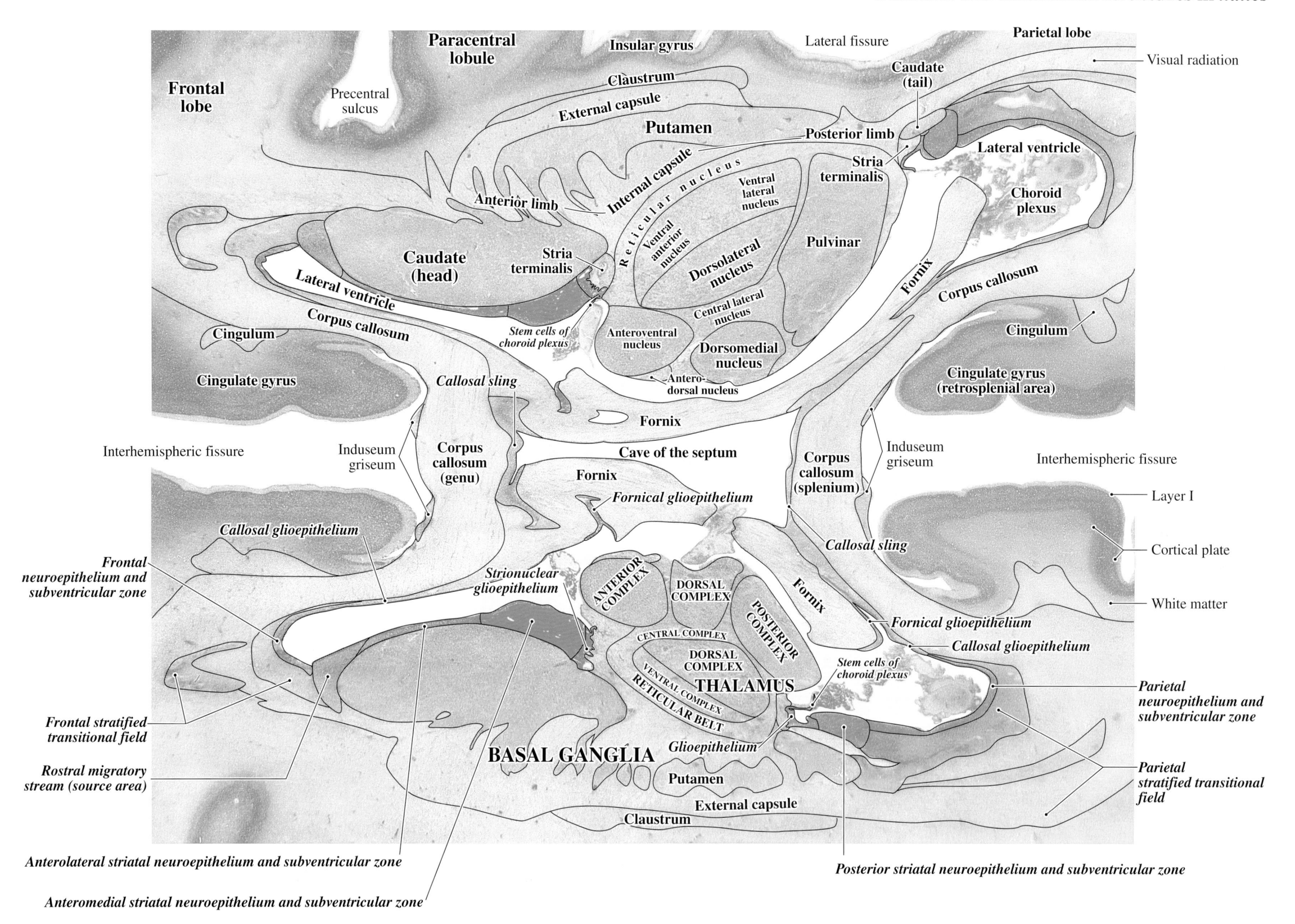

**Germinal and transitional structures in *italics***

## PLATE 82A

**GW37 Horizontal**
CR 320 mm
Y301-62
Level 4: Section 561

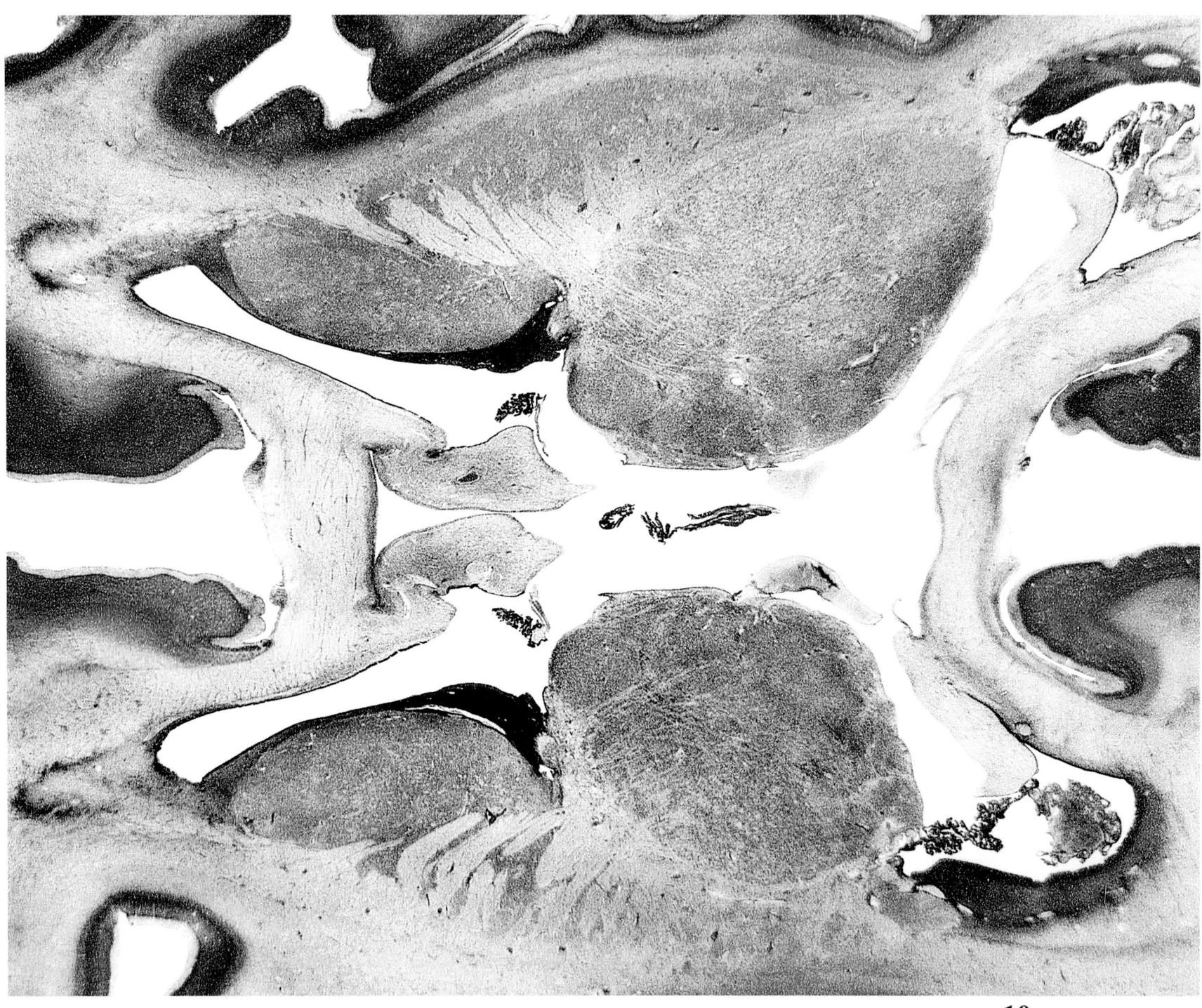

See the entire section in Plate 68.

10 mm

# PLATE 82B

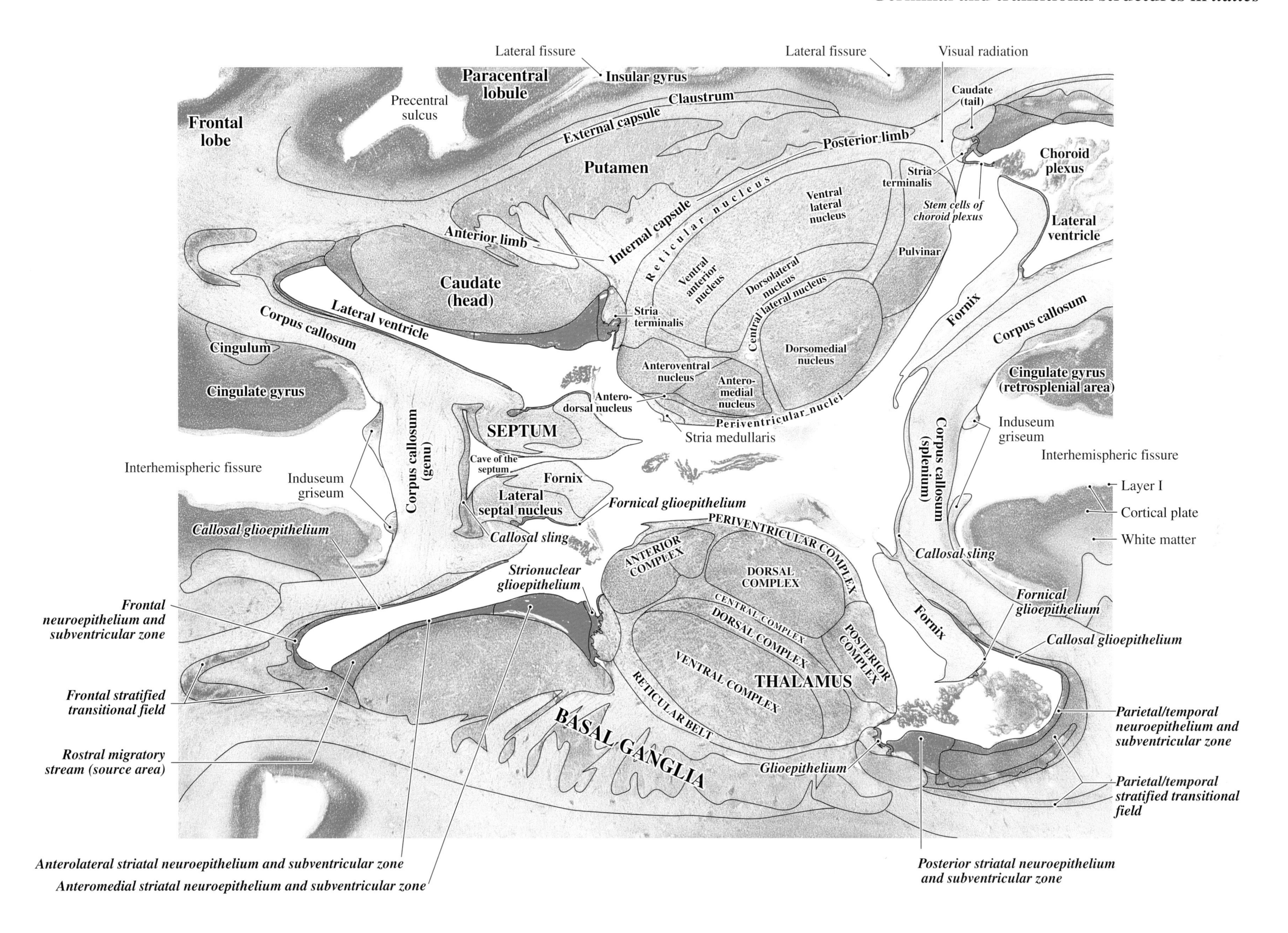

**Germinal and transitional structures in *italics***

# PLATE 83A

**GW37 Horizontal
CR 320 mm, Y301-62
Level 5: Section 631**

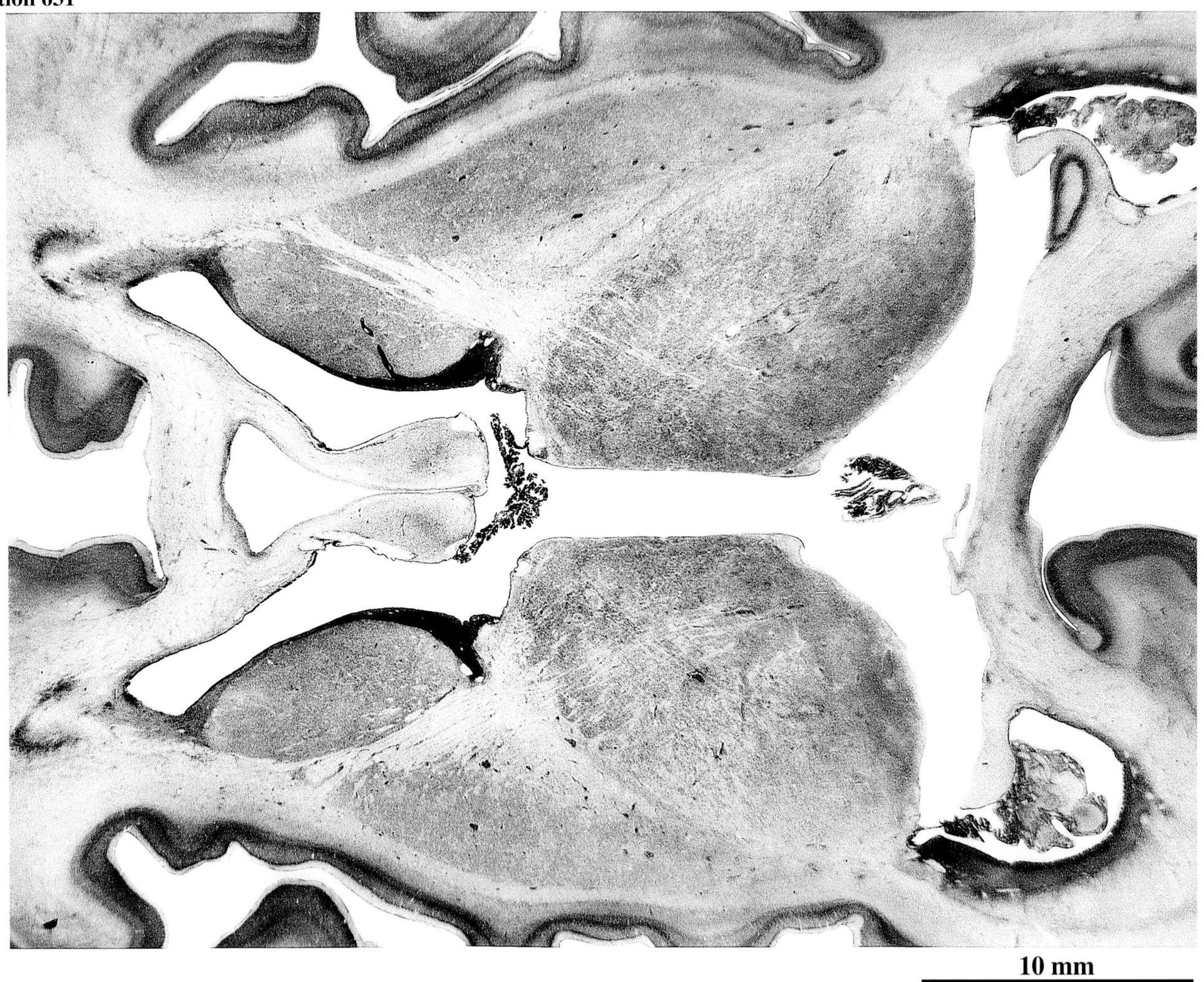

**10 mm**

See the entire section in Plate 69.

119

# PLATE 83B

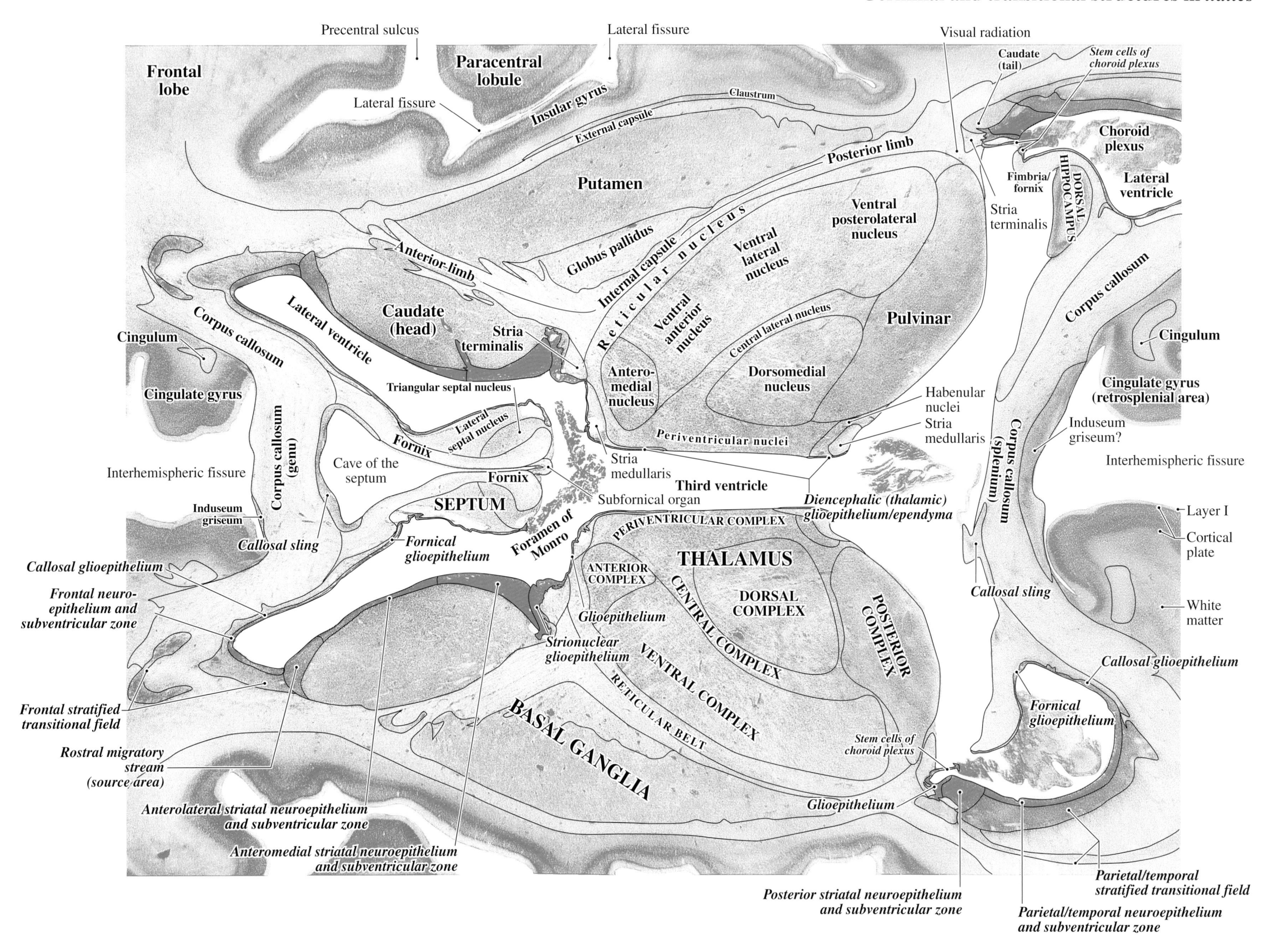

**Germinal and transitional structures in *italics***

## PLATE 84A

**GW37 Horizontal**
**CR 320 mm, Y301-62**
**Level 6: Section 731**

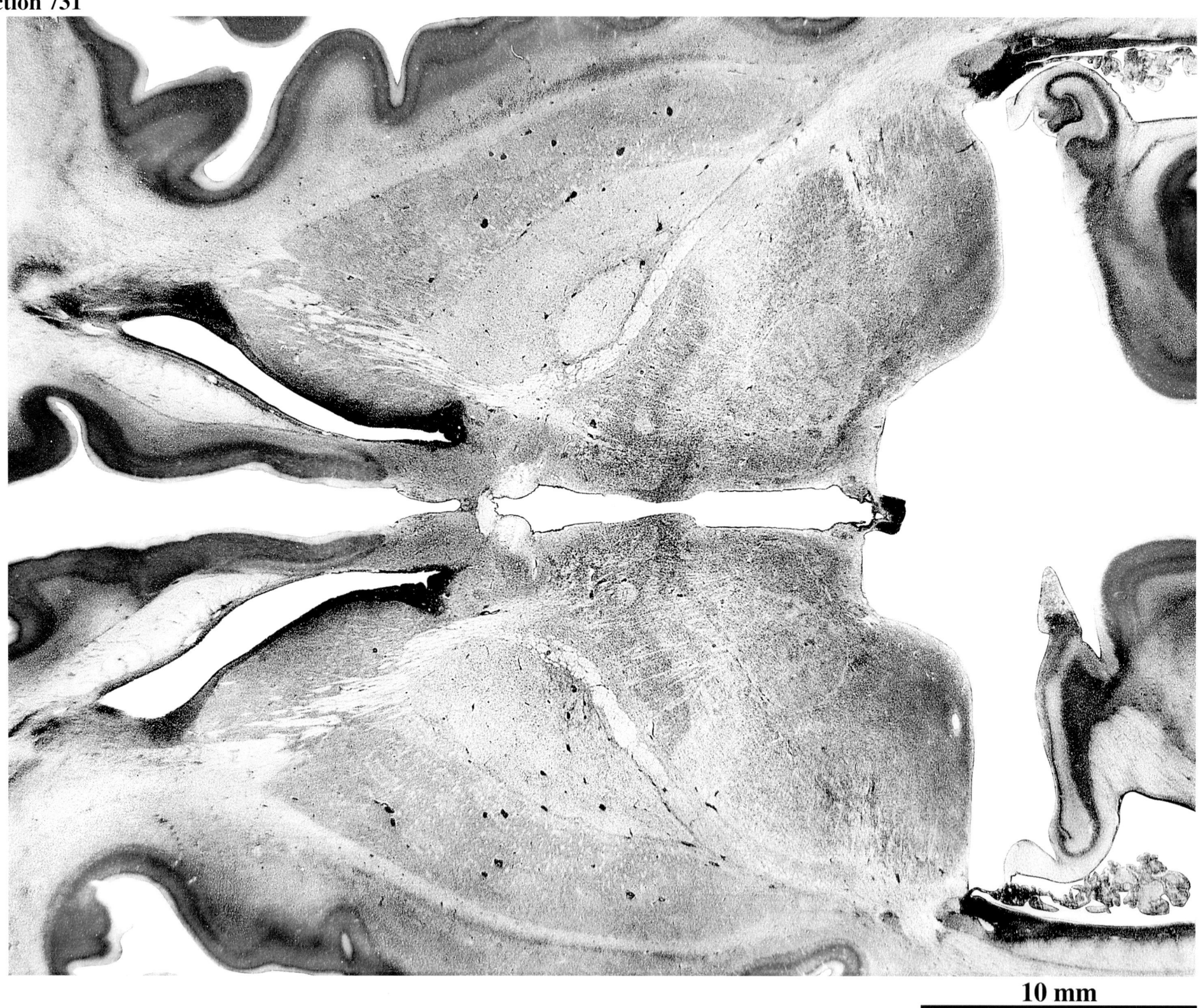

10 mm

See the entire section in Plate 70.

# PLATE 84B

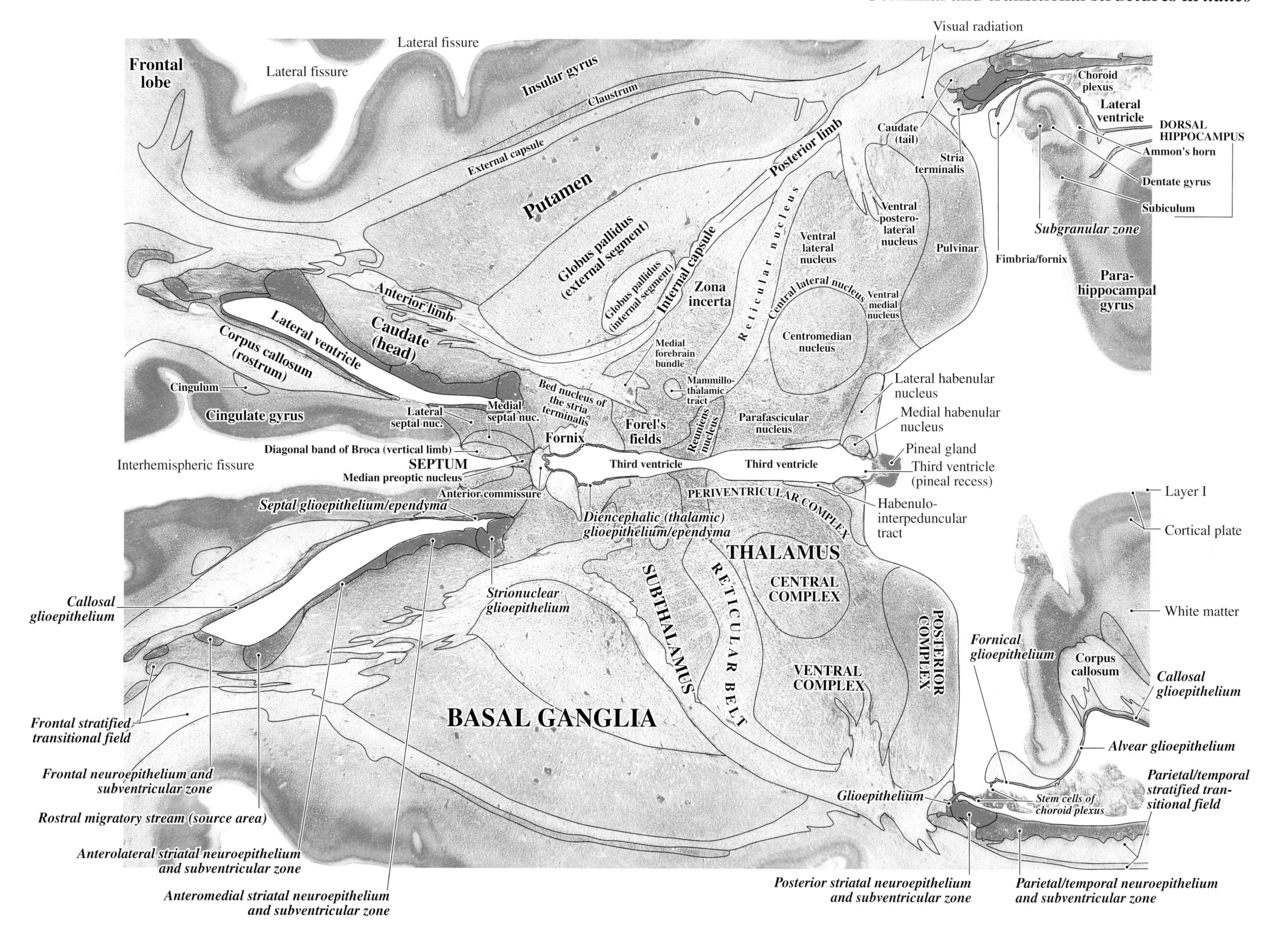

**Germinal and transitional structures in *italics***

## PLATE 85A

**GW37 Horizontal**
**CR 320 mm, Y301-62**
**Level 7: Section 761**

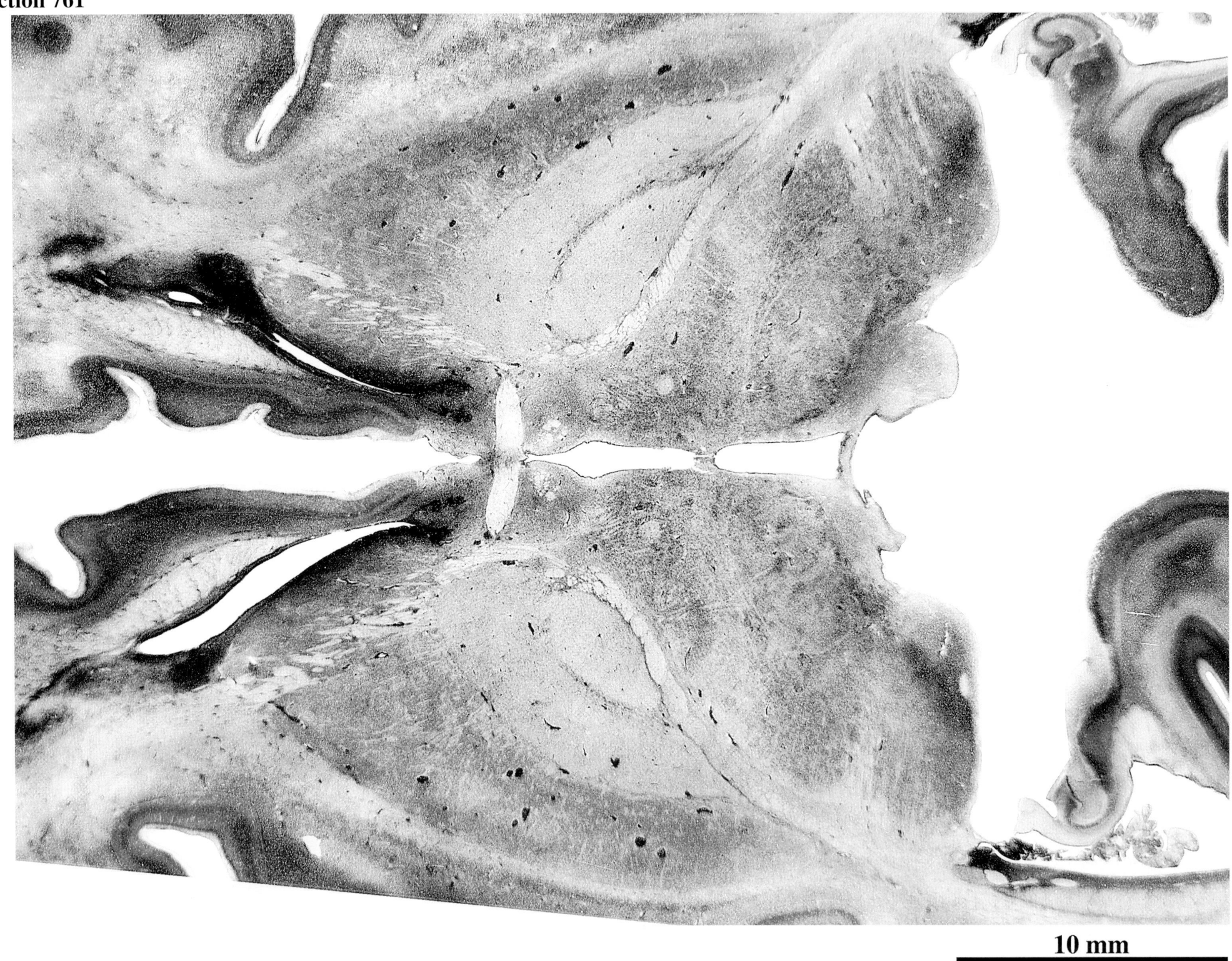

10 mm

See the entire section in Plate 71.

# PLATE 85B

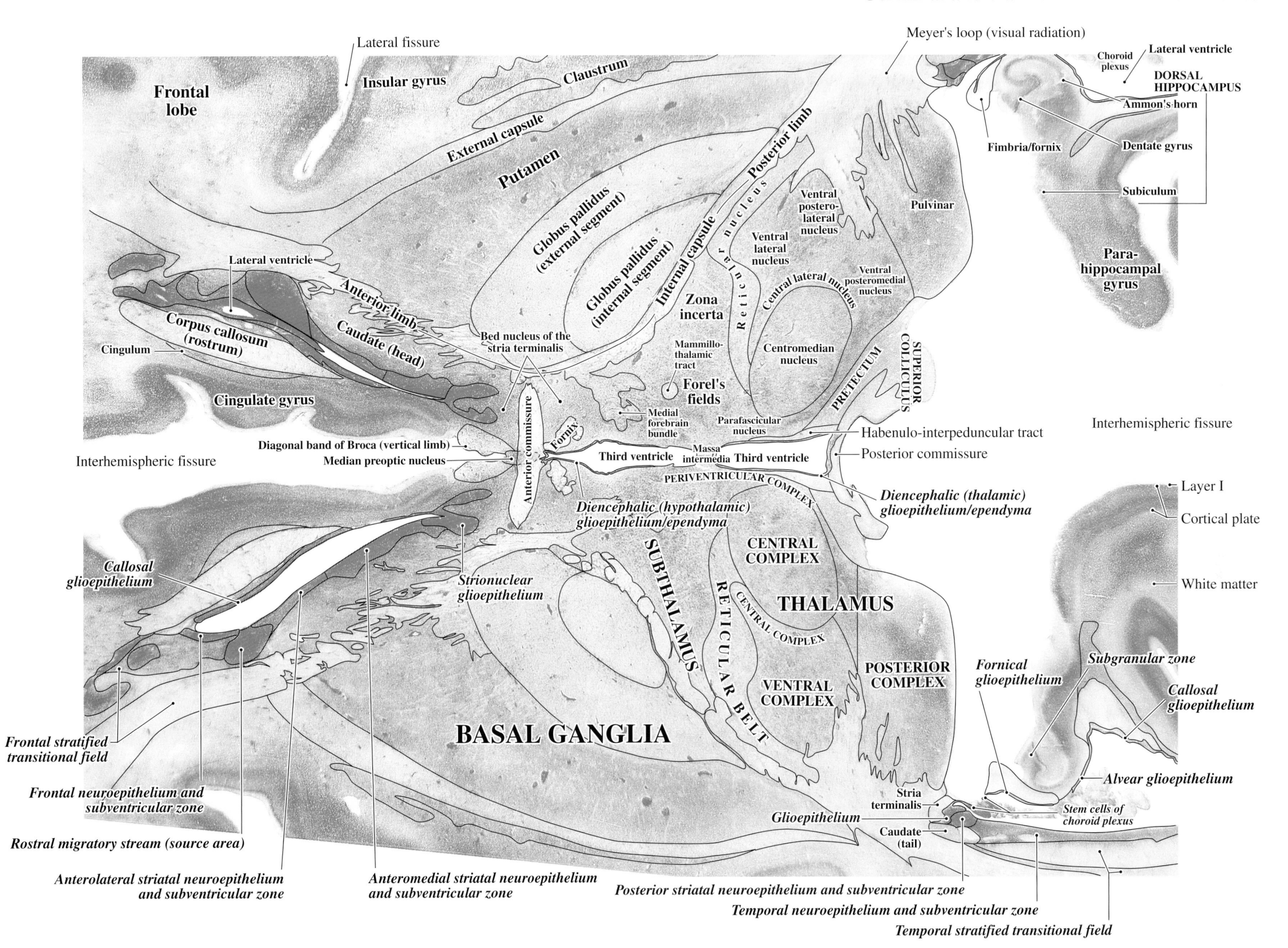

**Germinal and transitional structures in *italics***

## PLATE 86A

**GW37 Horizontal**
**CR 320 mm, Y301-62, Level 8: Section 801**

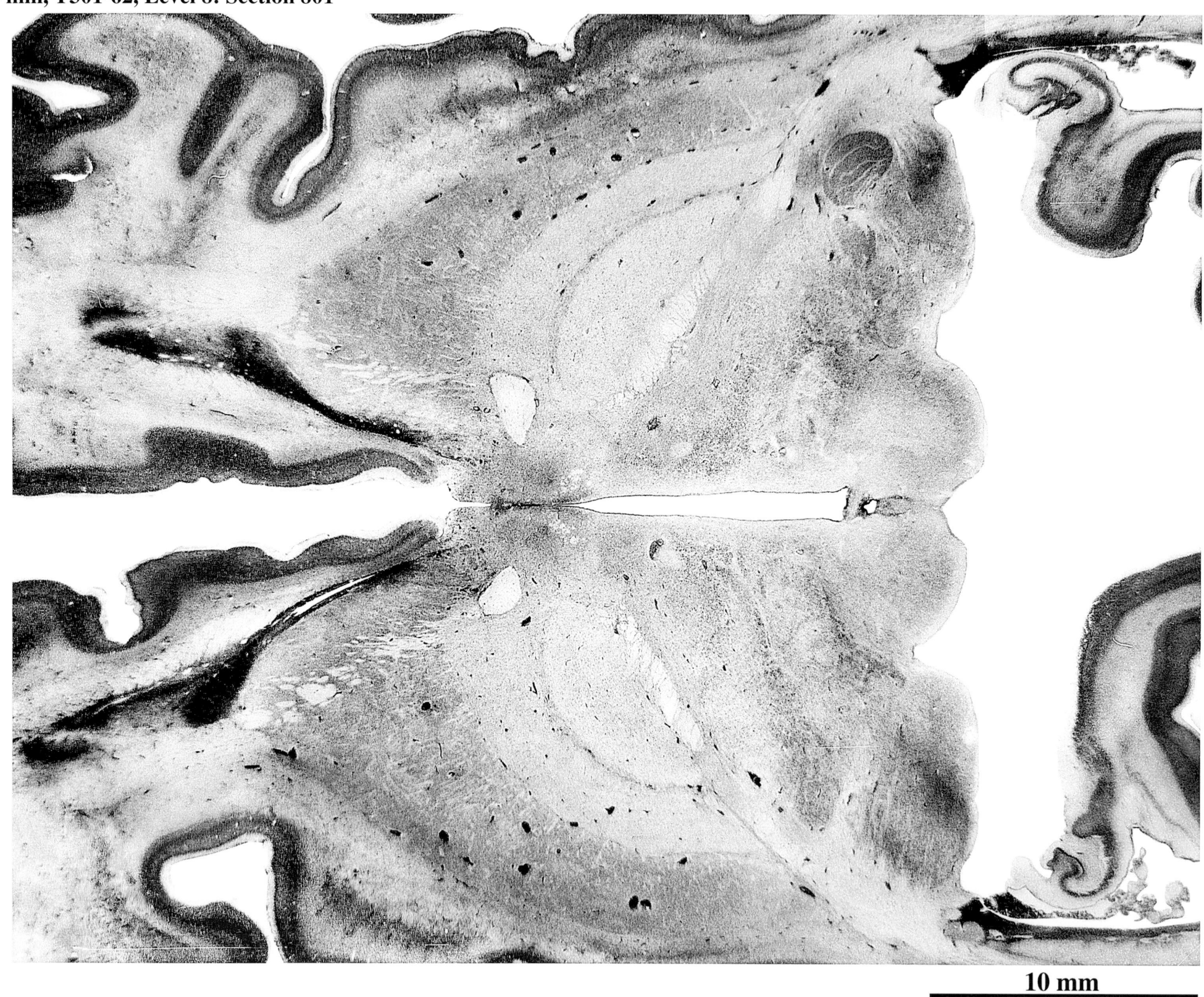

10 mm

See the entire section in Plate 72.

# PLATE 86B

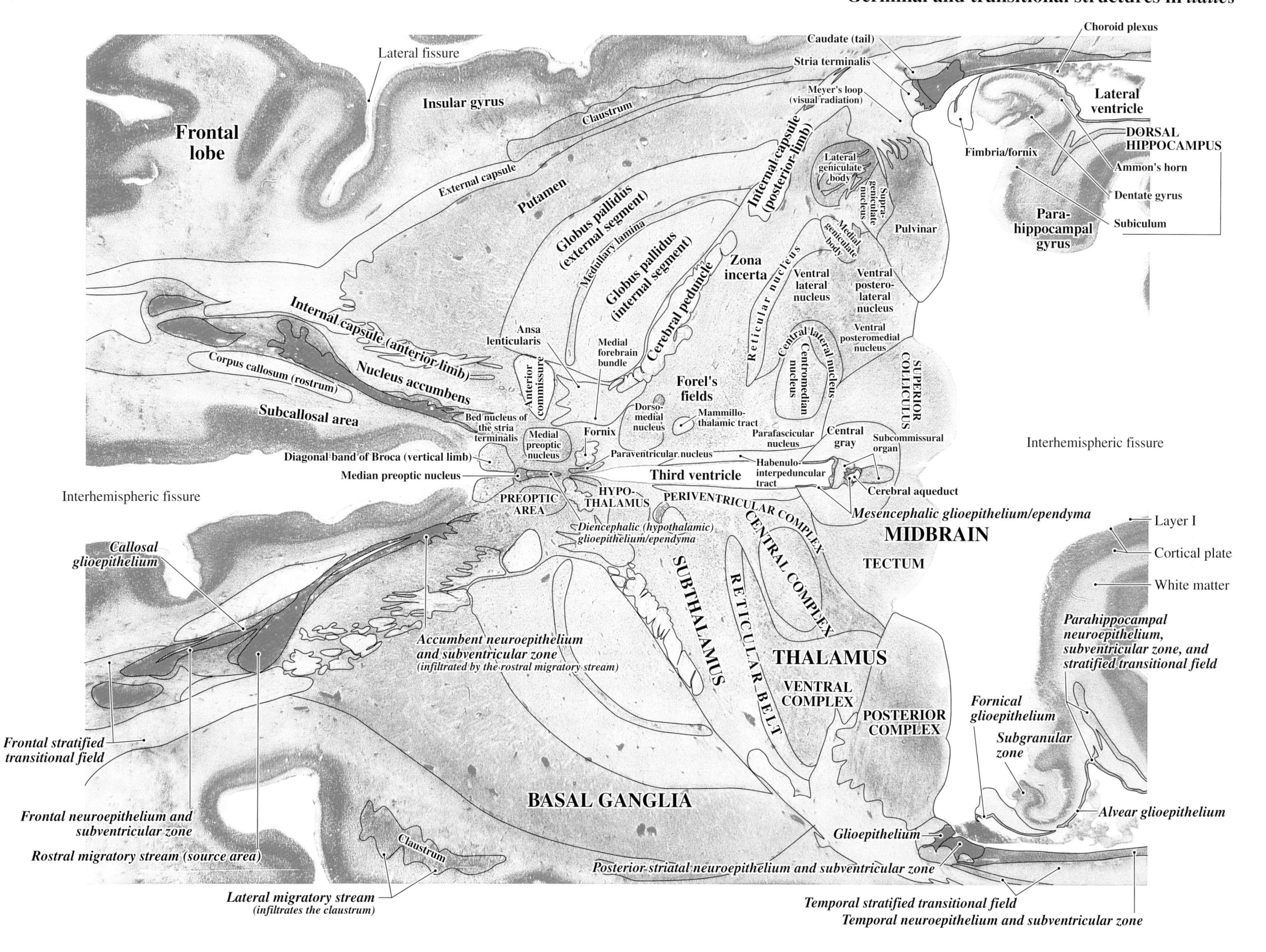

**Germinal and transitional structures in *italics***

# PLATE 87A

**GW37 Horizontal**
**CR 320 mm, Y301-62, Level 9: Section 851**

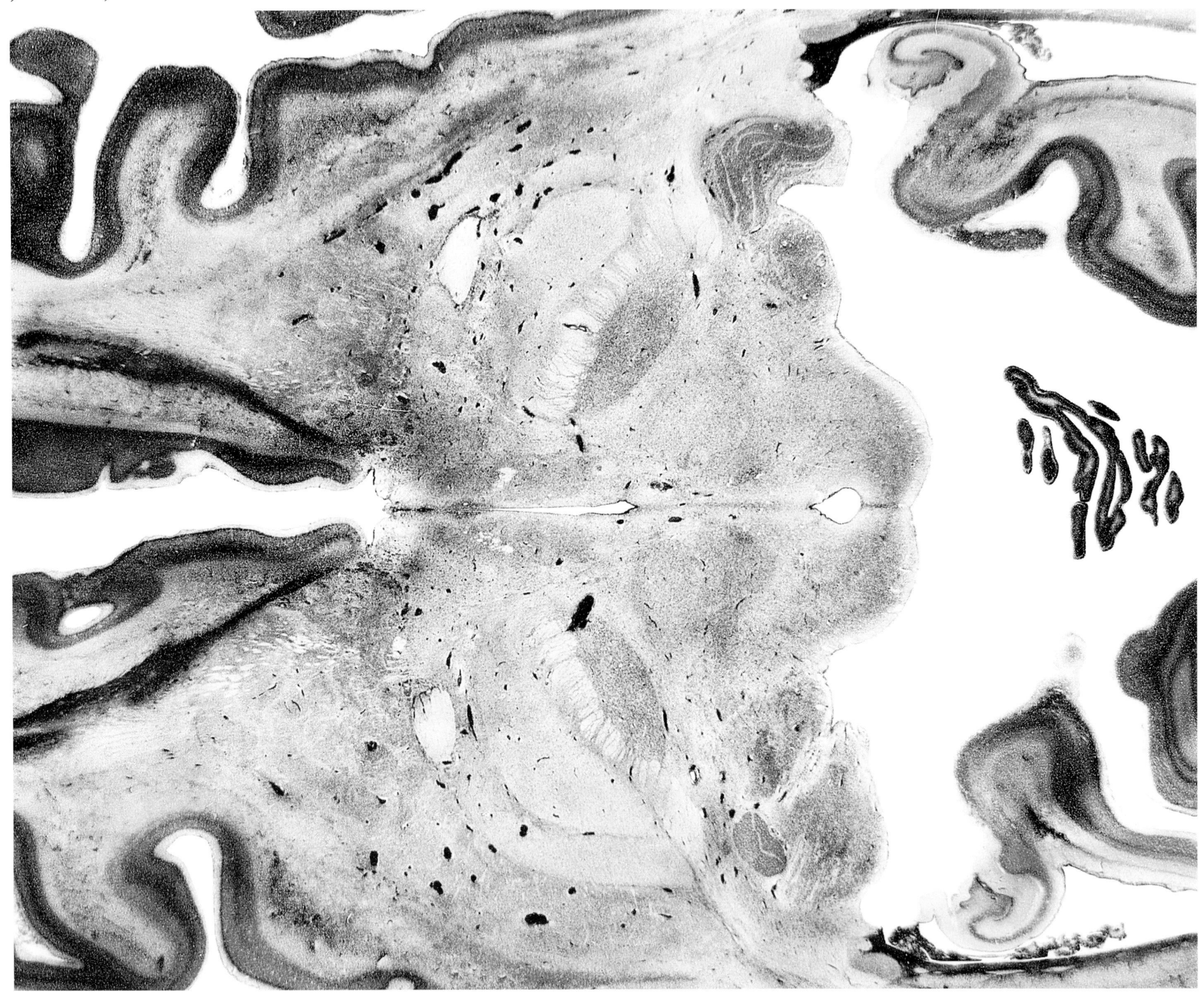

10 mm

See the entire section in Plate 73.

### PLATE 87B

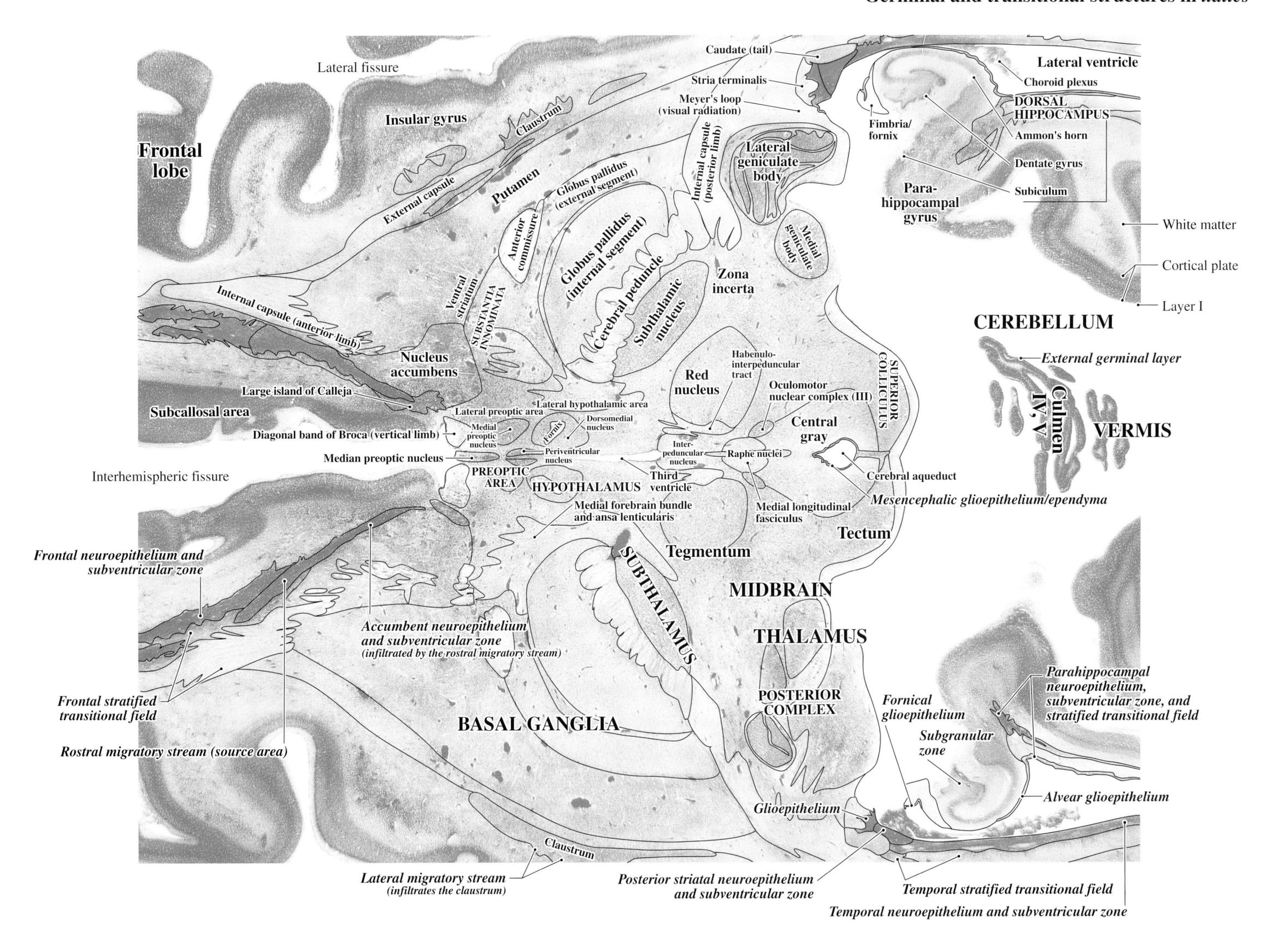

**Germinal and transitional structures in *italics***

## PLATE 88A

**GW37 Horizontal
CR 320 mm, Y301-62
Level 10: Section 921**

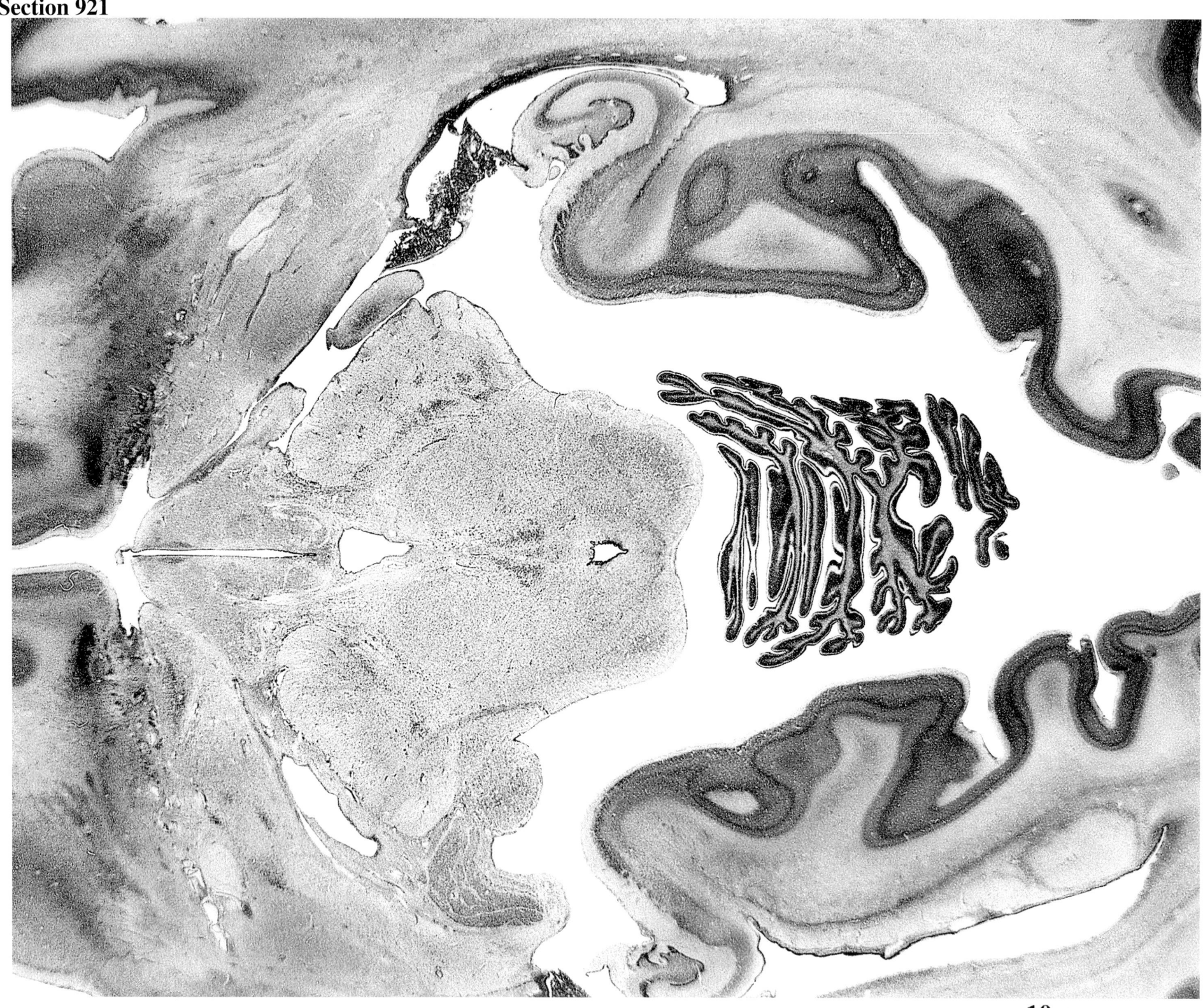

See the entire section in Plate 74.

10 mm

**PLATE 88B**

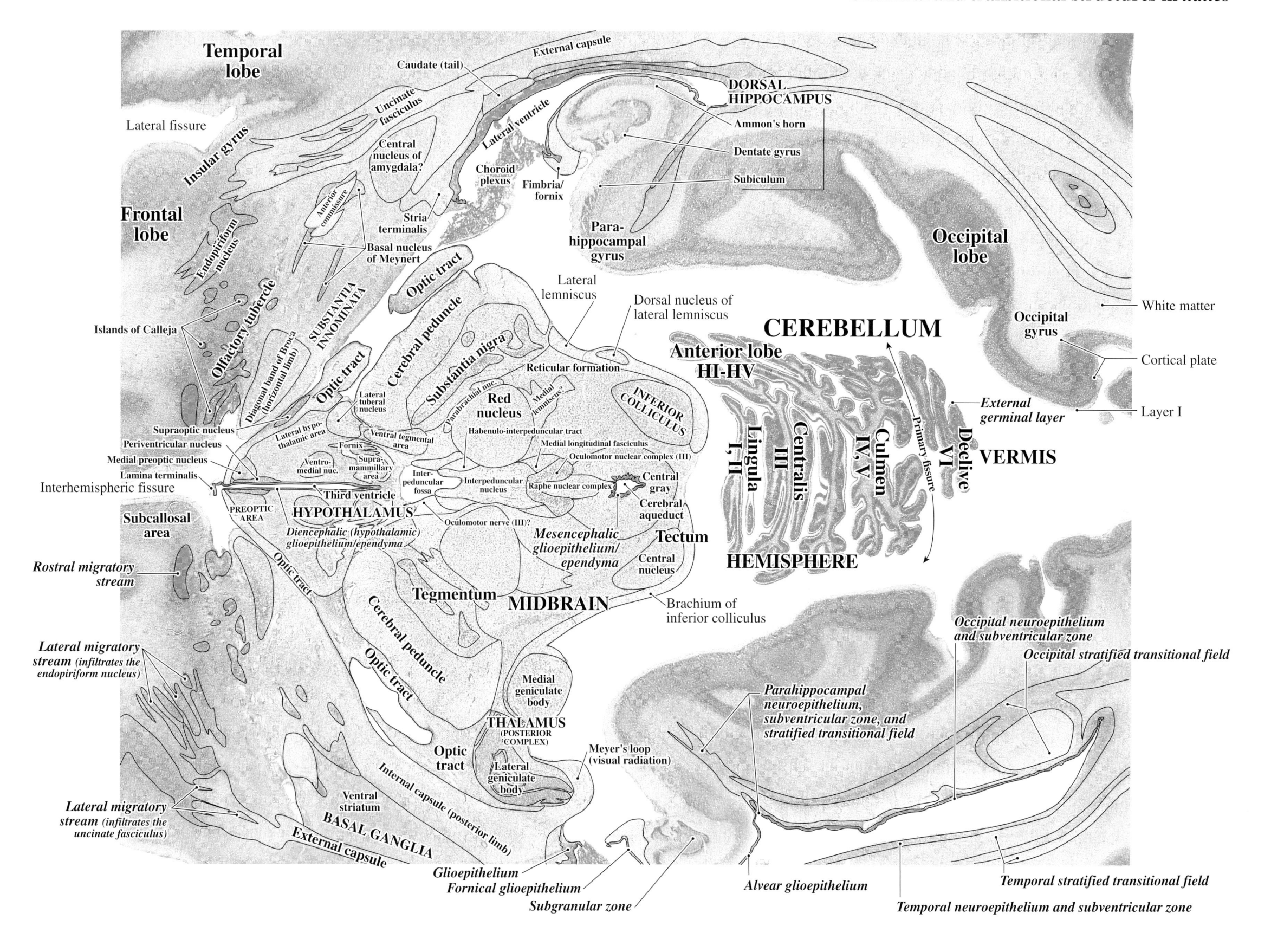

**Germinal and transitional structures in *italics***

## PLATE 89A

**GW37 Horizontal
CR 320 mm
Y301-62
Level 11:
Section 981**

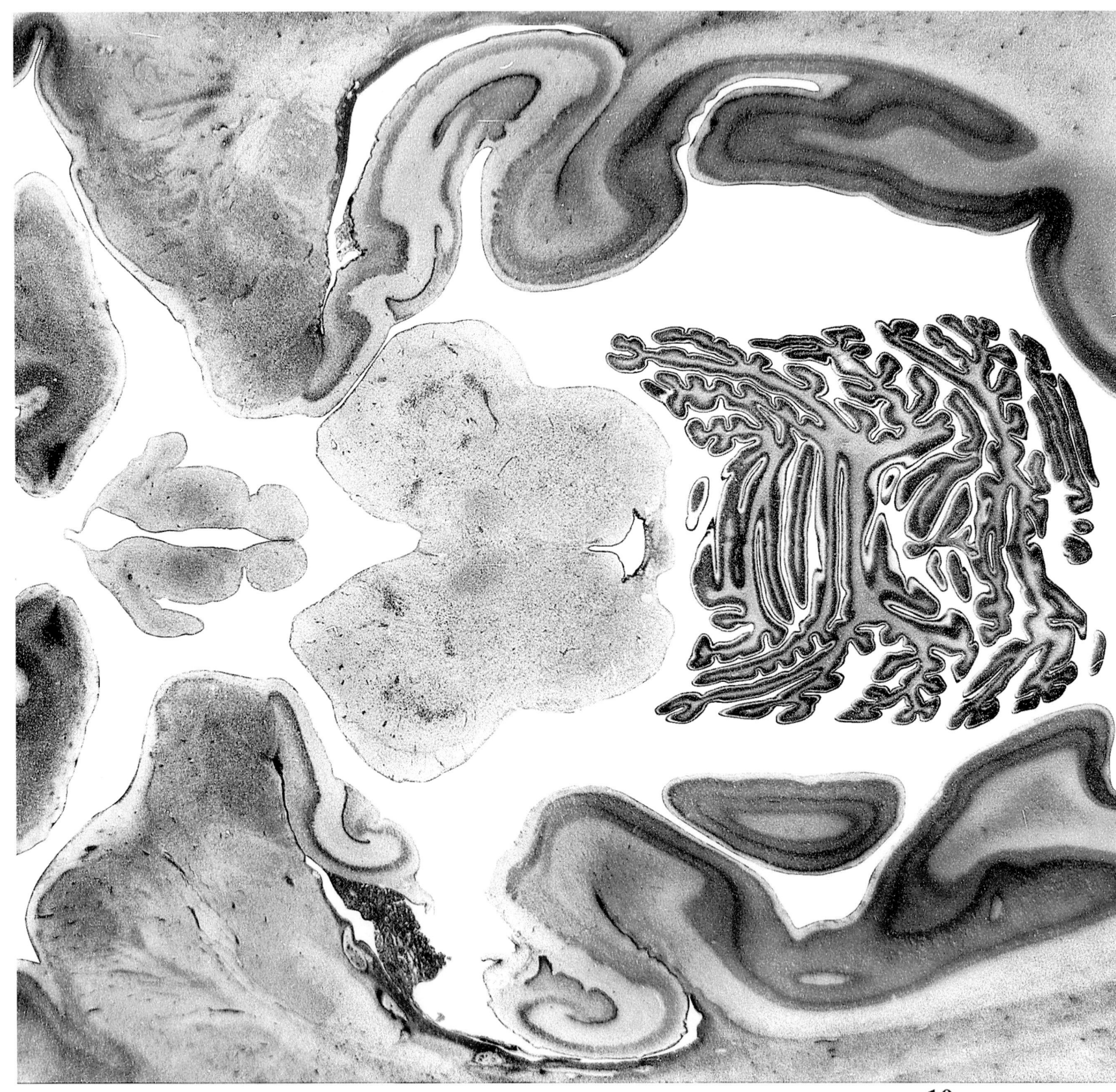

See the entire section in Plate 75.

10 mm

# PLATE 89B

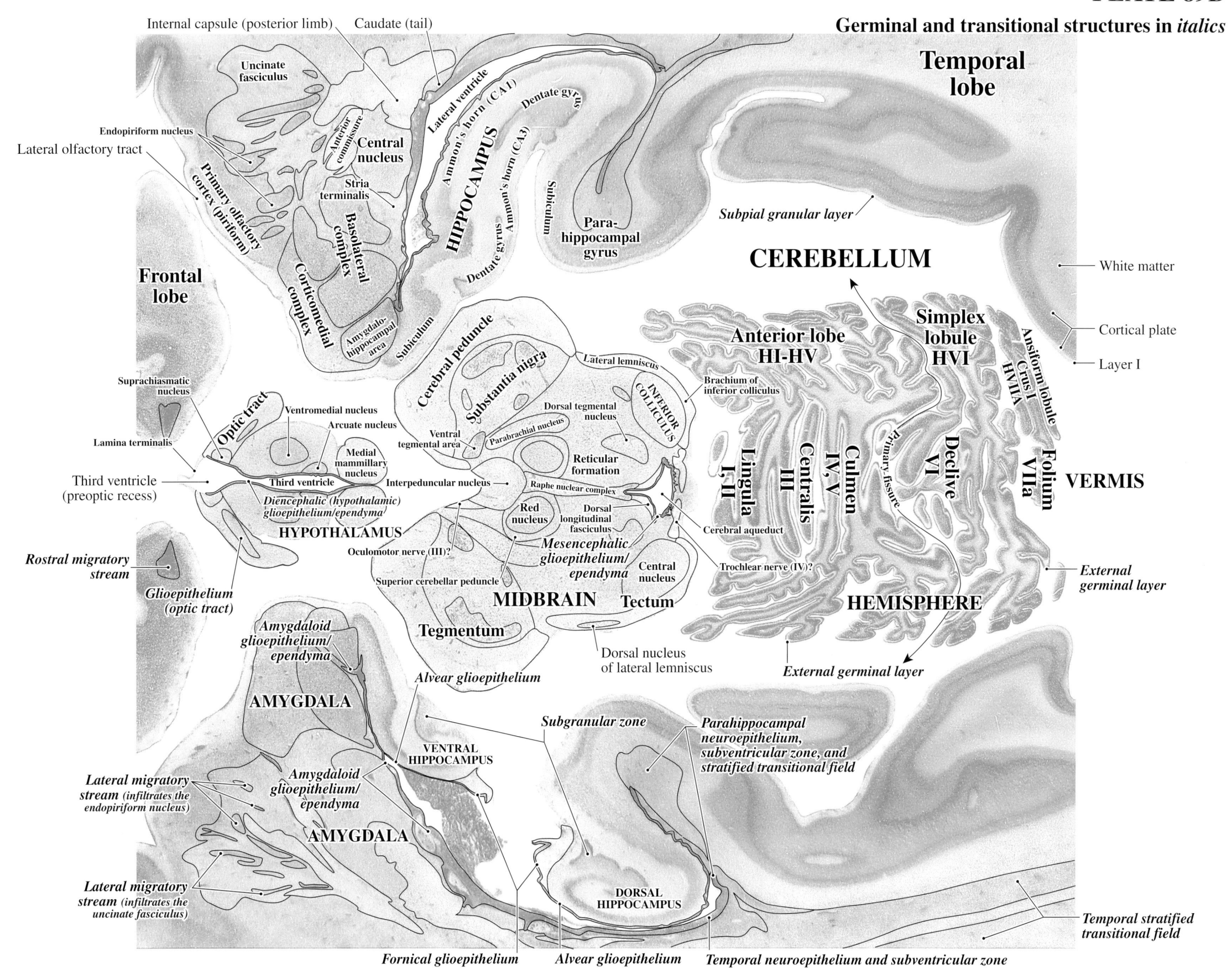

**Germinal and transitional structures in *italics***

## PLATE 90A

**GW37 Horizontal
CR 320 mm
Y301-62
Level 12:
Section 1051**

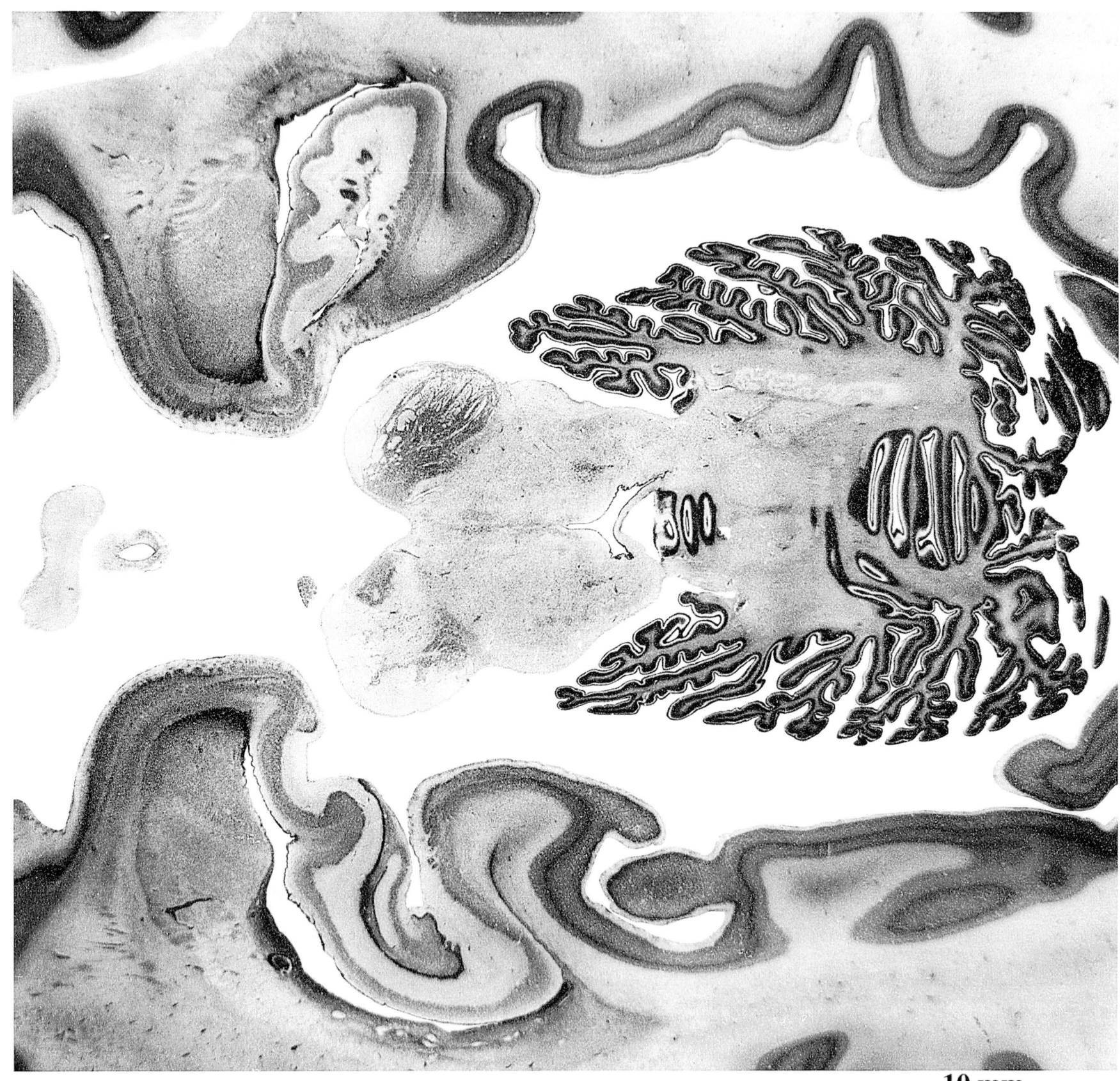

See the entire section in Plate 76.

10 mm

## PLATE 90B

**Germinal and transitional structures in *italics***

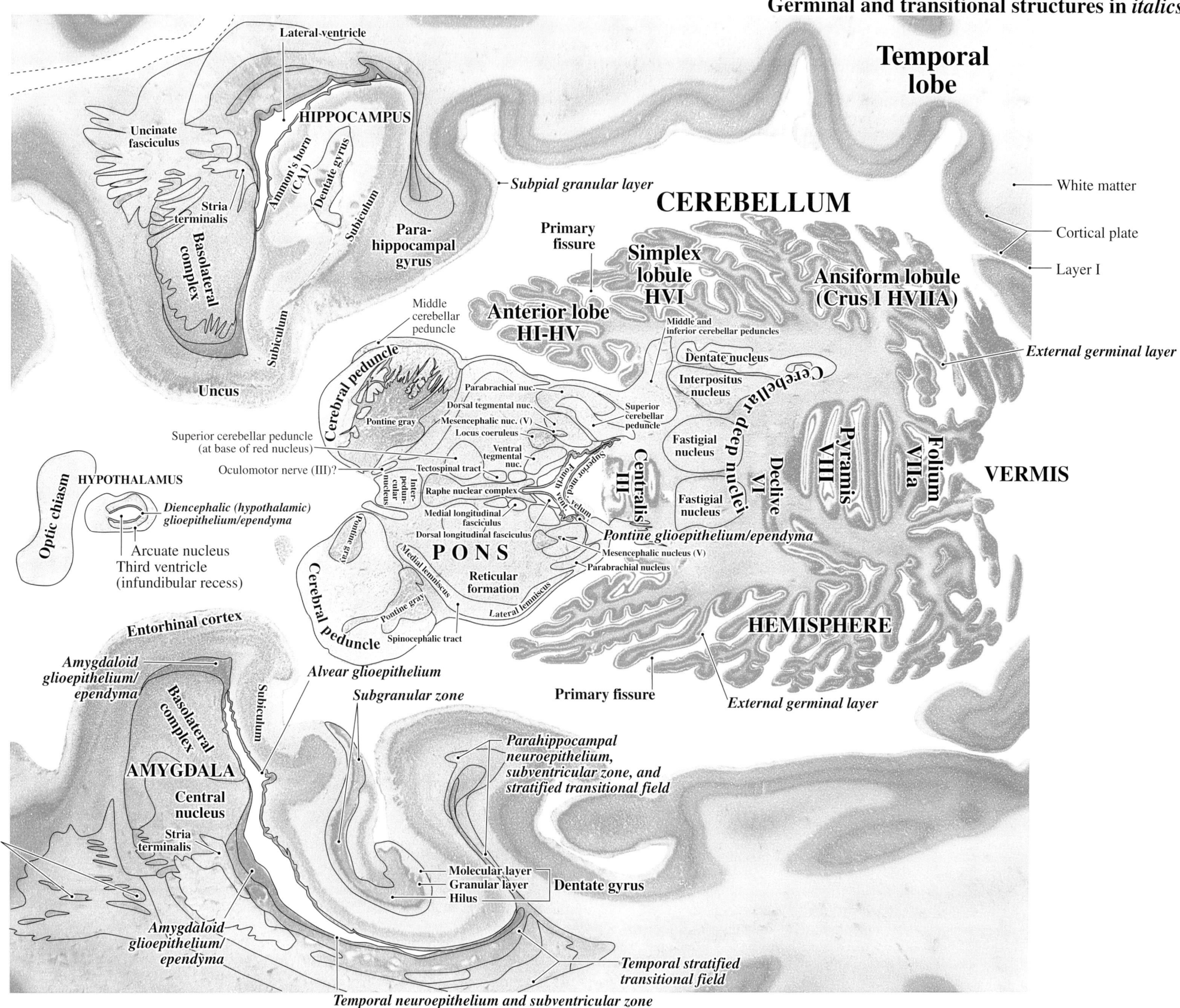

134

**PLATE 91A**

**GW37 Horizontal**
**CR 320 mm**
**Y301-62**
**Level 13: Section 1111**

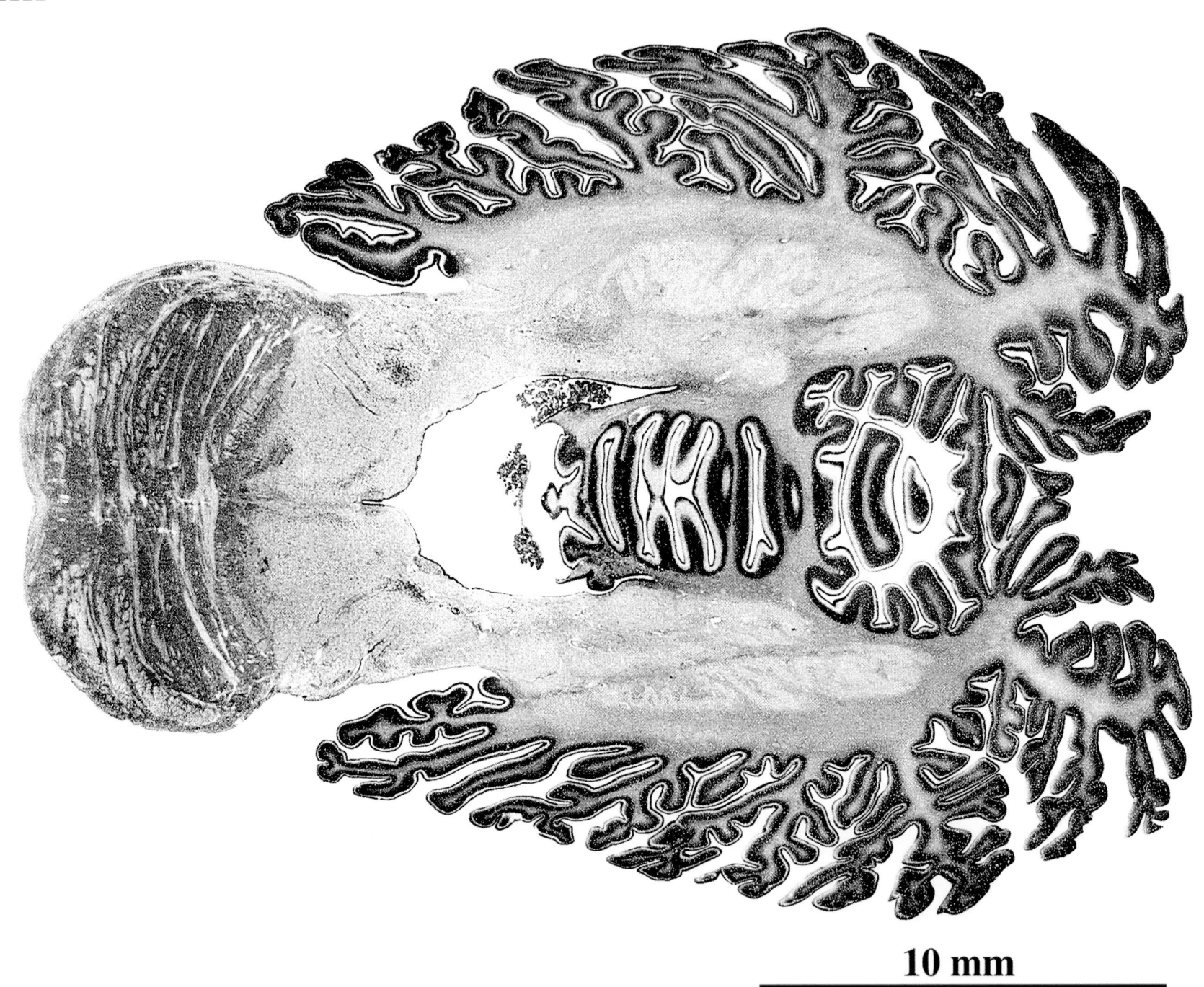

**10 mm**

See the entire section in **Plate 77.**

## PLATE 91B

**Germinal and transitional structures in *italics***

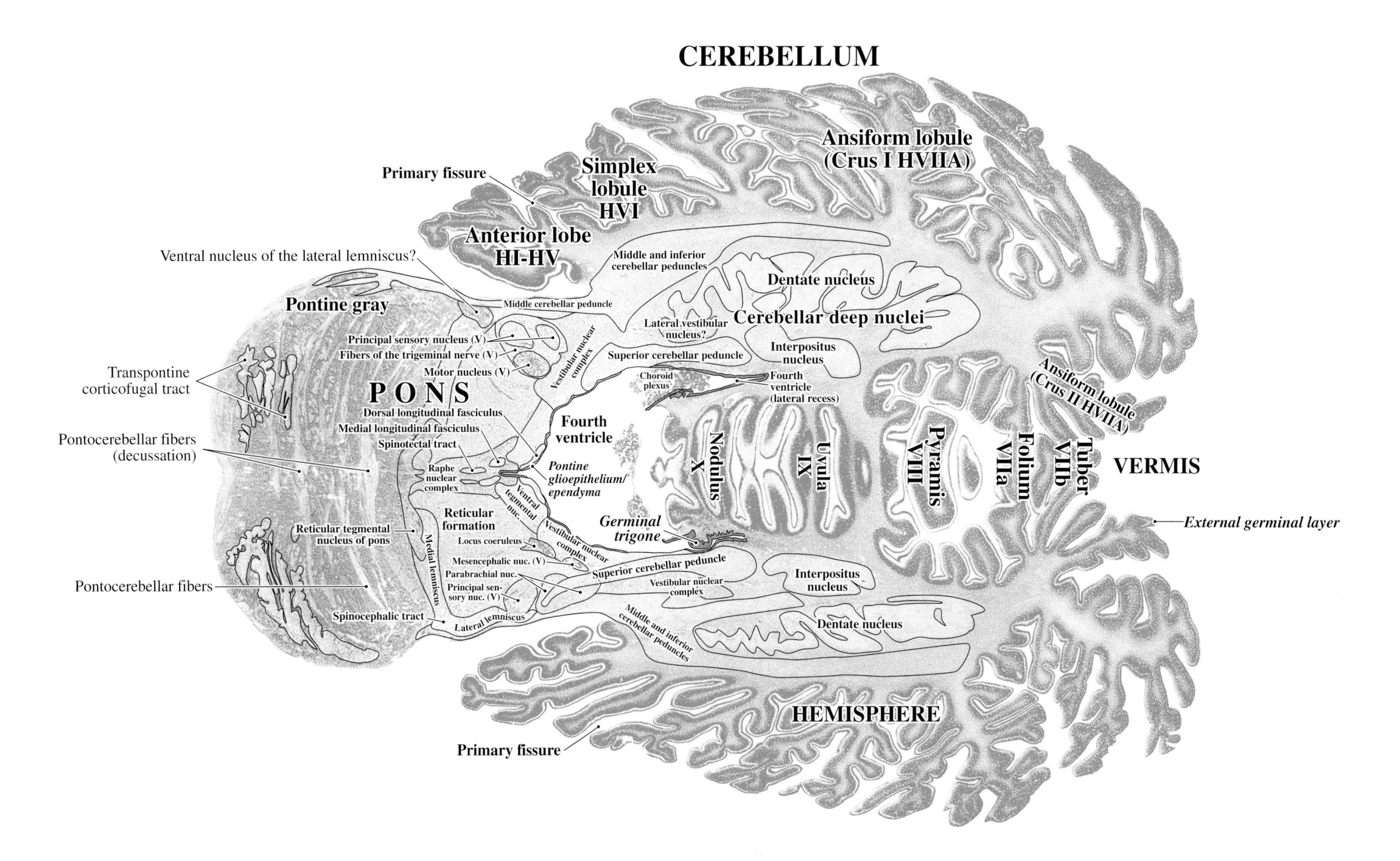

**PLATE 92A**

GW37 Horizontal
CR 320 mm
Y301-62
Level 14: Section 1161

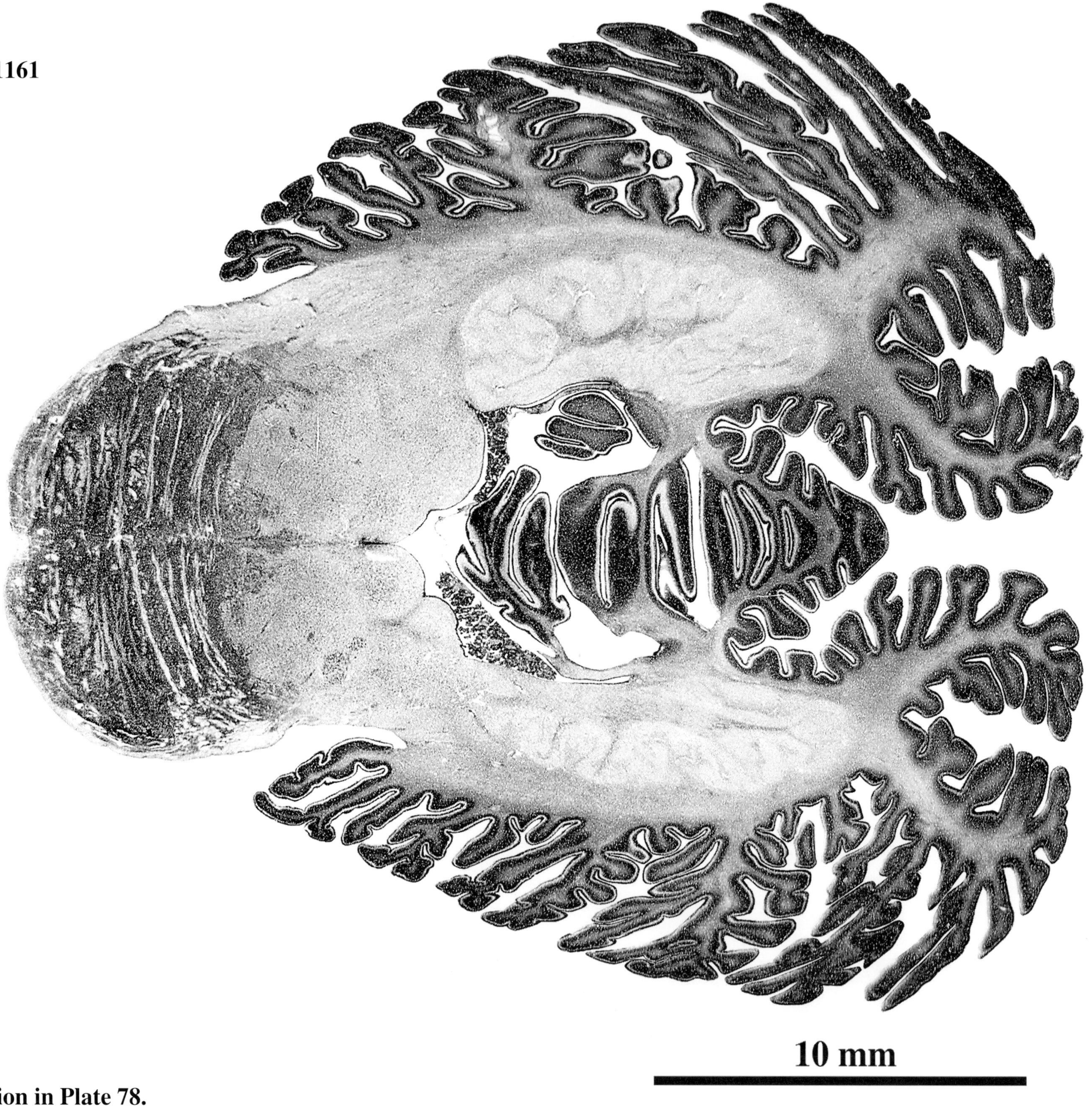

10 mm

See the entire section in Plate 78.

137

**PLATE 92B**

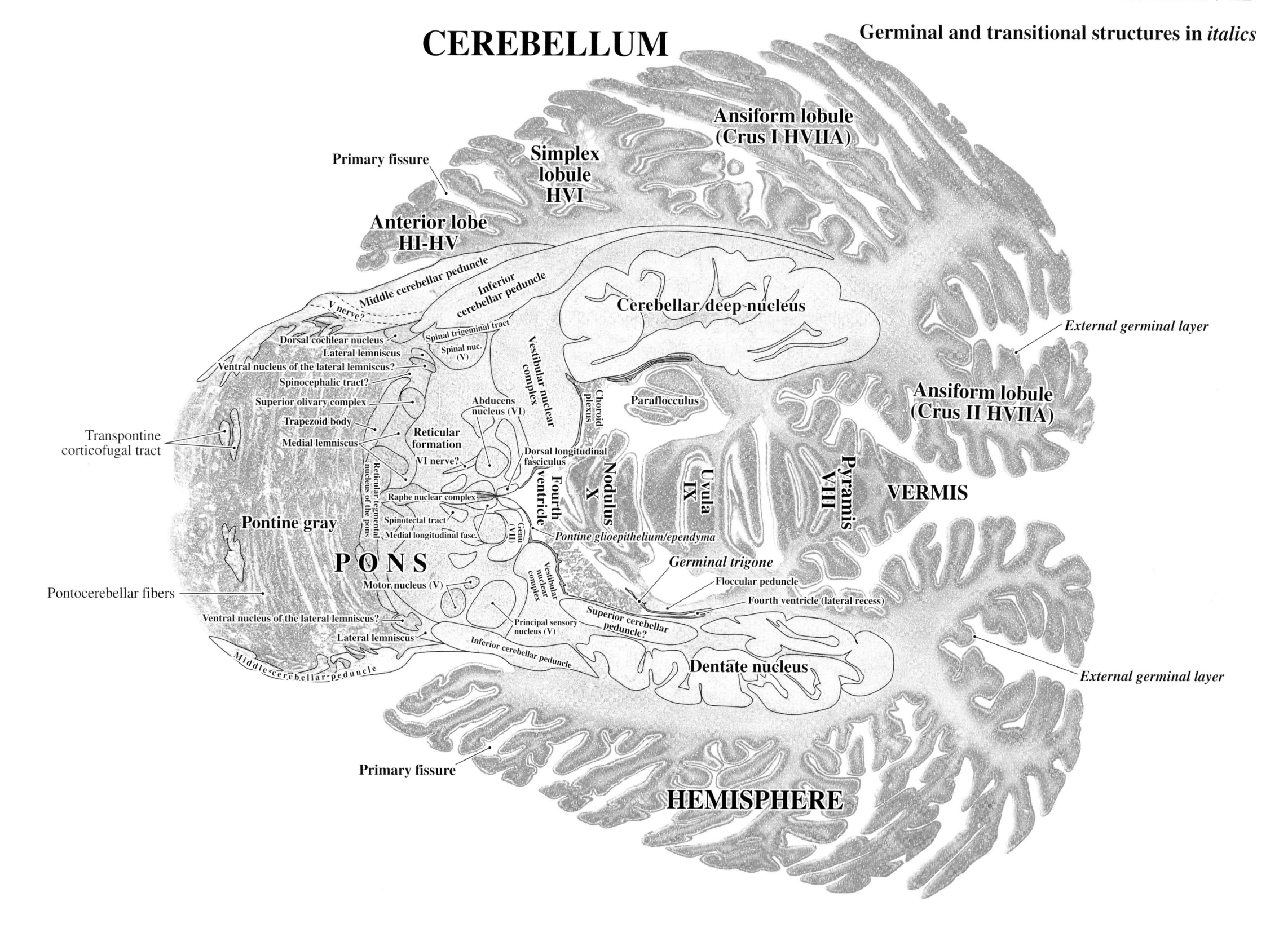

Germinal and transitional structures in *italics*

**CEREBELLUM**

138

**PLATE 93A**

**GW37 Horizontal
CR 320 mm
Y301-62
Level 15: Section 1251**

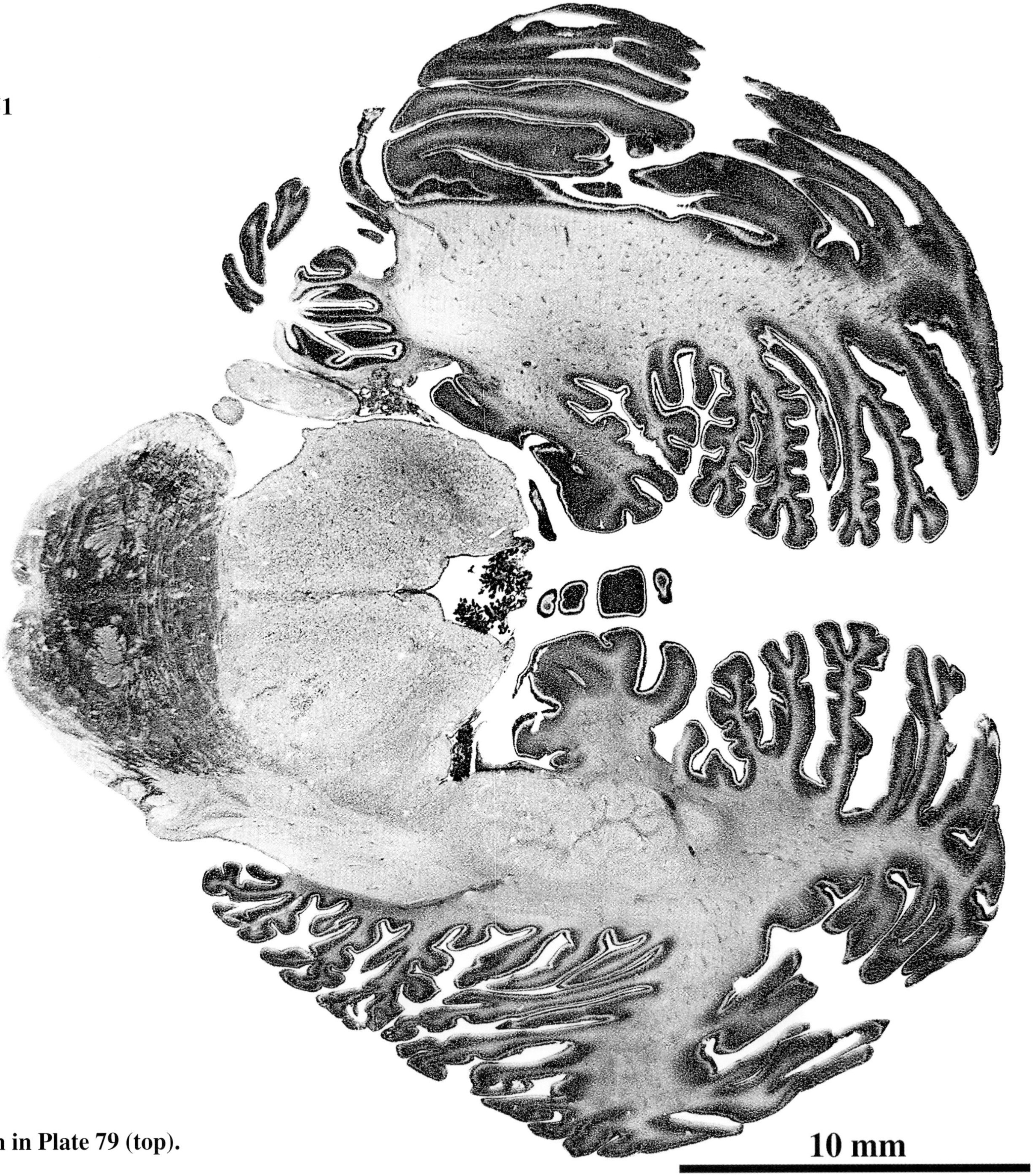

See the entire section in Plate 79 (top).

10 mm

**PLATE 93B**

Germinal and transitional structures in *italics*

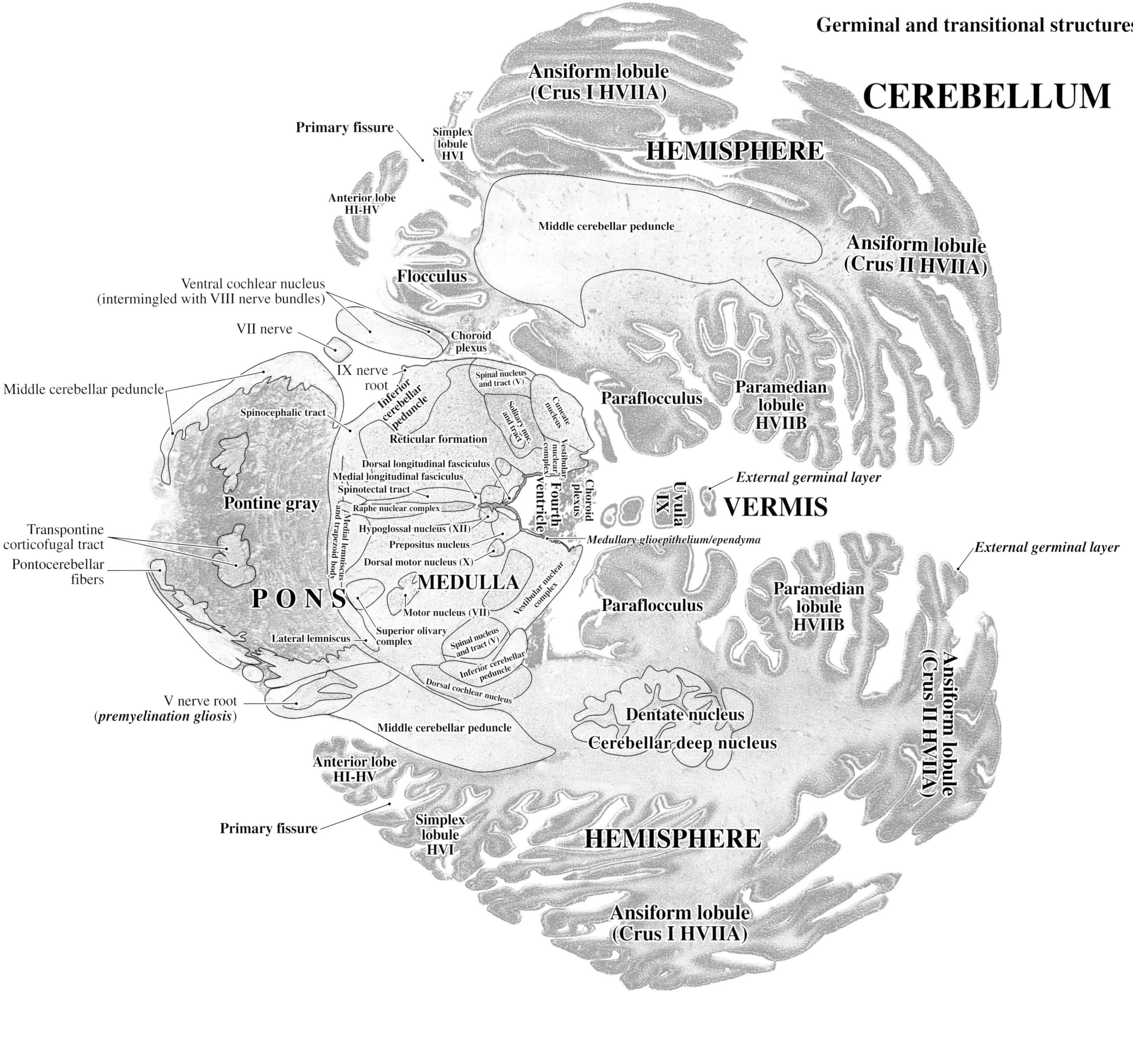

140

# PLATE 94A

**GW37 Horizontal**
**CR 320 mm**
**Y301-62**

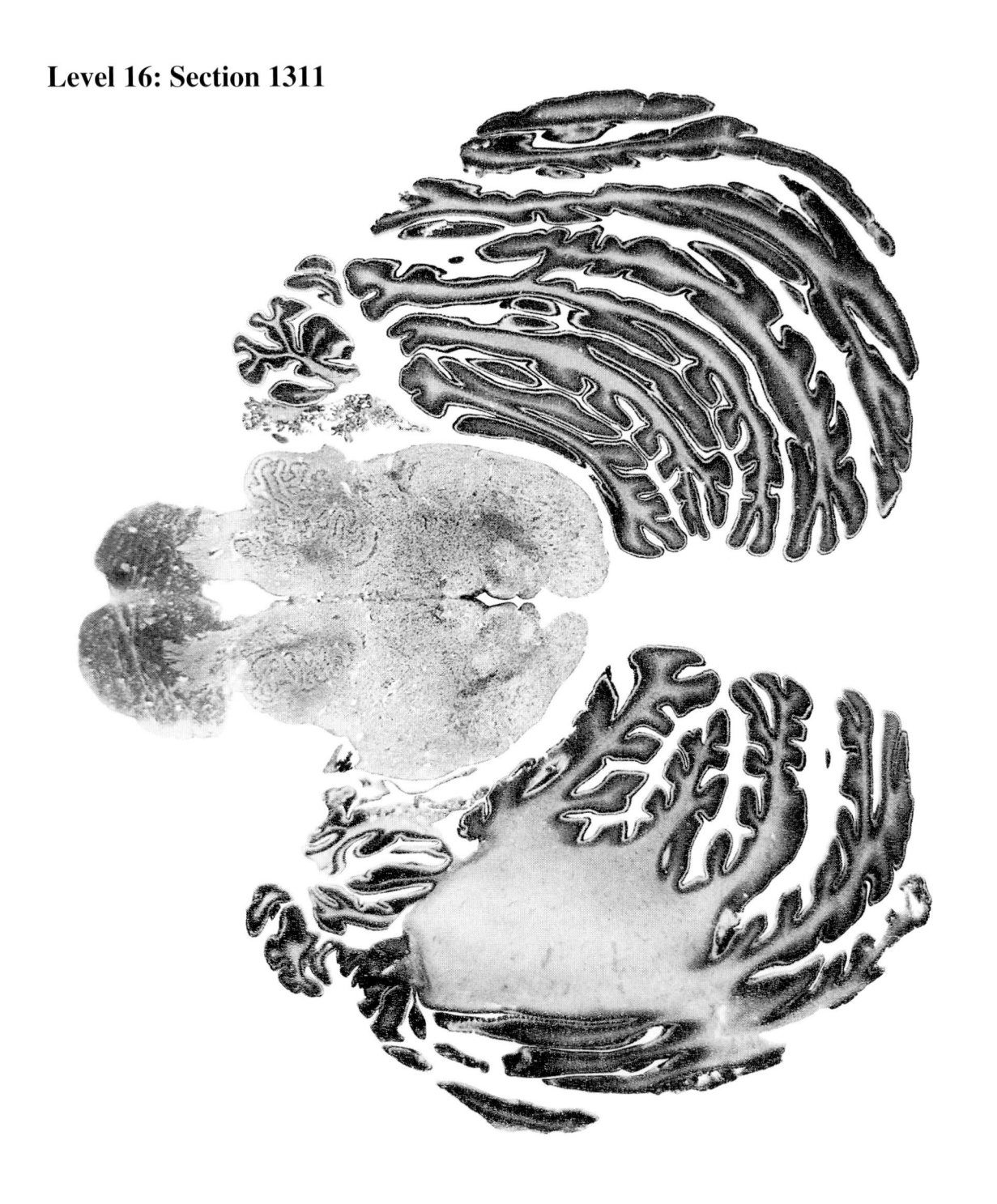

**Level 16: Section 1311**

**Level 17: Section 1331**

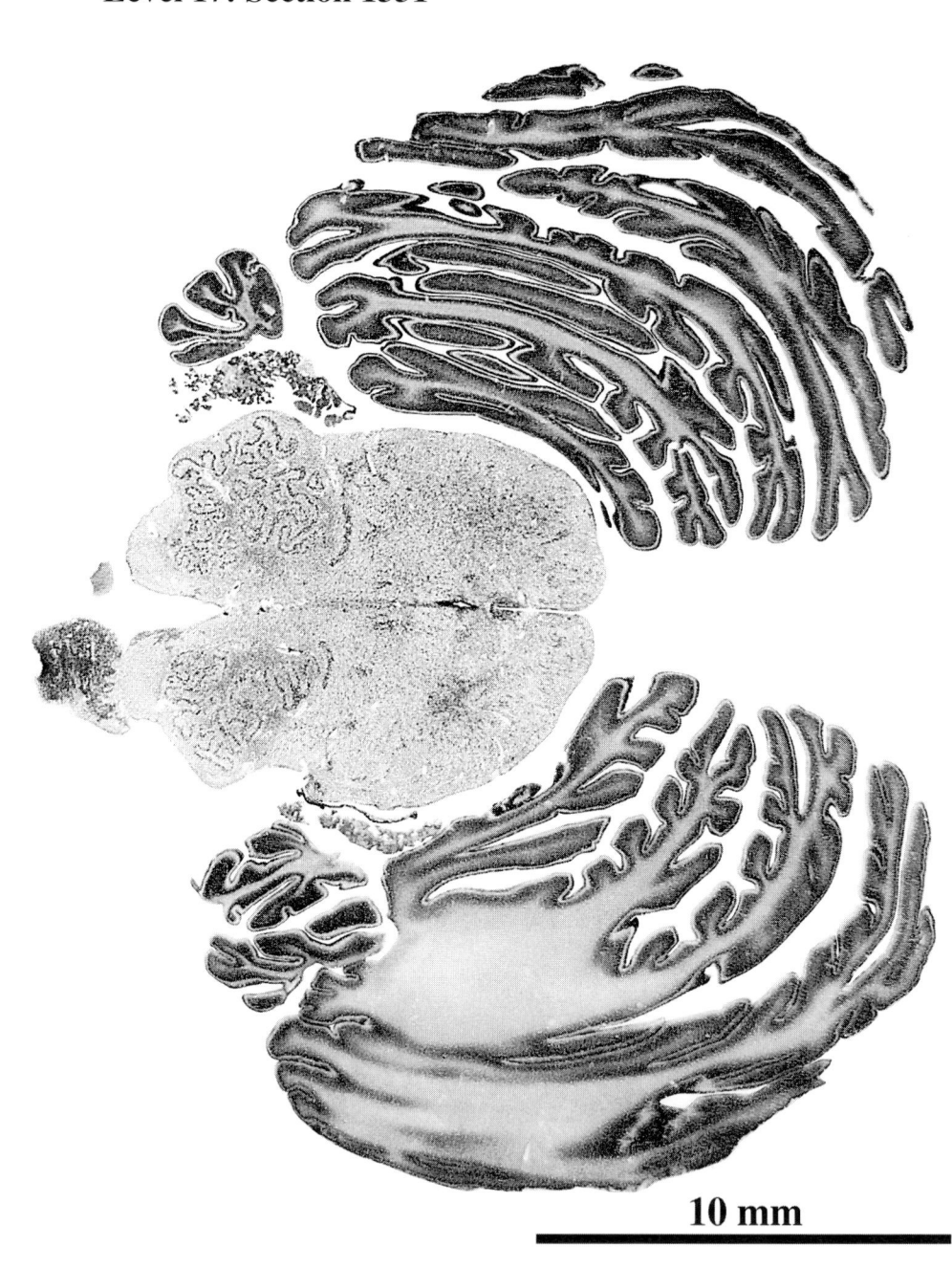

10 mm

See the entire section in Plate 79 (bottom).

See the entire section in Plate 80 (top).

# PLATE 94B

**Germinal and transitional structures in *italics***

## Section 1311

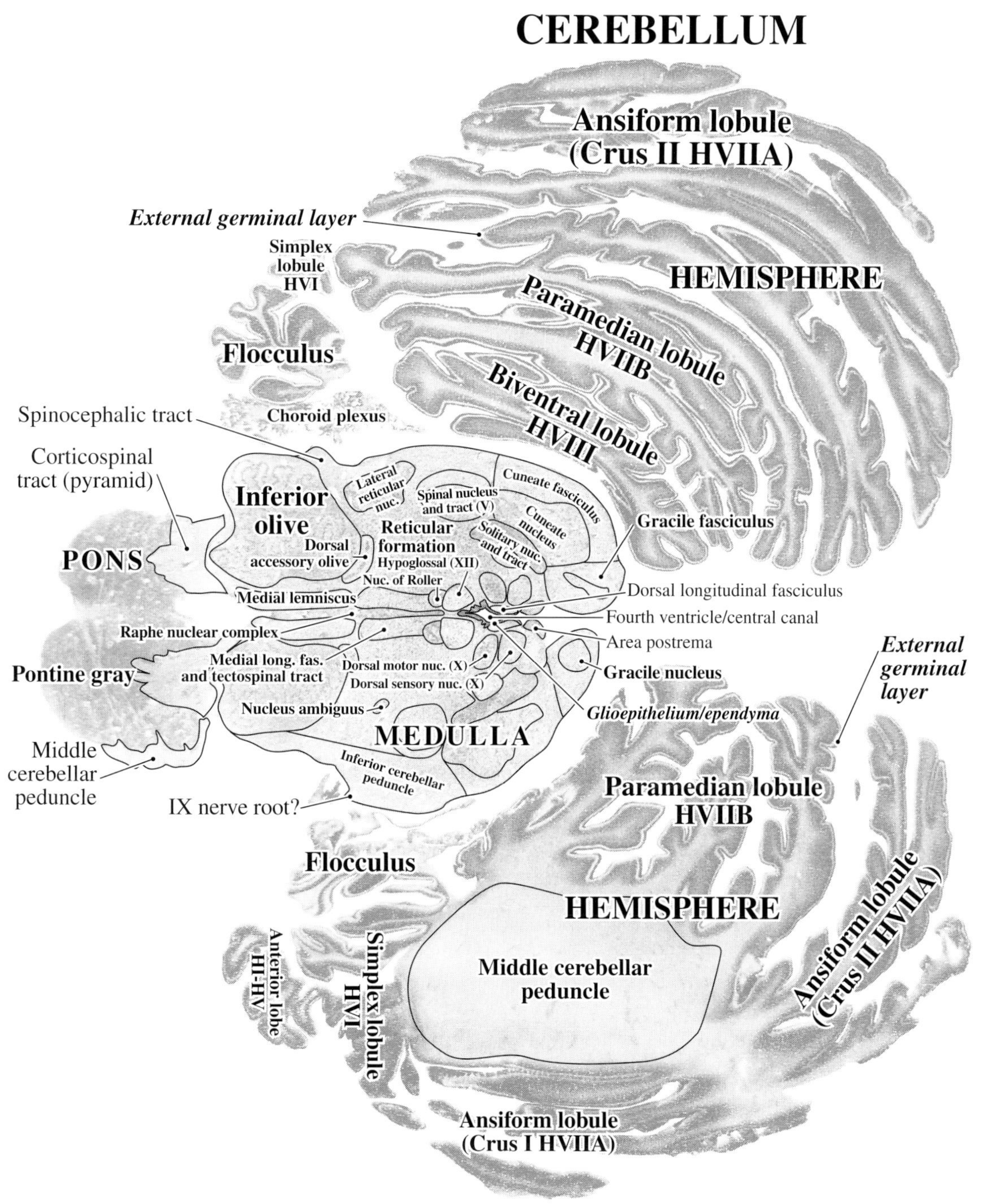


## Section 1331

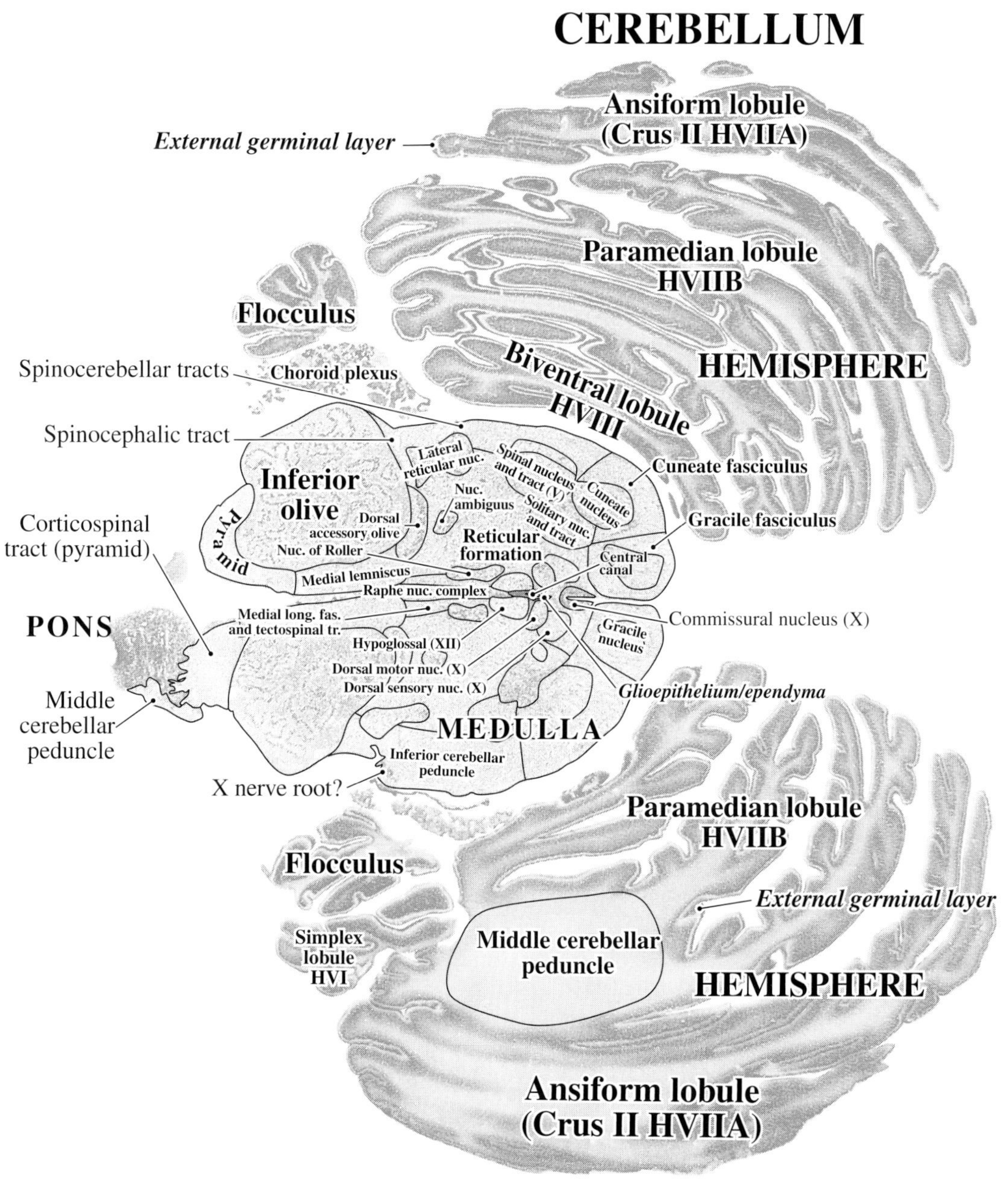

## PLATE 95A

GW37 Horizontal
CR 320 mm
Y301-62

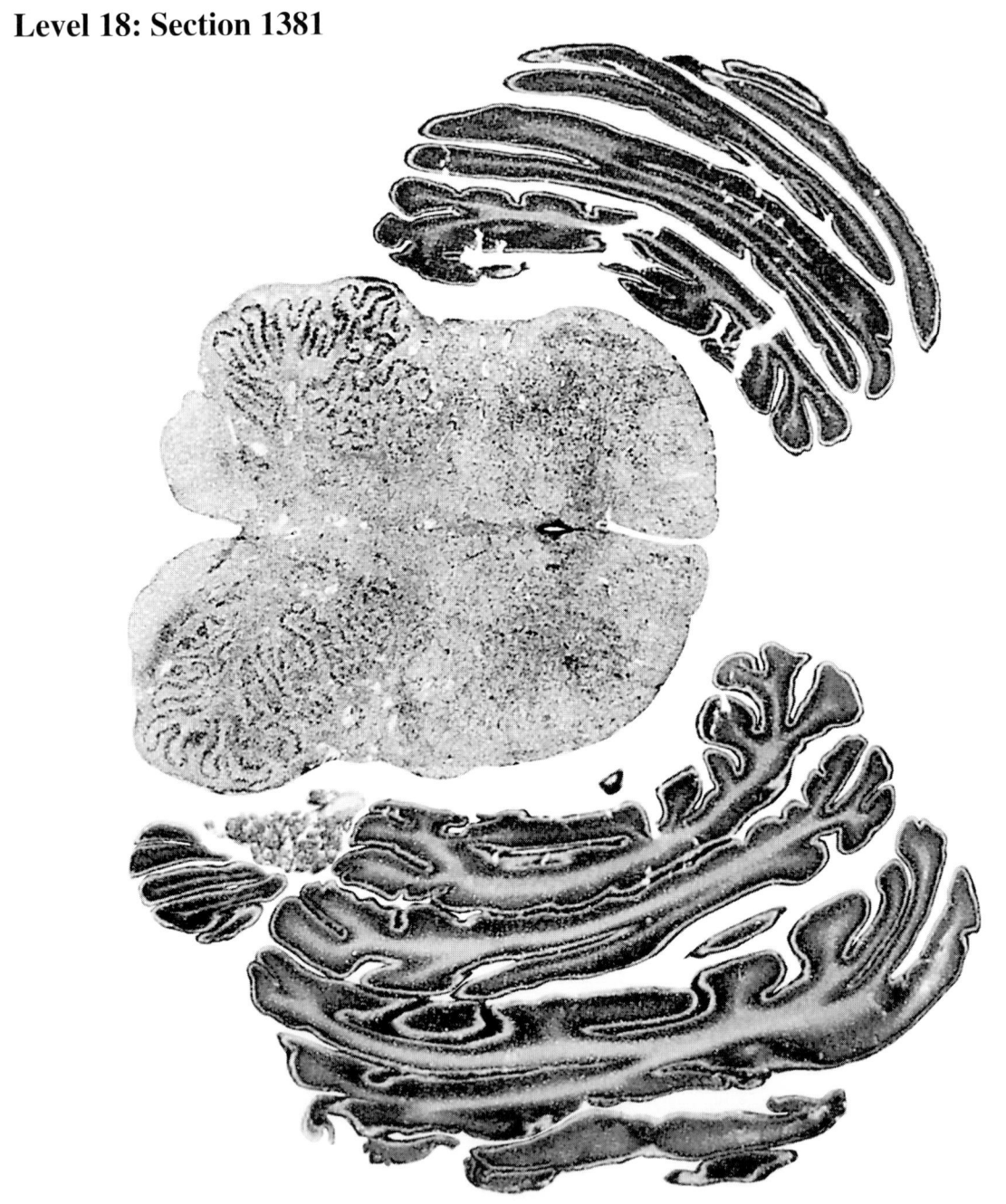

**Level 18: Section 1381**

**Level 19: Section 1441**

10 mm

See the entire section in Plate 80 (middle).

See the entire section in Plate 80 (bottom).

## PLATE 95B

**Germinal and transitional structures in *italics***

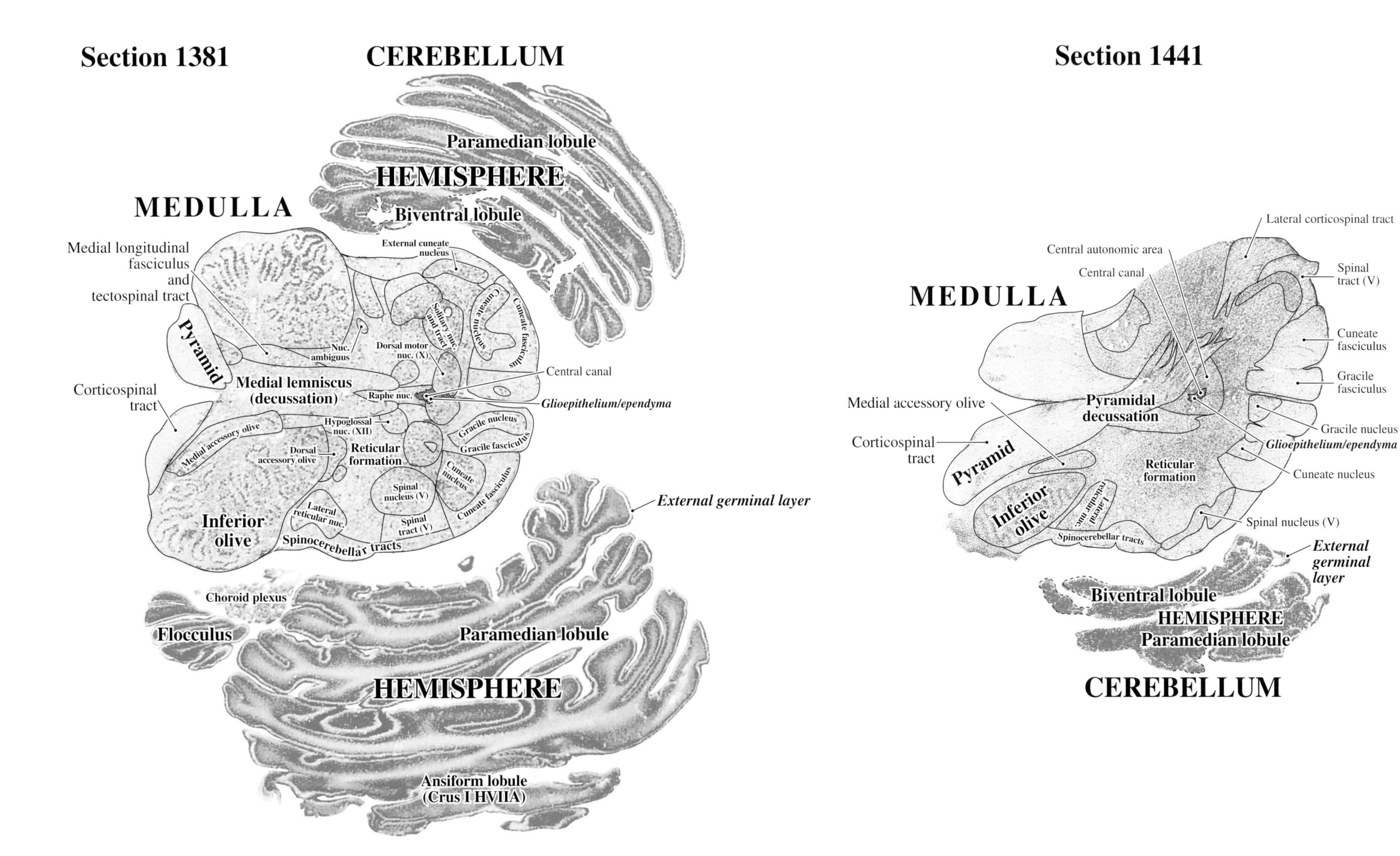

# PART V: GW32 SAGITTAL

This specimen is case number BX-15-60 (Perinatal RPSL) in the Yakovlev Collection. A female infant survived for one hour after a premature birth. Death occurred because a hyaline membrane obstructed the airway to the lungs. The brain was cut in the sagittal plane in 35-μm thick sections and is classified as a Normative Control in the Yakovlev Collection (Haleem, 1990). Since there is no photograph of this brain before it was embedded and cut, the photograph of the medial view of another GW32 brain that Larroche published in 1966 (**Figure 6**) is used.

Photographs of 6 different Nissl-stained sections (**Levels 1-6**) are shown at low magnification in **Plates 96-101**. The core of the brain and the cerebellum are shown at high magnification in companion **Plates 102AB-107AB** for **Levels 1-6**. Very high magnification views of different regions of the cerebellar cortex are shown in **Plates 108-111**. Because the section numbers decrease from Level 1 (most medial) to Level 6 (most lateral), they are from the left side of the brain; the right side has higher section numbers proceeding medial to lateral. The cutting plane of this brain is nearly parallel to the midline in anterior and posterior parts of each section. However, the occipital lobe has been displaced toward the left. For example, the occipital lobe in **Level 1** (**Plate 96**) is from the right side of the brain. There is no occipital lobe in **Level 2** (**Plate 97**), and the left occipital lobe first appears in **Level 3** (**Plate 98**). The sections chosen for illustration are spaced closer together near the midline to show small structures in the diencephalon, midbrain, pons, and medulla.

Y15-60 contains the same group of immature structures that are in the GW37 brains, except all of these structures are slightly more prominent. In the cortical regions of the telencephalon, remnants of the germinal matrices are present in all lobes of the cerebral cortex where the ***neuroepithelium/subventricular zone*** are presumably generating neocortical interneurons. Migrating and sojourning neurons and/or glia are visible in all lobes of the cerebral cortex as ***stratified transitional fields***, thin in the occipital lobe, and thicker in the frontal, parietal and temporal lobes. More neurons, glia, and their mitotic precursor cells are migrating through the olfactory peduncle toward the olfactory bulb (***rostral migratory stream***) from a presumed source area in the germinal matrix at the junction between the cerebral cortex, striatum, and nucleus accumbens. Within the lateral parts of the cerebral cortex, the ***lateral migratory stream*** contains neurons and glia that percolate through the claustrum, endopiriform nucleus, external capsule, and uncinate fasciculus. These cells appear to be heading toward the insular cortex, primary olfactory cortex, temporal cortex, and basolateral parts of the amygdaloid complex. In the basal ganglia, there is a thick ***neuroepithelium/subventricular zone*** overlying the striatum and nucleus accumbens where neurons and glia are being generated; some of these, especially from the accumbal area, will enter the ***rostral migratory stream***. Another region of active neurogenesis in the telencephalon is the ***subgranular zone*** in the hilus of the dentate gyrus that is the source of granule cells. Just as in the GW37 specimens, the septum, fornix, and Ammon's horn have only a thin, darkly staining layer at the ventricle, and these are presumed to be generating glia, cells of the choroid plexus, and the ependymal lining of the ventricle.

Most of the structures in the diencephalon appear to be settled and are maturing, and the third ventricle is lined by a thin ***glioepithelium/ependyma***. In the midbrain and anterior pons, there is a slightly thicker and more convoluted ***glioepithelium/ependyma*** lining the posterior cerebral aqueduct and anterior fourth ventricle. A thin ***glioepithelium/ependyma*** lines the fourth ventricle in the posterior pons and anterior medulla, but that thickens in the posterior medulla. The ***external germinal layer*** is prominent over the entire surface of the cerebellar cortex and is actively producing basket, stellate, and granule cells. The ***germinal trigone*** is visible at the base of the nodulus and along the floccular peduncle; choroid plexus cells and glia are originating here.

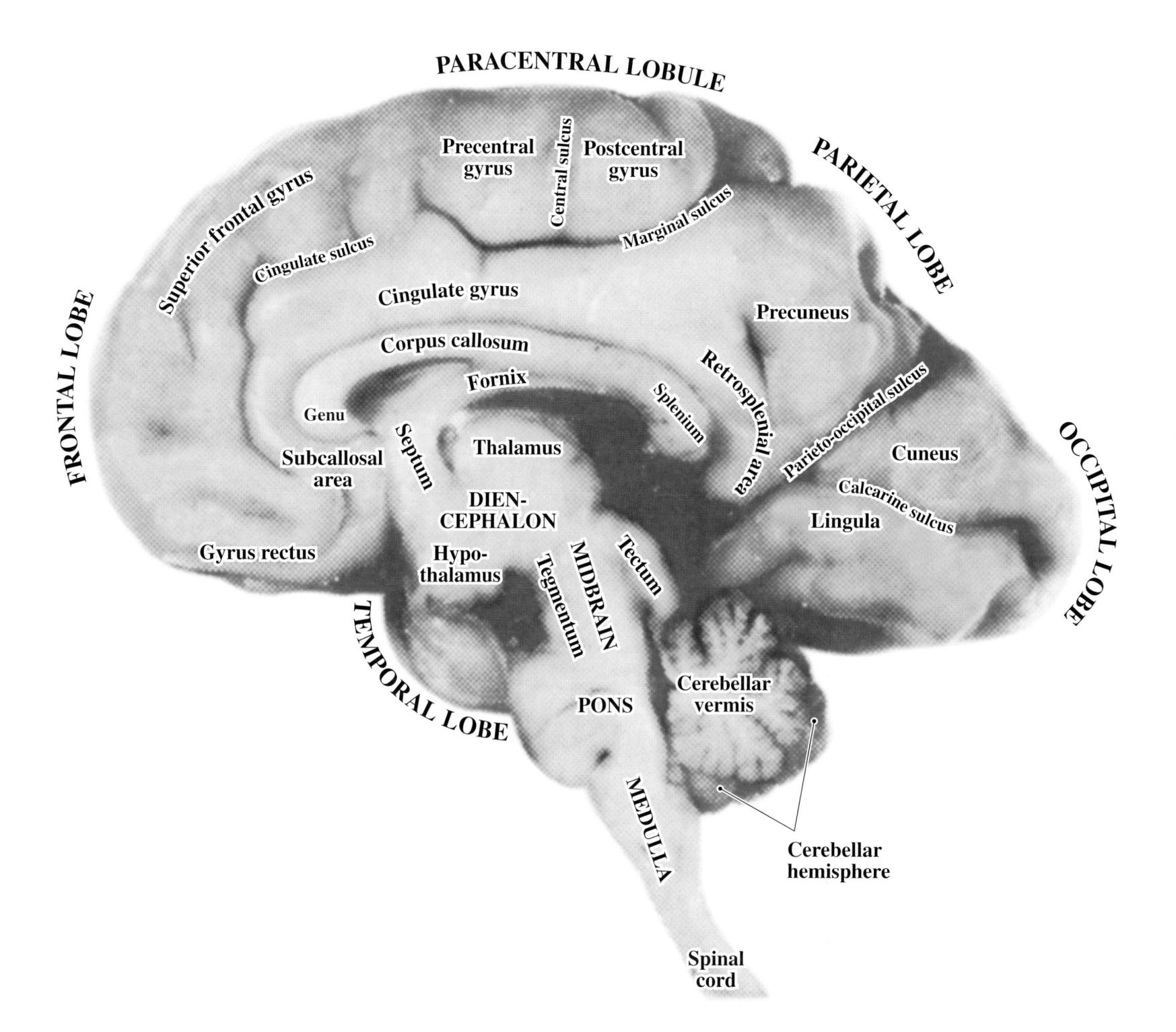

**Figure 6.** Midline sagittal view of a GW32 brain with major structures in the cerebral hemispheres and brainstem labeled. (This is part of Figure 2-9 on page 27 in B. A. Curtis, S. Jacobson, and E. M. Marcus (1972) *An Introduction to the Neurosciences*, Philadelphia: W. B. Saunders. The photograph was originally published by J. C. Larroche (1966) The development of the central nervous system during intrauterine life. In: *Human Development*, F. Falkner (ed.), Philadelphia: W. B. Saunders, page 259.)

## PLATE 96

GW32 Sagittal
CR 270 mm
Y15-60
Level 1: Section 801

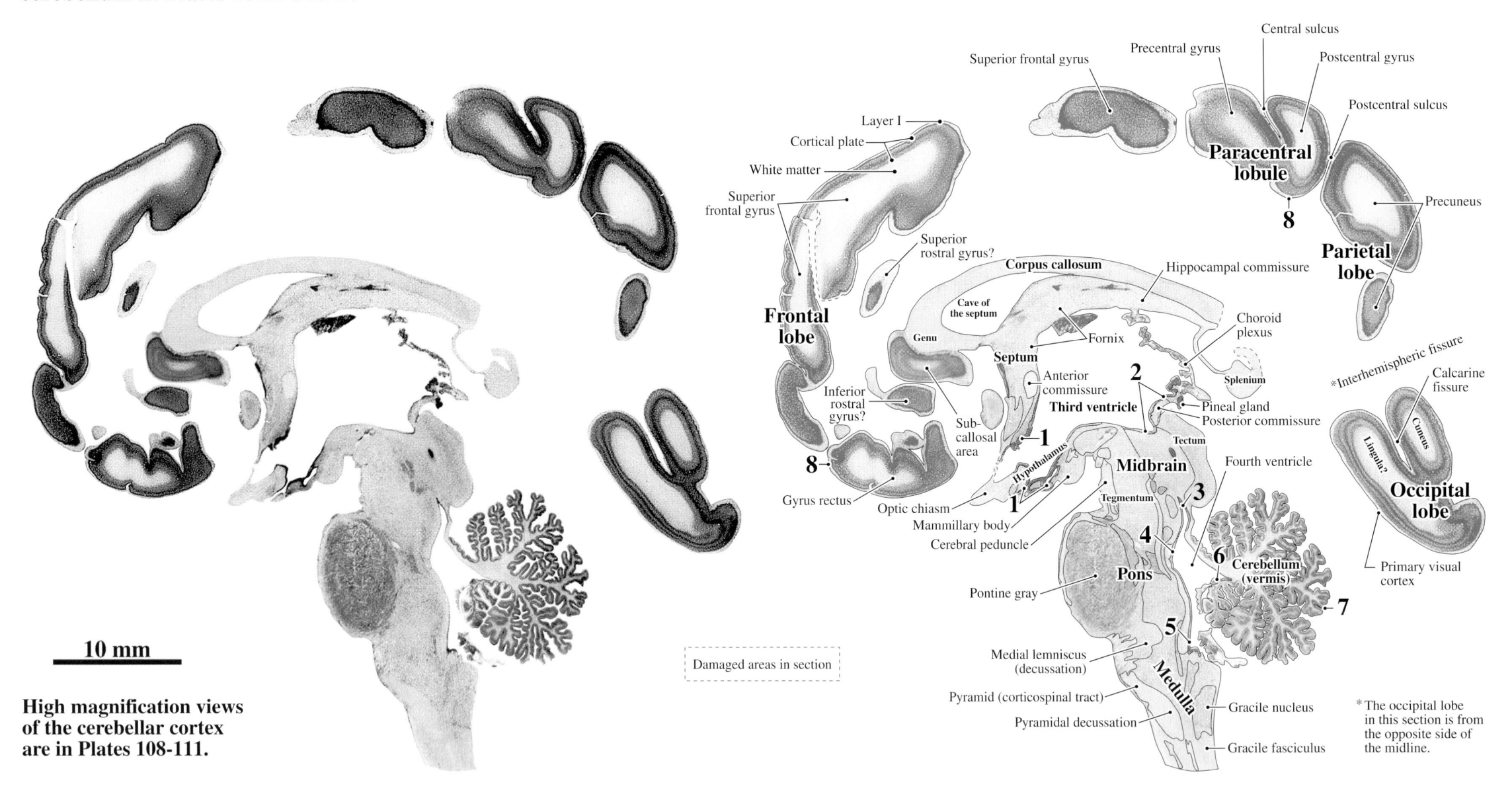

*Remnants of the germinal matrix, migratory streams, and transitional fields*

| 1 | *Diencephalic (hypothalamic) glioepithelium/ependyma* | 5 | *Medullary glioepithelium/ependyma* |
| --- | --- | --- | --- |
| 2 | *Diencephalic/mesencephalic glioepithelium/ependyma* | 6 | *Germinal trigone (cerebellum)* |
| 3 | *Mesencephalic glioepithelium/ependyma* | 7 | *External germinal layer (cerebellum)* |
| 4 | *Pontine glioepithelium/ependyma* | 8 | *Subpial granular layer (cortical)* |

See detail of brain core and cerebellum in Plates 102A and B.

High magnification views of the cerebellar cortex are in Plates 108-111.

*The occipital lobe in this section is from the opposite side of the midline.

# PLATE 97

**GW32 Sagittal**
**CR 270 mm**
**Y15-60**
**Level 2: Section 741**

## Remnants of the germinal matrix, migratory streams, and transitional fields

1. Rostral migratory stream
2. Accumbent neuroepithelium and subventricular zone (intermingled with the source of the rostral migratory stream)
3. Callosal sling
4. Callosal glioepithelium
5. Fornical glioepithelium
6. Septal glioepithelium/ependyma
7. Strionuclear glioepithelium
8. Pontine glioepithelium/ependyma
9. Germinal trigone (cerebellum)
10. External germinal layer (cerebellum)
11. Subpial granular layer (cortical)

**See detail of brain core and cerebellum in Plates 103A and B.**

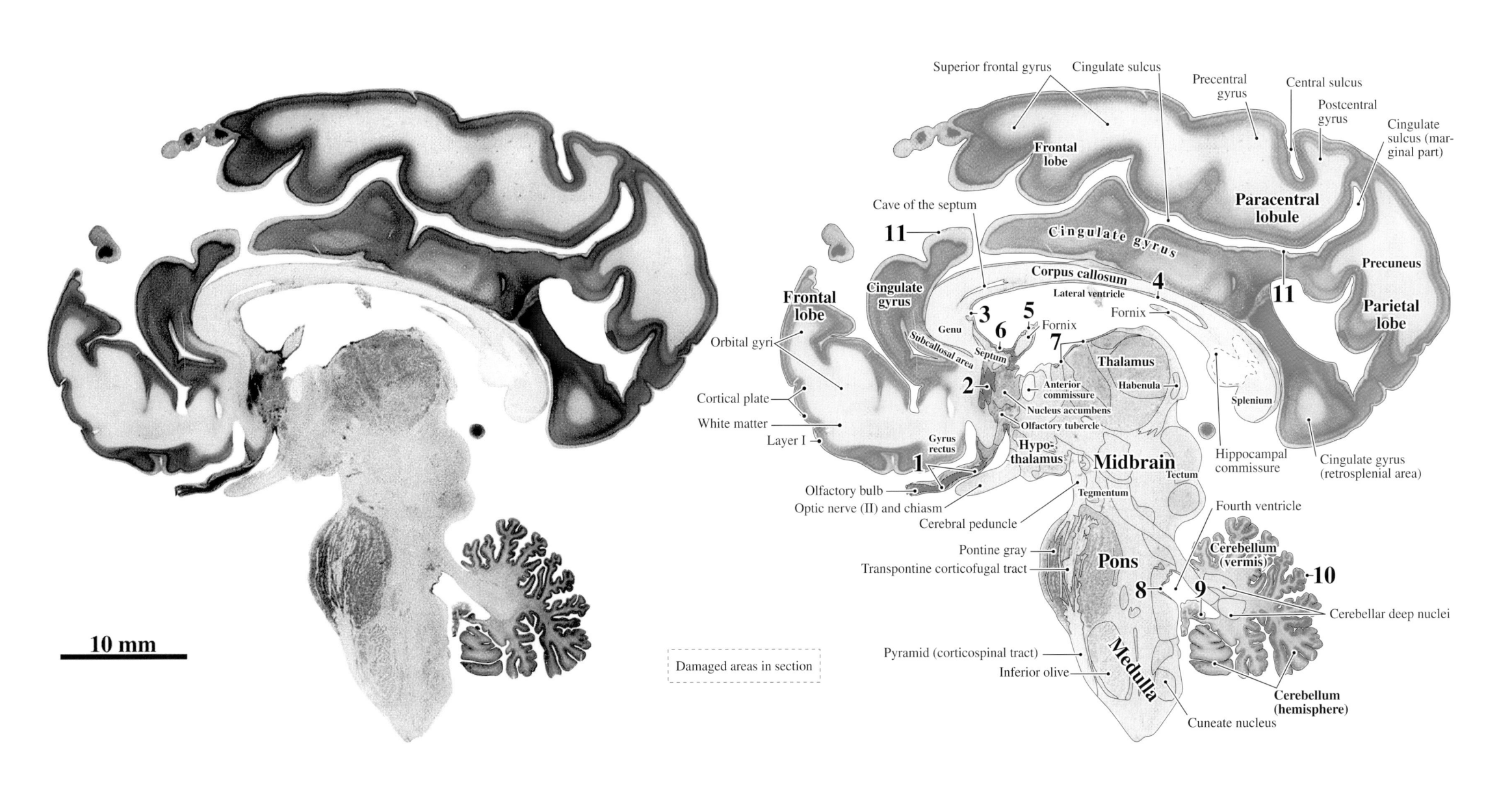

## PLATE 98

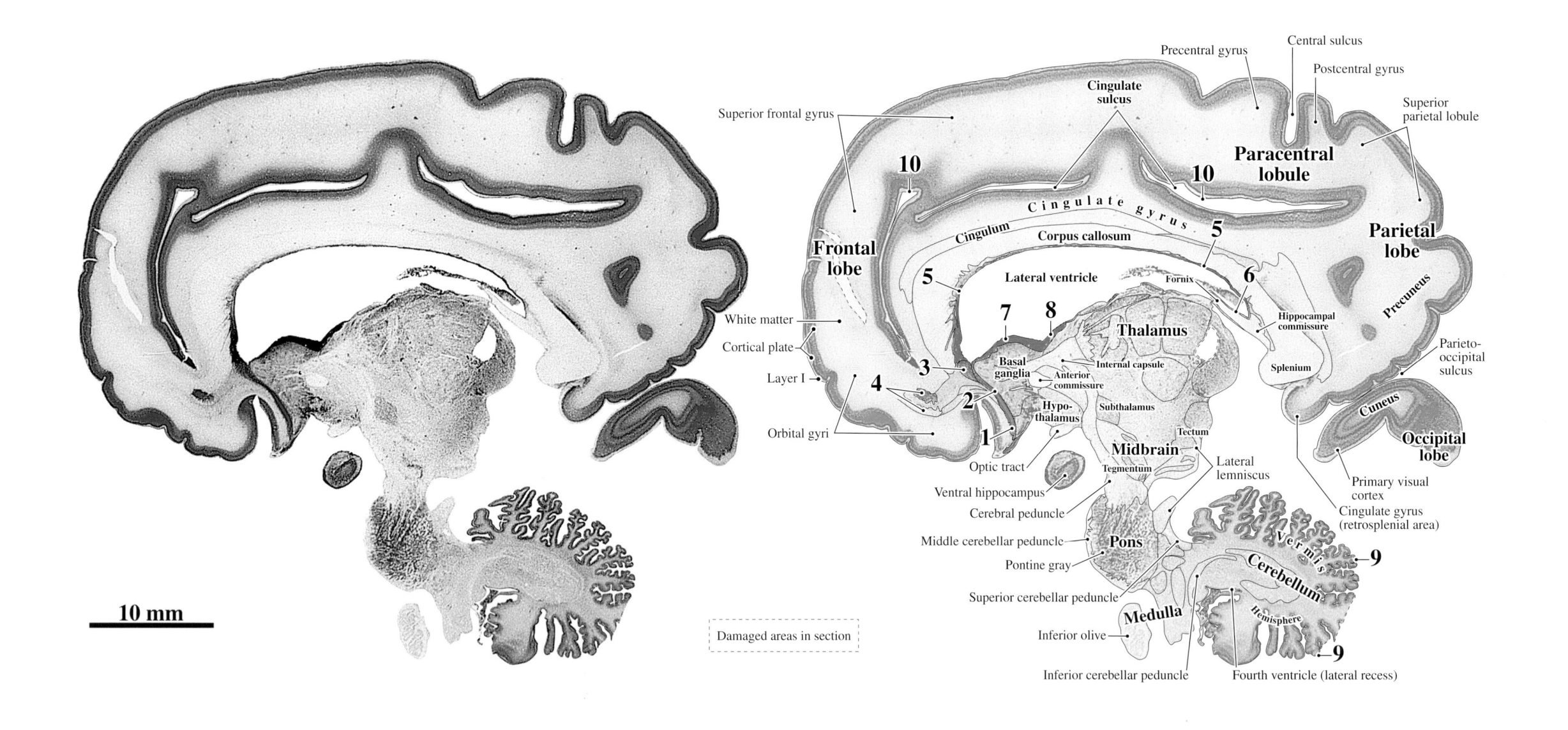

GW32 Sagittal
CR 270 mm
Y15-60
Level 3: Section 681

See detail of brain core and
cerebellum in Plates 104A and B.

*Remnants of the germinal matrix, migratory streams, and transitional fields*

| | | | |
|---|---|---|---|
| 1 | *Rostral migratory stream* | 6 | *Fornical glioepithelium* |
| 2 | *Accumbent neuroepithelium and subventricular zone (intermingled with the source of the rostral migratory stream)* | 7 | *Anteromedial striatal neuroepithelium and subventricular zone* |
| 3 | *Rostral migratory stream (source area)* | 8 | *Strionuclear glioepithelium* |
| 4 | *Frontal neuroepithelium, subventricular zone, and stratified transitional field* | 9 | *External germinal layer (cerebellum)* |
| 5 | *Callosal glioepithelium* | 10 | *Subpial granular layer (cortical)* |

10 mm

Damaged areas in section

# PLATE 99

**GW32 Sagittal**
**CR 270 mm**
**Y15-60**
**Level 4: Section 581**

## Remnants of the germinal matrix, migratory streams, and transitional fields

1. Frontal neuroepithelium, subventricular zone, and stratified transitional field
2. Paracentral neuroepithelium, subventricular zone, and stratified transitional field
3. Parahippocampal neuroepithelium, subventricular zone, and stratified transitional field
4. Callosal glioepithelium
5. Fornical glioepithelium
6. Alvear glioepithelium
7. Subgranular zone (dentate)
8. Amygdaloid glioepithelium/ependyma
9. Anterolateral striatal neuroepithelium and subventricular zone (intermingled with the source of the rostral migratory stream)
10. Anteromedial striatal neuroepithelium and subventricular zone
11. Strionuclear glioepithelium
12. External germinal layer (cerebellum)
13. Subpial granular layer (cortical)

See detail of brain core and cerebellum in Plates 105A and B.

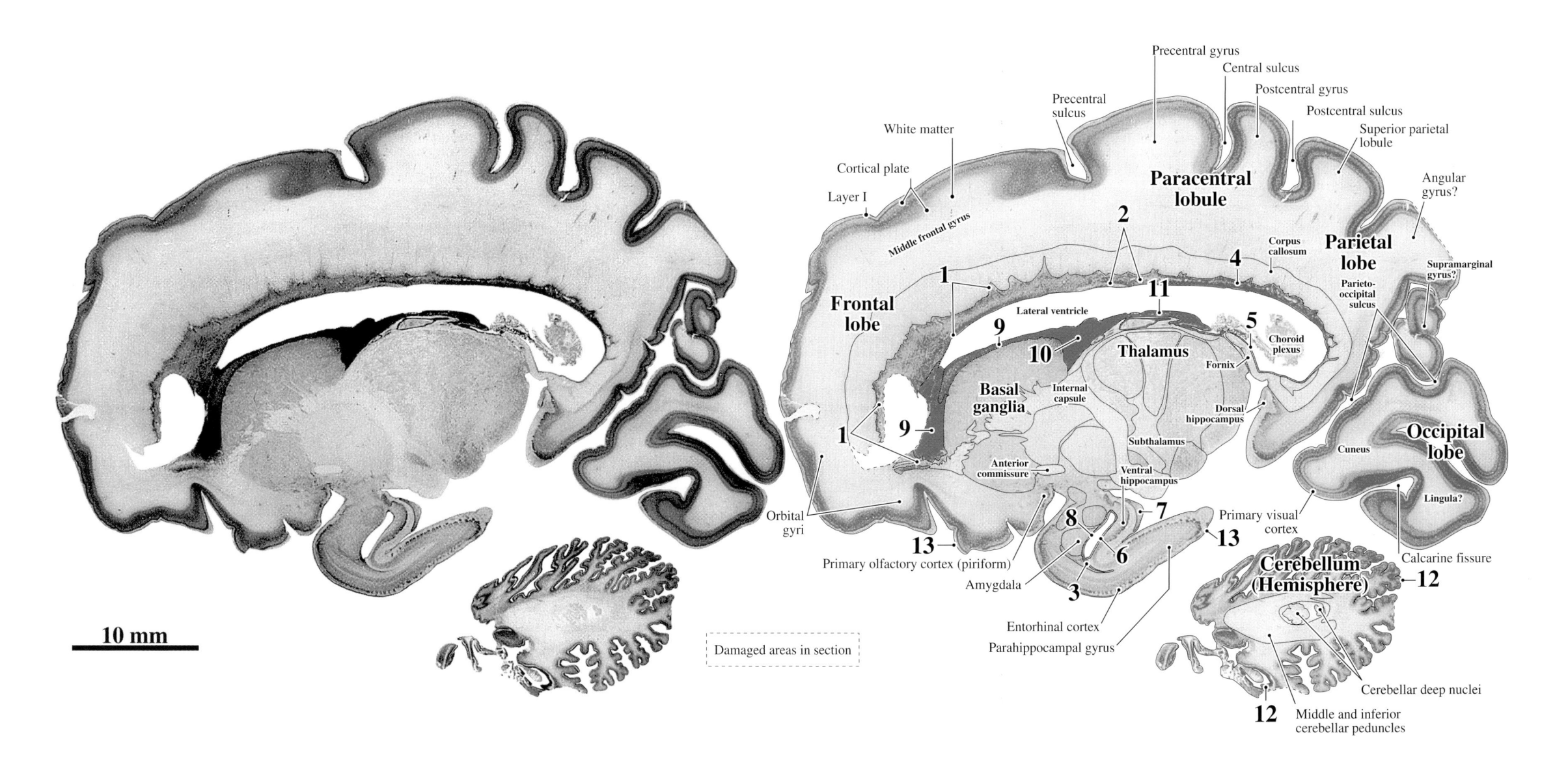

# PLATE 100

GW32 Sagittal
CR 270 mm
Y15-60
Level 5: Section 481

See detail of brain core and cerebellum in Plates 106A and B.

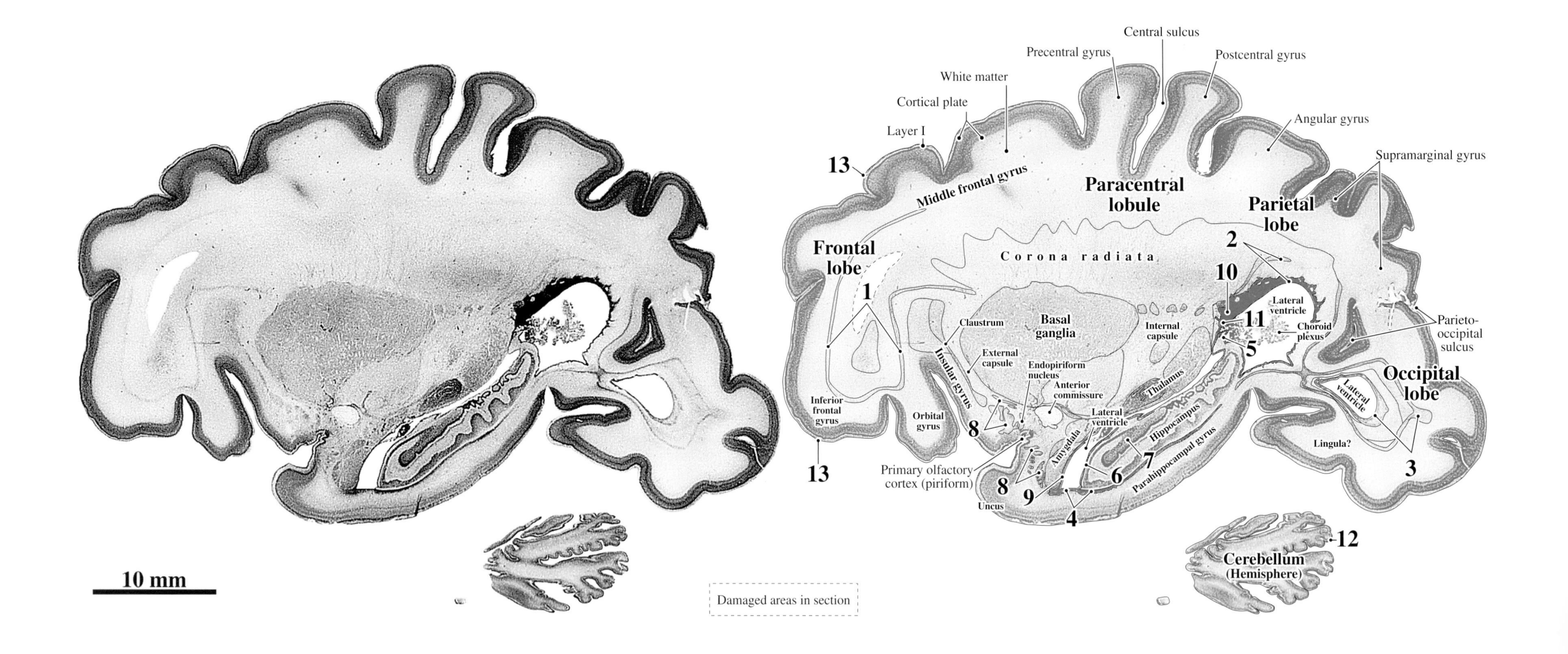

*Remnants of the germinal matrix, migratory streams, and transitional fields*

| | | | |
|---|---|---|---|
| 1 | Frontal stratified transitional field | 7 | Subgranular zone (dentate) |
| 2 | Parietal neuroepithelium, subventricular zone, and stratified transitional field | 8 | Lateral migratory stream (cortical) |
| 3 | Occipital neuroepithelium, subventricular zone, and stratified transitional field | 9 | Amygdaloid glioepithelium/ependyma |
| 4 | Parahippocampal neuroepithelium, subventricular zone, and stratified transitional field | 10 | Posterior striatal neuroepithelium and subventricular zone |
|   |  | 11 | Strionuclear glioepithelium |
| 5 | Fornical glioepithelium | 12 | External germinal layer (cerebellum) |
| 6 | Alvear glioepithelium | 13 | Subpial granular layer (cortical) |

# PLATE 101

**GW32 Sagittal**
**CR 270 mm**
**Y15-60**
**Level 6: Section 421**

## Remnants of the germinal matrix, migratory streams, and transitional fields

1. Parietal neuroepithelium, subventricular zone, and stratified transitional field
2. Occipital neuroepithelium, subventricular zone, and stratified transitional field
3. Parahippocampal neuroepithelium, subventricular zone, and stratified transitional field
4. Alvear glioepithelium
5. Lateral migratory stream (cortical)
6. Amygdaloid glioepithelium/ependyma
7. Posterior striatal neuroepithelium and subventricular zone
8. Strionuclear glioepithelium
9. External germinal layer (cerebellum)
10. Subpial granular layer (cortical)

**See detail of brain core and cerebellum in Plates 107A and B.**

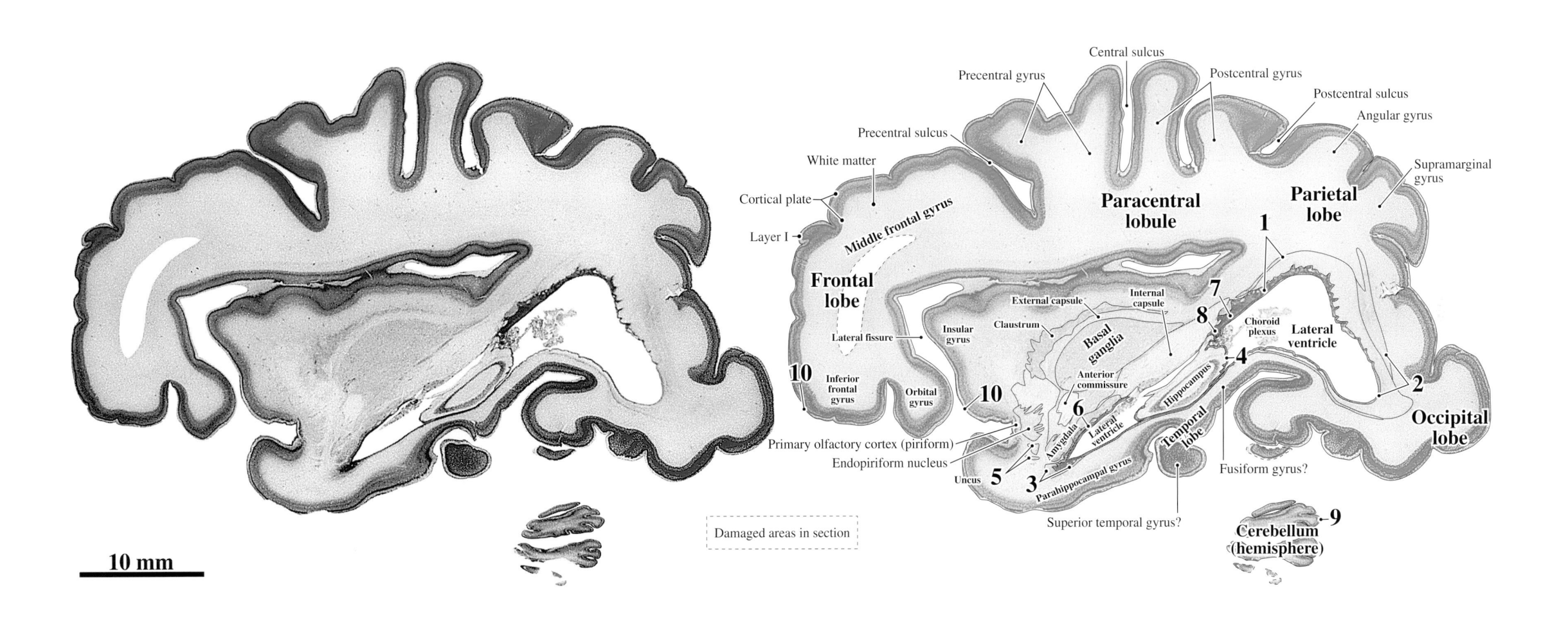

## PLATE 102A

GW32 Sagittal
CR 270 mm
Y15-60
Level 1: Section 801

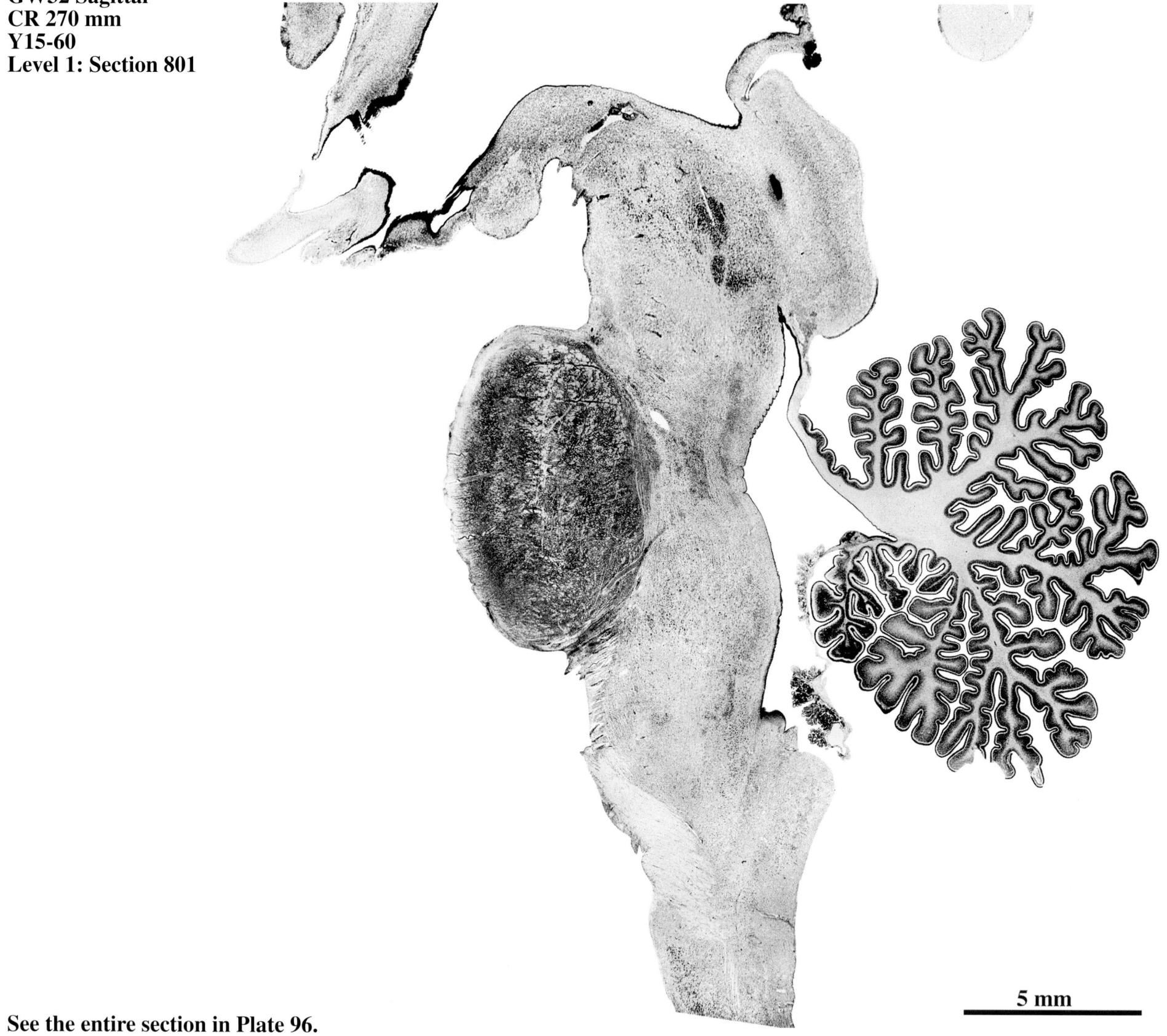

5 mm

See the entire section in Plate 96.

**PLATE 102B**

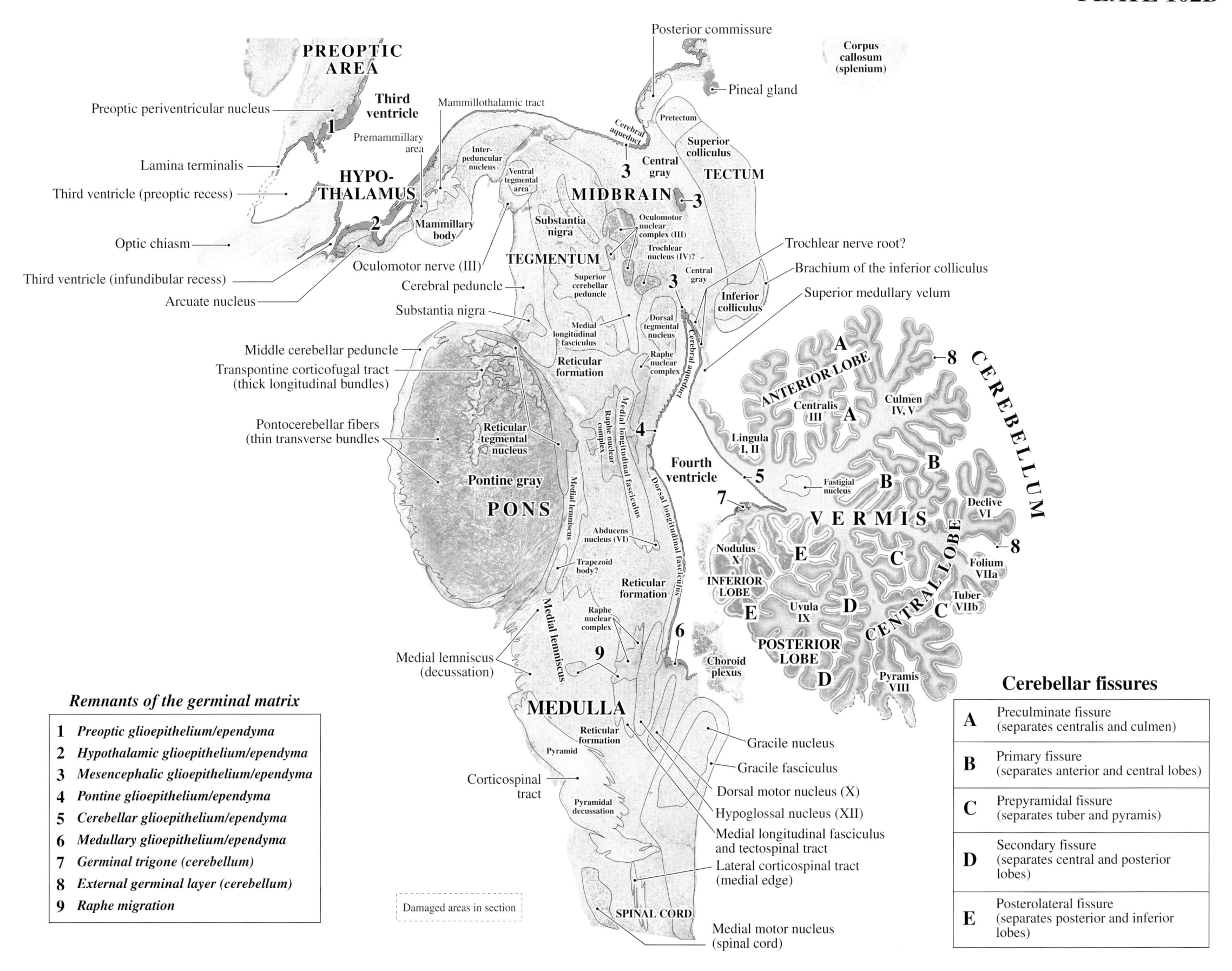

## PLATE 103A

**GW32 Sagittal
CR 270 mm
Y15-60
Level 2: Section 741**

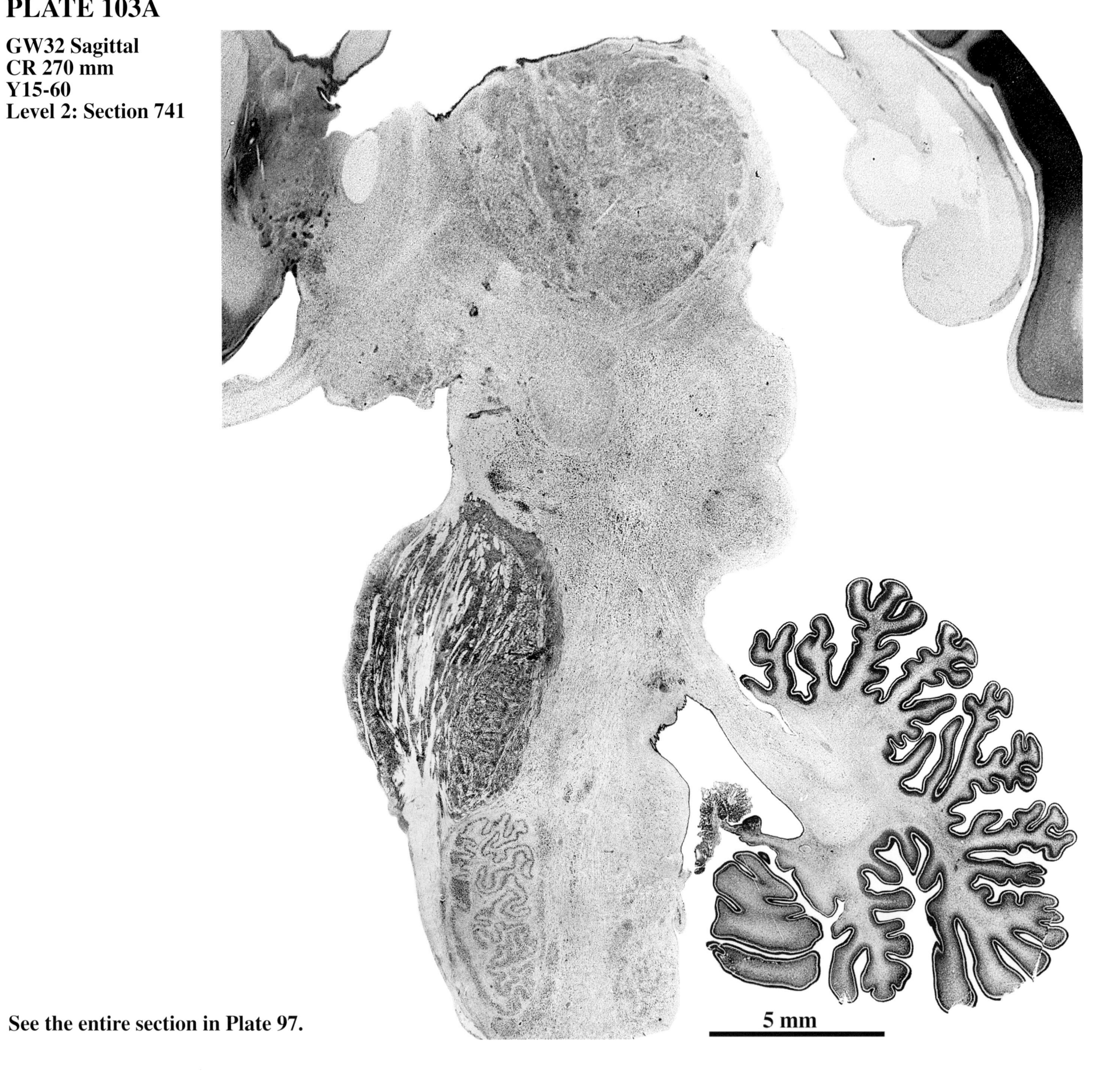

See the entire section in Plate 97.

5 mm

PLATE 103B

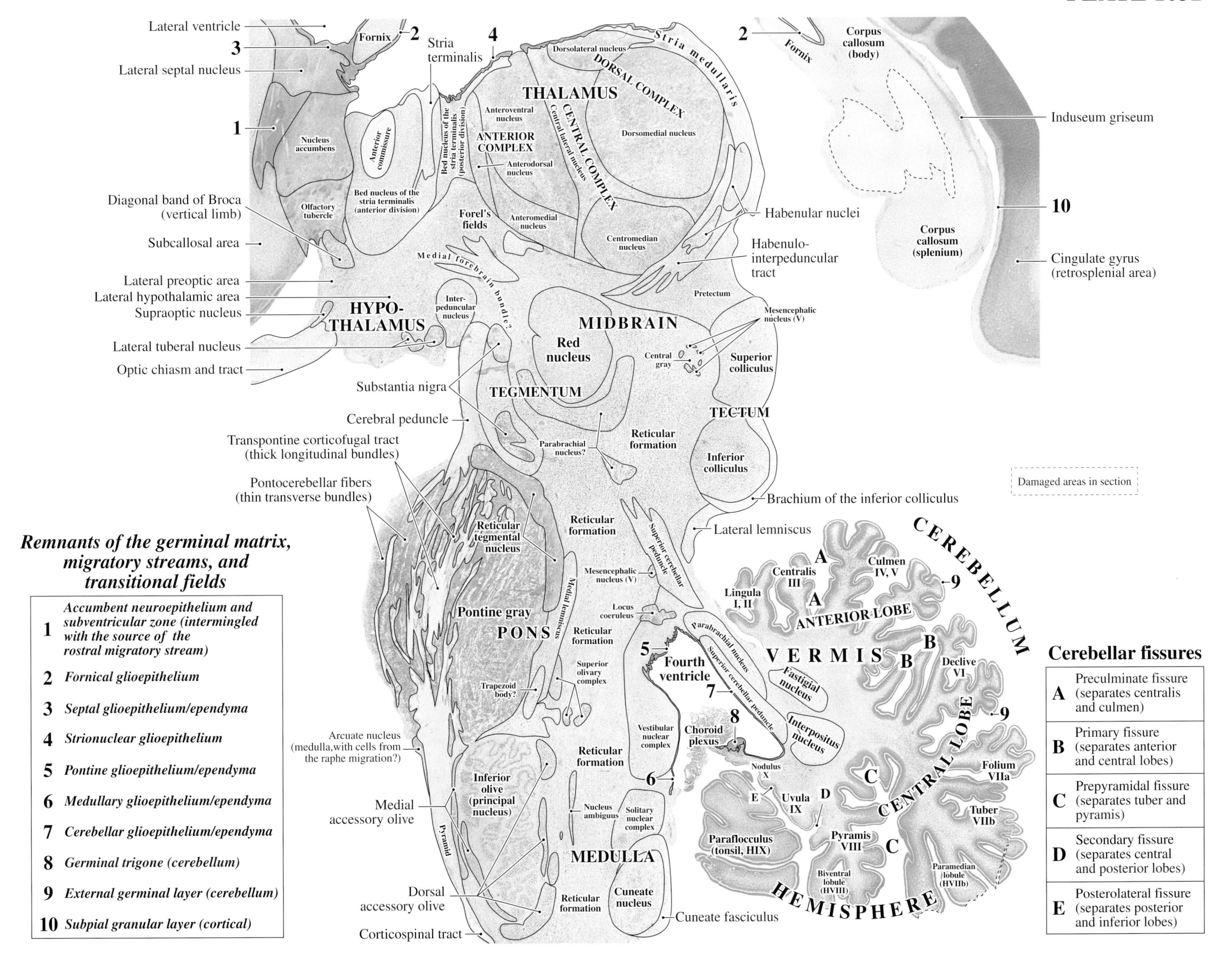

## PLATE 104A

GW32 Sagittal
CR 270 mm
Y15-60
Level 3: Section 681

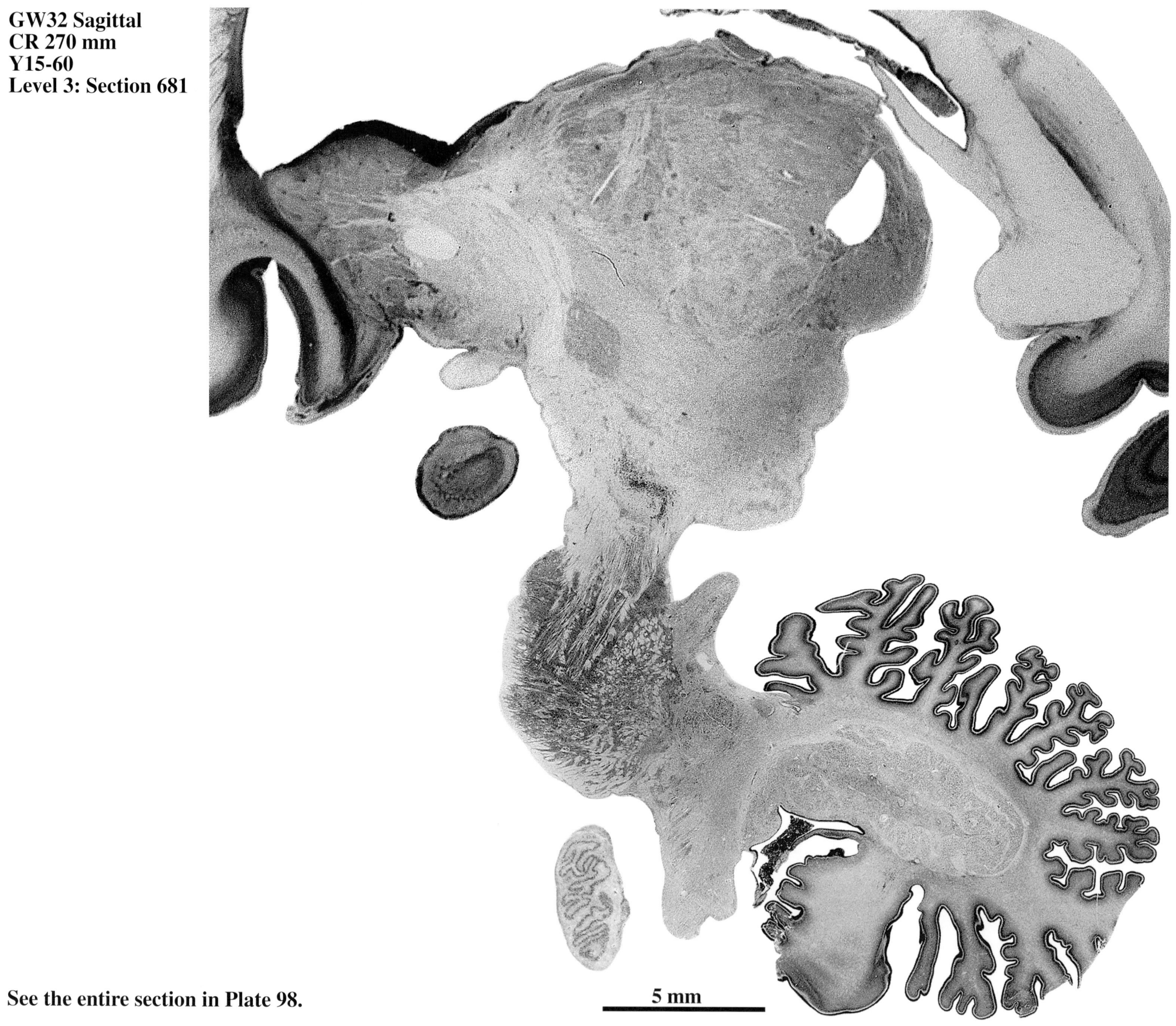

See the entire section in Plate 98.

5 mm

157

**PLATE 104B**

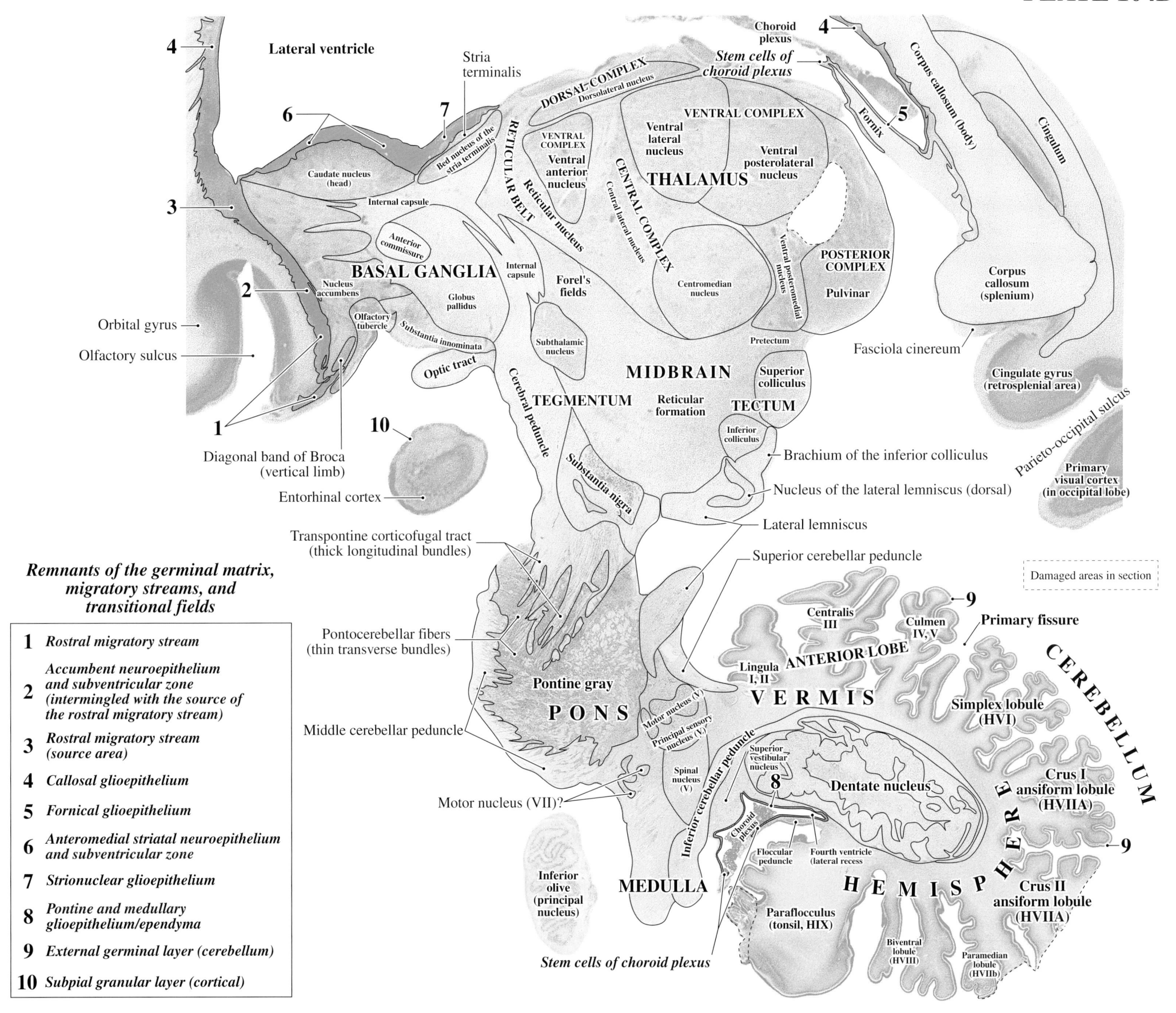

***Remnants of the germinal matrix, migratory streams, and transitional fields***

| | |
|---|---|
| **1** | *Rostral migratory stream* |
| **2** | *Accumbent neuroepithelium and subventricular zone (intermingled with the source of the rostral migratory stream)* |
| **3** | *Rostral migratory stream (source area)* |
| **4** | *Callosal glioepithelium* |
| **5** | *Fornical glioepithelium* |
| **6** | *Anteromedial striatal neuroepithelium and subventricular zone* |
| **7** | *Strionuclear glioepithelium* |
| **8** | *Pontine and medullary glioepithelium/ependyma* |
| **9** | *External germinal layer (cerebellum)* |
| **10** | *Subpial granular layer (cortical)* |

Labels on figure:
- Lateral ventricle
- Stria terminalis
- Choroid plexus
- *Stem cells of choroid plexus*
- Corpus callosum (body)
- Cingulum
- DORSAL COMPLEX — Dorsolateral nucleus
- VENTRAL COMPLEX — Ventral lateral nucleus, Ventral posterolateral nucleus
- Fornix
- Bed nucleus of the stria terminalis
- Caudate nucleus (head)
- RETICULAR BELT — Reticular nucleus
- VENTRAL COMPLEX — Ventral anterior nucleus
- THALAMUS
- CENTRAL COMPLEX — Central lateral nucleus
- Internal capsule
- Anterior commissure
- BASAL GANGLIA
- Nucleus accumbens
- Globus pallidus
- Internal capsule
- Forel's fields
- Centromedian nucleus
- Ventral posteromedial nucleus
- POSTERIOR COMPLEX — Pulvinar
- Corpus callosum (splenium)
- Orbital gyrus
- Olfactory sulcus
- Olfactory tubercle
- Substantia innominata
- Subthalamic nucleus
- Pretectum
- Fasciola cinereum
- Cingulate gyrus (retrosplenial area)
- Optic tract
- TEGMENTUM
- MIDBRAIN
- Reticular formation
- Superior colliculus
- TECTUM
- Diagonal band of Broca (vertical limb)
- Cerebral peduncle
- Substantia nigra
- Inferior colliculus
- Parieto-occipital sulcus
- Primary visual cortex (in occipital lobe)
- Entorhinal cortex
- Brachium of the inferior colliculus
- Nucleus of the lateral lemniscus (dorsal)
- Lateral lemniscus
- Transpontine corticofugal tract (thick longitudinal bundles)
- Superior cerebellar peduncle
- Damaged areas in section
- Pontocerebellar fibers (thin transverse bundles)
- Centralis III
- Culmen IV, V
- Primary fissure
- Lingula I, II
- ANTERIOR LOBE
- Simplex lobule (HVI)
- Pontine gray
- PONS
- VERMIS
- CEREBELLUM
- Middle cerebellar peduncle
- Motor nucleus (V)
- Principal sensory nucleus (V)
- Crus I ansiform lobule (HVIIA)
- Spinal nucleus (V)
- Superior vestibular nucleus
- Dentate nucleus
- HEMISPHERE
- Motor nucleus (VII)?
- Inferior cerebellar peduncle
- Choroid plexus
- Flocular peduncle
- Fourth ventricle (lateral recess)
- Crus II ansiform lobule (HVIIA)
- Inferior olive (principal nucleus)
- MEDULLA
- Paraflocculus (tonsil, HIX)
- Biventral lobule (HVIII)
- Paramedian lobule (HVIIb)
- *Stem cells of choroid plexus*

## PLATE 105A

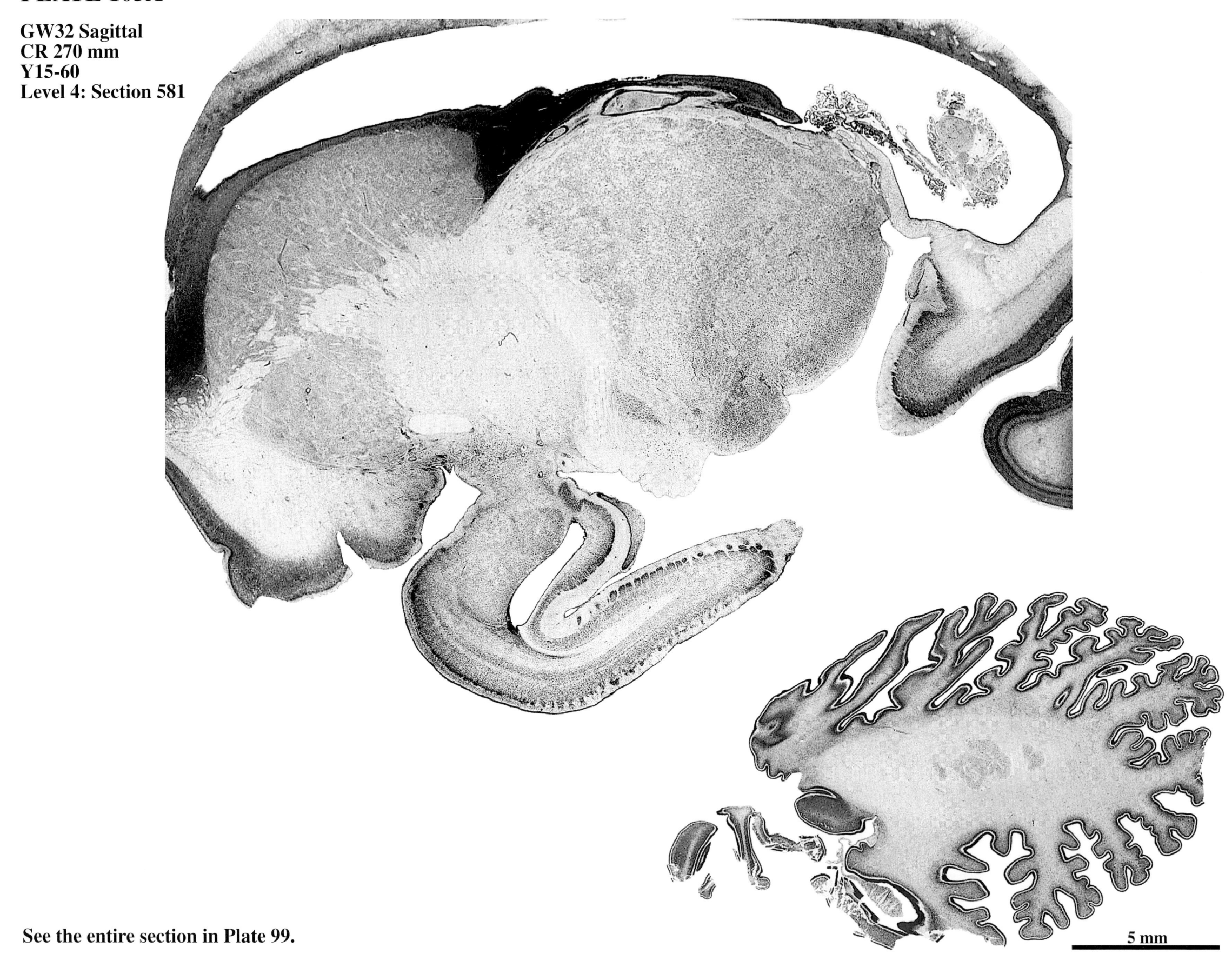

**GW32 Sagittal**
**CR 270 mm**
**Y15-60**
**Level 4: Section 581**

See the entire section in Plate 99.

5 mm

# PLATE 105B

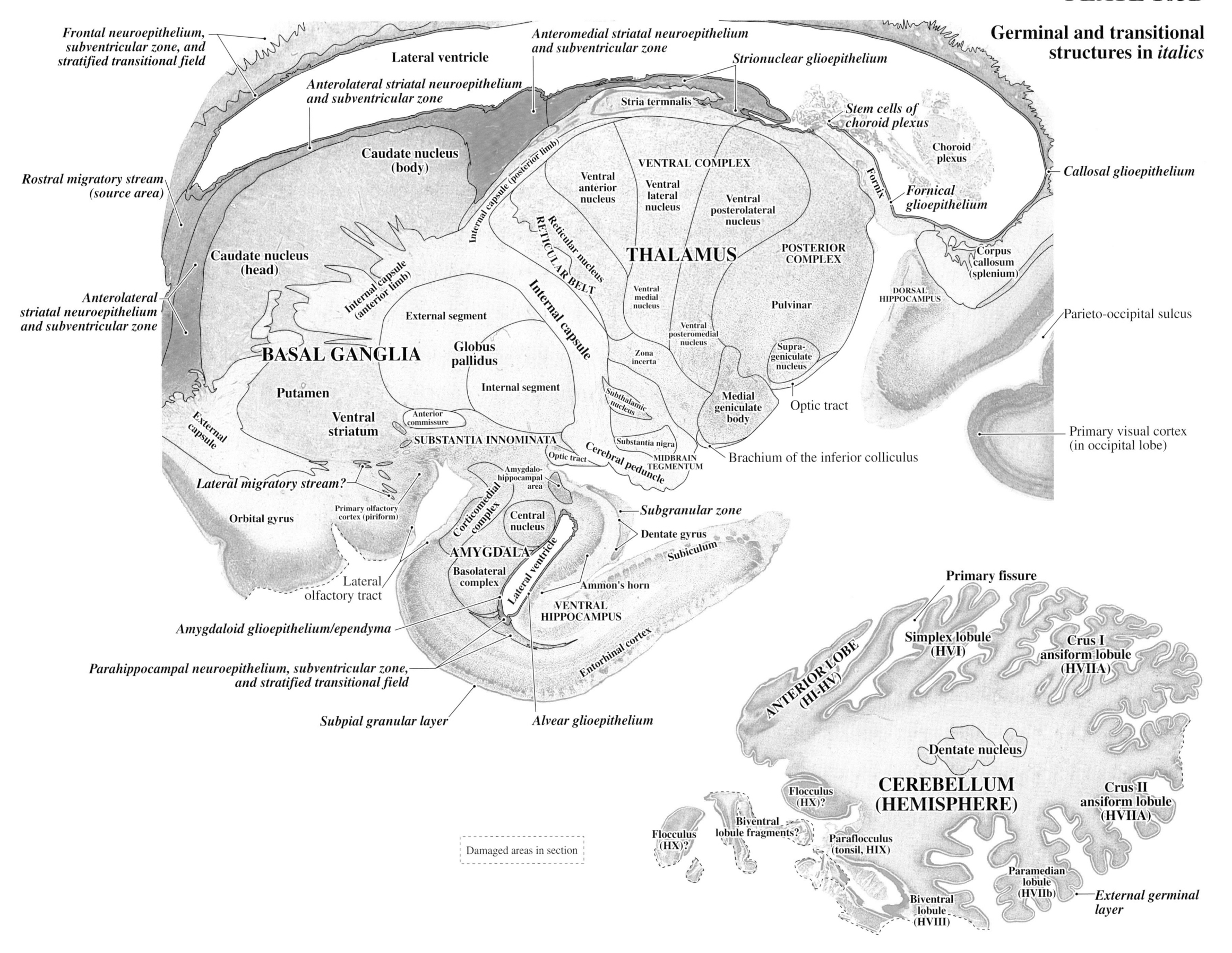

**Germinal and transitional structures in *italics***

*Frontal neuroepithelium, subventricular zone, and stratified transitional field*

Lateral ventricle

*Anteromedial striatal neuroepithelium and subventricular zone*

*Anterolateral striatal neuroepithelium and subventricular zone*

*Strionuclear glioepithelium*

*Stria terminalis*

Caudate nucleus (body)

*Stem cells of choroid plexus*

Choroid plexus

*Rostral migratory stream (source area)*

*Callosal glioepithelium*

VENTRAL COMPLEX

Ventral anterior nucleus

Ventral lateral nucleus

Ventral posterolateral nucleus

*Fornical glioepithelium*

Internal capsule (posterior limb)

Reticular nucleus

RETICULAR BELT

Caudate nucleus (head)

Corpus callosum (splenium)

THALAMUS

POSTERIOR COMPLEX

*Anterolateral striatal neuroepithelium and subventricular zone*

Internal capsule (anterior limb)

Ventral medial nucleus

Pulvinar

DORSAL HIPPOCAMPUS

External segment

Internal capsule

Zona incerta

Parieto-occipital sulcus

BASAL GANGLIA

Globus pallidus

Subthalamic nucleus

Supra-geniculate nucleus

Putamen

Internal segment

Medial geniculate body

Ventral striatum

Anterior commissure

Substantia nigra

Optic tract

*External capsule*

SUBSTANTIA INNOMINATA

Optic tract

Cerebral peduncle

MIDBRAIN TEGMENTUM

Brachium of the inferior colliculus

Primary visual cortex (in occipital lobe)

*Lateral migratory stream?*

Amygdalo-hippocampal area

Orbital gyrus

Primary olfactory cortex (piriform)

Corticomedial complex

Central nucleus

*Subgranular zone*

Dentate gyrus

AMYGDALA

Subiculum

Basolateral complex

**Primary fissure**

Lateral olfactory tract

Lateral ventricle

Ammon's horn

*Amygdaloid glioepithelium/ependyma*

VENTRAL HIPPOCAMPUS

Simplex lobule (HVI)

Crus I ansiform lobule (HVIIA)

*Parahippocampal neuroepithelium, subventricular zone, and stratified transitional field*

Entorhinal cortex

ANTERIOR LOBE (HI-HV)

*Subpial granular layer*

*Alvear glioepithelium*

Dentate nucleus

Crus II ansiform lobule (HVIIA)

CEREBELLUM (HEMISPHERE)

Flocculus (HX)?

Flocculus (HX)?

Biventral lobule fragments?

Paraflocculus (tonsil, HIX)

Paramedian lobule (HVIIb)

Damaged areas in section

Biventral lobule (HVIII)

*External germinal layer*

**PLATE 106A**

GW32 Sagittal
CR 270 mm
Y15-60
Level 5: Section 481

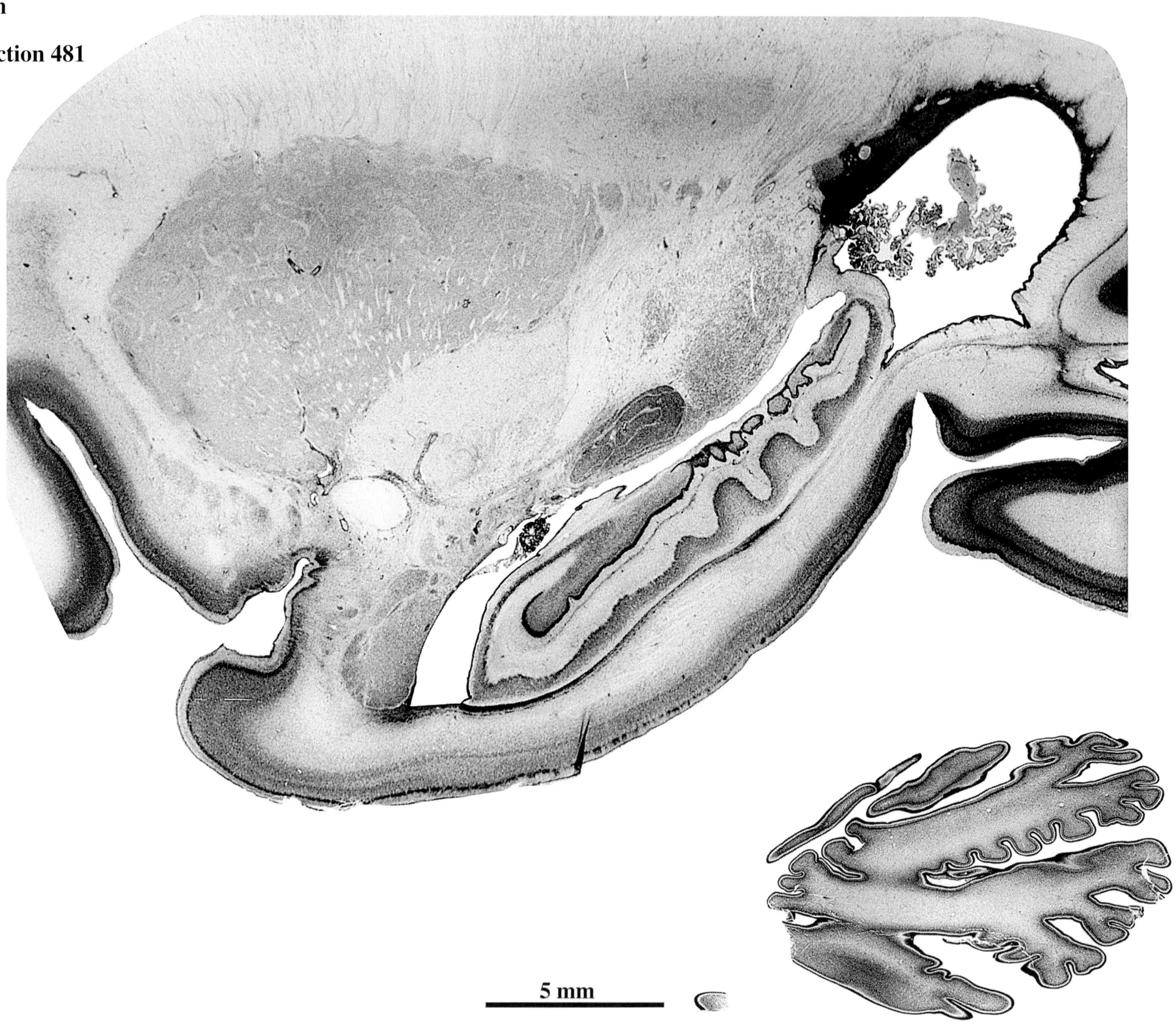

5 mm

See the entire section in Plate 100.

**PLATE 106B**

**Germinal and transitional structures in *italics***

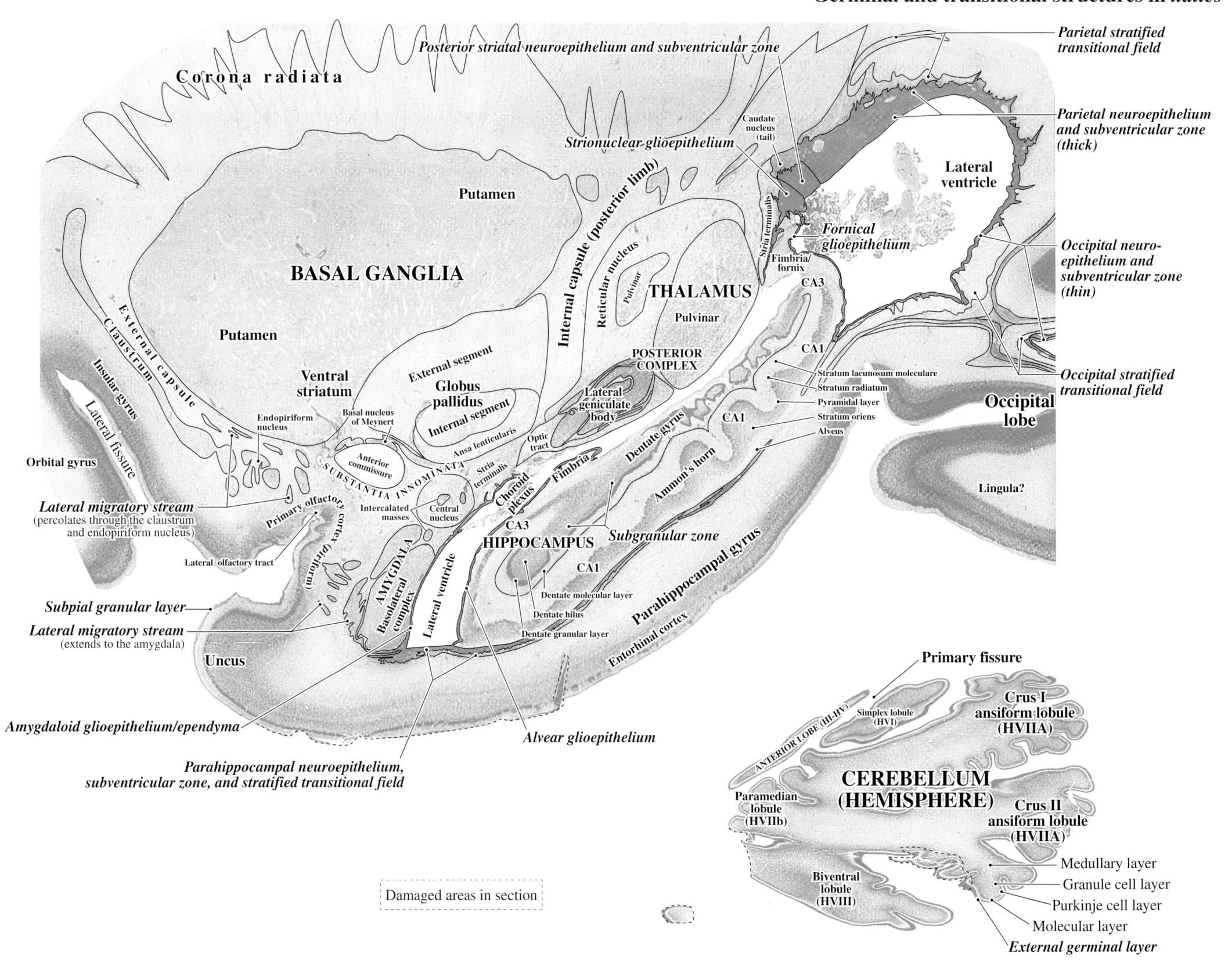

## PLATE 107A

GW32 Sagittal, CR 270 mm, Y15-60
Level 6: Section 421

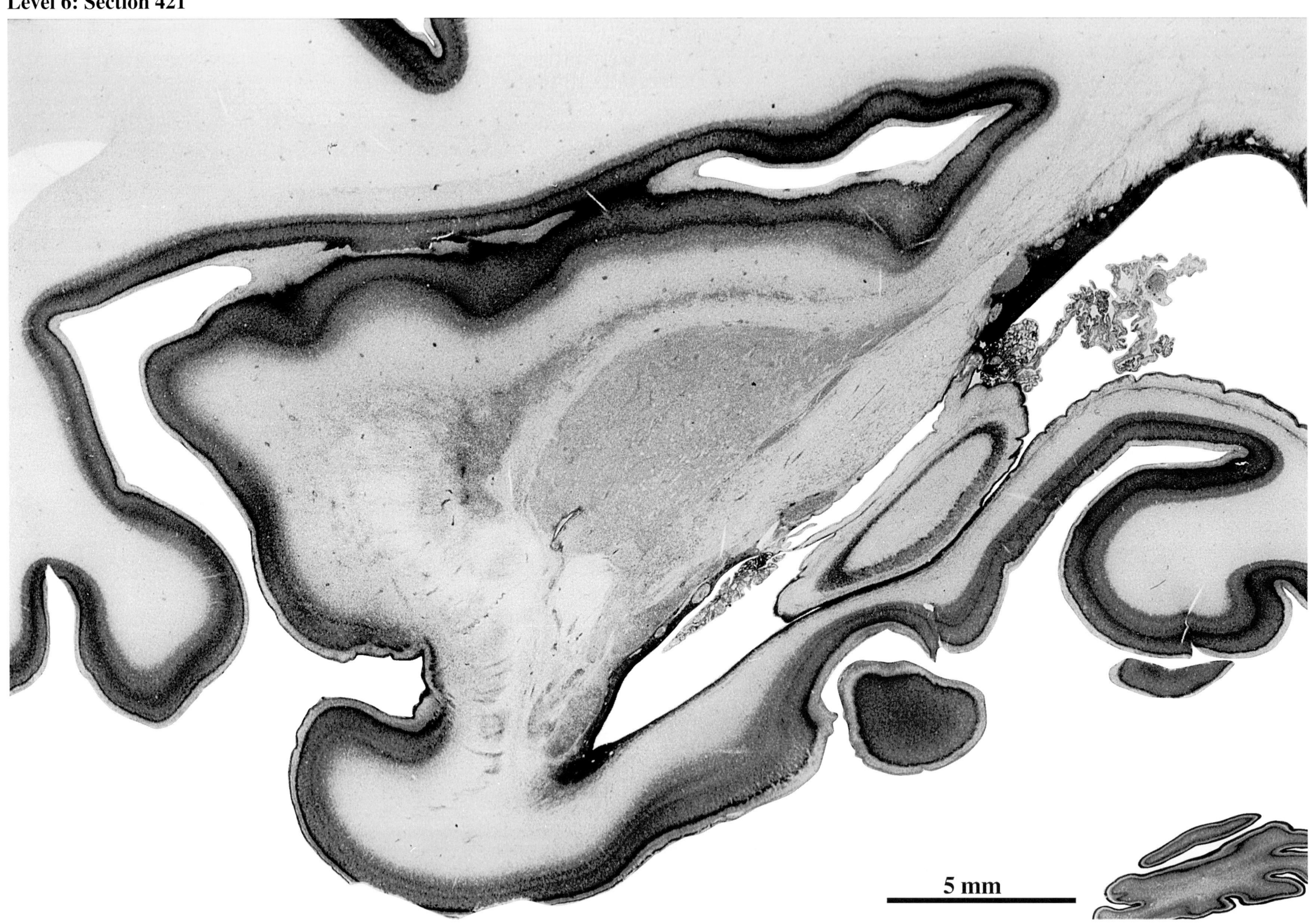

5 mm

See the entire section in Plate 101.

## PLATE 107B

**Germinal and transitional structures in *italics***

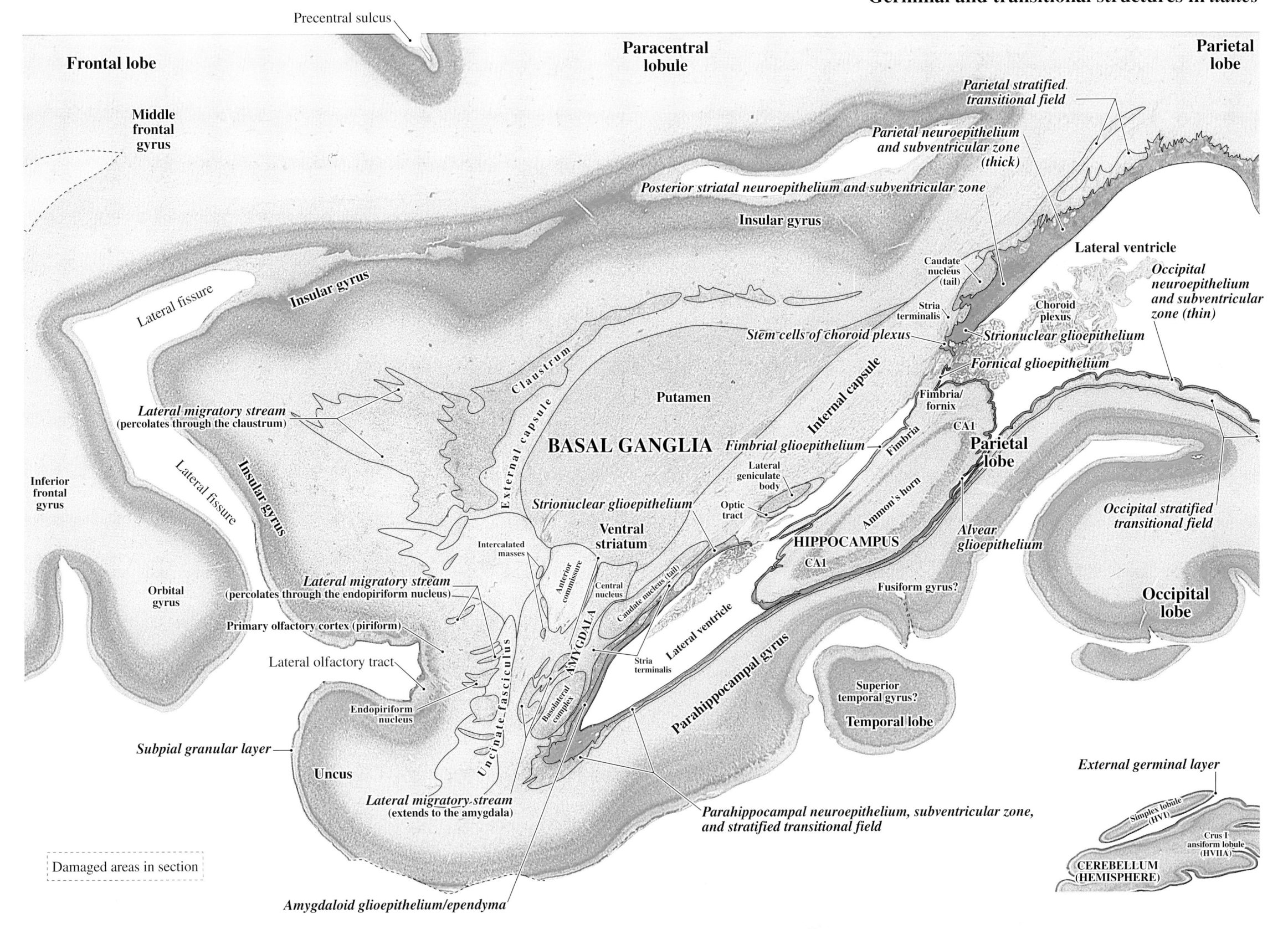

## PLATE 108

**GW32 Sagittal
CR 270 mm
Y15-60
Level 1:
Section 801
CEREBELLUM**

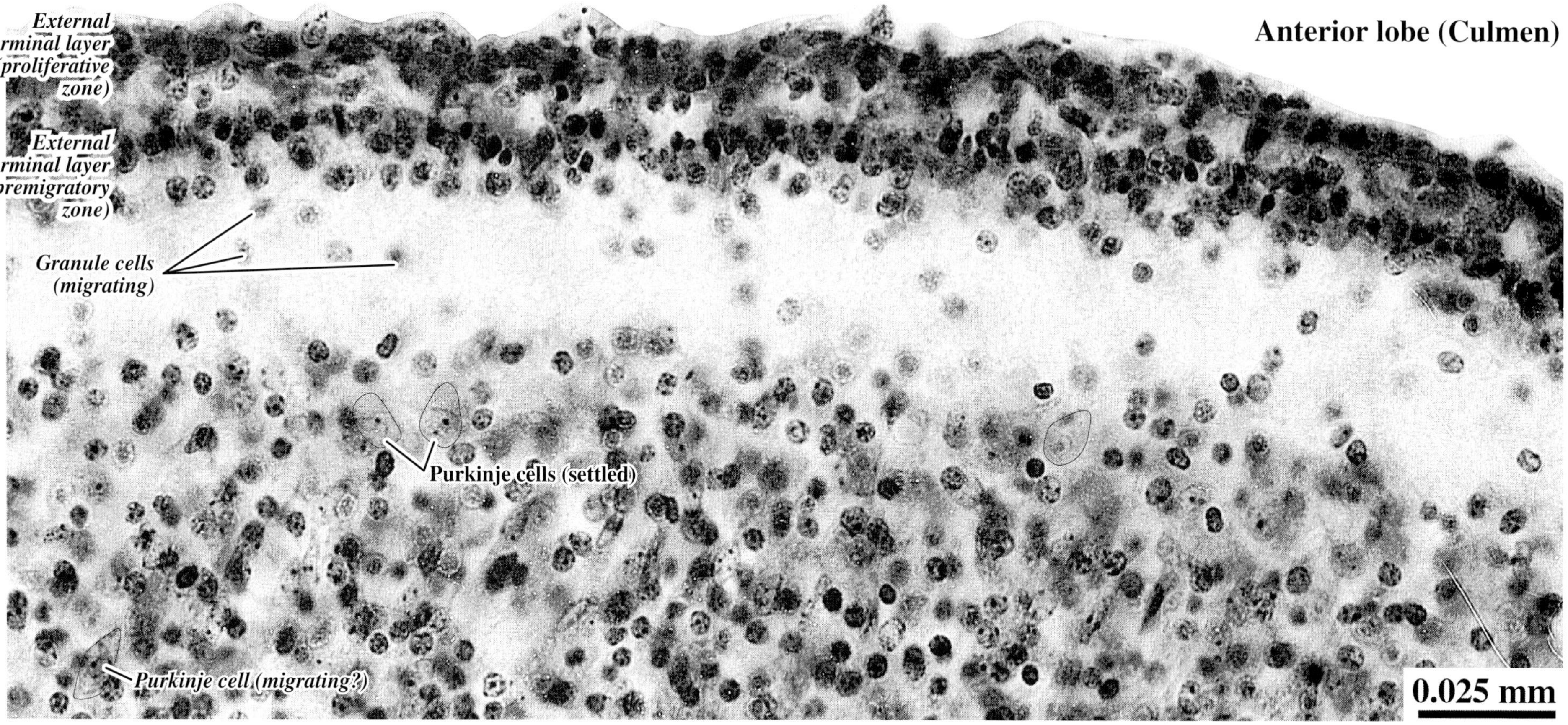

*External germinal layer (proliferative zone)*

*External germinal layer (premigratory zone)*

*Granule cells (migrating)*

Molecular layer

Purkinje cell layer:
*(immature Purkinje cells without apical caps migrating and settling)*

*Purkinje cells (settled)*

Granule cell layer

*Purkinje cell (migrating?)*

Anterior lobe (Culmen)

0.025 mm

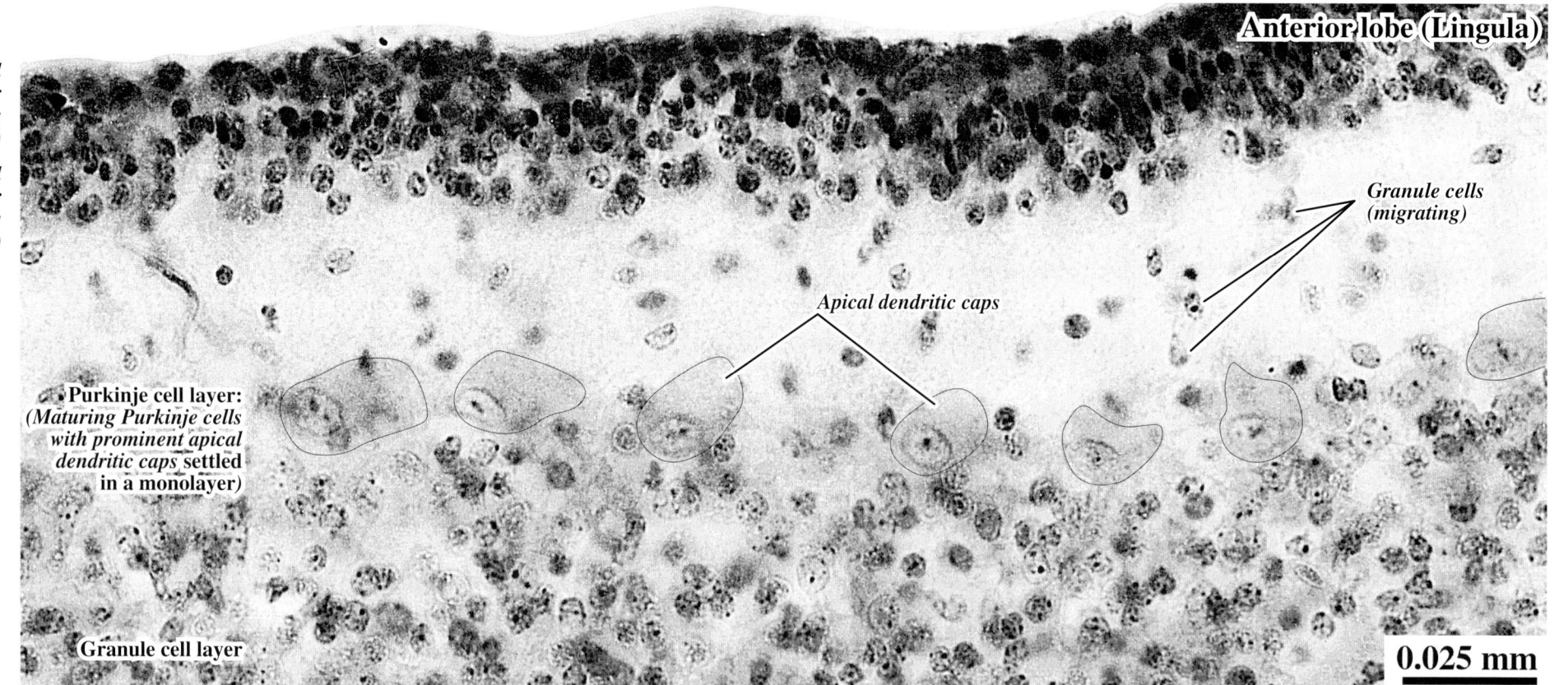

Anterior lobe (Lingula)

*External germinal layer (proliferative zone)*

*External germinal layer (premigratory zone)*

*Granule cells (migrating)*

Molecular layer

*Apical dendritic caps*

*Purkinje cell layer: (Maturing Purkinje cells with prominent apical dendritic caps settled in a monolayer)*

Granule cell layer

See the entire Section in Plate 96.

A medium-magnification view of the cerebellum is in Plates 102A and B.

0.025 mm

**PLATE 109**

GW32 Sagittal
CR 270 mm
Y15-60
Level 1:
Section 801
**CEREBELLUM**

Central lobe (Folium)

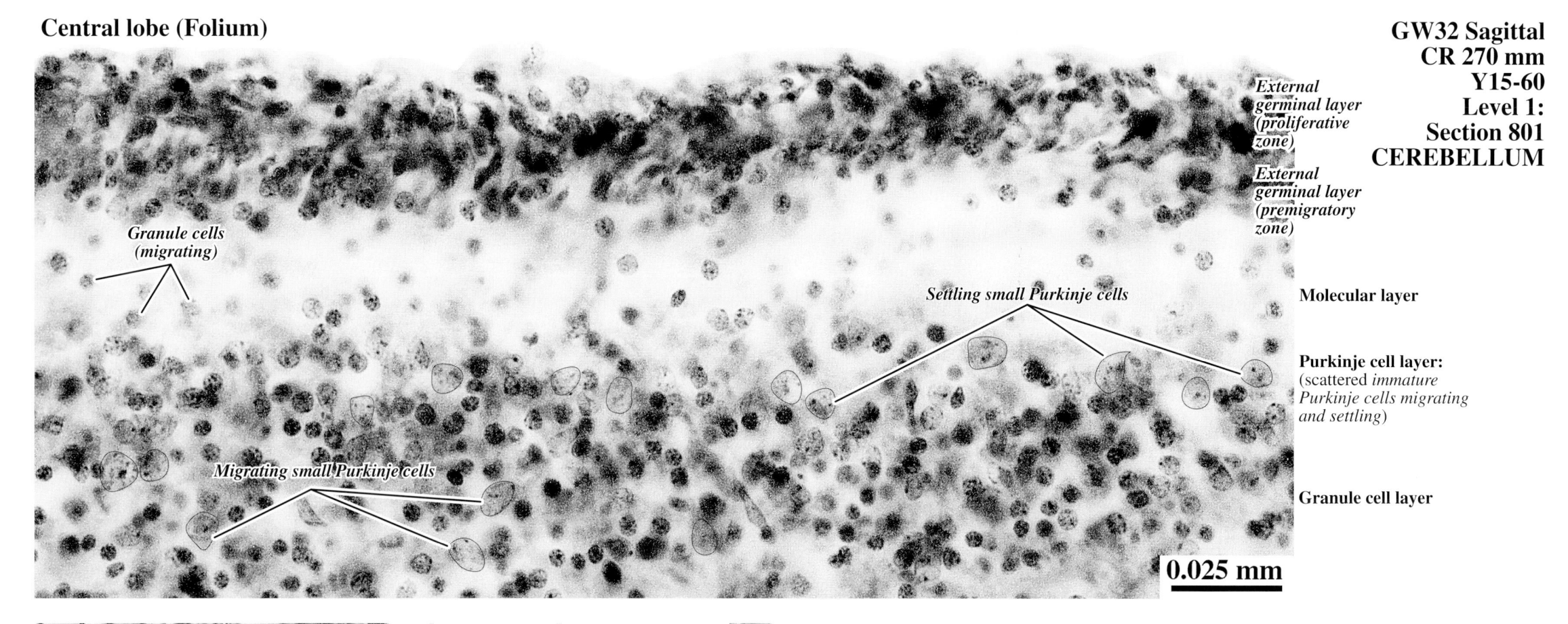

*Granule cells (migrating)*

*Settling small Purkinje cells*

*Migrating small Purkinje cells*

External germinal layer (proliferative zone)
External germinal layer (premigratory zone)

Molecular layer

Purkinje cell layer: (scattered *immature* Purkinje cells migrating and settling)

Granule cell layer

0.025 mm

Folium/Tuber

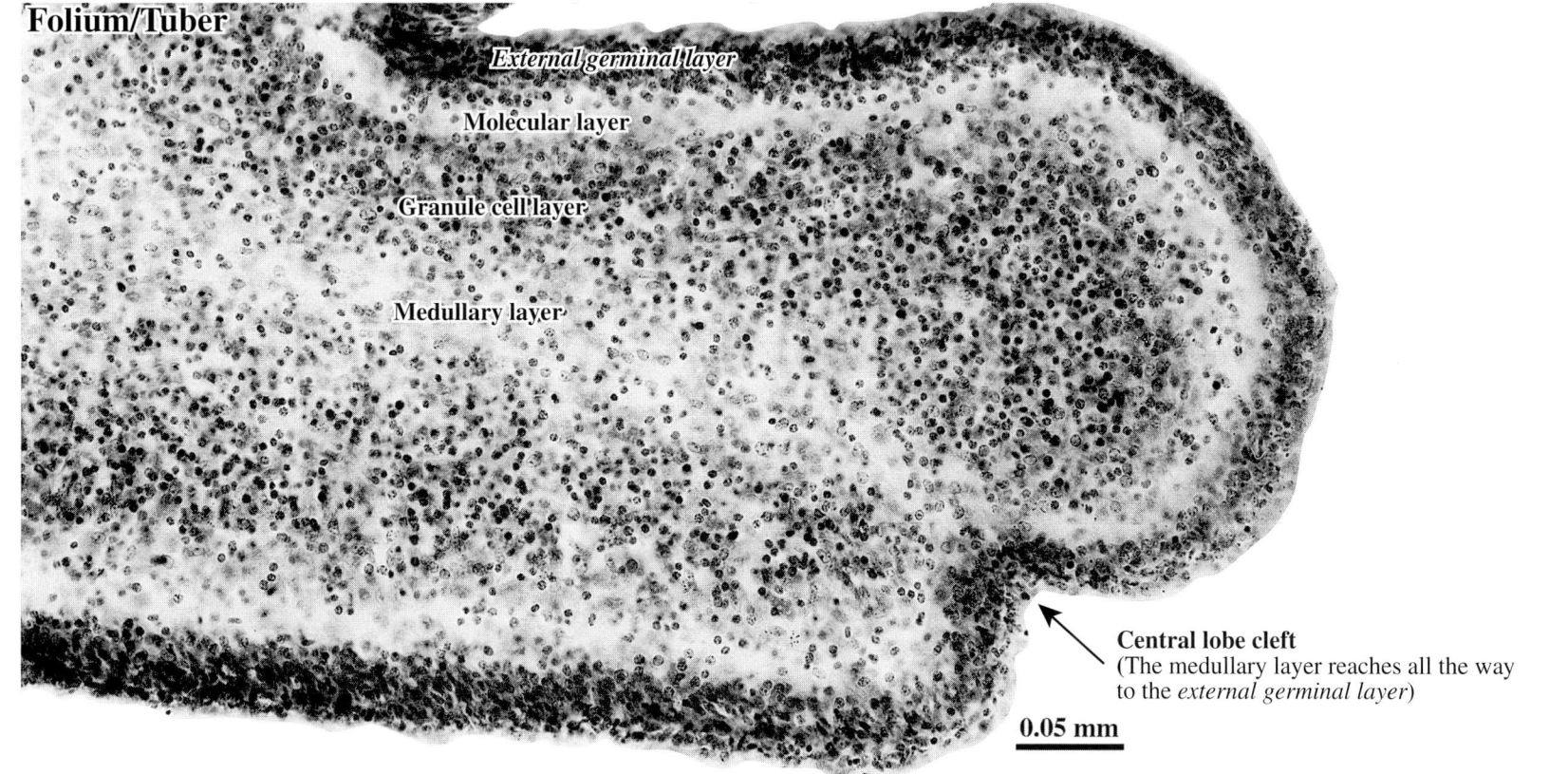

*External germinal layer*
Molecular layer
Granule cell layer
Medullary layer

Central lobe cleft
(The medullary layer reaches all the way to the *external germinal layer*)

0.05 mm

See the entire Section in Plate 96.

A medium-magnification view of the cerebellum is in Plates 102A and B.

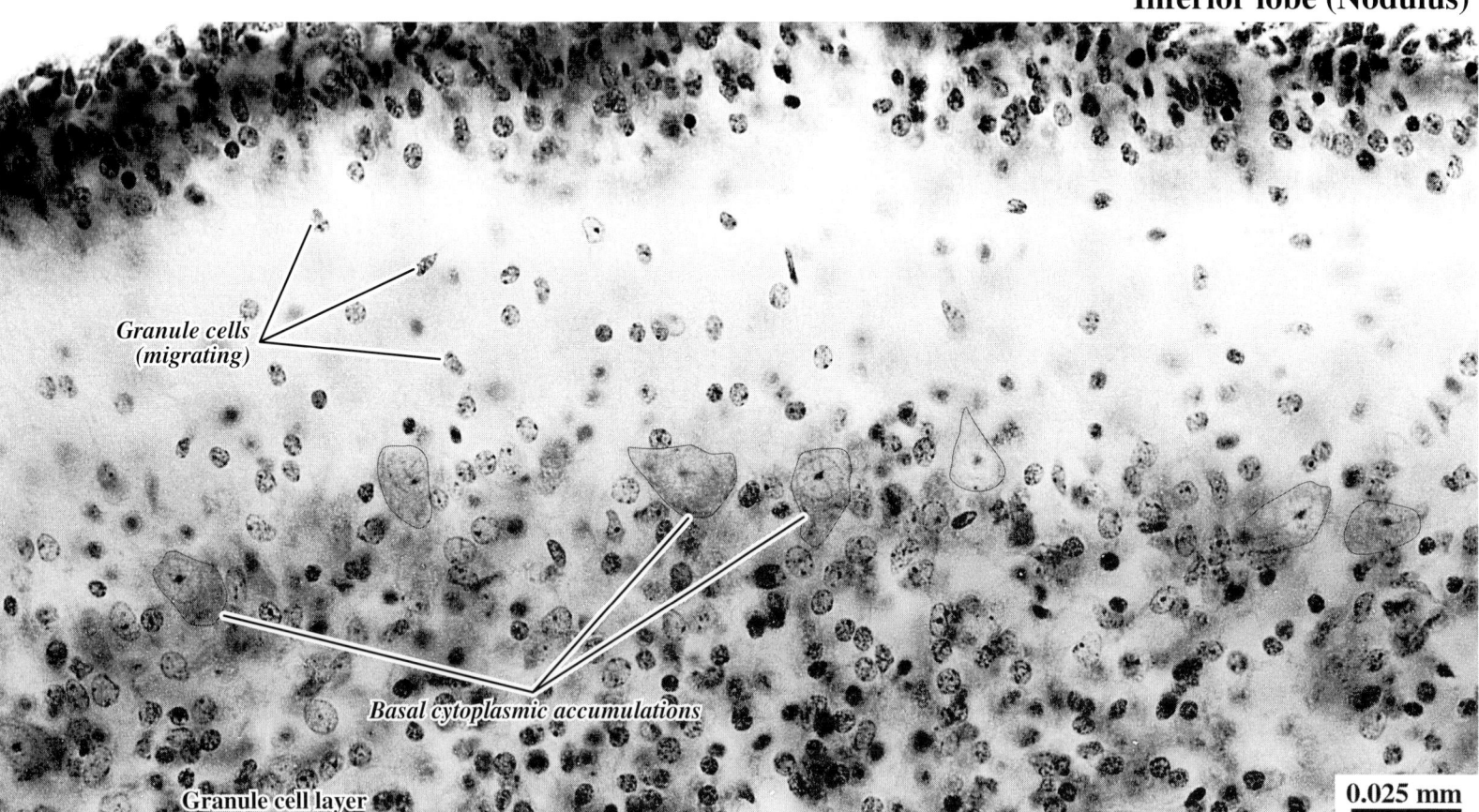

# PLATE 110

**GW32 Sagittal
CR 270 mm
Y15-60
Level 1: Section 801
CEREBELLUM**

See the entire Section in Plate 96.

A medium-magnification view of the cerebellum is in Plates 102A and B.

*Posterior lobe (Uvula)*

*External germinal layer (proliferative zone)*
*External germinal layer (premigratory zone)*
Molecular layer
Settled large Purkinje cells
Purkinje cell layer: (settled large Purkinje cells, widely scattered)
*Migrating small Purkinje cells?*
Granule cell layer

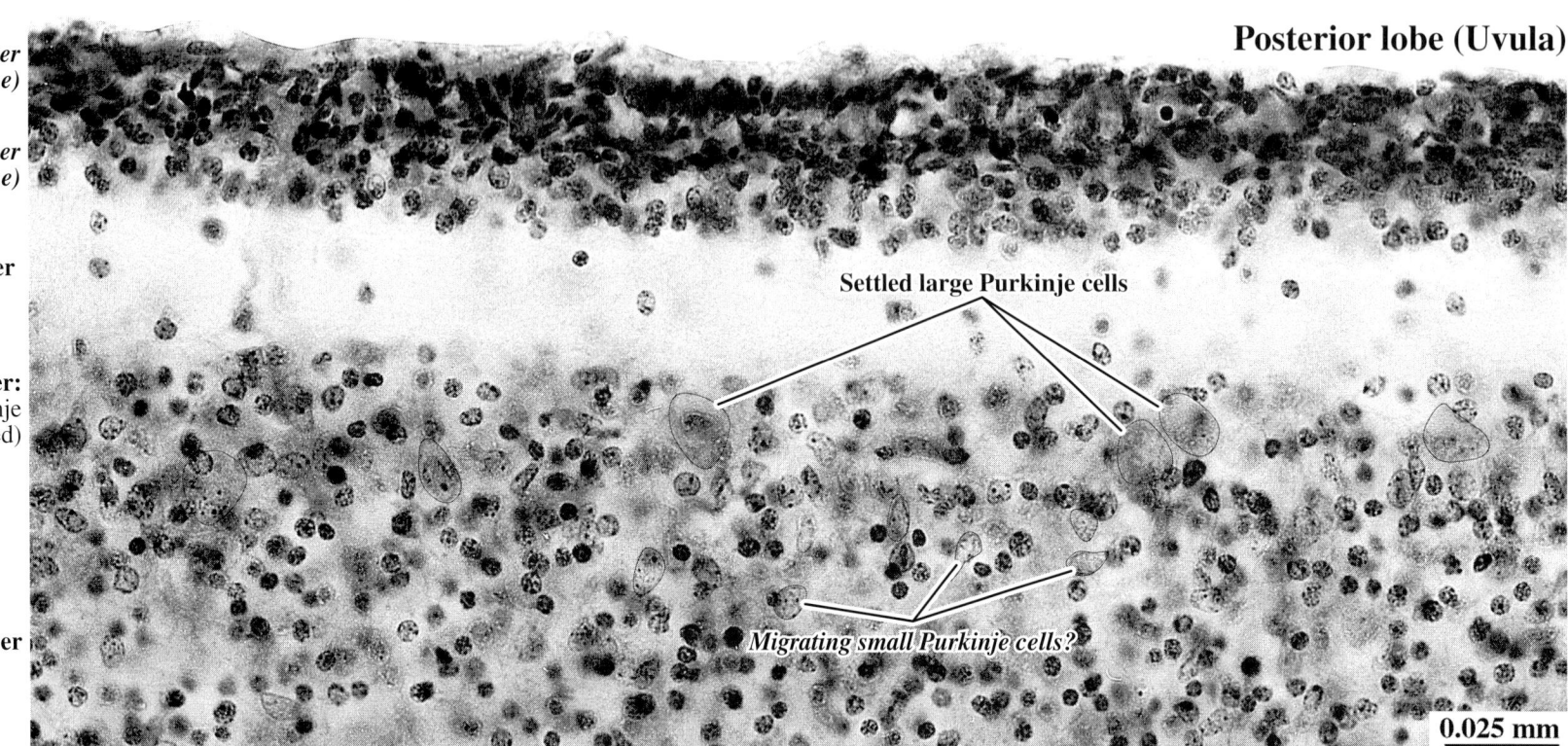

0.025 mm

*Inferior lobe (Nodulus)*

*External germinal layer (proliferative zone)*
*External germinal layer (premigratory zone)*
*Granule cells (migrating)*
Molecular layer
Purkinje cell layer: (Purkinje cells with *basal cytoplasmic accumulations* settled in a monolayer)
*Basal cytoplasmic accumulations*
Granule cell layer

0.025 mm

# PLATE 111

**GW32 Sagittal
CR 270 mm, Y15-60
Level 1: Section 801
CEREBELLUM**

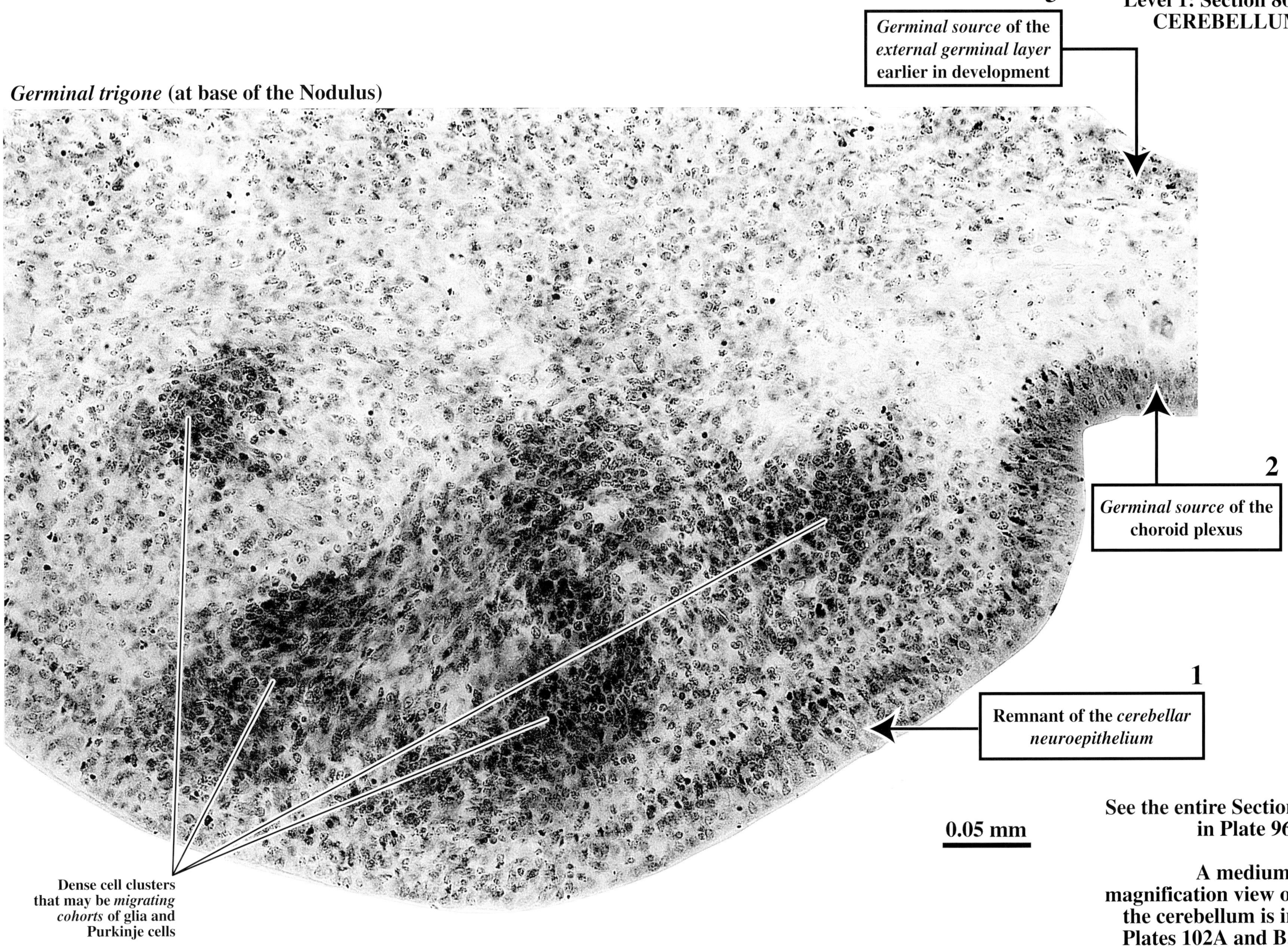

*Germinal trigone* (at base of the Nodulus)

**3** — *Germinal source* of the *external germinal layer* earlier in development

**2** — *Germinal source* of the *choroid plexus*

**1** — Remnant of the *cerebellar neuroepithelium*

Dense cell clusters that may be *migrating cohorts* of glia and Purkinje cells

0.05 mm

See the entire Section in Plate 96.

A medium-magnification view of the cerebellum is in Plates 102A and B.

# PART VI:
# GW30 HORIZONTAL

This specimen is case number W-187-65 (Perinatal RPSL) in the Yakovlev Collection. A male infant survived 17 hours after a premature birth. Death occurred because of a pulmonary hyaline membrane. Autopsy notes include a subarachnoid hemorrhage over the cerebrum, but the brain appears normal and is classified as a Normative Control in the Yakovlev Collection (Haleem, 1990). It was cut in the horizontal plane in 35-μm thick sections. Since there is no available photograph of this brain before it was embedded and cut, we use the lateral view of another GW30 brain that Larroche published in 1967 (**Figure 7**).

The approximate cutting plane of this brain is indicated in **Figure 8** (facing page) with lines superimposed on the GW30 brain from the Larroche (1967) series. The anterior part of each section is ventral to the posterior part. As in all other specimens, the sections chosen for illustration are more closely spaced to show small structures in the diencephalon, midbrain, pons, and medulla. Illustrated sections are spaced farther apart when they contain only large brain structures, such as the cerebral cortex, basal ganglia, and cerebellum. Photographs of 24 different Nissl-stained sections (**Levels 1-24**) are shown at low magnification in **Plates 112-131**. The core of the brain and the cerebellum are shown at high magnification in companion **Plates 132AB-150AB** for **Levels 3-24**.

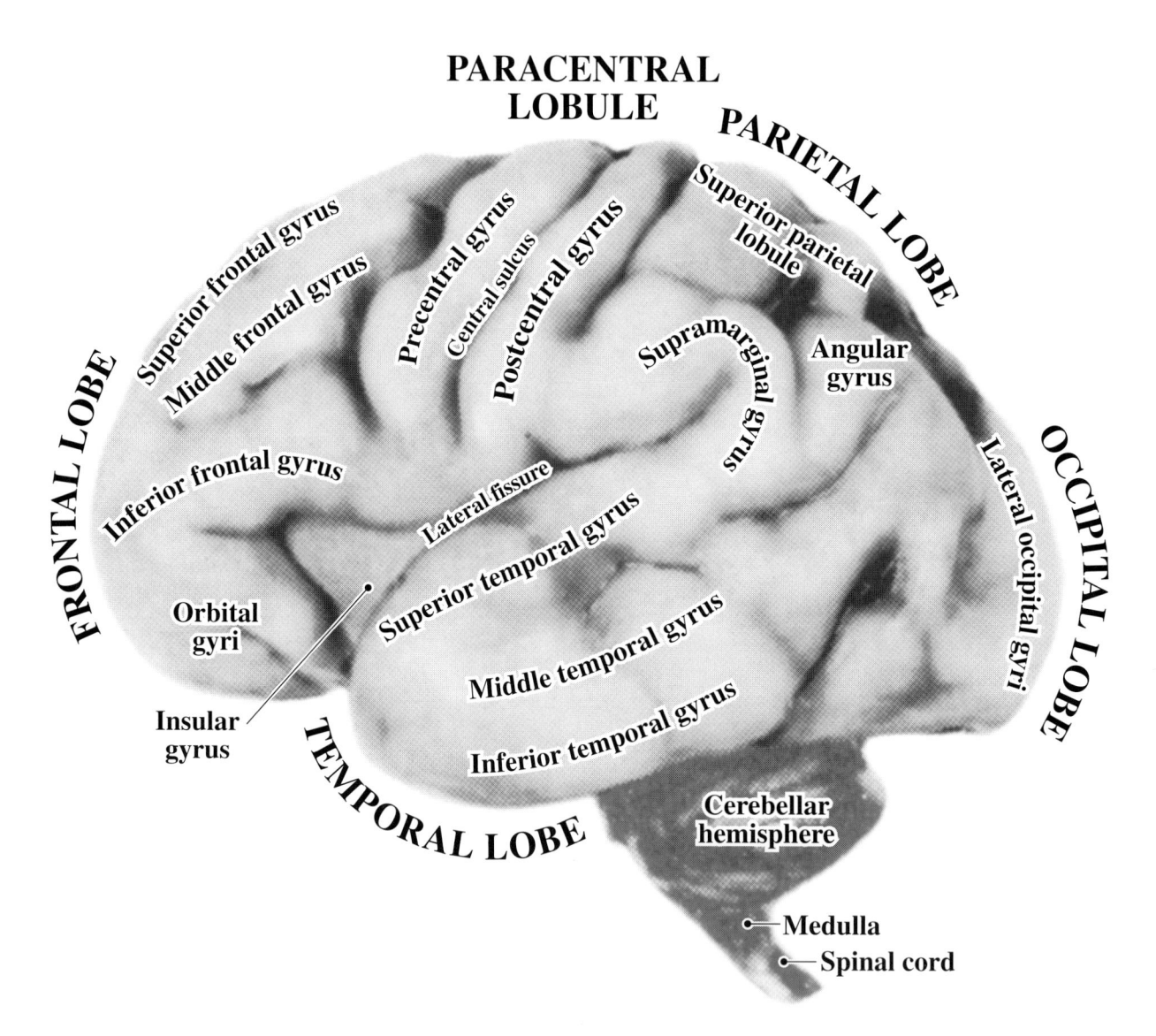

**Figure 7.** Lateral view of a GW30 brain with major structures in the cerebral hemispheres labeled. (From the photographic series of: J. C. Larroche (1967) Maturation morphologique du système nerveux central: ses rapports avec le développement pondéral du foetus et son age gestationnel. In: *Regional Development of the Brain in Early Life*, A. Minkowski (ed.), London: Blackwell, page 248.)

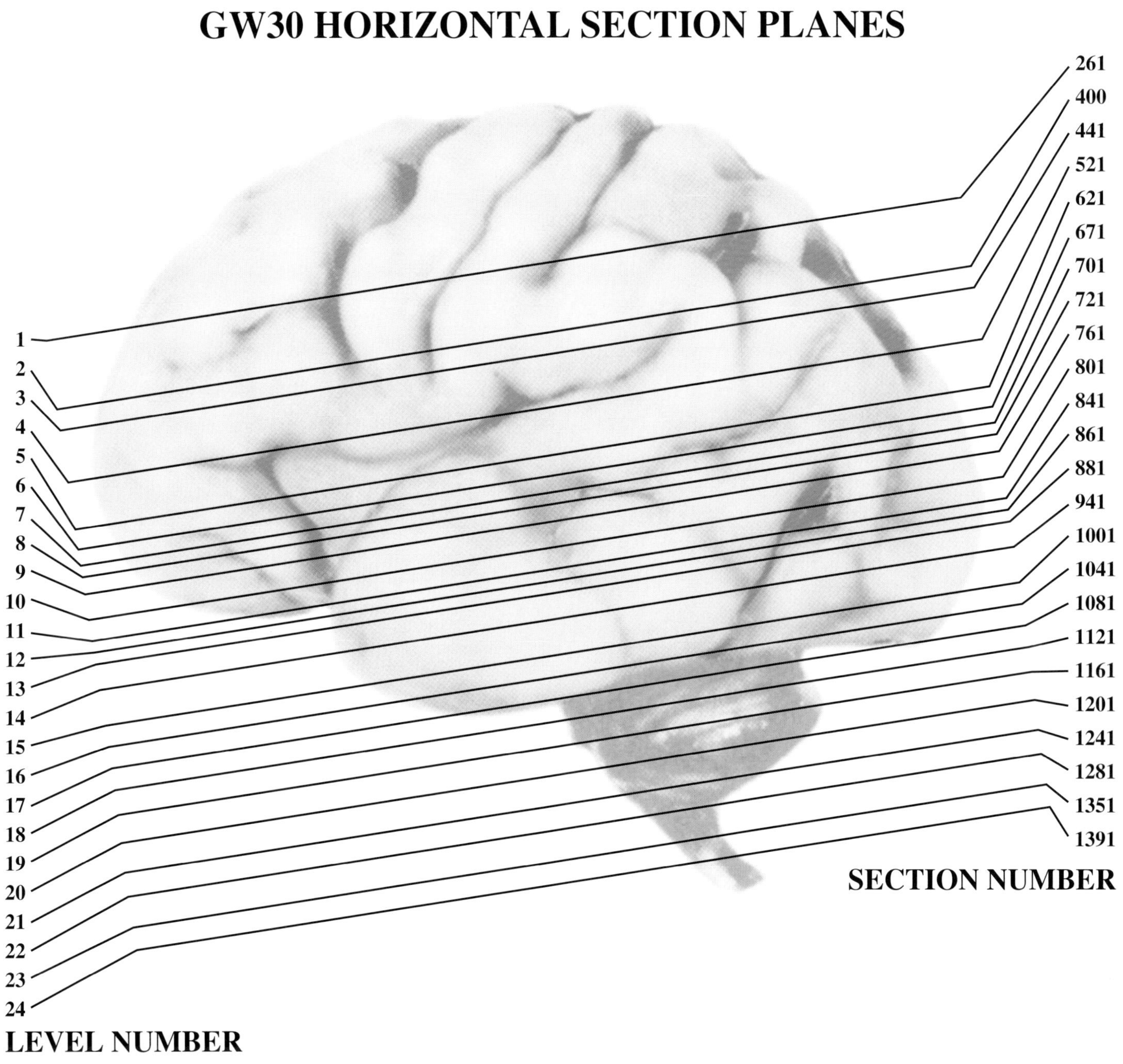

# GW30 HORIZONTAL SECTION PLANES

**SECTION NUMBER**

**LEVEL NUMBER**

**Figure 8.** Lateral view of the same GW30 brain shown in **Figure 7** with the approximate locations and cutting angle of the sections of Y187-65. (From the photographic series of: J. C. Larroche (1967) Maturation morphologique du système nerveux central: ses rapports avec le développement pondéral du foetus et son age gestationnel. In: *Regional Development of the Brain in Early Life*, A. Minkowski (ed.), London: Blackwell, page 248).

Y187-65 contains more prominent immature structures than in the older specimens. A densely staining *neuroepithelium/subventricular zone* is present and presumably generating neocortical interneurons in all lobes of the cerebral cortex. The same thickness variations between the occipital and other lobes of the cerebral cortex are still there. Remnants of migrating and sojourning neurons and/or glia are visible in all lobes of the cerebral cortex as *stratified transitional fields*. Many neurons, glia, and their mitotic precursor cells are still migrating through the olfactory peduncle toward the olfactory bulb (*rostral migratory stream*) from a presumed source area in the germinal matrix at the junction between the cerebral cortex, striatum, and nucleus accumbens. The *lateral migratory stream* percolates through the claustrum, endopiriform nucleus, external capsule, and uncinate fasciculus with dense streams of cells that appear to be heading toward the insular cortex, primary olfactory cortex, temporal cortex, and basolateral parts of the amygdaloid complex. In the basal ganglia, there is a large *neuroepithelium/subventricular zone* overlying the striatum and nucleus accumbens where neurons are presumably generated; at least three subdivisions (anteromedial, anterolateral, and posterior) can be distinguished in the striatal part. Another region of active neurogenesis in the telencephalon is the *subgranular zone* in the hilus of the dentate gyrus that is the source of granule cells. Other structures in the telencephalon, such as the septum, fornix, and Ammon's horn, have only a thin, darkly staining layer at the ventricle, and these are presumed to be generating glia, cells of the choroid plexus, and the ependymal lining of the ventricle.

Most of the structures in the diencephalon appear to be settled and are maturing, but the third ventricle is lined by a more densely staining *glioepithelium/ependyma* than in the GW37 specimens. A convoluted *glioepithelium/ependyma* lines the cerebral aqueduct in the midbrain that continues into the anterior fourth ventricle. A smooth *glioepithelium/ependyma* lines the fourth ventricle through the posterior pons. A slightly convoluted *glioepithelium/ependyma* lines the floor of the fourth ventricle through much of the medulla. The *external germinal layer* is prominent over the entire surface of the cerebellar cortex and is actively producing basket, stellate, and granule cells. The cerebellar cortex itself shows less definition between hemispheric lobules. The *germinal trigone* is at the base of the nodulus and along the floccular peduncle; choroid plexus cells and glia may still be originating here.

# PLATE 112

GW30 Horizontal
CR 260 mm
Y187-65
Level 1: Section 261

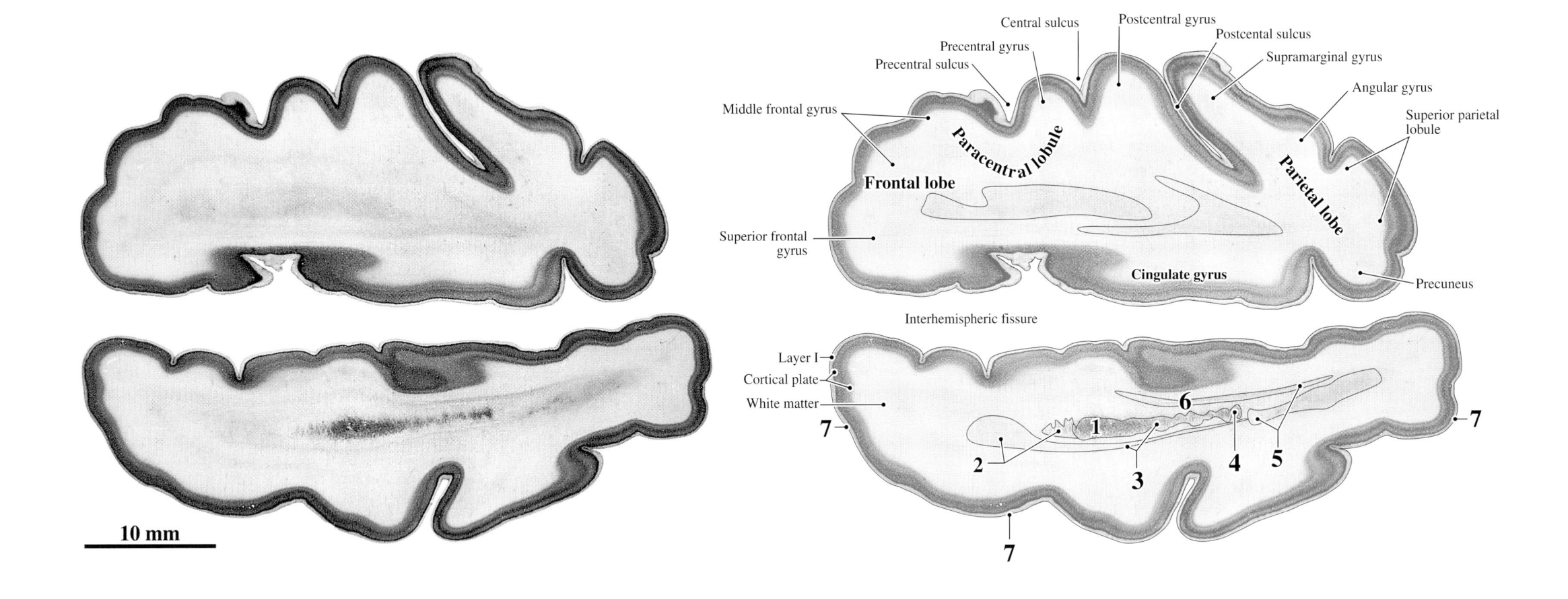

**Remnants of the germinal matrix, migratory streams, and transitional fields**

| | | | |
|---|---|---|---|
| 1 | Frontal subventricular zone | 5 | Parietal stratified transitional field |
| 2 | Frontal stratified transitional field | 6 | Cingulate stratified transitional field |
| 3 | Paracentral stratified transitional field | | |
| 4 | Parietal neuroepithelium and subventricular zone | 7 | Subpial granular layer (cortical) |

## PLATE 113

**GW30 Horizontal
CR 260 mm
Y187-65
Level 2: Section 400**

### Remnants of the germinal matrix, migratory streams, and transitional fields

| | | | |
|---|---|---|---|
| 1 | Frontal neuroepithelium and subventricular zone | 7 | Parietal stratified transitional field |
| 2 | Frontal stratified transitional field | 8 | Posterior striatal neuroepithelium and subventricular zone |
| 3 | Callosal glioepithelium | 9 | Anterolateral striatal neuroepithelium and subventricular zone |
| 4 | Callosal sling | 10 | Anteromedial striatal neuroepithelium and subventricular zone |
| 5 | Fornical glioepithelium | 11 | Strionuclear glioepithelium |
| 6 | Parietal neuroepithelium and subventricular zone | 12 | Subpial granular layer (cortical) |

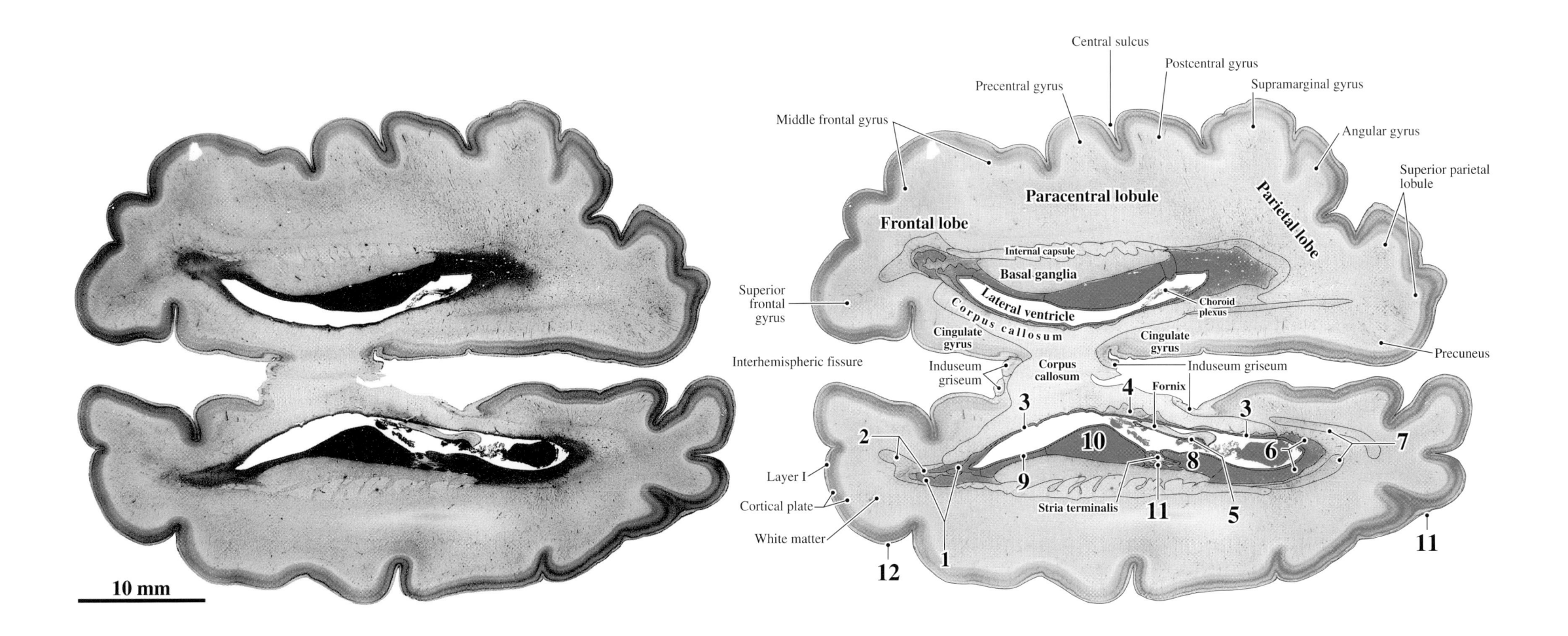

# PLATE 114

**GW30 Horizontal**
**CR 260 mm**
**Y187-65**
**Level 3: Section 441**

**See detail of brain core in Plates 132A and B.**

### *Remnants of the germinal matrix, migratory streams, and transitional fields*

| | | | |
|---|---|---|---|
| 1 | Frontal neuroepithelium and subventricular zone | 7 | Parietal stratified transitional field |
| 2 | Frontal stratified transitional field | 8 | Posterior striatal neuroepithelium and subventricular zone |
| 3 | Callosal glioepithelium | 9 | Anterolateral striatal neuroepithelium and subventricular zone |
| 4 | Callosal sling | 10 | Anteromedial striatal neuroepithelium and subventricular zone |
| 5 | Fornical glioepithelium | 11 | Strionuclear glioepithelium |
| 6 | Parietal neuroepithelium and subventricular zone | 12 | Subpial granular layer (cortical) |

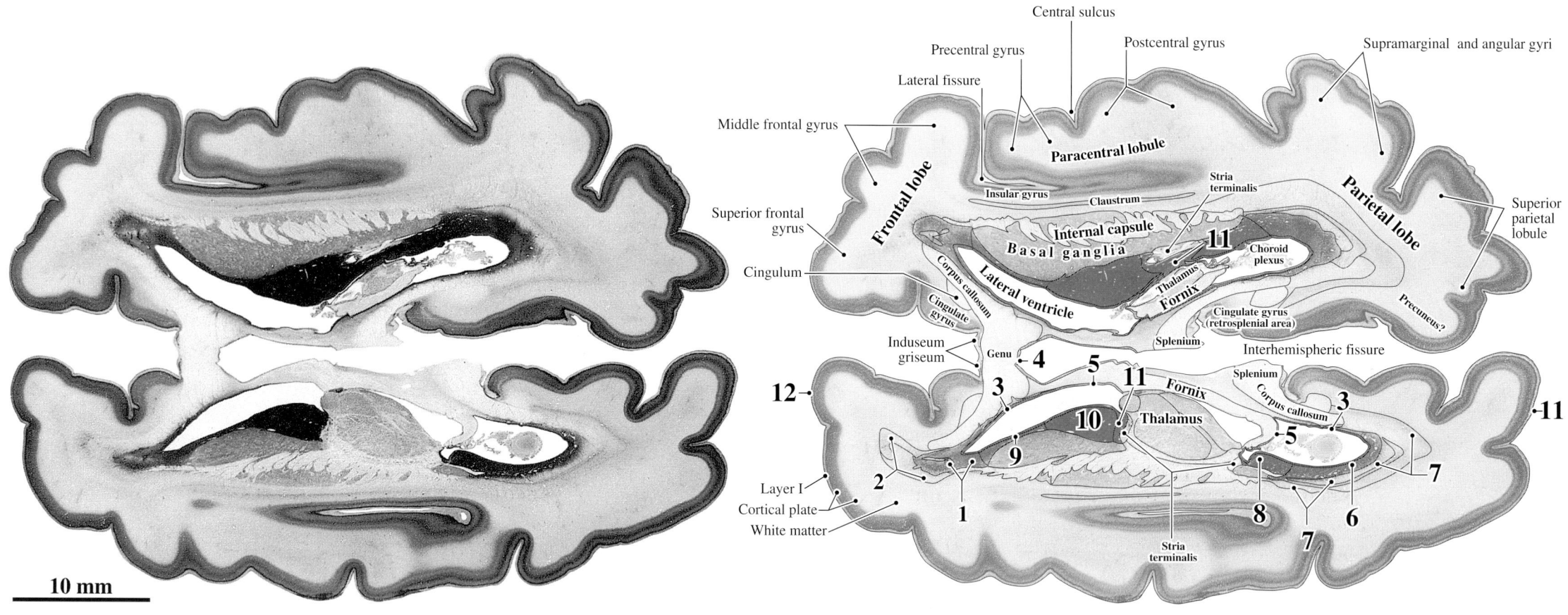

The splenium of the corpus callosum was cut during processing.

# PLATE 115

**GW30 Horizontal**
**CR 260 mm**
**Y187-65**
**Level 4: Section 521**

## Remnants of the germinal matrix, migratory streams, and transitional fields

| | | | |
|---|---|---|---|
| 1 | Rostral migratory stream (source area) | 8 | Parietal stratified transitional field |
| 2 | Frontal neuroepithelium and subventricular zone | 9 | Posterior sriatal neuroepithelium and subventricular zone |
| 3 | Frontal stratified transitional field | 10 | Anterolateral striatal neuroepithelium and subventricular zone |
| 4 | Callosal glioepithelium | 11 | Anteromedial striatal neuroepithelium and subventricular zone |
| 5 | Callosal sling | 12 | Strionuclear glioepithelium |
| 6 | Fornical glioepithelium | 13 | Septal glioepithelium/ependyma |
| 7 | Parietal neuroepithelium and subventricular zone | 14 | Subpial granular layer (cortical) |

**See detail of brain core in Plates 133A and B.**

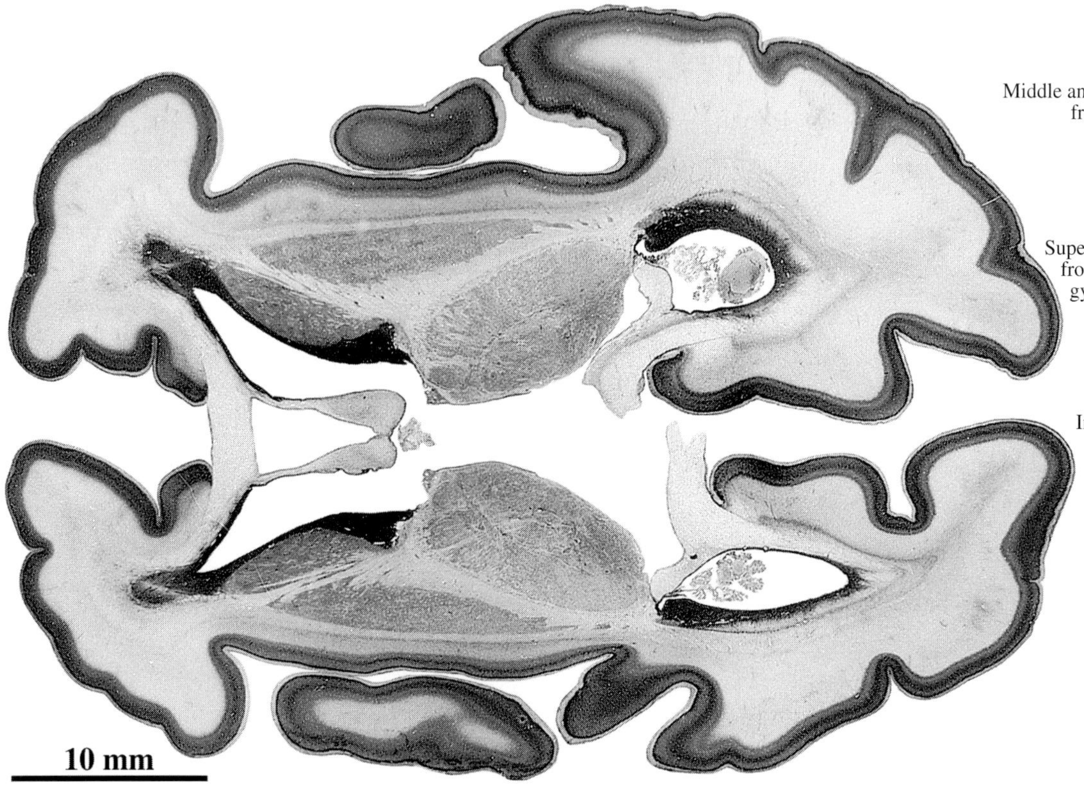

The splenium of the corpus callosum was cut during processing.

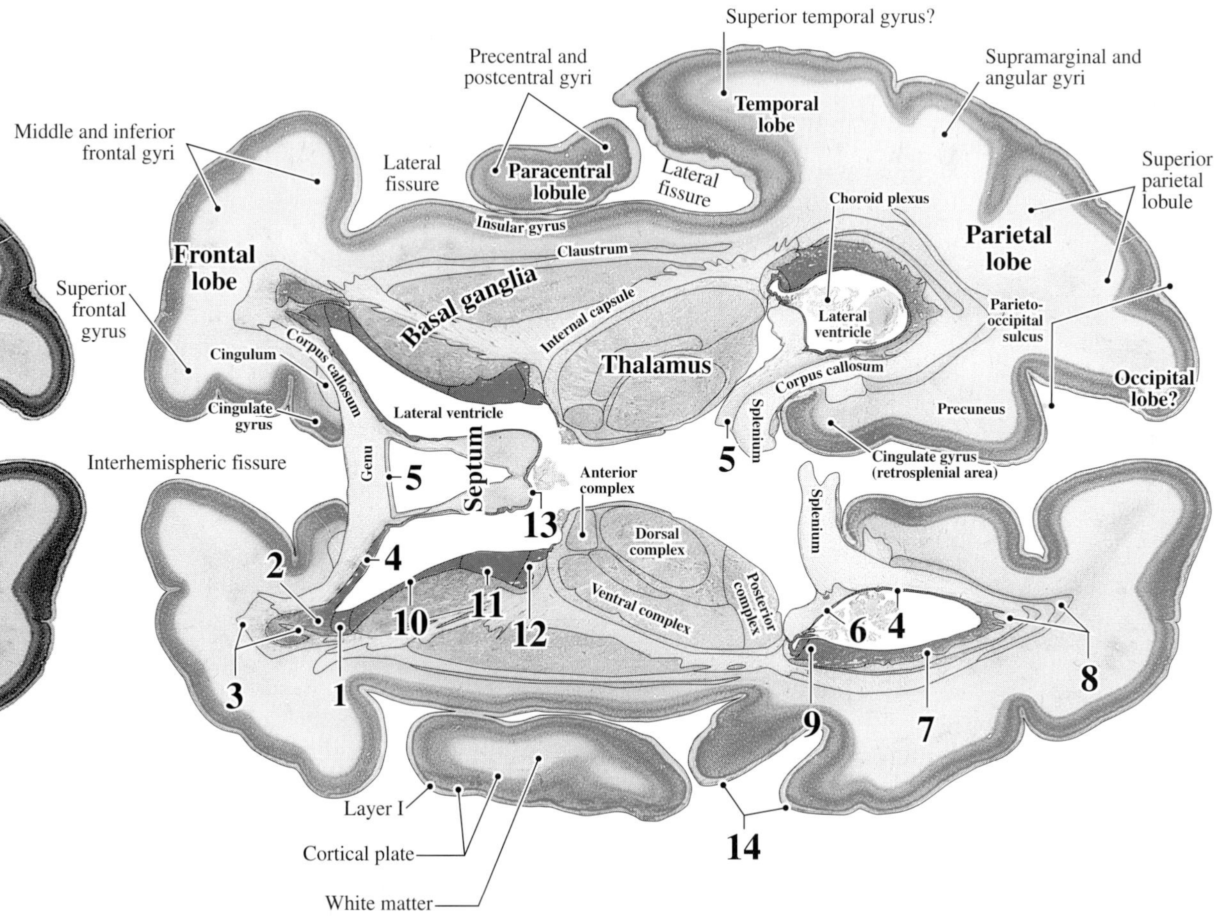

# PLATE 116

**GW30 Horizontal**
**CR 260 mm**
**Y187-65**
**Level 5: Section 621**

**See detail of brain core in Plates 134A and B.**

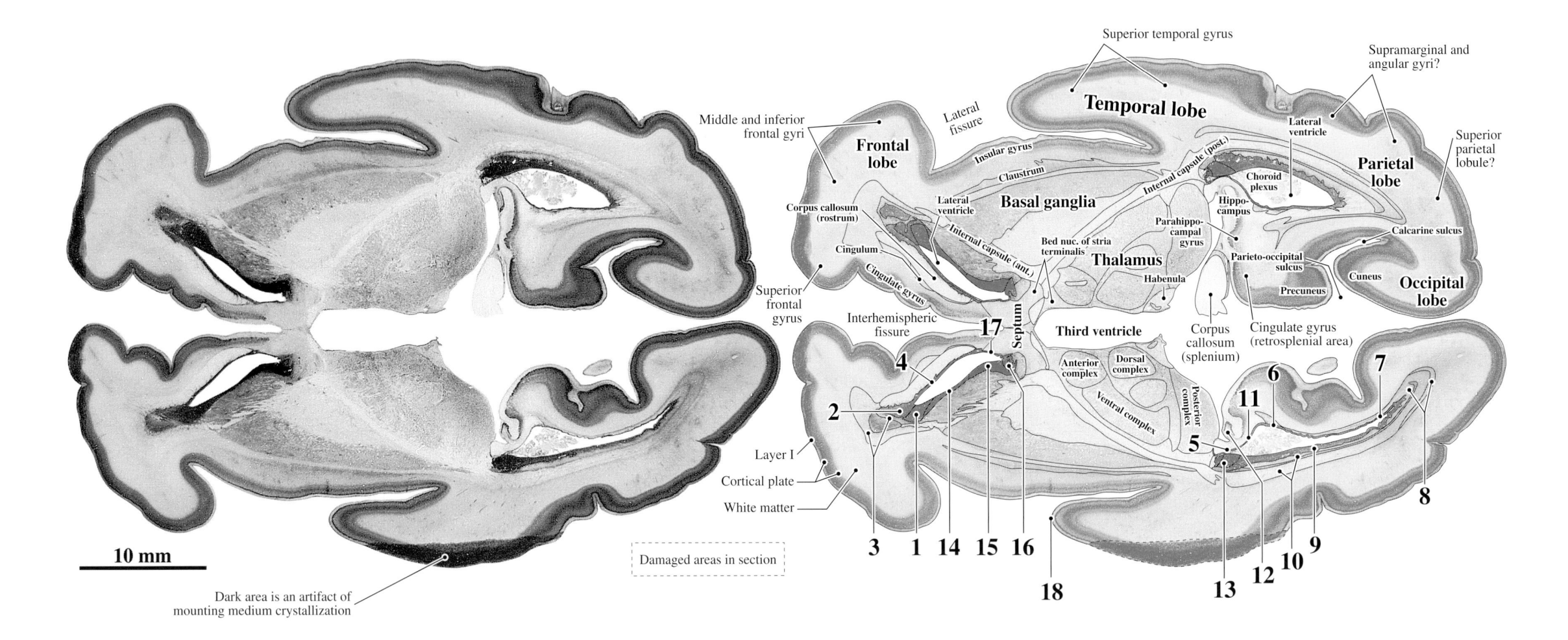

*Remnants of the germinal matrix, migratory streams, and transitional fields*

1. *Rostral migratory stream (source area)*
2. *Frontal neuroepithelium and subventricular zone*
3. *Frontal stratified transitional field*
4. *Callosal glioepithelium*
5. *Fornical glioepithelium*
6. *Parahippocampal neuroepithelium, subventricular zone, and stratified transitional field*
7. *Occipital neuroepithelium and subventricular zone*
8. *Occipital stratified transitional field*
9. *Parietal/temporal neuroepithelium and subventricular zone*
10. *Parietal/temporal stratified transitional field*
11. *Alvear glioepithelium*
12. *Subgranular zone (dentate)*
13. *Posterior striatal neuroepithelium and subventricular zone*
14. *Anterolateral striatal neuroepithelium and subventricular zone*
15. *Anteromedial striatal neuroepithelium and subventricular zone*
16. *Strionuclear neuroepithelium and subventricular zone*
17. *Septal glioepithelium/ependyma*
18. *Subpial granular layer (cortical)*

# PLATE 117

**GW30 Horizontal**
**CR 260 mm**
**Y187-65**
**Level 6: Section 671**

## *Remnants of the germinal matrix, migratory streams, and transitional fields*

1. *Rostral migratory stream (source area)*
2. *Frontal neuroepithelium and subventricular zone*
3. *Frontal stratified transitional field*
4. *Callosal glioepithelium*
5. *Fornical glioepithelium*
6. *Parahippocampal neuroepithelium, subventricular zone, and stratified transitional field*
7. *Occipital neuroepithelium and subventricular zone*
8. *Occipital stratified transitional field*
9. *Temporal neuroepithelium and subventricular zone*
10. *Temporal stratified transitional field*
11. *Alvear glioepithelium*
12. *Subgranular zone (dentate)*
13. *Lateral migratory stream (cortical)*
14. *Posterior striatal neuroepithelium and subventricular zone*
15. *Anterolateral striatal neuroepithelium and subventricular zone (infiltrated by the rostral migratory stream)*
16. *Accumbent neuroepithelium and subventricular zone (infiltrated by the rostral migratory stream)*
17. *Subpial granular layer (cortical)*

**See detail of brain core in Plates 135A and B.**

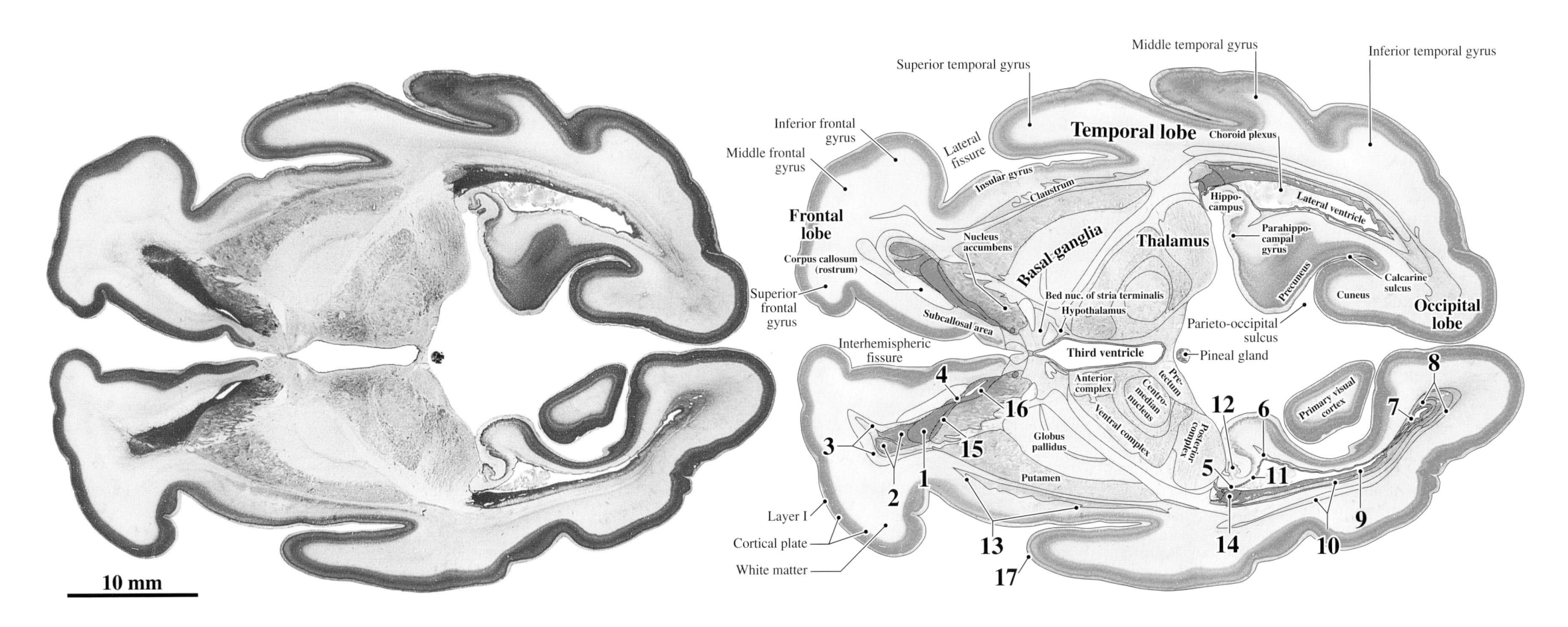

# PLATE 118

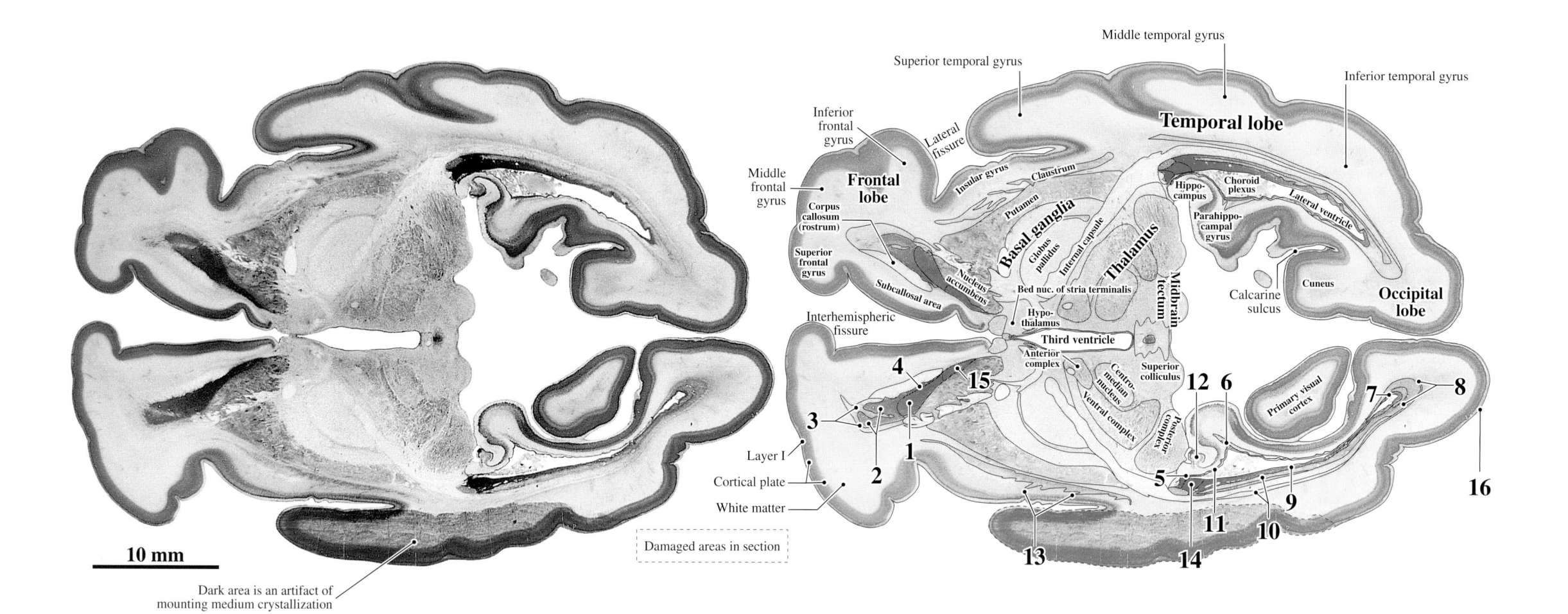

GW30 Horizontal
CR 260 mm
Y187-65
Level 7: Section 701

See detail of brain core
in Plates 136A and B.

### Remnants of the germinal matrix, migratory streams, and transitional fields

1. *Rostral migratory stream (source area)*
2. *Frontal neuroepithelium and subventricular zone*
3. *Frontal stratified transitional field*
4. *Callosal glioepithelium*
5. *Fornical glioepithelium*
6. *Parahippocampal neuroepithelium, subventricular zone, and stratified transitional field*
7. *Occipital neuroepithelium and subventricular zone*
8. *Occipital stratified transitional field*
9. *Temporal neuroepithelium and subventricular zone*
10. *Temporal stratified transitional field*
11. *Alvear glioepithelium*
12. *Subgranular zone (dentate)*
13. *Lateral migratory stream (cortical)*
14. *Posterior striatal neuroepithelium and subventricular zone*
15. *Accumbent neuroepithelium and subventricular zone (infiltrated by the rostral migratory stream)*
16. *Subpial granular layer (cortical)*

# PLATE 119

**GW30 Horizontal**
**CR 260 mm**
**Y187-65**
**Level 8: Section 721**

### Remnants of the germinal matrix, migratory streams, and transitional fields

1. Rostral migratory stream (source area)
2. Frontal neuroepithelium and subventricular zone
3. Frontal stratified transitional field
4. Callosal glioepithelium
5. Fornical glioepithelium
6. Parahippocampal neuroepithelium, subventricular zone, and stratified transitional field
7. Occipital neuroepithelium and subventricular zone
8. Occipital stratified transitional field
9. Temporal neuroepithelium and subventricular zone
10. Temporal stratified transitional field
11. Alvear glioepithelium
12. Subgranular zone (dentate)
13. Lateral migratory stream (cortical)
14. Posterior striatal neuroepithelium and subventricular zone
15. Accumbent neuroepithelium and subventricular zone (infiltrated by the rostral migratory stream)
16. Subpial granular layer (cortical)

**See detail of brain core in Plates 137A and B.**

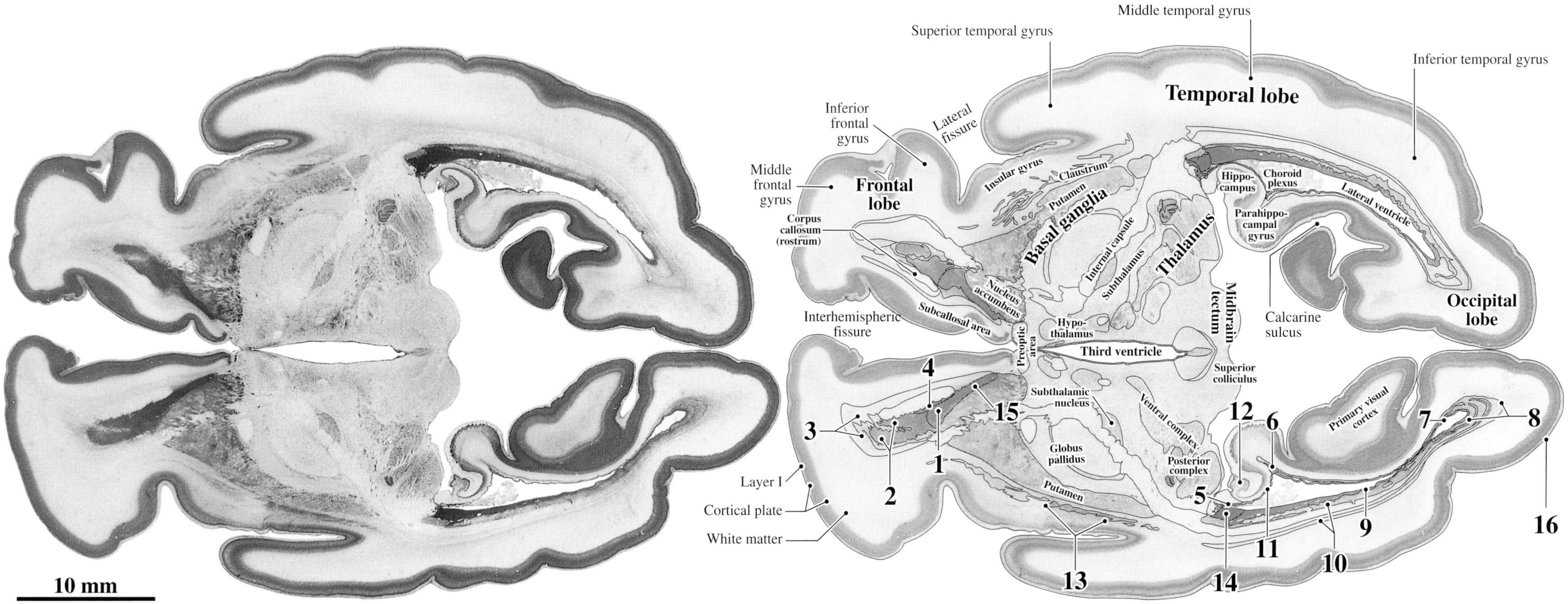

# PLATE 120

GW30 Horizontal
CR 260 mm
Y187-65
Level 9: Section 761

See detail of brain core
in Plates 138A and B.

### Remnants of the germinal matrix, migratory streams, and transitional fields

1. Rostral migratory stream (intermingled with the accumbent neuroepithelium and subventricular zone)
2. Frontal neuroepithelium and subventricular zone
3. Frontal stratified transitional field
4. Fornical glioepithelium
5. Parahippocampal neuroepithelium, subventricular zone, and stratified transitional field
6. Occipital neuroepithelium and subventricular zone
7. Occipital stratified transitional field
8. Temporal neuroepithelium and subventricular zone
9. Temporal stratified transitional field
10. Alvear glioepithelium
11. Subgranular zone (dentate)
12. Lateral migratory stream (cortical)
13. Posterior striatal neuroepithelium and subventricular zone
14. Subpial granular layer (cortical)

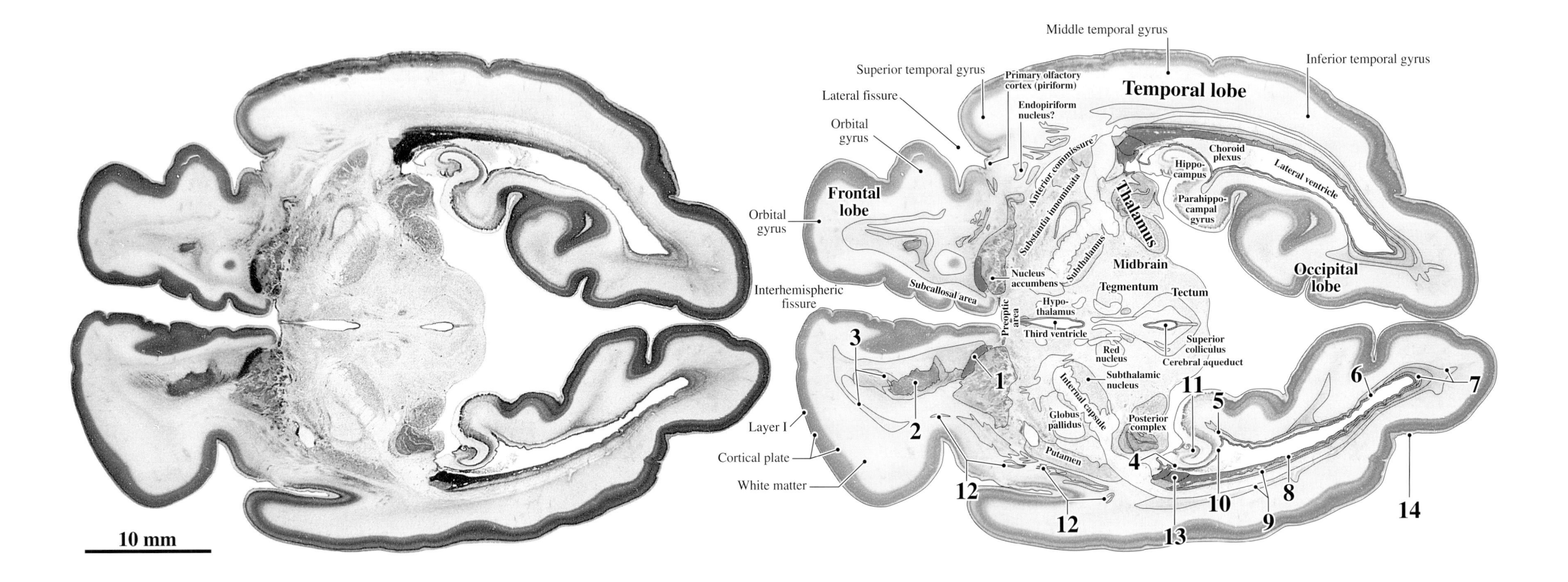

# PLATE 121

**GW30 Horizontal
CR 260 mm
Y187-65
Level 10: Section 801**

*Remnants of the germinal matrix, migratory streams, and transitional fields*

| | | | |
|---|---|---|---|
| 1 | Rostral migratory stream | 8 | Temporal neuroepithelium and subventricular zone |
| 2 | Frontal neuroepithelium and subventricular zone | 9 | Temporal stratified transitional field |
| 3 | Frontal stratified transitional field | 10 | Alvear glioepithelium |
| 4 | Fornical glioepithelium | 11 | Subgranular zone (dentate) |
| 5 | Parahippocampal neuroepithelium, subventricular zone, and stratified transitional field | 12 | Lateral migratory stream (cortical) |
| 6 | Occipital neuroepithelium and subventricular zone | 13 | Posterior striatal neuroepithelium and subventricular zone |
| 7 | Occipital stratified transitional field | 14 | Subpial granular layer (cortical) |

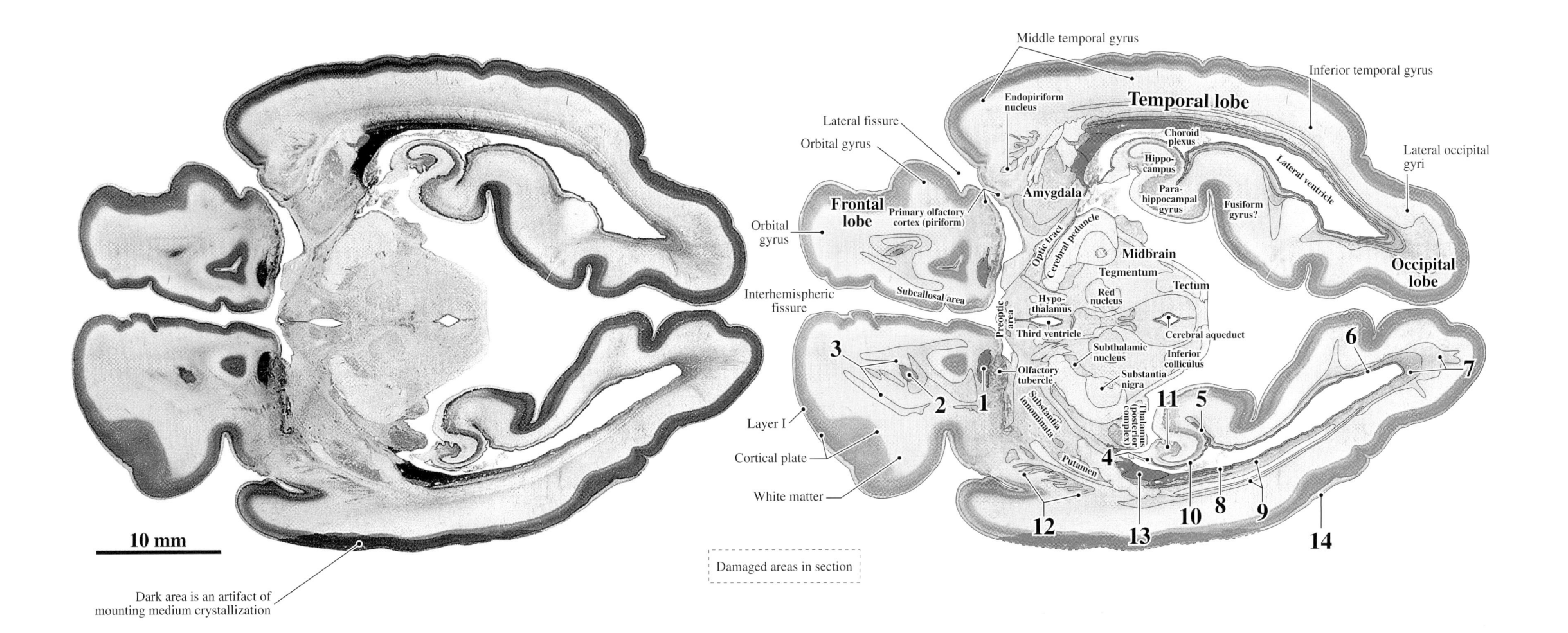

**See detail of brain core in Plates 139A and B.**

# PLATE 122

GW30 Horizontal
CR 260 mm
Y187-65
Level 11: Section 841

See detail of brain core
in Plates 140A and B.

### Remnants of the germinal matrix, migratory streams, and transitional fields

| | | | |
|---|---|---|---|
| 1 | Rostral migratory stream | 8 | Temporal stratified transitional field |
| 2 | Frontal stratified transitional field | 9 | Alvear glioepithelium |
| 3 | Fornical glioepithelium | 10 | Subgranular zone (dentate) |
| 4 | Parahippocampal neuroepithelium, subventricular zone, and stratified transitional field | 11 | Lateral migratory stream (cortical) |
| 5 | Occipital neuroepithelium and subventricular zone | 12 | Posterior striatal neuroepithelium and subventricular zone |
| 6 | Occipital stratified transitional field | 13 | Amygdaloid glioepithelium/ependyma |
| 7 | Temporal neuroepithelium and subventricular zone | 14 | Subpial granular layer (cortical) |

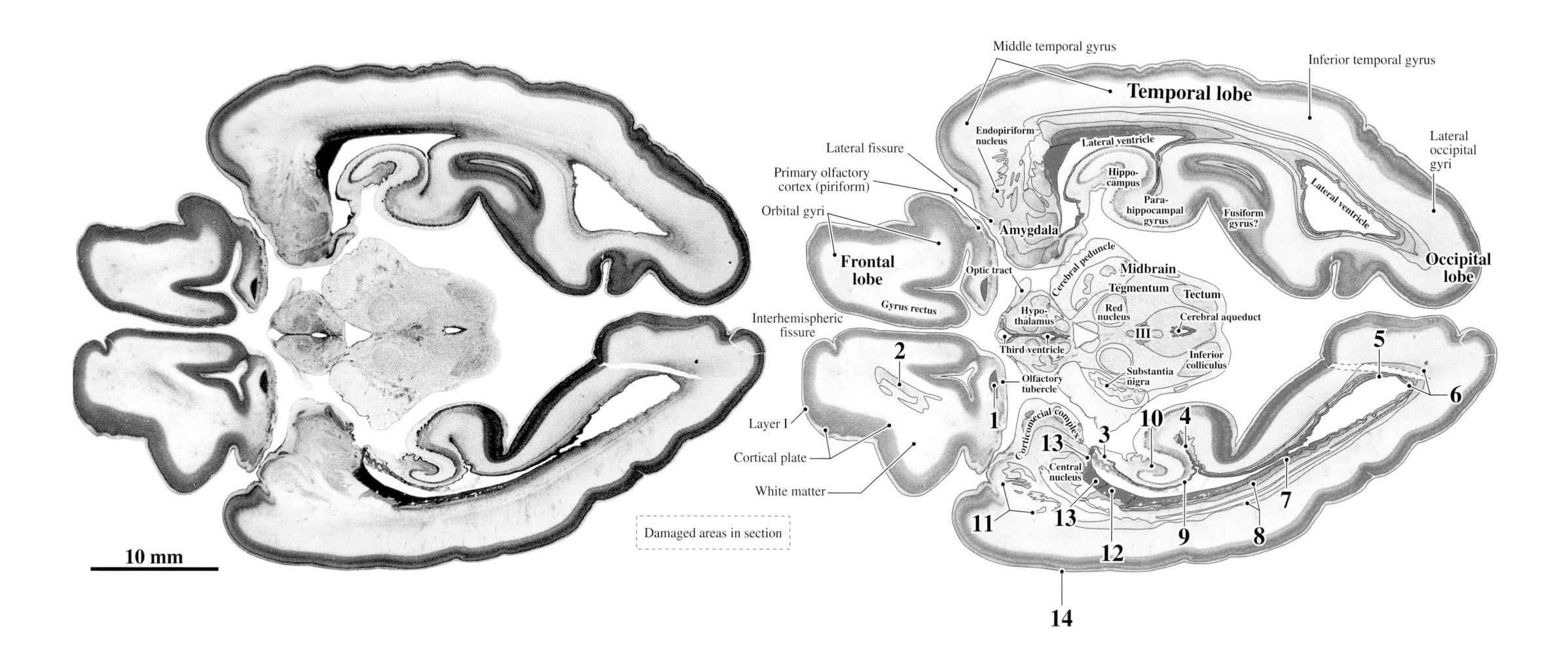

# PLATE 123

**GW30 Horizontal**
**CR 260 mm**
**Y187-65**
**Level 12: Section 861**

## Remnants of the germinal matrix, migratory streams, and transitional fields

| | | | |
|---|---|---|---|
| 1 | Rostral migratory stream | 8 | Alvear glioepithelium |
| 2 | Fornical glioepithelium | 9 | Subgranular zone (dentate) |
| 3 | Parahippocampal neuroepithelium, subventricular zone, and stratified transitional field | 10 | Lateral migratory stream (cortical) |
| 4 | Occipital neuroepithelium and subventricular zone | 11 | Posterior striatal neuroepithelium and subventricular zone |
| 5 | Occipital stratified transitional field | 12 | Amygdaloid glioepithelium/ependyma |
| 6 | Temporal neuroepithelium and subventricular zone | 13 | External germinal layer (cerebellum) |
| 7 | Temporal stratified transitional field | 14 | Subpial granular layer (cortical) |

**See detail of brain core and cerebellum in Plates 141A and B.**

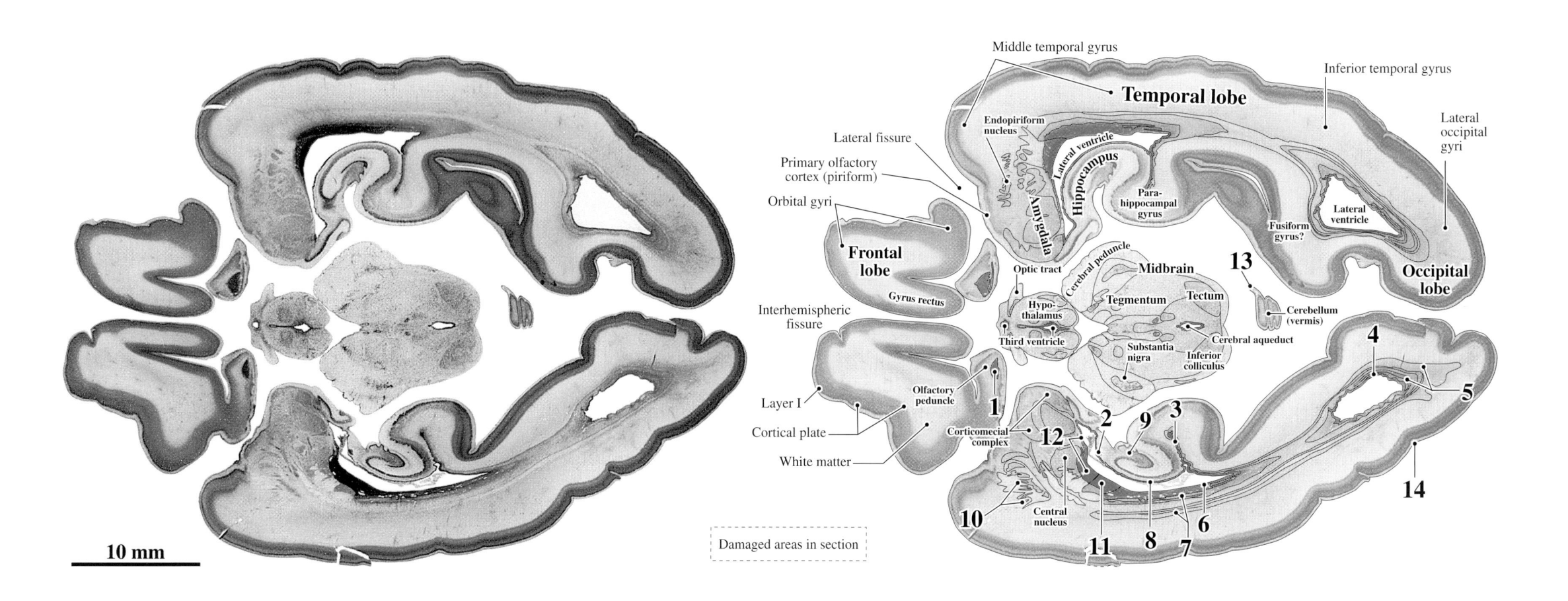

# PLATE 124

**GW30 Horizontal**
**CR 260 mm**
**Y187-65**
**Level 13: Section 881**

**See detail of brain core and cerebellum in Plates 142A and B.**

*Remnants of the germinal matrix, migratory streams, and transitional fields*

| | | | |
|---|---|---|---|
| 1 | Rostral migratory stream | 7 | Alvear glioepithelium |
| 2 | Parahippocampal neuroepithelium, subventricular zone, and stratified transitional field | 8 | Subgranular zone (dentate gyrus) |
| 3 | Occipital neuroepithelium and subventricular zone | 9 | Lateral migratory stream (cortical) |
| 4 | Occipital stratified transitional field | 10 | Amygdaloid glioepithelium/ependyma |
| 5 | Temporal neuroepithelium and subventricular zone | 11 | External germinal layer (cerebellum) |
| 6 | Temporal stratified transitional field | 12 | Subpial granular layer (cortical) |

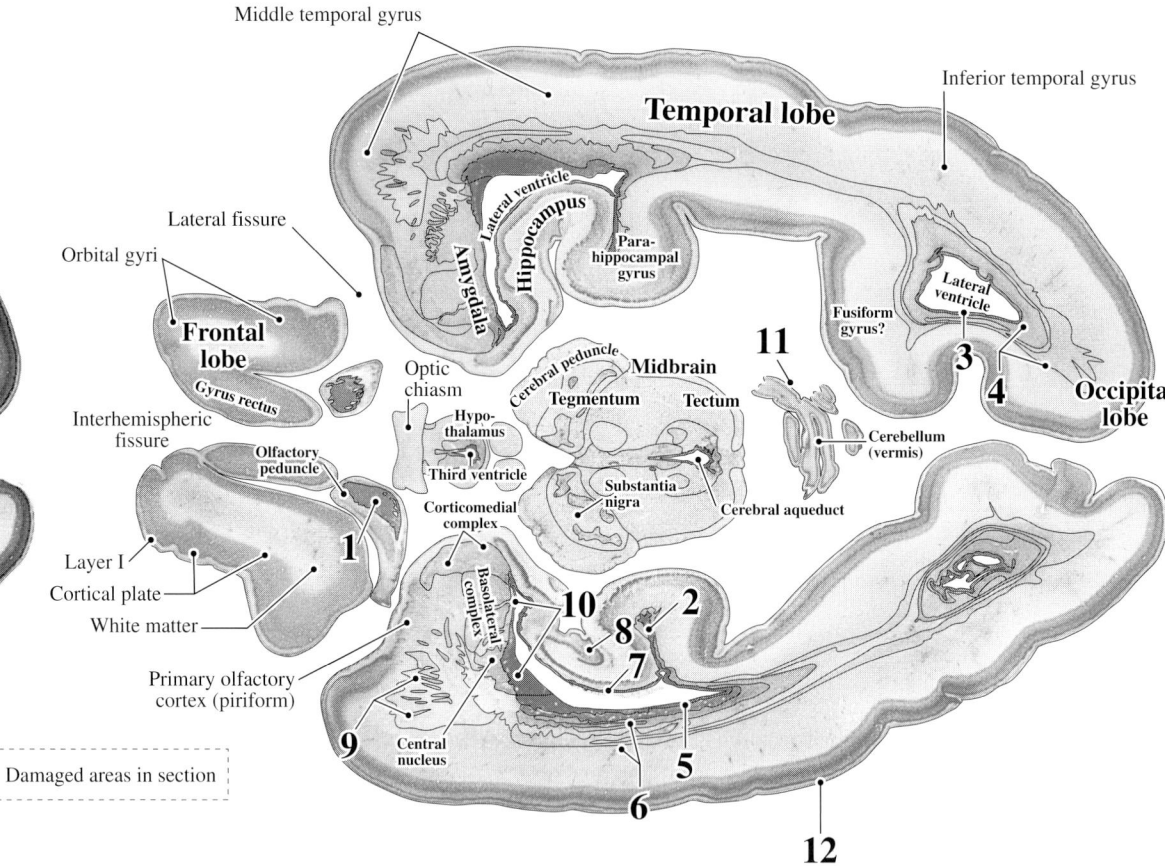

# PLATE 125

**GW30 Horizontal**
**CR 260 mm**
**Y187-65**
**Level 14: Section 941**

*Remnants of the germinal matrix, migratory streams, and transitional fields*

| | | | |
|---|---|---|---|
| 1 | Rostral migratory stream | 6 | Alvear glioepithelium |
| 2 | Parahippocampal neuroepithelium, subventricular zone, and stratified transitional field | 7 | Subgranular zone (dentate) |
| | | 8 | Lateral migratory stream (cortical) |
| 3 | Occipital stratified transitional field | 9 | Amygdaloid glioepithelium/ependyma |
| 4 | Temporal neuroepithelium and subventricular zone | 10 | External germinal layer (cerebellum) |
| 5 | Temporal stratified transitional field | 11 | Subpial granular layer (cortical) |

**See detail of brain core and cerebellum in Plates 143A and B.**

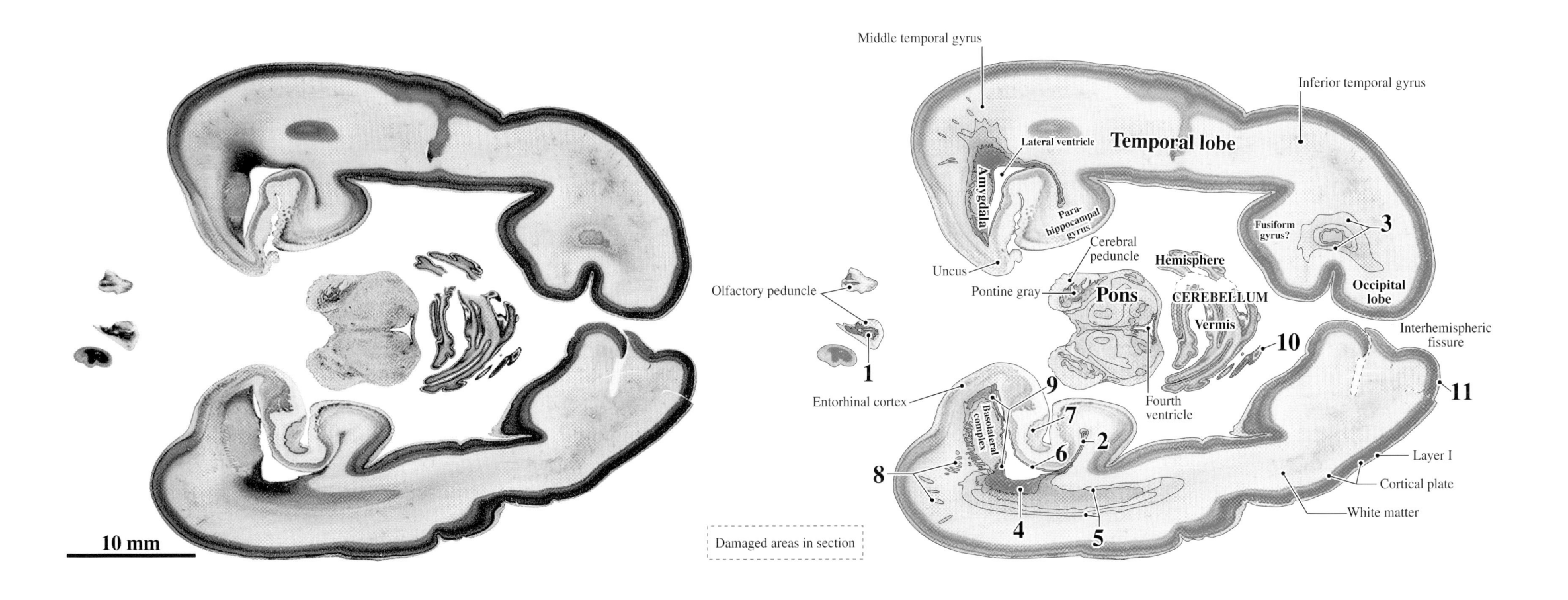

## PLATE 126

**GW30 Horizontal
CR 260 mm
Y187-65
Level 15: Section 1001**

*Remnants of the germinal matrix, migratory streams, and transitional fields*

| | | | |
|---|---|---|---|
| 1 | *Parahippocampal neuroepithelium and subventricular zone* | 4 | *Lateral migratory stream (cortical)* |
| 2 | *Parahippocampal stratified transitional field* | 5 | *Amygdaloid glioepithelium/ependyma* |
| | | 6 | *External germinal layer (cerebellum)* |
| 3 | *Alvear glioepithelium* | 7 | *Subpial granular layer (cortical)* |

**See detail of brain core and cerebellum in Plates 144A and B.**

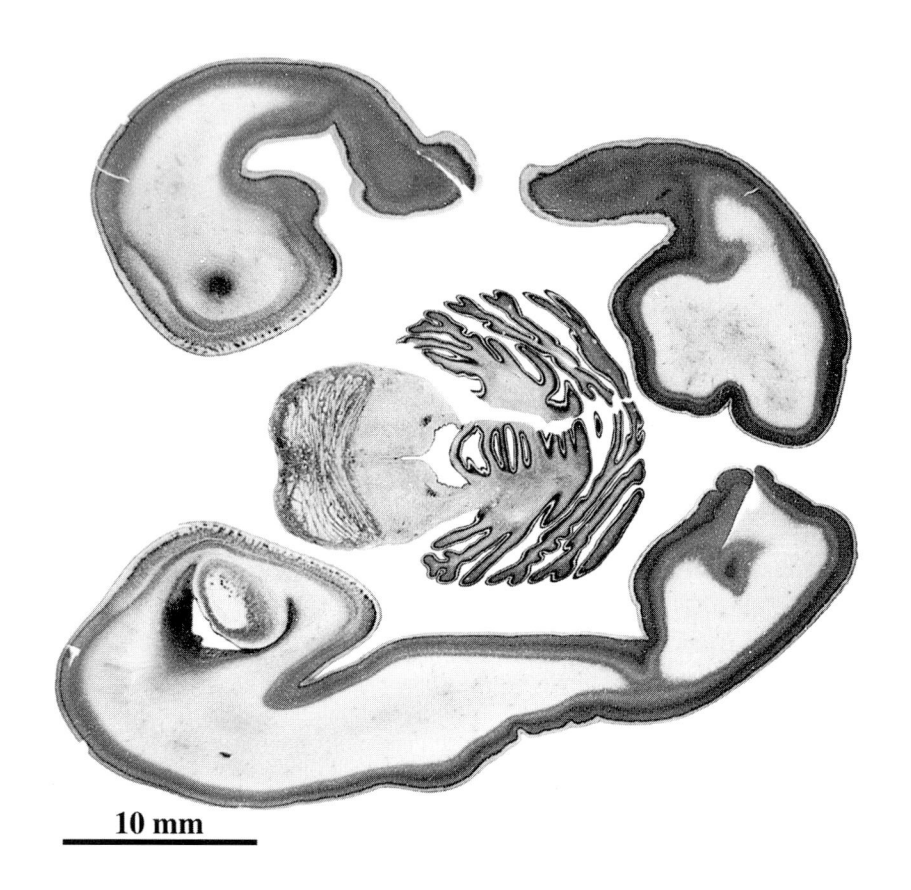

10 mm

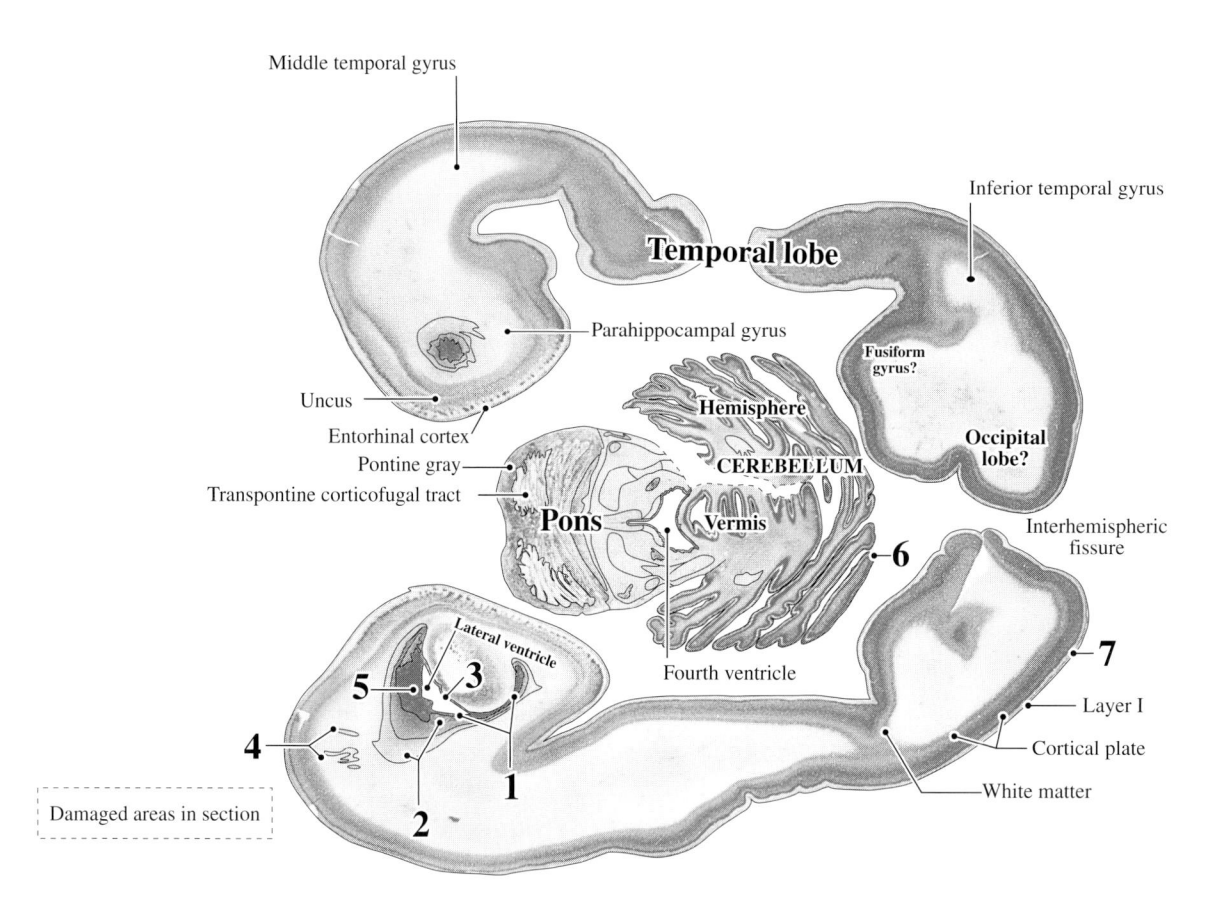

## PLATE 127

**GW30 Horizontal
CR 260 mm
Y187-65
Level 16: Section 1041**

*Remnants of the germinal matrix, migratory streams, and transitional fields*

1 *External germinal layer (cerebellum)*
2 *Subpial granular layer (cortical)*

See detail of brain core and cerebellum in Plates 145A and B.

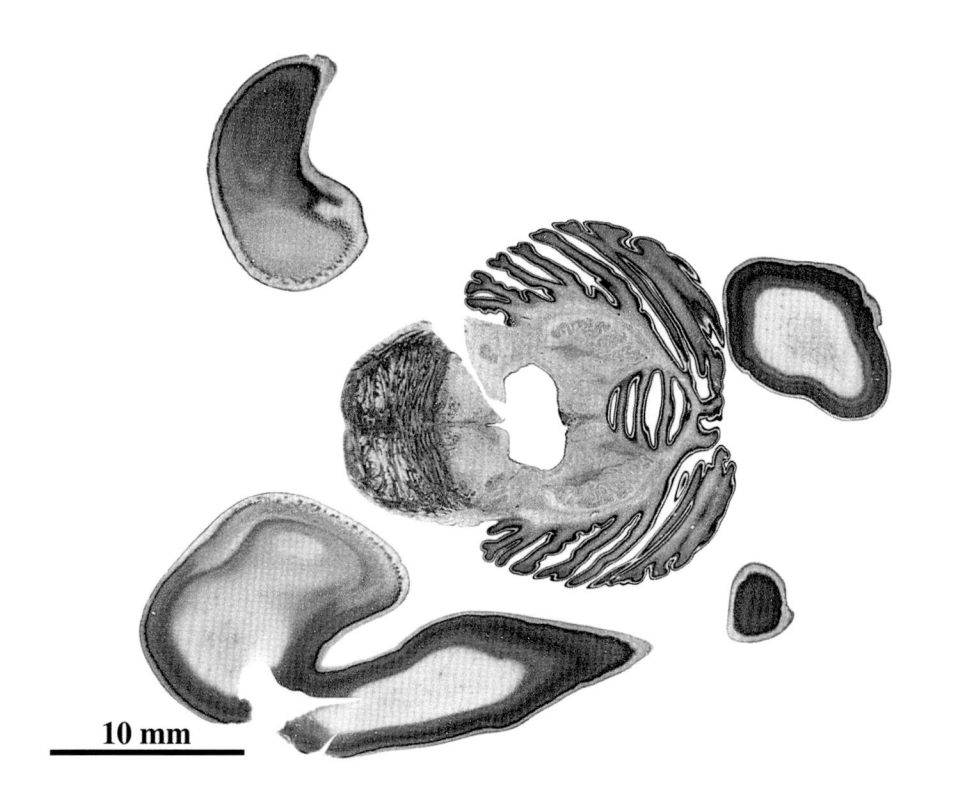

10 mm

Uncus
Middle cerebellar peduncle
Pontine gray
Transpontine corticofugal tract
Entorhinal cortex
Pons
Hemisphere
Deep nuclei
Fourth ventricle
Vermis
Occipital lobe
CEREBELLUM
Temporal lobe
Inferior temporal gyrus?

Damaged areas in section

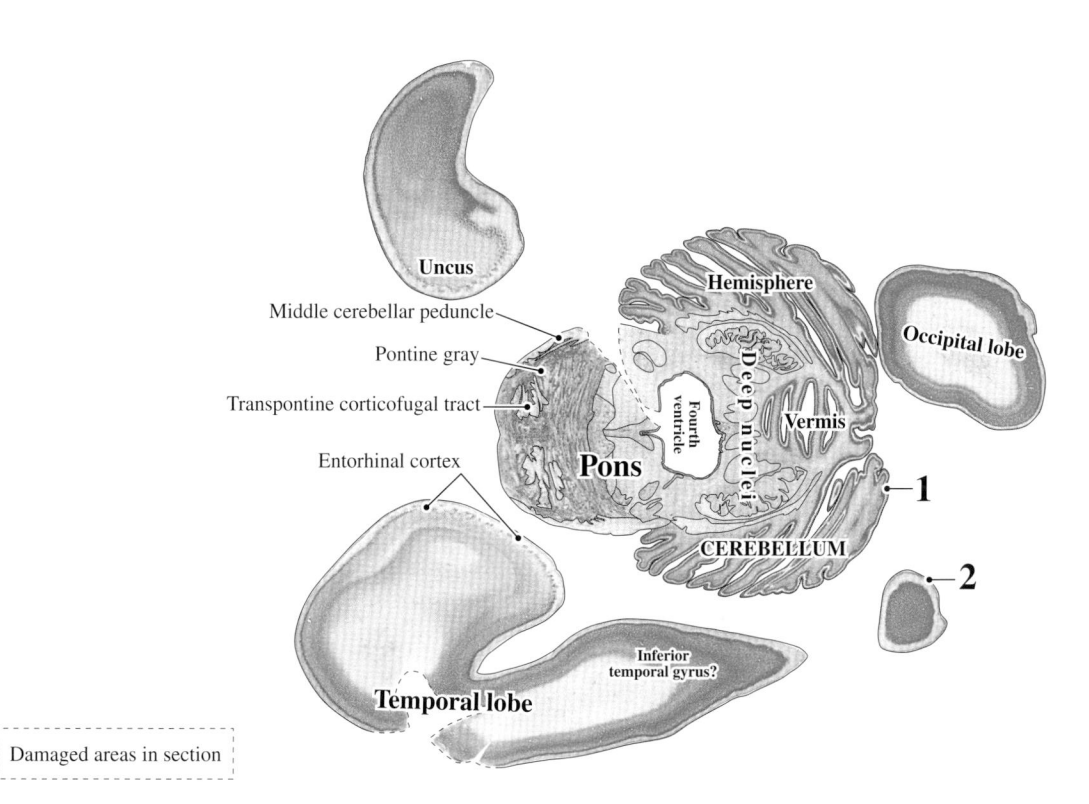

# PLATE 128

**GW30 Horizontal, CR 260 mm, Y187-65**

*Remnants of the germinal matrix, migratory streams, and transitional fields*

1. *External germinal layer (cerebellum)*
2. *Germinal trigone*
3. *Subpial granular layer (cortical)*

**Level 17: Section 1081**

See detail of brain core and cerebellum in Plates 146A and B.

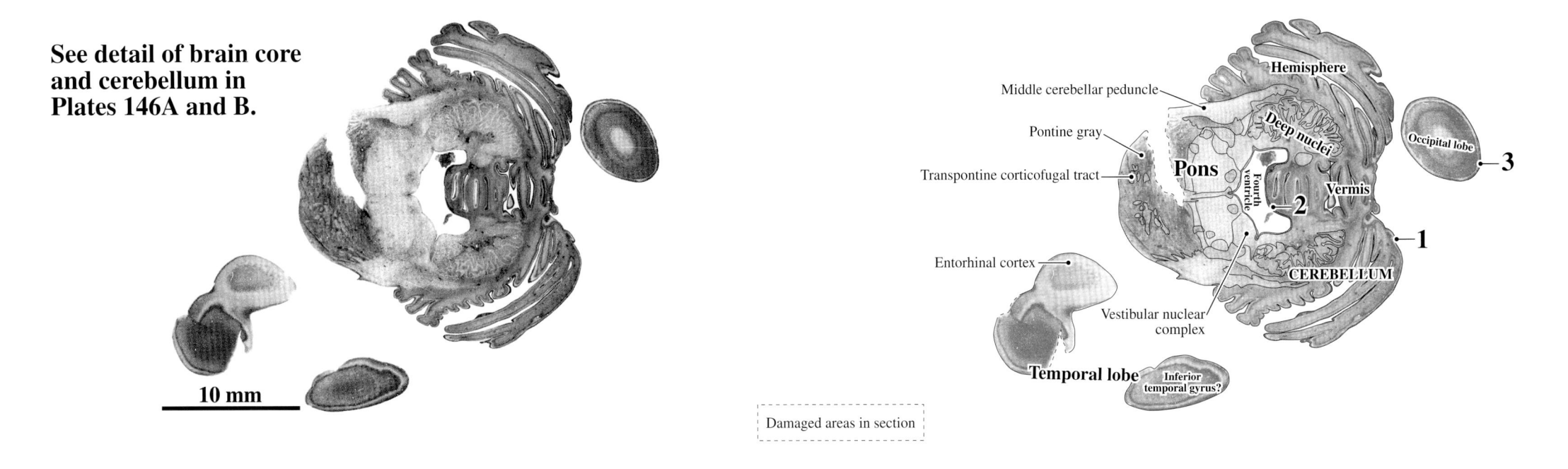

10 mm

Damaged areas in section

Middle cerebellar peduncle — Hemisphere — Deep nuclei — Occipital lobe — 3
Pontine gray — Pons — Fourth ventricle — Vermis
Transpontine corticofugal tract — 2 — 1
Entorhinal cortex — CEREBELLUM
Vestibular nuclear complex
Temporal lobe — Inferior temporal gyrus?

**Level 18: Section 1121**

See high magnification view of this section in Plates 147A and B.

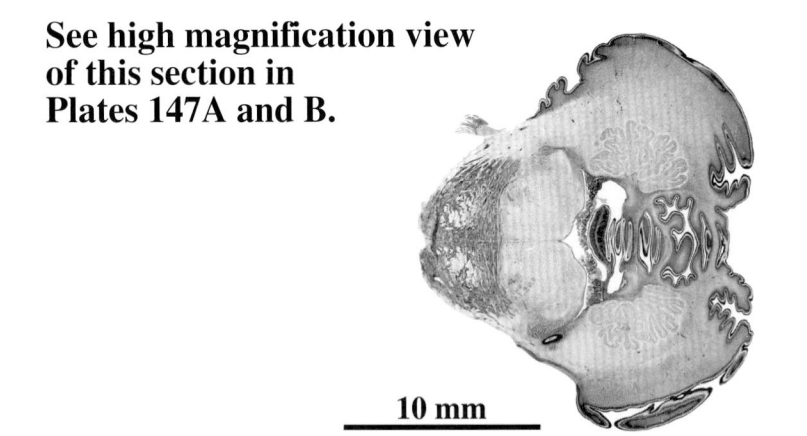

10 mm

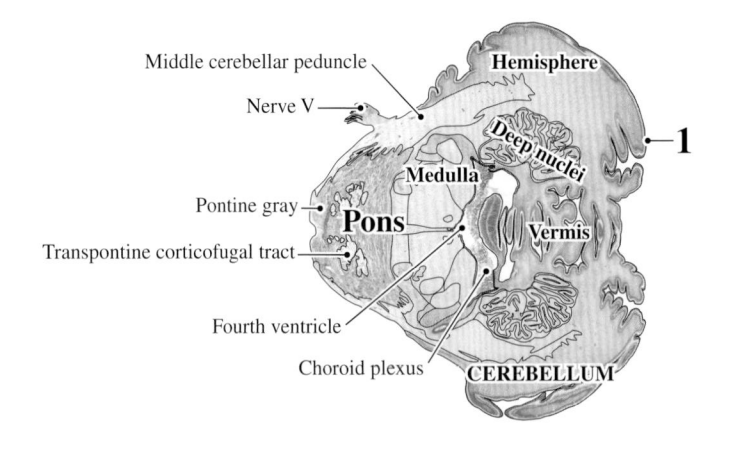

Middle cerebellar peduncle — Hemisphere
Nerve V — Deep nuclei — 1
Medulla
Pontine gray — Pons — Vermis
Transpontine corticofugal tract
Fourth ventricle
Choroid plexus — CEREBELLUM

## PLATE 129

***Remnants of the germinal matrix, migratory streams, and transitional fields***

| 1 | *External germinal layer (cerebellum)* |

GW30 Horizontal, CR 260 mm, Y187-65

**Level 19: Section 1161**

See high magnification view in Plates 148A and B.

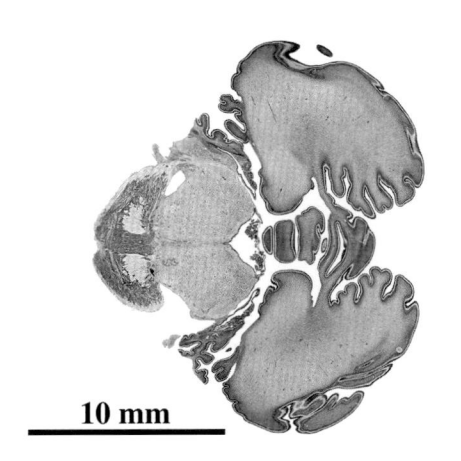

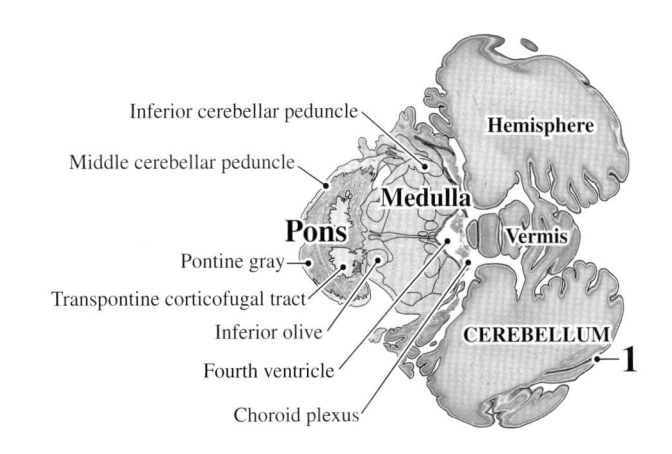

**Level 20: Section 1201**

See high magnification view in Plates 149A and B (left panel).

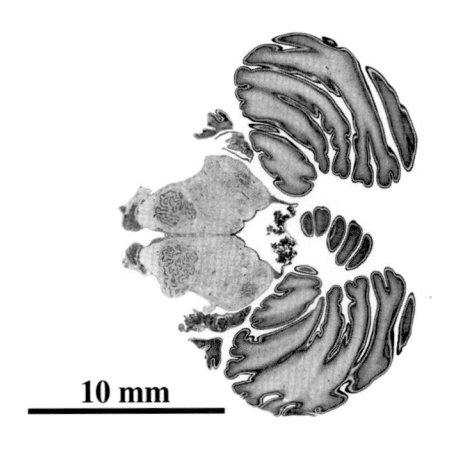

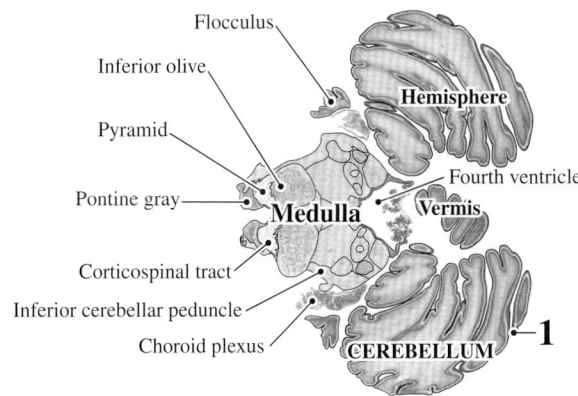

# PLATE 130

**GW30 Horizontal, CR 260 mm, Y187-65**

*Remnants of the germinal matrix, migratory streams, and transitional fields*

| 1 | External germinal layer (cerebellum) |

---

**Level 21: Section 1241**

See high magnification view in Plates 149A and B (right panel).

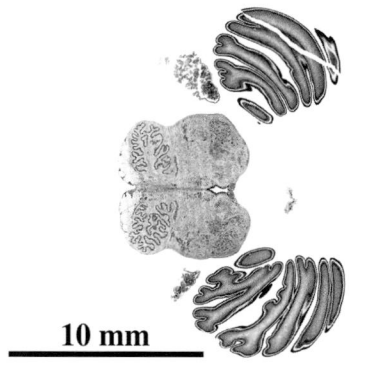

10 mm

Choroid plexus
Inferior cerebellar peduncle and spinocerebellar tracts
Inferior olive
Pyramid
Corticospinal tract
Solitary nuclear complex
**Hemisphere**
**Medulla**
**CEREBELLUM** —1

Damaged areas in section

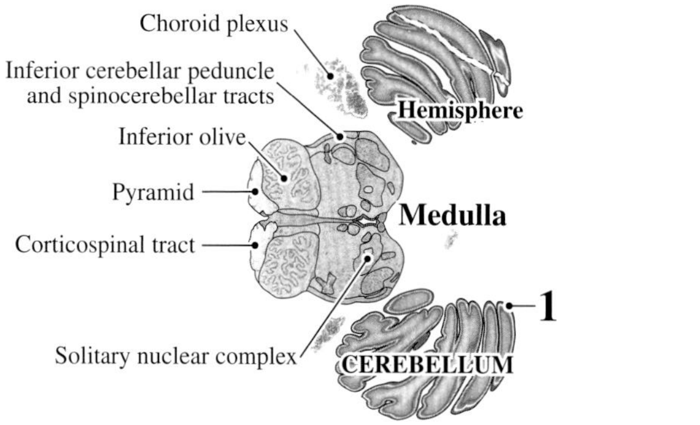

---

**Level 22: Section 1281**

See high magnification view in Plates 150A and B (left panel).

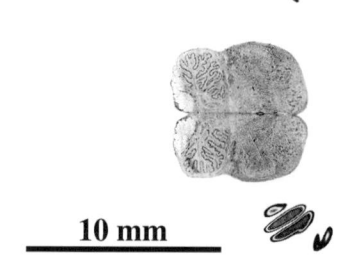

10 mm

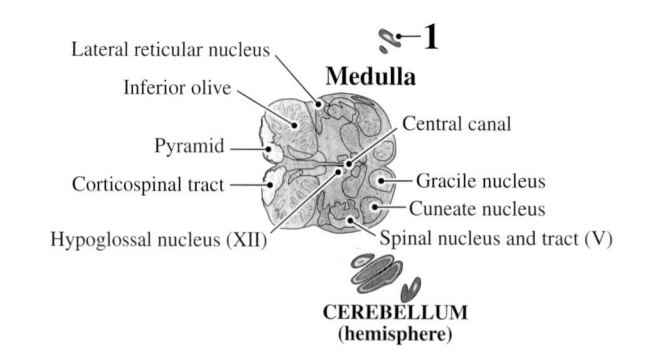

Lateral reticular nucleus
Inferior olive
Pyramid
Corticospinal tract
Hypoglossal nucleus (XII)
**Medulla** —1
Central canal
Gracile nucleus
Cuneate nucleus
Spinal nucleus and tract (V)
**CEREBELLUM** (hemisphere)

# PLATE 131

*Remnants of the germinal matrix, migratory streams, and transitional fields*

GW30 Horizontal, CR 260 mm, Y187-65

| 1 | *Dorsal funicular myelination gliosis* |
|---|---|
| 2 | *Ventral funicular myelination gliosis* |

**Level 23: Section 1351**

**See high magnification view in Plates 150A and B (middle panel).**

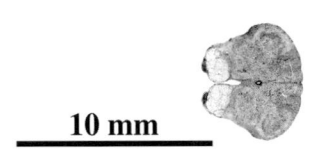

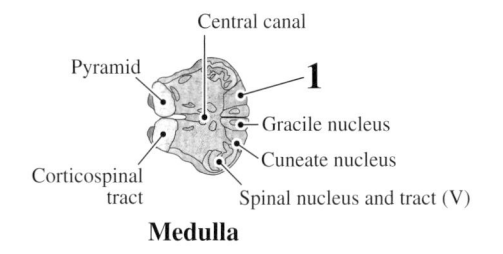

**Medulla**

**Level 24: Section 1391**

**See high magnification view in Plates 150A and B (right panel).**

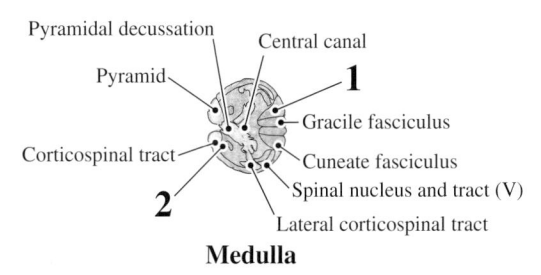

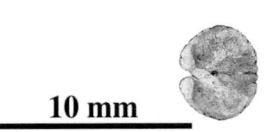

**Medulla**

## PLATE 132A

**GW30 Horizontal
CR 260 mm
Y187-65
Level 3: Section 441**

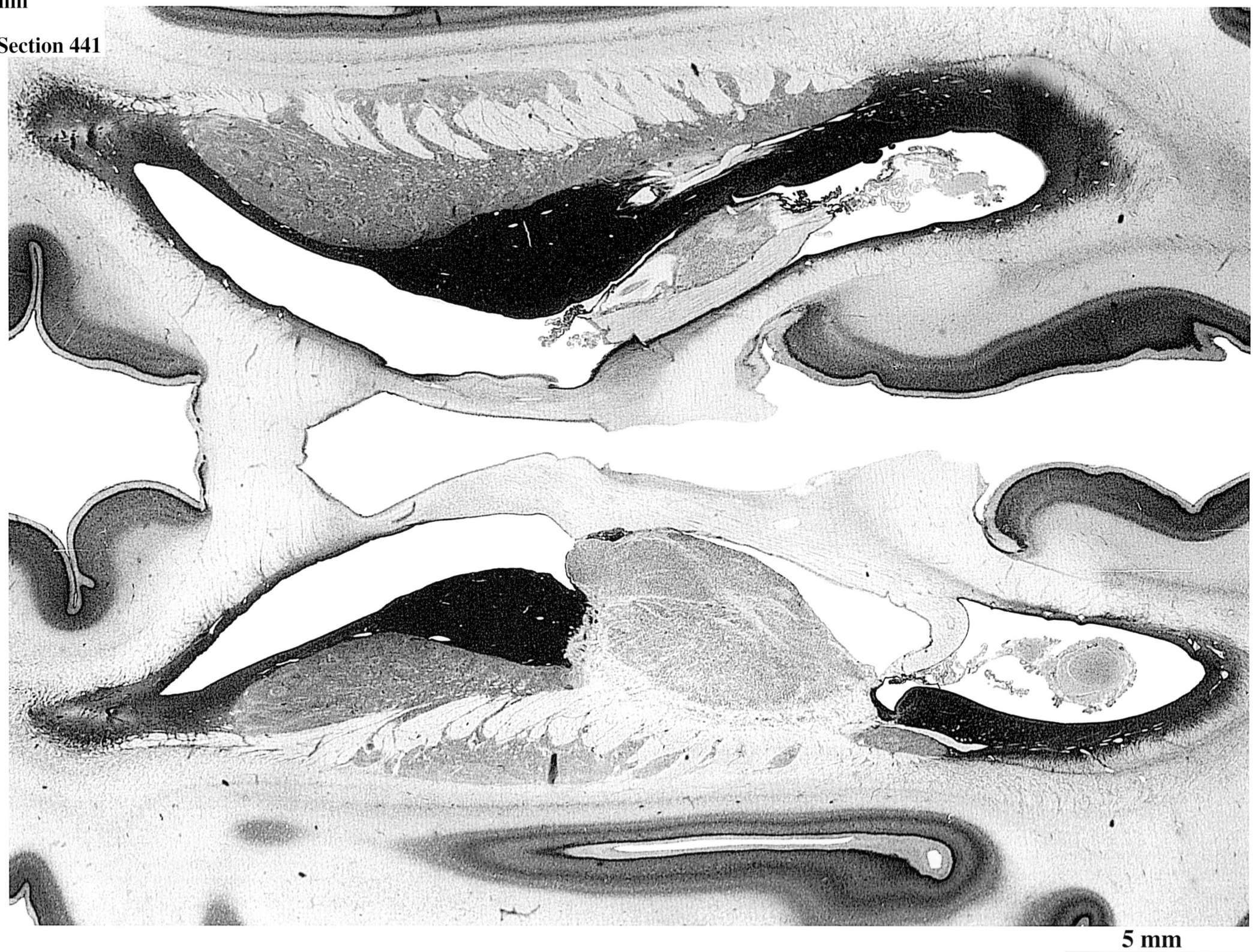

See the entire section
in Plate 114.

5 mm

## PLATE 132B

**Germinal and transitional structures in *italics***

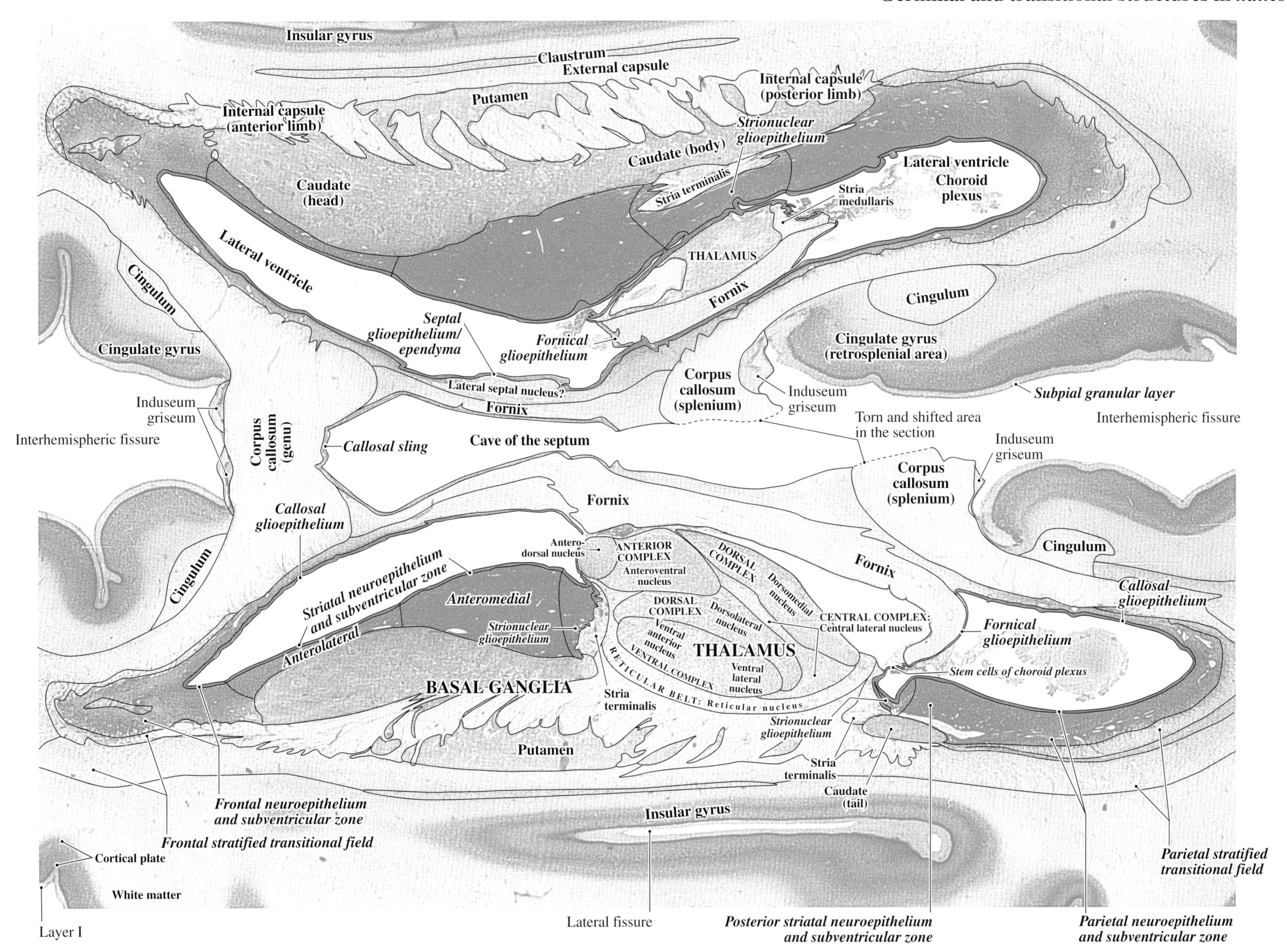

# PLATE 133A

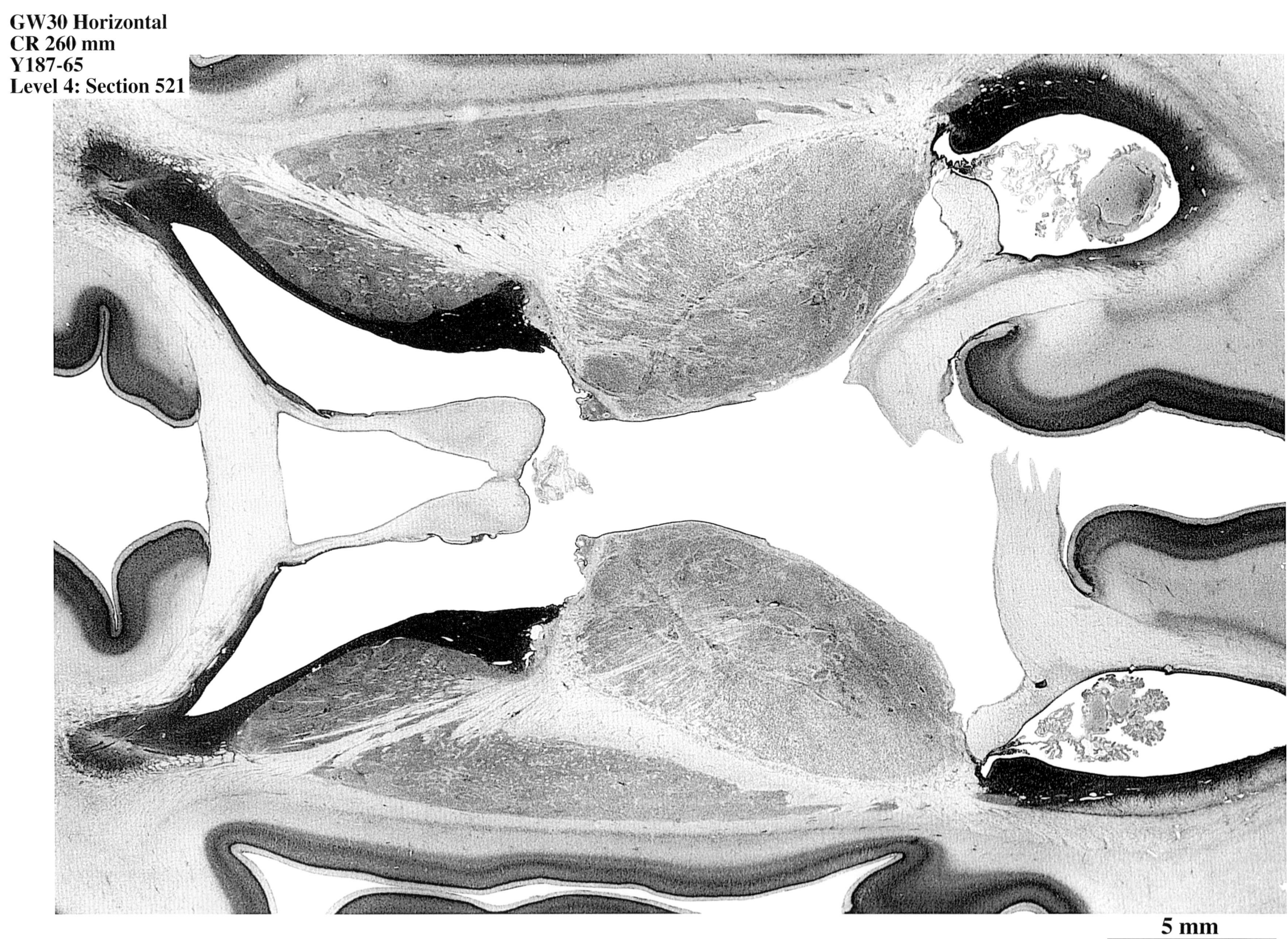

**GW30 Horizontal
CR 260 mm
Y187-65
Level 4: Section 521**

5 mm

See the entire section in Plate 115.

## PLATE 133B

**Germinal and transitional structures in *italics***

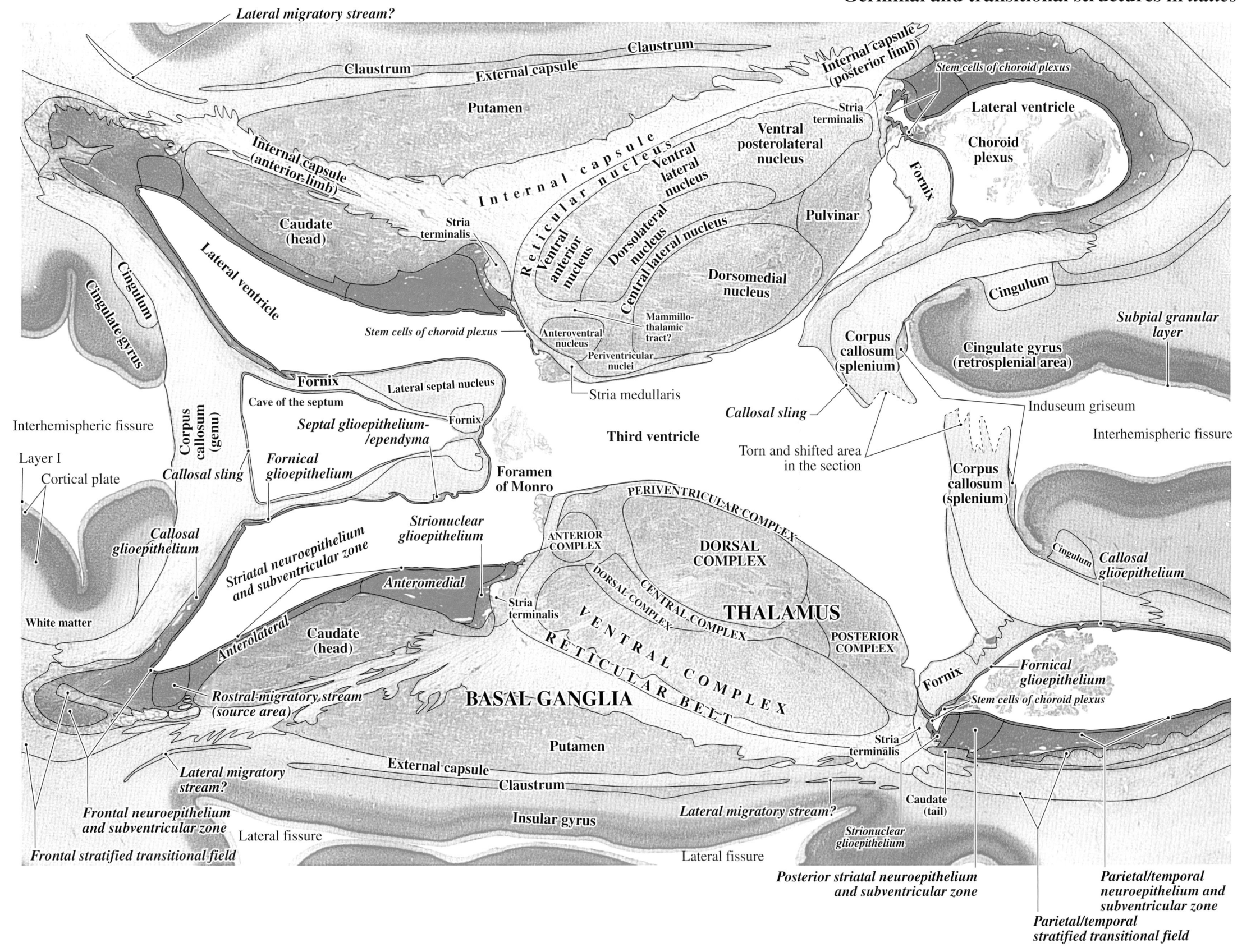

## PLATE 134A

GW30 Horizontal
CR 260 mm
Y187-65
Level 5: Section 621

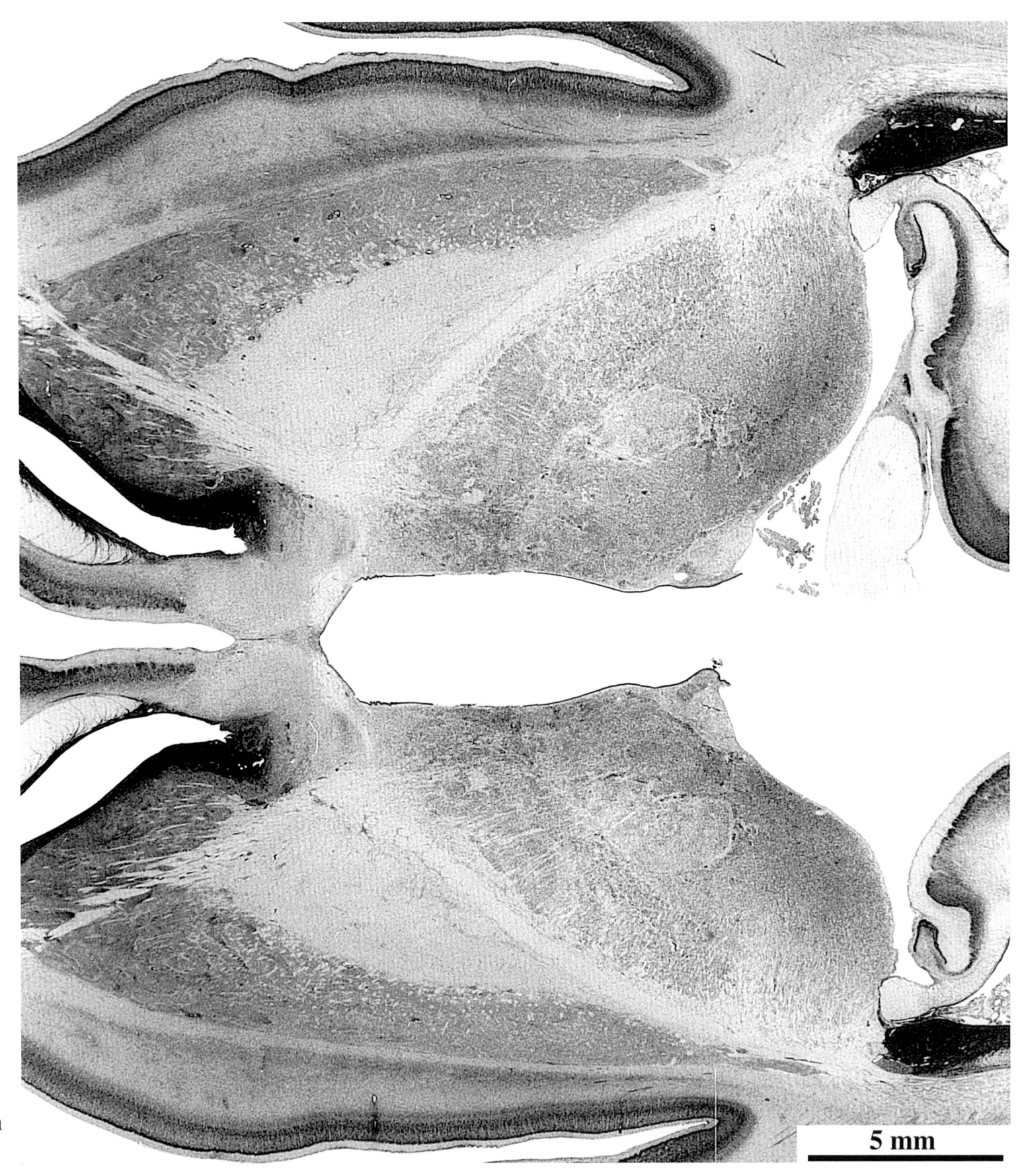

See the entire section in Plate 116.

5 mm

## PLATE 134B

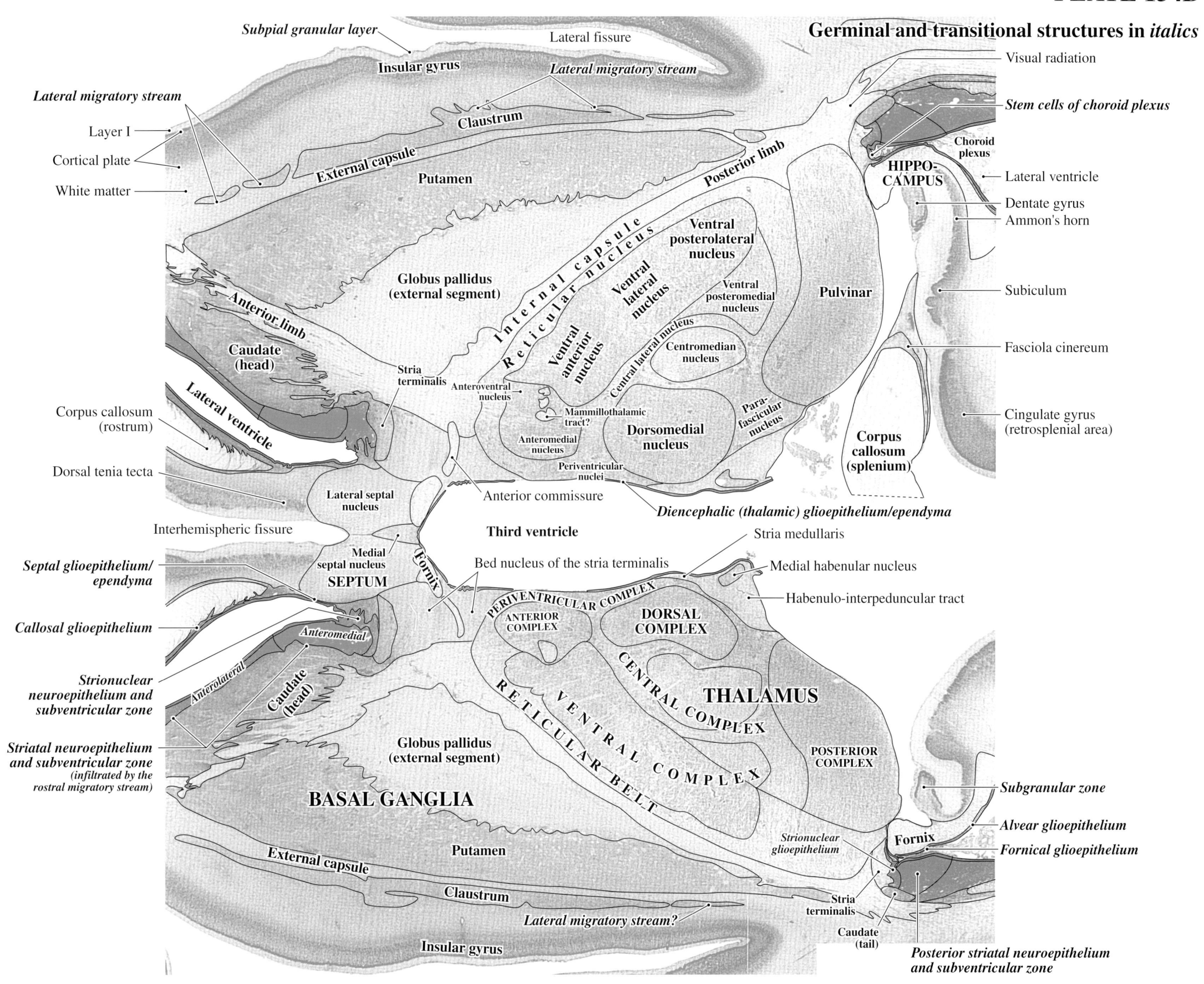

**Germinal and transitional structures in *italics***

## PLATE 135A

GW30 Horizontal
CR 260 mm
Y187-65
Level 6: Section 671

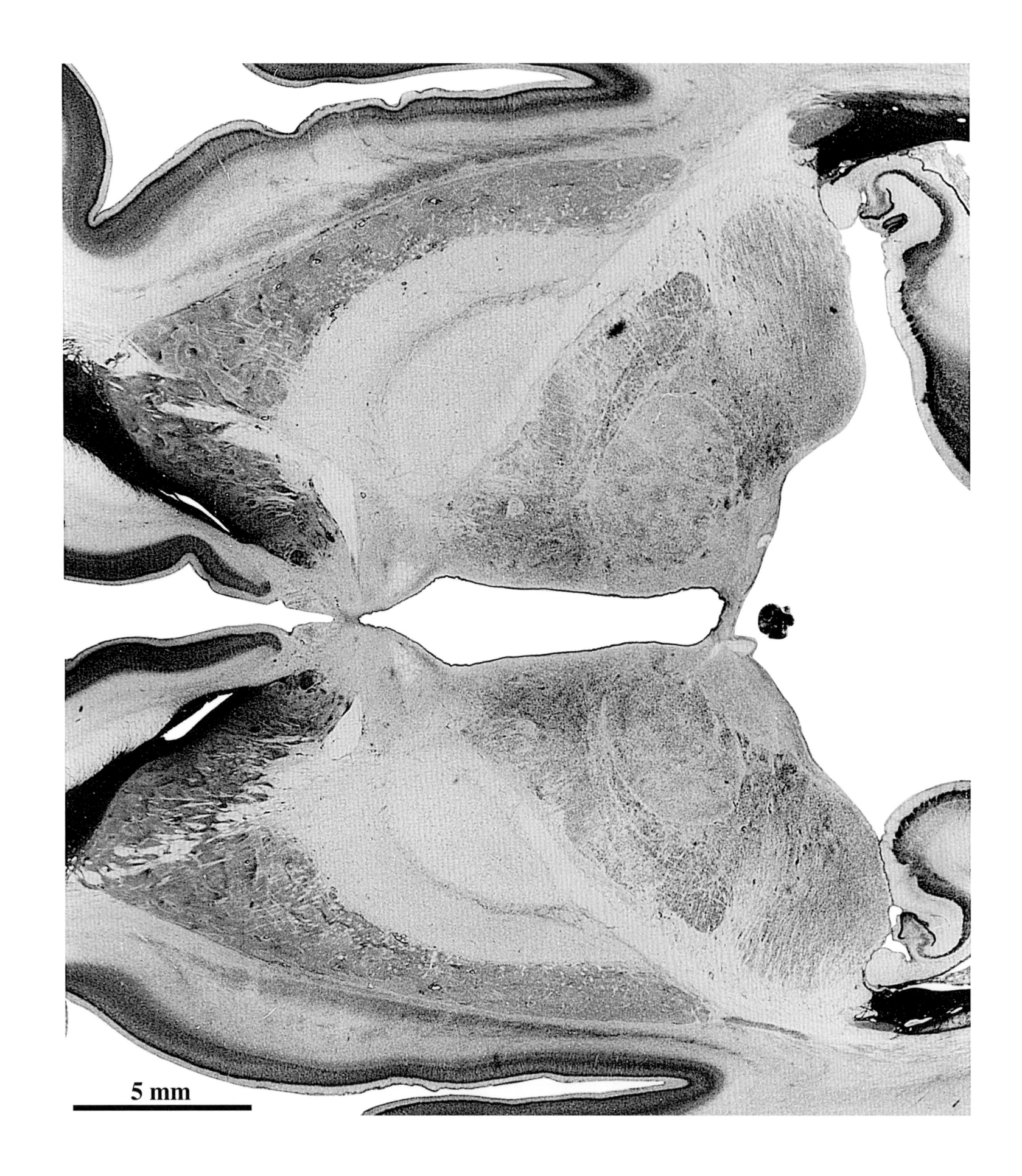

See the entire section in Plate 117.

5 mm

PLATE 135B

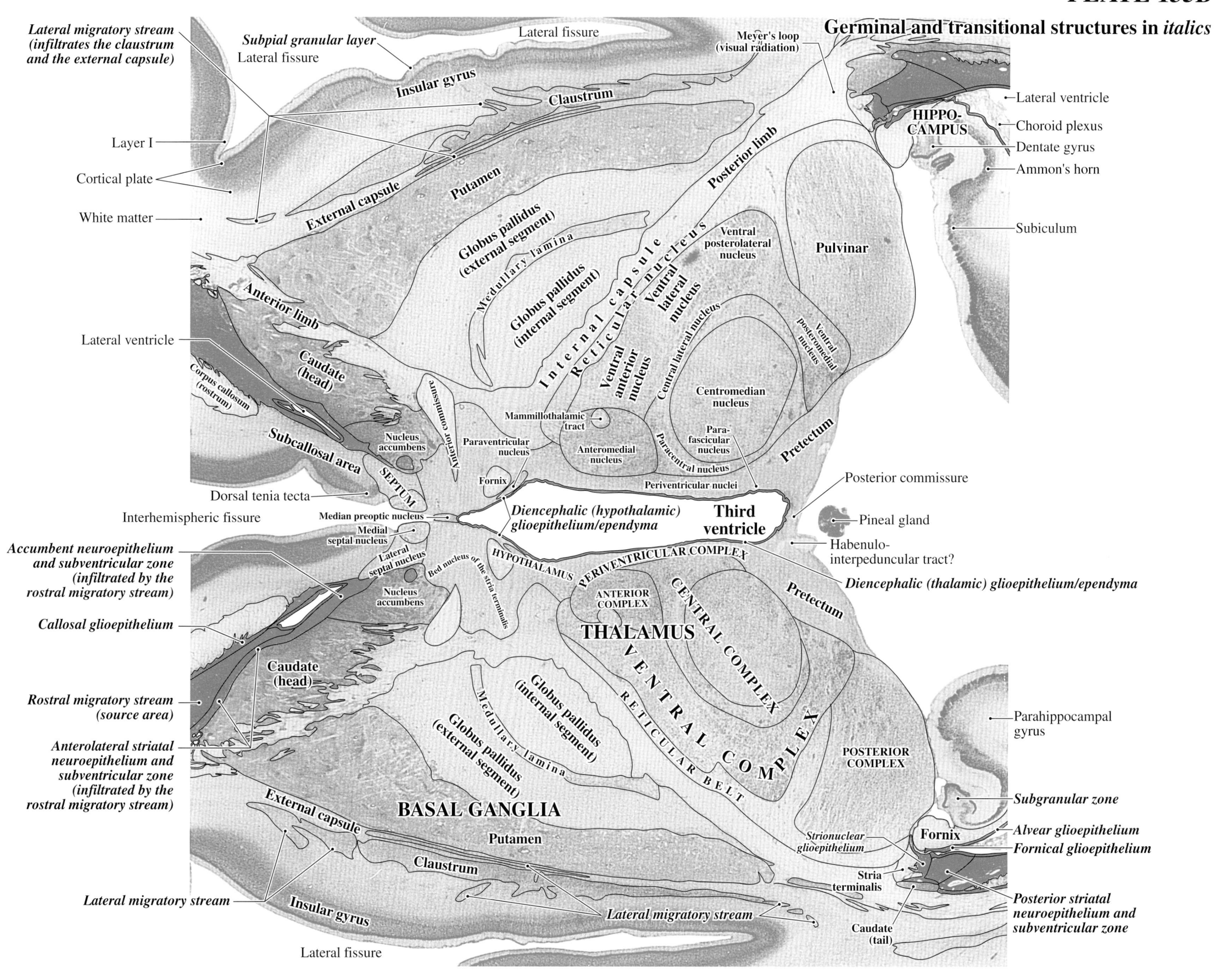

**Germinal and transitional structures in *italics***

**PLATE 136A**

GW30 Horizontal
CR 260 mm
Y187-65
Level 7: Section 701

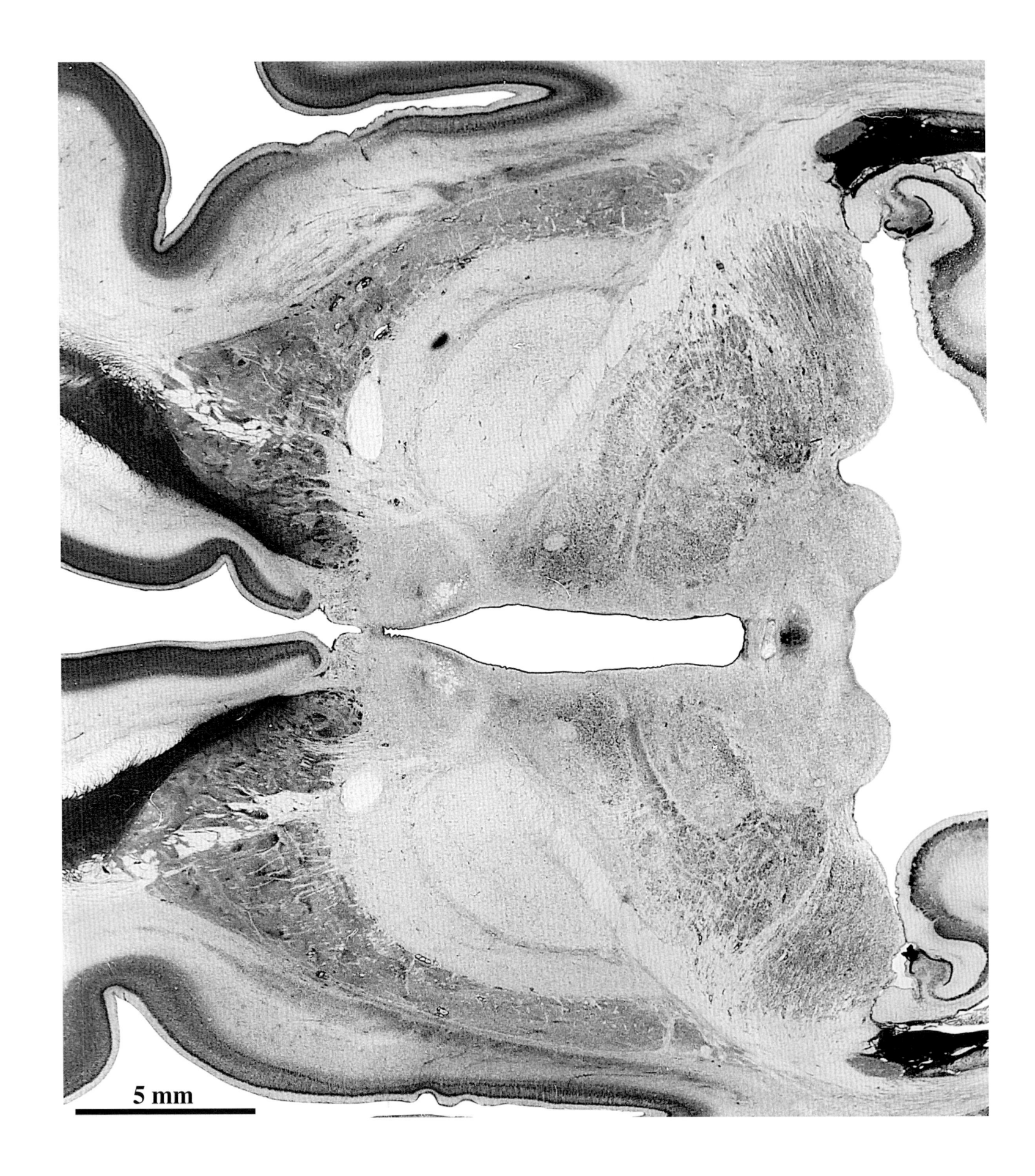

See the entire section in Plate 118.

5 mm

**PLATE 136B**

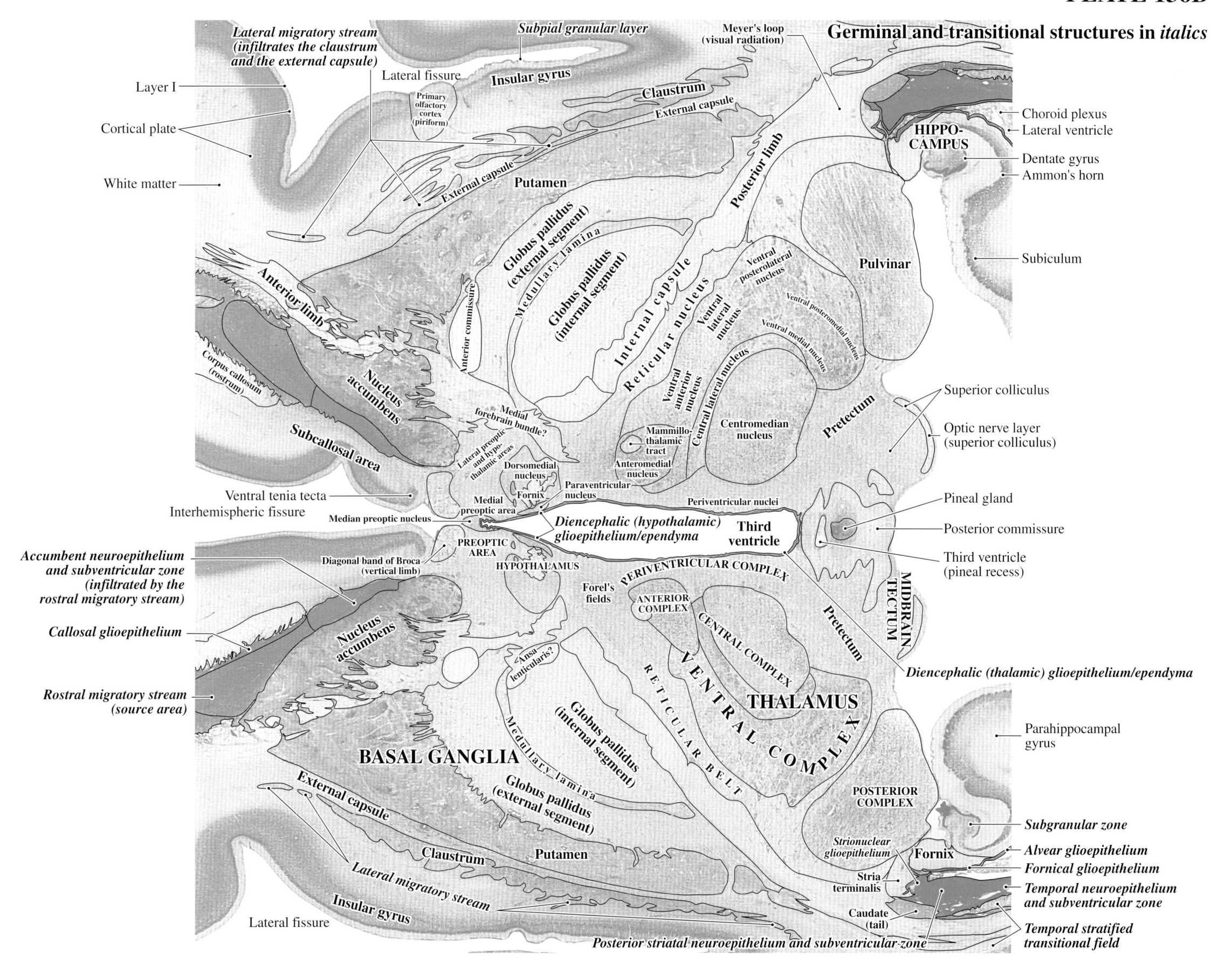

**PLATE 137A**

GW30 Horizontal
CR 260 mm
Y187-65
Level 8: Section 721

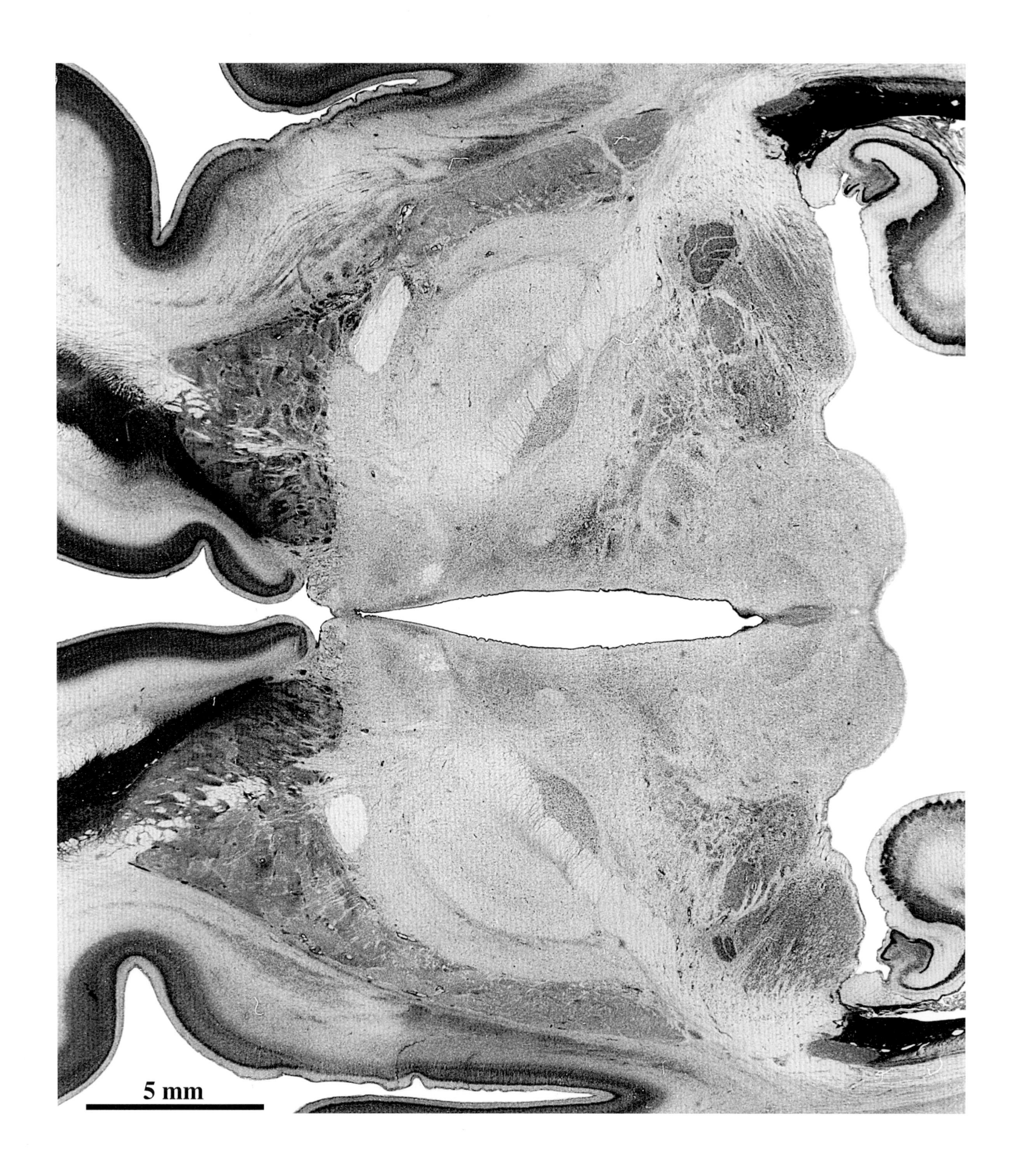

See the entire section in Plate 119.

5 mm

## PLATE 137B

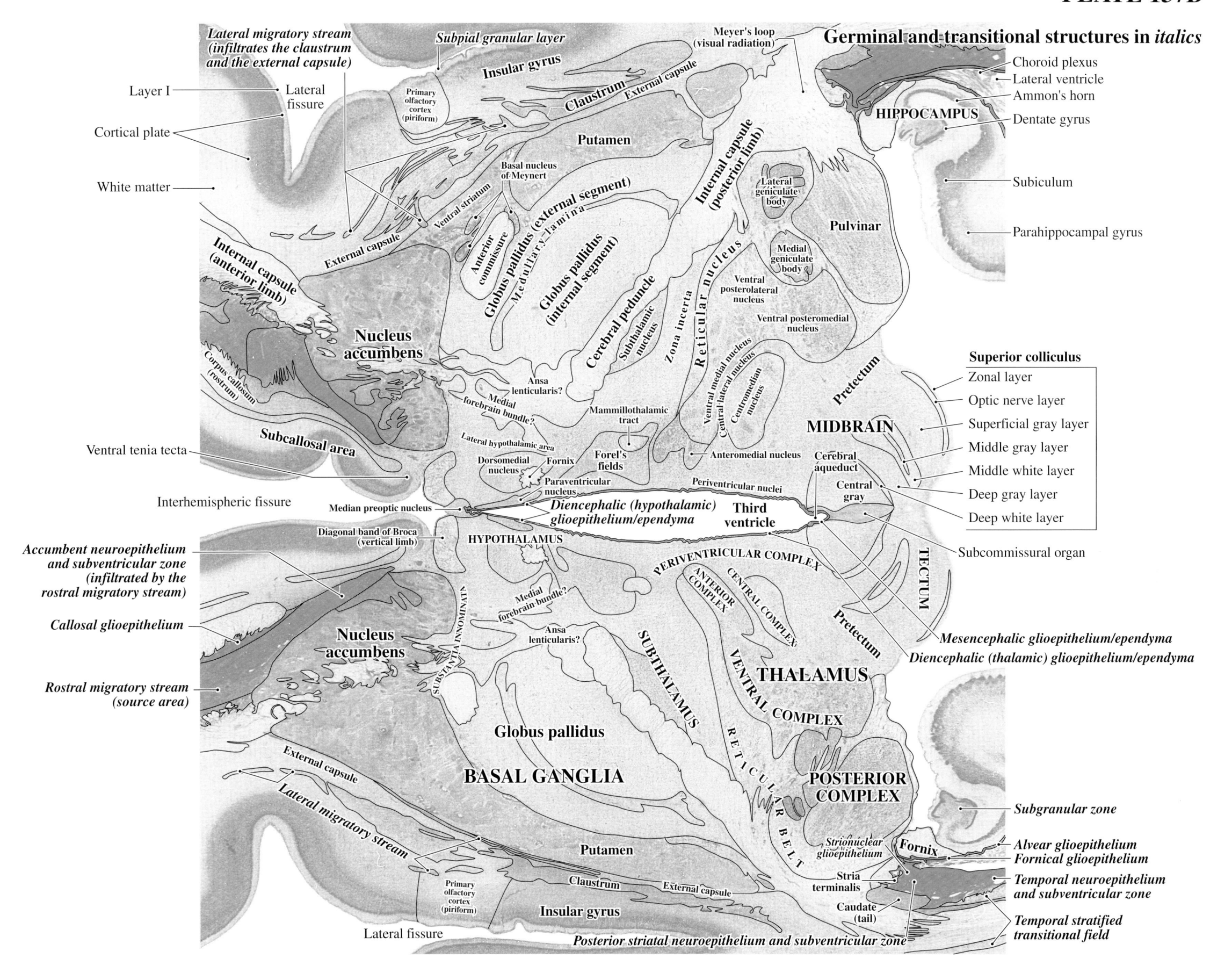

## PLATE 138A

**GW30 Horizontal
CR 260 mm
Y187-65
Level 9: Section 761**

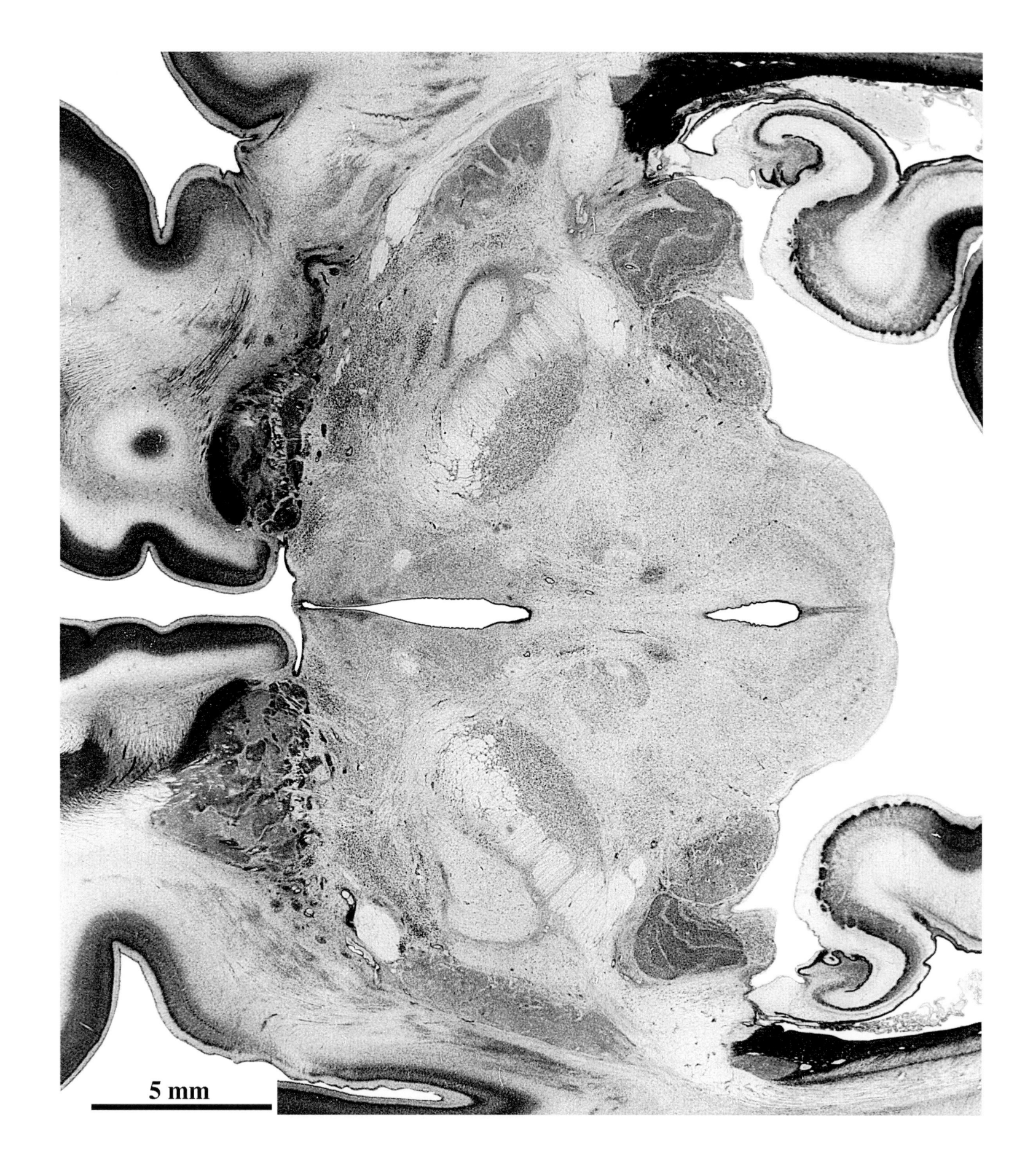

See the entire section in Plate 120.

5 mm

**PLATE 138B**

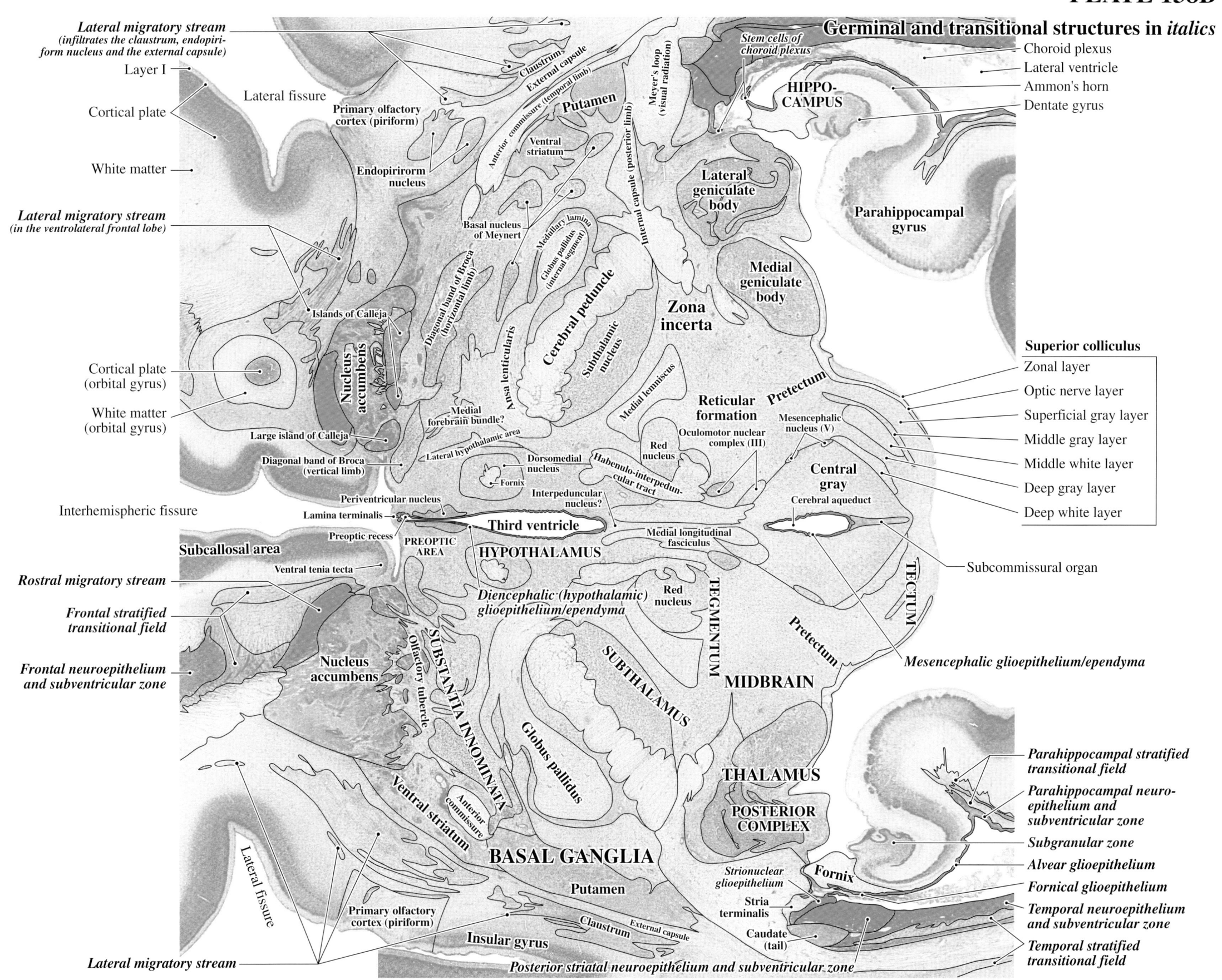

# PLATE 139A

**GW30 Horizontal
CR 260 mm
Y187-65
Level 10: Section 801**

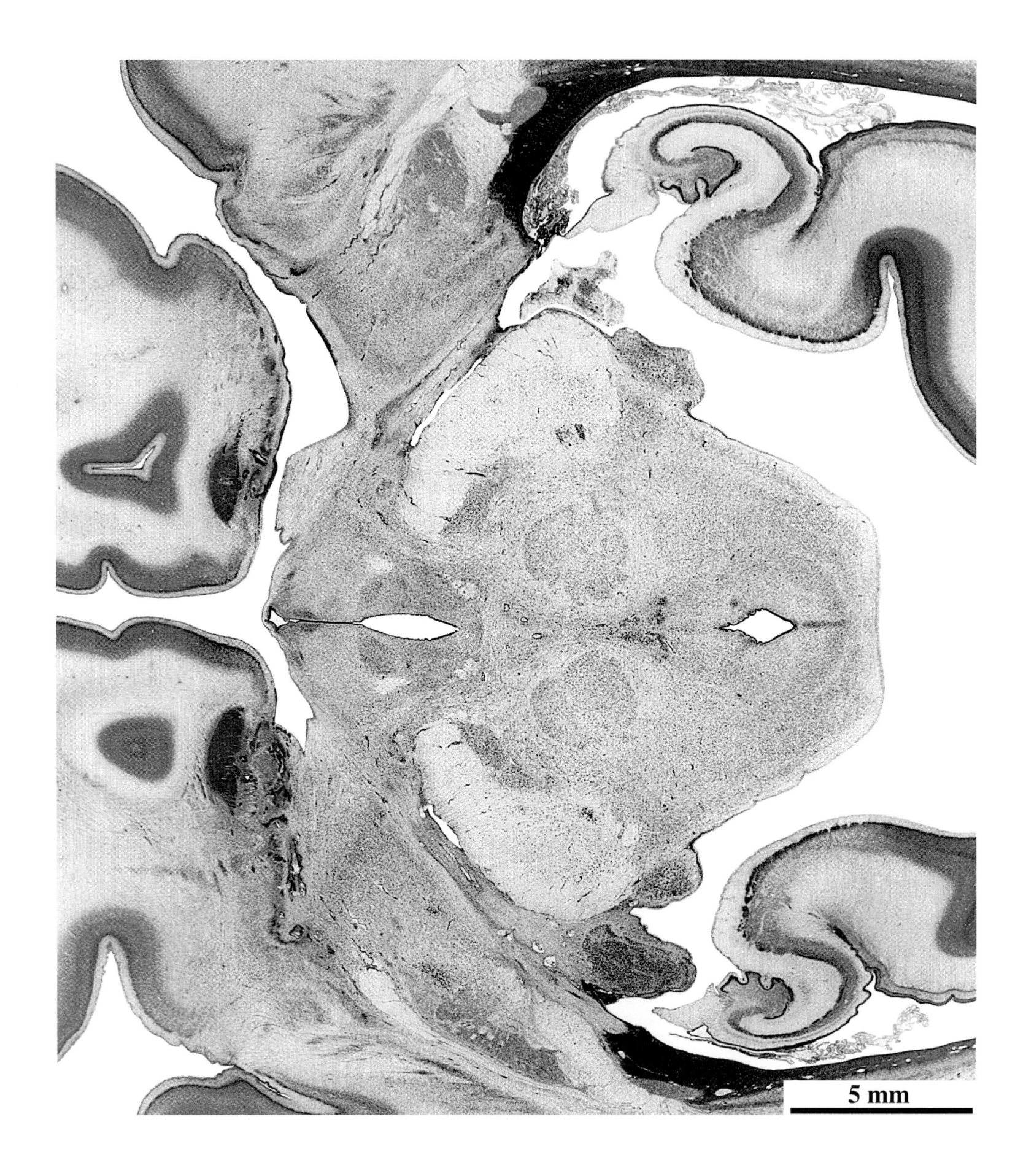

See the entire section in Plate 121.

5 mm

**PLATE 139B**

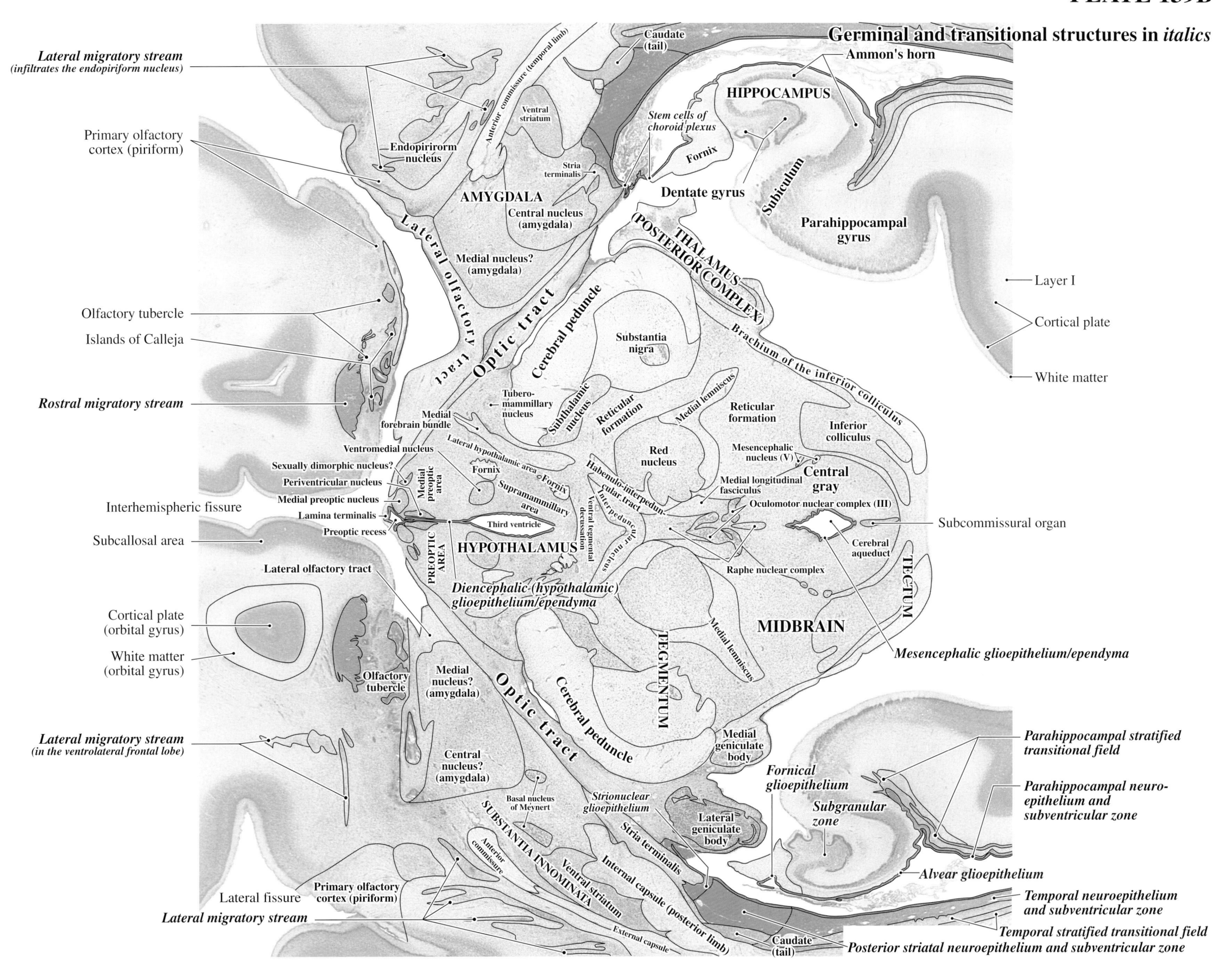

**Germinal and transitional structures in *italics***

… 206

**PLATE 140A**

**GW30 Horizontal
CR 260 mm
Y187-65
Level 11: Section 841**

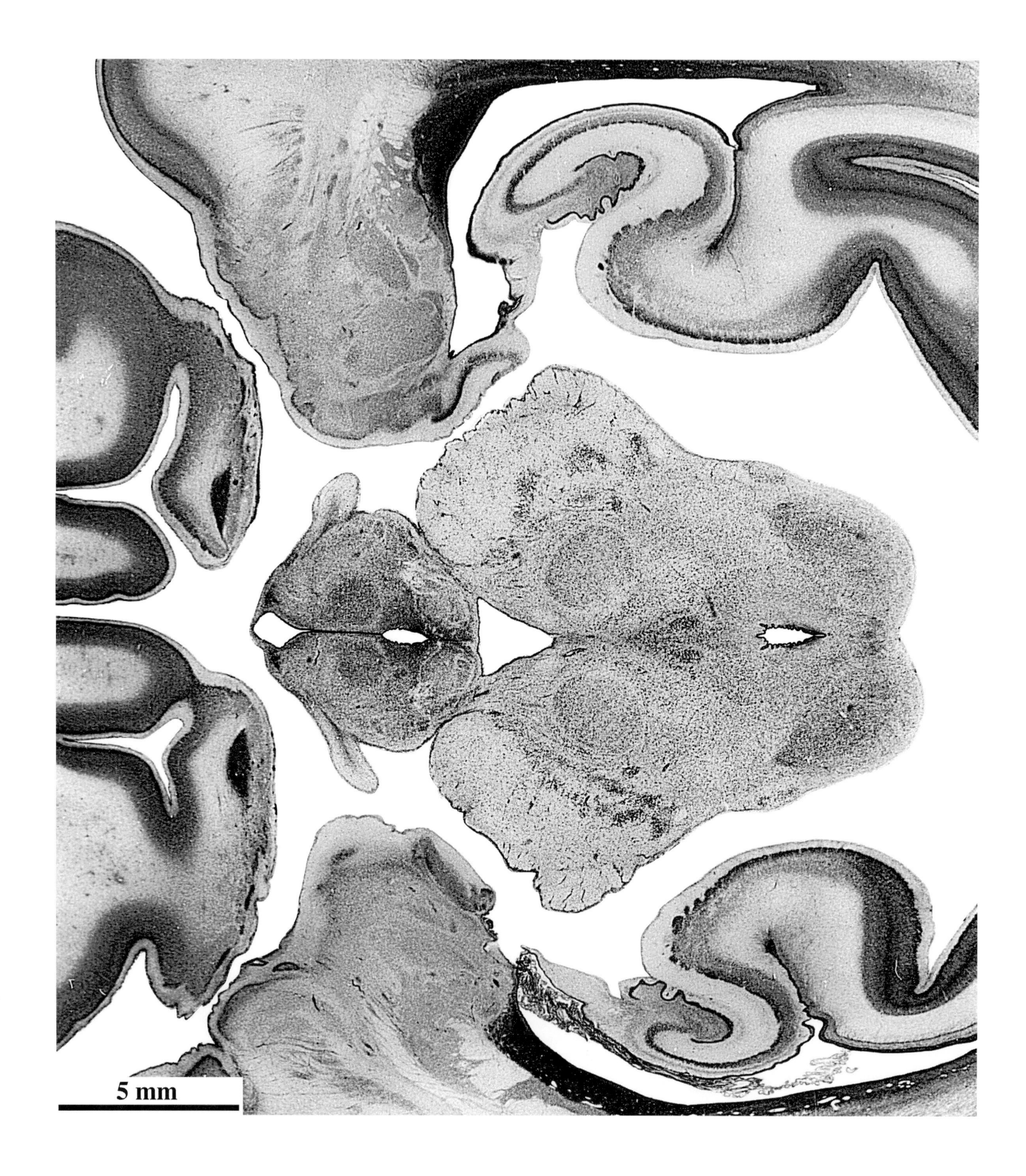

See the entire section in Plate 122.

5 mm

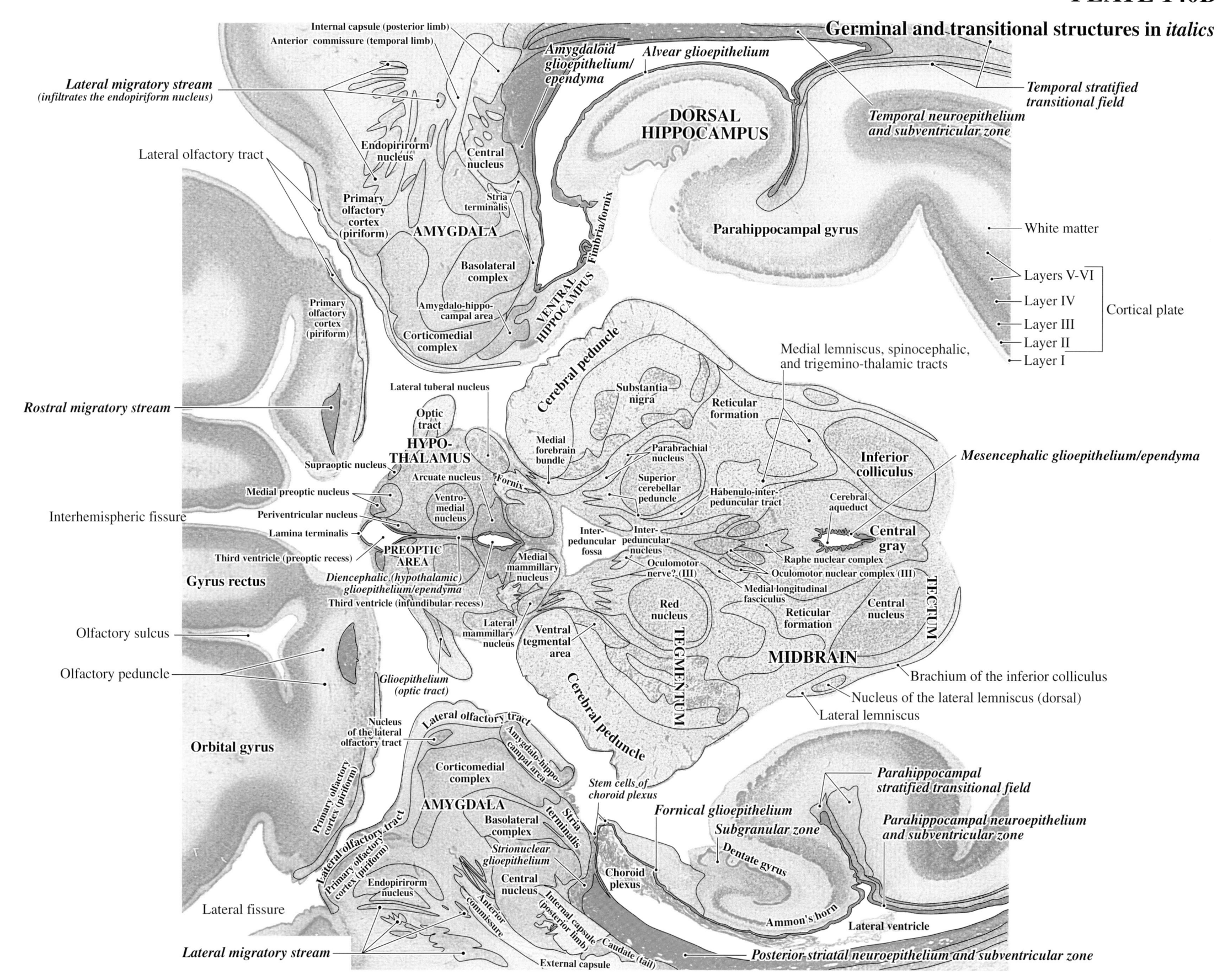

**PLATE 140B**

**Germinal and transitional structures in *italics***

## PLATE 141A

**GW30 Horizontal
CR 260 mm, Y187-65
Level 12: Section 861**

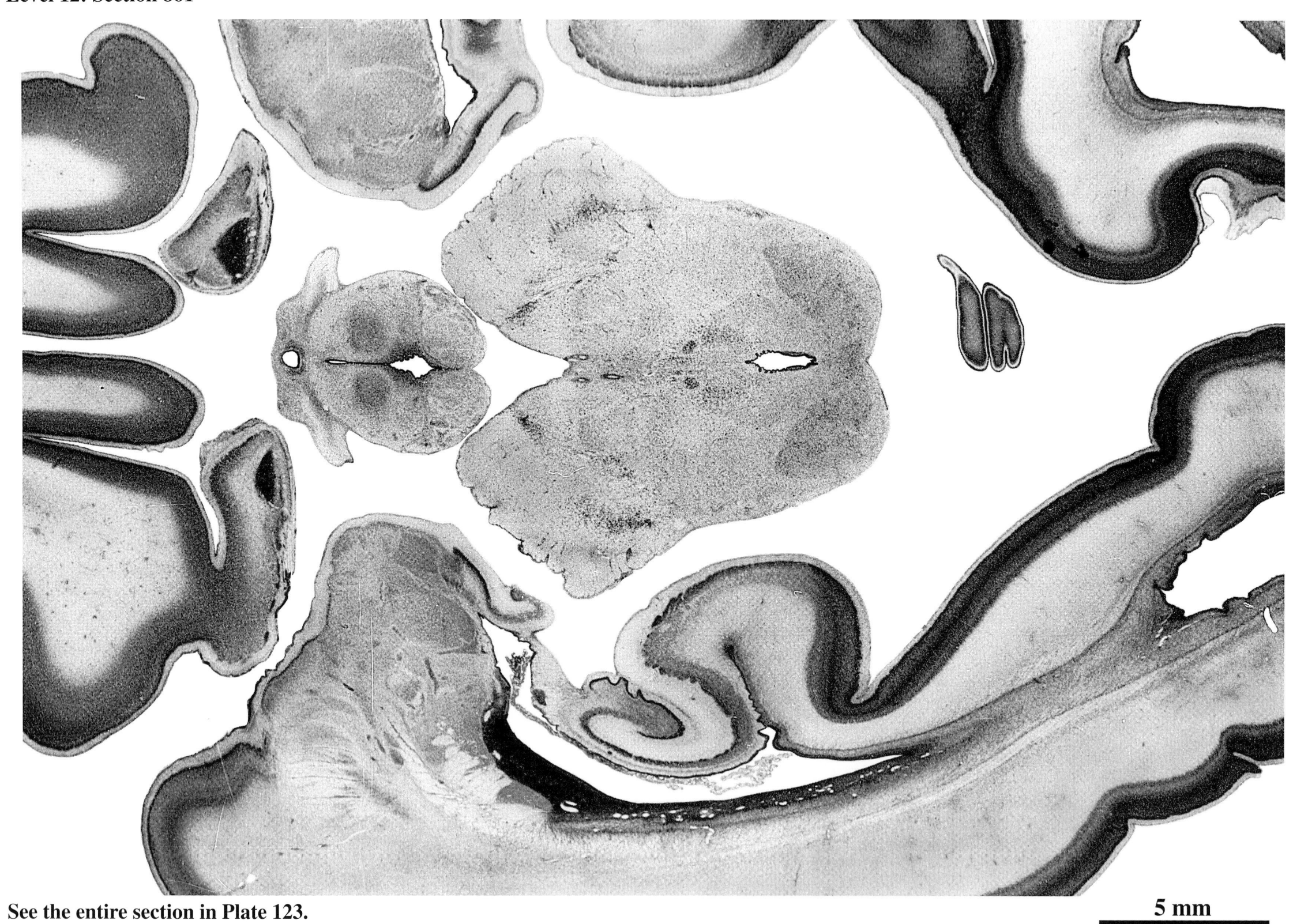

See the entire section in Plate 123.

5 mm

## PLATE 141B

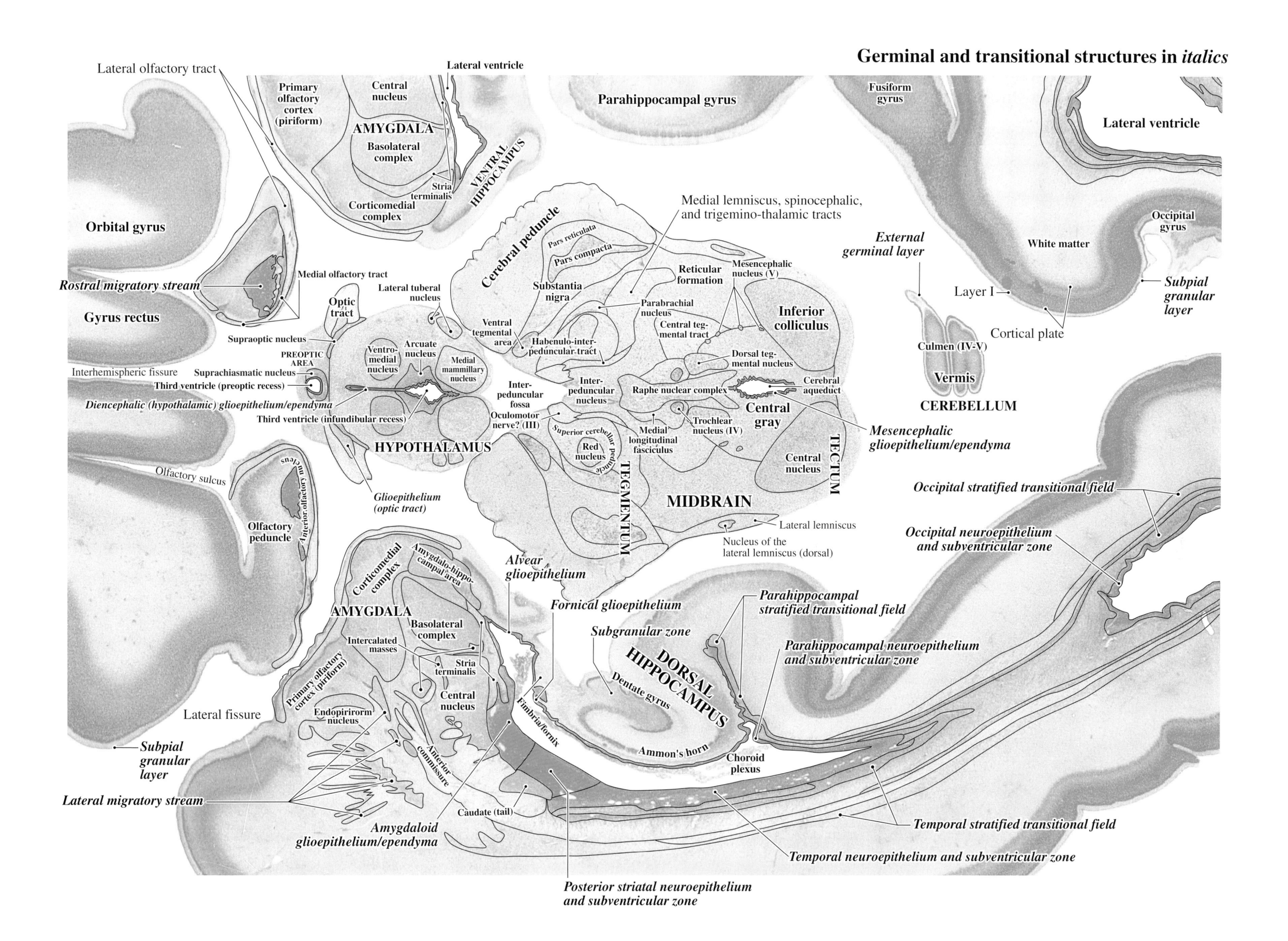

**Germinal and transitional structures in *italics***

**PLATE 142A**

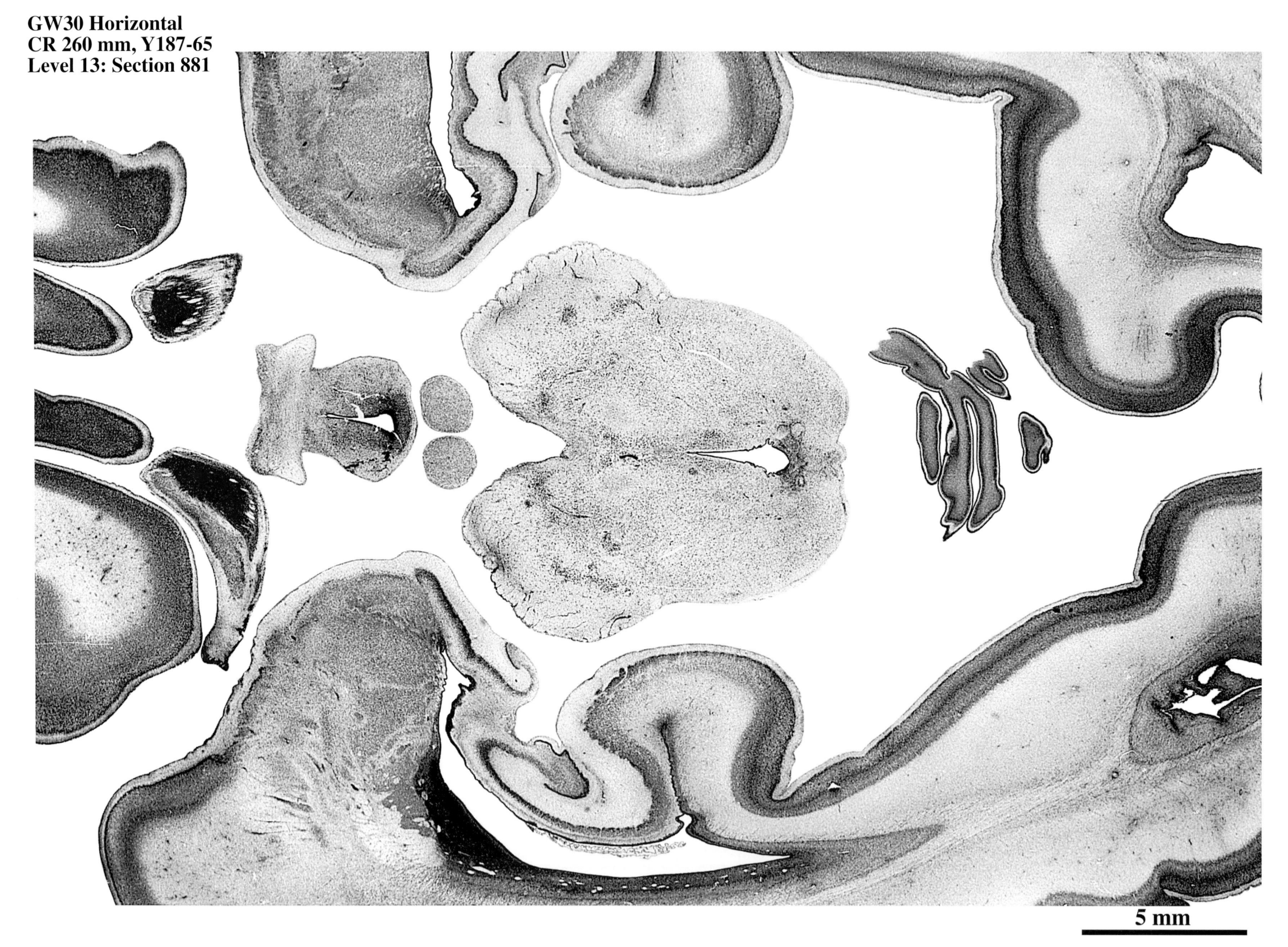

GW30 Horizontal
CR 260 mm, Y187-65
Level 13: Section 881

5 mm

See the entire section in Plate 124.

# PLATE 142B

**Germinal and transitional structures in *italics***

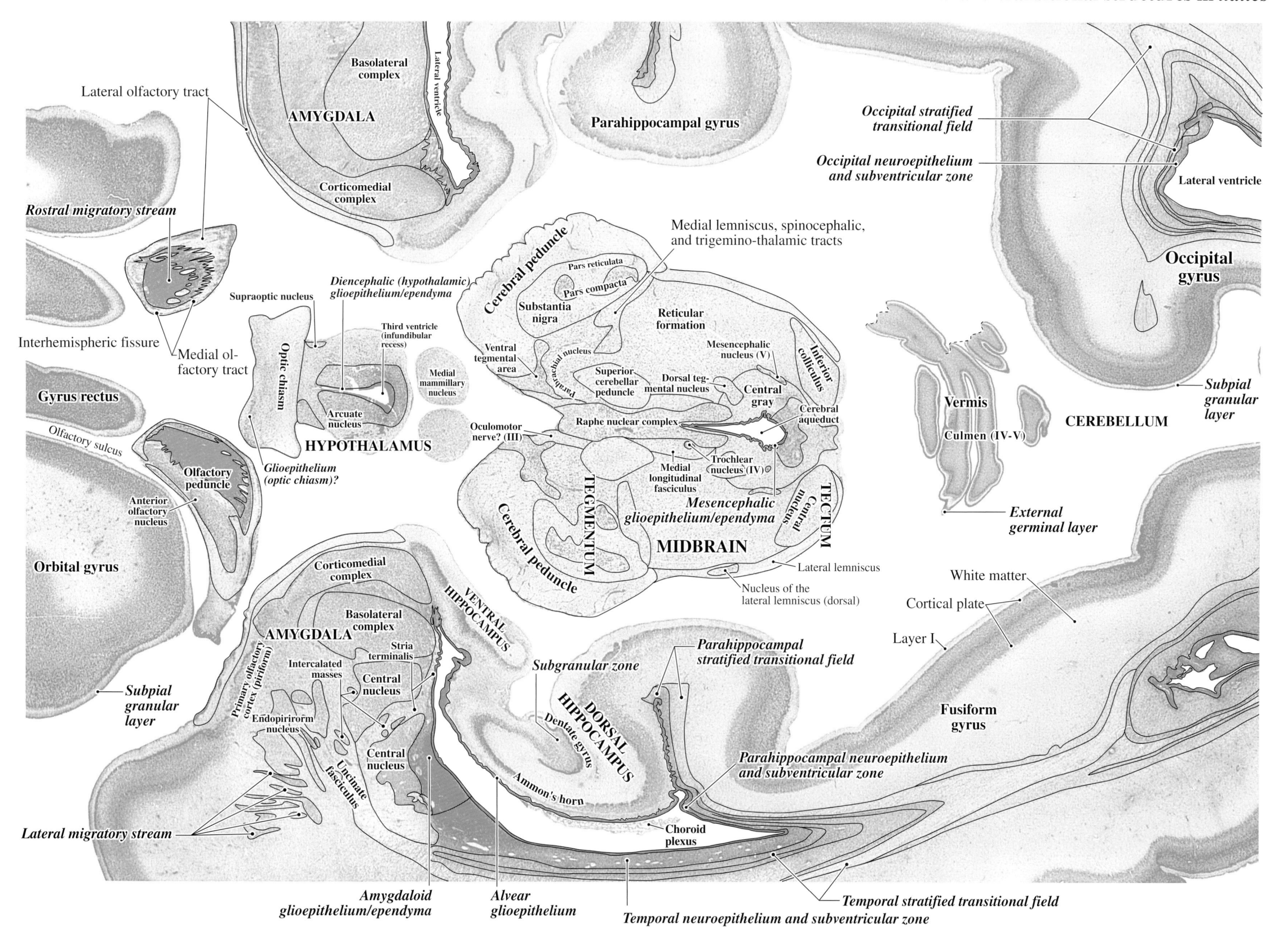

## PLATE 143A

**GW30 Horizontal
CR 260 mm, Y187-65
Level 14: Section 941**

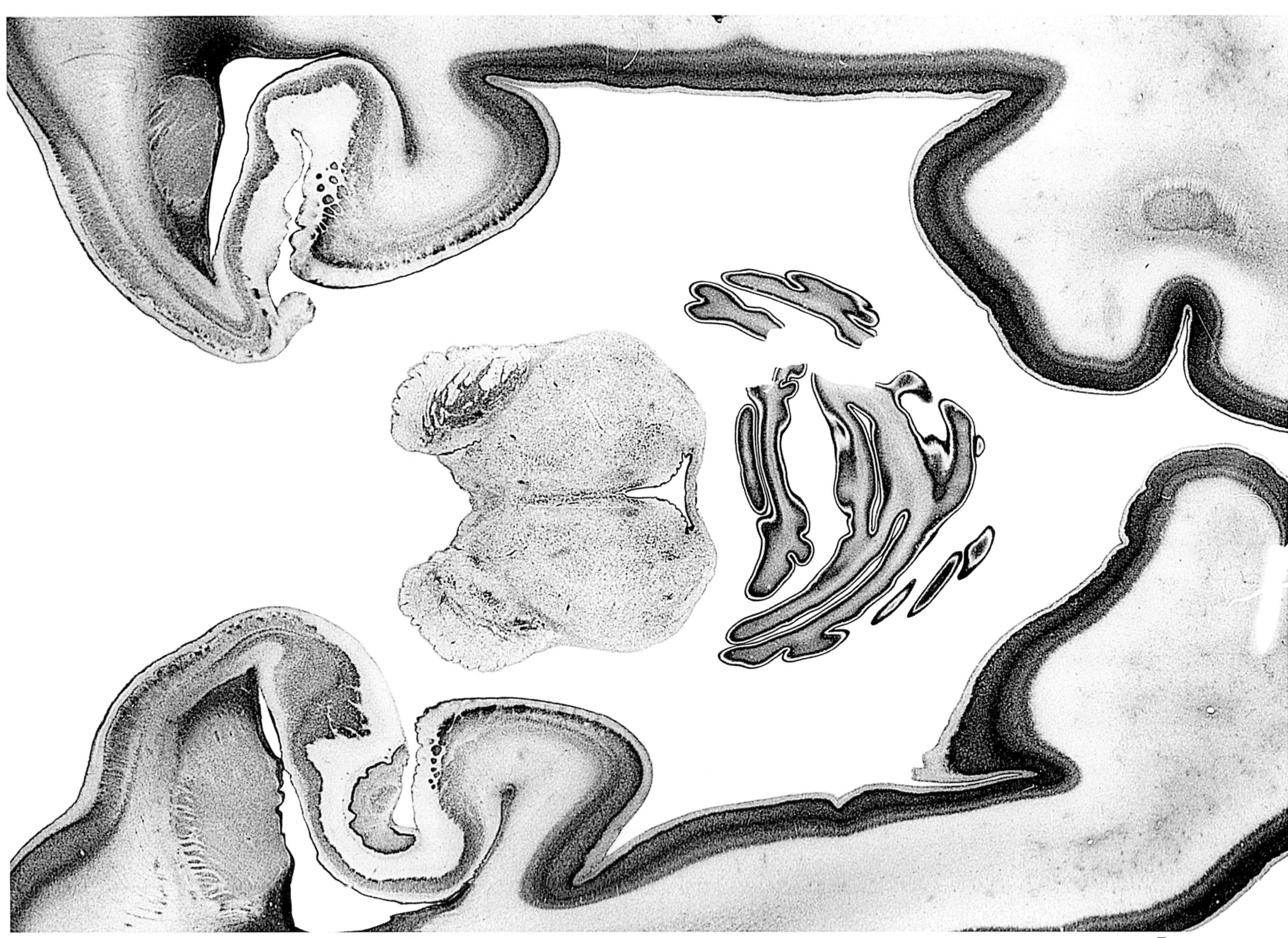

See the entire section in Plate 125.

5 mm

**PLATE 143B**

**Germinal and transitional structures in *italics***

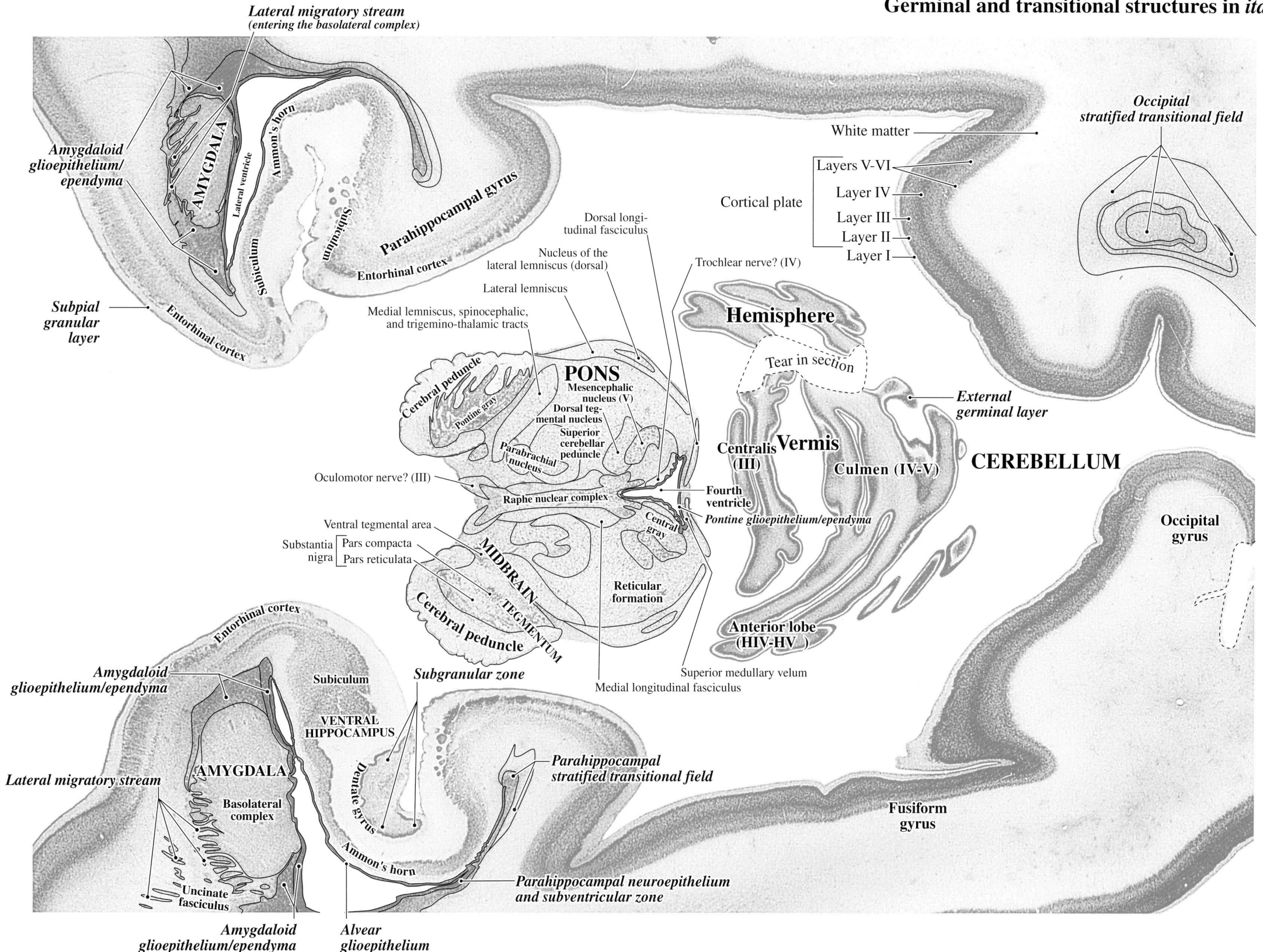

## PLATE 144A

**GW30 Horizontal**
**CR 260 mm, Y187-65**
**Level 15: Section 1001**

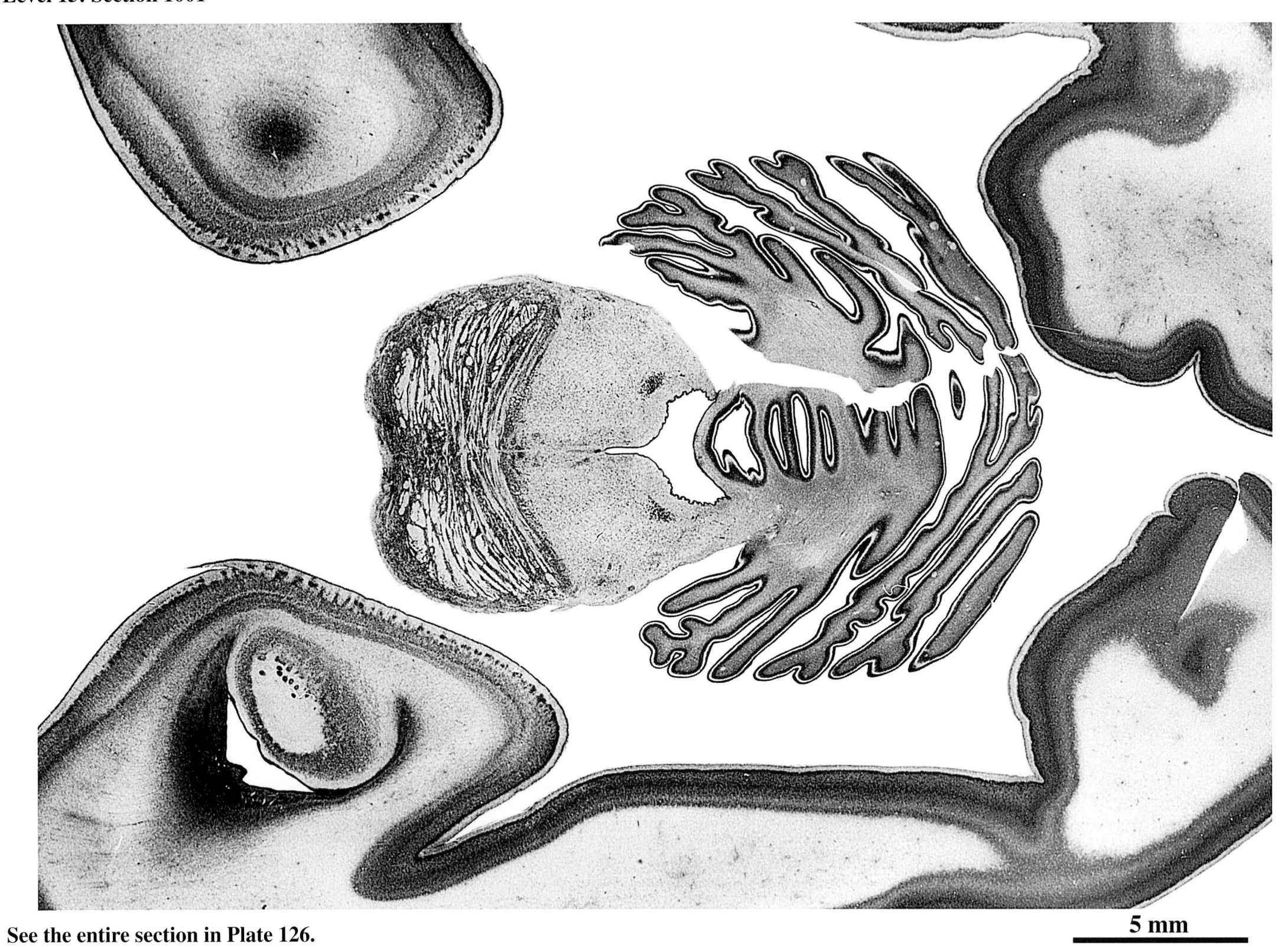

See the entire section in Plate 126.

5 mm

## PLATE 144B
**Germinal and transitional structures in *italics***

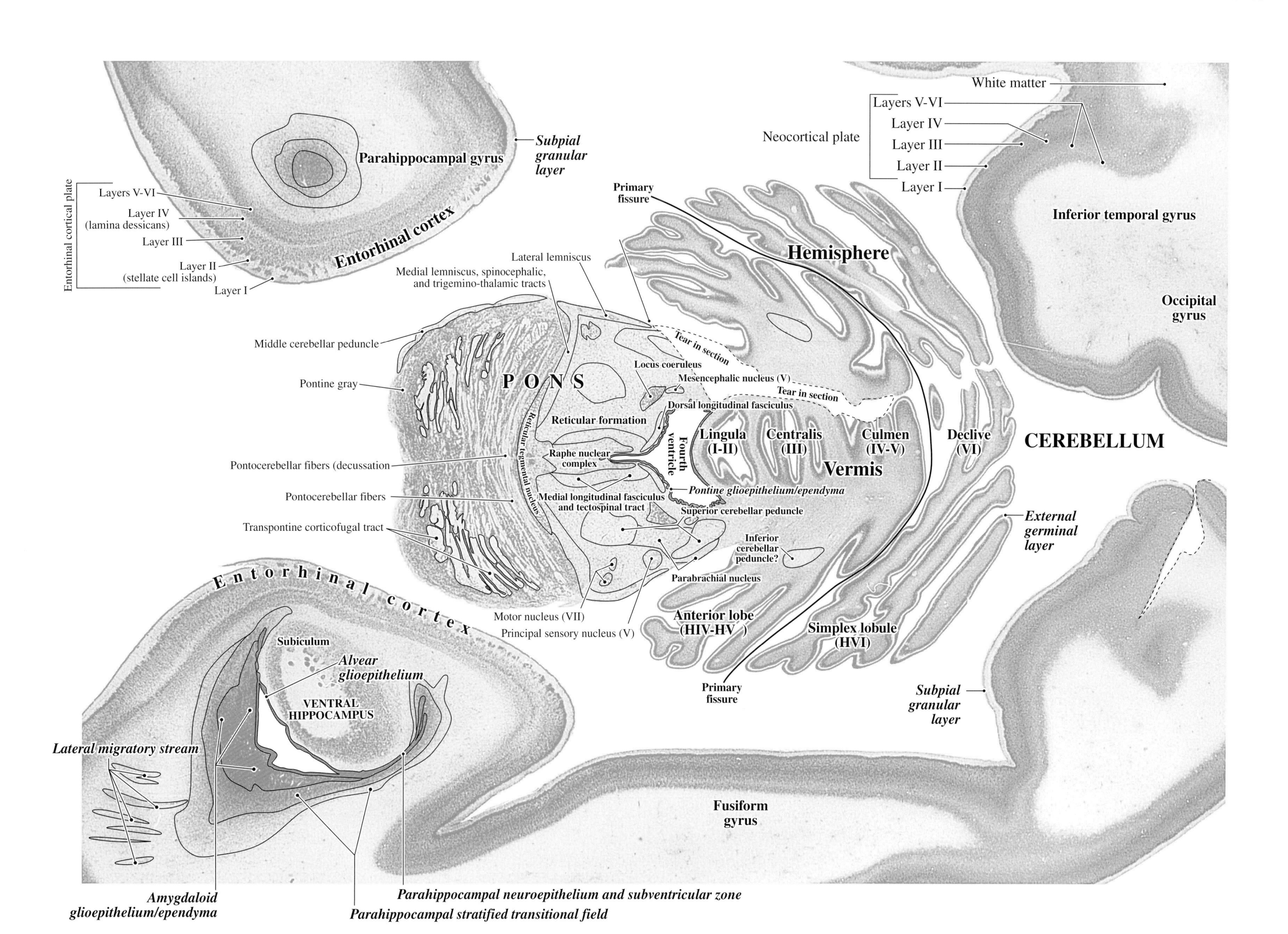

**PLATE 145A**

**GW30 Horizontal**
**CR 260 mm**
**Y187-65**
**Level 16: Section 1041**

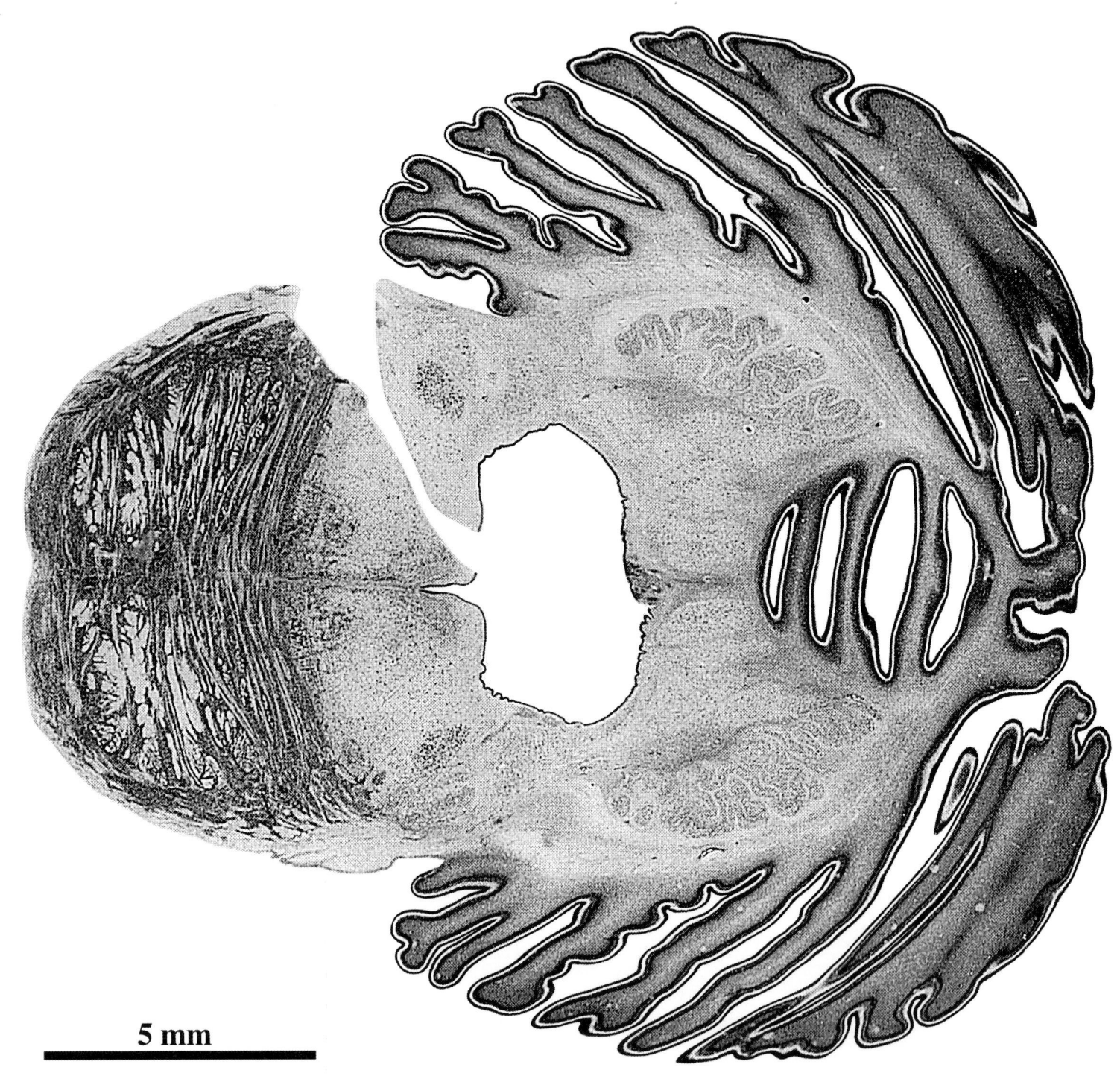

5 mm

See the entire section in Plate 127.

**PLATE 145B**

**Germinal and transitional structures in *italics***

# CEREBELLUM

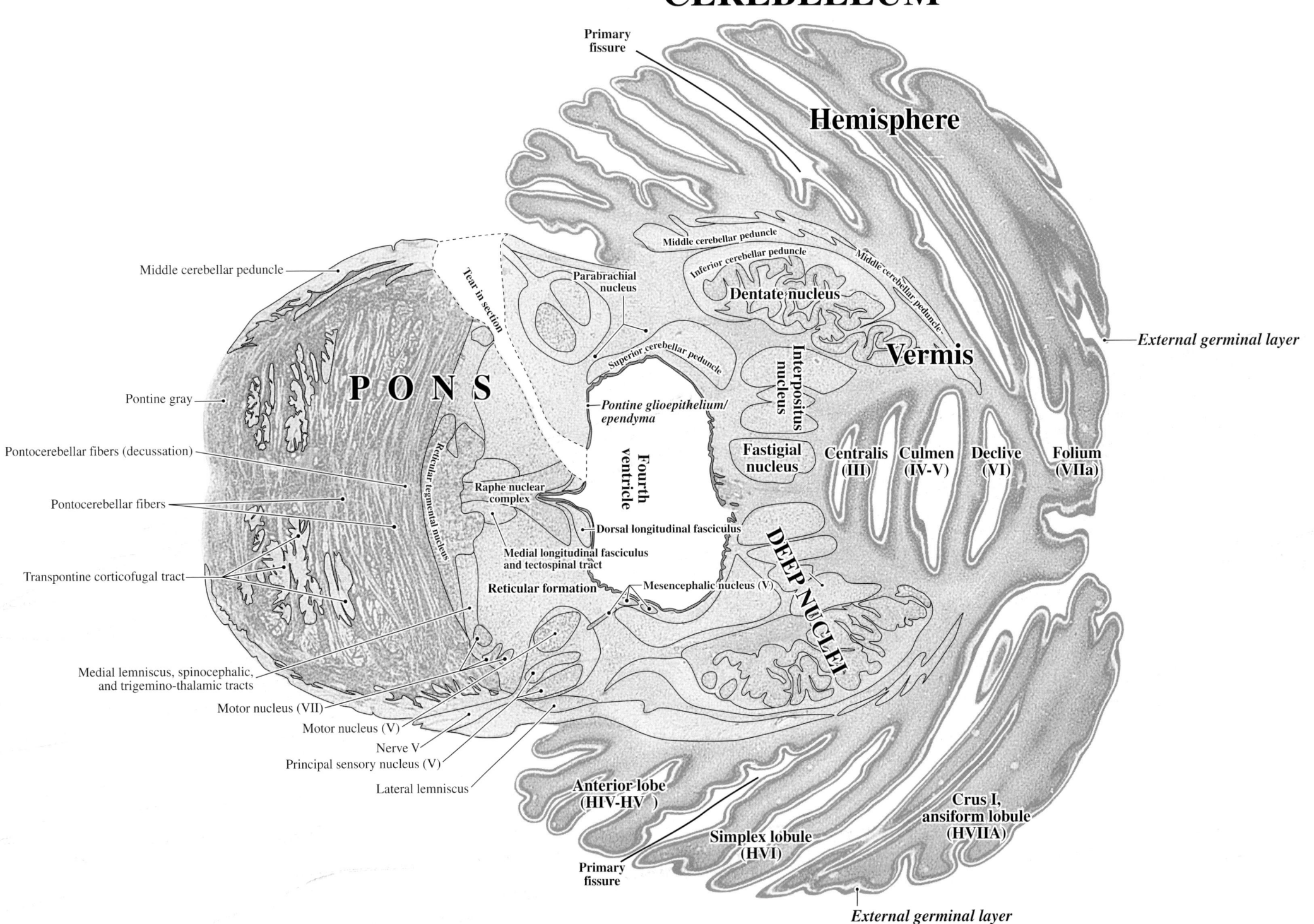

**PLATE 146A**

**GW30 Horizontal
CR 260 mm
Y187-65
Level 17: Section 1081**

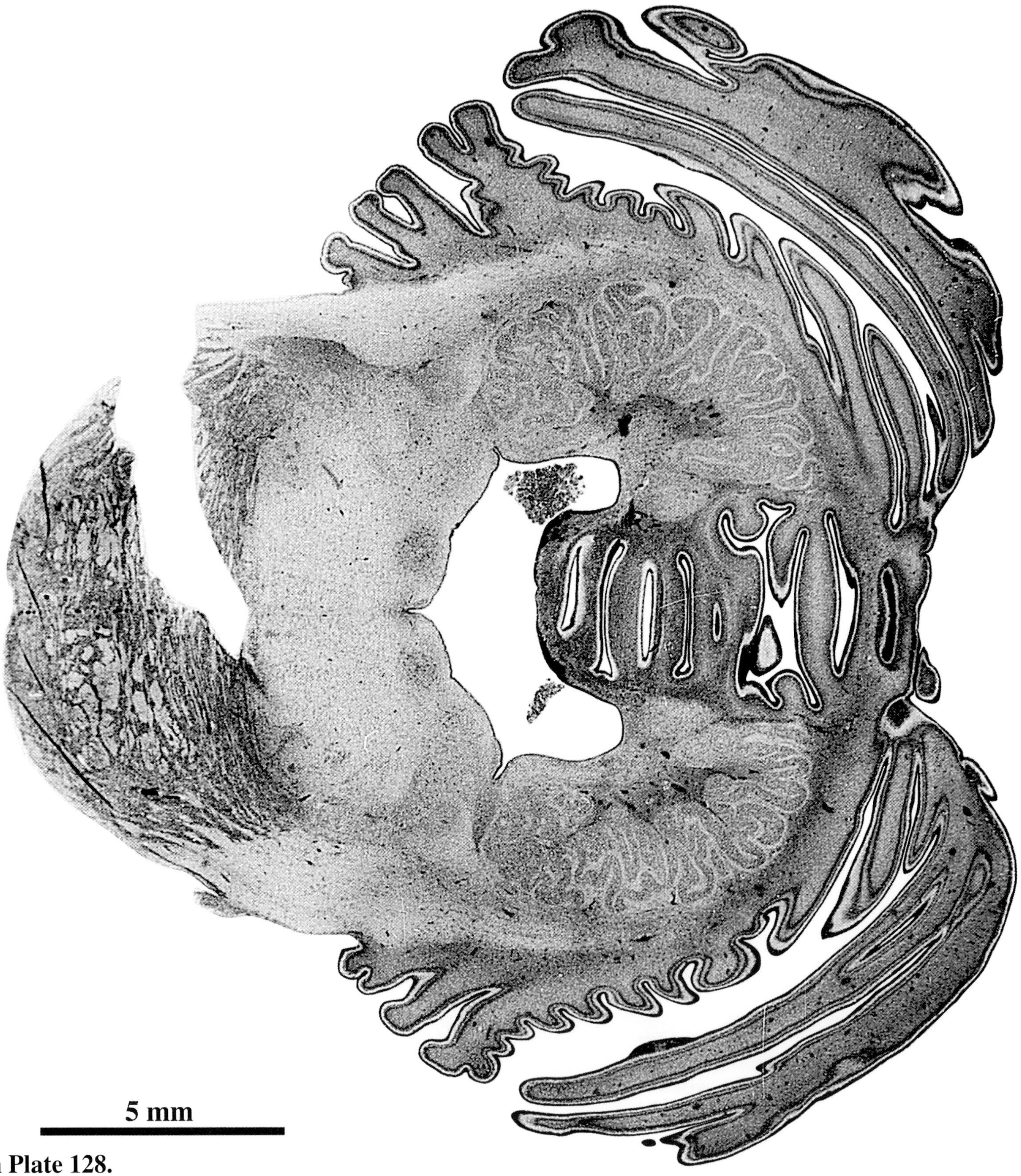

5 mm

See the entire section in Plate 128.

## PLATE 146B

**Germinal and transitional structures in *italics***

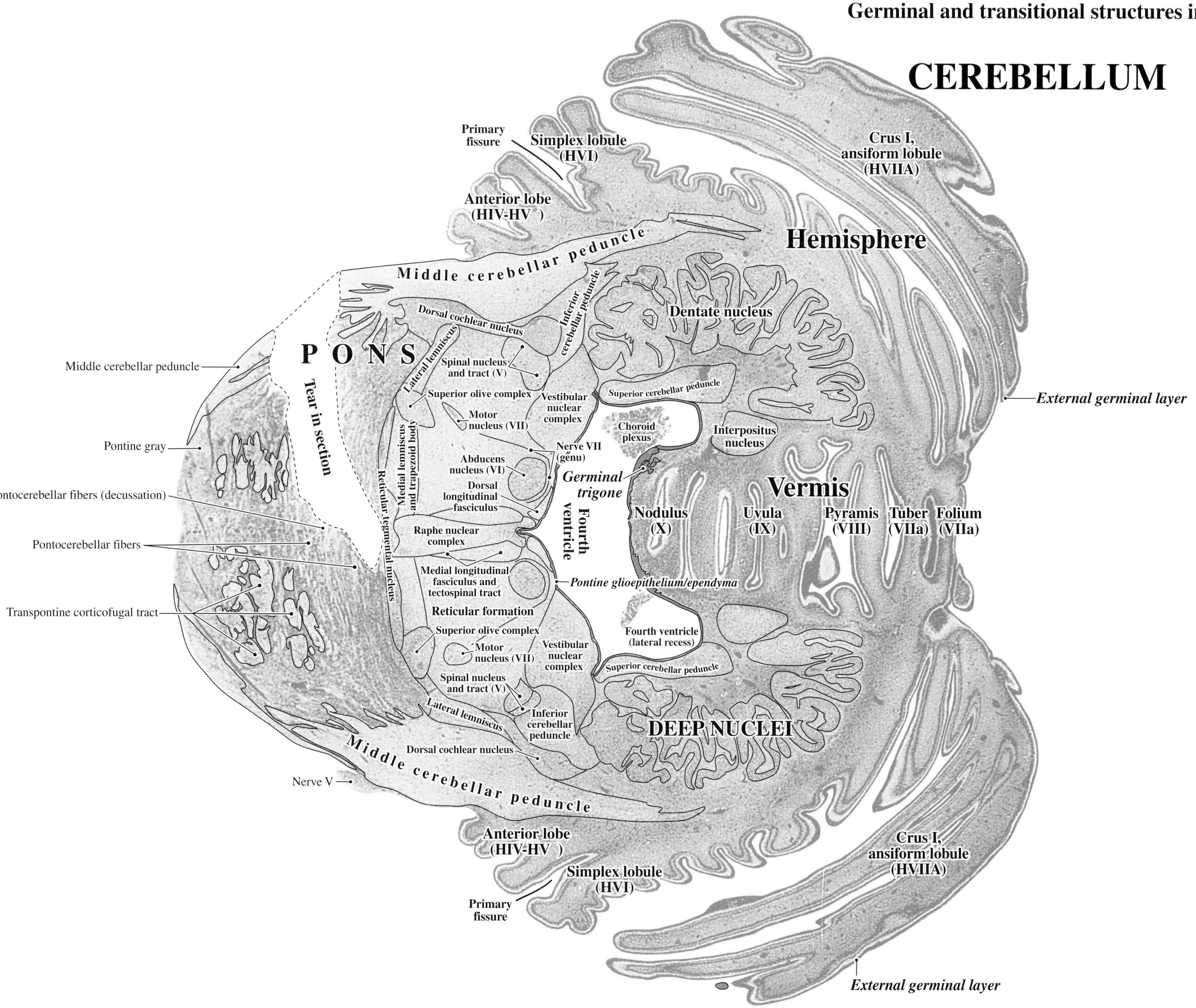

# CEREBELLUM

Labels on the figure:
- Primary fissure
- Simplex lobule (HVI)
- Crus I, ansiform lobule (HVIIA)
- Anterior lobe (HIV-HV)
- Hemisphere
- Middle cerebellar peduncle
- Dorsal cochlear nucleus
- Inferior cerebellar peduncle
- Dentate nucleus
- PONS
- Middle cerebellar peduncle
- Tear in section
- Lateral lemniscus
- Spinal nucleus and tract (V)
- Vestibular nuclear complex
- Superior olive complex
- Superior cerebellar peduncle
- Pontine gray
- Motor nucleus (VII)
- Interpositus nucleus
- External germinal layer
- Pontocerebellar fibers (decussation)
- Medial lemniscus and trapezoid body
- Nerve VII (genu)
- Choroid plexus
- Abducens nucleus (VI)
- *Germinal trigone*
- Vermis
- Pontocerebellar fibers
- Dorsal longitudinal fasciculus
- Reticular tegmental nucleus
- Raphe nuclear complex
- Nodulus (X)
- Uvula (IX)
- Pyramis (VIII)
- Tuber (VIIa)
- Folium (VIIa)
- Fourth ventricle
- Transpontine corticofugal tract
- Medial longitudinal fasciculus and tectospinal tract
- *Pontine glioepithelium/ependyma*
- Reticular formation
- Superior olive complex
- Fourth ventricle (lateral recess)
- Motor nucleus (VII)
- Vestibular nuclear complex
- Superior cerebellar peduncle
- Spinal nucleus and tract (V)
- Inferior cerebellar peduncle
- Lateral lemniscus
- Dorsal cochlear nucleus
- DEEP NUCLEI
- Nerve V
- Middle cerebellar peduncle
- Anterior lobe (HIV-HV)
- Simplex lobule (HVI)
- Primary fissure
- Crus I, ansiform lobule (HVIIA)
- *External germinal layer*

219

**PLATE 147A**

GW30 Horizontal
CR 260 mm
Y187-65
Level 18: Section 1121

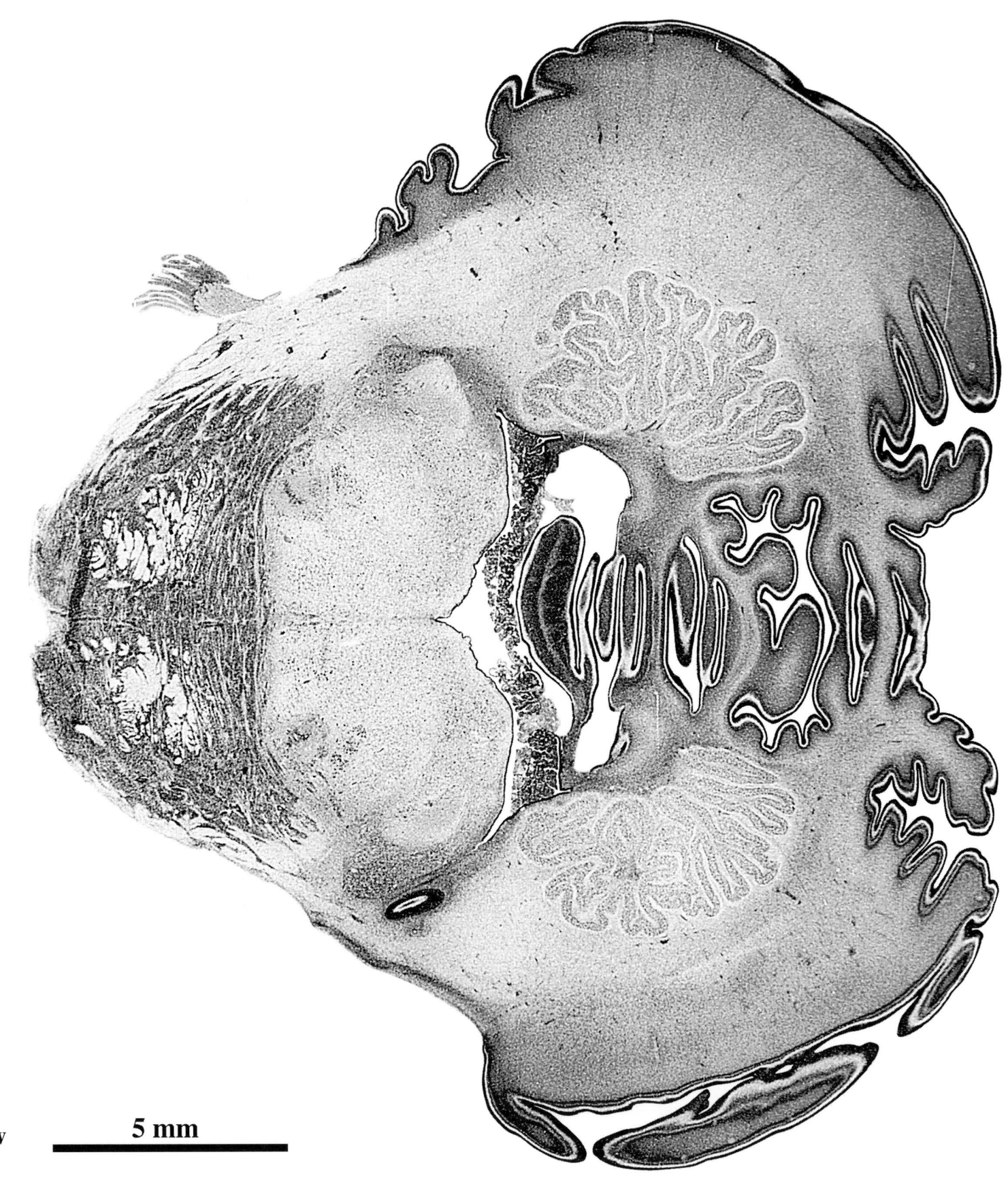

See the low magnification view in Plate 128.

5 mm

PLATE 147B

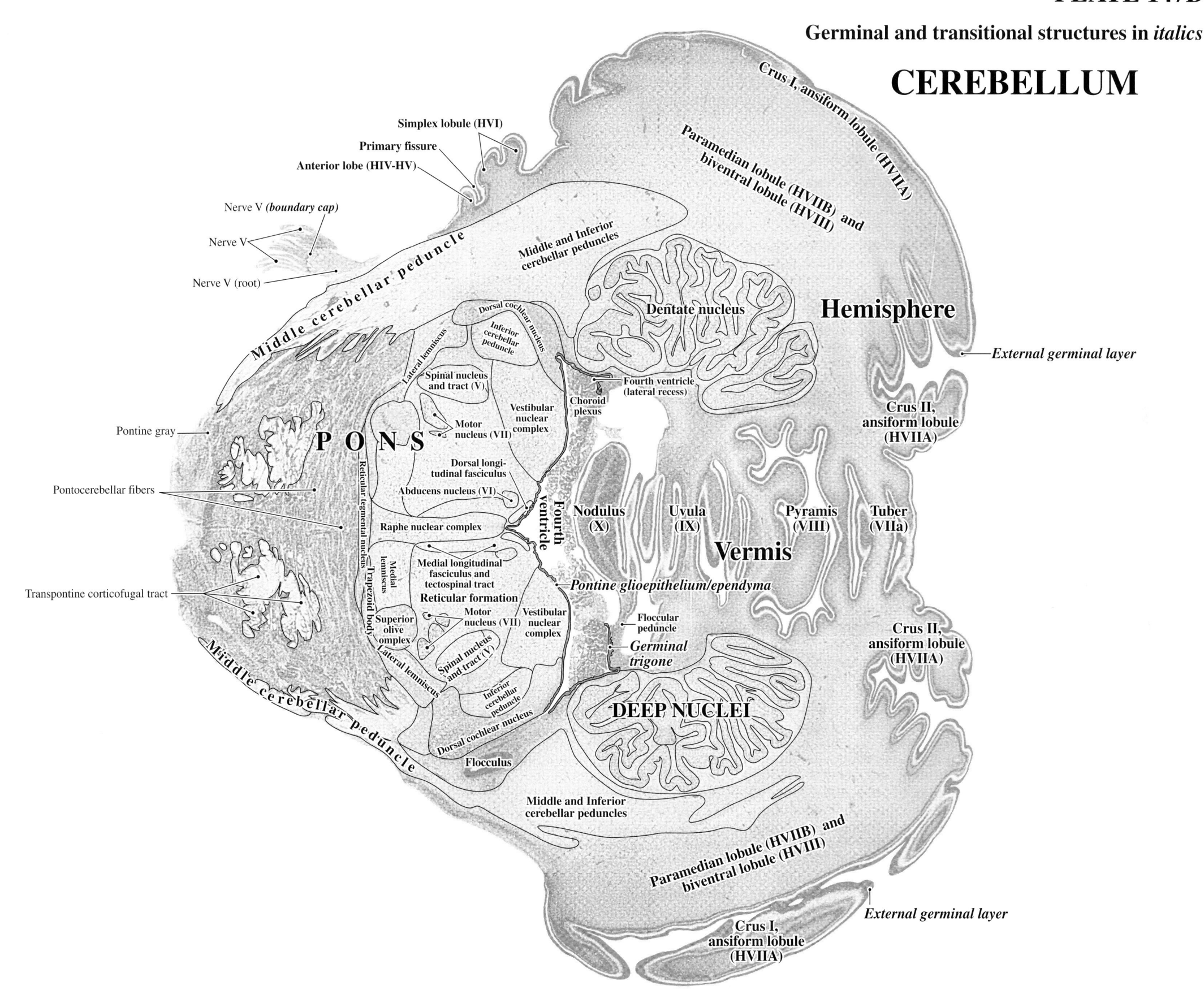

Germinal and transitional structures in *italics*

# CEREBELLUM

**PLATE 148A**

GW30 Horizontal
CR 260 mm
Y187-65
Level 19: Section 1161

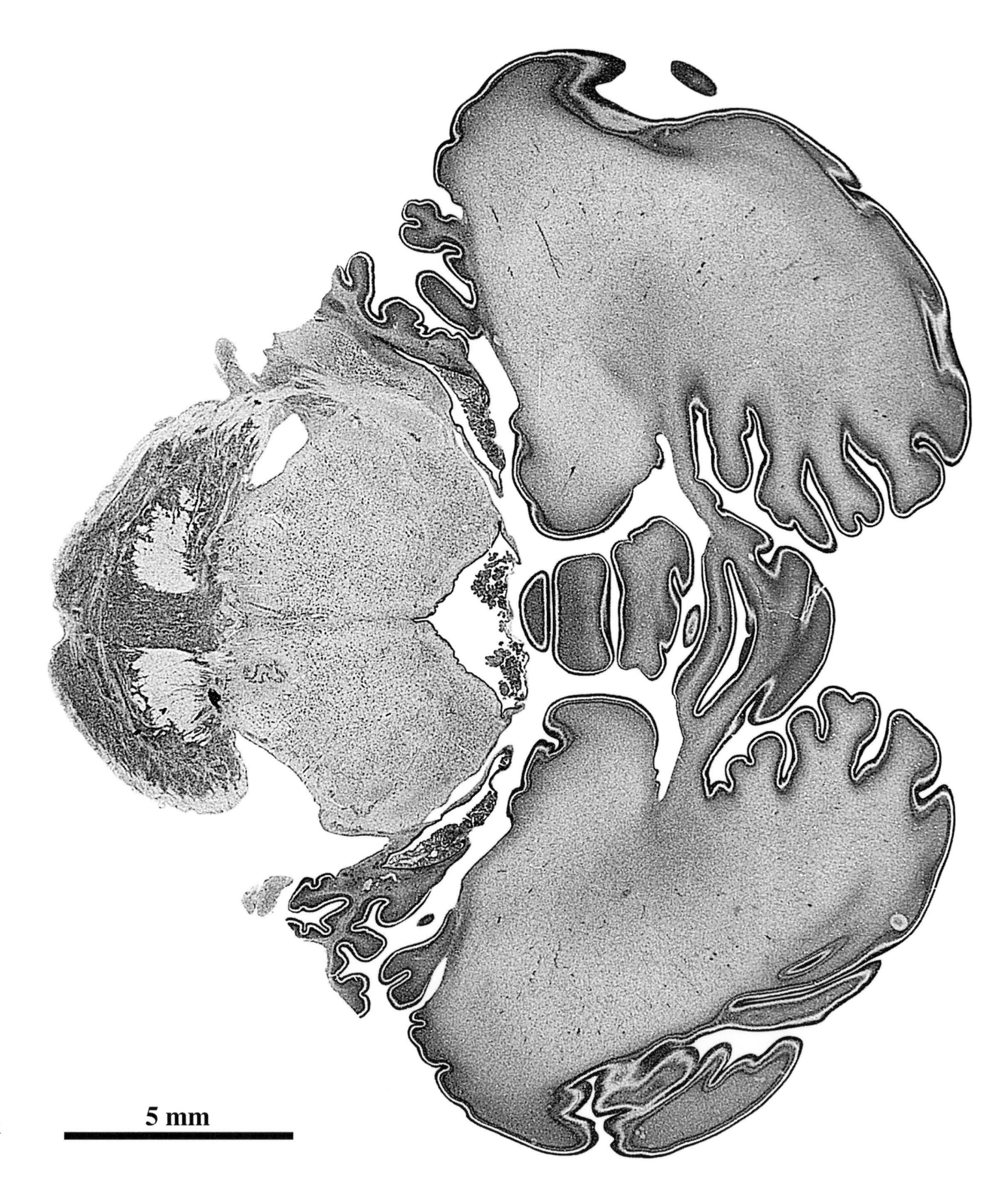

See the low magnification view in Plate 129.

5 mm

## PLATE 148B

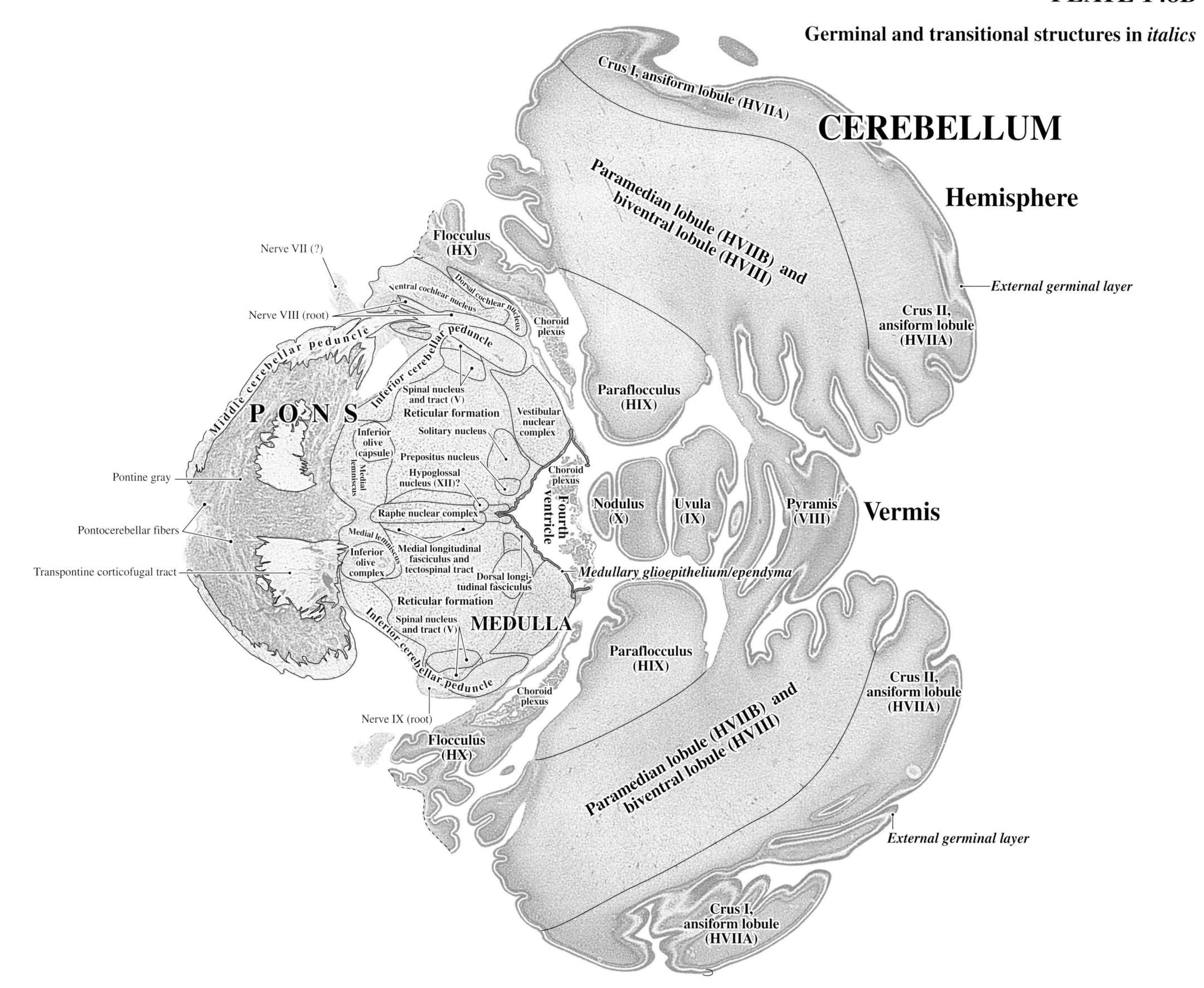

**Germinal and transitional structures in *italics***

## PLATE 149A

**GW30 Horizontal**
CR 260 mm
Y187-65

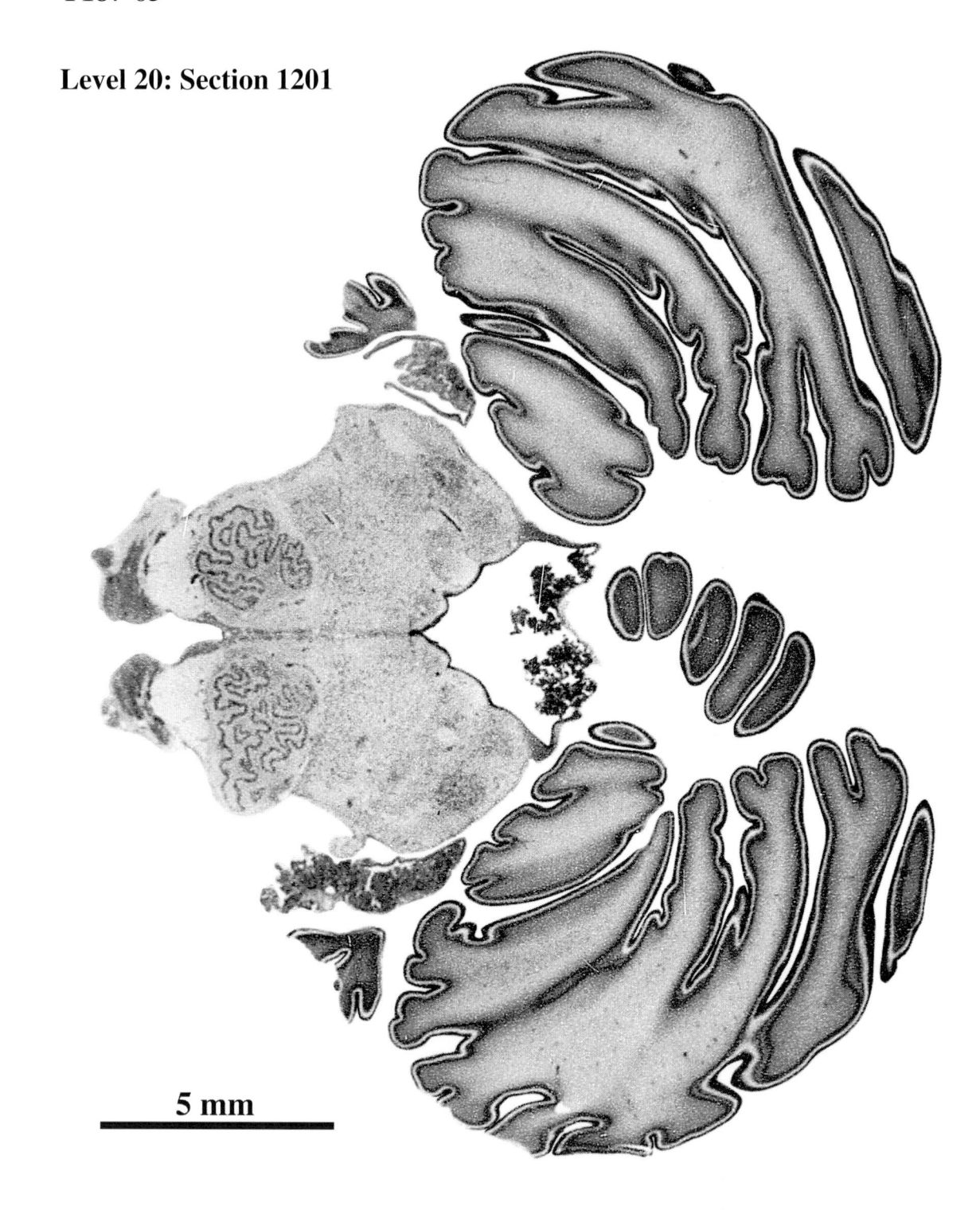

**Level 20: Section 1201**

**Level 21: Section 1241**

5 mm

5 mm

See the low magnification view in Plate 129.

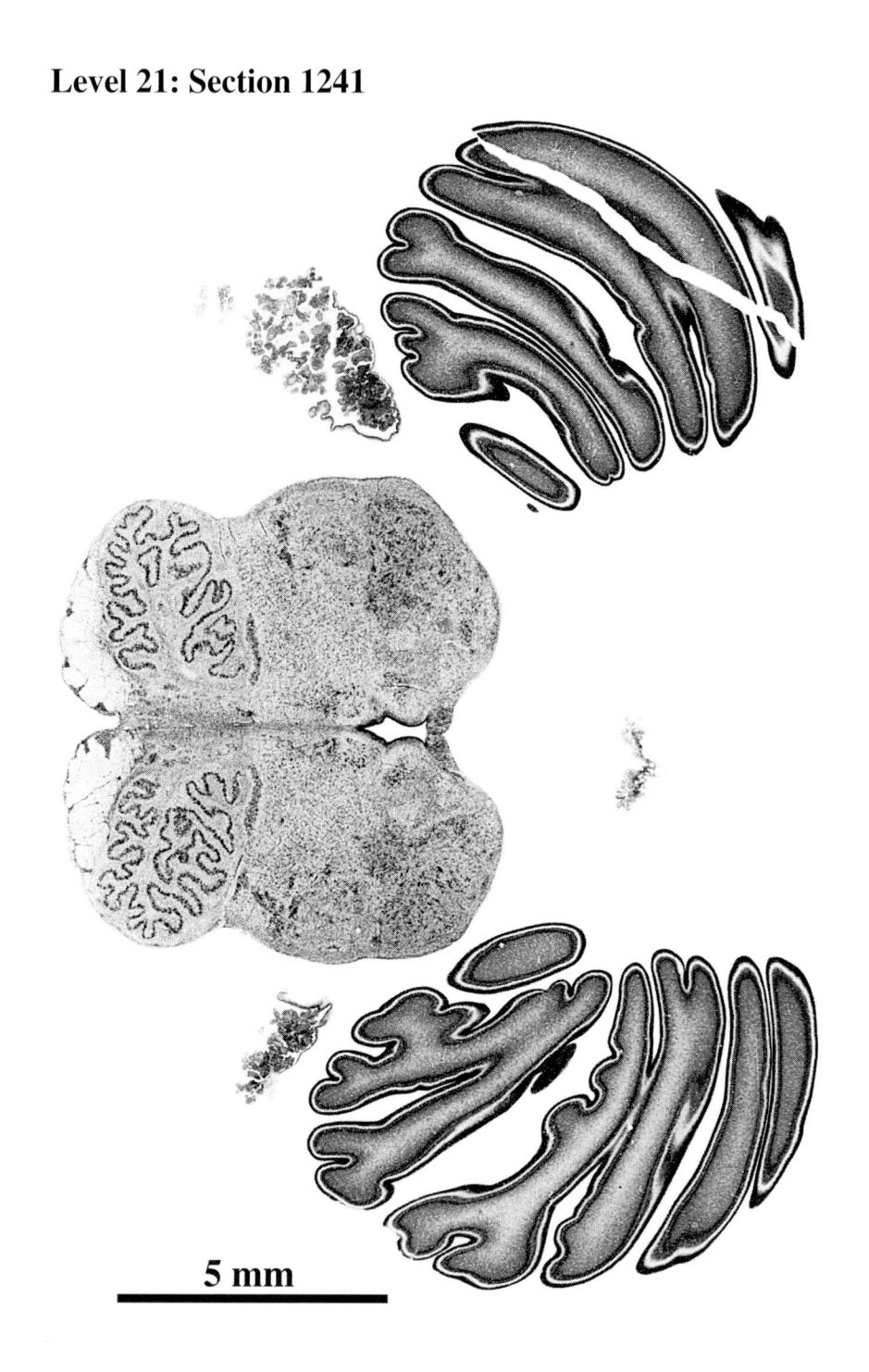

See the low magnification view in Plate 130.

**PLATE 149B**

Germinal and transitional structures in *italics*

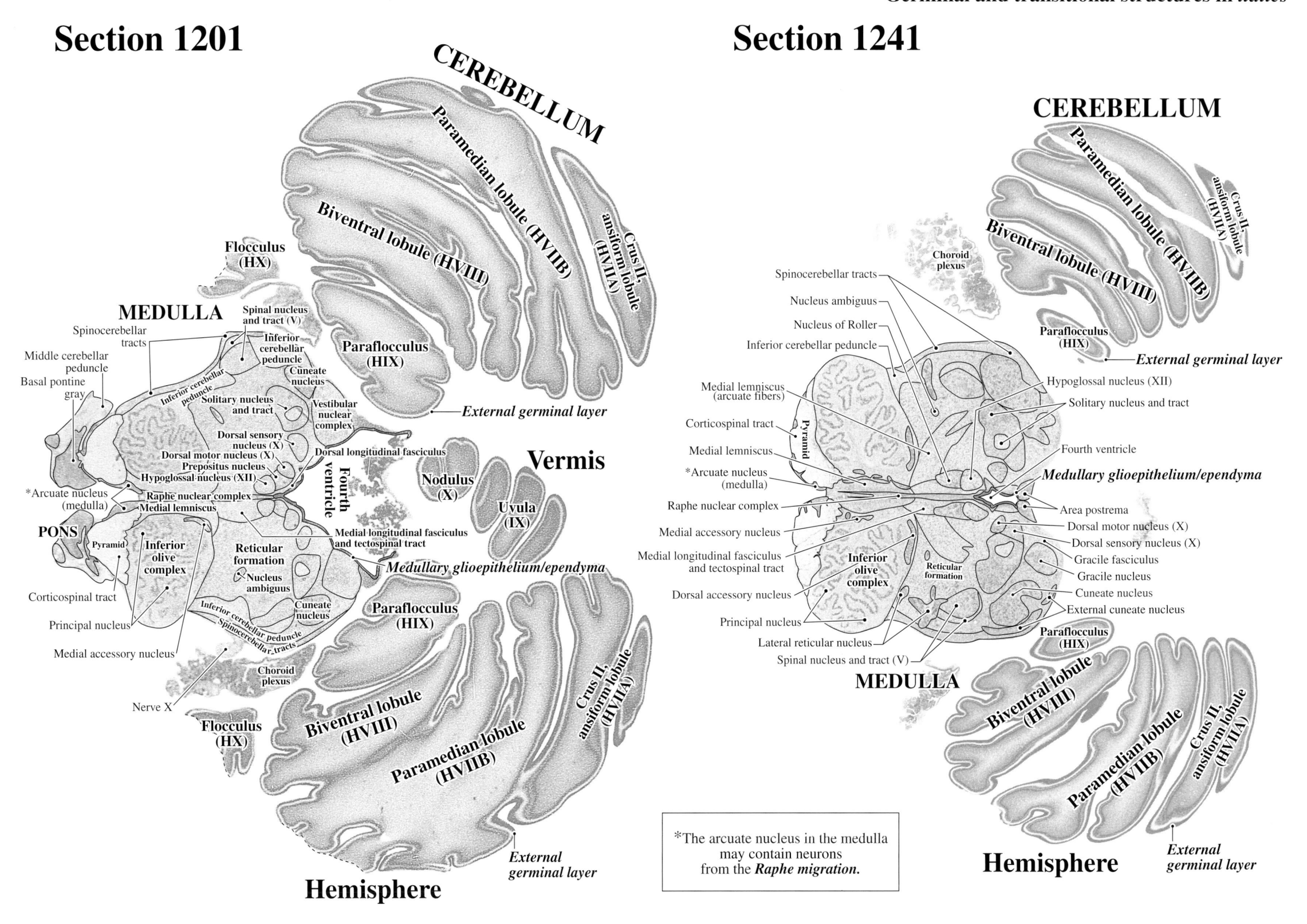

## PLATE 150A

GW30 Horizontal
CR 260 mm
Y187-65

**Level 22: Section 1281**

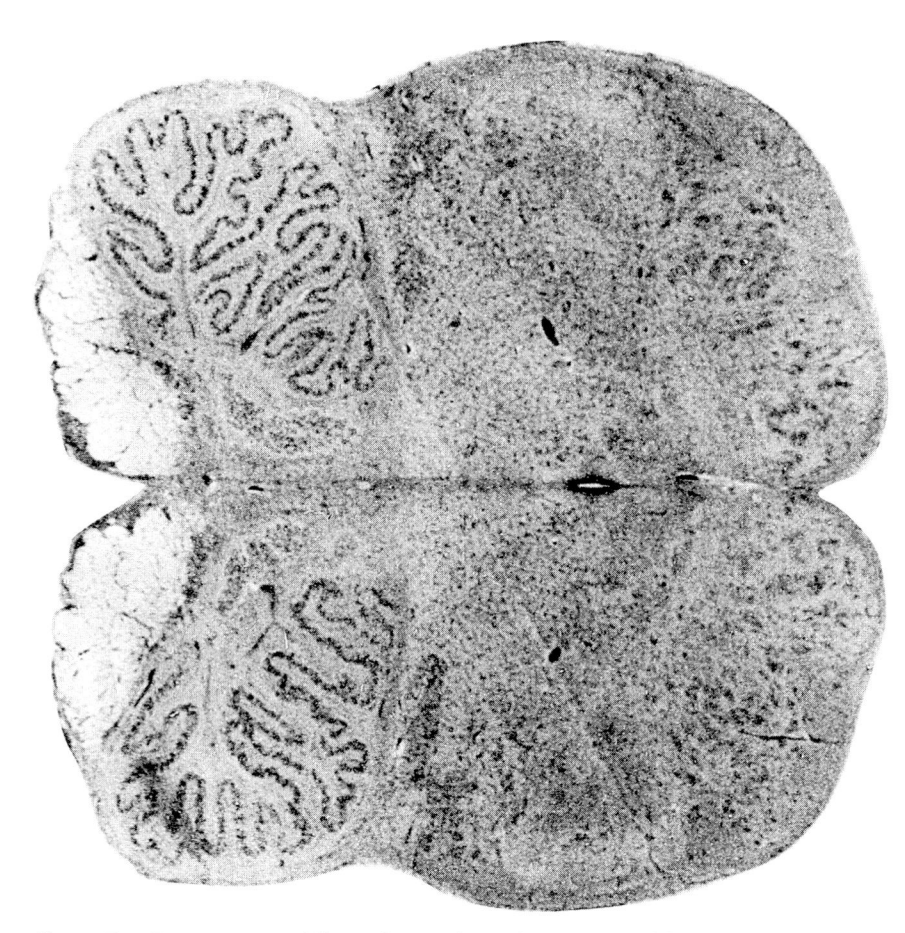

**Level 23: Section 1351**

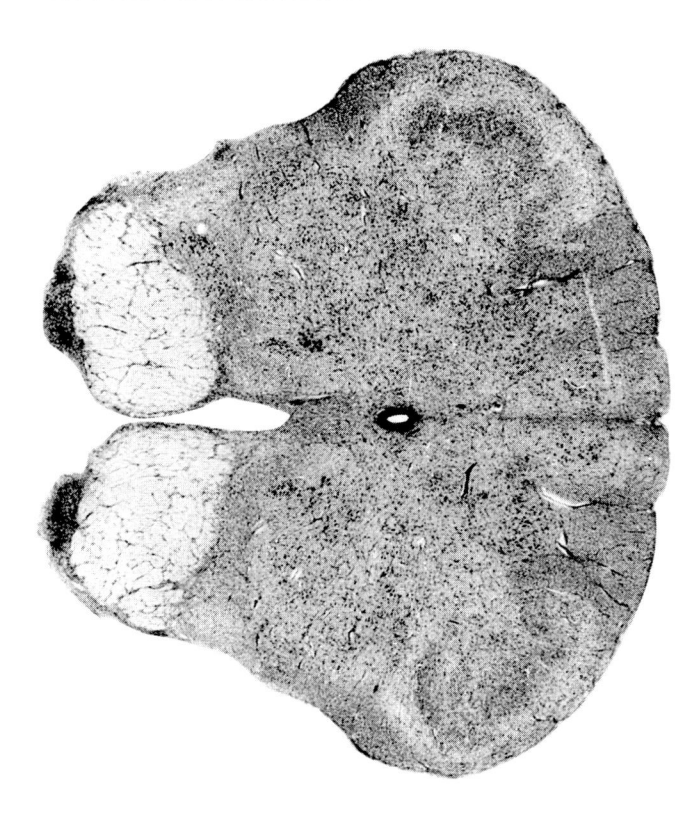

**Level 24: Section 1391**

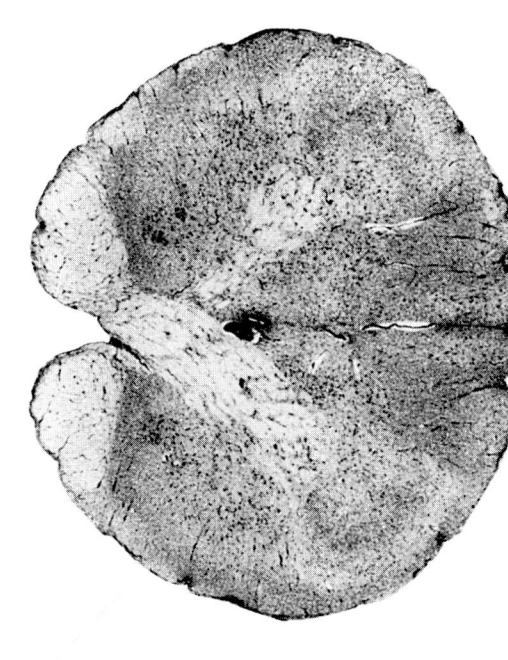

See the low magnification view in Plate 130.

See the low magnification view in Plate 131.

See the low magnification view in Plate 131.

5 mm

**PLATE 150B**

Germinal and transitional structures in *italics*

**Section 1281**
**LOWER MEDULLA**

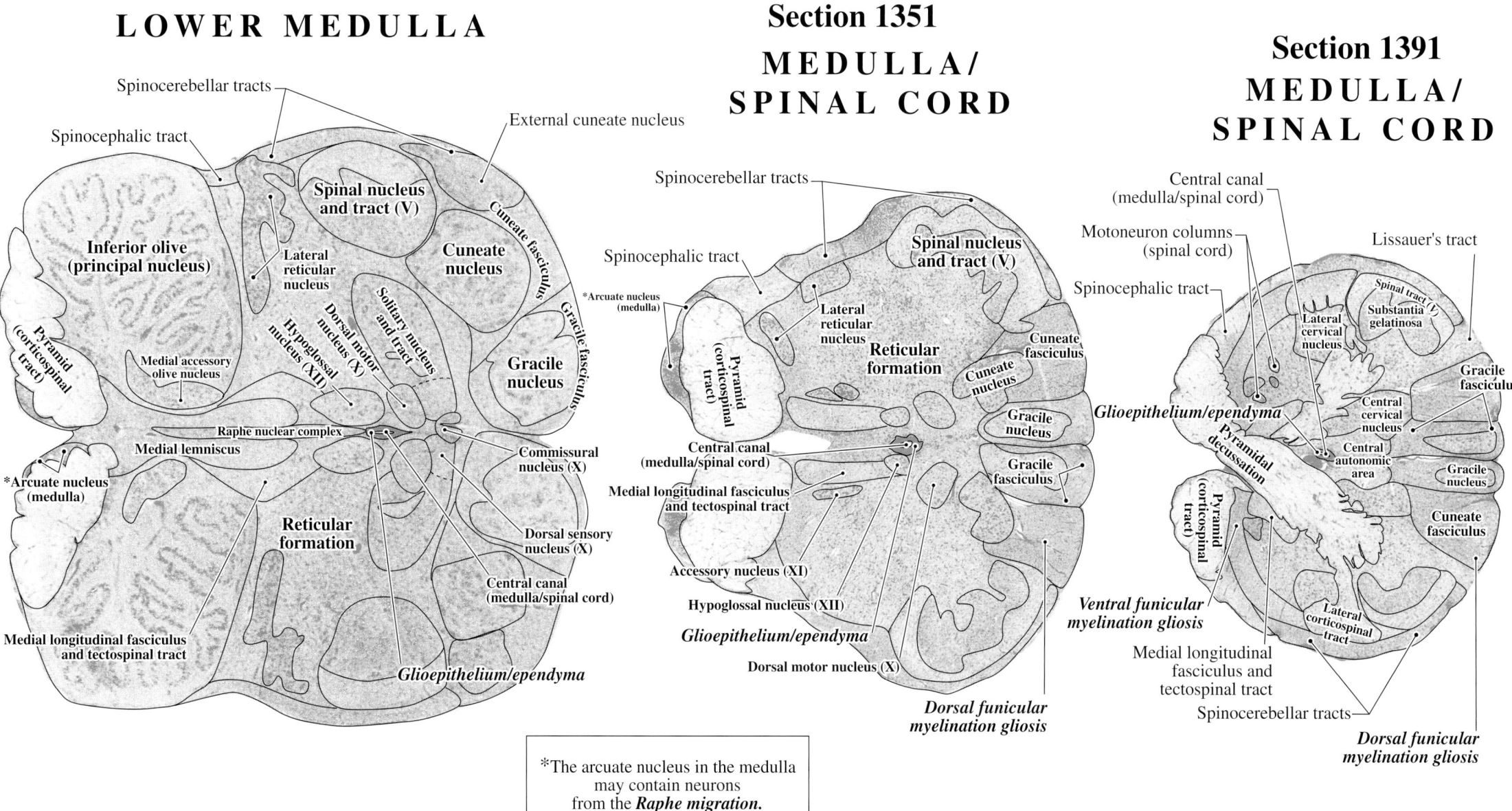

**Section 1351**
**MEDULLA/ SPINAL CORD**

**Section 1391**
**MEDULLA/ SPINAL CORD**

*The arcuate nucleus in the medulla may contain neurons from the **Raphe migration**.

# PART VII: GW29 CORONAL

This specimen is case number W-14-59 (Perinatal RPSL) in the Yakovlev Collection. A female fetus was prematurely stillborn after intrauterine asphyxia. Autopsy notes include a subdural hemorrhage, and there is a hemorrhage in the striatal germinal matrix on the left side of many sections. However, the remainder of the brain appears normal and is classified as a Normative Control in the Yakovlev Collection (Haleem, 1990). It was cut in the coronal plane in 35-μm and 15-μm thick sections. Since there is no available photograph of this brain before it was embedded and cut, the photograph of the lateral view of a GW30 brain that Larroche published in 1967 (**Figure 9**) is used.

The approximate cutting plane of this brain is indicated in **Figure 10** (facing page) with lines superimposed on the GW30 brain from the Larroche (1967) series. This brain is cut perpendicular to the longitudinal axis of the cerebral hemispheres between the frontal and occipital poles and is remarkably even in the medial/lateral plane (both temporal poles appear in **Level 3, Section 621**). As in all other specimens, the sections chosen for illustration are spaced closer together to show small structures in the diencephalon, midbrain, pons, and medulla. Illustrated sections are spaced farther apart when they contain only large brain structures, such as the cerebral cortex, basal ganglia, and cerebellum. Photographs of 20 different Nissl-stained sections (**Levels 1-20**) are shown at low magnification in **Plates 151-170**. Different areas of the cerebral cortex are shown at very high magnification in **Plates 171-174**. The core of the brain and the cerebellum are shown at high magnification in companion **Plates 175AB-189AB** for **Levels 3-18**.

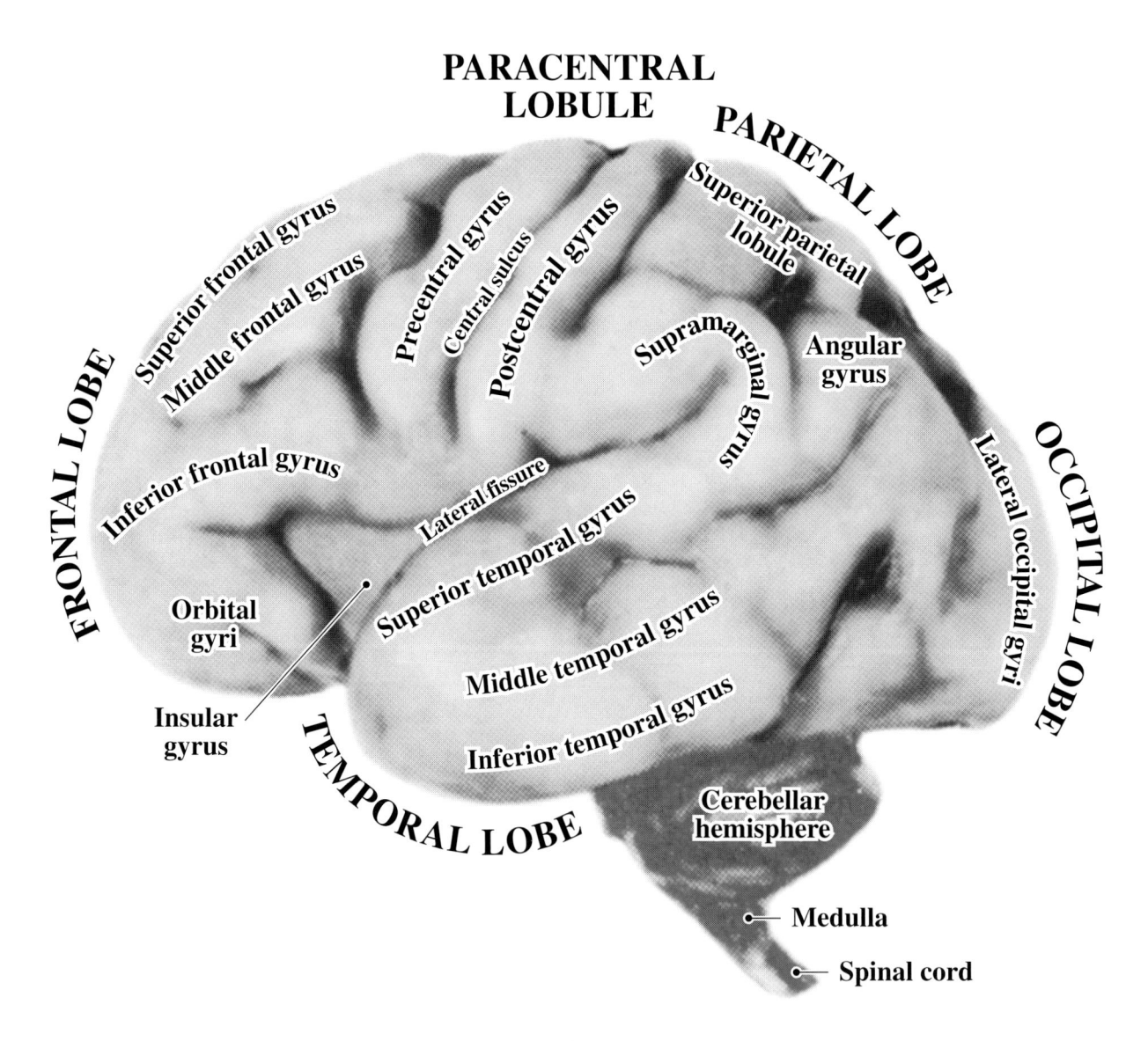

**Figure 9.** Lateral view of a GW30 brain with major structures in the cerebral hemispheres labeled. This is the same as **Figure 7** repeated here for convenience. (From the photographic series of J. C. Larroche (1967) Maturation morphologique du système nerveux central: ses rapports avec le développement pondéral du foetus et son age gestationnel. In: *Regional Development of the Brain in Early Life*, A. Minkowski (ed.), London: Blackwell, page 248.)

# GW29 CORONAL SECTION PLANES

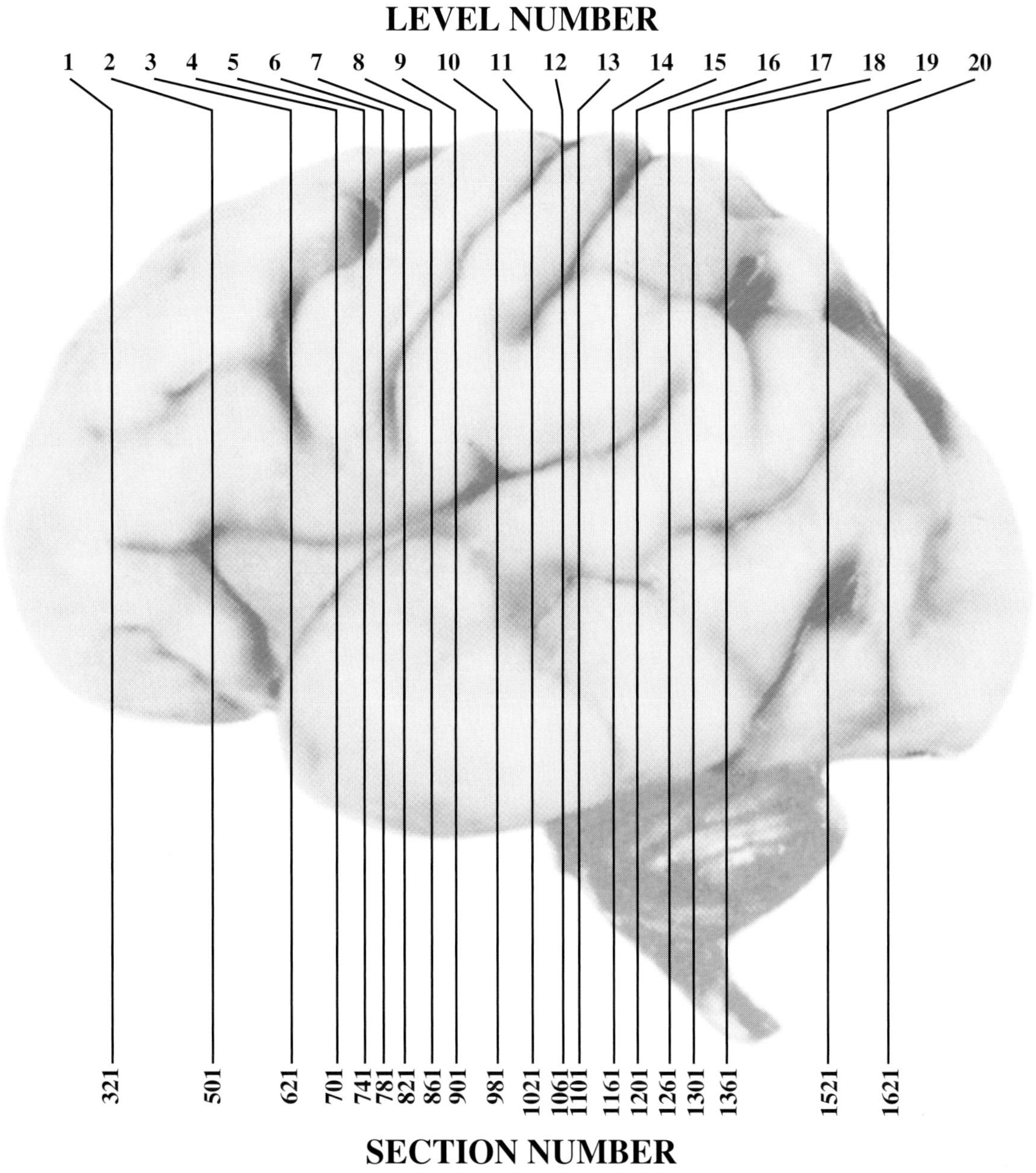

## LEVEL NUMBER

1  2  3  4  5  6  7  8  9  10  11  12  13  14  15  16  17  18  19  20

321  501  621  701  741  781  821  861  901  981  1021  1061  1101  1161  1201  1261  1301  1361  1521  1621

## SECTION NUMBER

**Figure 10.** Lateral view of the same GW30 brain shown in **Figure 9** with the approximate locations and cutting angle of the sections of Y14-59. (From the photographic series of J. C. Larroche (1967) Maturation morphologique du système nerveux central: ses rapports avec le développement pondéral du foetus et son age gestationnel. In: *Regional Development of the Brain in Early Life*, A. Minkowski (ed.), London: Blackwell, page 248.)

Y14-59 contains the same immature structures shown in the GW30 Horizontal specimen. A densely staining ***neuroepithelium/ subventricular zone*** is present and presumably generating neocortical interneurons in all lobes of the cerebral cortex. The same thickness variations between the occipital and other lobes of the cerebral cortex are still there. Remnants of migrating and sojourning neurons and/or glia are visible in all lobes of the cerebral cortex as ***stratified transitional fields***. Many neurons, glia, and their mitotic precursor cells are still migrating through the olfactory peduncle toward the olfactory bulb (***rostral migratory stream***) from a presumed source area in the germinal matrix at the junction between the cerebral cortex, striatum, and nucleus accumbens. Within the lateral parts of the cerebral cortex, definite streams of neurons and glia are in the ***lateral migratory stream*** that percolates through the claustrum, endopiriform nucleus, external capsule, and uncinate fasciculus. These cells appear to be heading toward the insular cortex, primary olfactory cortex, temporal cortex, and basolateral parts of the amygdaloid complex. In the basal ganglia, there is a large ***neuroepithelium/subventricular zone*** overlying the striatum and nucleus accumbens where neurons are being generated; at least three subdivisions (anteromedial, anterolateral, and posterior) can be distinguished in the striatal part. Another region of active neurogenesis in the telencephalon is the ***subgranular zone*** in the hilus of the dentate gyrus that is the source of granule cells. Other structures in the telencephalon, such as the septum, fornix, and Ammon's horn have only a thin, darkly staining layer at the ventricle, and these are presumed to be generating glia, cells of the choroid plexus, and the ependymal lining of the ventricle.

Most of the structures in the diencephalon appear to be settled and are maturing, but the third ventricle is lined by a more densely staining ***glioepithelium/ependyma*** than in the GW37 specimens. A convoluted ***glioepithelium/ependyma*** lines the cerebral aqueduct in the midbrain that continues into the anterior fourth ventricle. A smooth ***glioepithelium/ependyma*** lines the fourth ventricle through the posterior pons. A slightly convoluted ***glioepithelium/ependyma*** lines the floor of the fourth ventricle near the midline in the medulla. The ***external germinal layer*** is prominent over the entire surface of the cerebellar cortex and is actively producing basket, stellate, and granule cells. The cerebellar cortex itself shows less definition between hemispheric lobules. The ***germinal trigone*** is at the base of the nodulus and along the floccular peduncle; choroid plexus cells and glia may still be originating here.

# PLATE 151

GW29 Coronal
CR 260 mm
Y14-59
Level 1: Section 321

*Remnants of the germinal matrix, migratory streams, and transitional fields*

1 *Frontal neuroepithelium and subventricular zone*
2 *Frontal stratified transitional field*

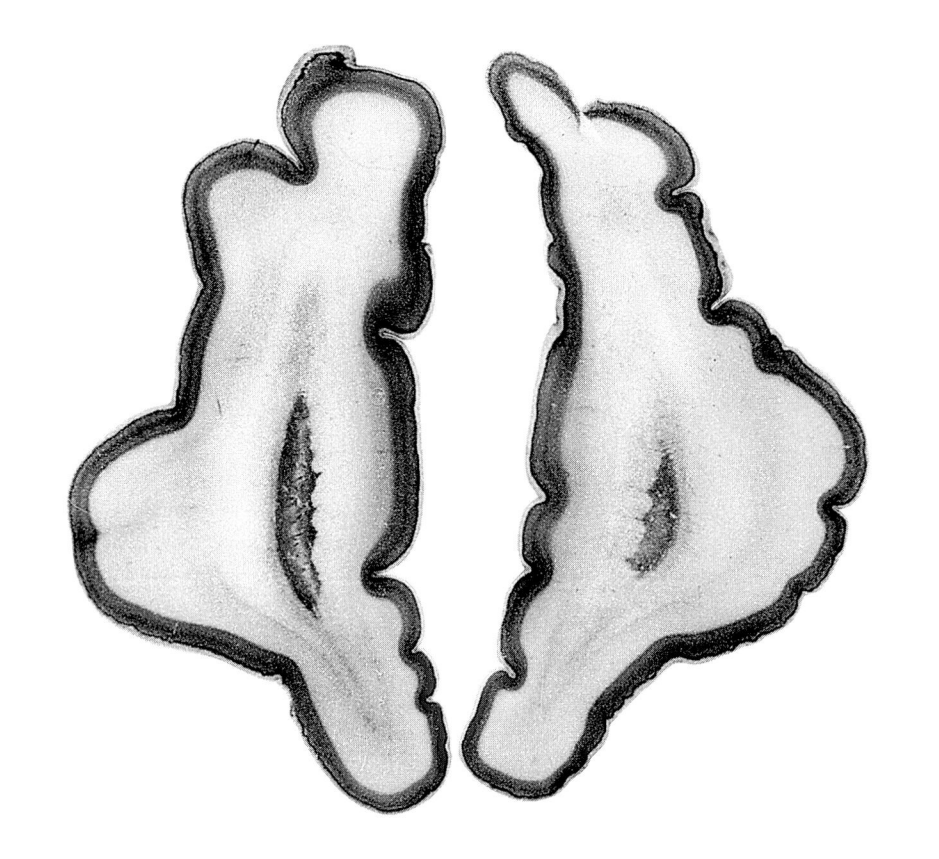

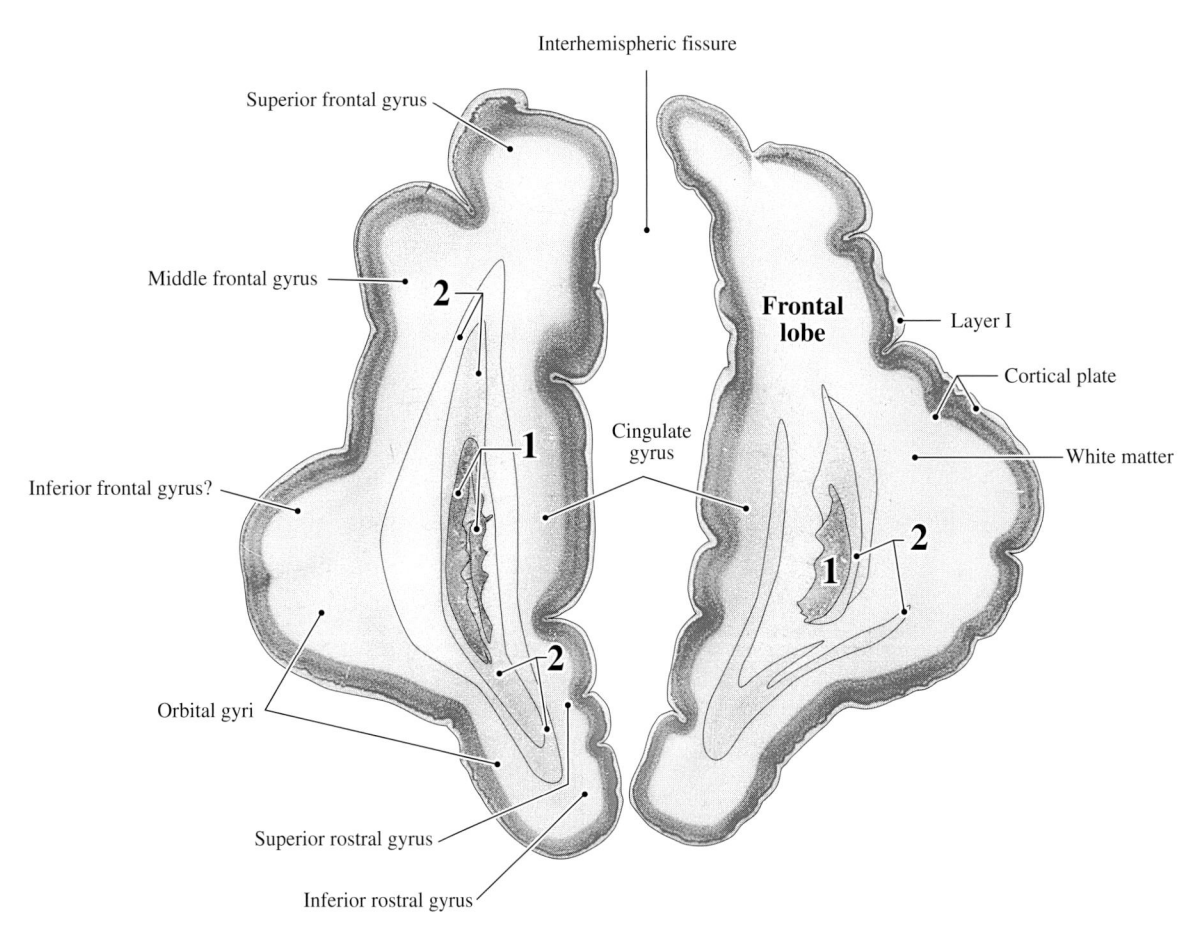

## PLATE 152

**GW29 Coronal**
**CR 260 mm**
**Y14-59**
**Level 2: Section 501**

*Remnants of the germinal matrix, migratory streams, and transitional fields*

1. Frontal neuroepithelium and subventricular zone
2. Frontal stratified transitional field
3. Callosal glioepithelium
4. Frontal neuroepithelium and subventricular zone (intermingled with the source of the rostral migratory stream)
5. Rostral migratory stream
6. Anterolateral striatal neuroepithelium and subventricular zone

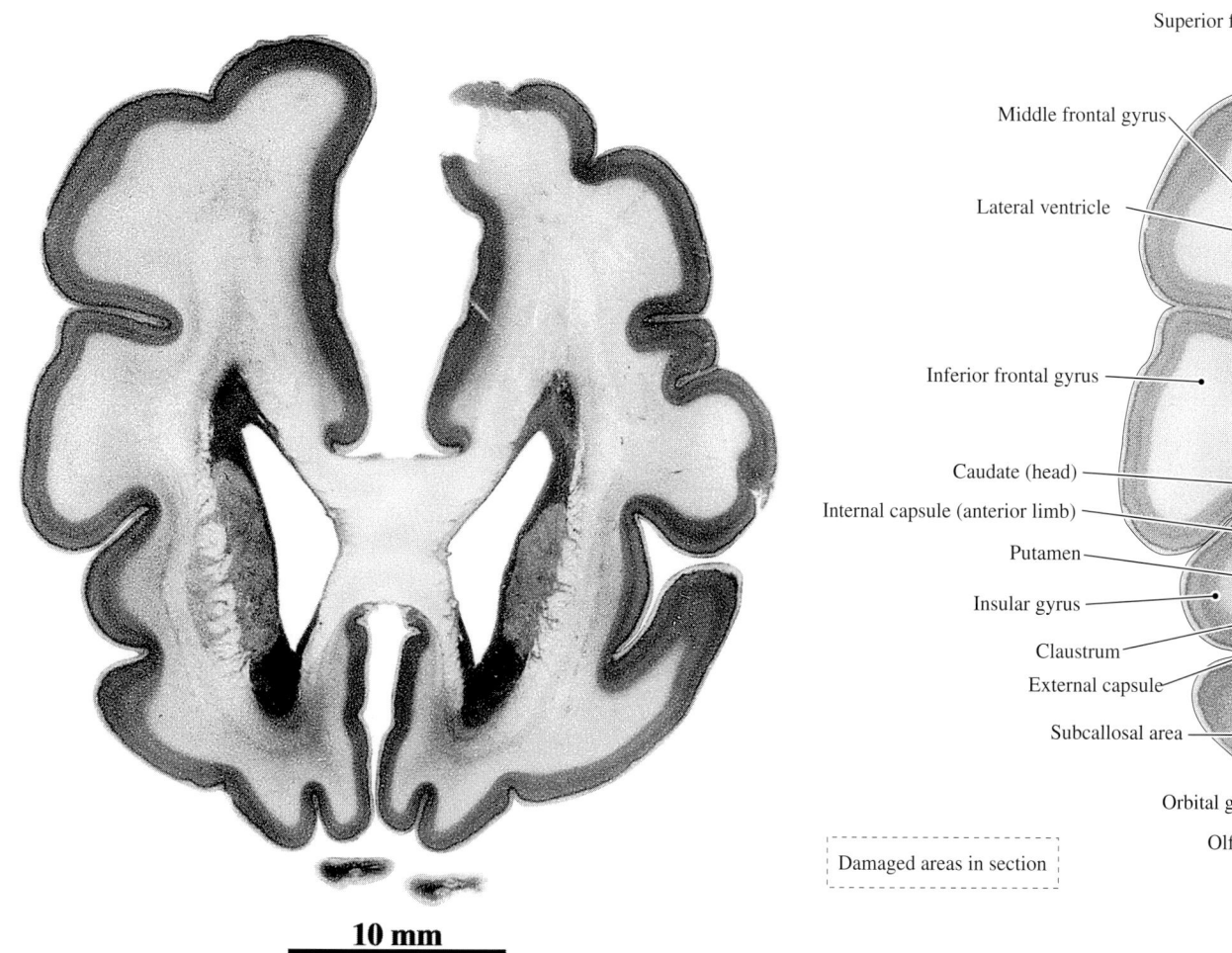

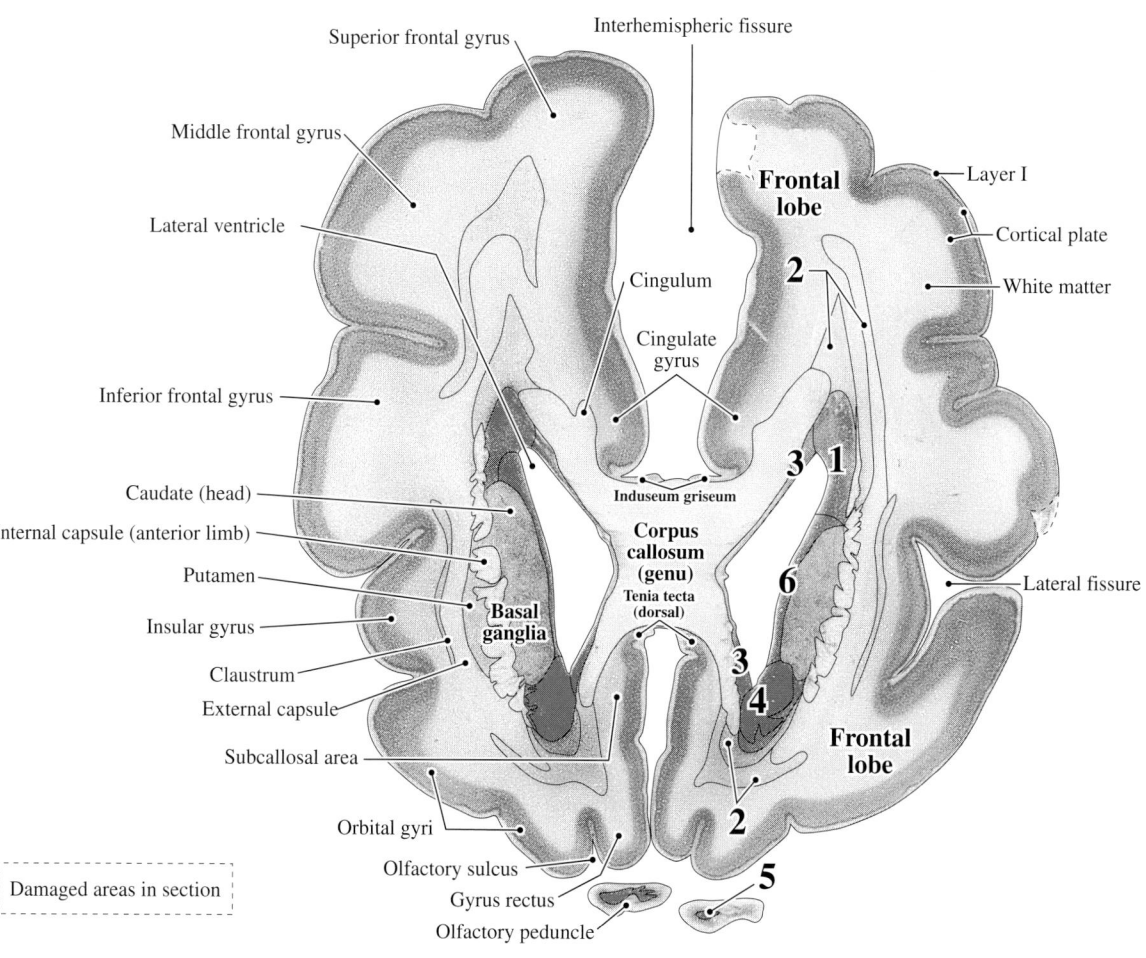

PLATE 153

GW29 Coronal
CR 260 mm
Y14-59
Level 3: Section 621

*Remnants of the germinal matrix, migratory streams, and transitional fields*

| 1 | Frontal neuroepithelium and subventricular zone | 7 | Accumbent neuroepithelium and subventricular zone (intermingled with the rostral migratory stream and the frontal neuroepithelium/subventricular zone) |
|---|---|---|---|
| 2 | Frontal stratified transitional field | | |
| 3 | Callosal glioepithelium | 8 | Anteromedial striatal neuroepithelium and subventricular zone |
| 4 | Callosal sling | 9 | Anterolateral striatal neuroepithelium and subventricular zone |
| 5 | Fornical glioepithelium | 10 | Lateral migratory stream (cortical) |
| 6 | Rostral migratory stream | 11 | Subpial granular layer (cortical) |

10 mm

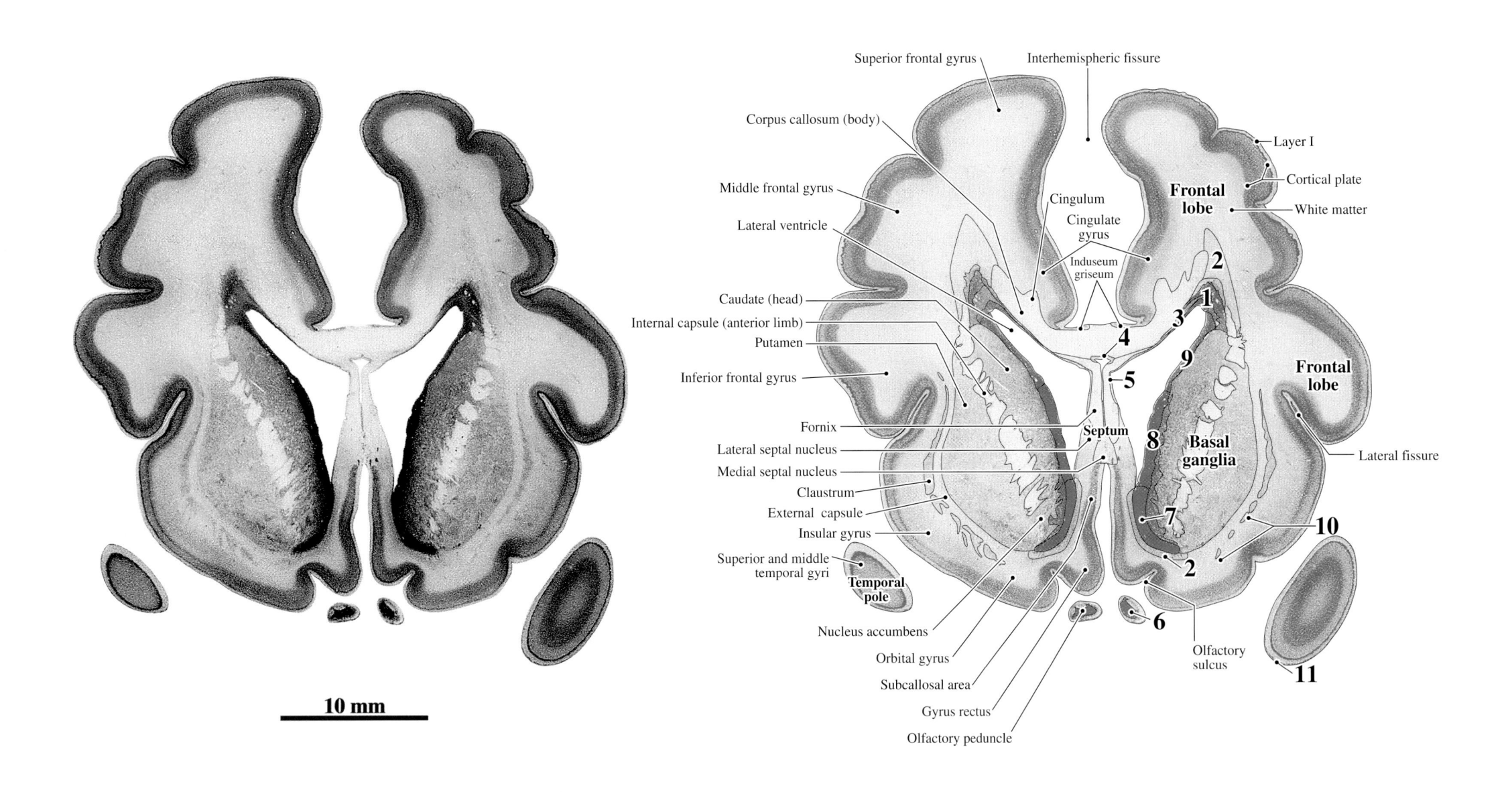

## PLATE 154

**GW29 Coronal**
**CR 260 mm**
**Y14-59**
**Level 4: Section 701**

### Remnants of the germinal matrix, migratory streams, and transitional fields

1. Frontal neuroepithelium and subventricular zone
2. Frontal stratified transitional field
3. Callosal glioepithelium
4. Callosal sling
5. Fornical glioepithelium
6. Rostral migratory stream
7. Accumbent neuroepithelium and subventricular zone
8. Anteromedial striatal neuroepithelium and subventricular zone
9. Anterolateral striatal neuroepithelium and subventricular zone
10. Lateral migratory stream (cortical)
11. Subpial granular layer (cortical)

**See detail of brain core in Plates 175A and B.**

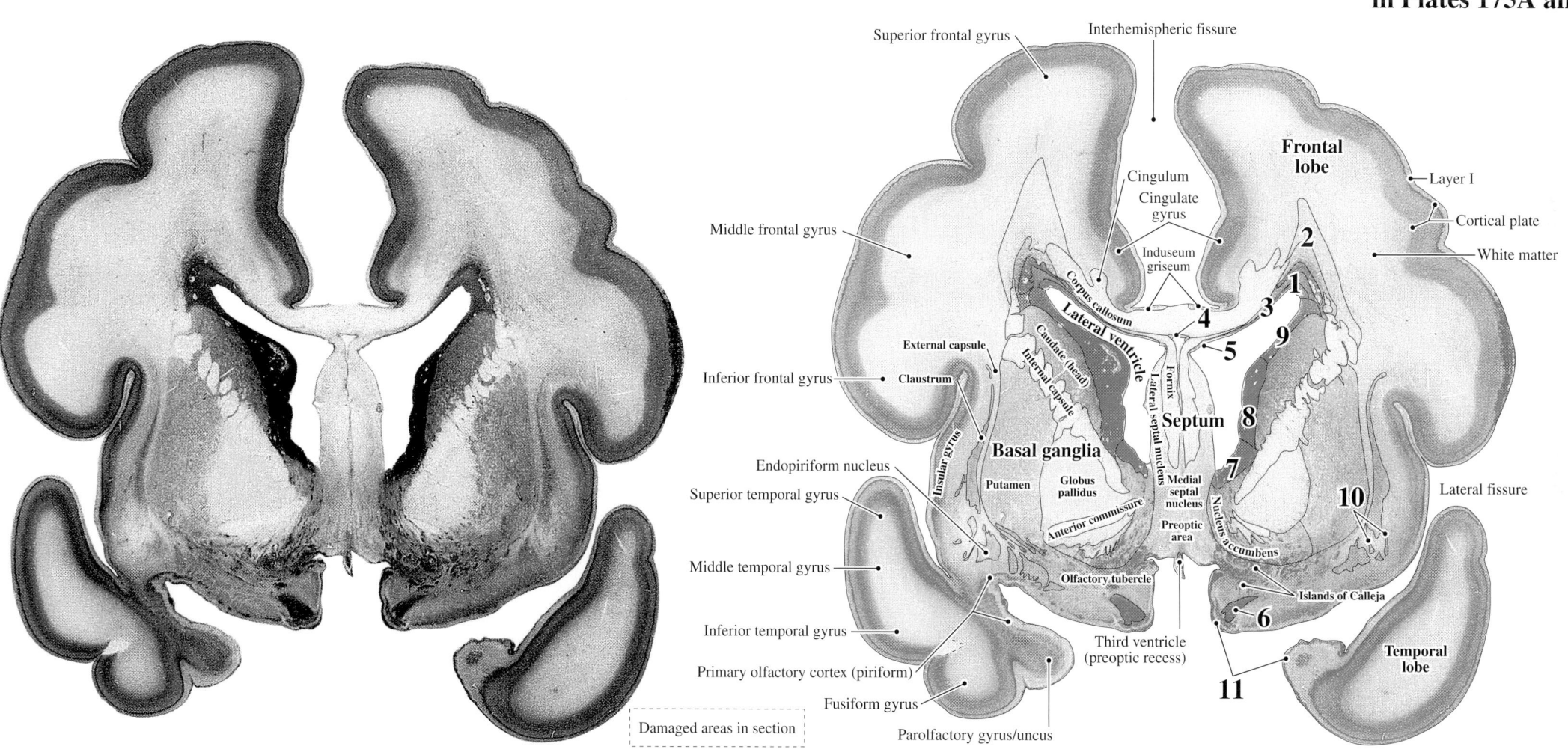

# PLATE 155

GW29 Coronal
CR 260 mm
Y14-59
Level 5: Section 741

*Remnants of the germinal matrix, migratory streams, and transitional fields*

| | | | |
|---|---|---|---|
| 1 | Frontal neuroepithelium and subventricular zone | 6 | Strionuclear glioepithelium |
| 2 | Frontal stratified transitional field | 7 | Anteromedial striatal neuroepithelium and subventricular zone |
| 3 | Callosal glioepithelium | 8 | Anterolateral striatal neuroepithelium and subventricular zone |
| 4 | Callosal sling | 9 | Lateral migratory stream (cortical) |
| 5 | Fornical glioepithelium | 10 | Subpial granular layer (cortical) |

**See detail of brain core in Plates 176A and B.**

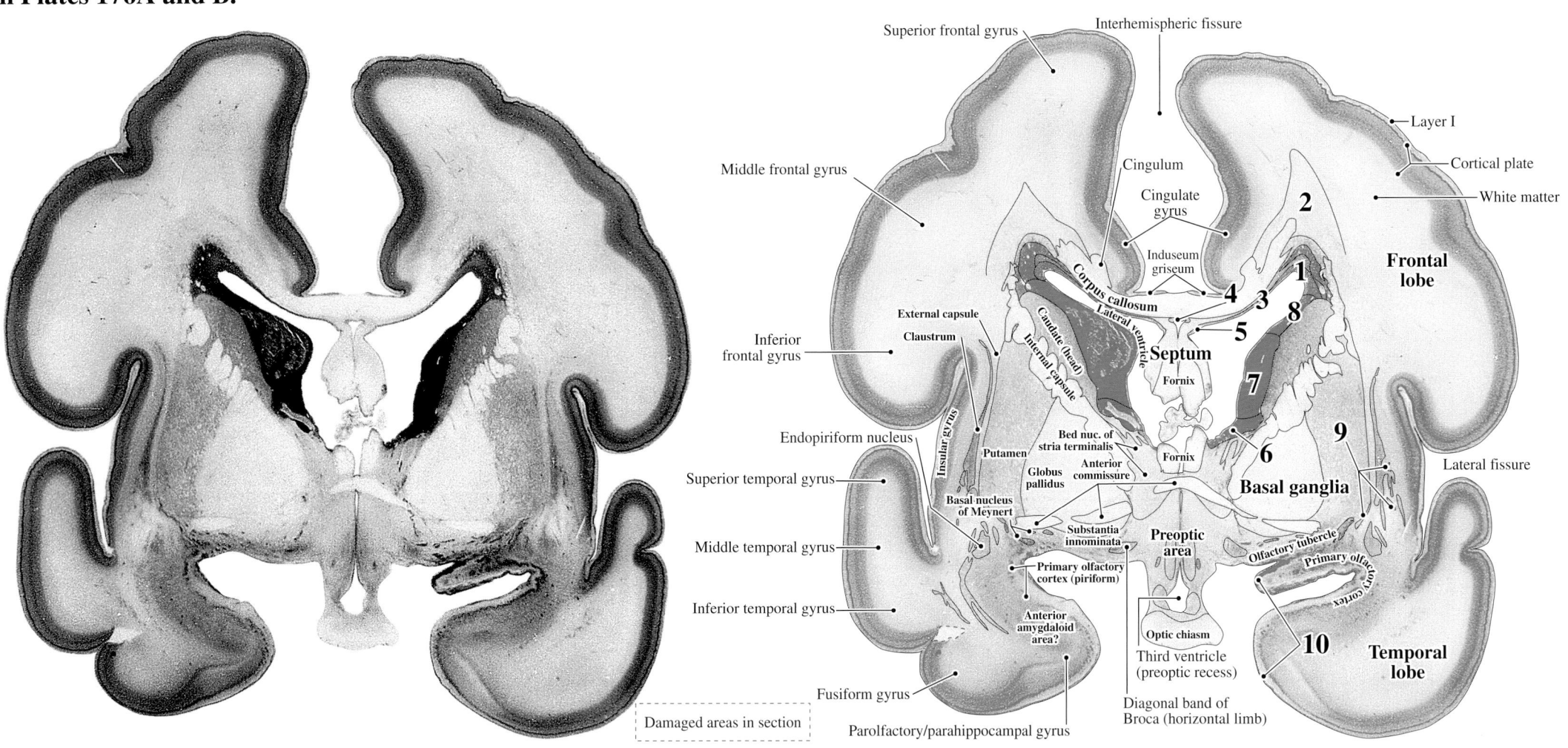

10 mm

PLATE 156

GW29 Coronal
CR 260 mm
Y14-59
Level 6: Section 781

### Remnants of the germinal matrix, migratory streams, and transitional fields

| | | | |
|---|---|---|---|
| 1 | Frontal neuroepithelium and subventricular zone | 6 | Strionuclear glioepithelium |
| 2 | Frontal stratified transitional field | 7 | Anteromedial striatal neuroepithelium and subventricular zone |
| 3 | Callosal glioepithelium | 8 | Anterolateral striatal neuroepithelium and subventricular zone |
| 4 | Callosal sling | 9 | Lateral migratory stream (cortical) |
| 5 | Fornical glioepithelium | 10 | Parahippocampal stratified transitional field (intermingled with the amygdaloid glioepithelium/ependyma) |

See detail of brain core in Plates 177A and B.

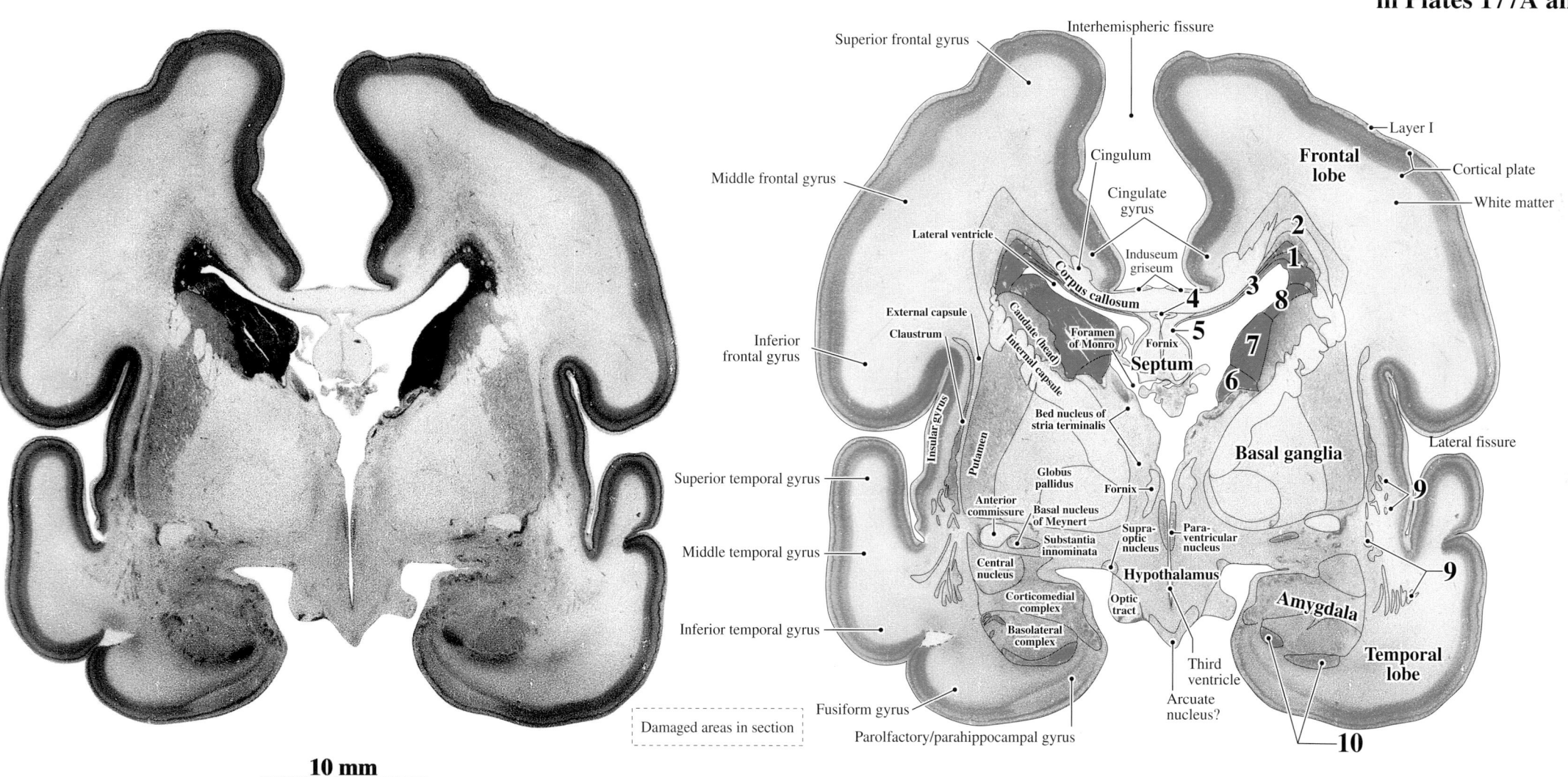

# PLATE 157

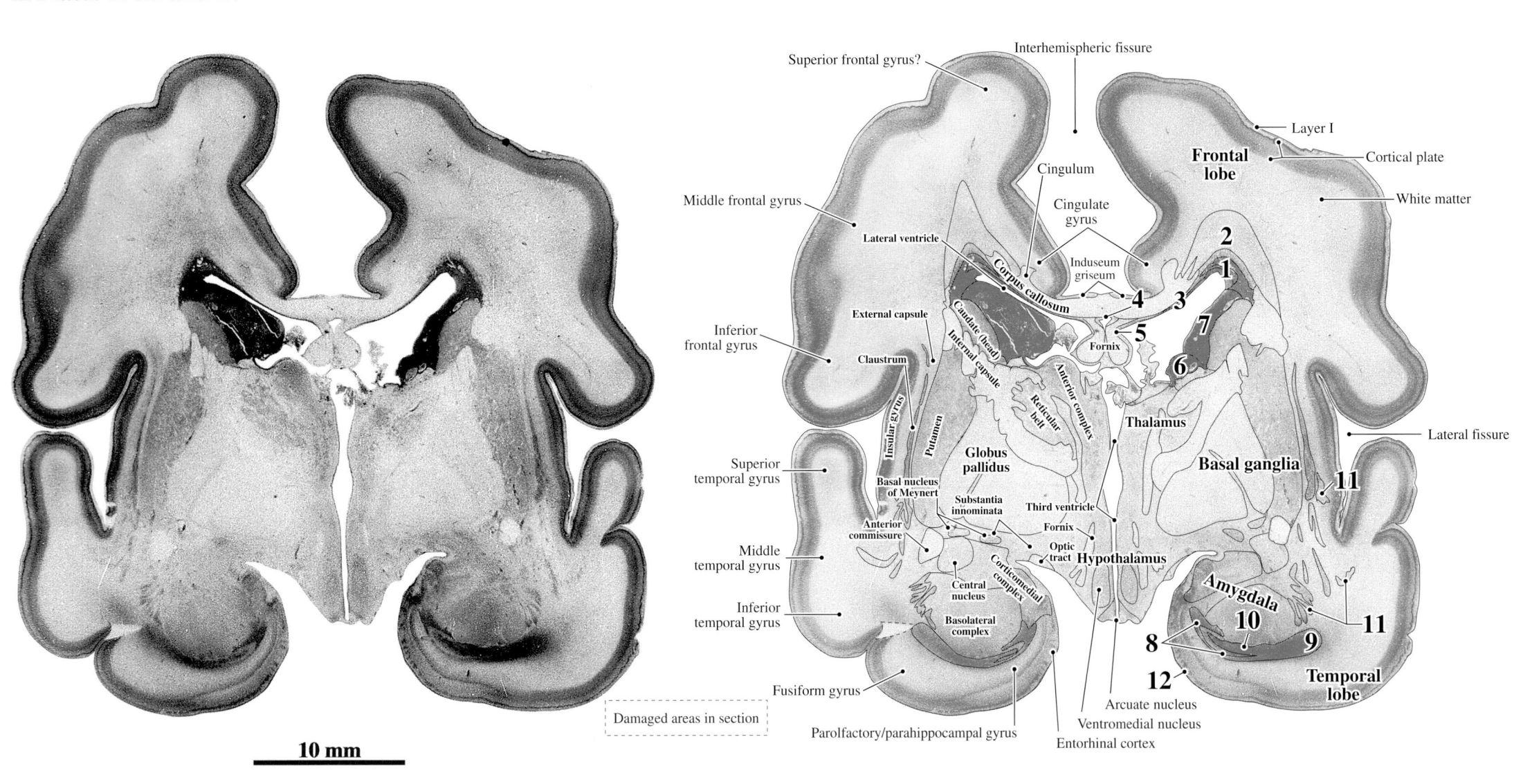

GW29 Coronal
CR 260 mm
Y14-59
Level 7: Section 821

**See detail of brain core in Plates 178A and B.**

*Remnants of the germinal matrix, migratory streams, and transitional fields*

| | | | |
|---|---|---|---|
| 1 | Frontal neuroepithelium and subventricular zone | 7 | Anteromedial striatal neuroepithelium and subventricular zone |
| 2 | Frontal stratified transitional field | 8 | Parahippocampal neuroepithelium, subventricular zone, and stratified transitional field |
| 3 | Callosal glioepithelium | 9 | Temporal neuroepithelium and subventricular zone (intermingled with the amygdaloid glioepithelium/ependyma) |
| 4 | Callosal sling | 10 | Amygdaloid glioepithelium/ependyma |
| 5 | Fornical glioepithelium | 11 | Lateral migratory stream (cortical) |
| 6 | Strionuclear glioepithelium | 12 | Subpial granular layer (cortical) |

10 mm

# PLATE 158

**GW29 Coronal**
**CR 260 mm**
**Y14-59**
**Level 8: Section 861**

## Remnants of the germinal matrix, migratory streams, and transitional fields

1. Frontal/paracentral neuroepithelium and subventricular zone
2. Frontal/paracentral stratified transitional field
3. Callosal glioepithelium
4. Strionuclear glioepithelium
5. Anteromedial/posterior striatal neuroepithelium and subventricular zone
6. Amygdaloid glioepithelium/ependyma
7. Alvear glioepithelium
8. Parahippocampal neuroepithelium, subventricular zone, and stratified transitional field
9. Temporal neuroepithelium and subventricular zone
10. Temporal stratified transitional field
11. Lateral migratory stream (cortical)
12. Subpial granular layer (cortical)

See this area of cortex in Plate 171.

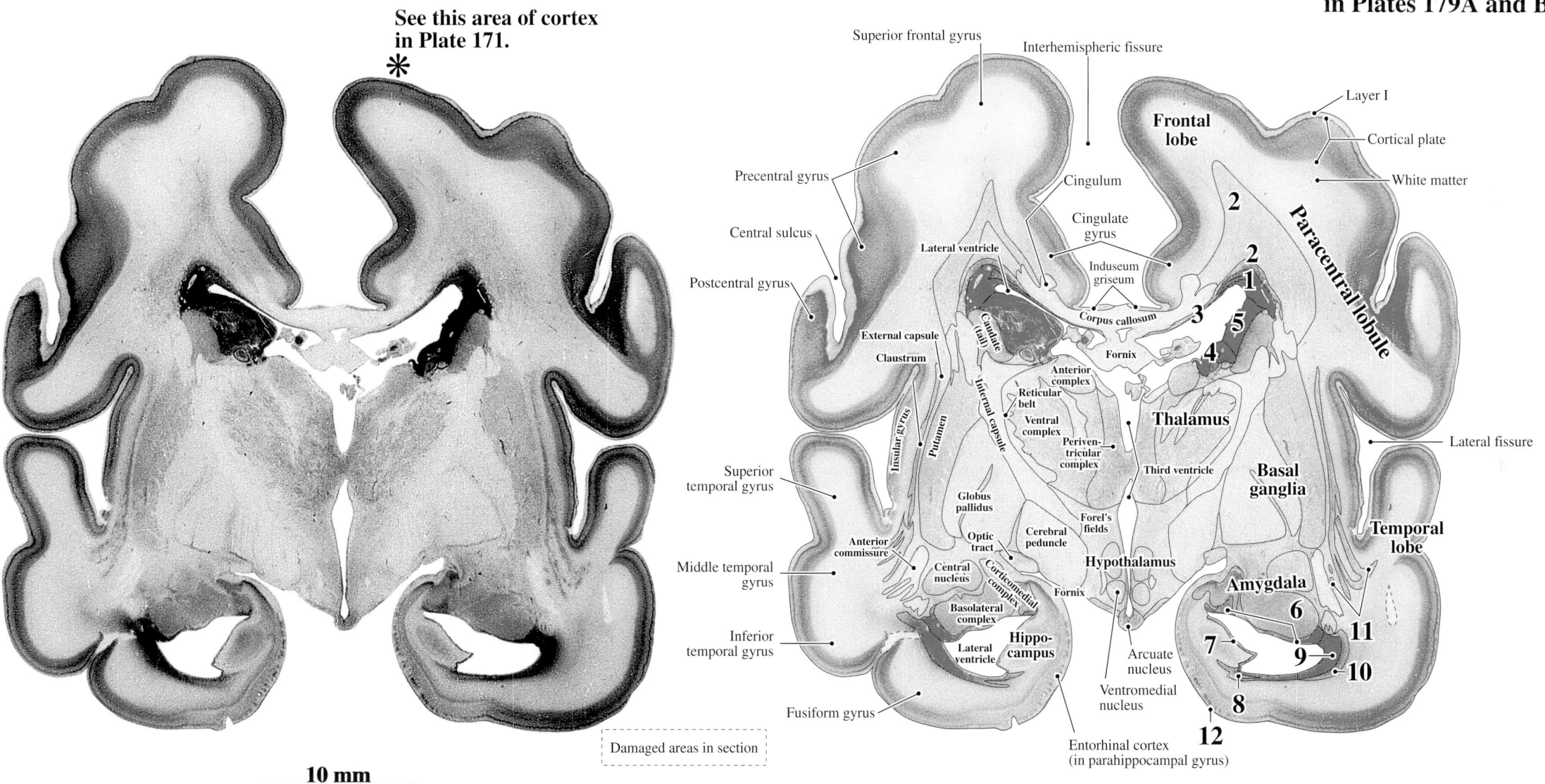

See detail of brain core in Plates 179A and B.

# PLATE 159

**GW29 Coronal**
**CR 260 mm**
**Y14-59**
**Level 9: Section 901**

*Remnants of the germinal matrix, migratory streams, and transitional fields*

| | | | |
|---|---|---|---|
| 1 | Paracentral neuroepithelium and subventricular zone | 7 | Alvear glioepithelium |
| 2 | Paracentral stratified transitional field | 8 | Parahippocampal neuroepithelium, subventricular zone, and stratified transitional field |
| 3 | Callosal glioepithelium | 9 | Temporal neuroepithelium and subventricular zone |
| 4 | Posterior striatal neuroepithelium and subventricular zone | 10 | Temporal stratified transitional field |
| 5 | Strionuclear glioepithelium | 11 | Lateral migratory stream (cortical) |
| 6 | Amygdaloid glioepithelium/ependyma | 12 | Subpial granular layer (cortical) |

See detail of brain core in Plates 180A and B.

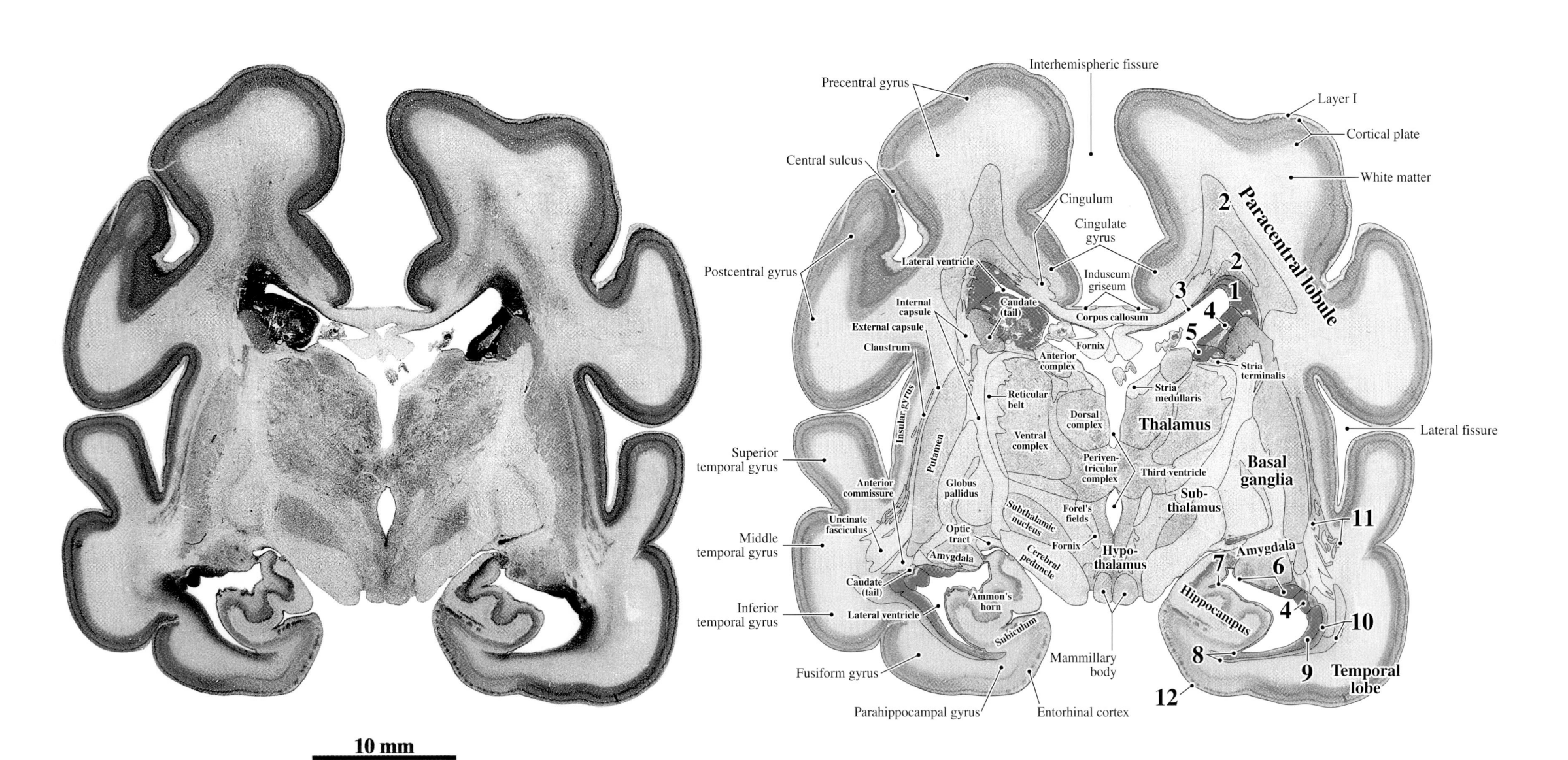

## PLATE 160

**GW29 Coronal**
**CR 260 mm**
**Y14-59**
**Level 10: Section 981**

### Remnants of the germinal matrix, migratory streams, and transitional fields

| | | | |
|---|---|---|---|
| 1 | Paracentral neuroepithelium and subventricular zone | 7 | Subgranular zone (dentate) |
| 2 | Paracentral stratified transitional field | 8 | Parahippocampal neuroepithelium, subventricular zone, and stratified transitional field |
| 3 | Callosal glioepithelium | 9 | Temporal neuroepithelium and subventricular zone |
| 4 | Posterior striatal neuroepithelium and subventricular zone | 10 | Temporal stratified transitional field |
| 5 | Strionuclear glioepithelium | 11 | Lateral migratory stream (cortical) |
| 6 | Alvear glioepithelium | 12 | Subpial granular layer (cortical) |

**See detail of brain core in Plates 181A and B.**

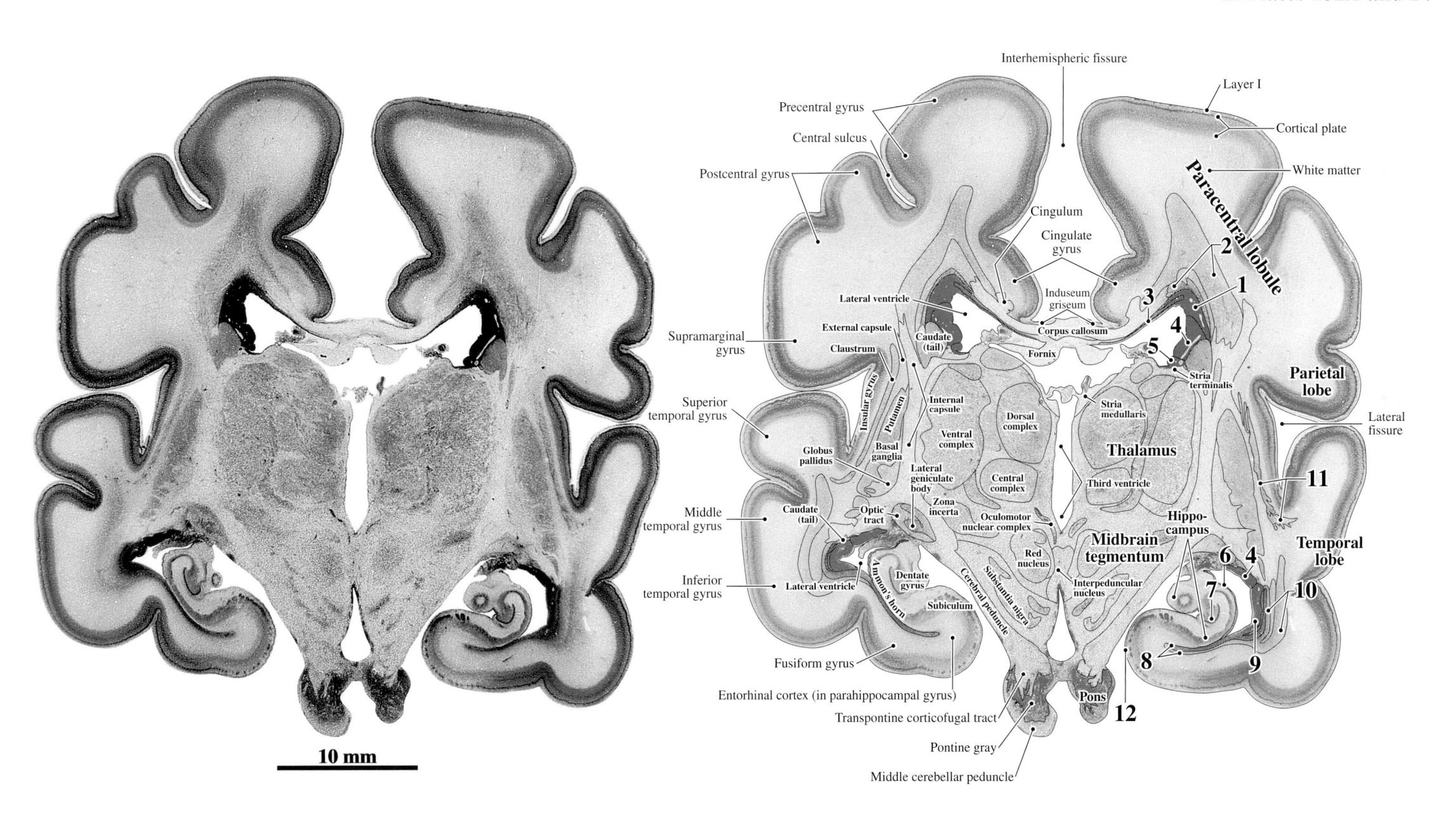

10 mm

## PLATE 161

GW29 Coronal
CR 260 mm
Y14-59
Level 11: Section 1021

See detail of brain core
in Plates 182A and B.

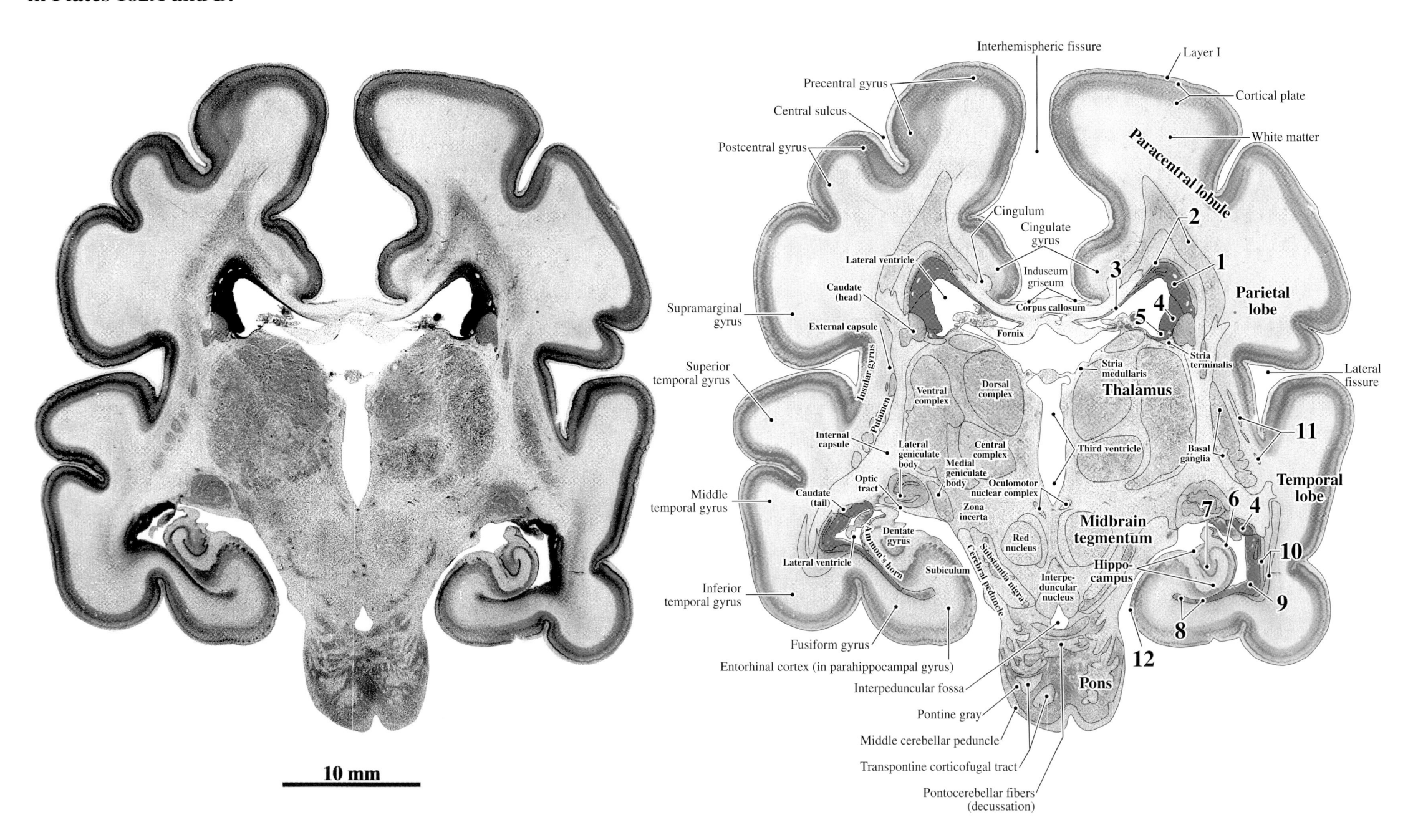

### Remnants of the germinal matrix, migratory streams, and transitional fields

1. *Paracentral/parietal neuroepithelium and subventricular zone*
2. *Paracentral/parietal stratified transitional field*
3. *Callosal glioepithelium*
4. *Posterior striatal neuroepithelium and subventricular zone*
5. *Strionuclear glioepithelium*
6. *Alvear glioepithelium*
7. *Subgranular zone (dentate)*
8. *Parahippocampal neuroepithelium, subventricular zone, and stratified transitional field*
9. *Temporal neuroepithelium and subventricular zone*
10. *Temporal stratified transitional field*
11. *Lateral migratory stream (cortical)*
12. *Subpial granular layer (cortical)*

# PLATE 162

**GW29 Coronal**
**CR 260 mm**
**Y14-59**
**Level 12: Section 1061**

## Remnants of the germinal matrix, migratory streams, and transitional fields

1. Paracentral/parietal neuroepithelium and subventricular zone
2. Paracentral/parietal stratified transitional field
3. Callosal glioepithelium
4. Posterior striatal neuroepithelium and subventricular zone
5. Strionuclear glioepithelium
6. Alvear glioepithelium
7. Subgranular zone (dentate)
8. Parahippocampal neuroepithelium, subventricular zone, and stratified transitional field
9. Temporal neuroepithelium and subventricular zone
10. Temporal stratified transitional field
11. Mesencephalic glioepithelium/ependyma
12. Subpial granular layer (cortical)

**See detail of brain core in Plates 183A and B.**

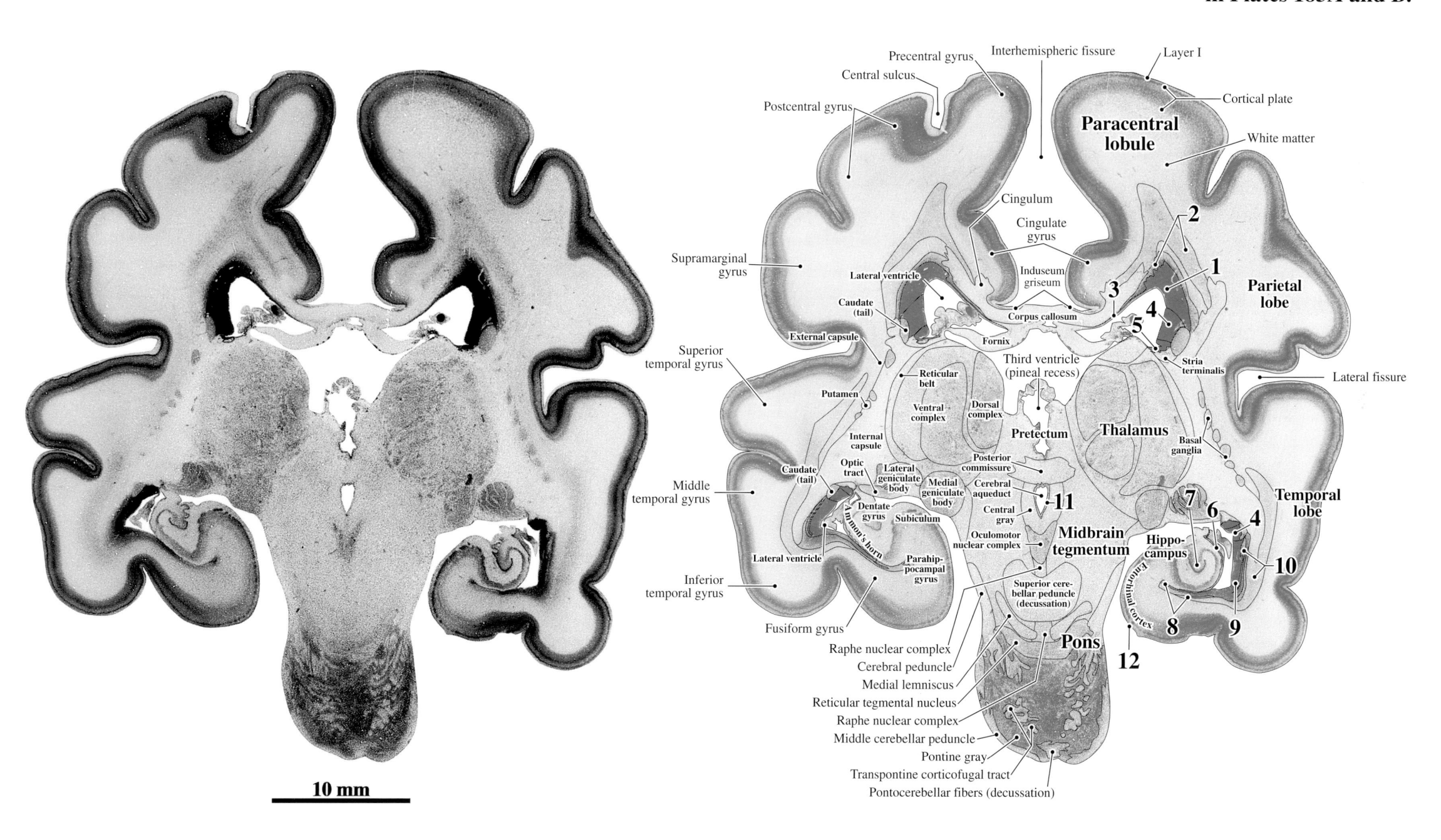

# PLATE 163

GW29 Coronal
CR 260 mm
Y14-59
Level 13: Section 1101

*Remnants of the germinal matrix, migratory streams, and transitional fields*

1 *Paracentral/parietal neuroepithelium and subventricular zone*
2 *Paracentral/parietal stratified transitional field*
3 *Callosal glioepithelium*
4 *Posterior striatal neuroepithelium and subventricular zone*
5 *Strionuclear glioepithelium*
6 *Alvear glioepithelium*
7 *Subgranular zone (dentate)*
8 *Parahippocampal neuroepithelium, subventricular zone, and stratified transitional field*
9 *Temporal neuroepithelium and subventricular zone*
10 *Temporal stratified transitional field*
11 *Mesencephalic glioepithelium/ependyma*
12 *Subpial granular layer (cortical)*

**See detail of brain core in Plates 184A and B.**

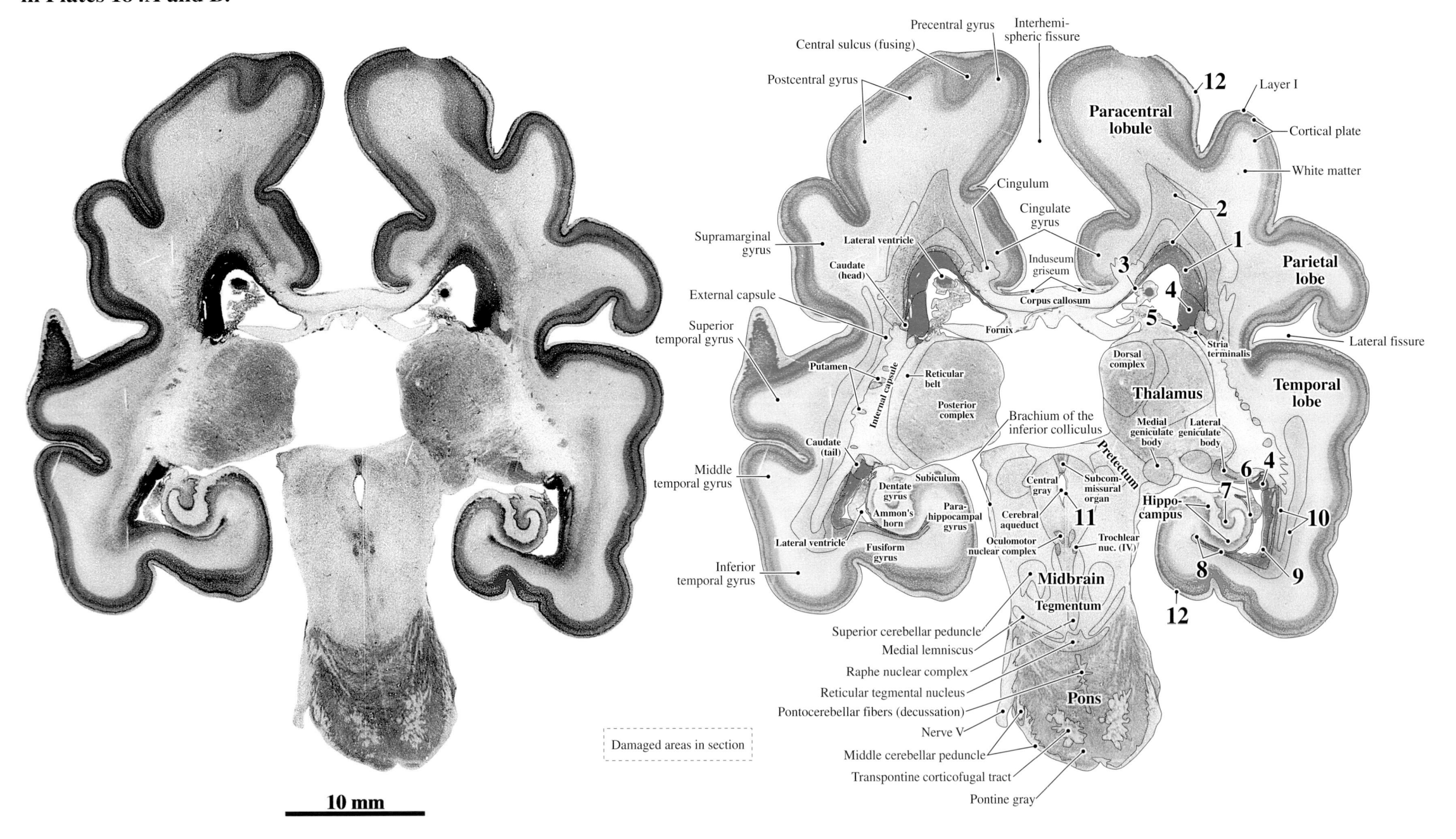

# PLATE 164

**GW29 Coronal**
**CR 260 mm**
**Y14-59**
**Level 14: Section 1161**

### Remnants of the germinal matrix, migratory streams, and transitional fields

1. *Paracentral/parietal neuroepithelium and subventricular zone*
2. *Paracentral/parietal stratified transitional field*
3. *Callosal glioepithelium*
4. *Posterior striatal neuroepithelium and subventricular zone*
5. *Strionuclear glioepithelium*
6. *Alvear glioepithelium*
7. *Subgranular zone (dentate)*
8. *Parahippocampal neuroepithelium, subventricular zone, and stratified transitional field*
9. *Temporal neuroepithelium and subventricular zone*
10. *Temporal stratified transitional field*
11. *Mesencephalic glioepithelium/ependyma*
12. *Raphe migration*
13. *External germinal layer (cerebellum)*
14. *Subpial granular layer (cortical)*

See this area of cortex ✱ in Plate 172.

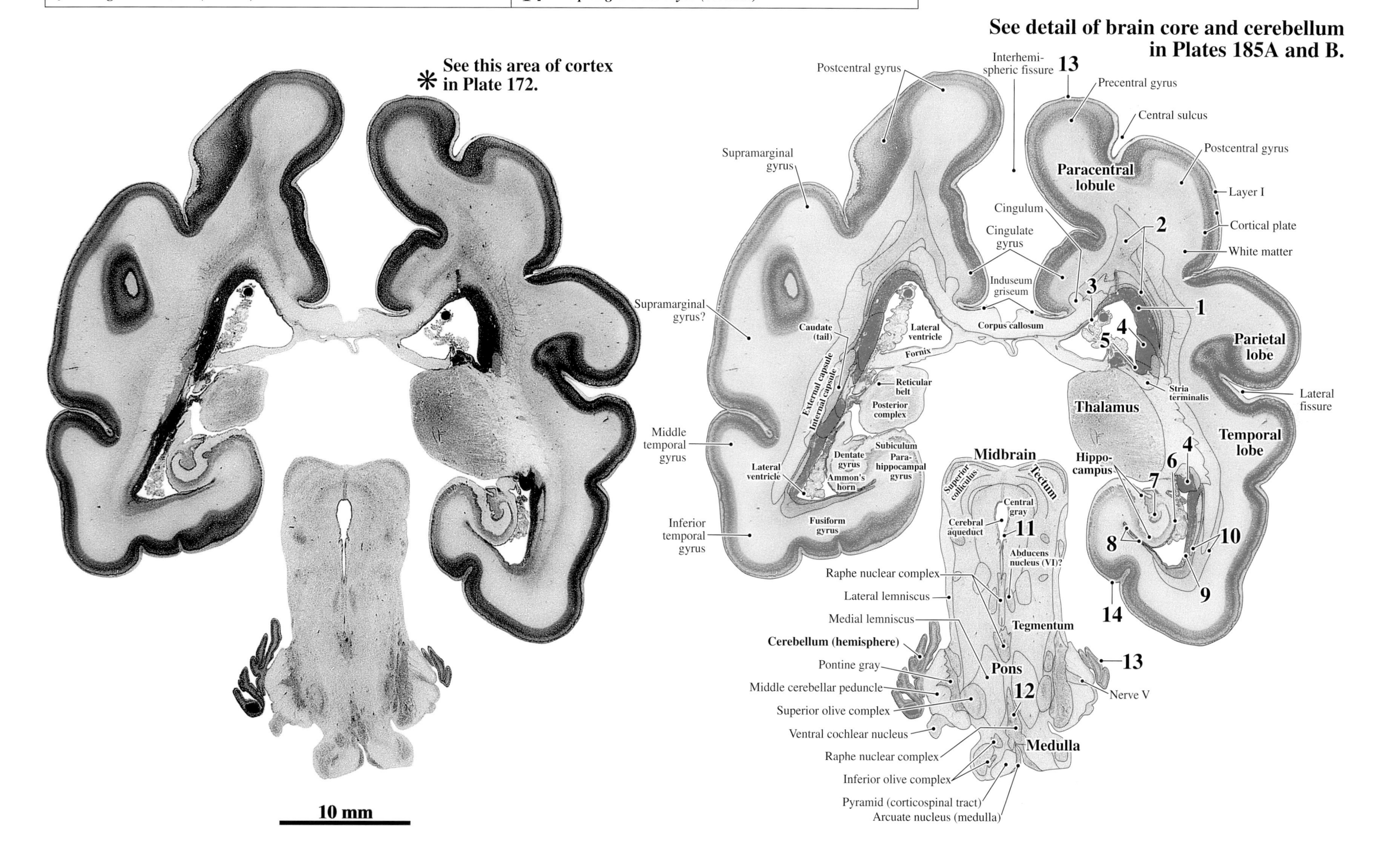

See detail of brain core and cerebellum in Plates 185A and B.

# PLATE 165

GW29 Coronal
CR 260 mm
Y14-59
Level 15: Section 1201

*Remnants of the germinal matrix, migratory streams, and transitional fields*

1 *Paracentral/parietal neuroepithelium and subventricular zone*
2 *Paracentral/parietal stratified transitional field*
3 *Callosal glioepithelium*
4 *Posterior striatal neuroepithelium and subventricular zone*
5 *Strionuclear neuroepithelium/glioepithelium*
6 *Alvear glioepithelium*
7 *Subgranular zone (dentate)*
8 *Parahippocampal neuroepithelium, subventricular zone, and stratified transitional field*
9 *Temporal neuroepithelium and subventricular zone*
10 *Temporal stratified transitional field*
11 *Mesencephalic glioepithelium/ependyma*
12 *Isthmal and pontine glioepithelium/ependyma*
13 *Raphe migration*
14 *External germinal layer (cerebellum)*
15 *Subpial granular layer (cortical)*

**See detail of brain core and cerebellum in Plates 186A and B.**

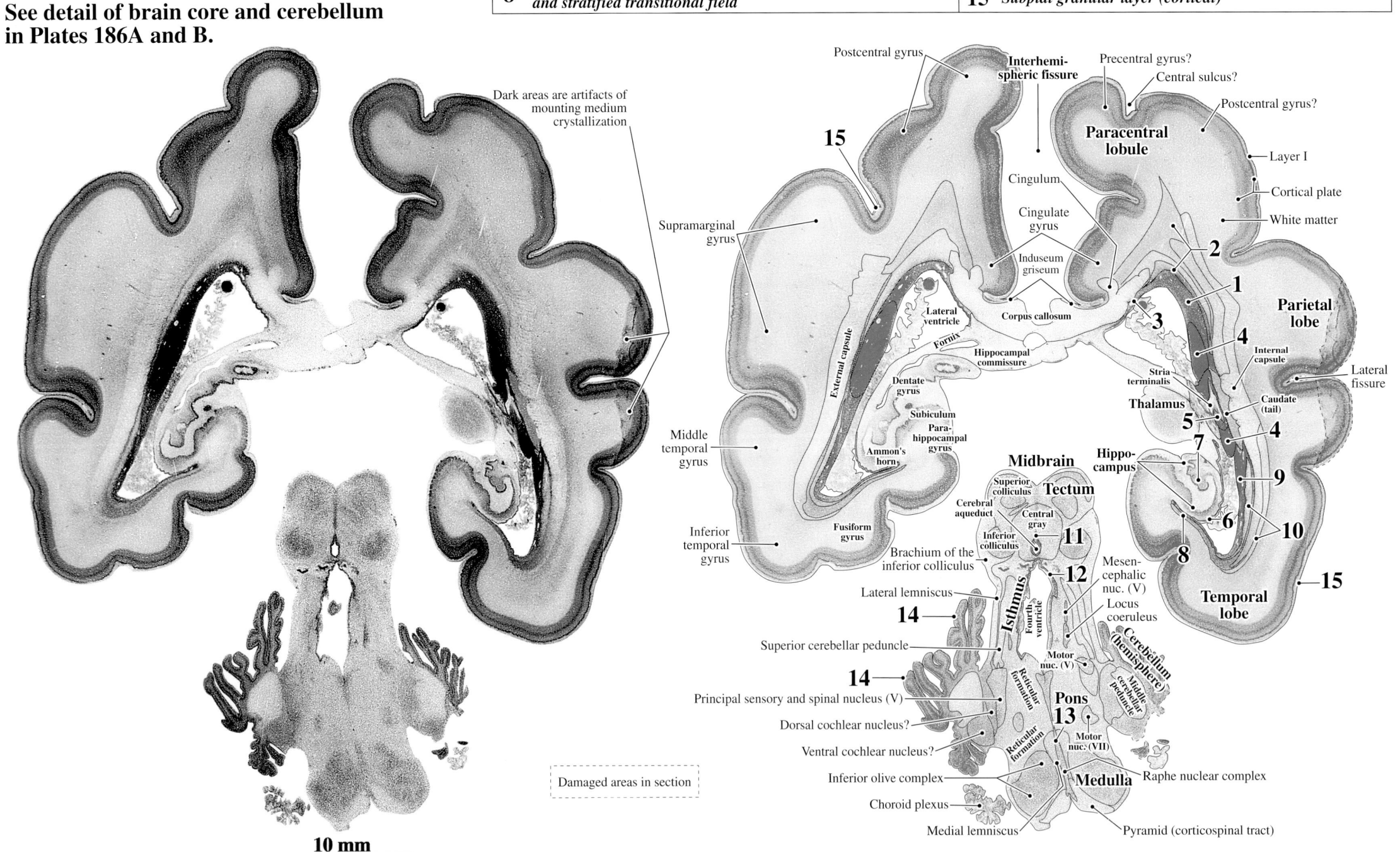

# PLATE 166

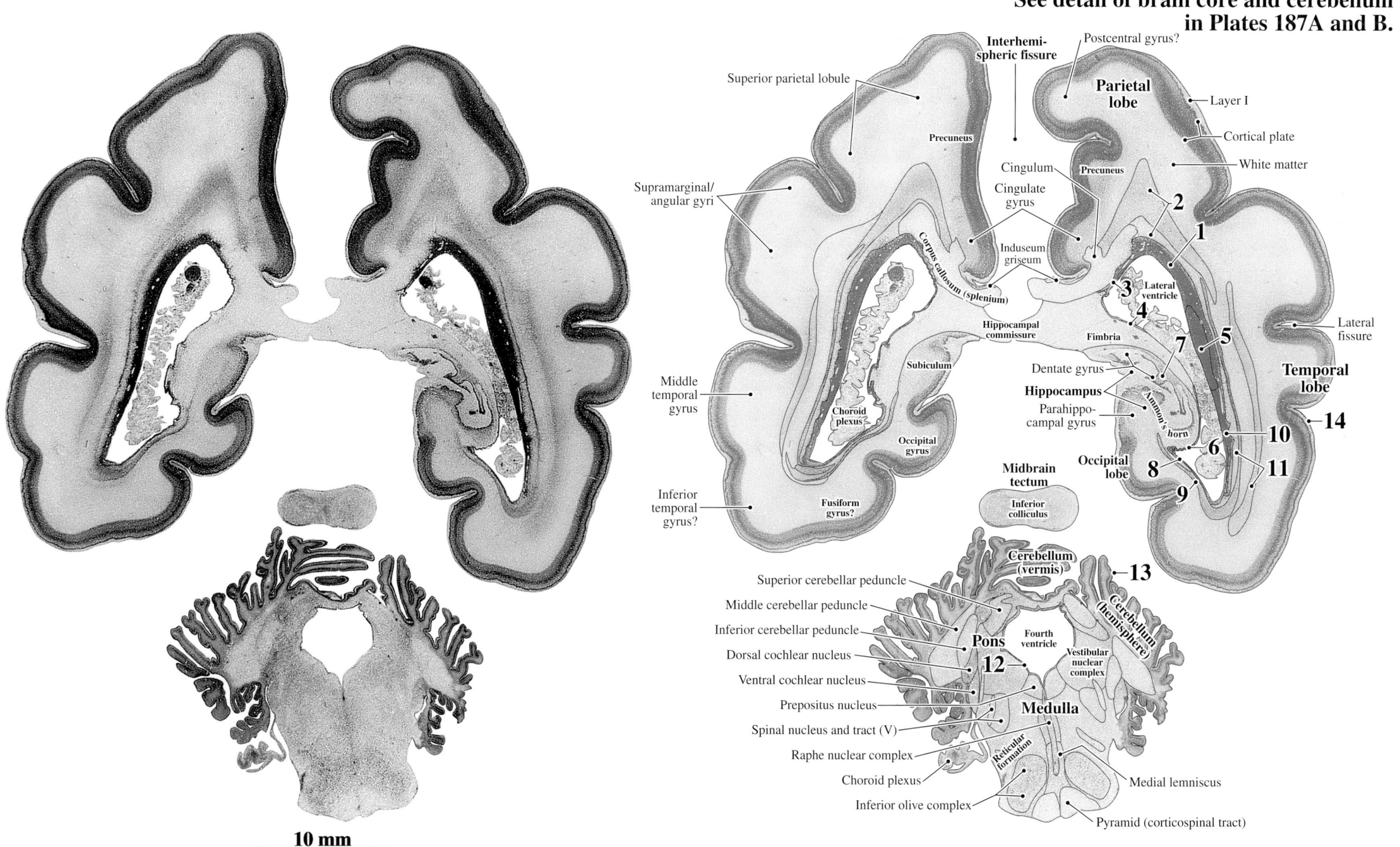

**GW29 Coronal**
**CR 260 mm**
**Y14-59**
**Level 16: Section 1261**

### Remnants of the germinal matrix, migratory streams, and transitional fields

1. *Parietal neuroepithelium and subventricular zone*
2. *Parietal stratified transitional field*
3. *Callosal glioepithelium*
4. *Fimbrial glioepithelium*
5. *Posterior striatal neuroepithelium and subventricular zone*
6. *Alvear glioepithelium*
7. *Subgranular zone (dentate)*
8. *Occipital neuroepithelium and subventricular zone*
9. *Occipital stratified transitional field*
10. *Temporal neuroepithelium and subventricular zone*
11. *Temporal stratified transitional field*
12. *Pontine and medullary glioepithelium/ependyma*
13. *External germinal layer (cerebellum)*
14. *Subpial granular layer (cortical)*

**See detail of brain core and cerebellum in Plates 187A and B.**

# PLATE 167

GW29 Coronal
CR 260 mm
Y14-59
Level 17: Section 1301

### Remnants of the germinal matrix, migratory streams, and transitional fields

1  Parietal neuroepithelium and subventricular zone
2  Parietal stratified transitional field
3  Callosal glioepithelium
4  Fimbrial glioepithelium
5  Occipital neuroepithelium and subventricular zone
6  Occipital stratified transitional field
7  Temporal neuroepithelium and subventricular zone
8  Temporal stratified transitional field
9  Medullary glioepithelium/ependyma
10  Germinal trigone (cerebellum)
11  External germinal layer (cerebellum)
12  Subpial granular layer (cortical)

**See detail of brain core and cerebellum in Plates 188A and B.**

∗ See this area of cortex in Plate 173.

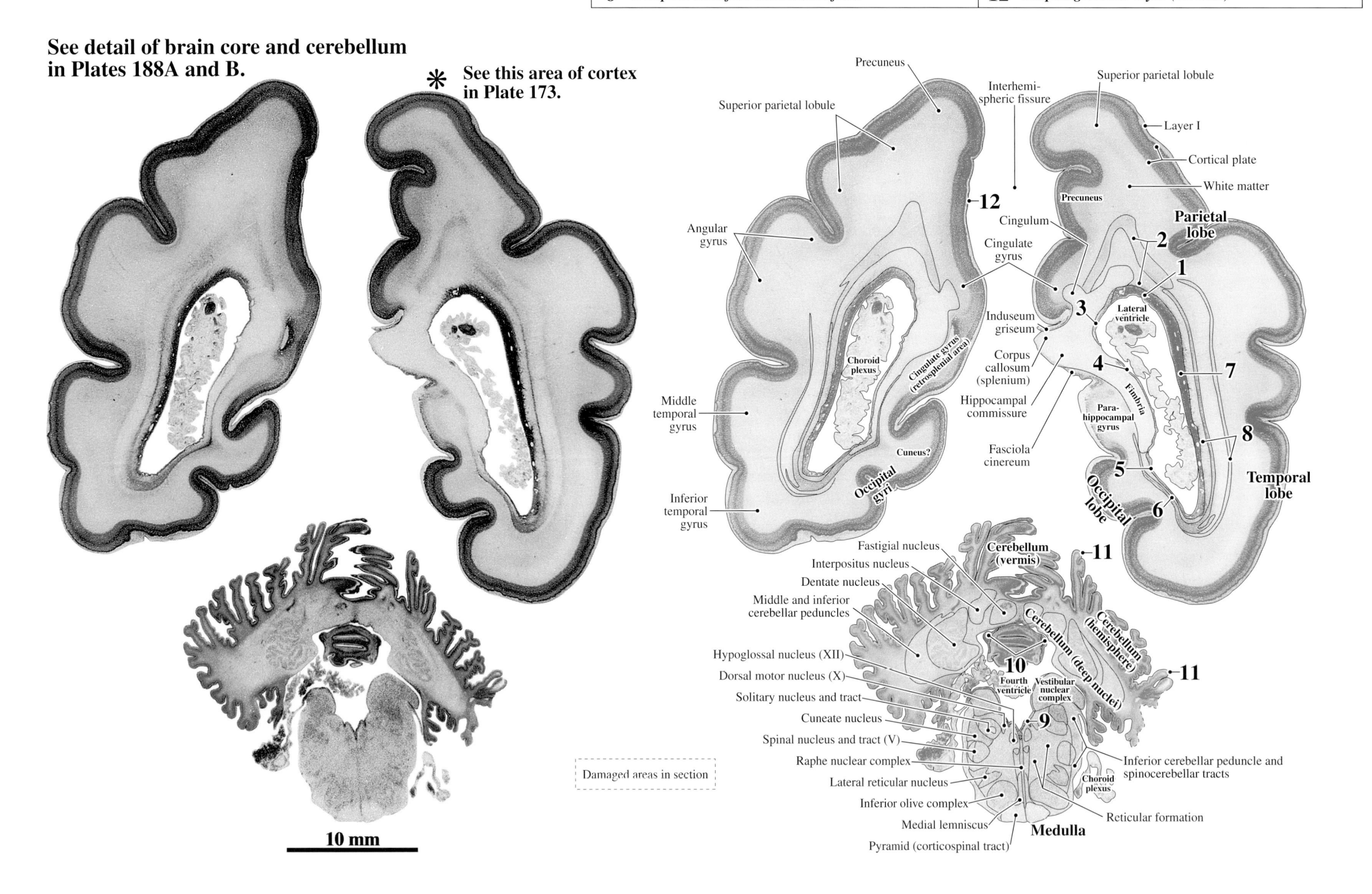

PLATE 168

GW29 Coronal
CR 260 mm
Y14-59
Level 18: Section 1361

**Remnants of the germinal matrix, migratory streams, and transitional fields**

1  *Parietal neuroepithelium and subventricular zone*
2  *Parietal stratified transitional field*
3  *Callosal glioepithelium*
4  *Occipital neuroepithelium and subventricular zone*
5  *Occipital stratified transitional field*
6  *Temporal neuroepithelium and subventricular zone*
7  *Temporal stratified transitional field*
8  *External germinal layer (cerebellum)*
9  *Spinomedullary glioepithelium/ependyma*
10 *Subpial granular layer (cortical)*

See detail of brain core and cerebellum in Plates 189A and B.

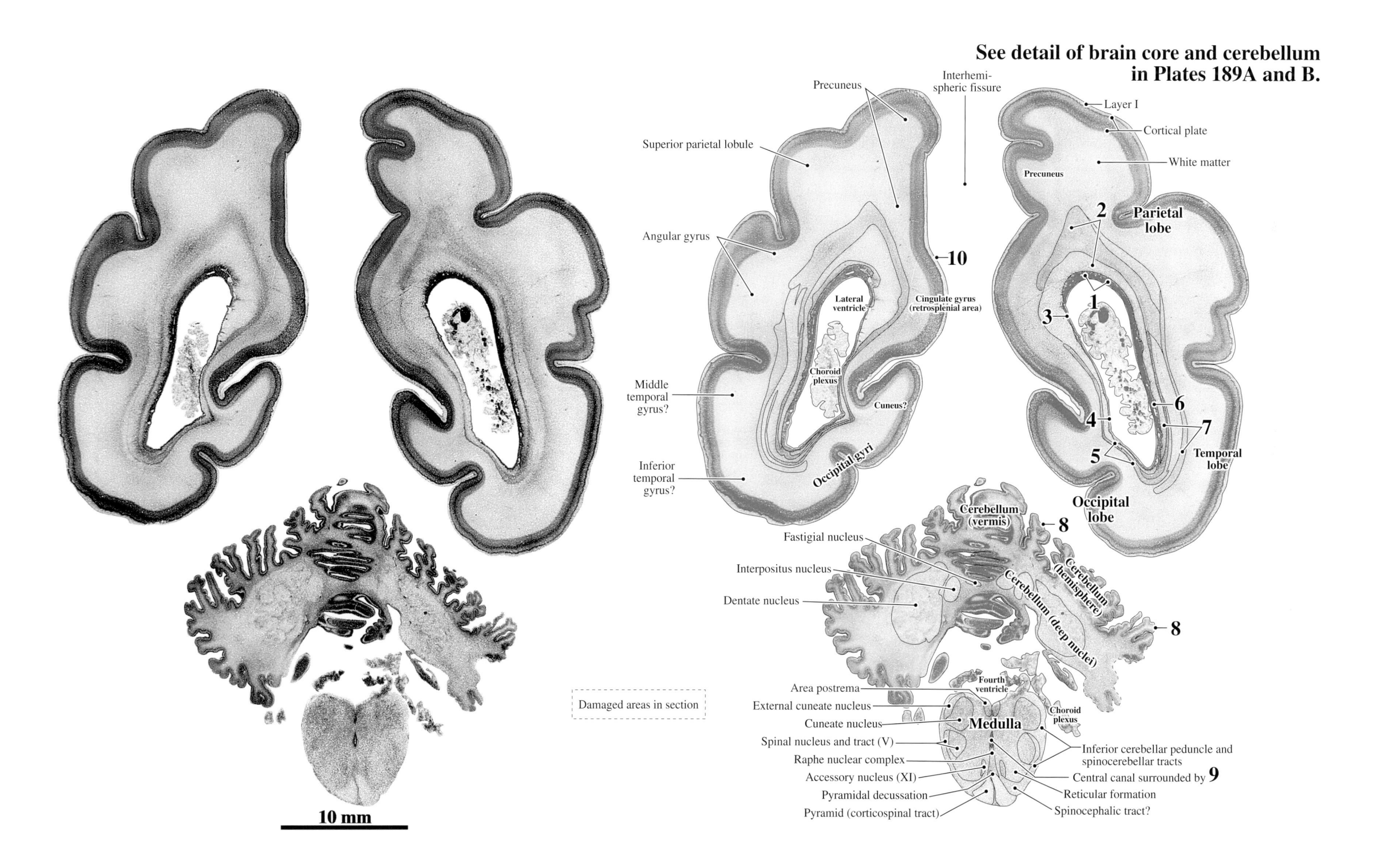

# PLATE 169

**GW29 Coronal**
**CR 260 mm**
**Y14-59**
**Level 19: Section 1521**

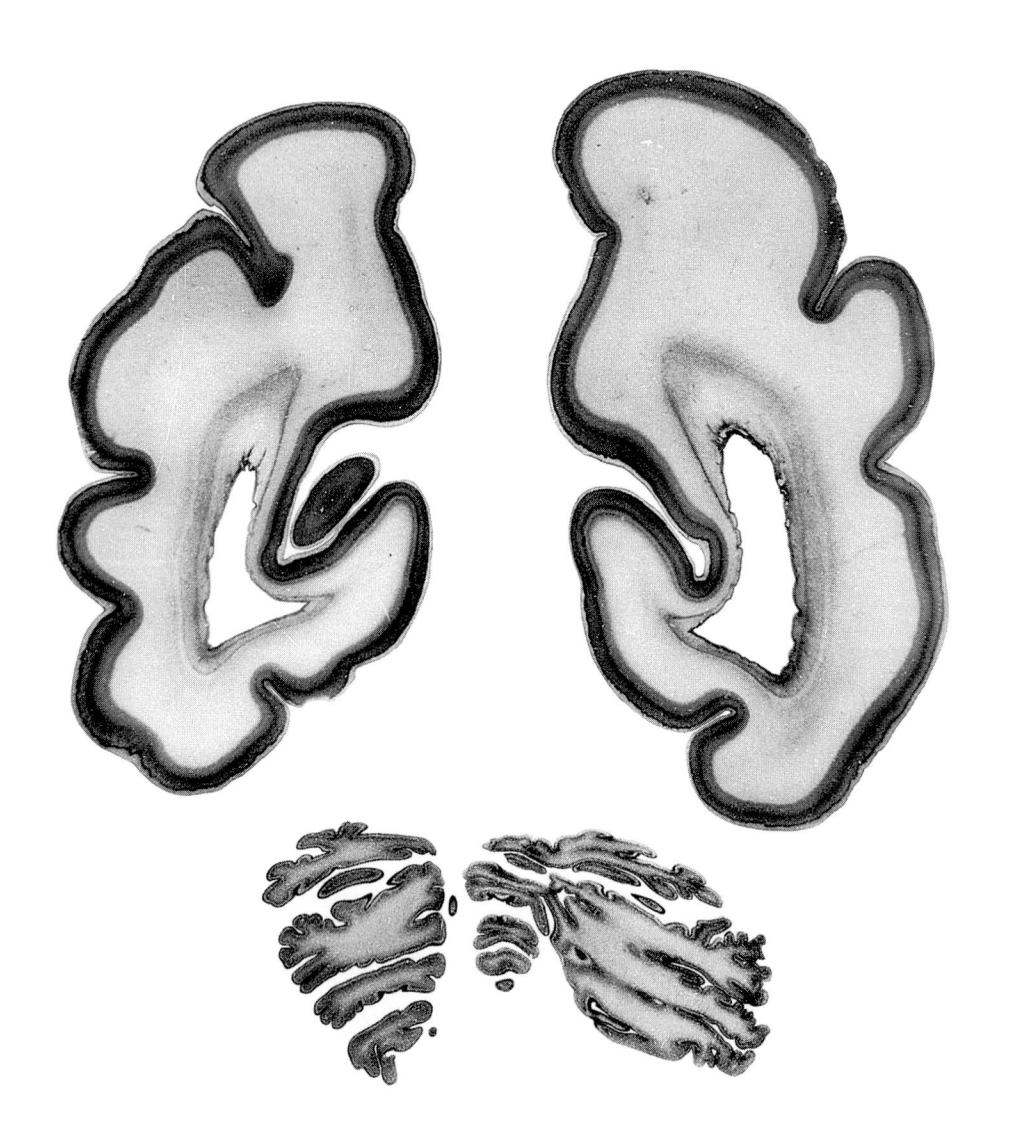

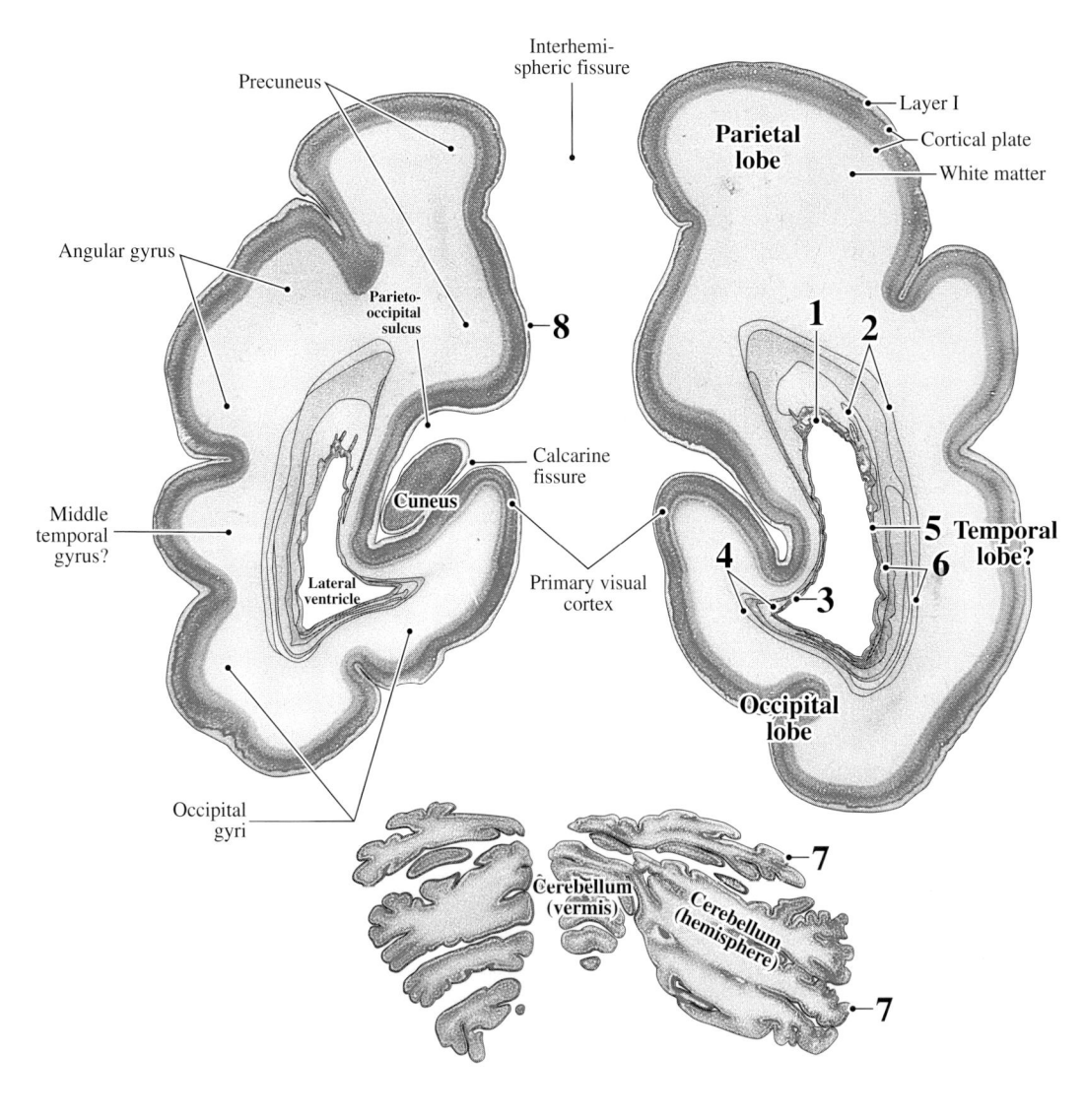

*Remnants of the germinal matrix, migratory streams, and transitional fields*

| | | | |
|---|---|---|---|
| 1 | *Parietal neuroepithelium and subventricular zone* | 5 | *Temporal neuroepithelium and subventricular zone?* |
| 2 | *Parietal stratified transitional field* | 6 | *Temporal stratified transitional field?* |
| 3 | *Occipital neuroepithelium and subventricular zone* | 7 | *External germinal layer (cerebellum)* |
| 4 | *Occipital stratified transitional field* | 8 | *Subpial granular layer (cortical)* |

**PLATE 170**

GW29 Coronal
CR 260 mm
Y14-59
Level 20: Section 1621

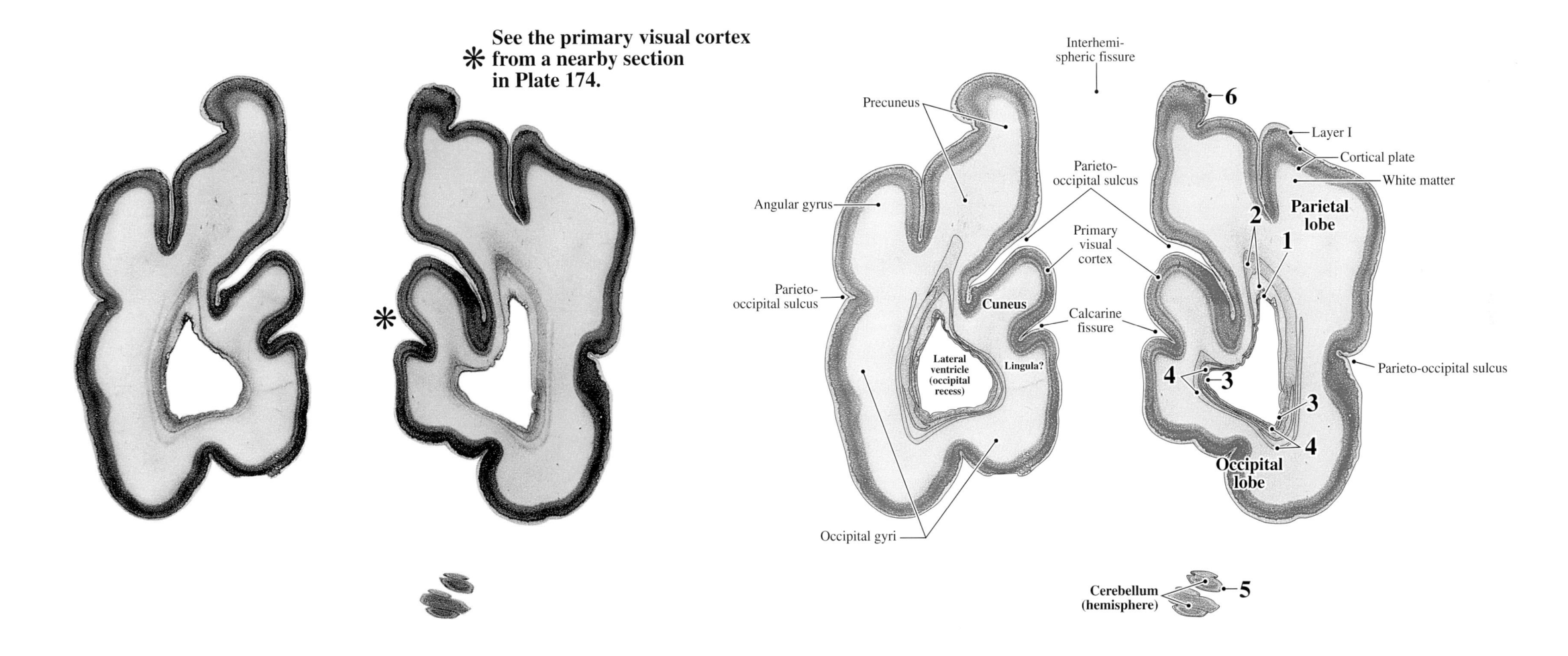

*Remnants of the germinal matrix, migratory streams, and transitional fields*

1  Parietal neuroepithelium and subventricular zone
2  Parietal stratified transitional field
3  Occipital neuroepithelium and subventricular zone
4  Occipital stratified transitional field
5  External germinal layer (cerebellum)
6  Subpial granular layer (cortical)

See the primary visual cortex from a nearby section in Plate 174.

10 mm

# PLATE 171

**GW29 Coronal**
**CR 260 mm, Y14-59**
**Level 8: Section 861**

**FRONTAL CORTEX**

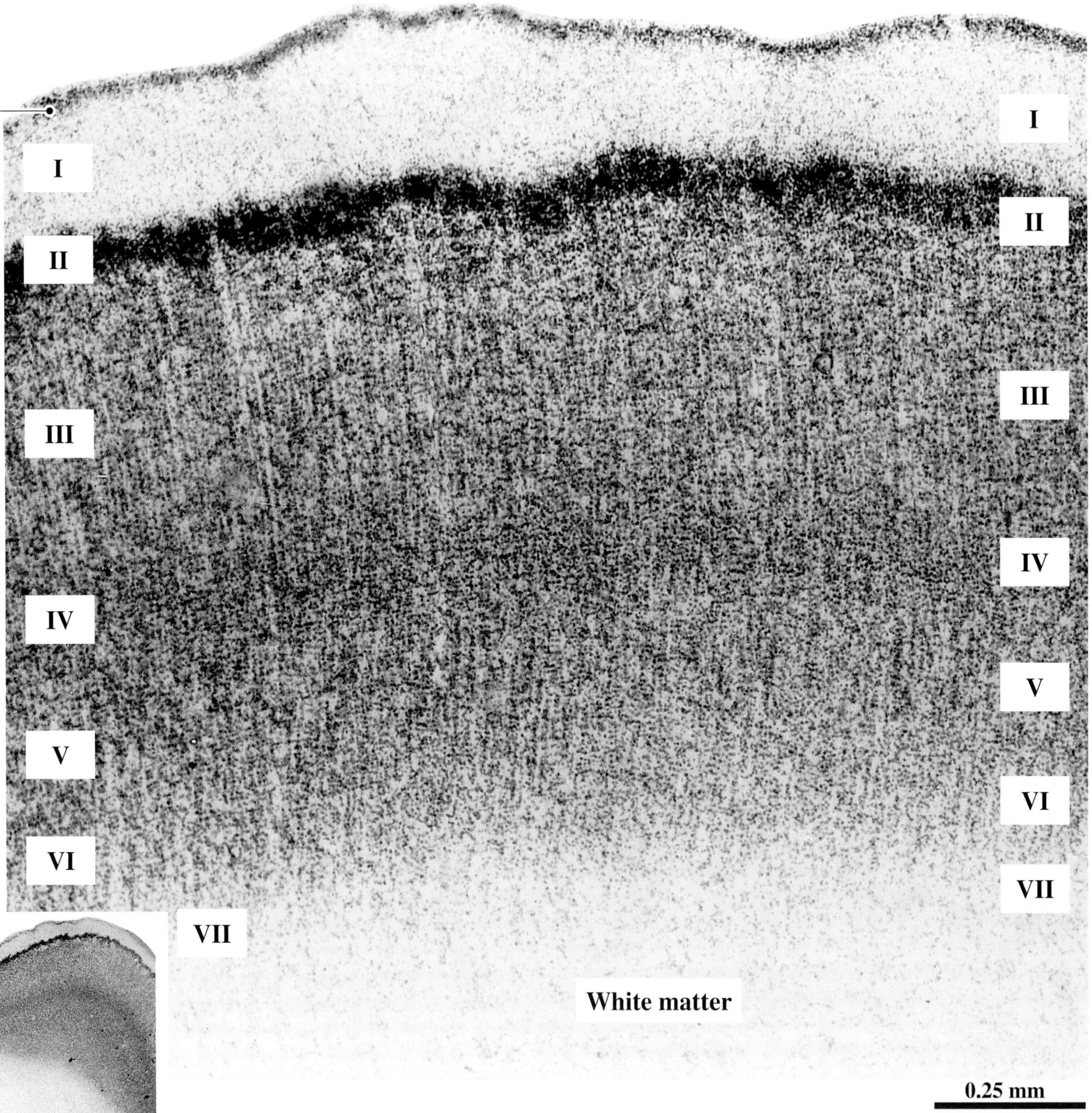

*Subpial granular layer*
*(transient proliferative glial matrix)*
I
II
*Latest arriving cells*
III
IV
V
VI
VII
I
II
III
IV
V
VI
VII
White matter
0.25 mm


**See the entire section in Plate 158.**

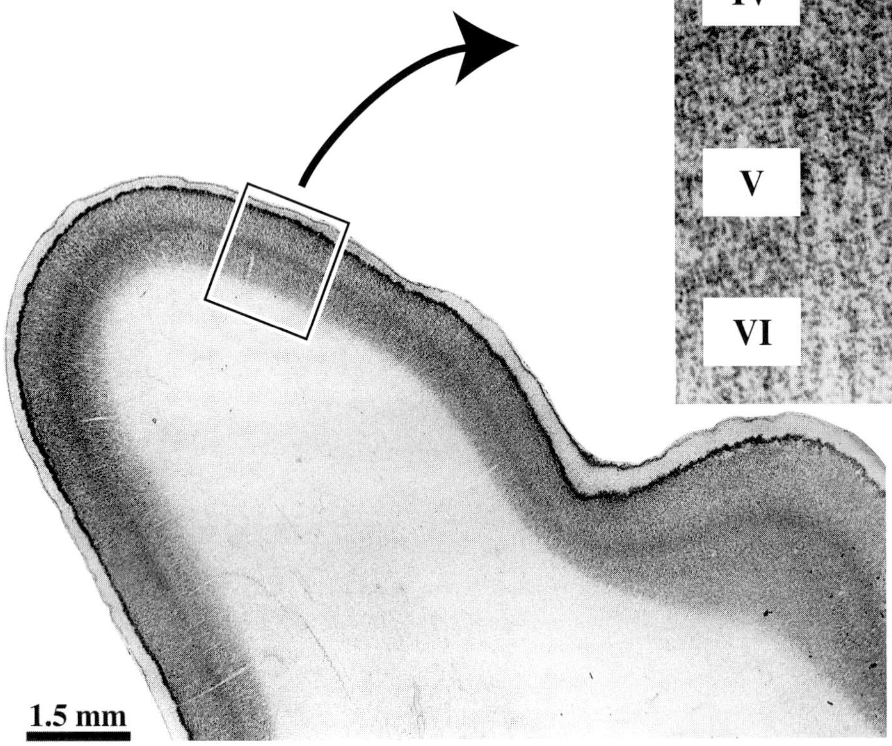

1.5 mm

# PLATE 172

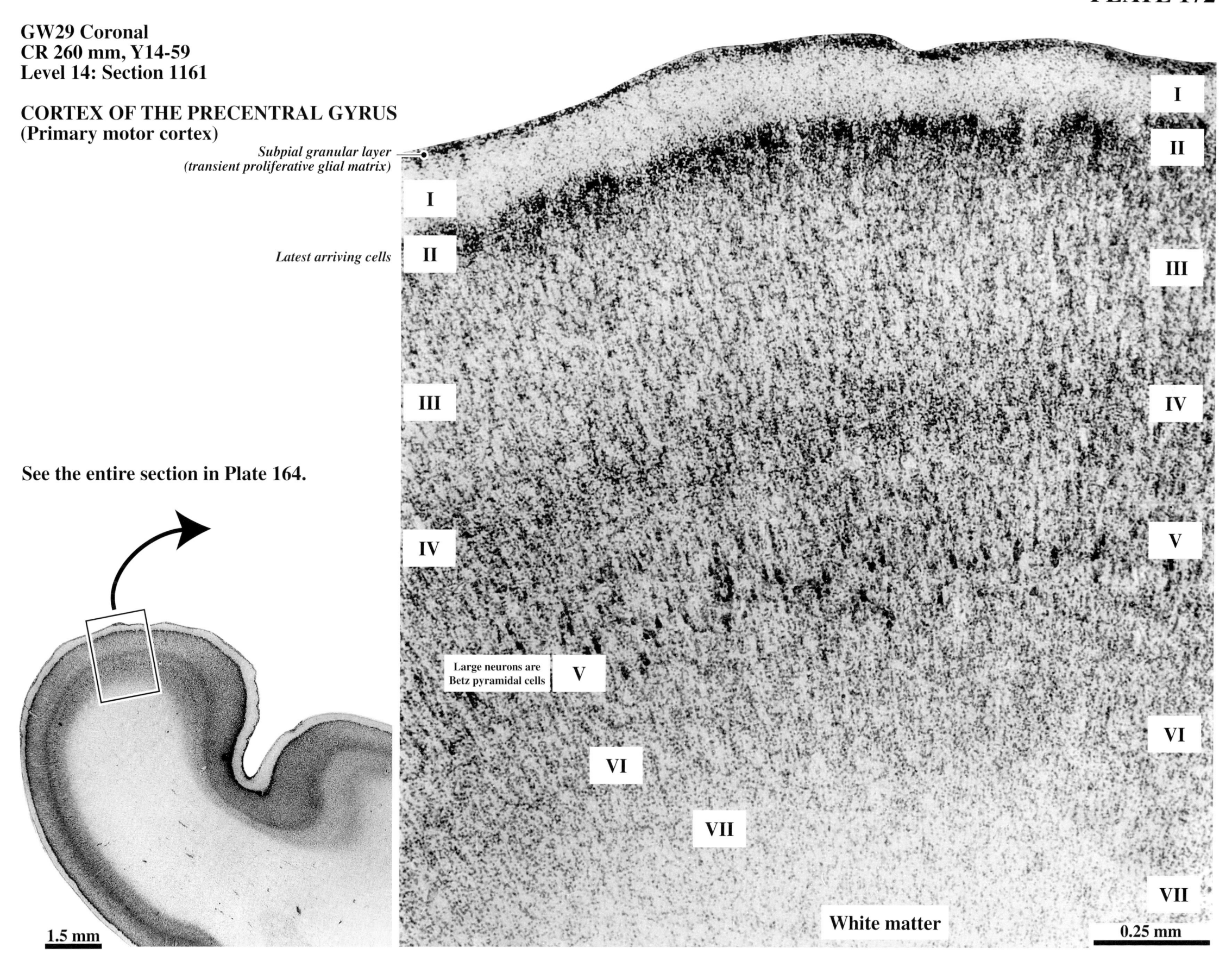

**GW29 Coronal**
**CR 260 mm, Y14-59**
**Level 14: Section 1161**

**CORTEX OF THE PRECENTRAL GYRUS**
(Primary motor cortex)

*Subpial granular layer*
*(transient proliferative glial matrix)*

*Latest arriving cells*

See the entire section in Plate 164.

1.5 mm

I
II
III
IV
V — Large neurons are Betz pyramidal cells
VI
VII
White matter

0.25 mm

# PLATE 173

**GW29 Coronal**
**CR 260 mm, Y14-59**
**Level 17: Section 1301**

**PARIETAL CORTEX**

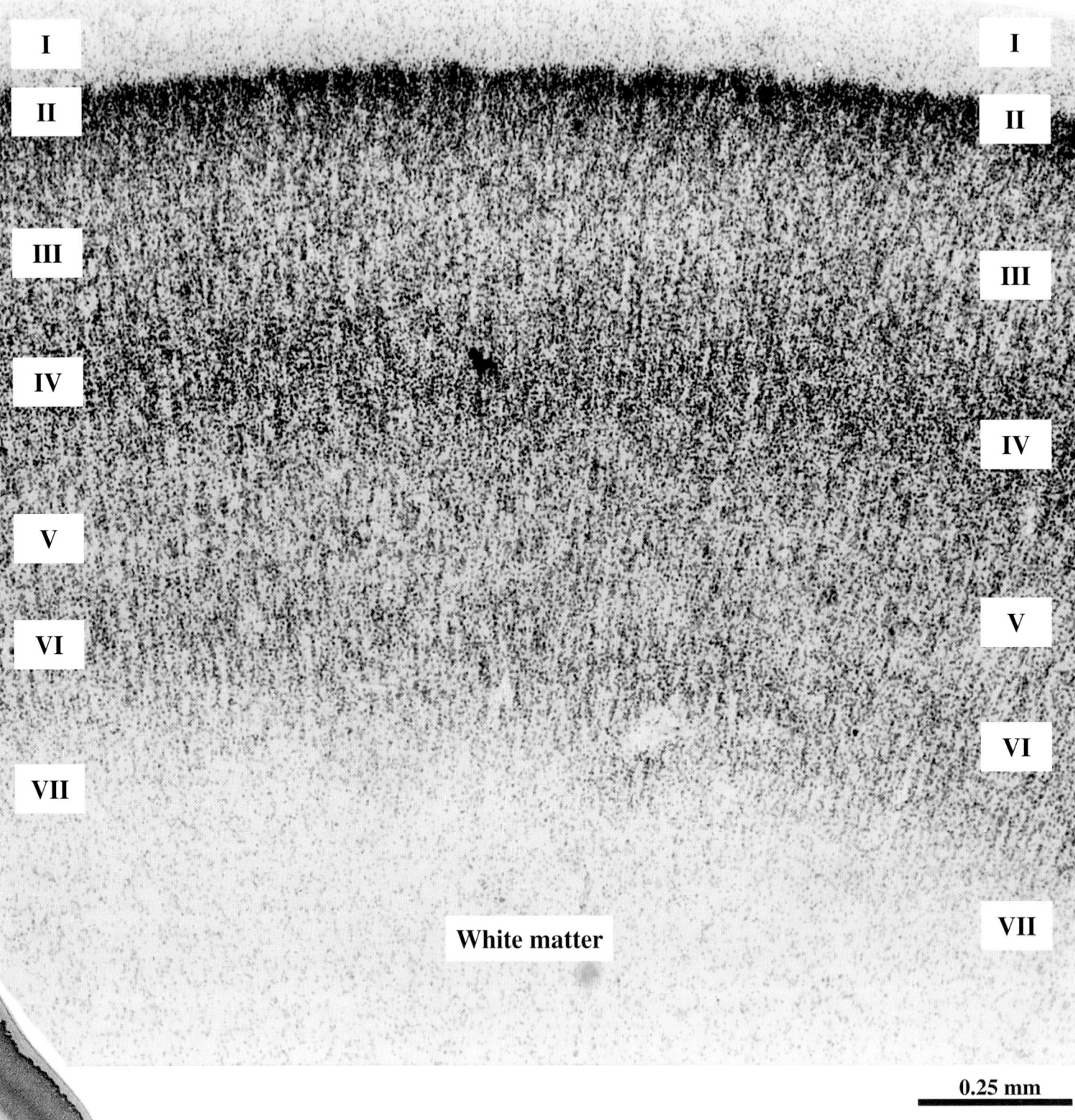

*Latest arriving cells*

See the entire section in Plate 167.

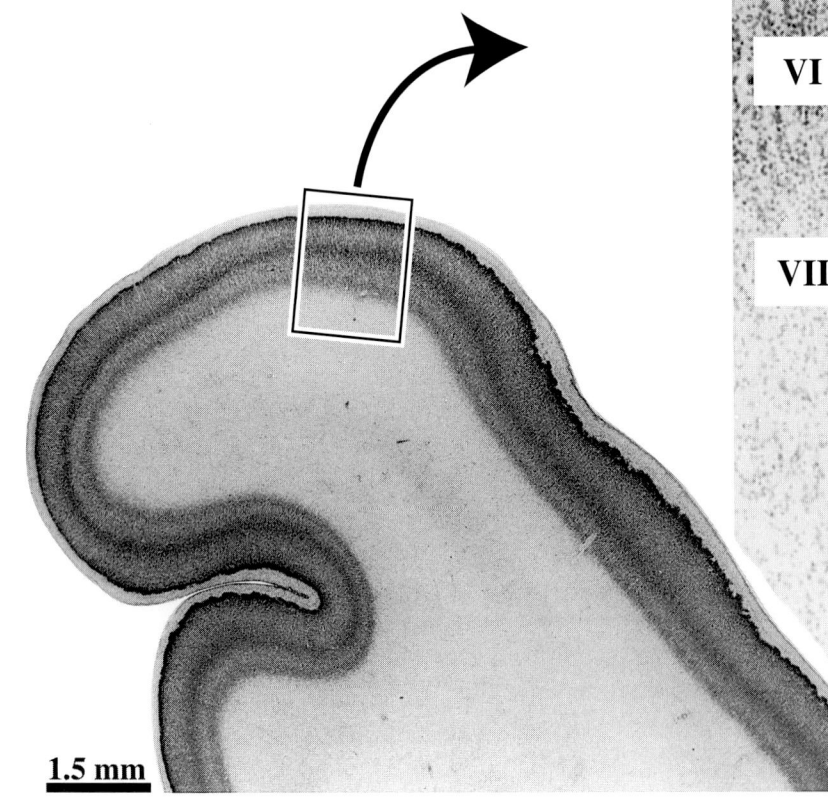

1.5 mm

I
II
III
IV
V
VI
VII

I
II
III
IV
V
VI
VII

White matter

0.25 mm

# PLATE 174

GW29 Coronal
CR 260 mm, Y14-59
Posterior to Level 20: Section 1681

STRIATE (Primary visual) CORTEX

See level 20 in Plate 170.

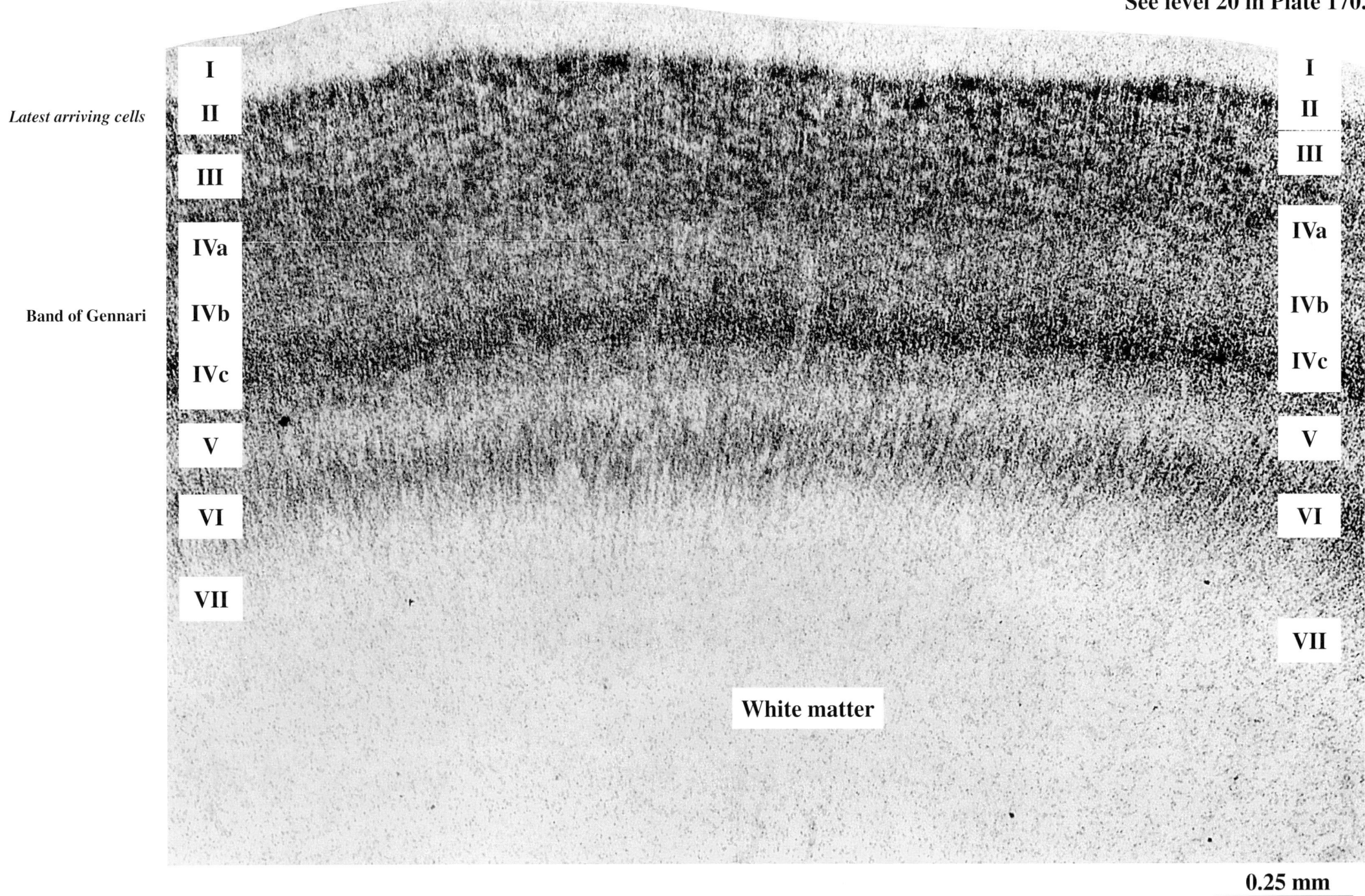

*Latest arriving cells*

Band of Gennari

I
II
III
IVa
IVb
IVc
V
VI
VII

White matter

0.25 mm

# PLATE 175A

GW29 Coronal
CR 260 mm
Y14-59
Level 4: Section 701

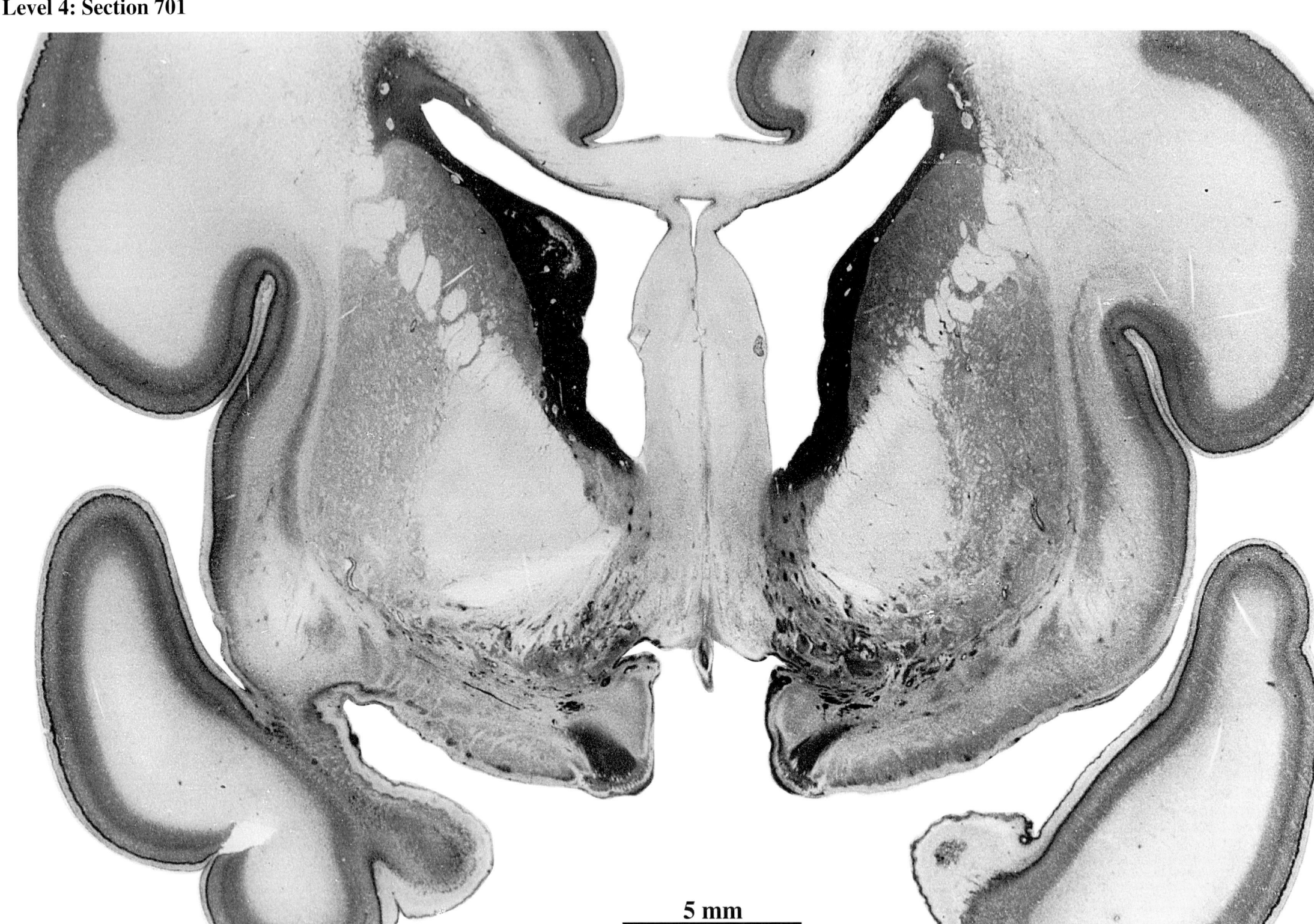

5 mm

See the entire section in Plate 154.

## PLATE 175B

**Germinal and transitional structures in *italics***

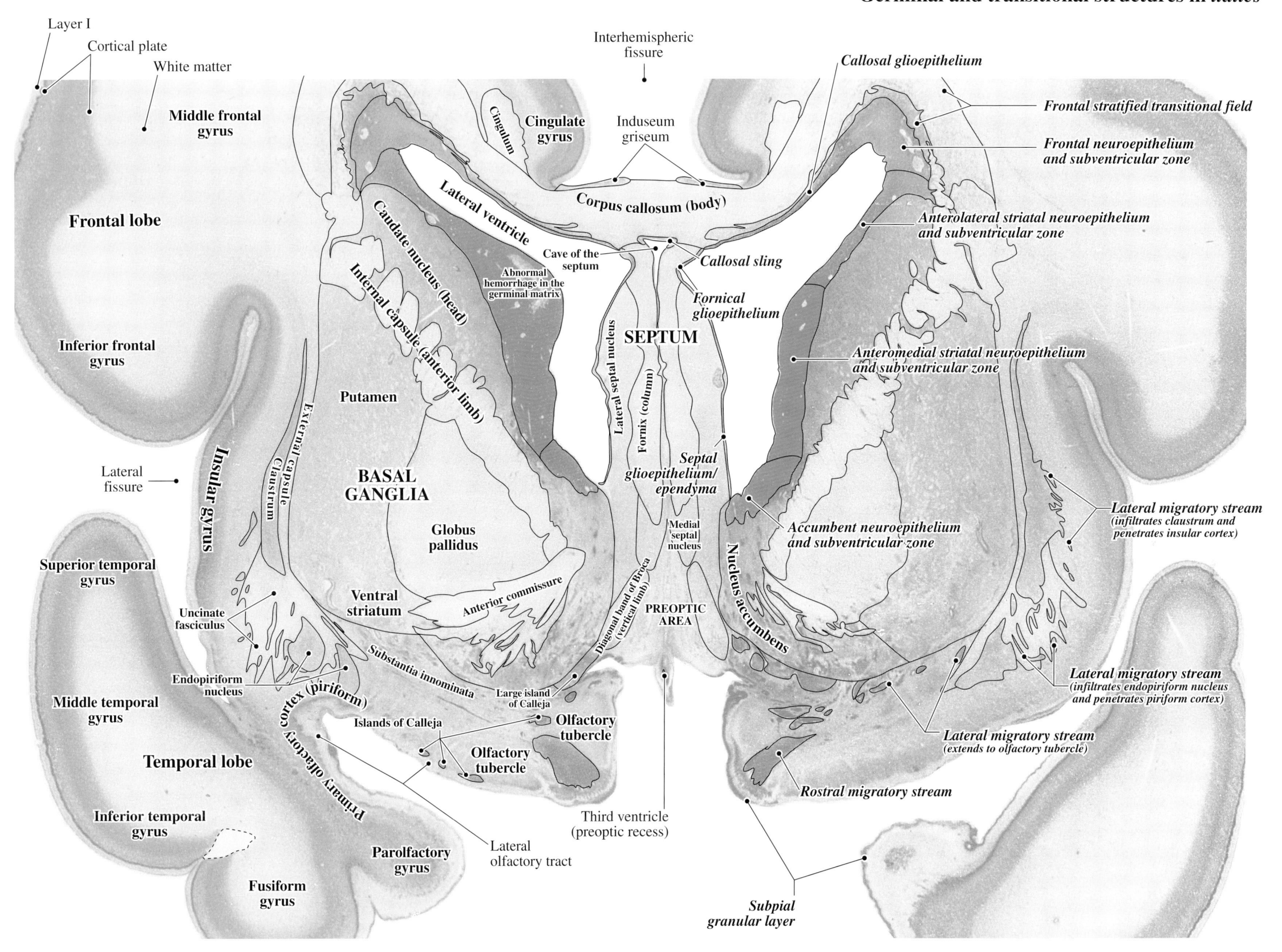

## PLATE 176A

**GW29 Coronal**
CR 260 mm
Y14-59
Level 5: Section 741

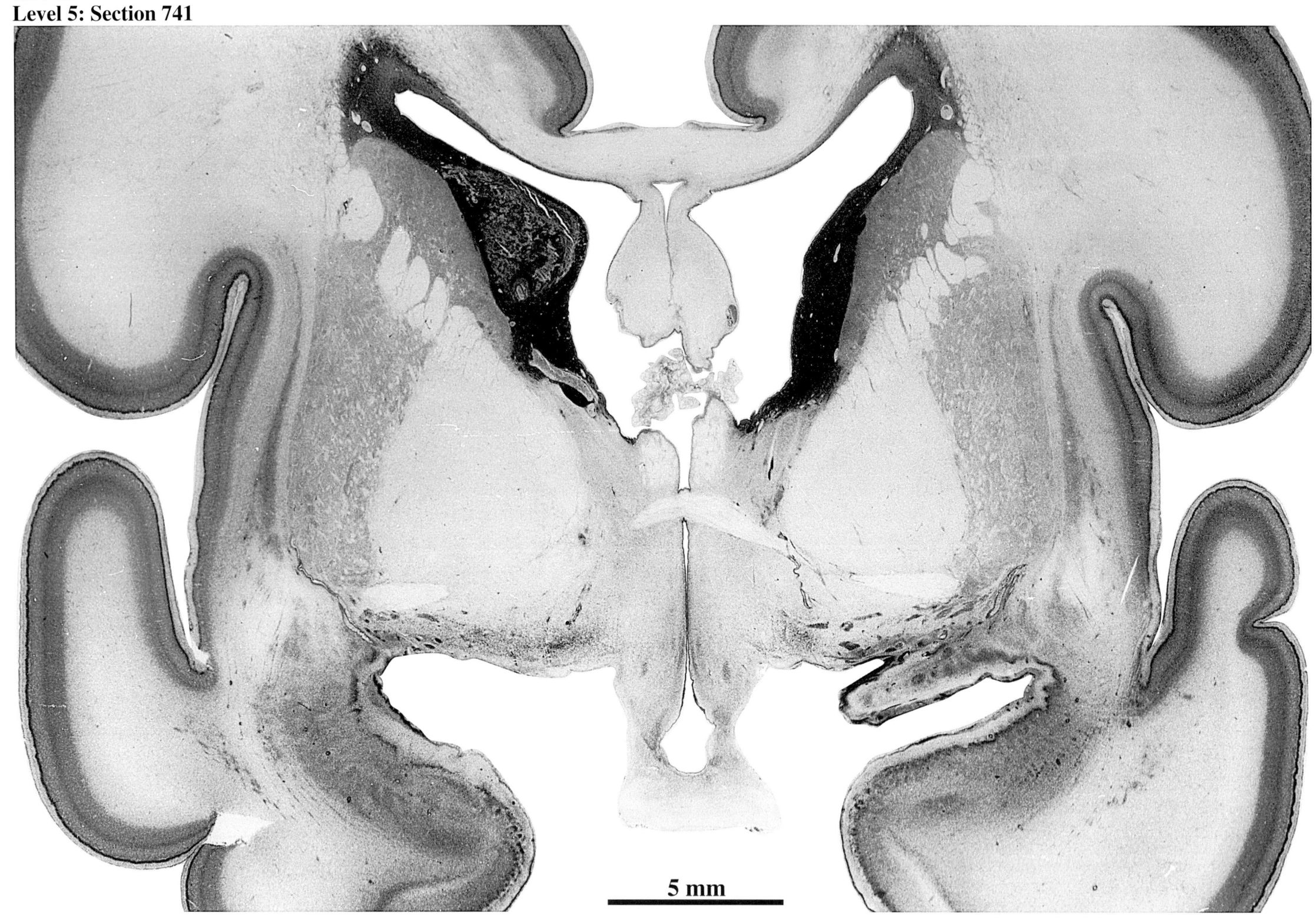

See the entire section in Plate 155.

## PLATE 176B

**Germinal and transitional structures in *italics***

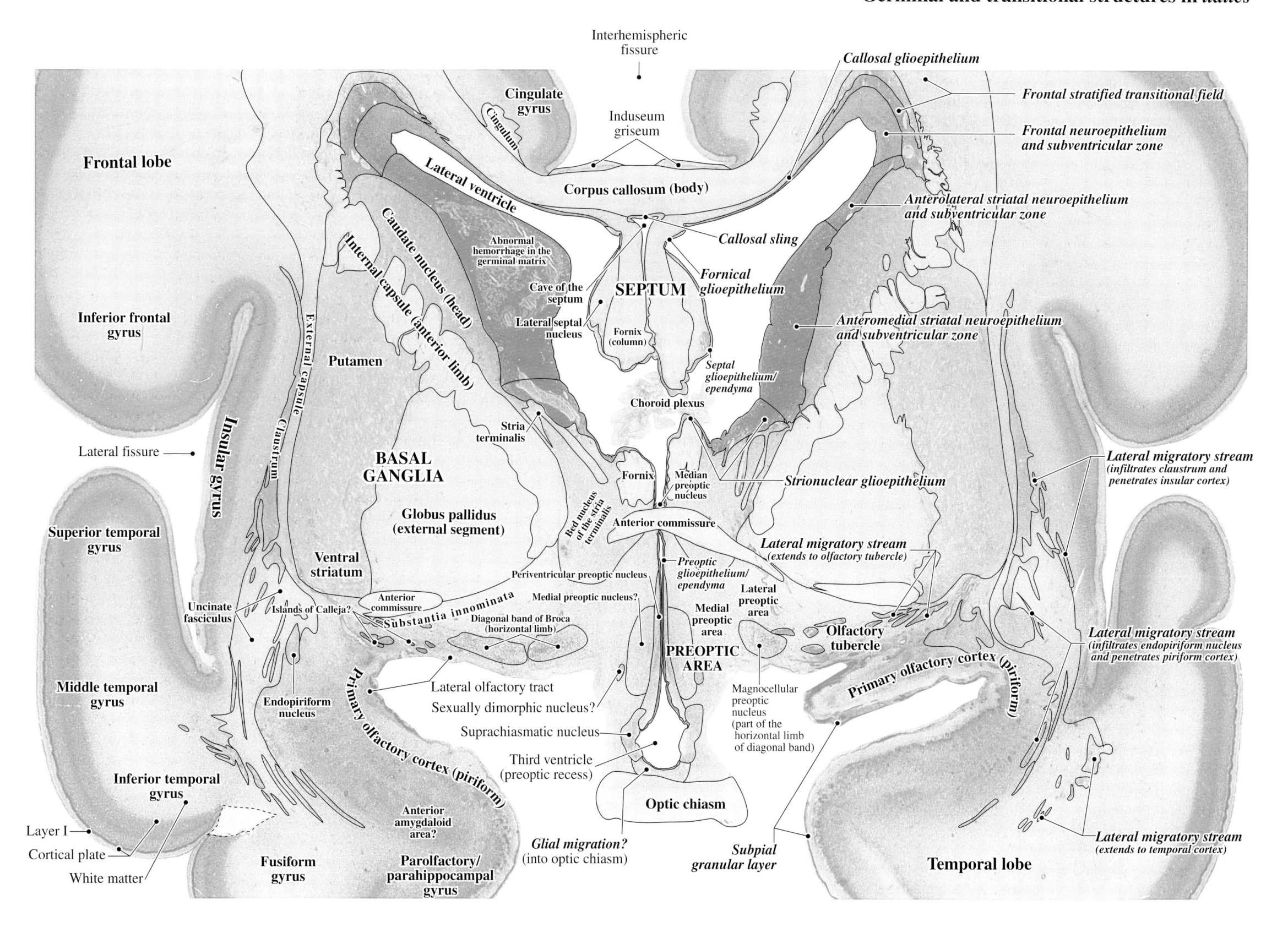

**PLATE 177A**

GW29 Coronal
CR 260 mm
Y14-59
Level 6: Section 781

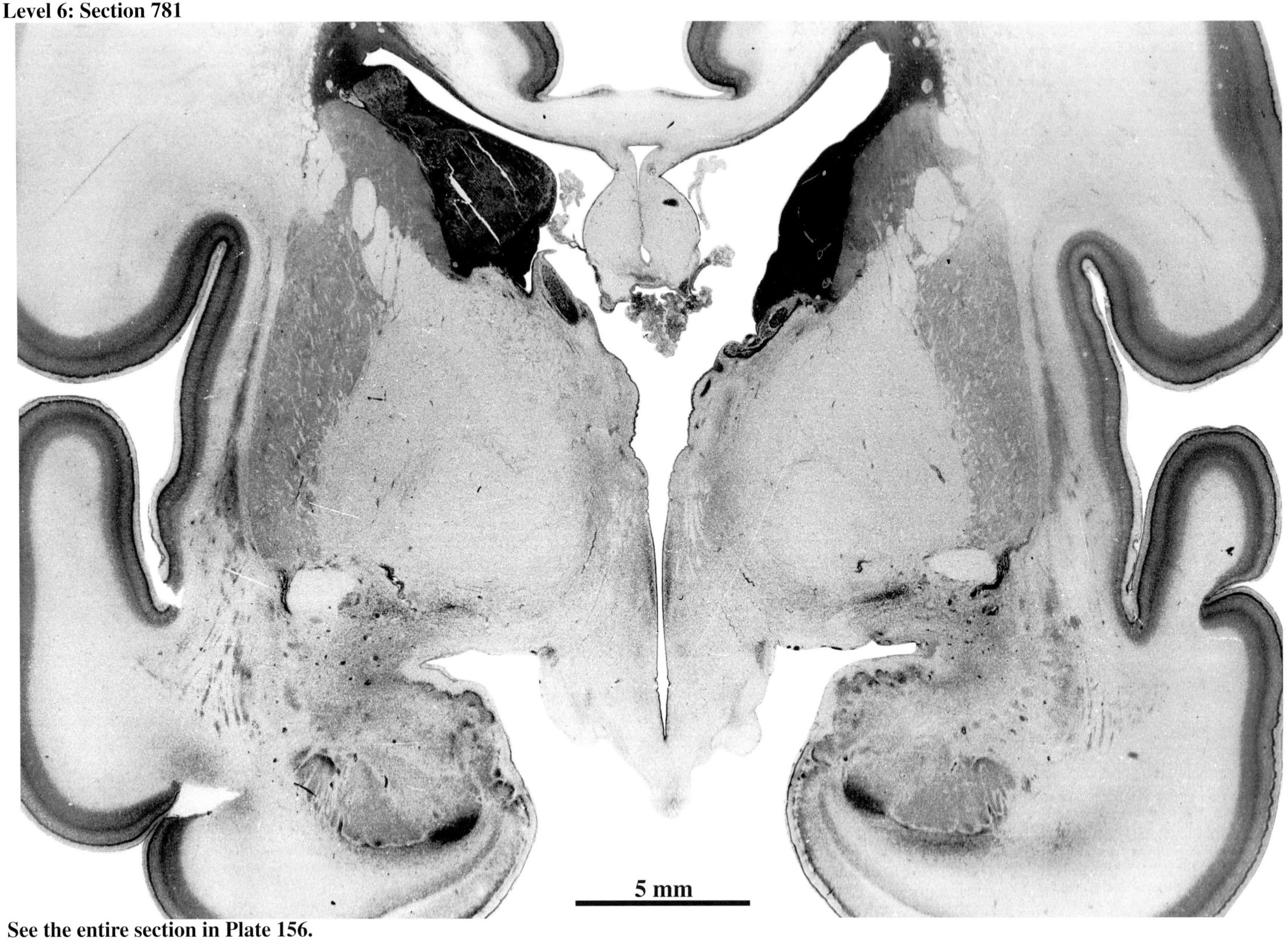

See the entire section in Plate 156.

**PLATE 177B**

**Germinal and transitional structures in *italics***

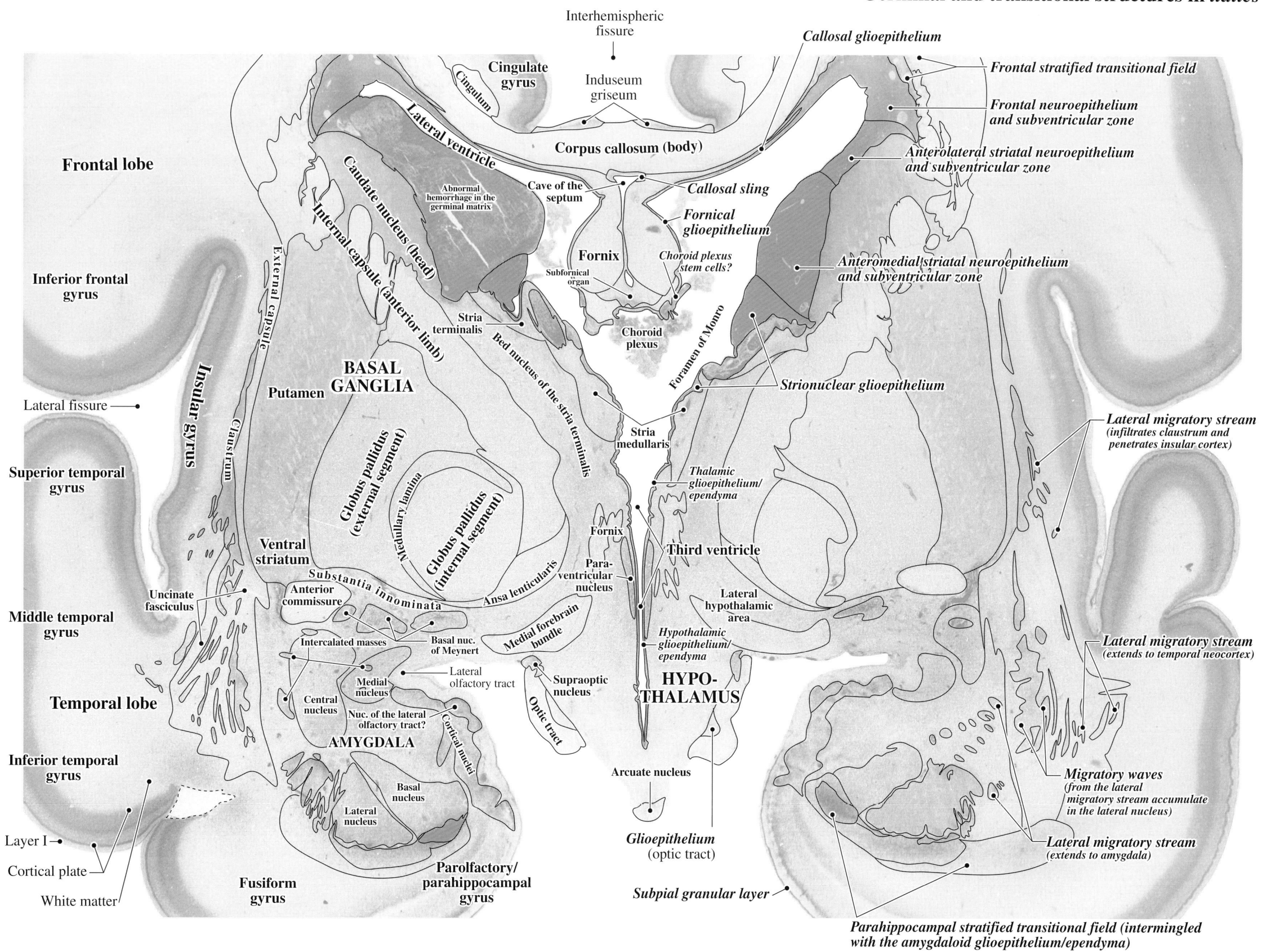

**PLATE 178A**

GW29 Coronal
CR 260 mm
Y14-59
Level 7: Section 821

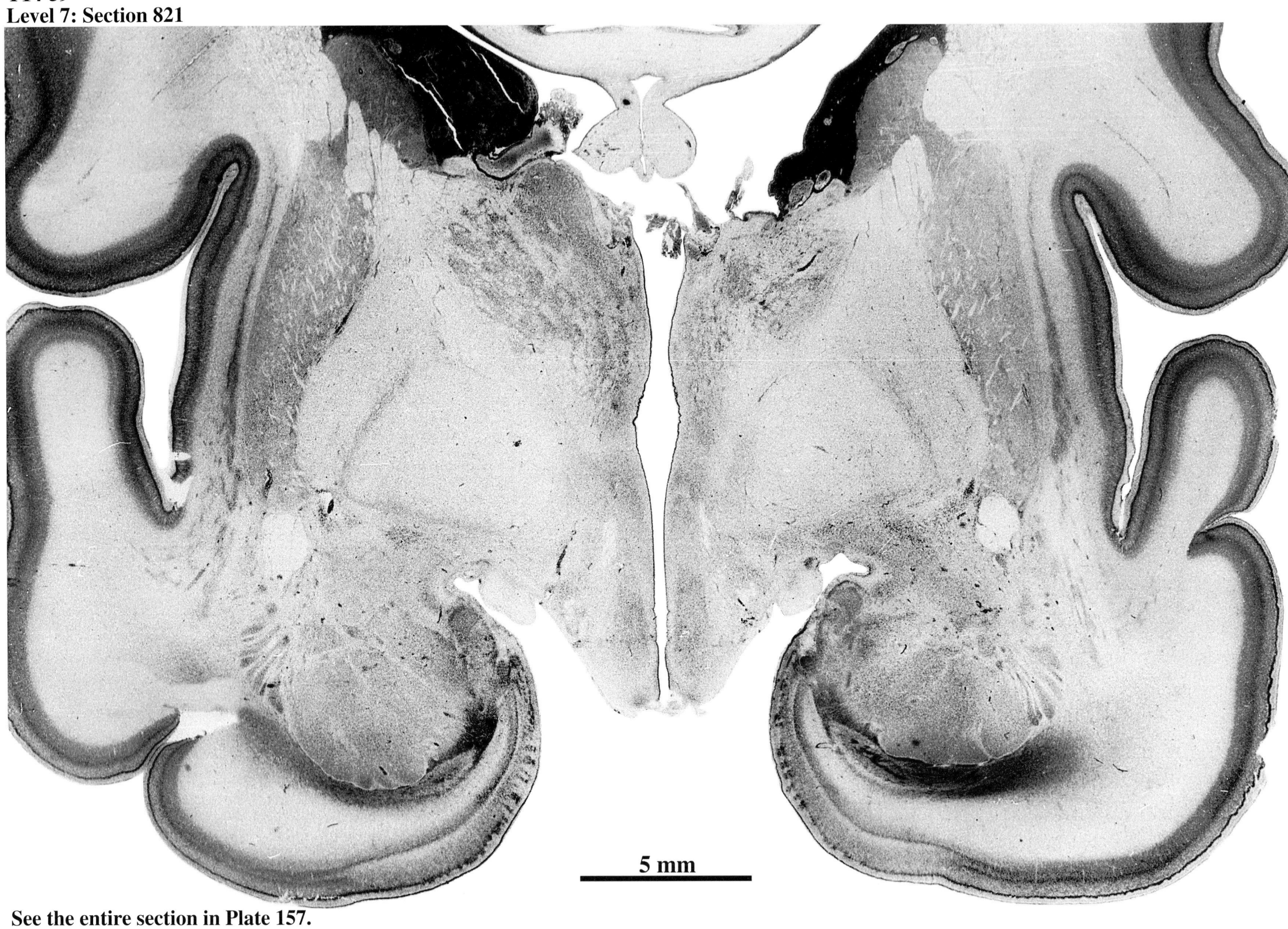

5 mm

See the entire section in Plate 157.

**PLATE 178B**

**Germinal and transitional structures in *italics***

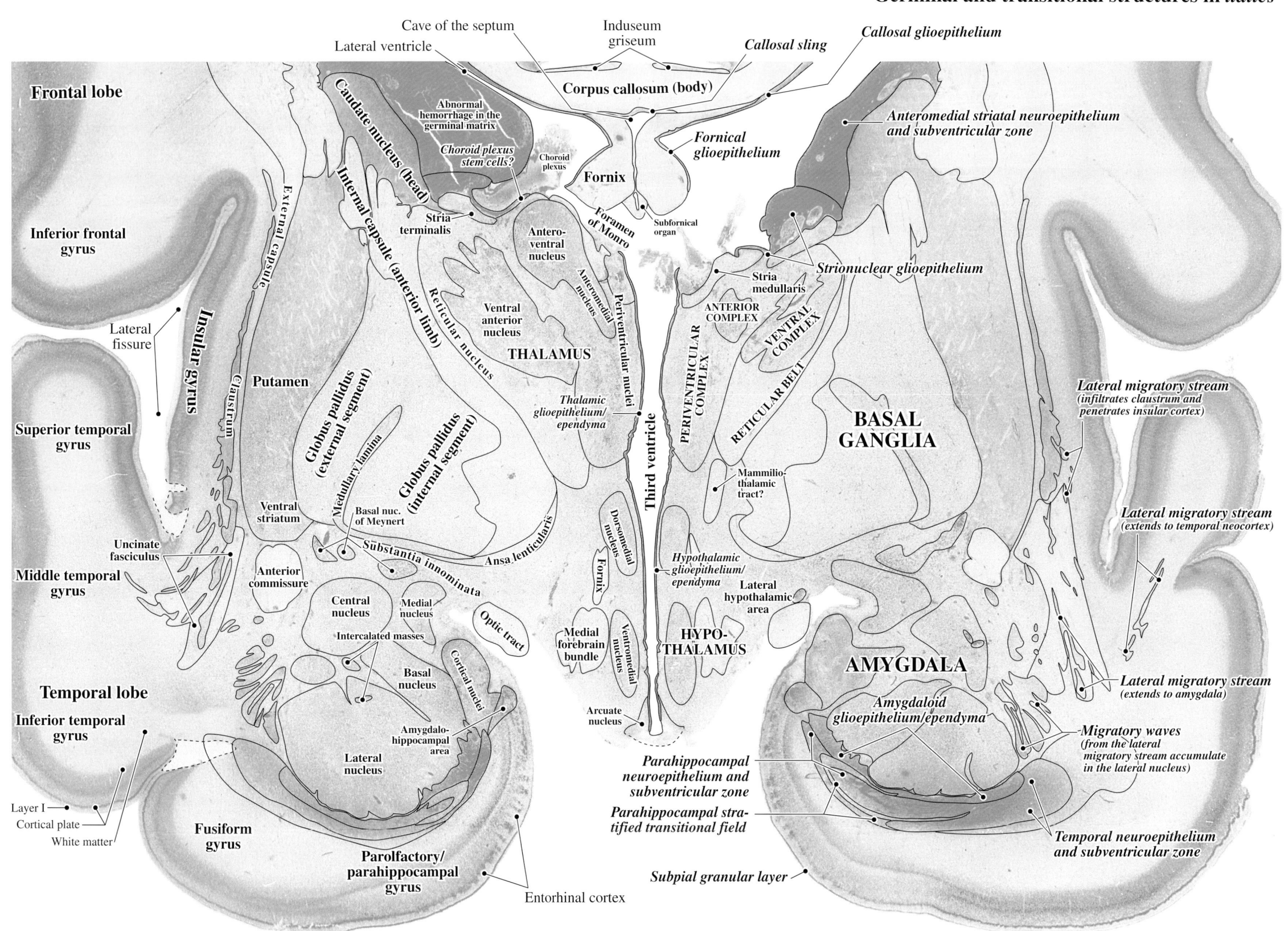

## PLATE 179A

GW29 Coronal
CR 260 mm
Y14-59
Level 8: Section 861

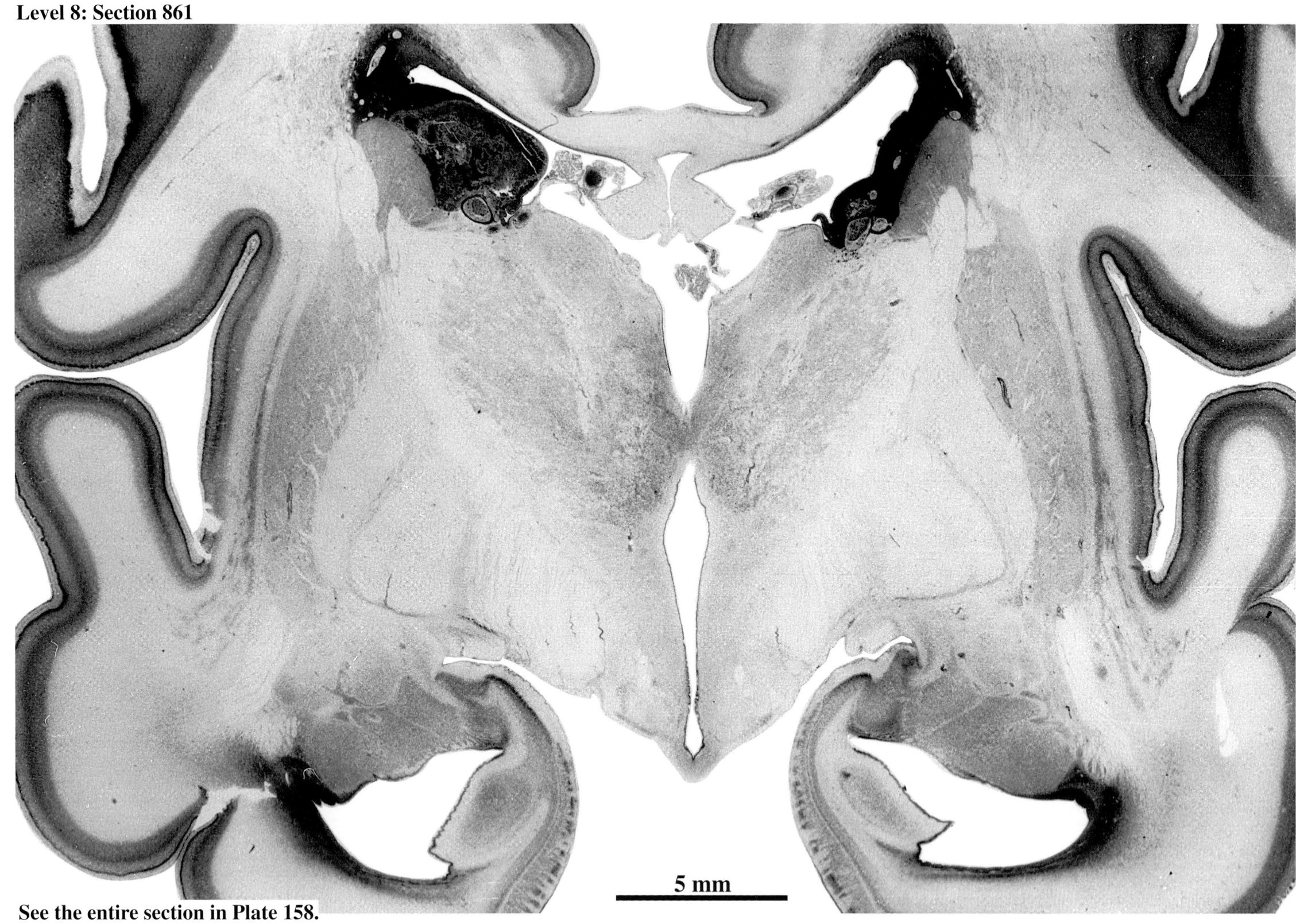

See the entire section in Plate 158.

# PLATE 179B

**Germinal and transitional structures in *italics***

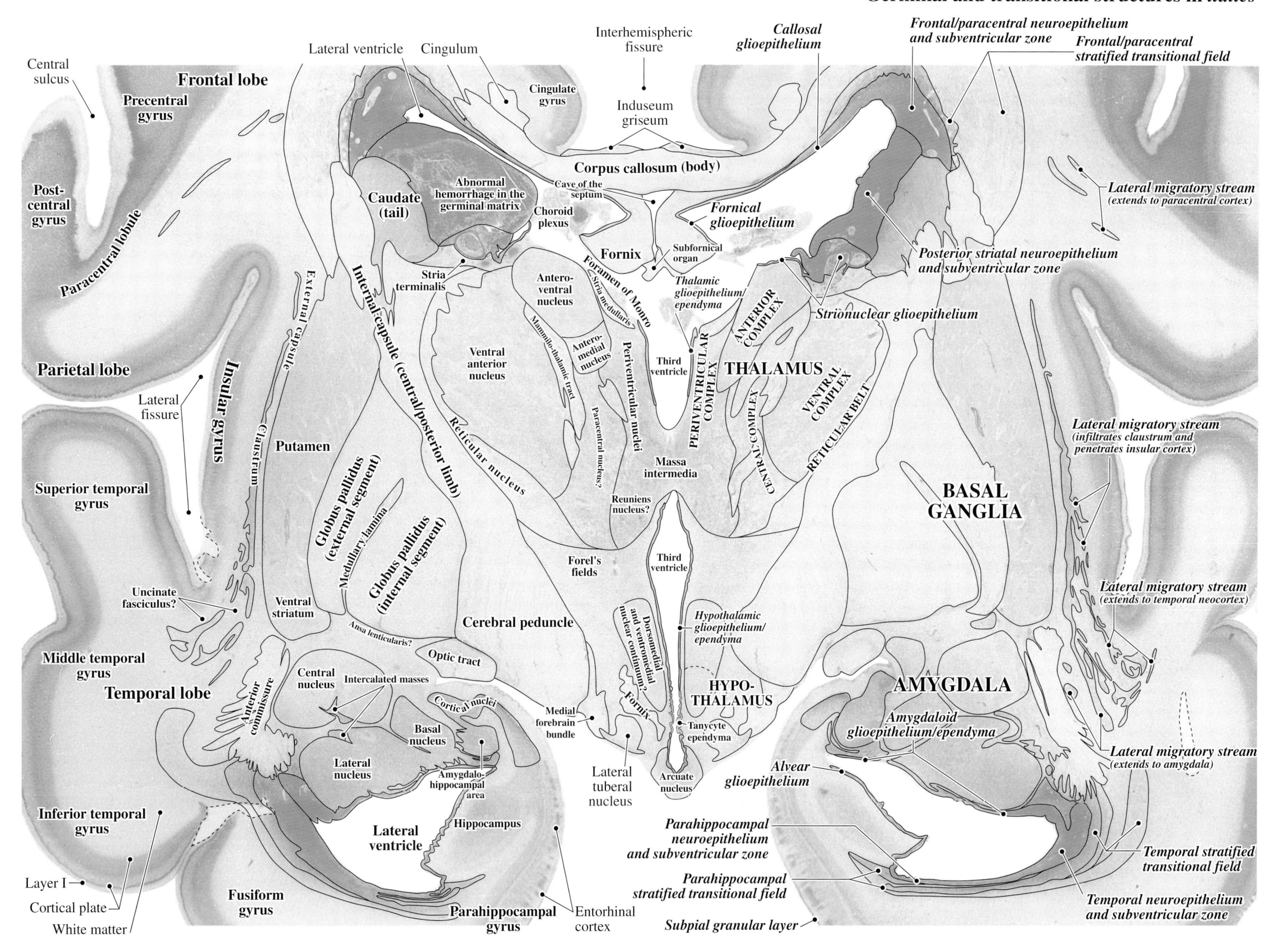

**PLATE 180A**

GW29 Coronal
CR 260 mm
Y14-59
Level 9: Section 901

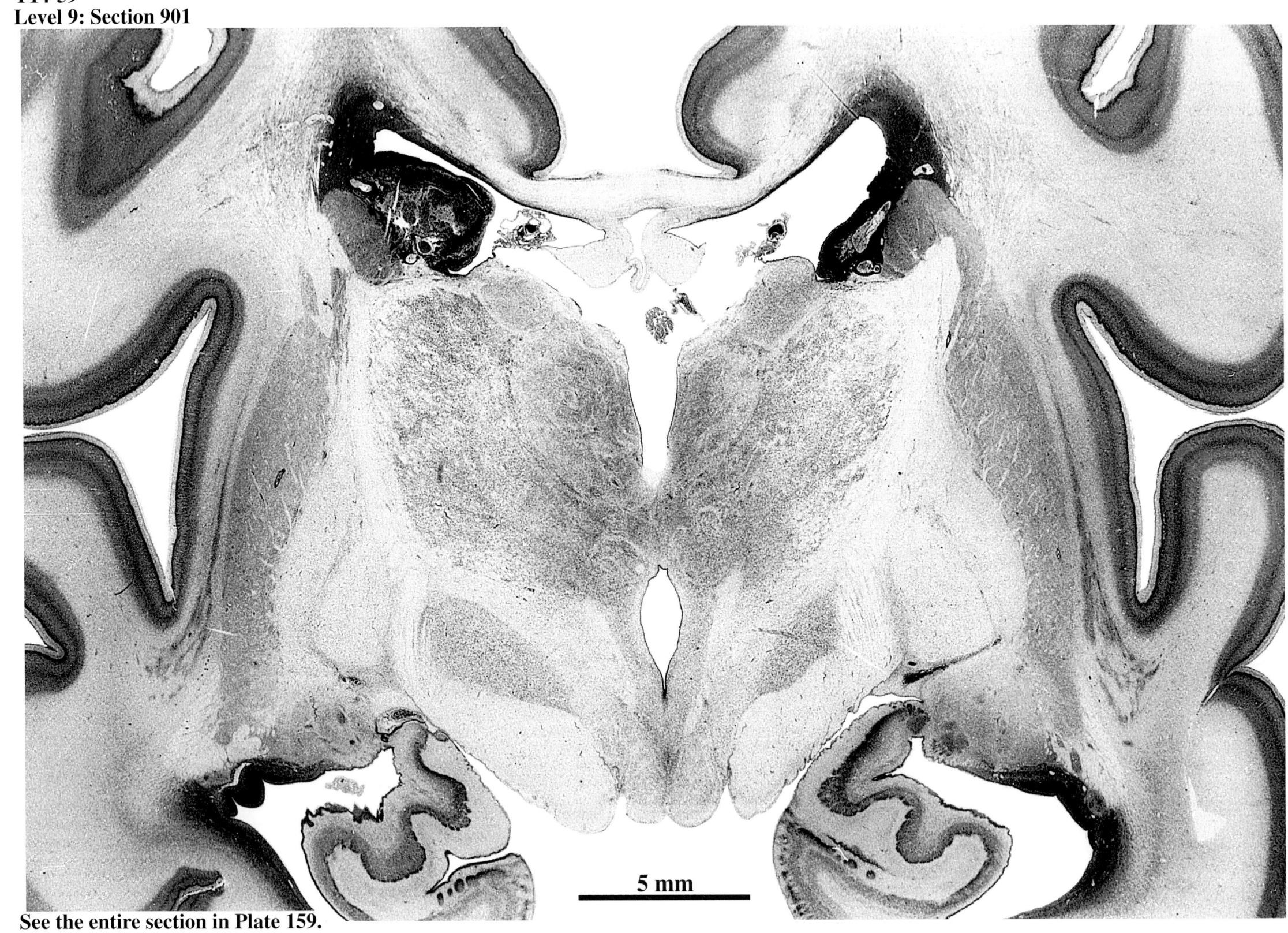

5 mm

See the entire section in Plate 159.

## PLATE 180B

**Germinal and transitional structures in *italics***

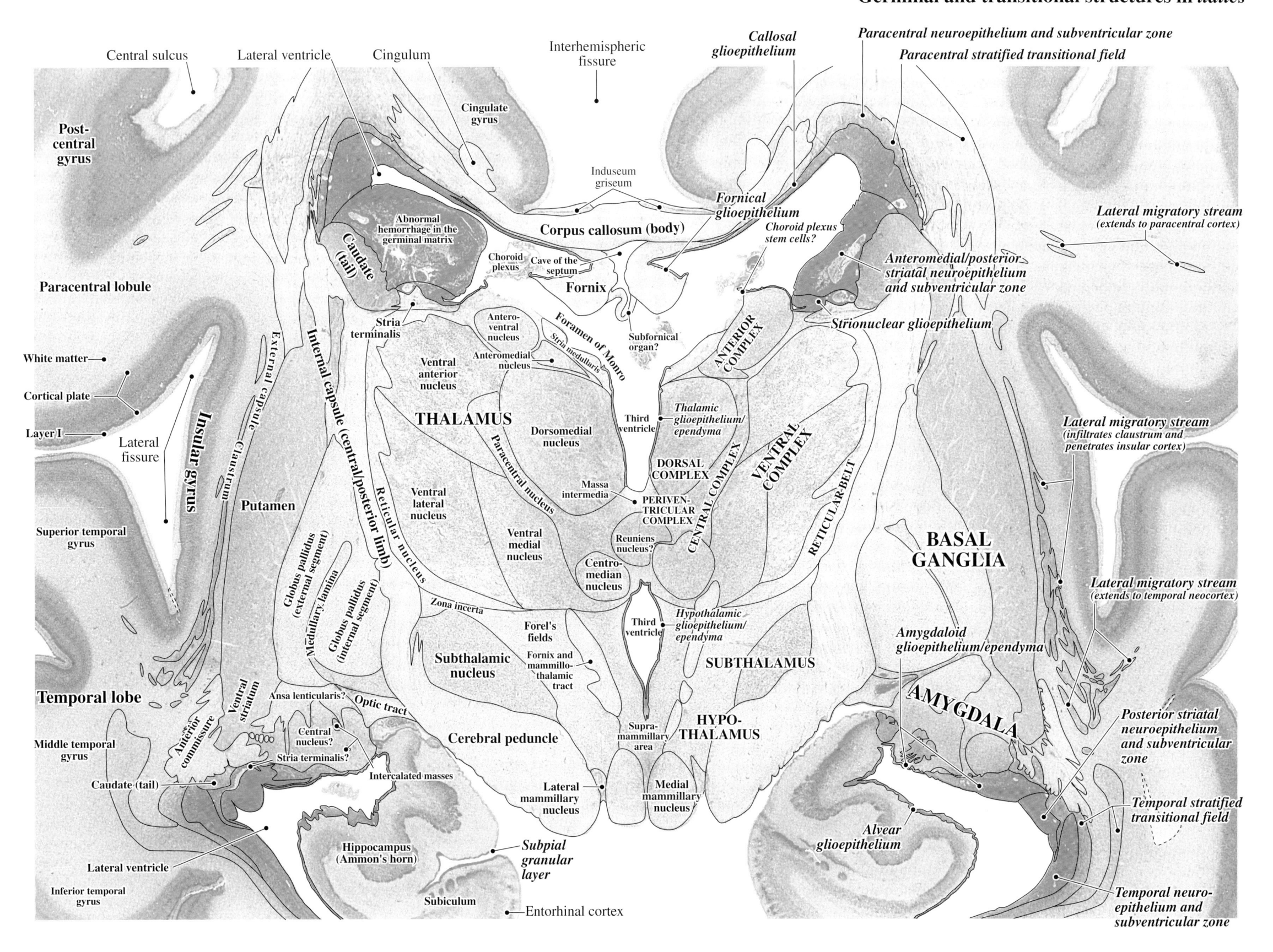

**PLATE 181A**

GW29 Coronal
CR 260 mm
Y14-59
Level 10: Section 981

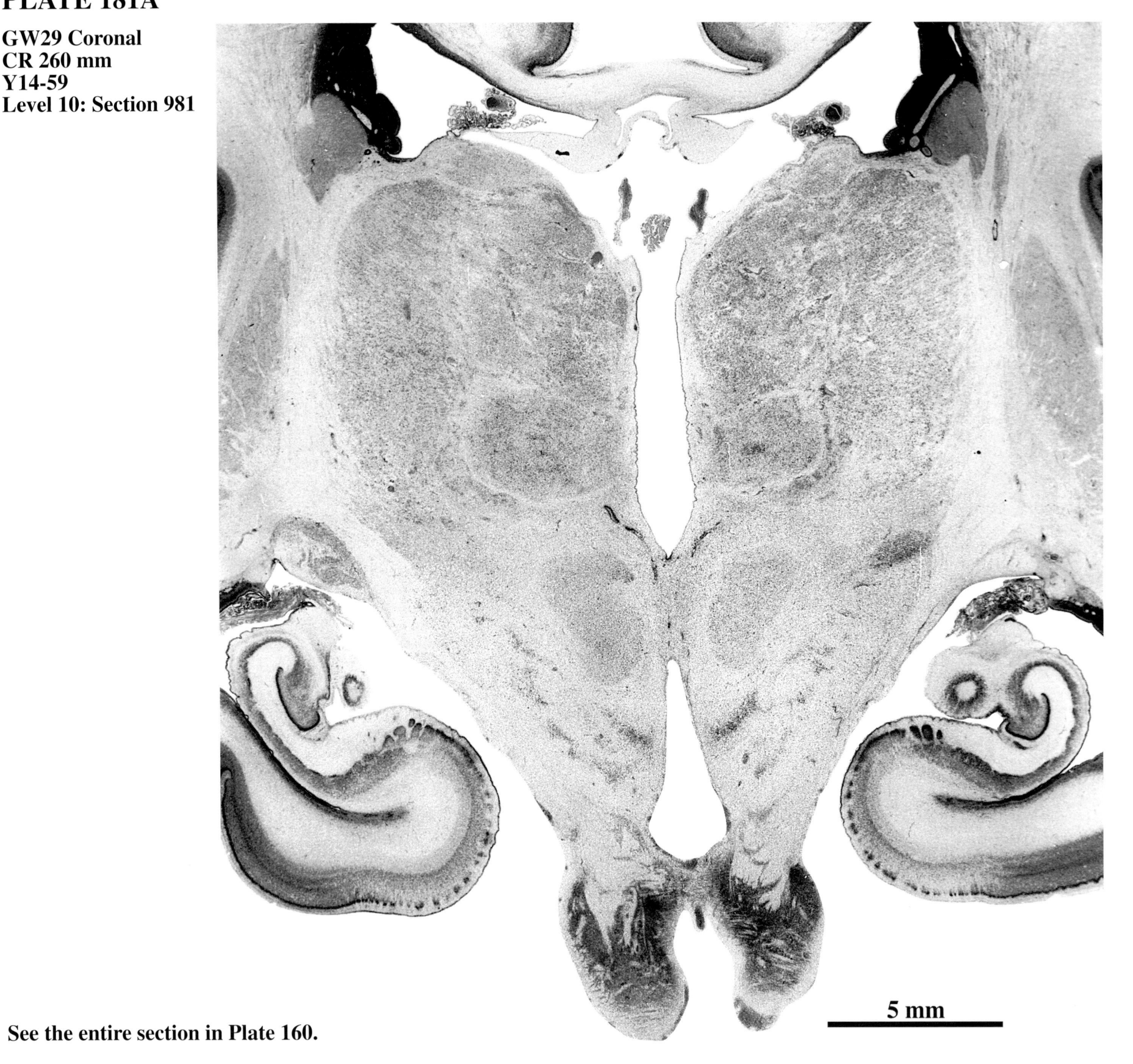

5 mm

See the entire section in Plate 160.

# PLATE 181B

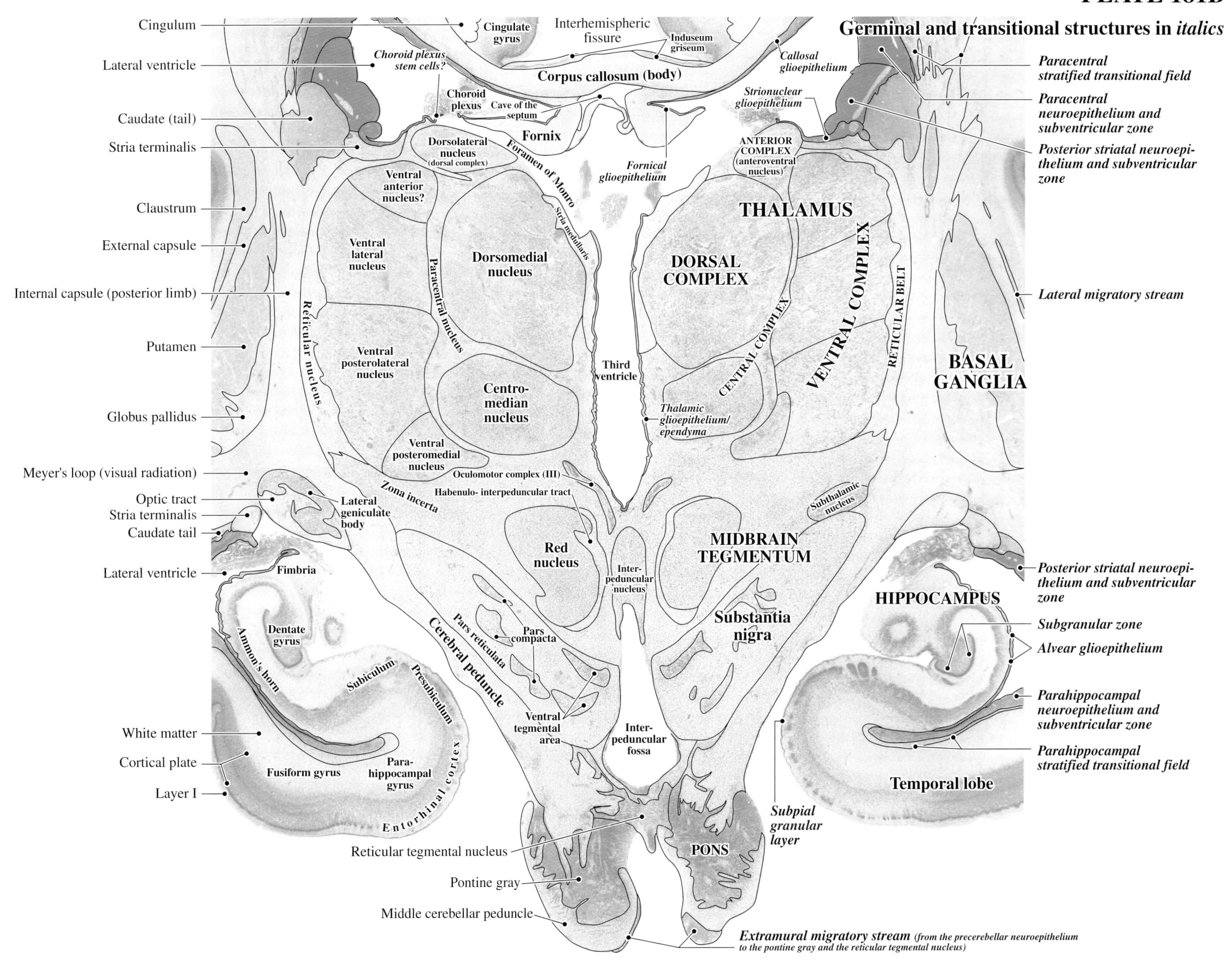

**Germinal and transitional structures in *italics***

## PLATE 182A

GW29 Coronal
CR 260 mm
Y14-59
Level 11: Section 1021

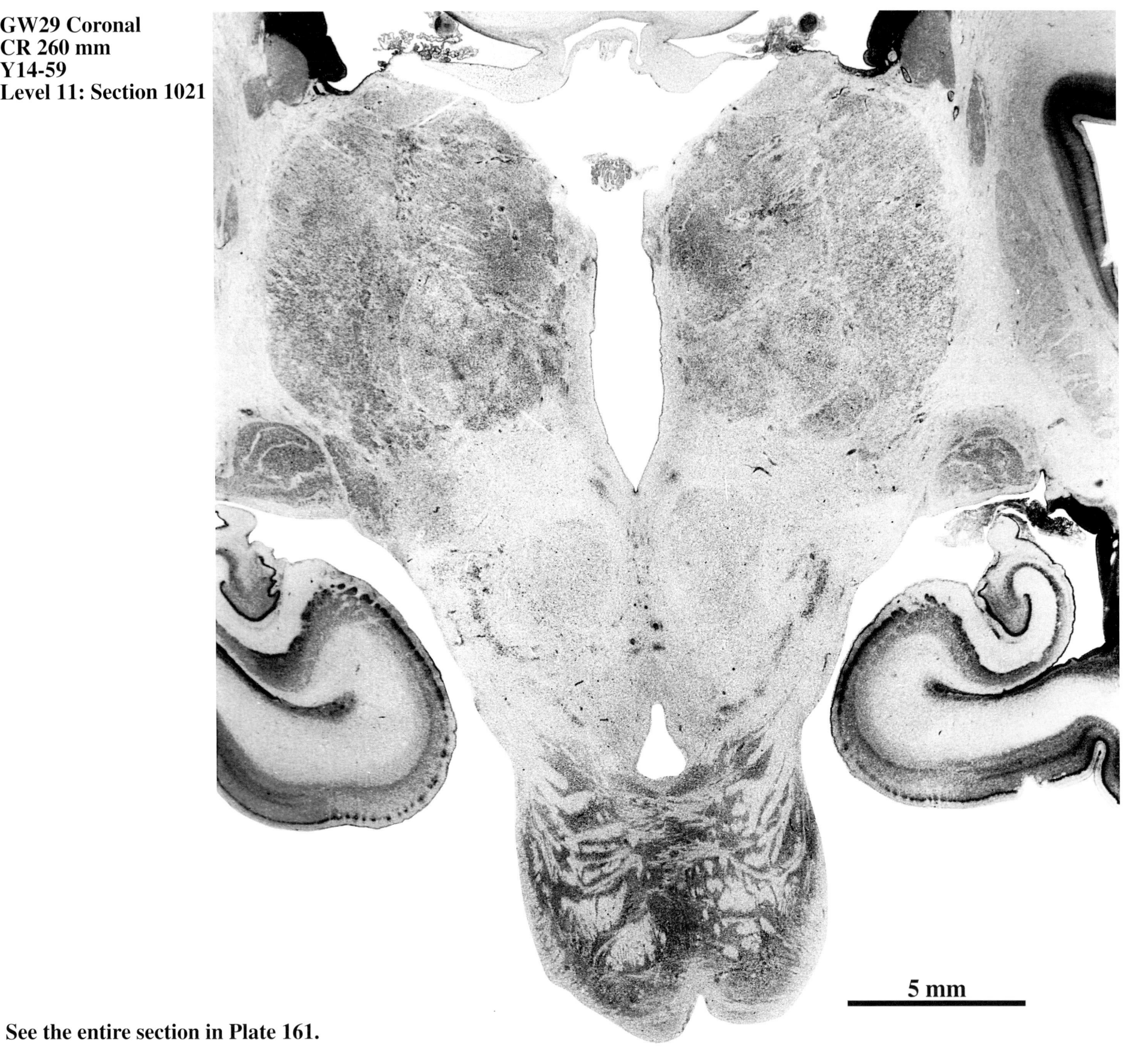

5 mm

See the entire section in Plate 161.

**PLATE 182B**

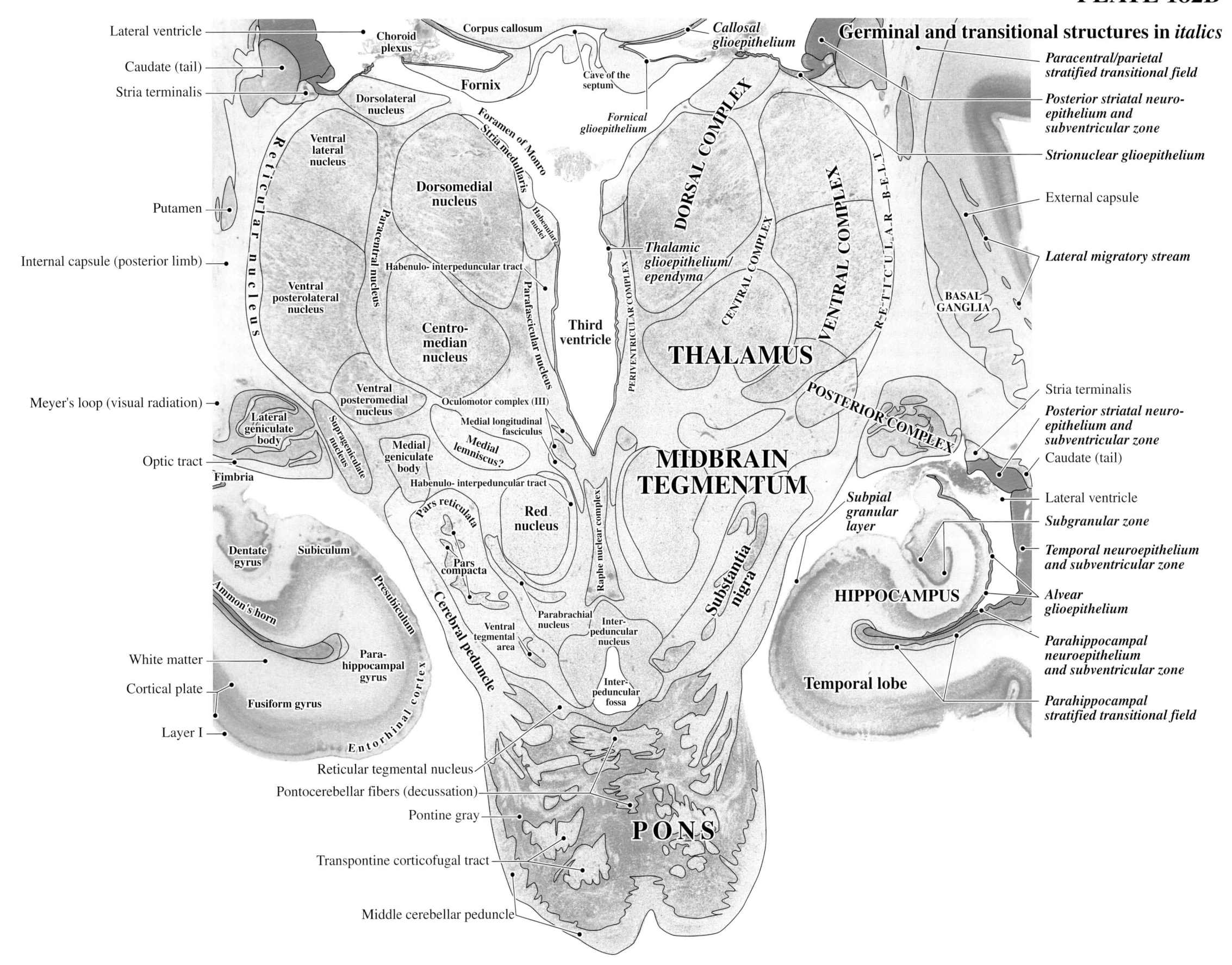

## PLATE 183A

GW29 Coronal
CR 260 mm
Y14-59
Level 12: Section 1061

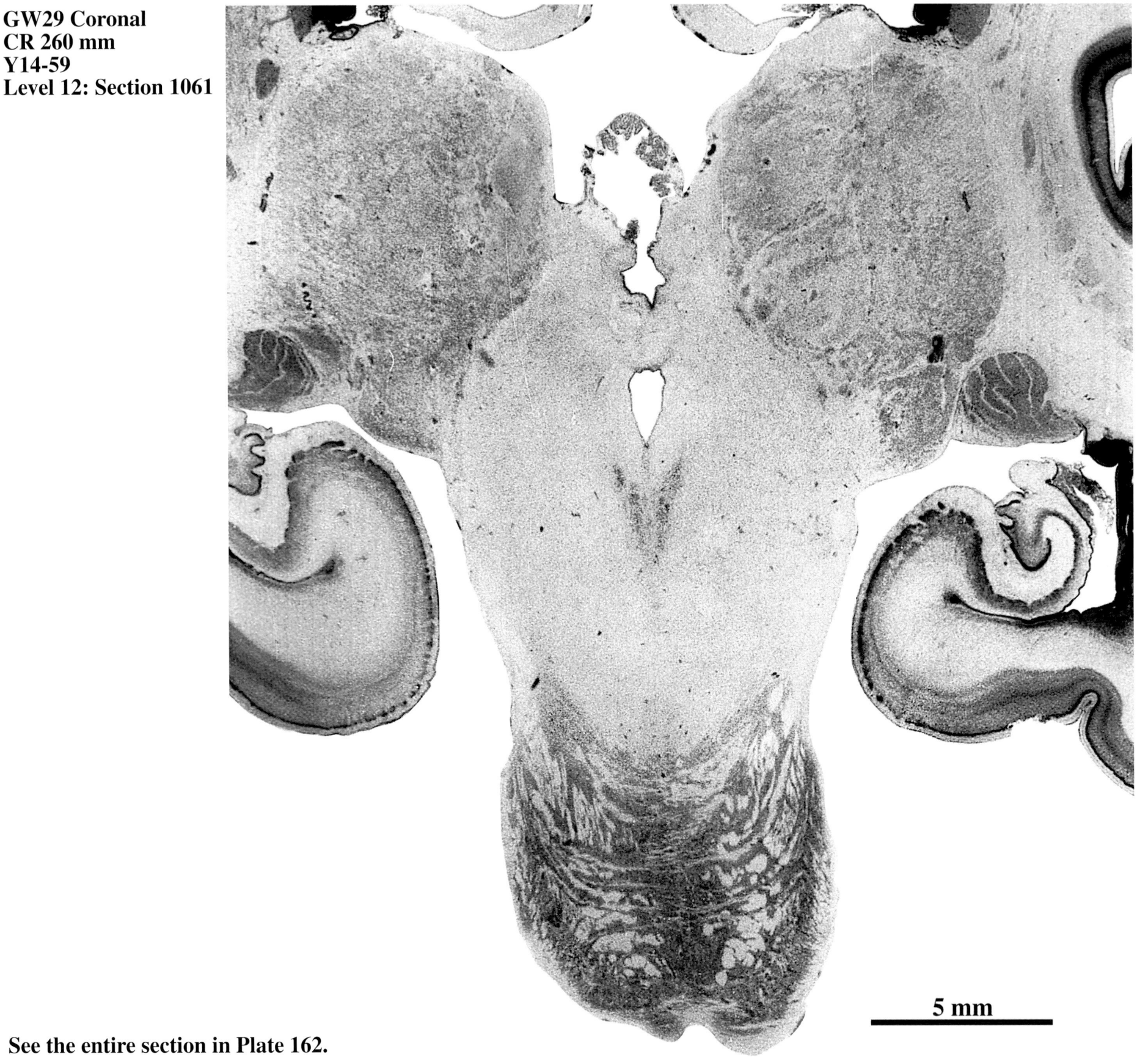

5 mm

See the entire section in Plate 162.

PLATE 183B

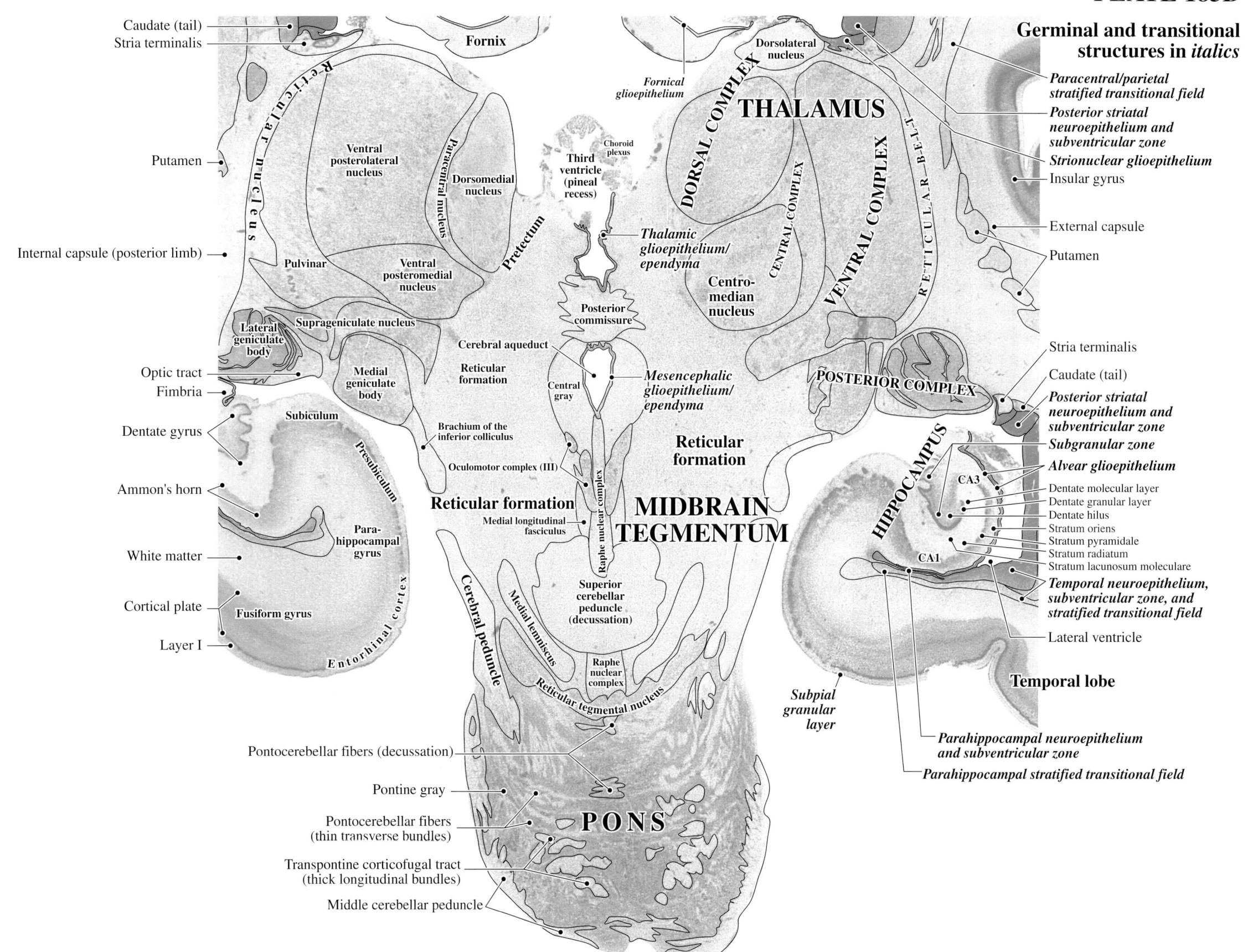

272

**PLATE 184A**

GW29 Coronal
CR 260 mm
Y14-59
Level 13: Section 1101

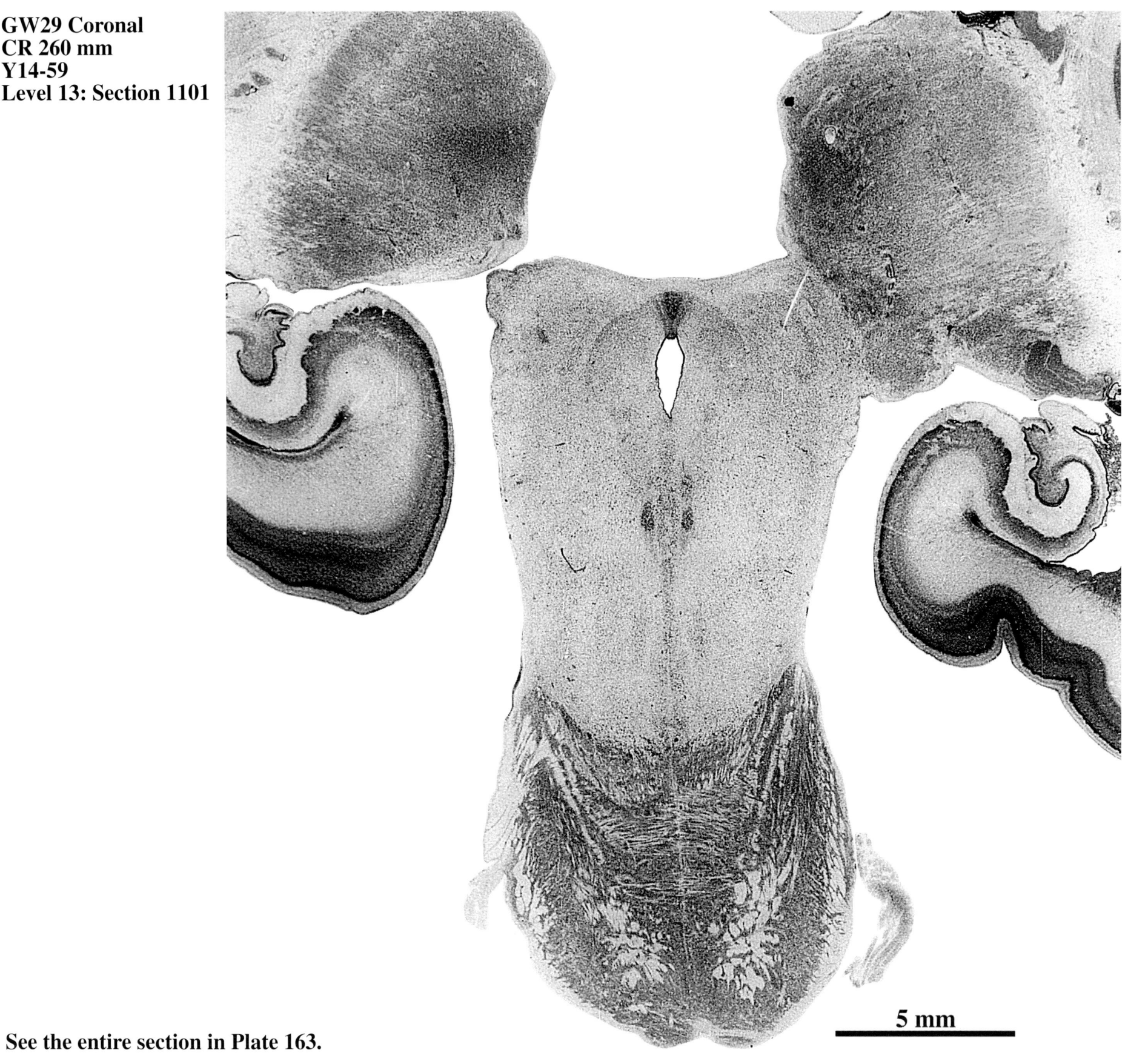

5 mm

See the entire section in Plate 163.

## PLATE 184B

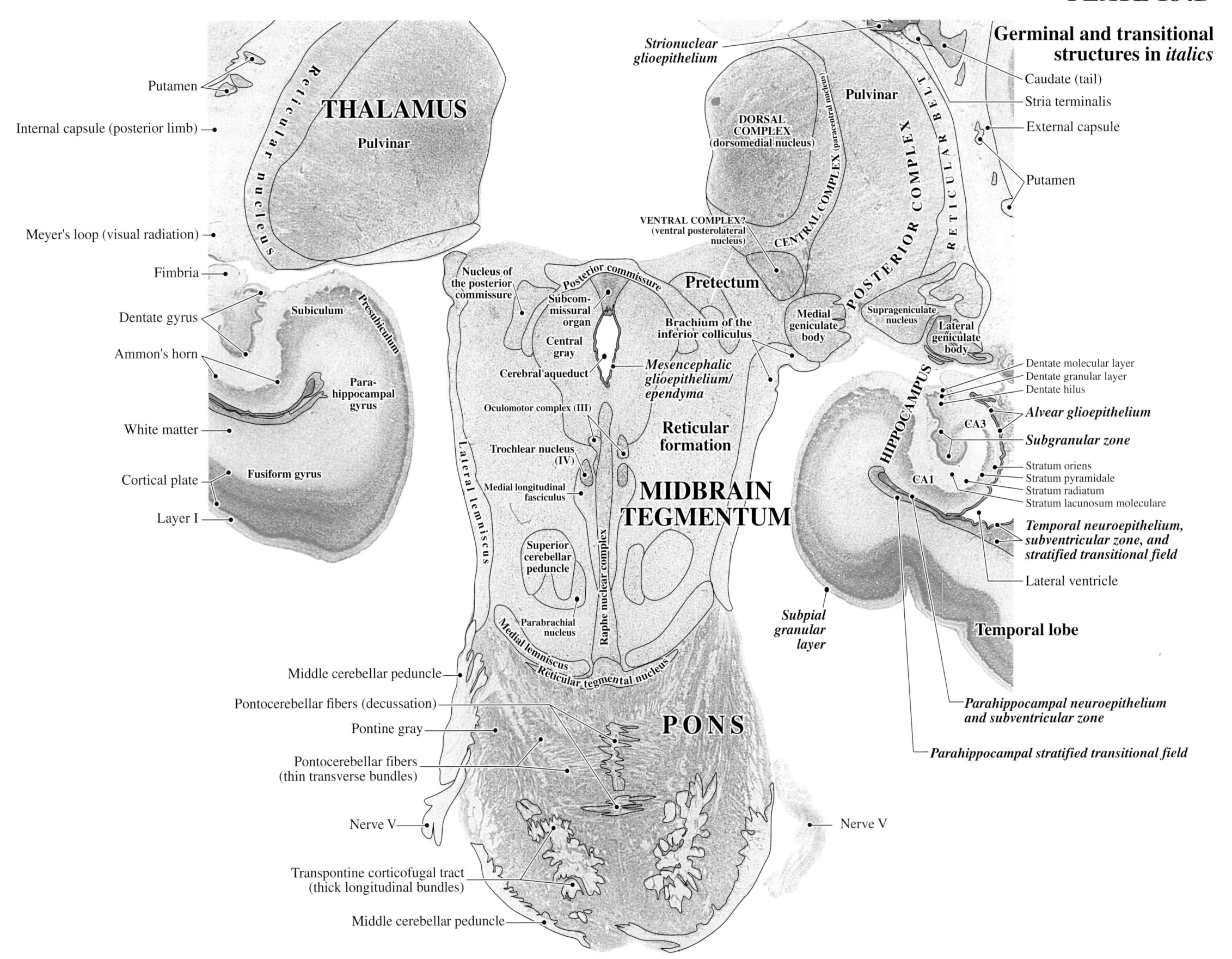

# PLATE 185A

GW29 Coronal
CR 260 mm
Y14-59
Level 14: Section 1161

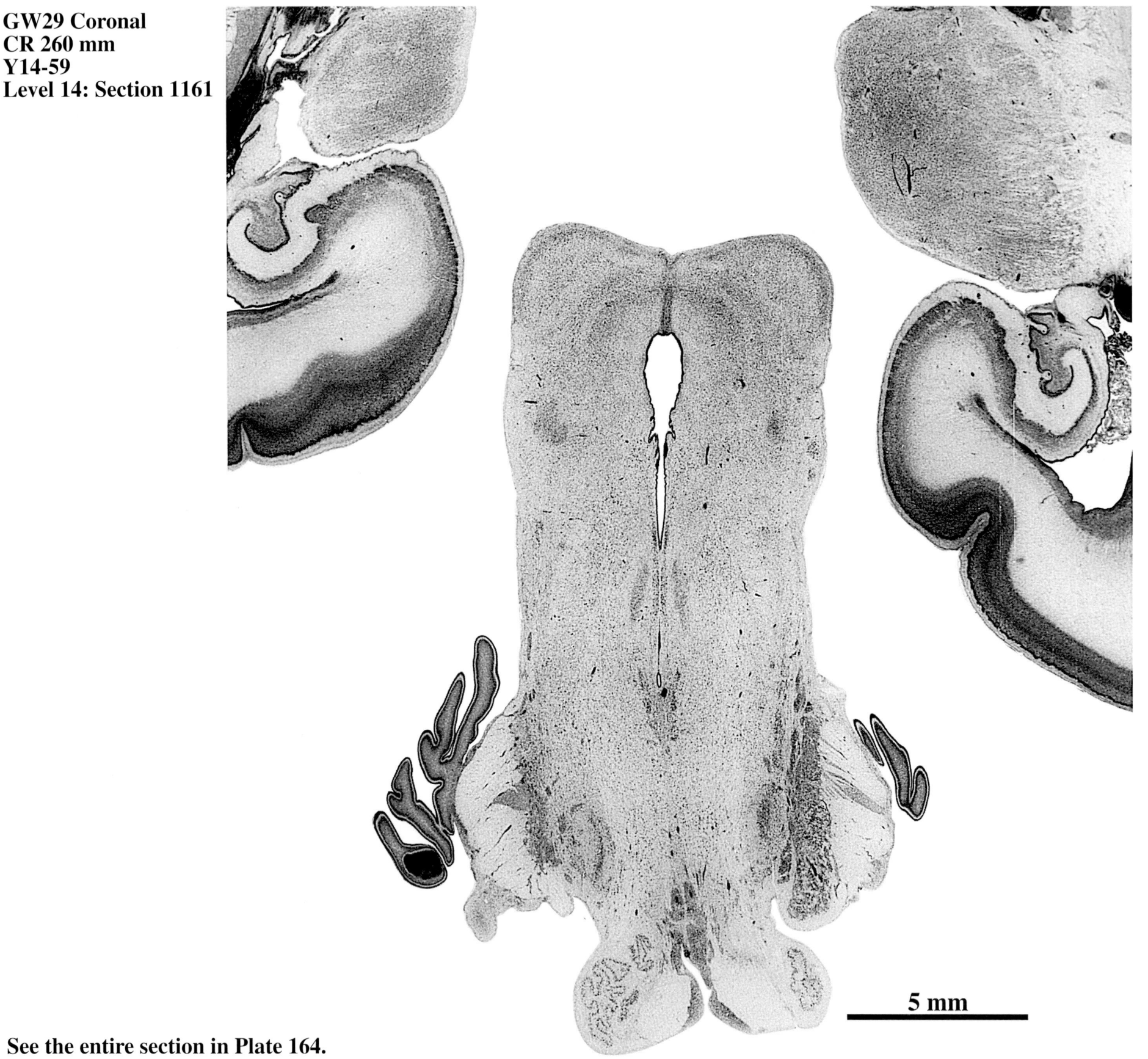

5 mm

See the entire section in Plate 164.

275

**PLATE 185B**

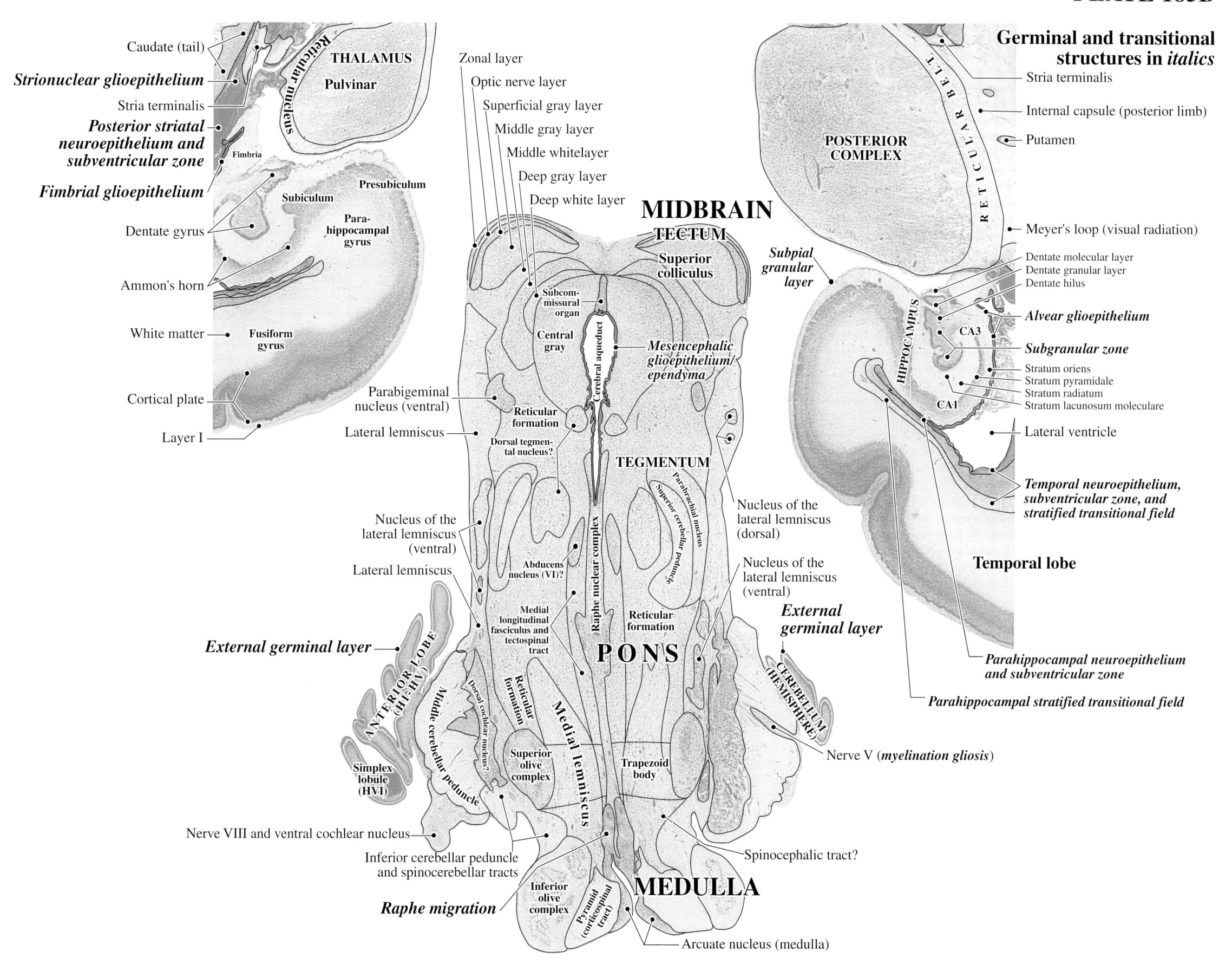

**PLATE 186A**

GW29 Coronal
CR 260 mm
Y14-59
Level 15: Section 1201

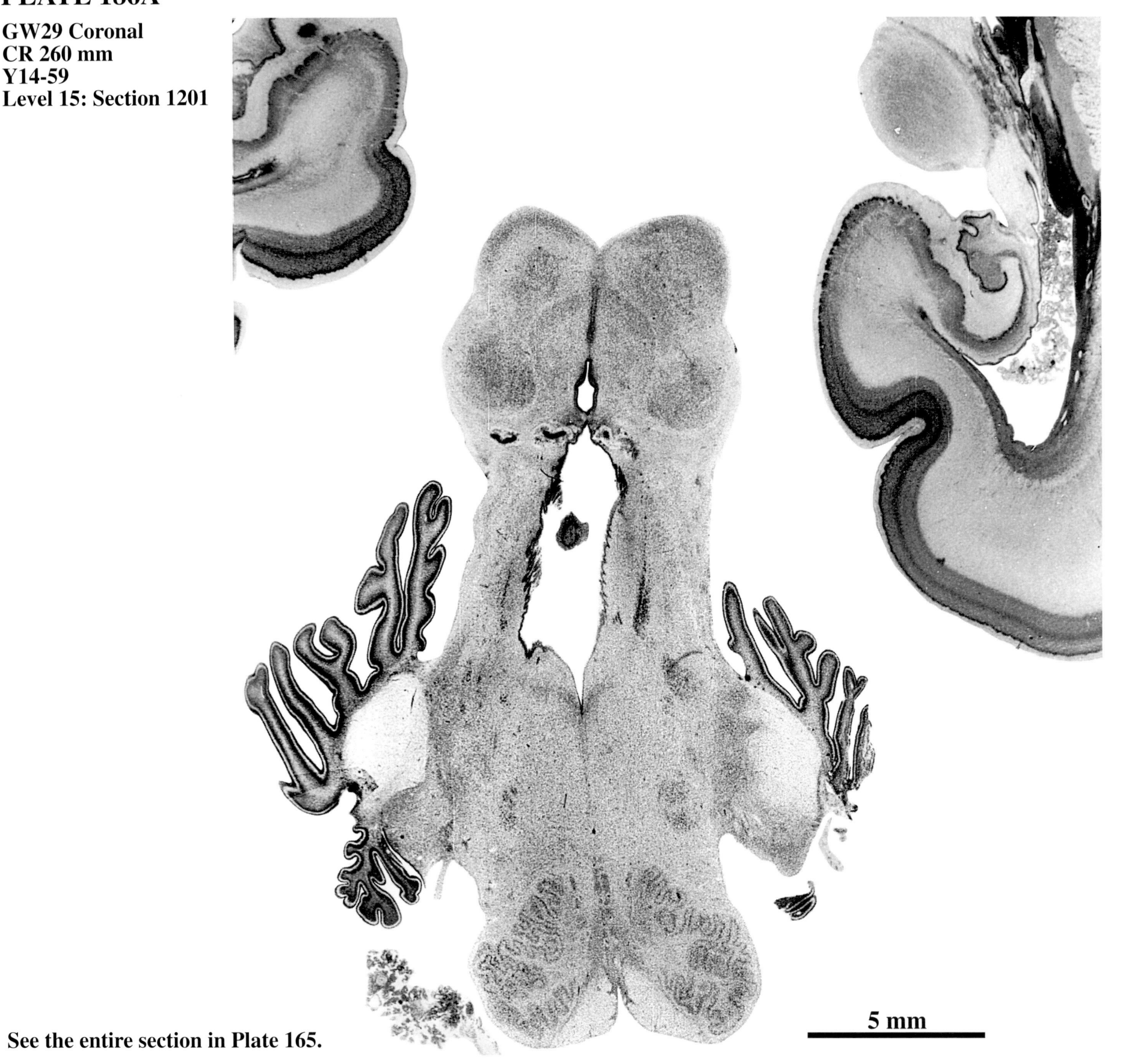

See the entire section in Plate 165.

5 mm

## PLATE 186B

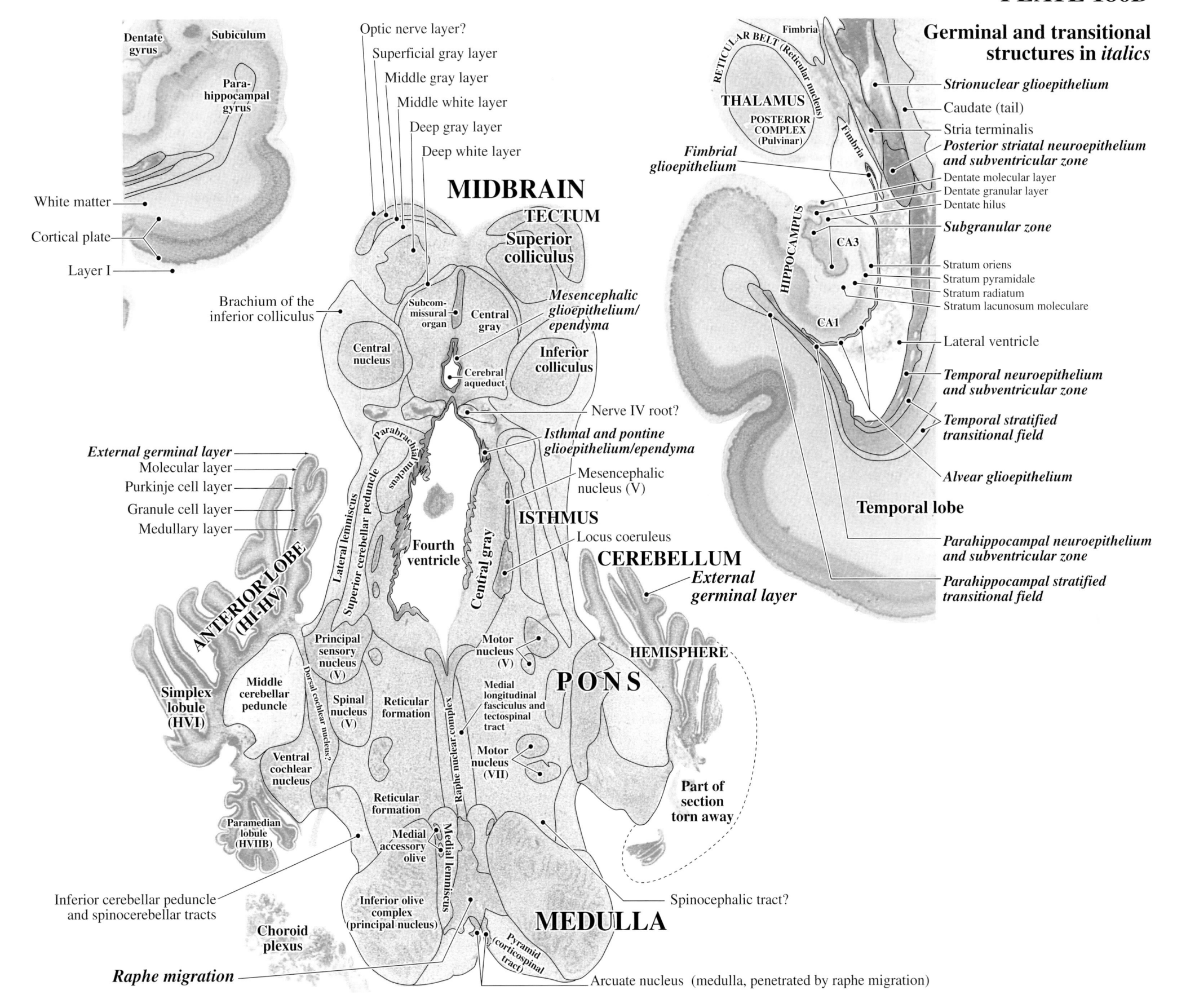

**Germinal and transitional structures in *italics***

**PLATE 187A**

GW29 Coronal
CR 260 mm
Y14-59
Level 16: Section 1261

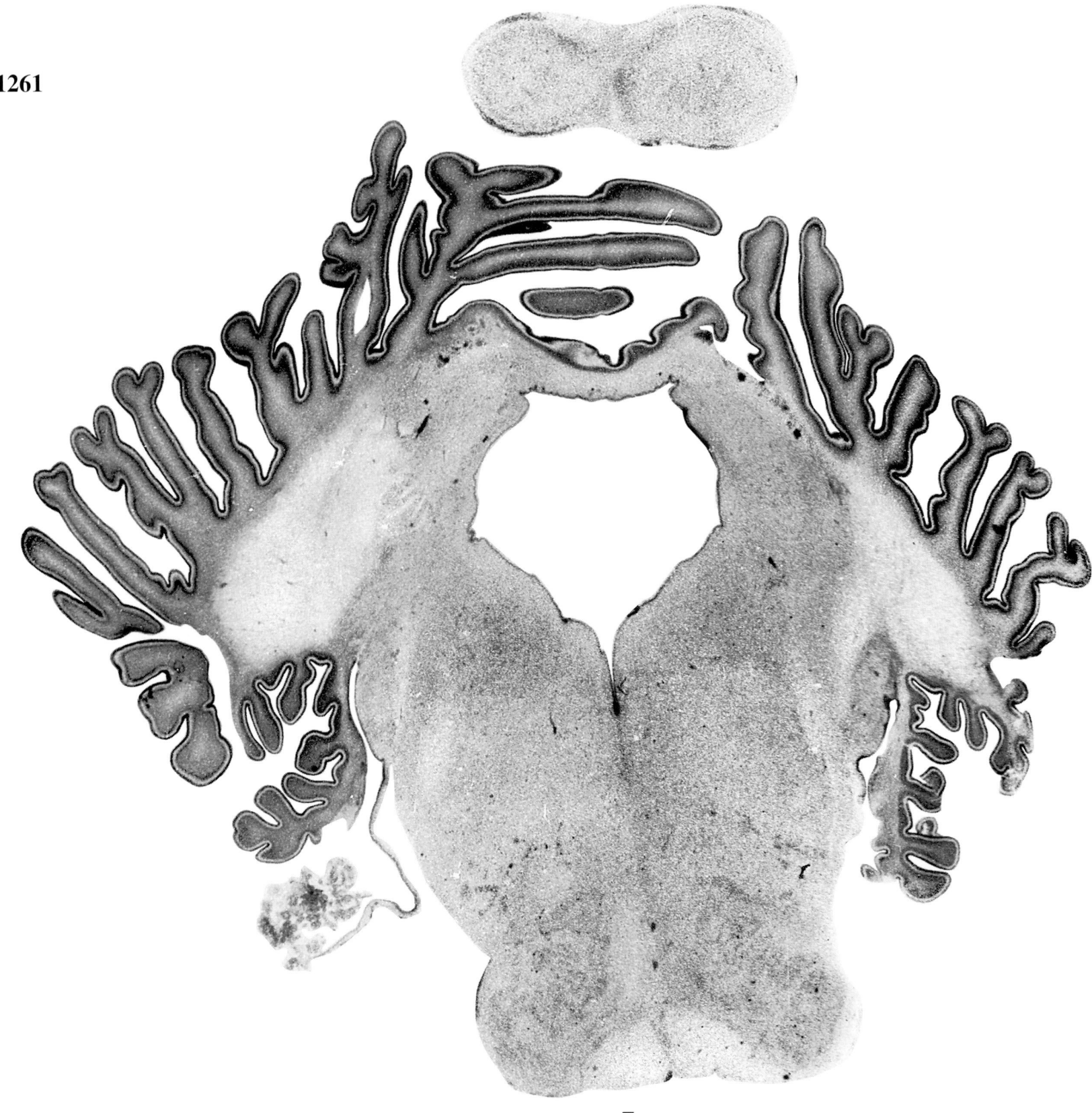

See the entire section in Plate 166.

5 mm

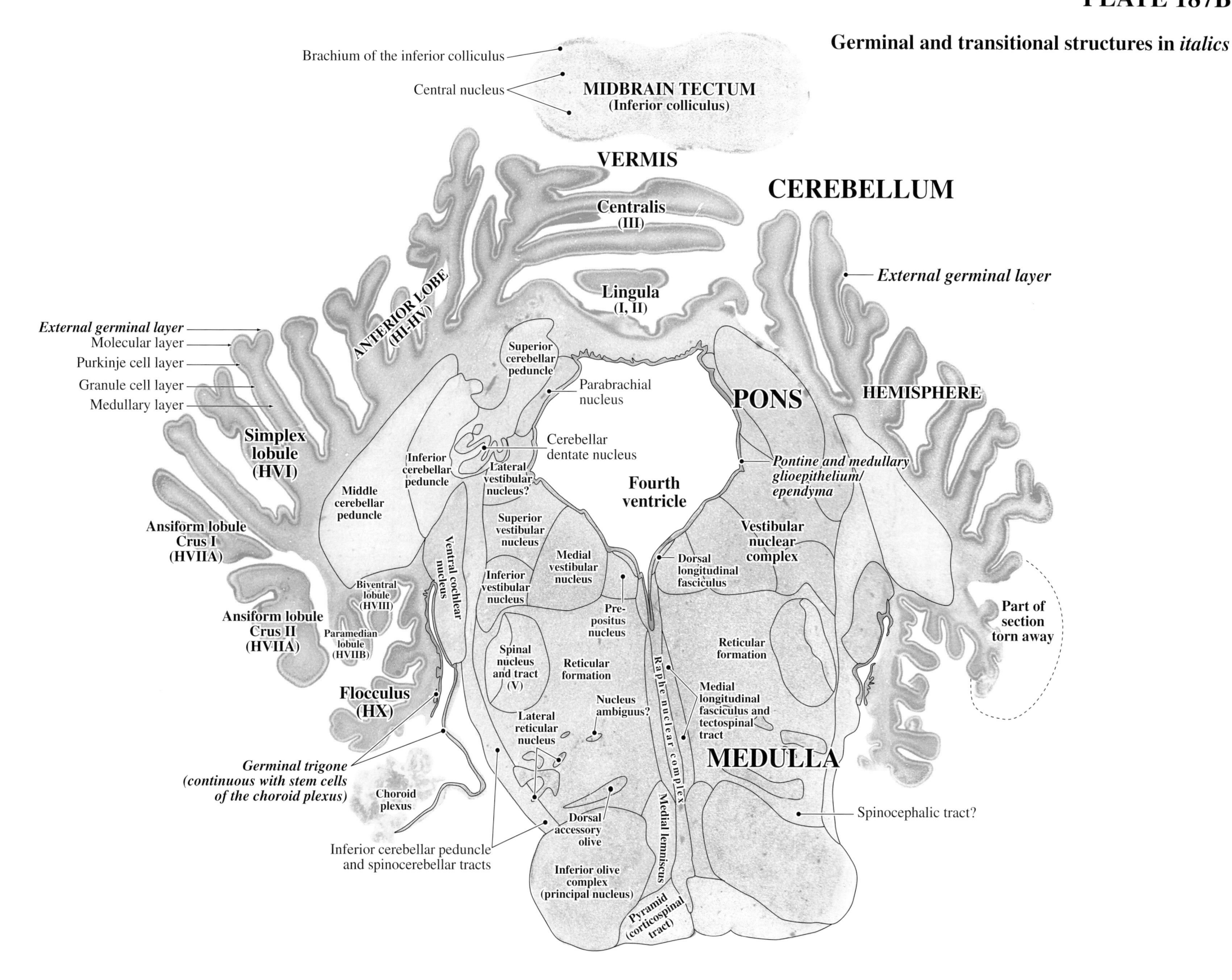

**PLATE 187B**

**Germinal and transitional structures in *italics***

**PLATE 188A**

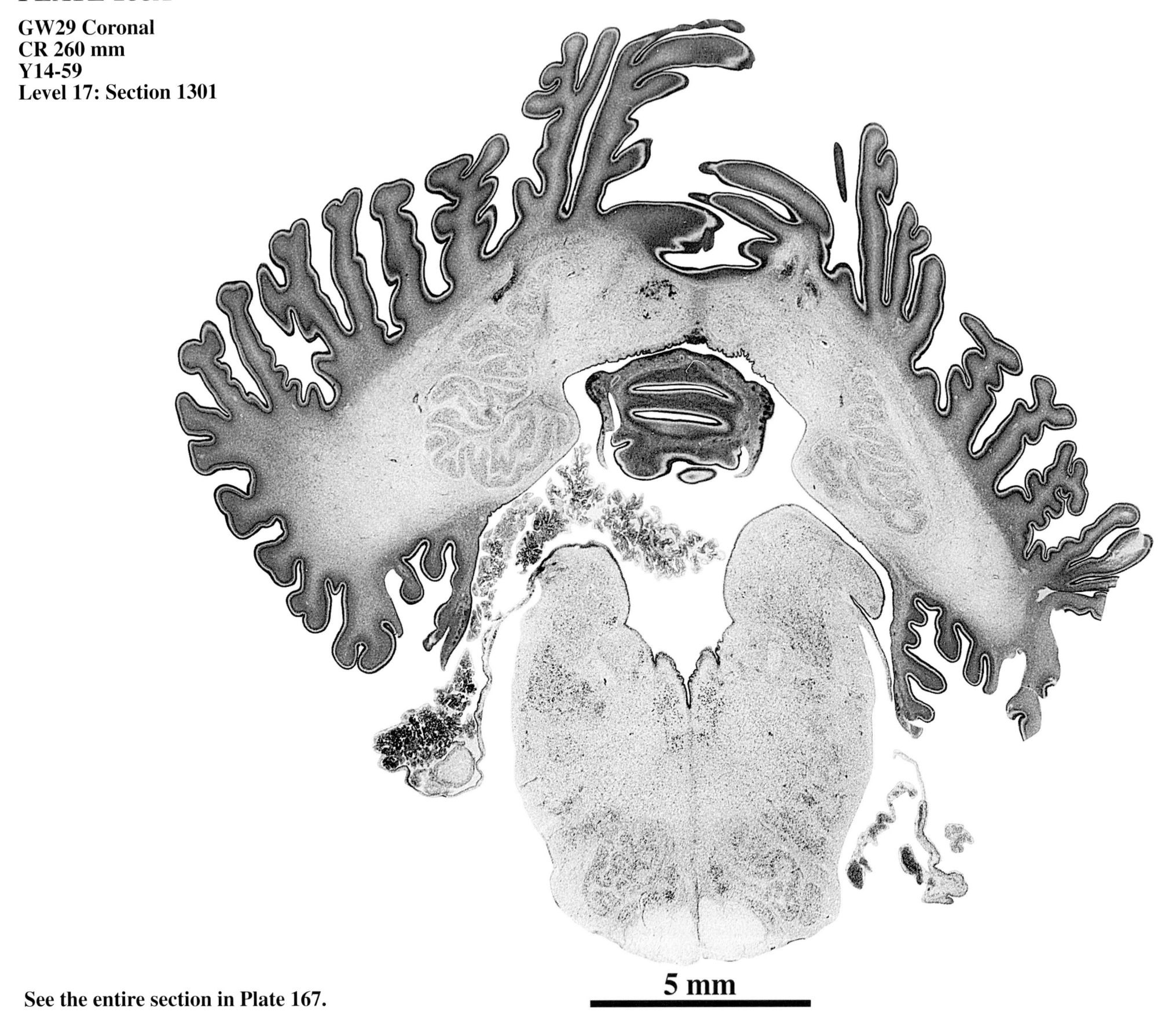

GW29 Coronal
CR 260 mm
Y14-59
Level 17: Section 1301

See the entire section in Plate 167.

5 mm

# PLATE 188B

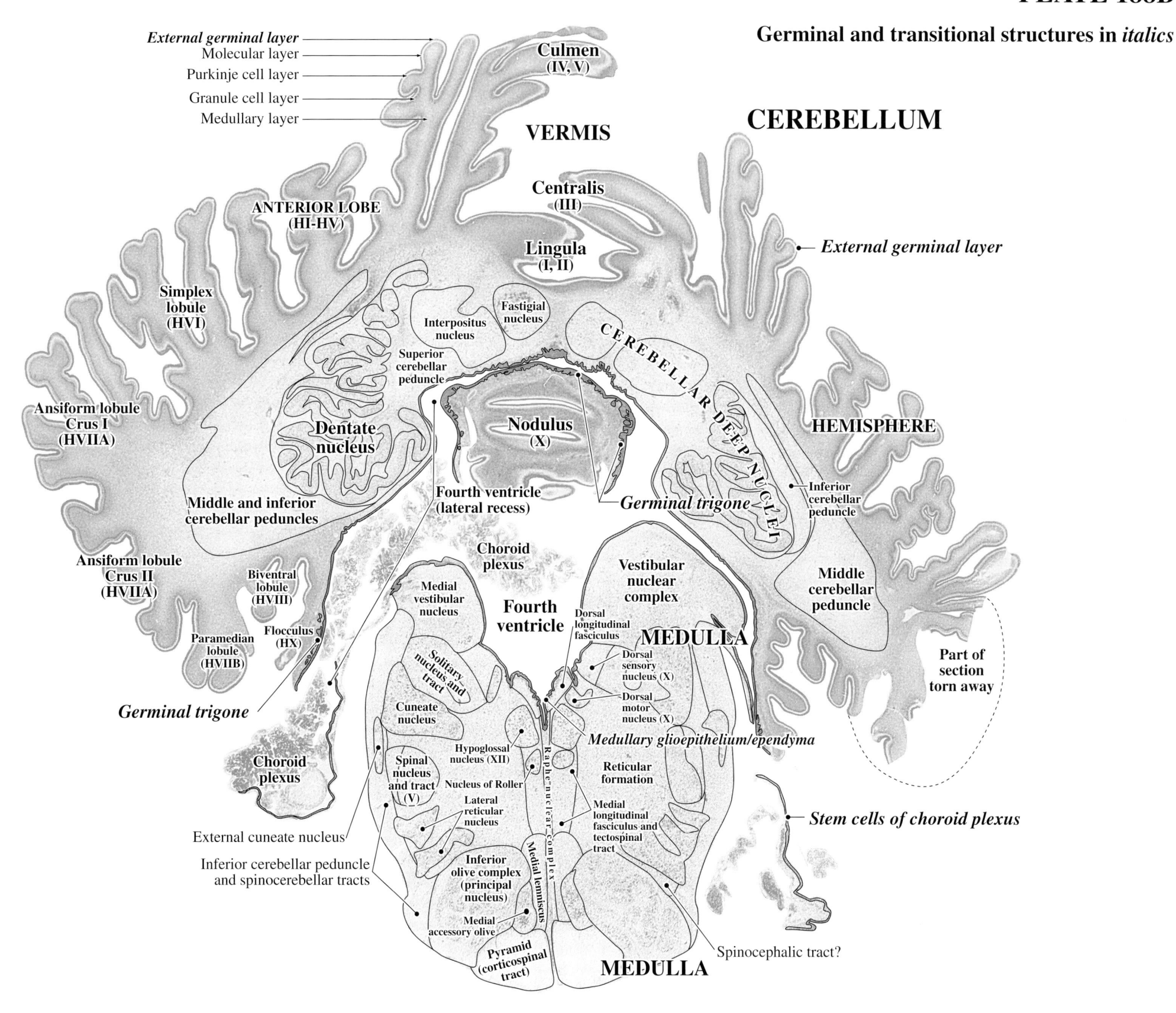

**Germinal and transitional structures in *italics***

## PLATE 189A

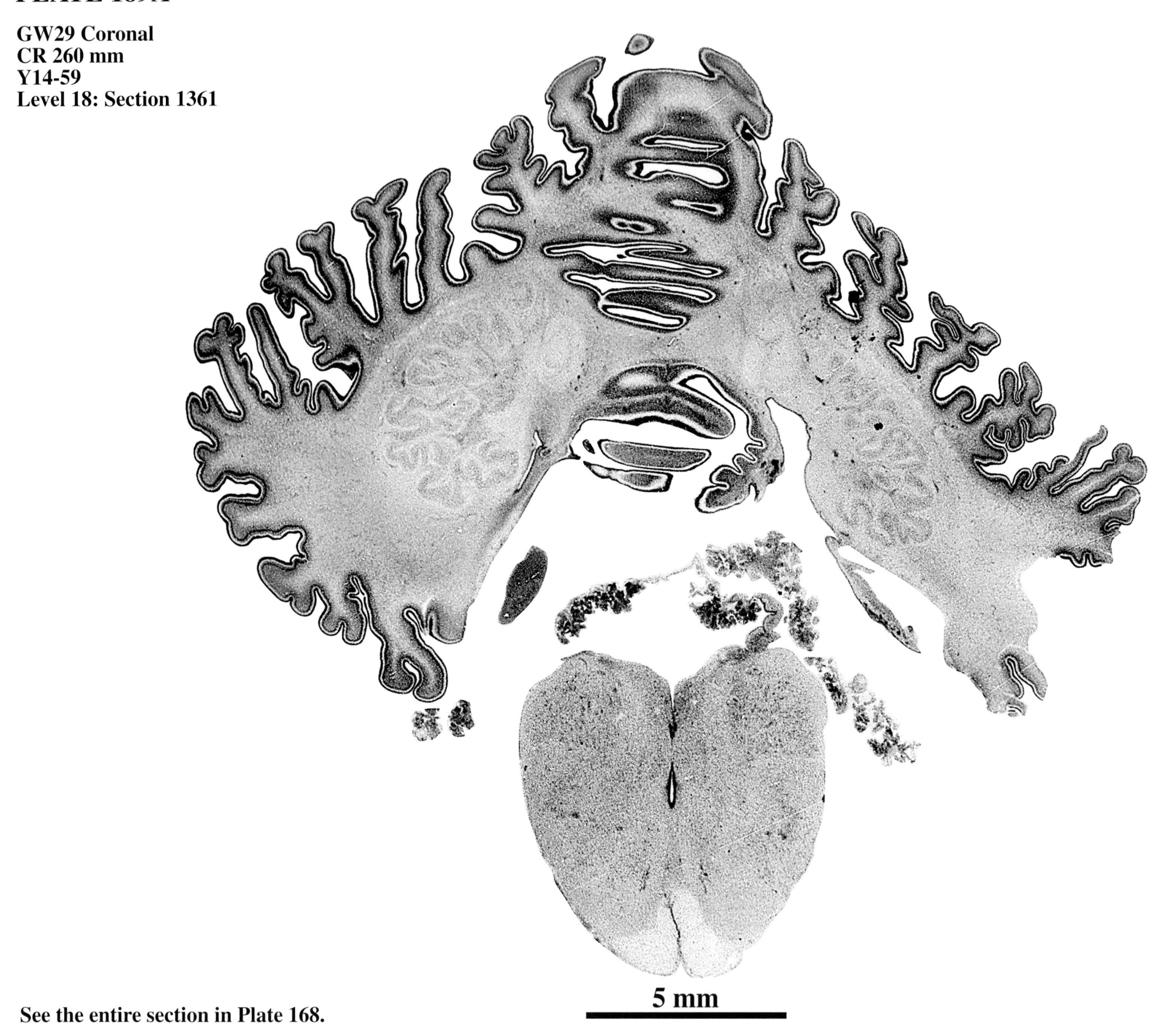

GW29 Coronal
CR 260 mm
Y14-59
Level 18: Section 1361

5 mm

See the entire section in Plate 168.

## PLATE 189B

**Germinal and transitional structures in *italics***

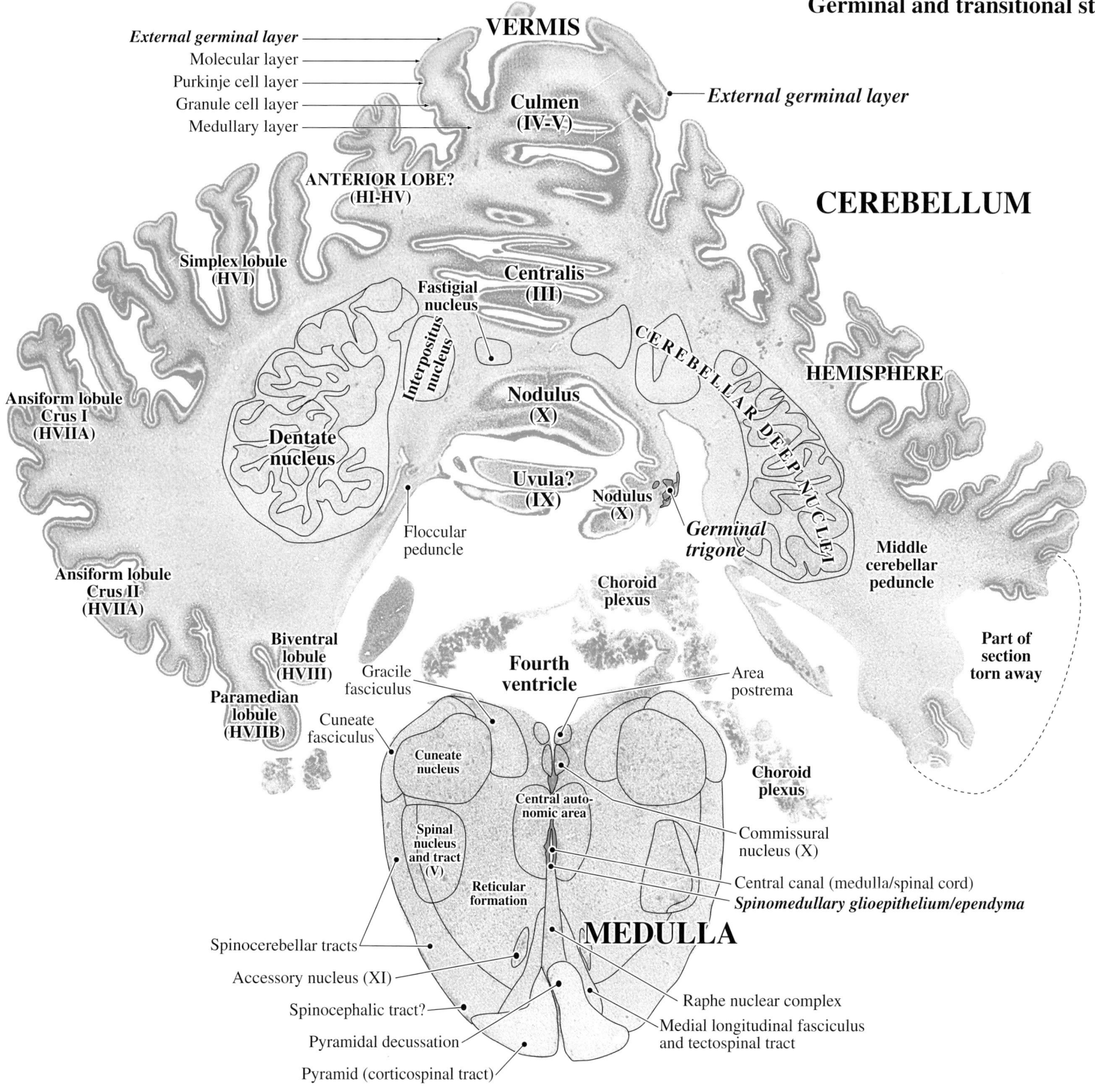

# PART VIII: GW26 SAGITTAL

This specimen is case number W-147-63 (Perinatal RPSL) in the Yakovlev Collection. A female infant survived for 27 days after a premature birth. Death occurred from a hemorrhage in the abdominal cavity. The brain was cut in the sagittal plane in 35-μm thick sections and is classified as a Normative Control in the Yakovlev Collection (Haleem, 1990). Since there is no photograph of this brain before it was embedded and cut, the photograph of the medial view of a GW25 brain that Larroche published in 1966 (**Figure 6**) is used to show gross anatomical features.

Photographs of 10 different Nissl-stained sections (**Levels 1-10**) are shown at low magnification in **Plates 190-199**. The core of the brain and the cerebellum are shown at high magnification in companion **Plates 200AB-209AB** for **Levels 1-6**. Very high magnification views of different regions of the cerebellar cortex are shown in **Plates 210-215**. Because the section numbers decrease from Level 1 (most medial) to Level 10 (most lateral), they are from the left side of the brain; the right side has higher section numbers proceeding medial to lateral. The cutting plane of this brain is nearly parallel to the midline in anterior and posterior parts of each section, including different parts of the cortex. The sections chosen for illustration are spaced closer together near the midline to show small structures in the diencephalon, midbrain, pons, and medulla.

Y15-60 contains the same group of immature structures as in the older specimens, except that all of these structures are more prominent. In the telencephalon, thicker remnants of the germinal matrices are present in all lobes of the cerebral cortex where the ***neuroepithelium/subventricular zone*** are generating neocortical interneurons. Migrating and sojourning neurons and/or glia are visible in all lobes of the cerebral cortex as ***stratified transitional fields***, thin in the occipital lobe, and thicker in the frontal, parietal and temporal lobes. More neurons, glia, and their mitotic precursor cells are migrating through the olfactory peduncle toward the olfactory bulb (***rostral migratory stream***) from a presumed source area in the germinal matrix at the junction between the cerebral cortex, striatum, and nucleus accumbens. Streams of neurons and glia percolate through the claustrum, endopiriform nucleus, external capsule, and uncinate fasciculus in the ***lateral migratory stream***. These cells appear to be heading toward the insular cortex, primary olfactory cortex, temporal cortex, and basolateral parts of the amygdaloid complex. In the basal ganglia, there is a thick ***neuroepithelium/subventricular zone*** overlying the striatum and nucleus accumbens where neurons and glia are being generated; some of these, especially from the accumbal area, will enter the ***rostral migratory stream***. Another region of active neurogenesis in the telencephalon is the ***subgranular zone*** in the hilus of the dentate gyrus that is the source of granule cells. Just as in the GW37 specimens, the septum, fornix, and Ammon's horn have only a thin, darkly staining layer at the ventricle, and these are presumed to be generating glia, cells of the choroid plexus, and the ependymal lining of the ventricle.

Most of the structures in the diencephalon appear to be settled and are maturing, but the ***glioepithelium/ependyma*** lining the third ventricle is more thick than in the older specimens. A convoluted ***glioepithelium/ependyma*** lines the cerebral aqueduct in the midbrain. A smooth ***glioepithelium/ependyma*** lines the fourth ventricle through much of the pons. Another convoluted ***glioepithelium/ependyma*** lines the fourth ventricle through much of the medulla, part of that may be a remnant of the germinal source of the precerebellar nuclei. The ***external germinal layer*** is prominent over the entire surface of the cerebellar cortex and is actively producing basket, stellate, and granule cells. The ***germinal trigone*** is visible at the base of the nodulus and along the floccular peduncle; choroid plexus cells and glia are originating here and some migrating neurons may still be in the trigone.

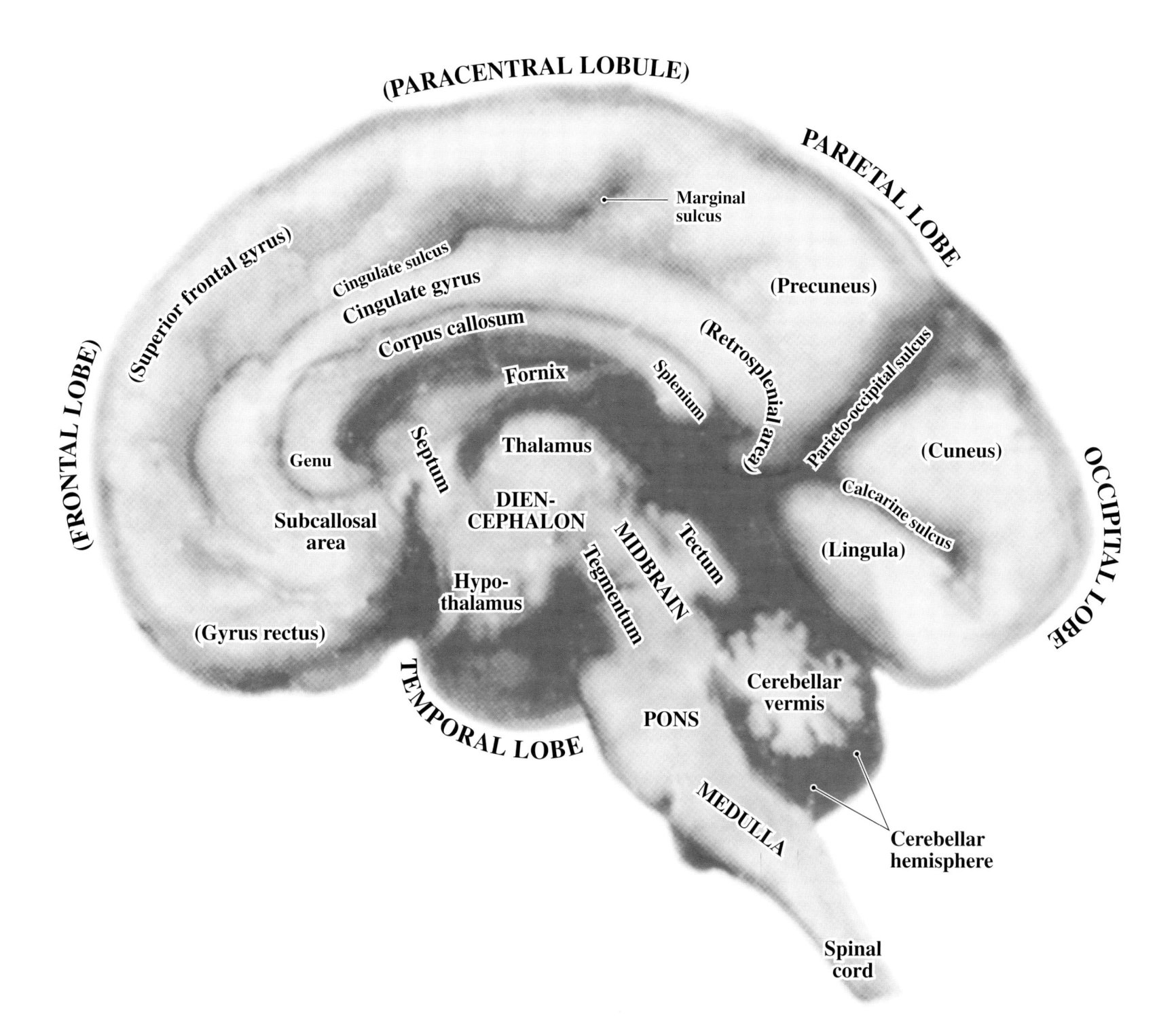

**Figure 11.** Midline sagittal view of a GW25 brain with major structures in the cerebral hemispheres and brainstem labeled. (This is part of Figure 2-9 on page 27 in B. A. Curtis, S. Jacobson, and E. M. Marcus (1972) *An Introduction to the Neurosciences*, Philadelphia: W. B. Saunders. The photograph was originally published by J. C. Larroche (1966) The development of the central nervous system during intrauterine life. In: *Human Development*, F. Falkner (ed.), Philadelphia: W. B. Saunders, page 259.)

# PLATE 190

GW26 Sagittal
CR 225 mm
Y147-63
Level 1: Section 501

*Remnants of the germinal matrix, migratory streams, and transitional fields*

| | | | |
|---|---|---|---|
| 1 | *Rostral migratory stream* | 6 | *Precerebellar glioepithelium* |
| 2 | *Diencephalic glioepithelium/ependyma* | 7 | *Medullary glioepithelium/ependyma* |
| 3 | *Diencephalic/mesencephalic glioepithelium/ependyma* | 8 | *Raphe migration* |
| 4 | *Mesencephalic glioepithelium/ependyma* | 9 | *Germinal trigone (cerebellum)* |
| 5 | *Pontine glioepithelium/ependyma* | 10 | *External germinal layer (cerebellum)* |
| | | 11 | *Subpial granular layer (cortical)* |

**See detail of brain core and cerebellum in Plates 200A and B.**

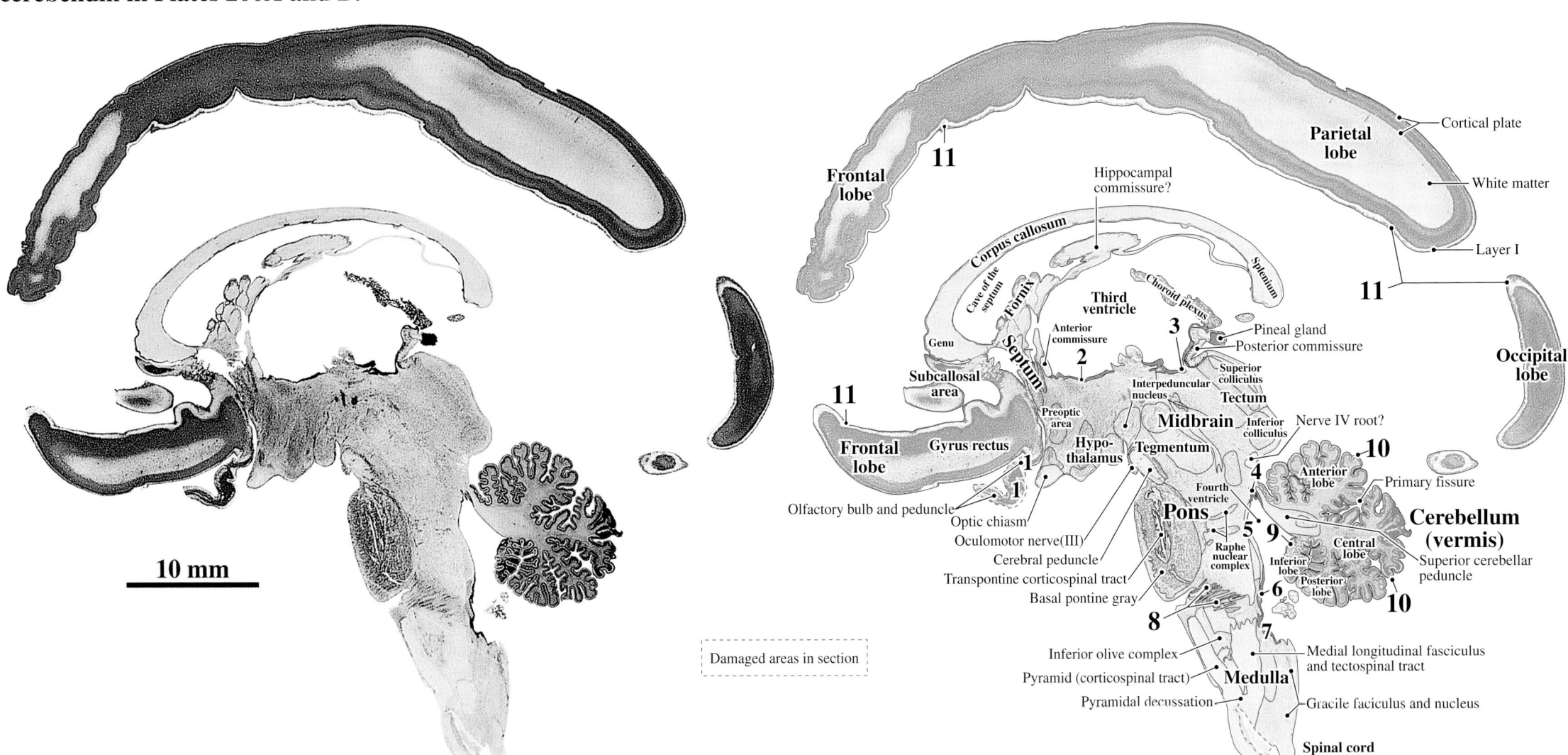

10 mm

Damaged areas in section

# PLATE 191

**GW26 Sagittal**
**CR 225 mm**
**Y147-63**
**Level 2: Section 481**

*Remnants of the germinal matrix, migratory streams, and transitional fields*

| | | | |
|---|---|---|---|
| 1 | *Rostral migratory stream* | 7 | *Mesencephalic glioepithelium/ependyma* |
| 2 | *Callosal sling* | 8 | *Pontine and medullary glioepithelium/ependyma* |
| 3 | *Callosal glioepithelium* | 9 | *Raphe migration* |
| 4 | *Fornical glioepithelium* | 10 | *Spinal glioepithelium/ependyma* |
| 5 | *Strionuclear glioepithelium* | 11 | *Germinal trigone (cerebellum)* |
| 6 | *Diencephalic (thalamic) glioepithelium/ependyma* | 12 | *External germinal layer (cerebellum)* |
| | | 13 | *Subpial granular layer (cortical)* |

**See detail of brain core and cerebellum in Plates 201A and B.**

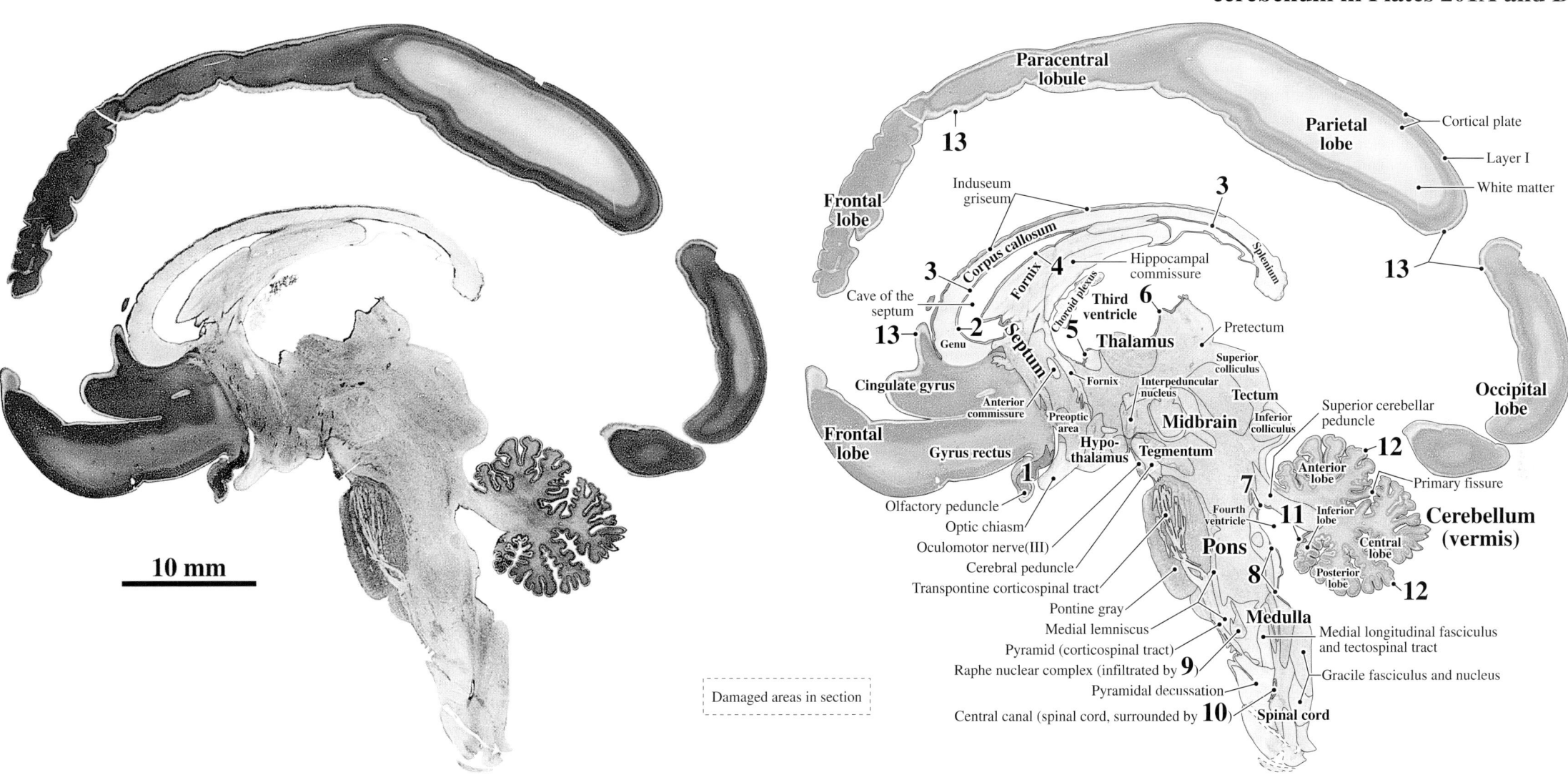

**See high magnification views of the cerebellar cortex and germinal trigone in Plates 210-215.**

# PLATE 192

GW26 Sagittal
CR 225 mm
Y147-63
Level 3: Section 461

**Remnants of the germinal matrix, migratory streams, and transitional fields**

| | | | |
|---|---|---|---|
| 1 | Rostral migratory stream | 6 | Mesencephalic glioepithelium/ependyma |
| 2 | Callosal glioepithelium | 7 | Pontine and medullary glioepithelium/ependyma |
| 3 | Fornical glioepithelium | 8 | Germinal trigone (cerebellum) |
| 4 | Strionuclear glioepithelium | 9 | External germinal layer (cerebellum) |
| 5 | Diencephalic (thalamic) glioepithelium/ependyma | 10 | Subpial granular layer (cortical) |

See detail of brain core and cerebellum in Plates 202A and B.

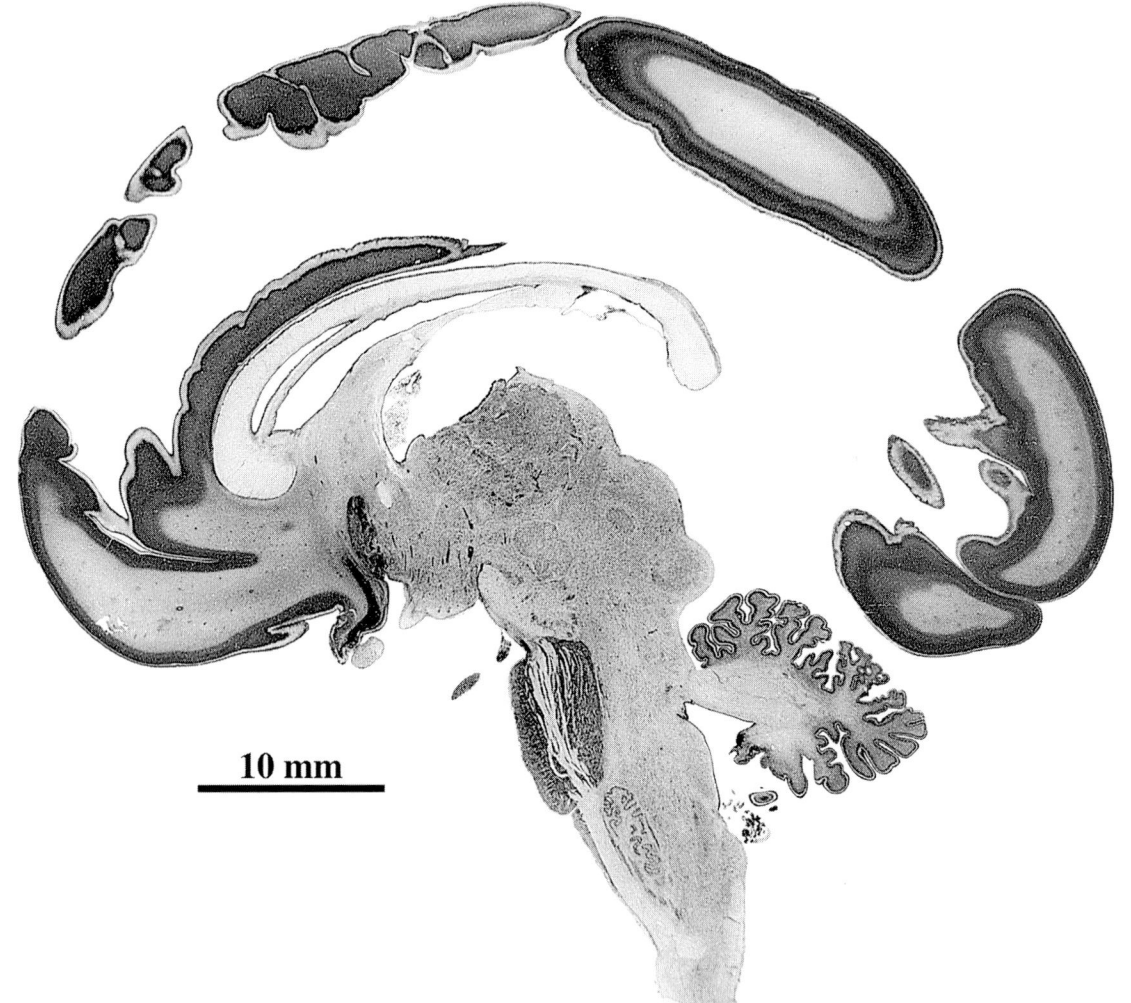

10 mm

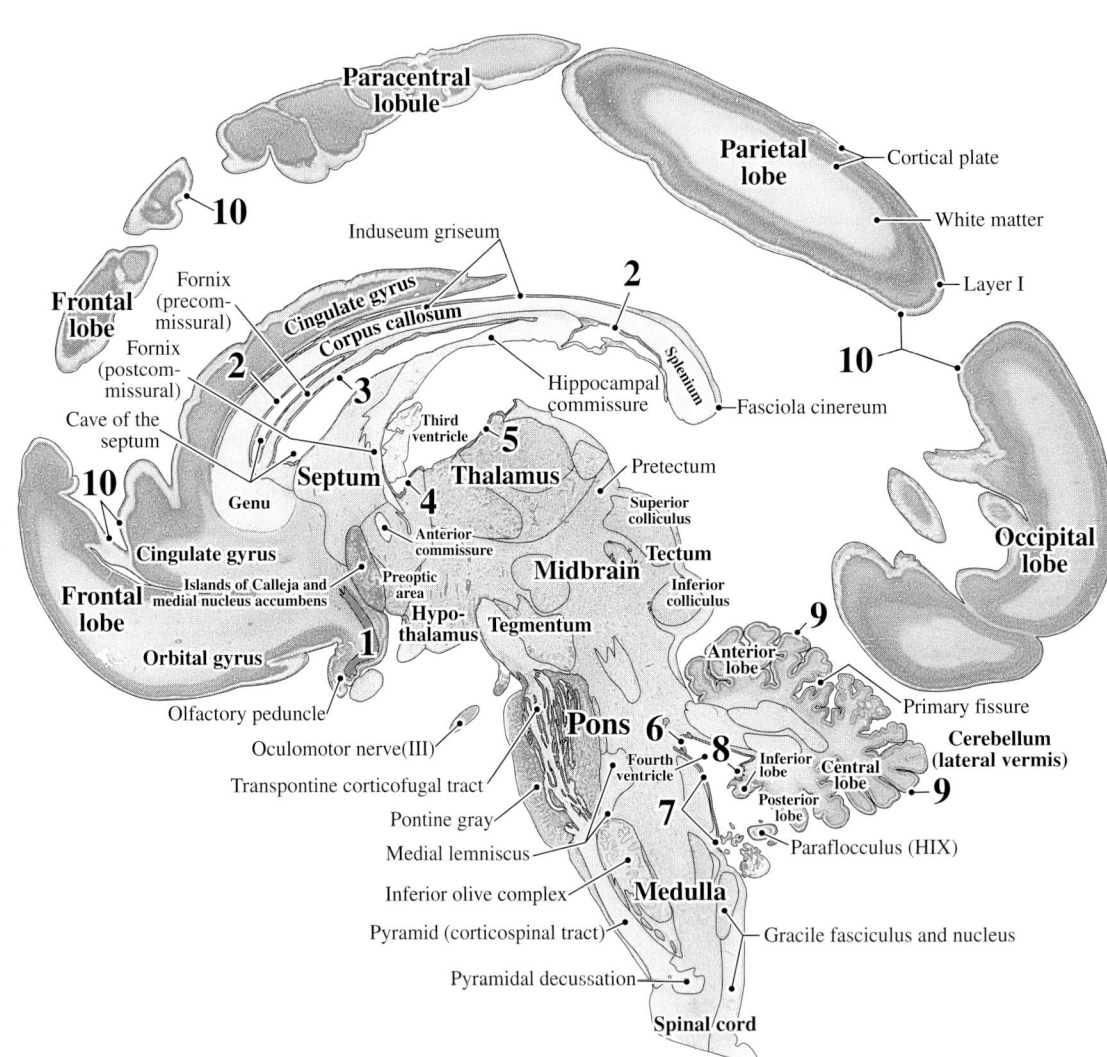

PLATE 193

GW26 Sagittal
CR 225 mm
Y147-63
Level 4: Section 421

### Remnants of the germinal matrix, migratory streams, and transitional fields

1. *Rostral migratory stream*
2. *Frontal neuroepithelium and subventricular zone*
3. *Frontal stratified transitional field*
4. *Callosal glioepithelium*
5. *Fornical glioepithelium*
6. *Accumbent neuroepithelium and subventricular zone (intermingled with the rostral migratory stream)*
7. *Strionuclear neuroepithelium and subventricular zone*
8. *Pontine glioepithelium/ependyma*
9. *External germinal layer (cerebellum)*
10. *Subpial granular layer (cortical)*

See detail of brain core and cerebellum in Plates 203A and B.

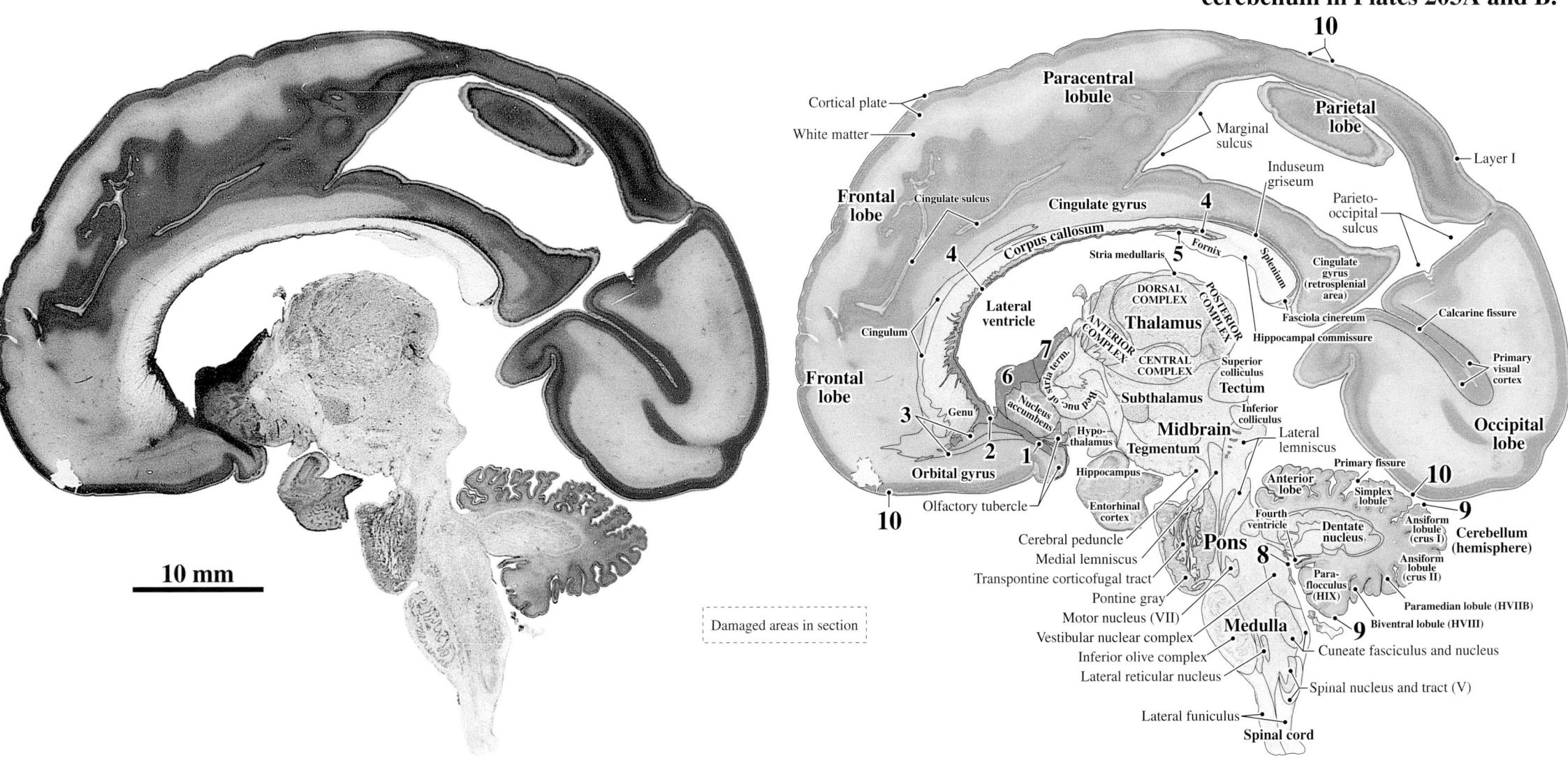

# PLATE 194

**GW26 Sagittal**
**CR 225 mm**
**Y147-63**
**Level 5: Section 381**

See detail of brain core and cerebellum in Plates 204A and B.

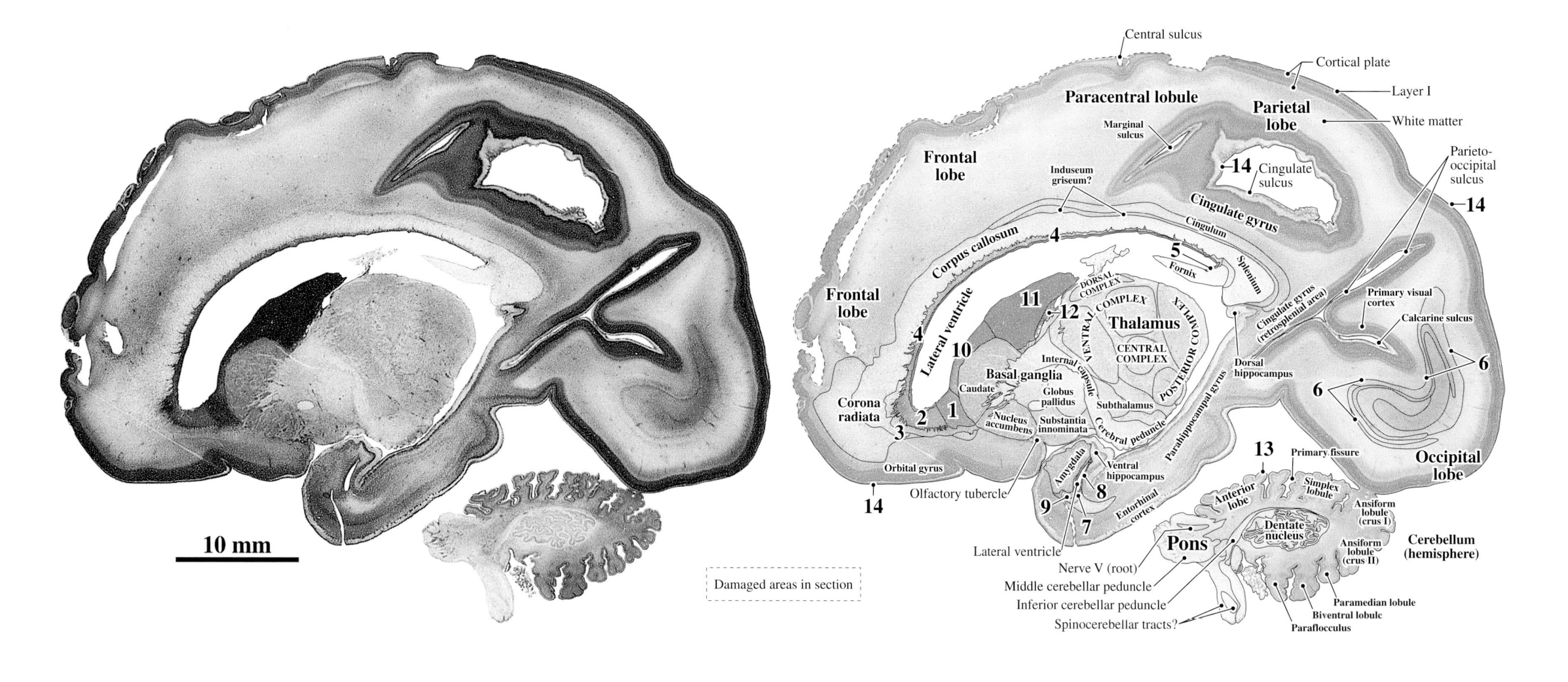

### Remnants of the germinal matrix, migratory streams, and transitional fields

| | | | |
|---|---|---|---|
| 1 | *Rostral migratory stream (source area)* | 8 | *Alvear glioepithelium* |
| 2 | *Frontal neuroepithelium and subventricular zone* | 9 | *Amygdaloid glioepithelium/ependyma* |
| 3 | *Frontal stratified transitional field* | 10 | *Anterolateral striatal neuroepithelium and subventricular zone* |
| 4 | *Callosal glioepithelium* | 11 | *Anteromedial striatal neuroepithelium and subventricular zone* |
| 5 | *Fornical glioepithelium* | 12 | *Strionuclear glioepithelium* |
| 6 | *Occipital stratified transitional field* | 13 | *External germinal layer (cerebellum)* |
| 7 | *Parahippocampal neuroepithelium, subventricular zone, and stratified transitional field* | 14 | *Subpial granular layer (cortical)* |

**PLATE 195**

GW26 Sagittal
CR 225 mm
Y147-63
Level 6: Section 361

### Remnants of the germinal matrix, migratory streams, and transitional fields

| | | | |
|---|---|---|---|
| 1 | Rostral migratory stream (source area) | 9 | Alvear glioepithelium |
| 2 | Frontal neuroepithelium and subventricular zone | 10 | Subgranular zone (dentate) |
| 3 | Frontal stratified transitional field | 11 | Amygdaloid glioepithelium/ependyma |
| 4 | Callosal glioepithelium | 12 | Anterolateral striatal neuroepithelium and subventricular zone |
| 5 | Fornical glioepithelium | 13 | Anteromedial striatal neuroepithelium and subventricular zone |
| 6 | Occipital neuroepithelium and subventricular zone | 14 | Strionuclear glioepithelium |
| 7 | Occipital stratified transitional field | 15 | External germinal layer (cerebellum) |
| 8 | Parahippocampal neuroepithelium, subventricular zone, and stratified transitional field | 16 | Subpial granular layer (cortical) |

See detail of brain core and cerebellum in Plates 205A and B.

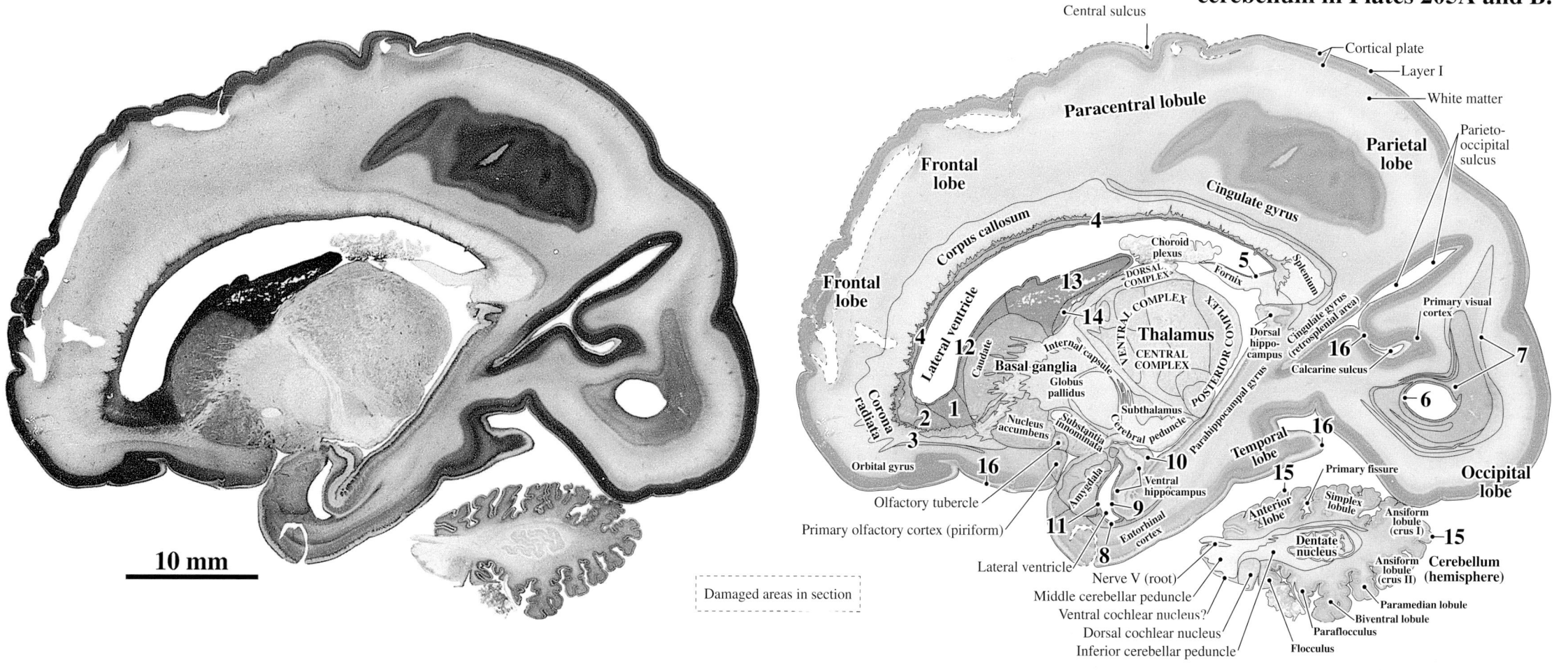

# PLATE 196

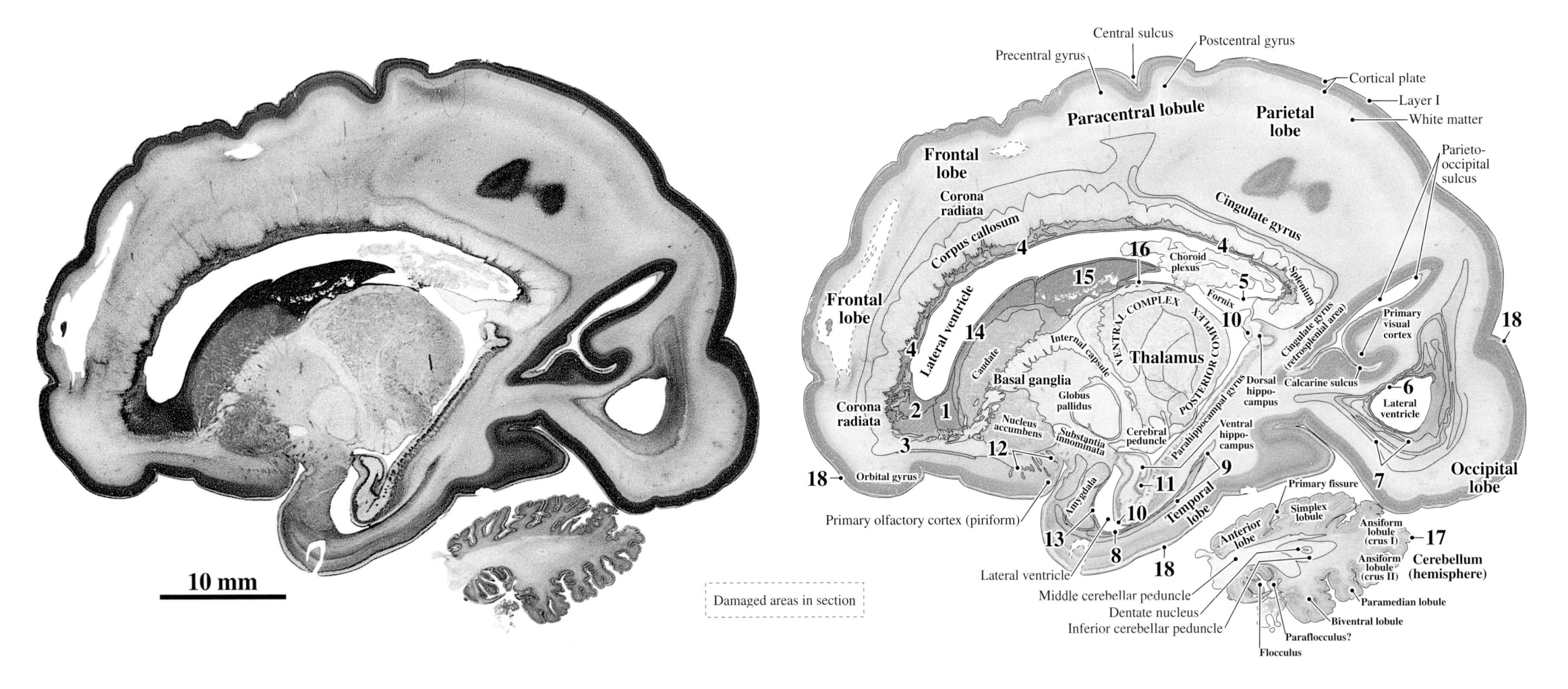

GW26 Sagittal
CR 225 mm
Y147-63
Level 7: Section 341

*Remnants of the germinal matrix, migratory streams, and transitional fields*

| | | | |
|---|---|---|---|
| 1 | Rostral migratory stream (source area) | 10 | Alvear glioepithelium |
| 2 | Frontal neuroepithelium and subventricular zone | 11 | Subgranular zone (dentate) |
| 3 | Frontal stratified transitional field | 12 | Lateral migratory stream (cortical) |
| 4 | Callosal glioepithelium | 13 | Amygdaloid glioepithelium/ependyma |
| 5 | Fornical glioepithelium | 14 | Anterolateral striatal neuroepithelium and subventricular zone |
| 6 | Occipital neuroepithelium and subventricular zone | 15 | Anteromedial striatal neuroepithelium and subventricular zone |
| 7 | Occipital stratified transitional field | 16 | Strionuclear glioepithelium |
| 8 | Temporal neuroepithelium and subventricular zone | 17 | External germinal layer (cerebellum) |
| 9 | Parahippocampal/temporal stratified transitional field | 18 | Subpial granular layer (cortical) |

**See detail of brain core and cerebellum in Plates 206A and B.**

# PLATE 197

**GW26 Sagittal**
**CR 225 mm**
**Y147-63**
**Level 8: Section 321**

*Remnants of the germinal matrix, migratory streams, and transitional fields*

1. *Rostral migratory stream (source area)*
2. *Frontal neuroepithelium and subventricular zone*
3. *Frontal stratified transitional field*
4. *Paracentral neuroepithelium and subventricular zone*
5. *Paracentral stratified transitional field*
6. *Callosal glioepithelium*
7. *Fornical glioepithelium*
8. *Occipital neuroepithelium and subventricular zone*
9. *Occipital stratified transitional field*
10. *Temporal neuroepithelium and subventricular zone*
11. *Temporal stratified transitional field*
12. *Alvear glioepithelium*
13. *Subgranular zone (dentate)*
14. *Lateral migratory stream (cortical)*
15. *Amygdaloid glioepithelium/ependyma*
16. *Anterolateral striatal neuroepithelium and subventricular zone*
17. *Anteromedial/posterior striatal neuroepithelium and subventricular zone*
18. *Strionuclear glioepithelium*
19. *External germinal layer (cerebellum)*
20. *Subpial granular layer (cortical)*

**See detail of brain core and cerebellum in Plates 207A and B.**

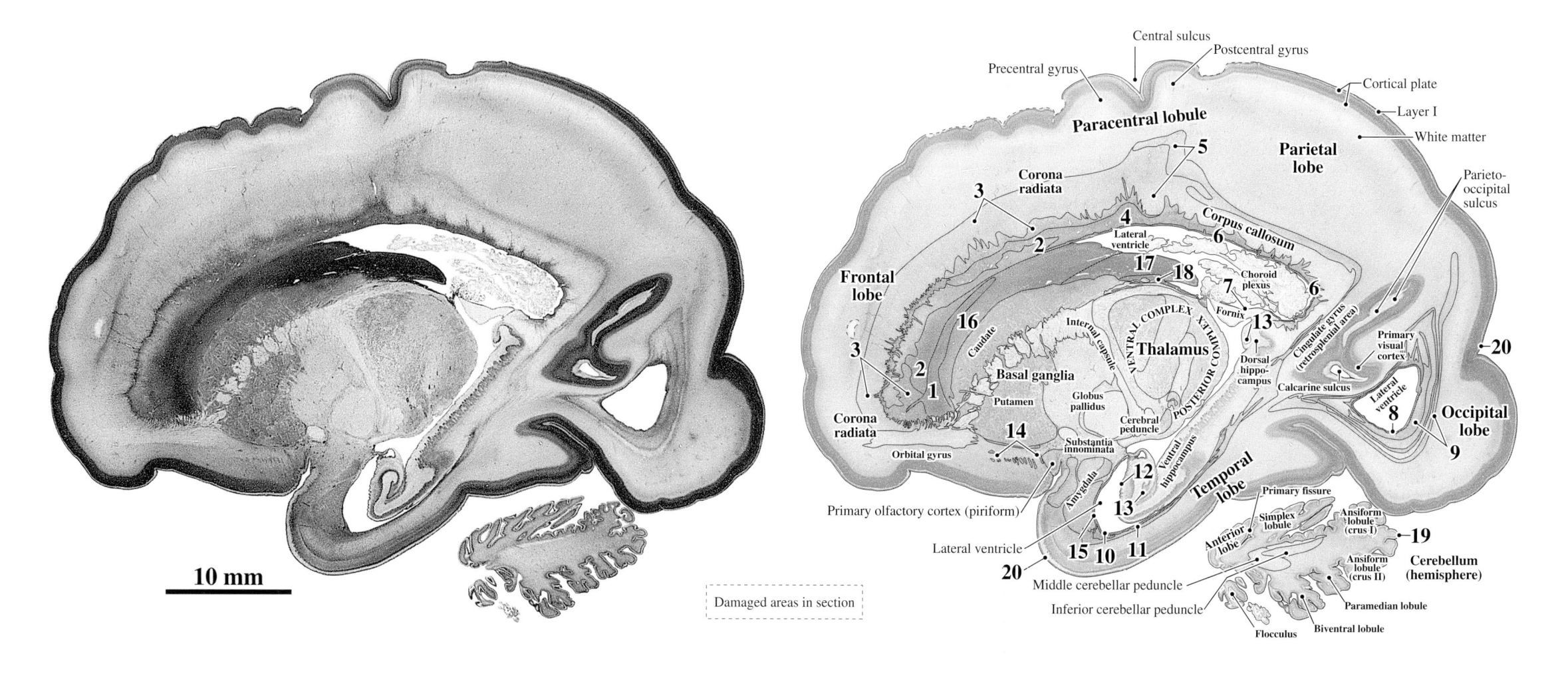

# PLATE 198

GW26 Sagittal
CR 225 mm
Y147-63
Level 9: Section 261

**Remnants of the germinal matrix, migratory streams, and transitional fields**

| | | | |
|---|---|---|---|
| 1 | Frontal stratified transitional field | 9 | Alvear glioepithelium |
| 2 | Paracentral stratified transitional field | 10 | Subgranular zone (dentate) |
| 3 | Parietal neuroepithelium and subventricular zone | 11 | Lateral migratory stream (cortical) |
| 4 | Parietal stratified transitional field | 12 | Amygdaloid glioepithelium/ependyma |
| 5 | Occipital neuroepithelium and subventricular zone | 13 | Posterior striatal neuroepithelium and subventricular zone |
| 6 | Occipital stratified transitional field | 14 | Strionuclear glioepithelium |
| 7 | Temporal neuroepithelium and subventricular zone | 15 | External germinal layer (cerebellum) |
| 8 | Temporal stratified transitional field | 16 | Subpial granular layer (cortical) |

**See detail of brain core in Plates 208A and B.**

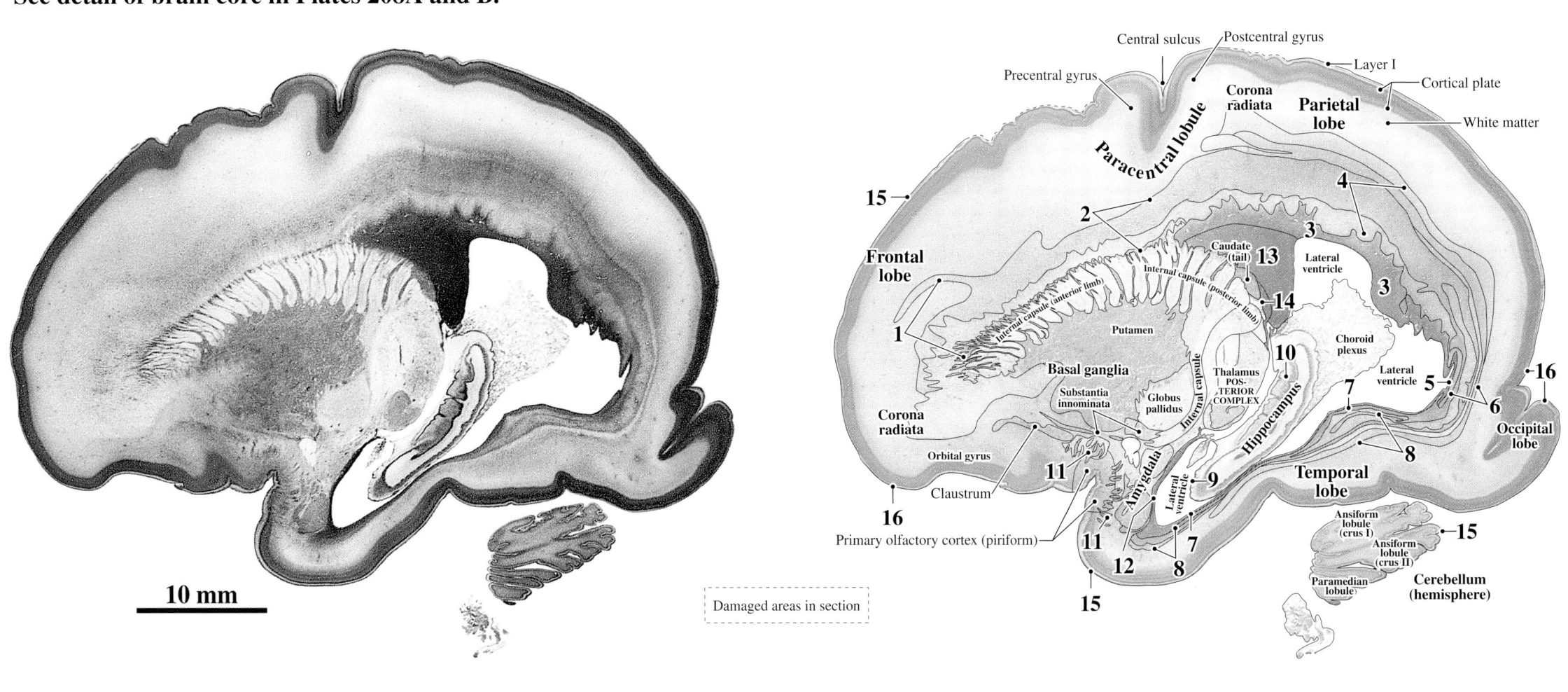

## PLATE 199

**GW26 Sagittal**
**CR 225 mm**
**Y147-63**
**Level 10: Section 221**

### Remnants of the germinal matrix, migratory streams, and transitional fields

| | | | |
|---|---|---|---|
| 1 | Parietal neuroepithelium and subventricular zone | 6 | Amygdaloid glioepithelium/ependyma |
| 2 | Parietal stratified transitional field | 7 | Posterior striatal neuroepithelium and subventricular zone |
| 3 | Temporal neuroepithelium and subventricular zone | 8 | Strionuclear glioepithelium |
| 4 | Temporal stratified transitional field | 9 | External germinal layer (cerebellum) |
| 5 | Lateral migratory stream (cortical) | 10 | Subpial granular layer (cortical) |

See detail of brain core in Plates 209A and B.

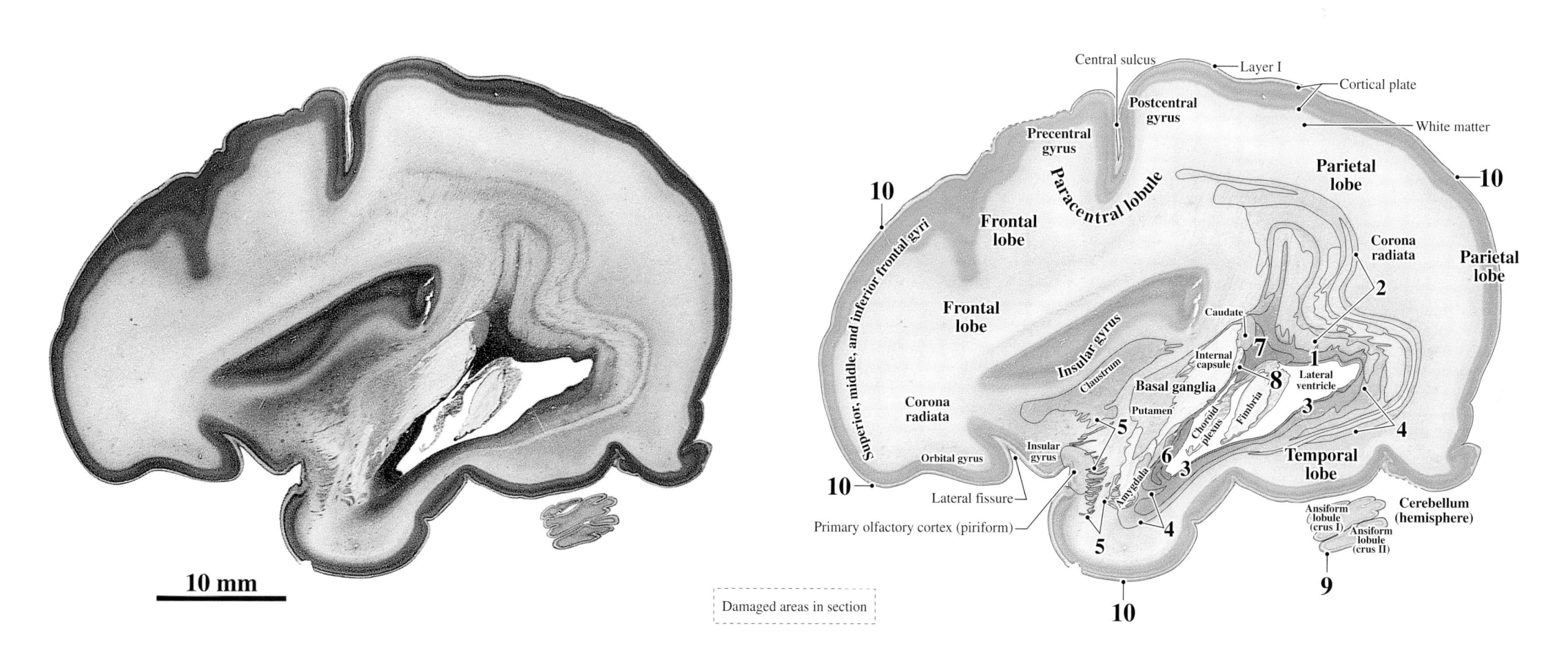

## PLATE 200A

GW26 Sagittal
CR 225 mm
Y147-63
Level 1: Section 501

See the entire section in Plate 190.

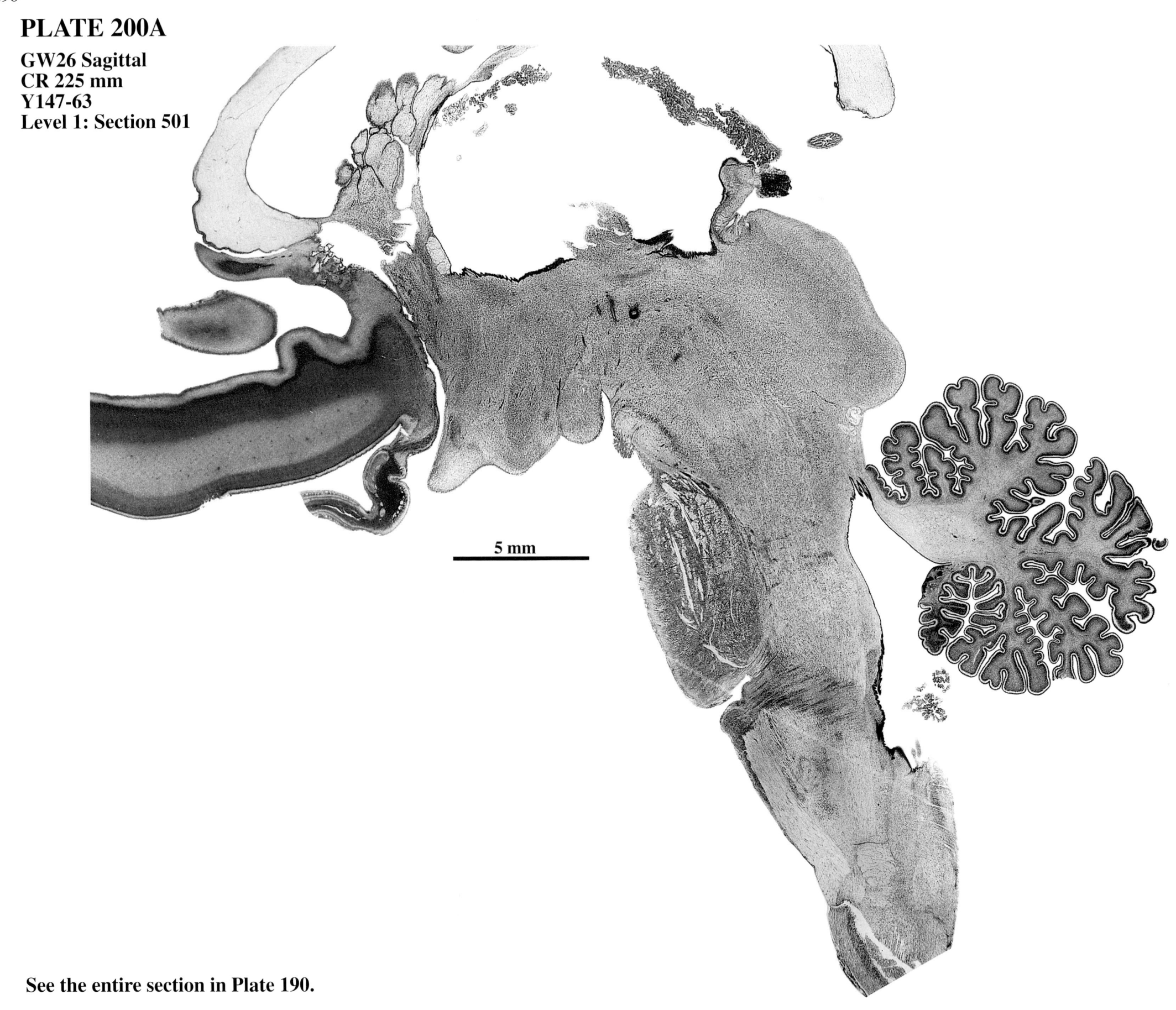

PLATE 200B

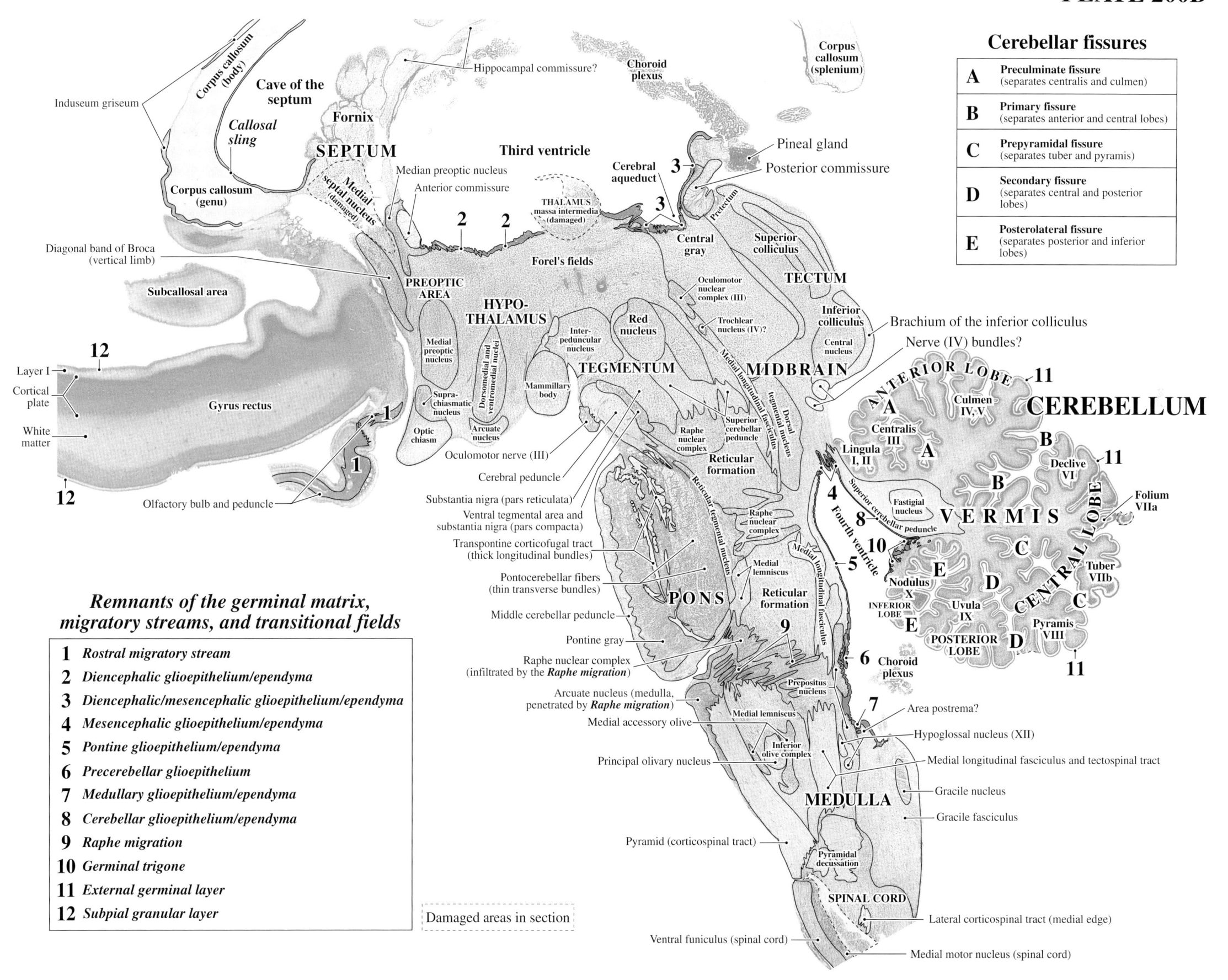

## Cerebellar fissures

| | | |
|---|---|---|
| A | Preculminate fissure | (separates centralis and culmen) |
| B | Primary fissure | (separates anterior and central lobes) |
| C | Prepyramidal fissure | (separates tuber and pyramis) |
| D | Secondary fissure | (separates central and posterior lobes) |
| E | Posterolateral fissure | (separates posterior and inferior lobes) |

### Remnants of the germinal matrix, migratory streams, and transitional fields

1. *Rostral migratory stream*
2. *Diencephalic glioepithelium/ependyma*
3. *Diencephalic/mesencephalic glioepithelium/ependyma*
4. *Mesencephalic glioepithelium/ependyma*
5. *Pontine glioepithelium/ependyma*
6. *Precerebellar glioepithelium*
7. *Medullary glioepithelium/ependyma*
8. *Cerebellar glioepithelium/ependyma*
9. *Raphe migration*
10. *Germinal trigone*
11. *External germinal layer*
12. *Subpial granular layer*

Damaged areas in section

Labels on figure:
Induseum griseum; Corpus callosum (body); Cave of the septum; Callosal sling; Corpus callosum (genu); Hippocampal commissure?; Choroid plexus; Corpus callosum (splenium); Fornix; SEPTUM; Median preoptic nucleus; Anterior commissure; Third ventricle; Cerebral aqueduct; Pineal gland; Posterior commissure; Medial septal nucleus (damaged); THALAMUS massa intermedia (damaged); Pretectum; Diagonal band of Broca (vertical limb); Forel's fields; Central gray; Superior colliculus; Subcallosal area; PREOPTIC AREA; HYPO-THALAMUS; Oculomotor nuclear complex (III); TECTUM; Medial preoptic nucleus; Inter-peduncular nucleus; Red nucleus; Trochlear nucleus (IV)?; Inferior colliculus; Central nucleus; Brachium of the inferior colliculus; Layer I; Cortical plate; Gyrus rectus; Supra-chiasmatic nucleus; Dorsomedial and ventromedial nuclei; Mammillary body; TEGMENTUM; Medial longitudinal fasciculus; MIDBRAIN; Nerve (IV) bundles?; ANTERIOR LOBE; Culmen IV,V; CEREBELLUM; White matter; Arcuate nucleus; Optic chiasm; Oculomotor nerve (III); Raphe nuclear complex; Superior cerebellar peduncle; Dorsal tegmental nucleus; Centralis III; Lingula I, II; Declive VI; Olfactory bulb and peduncle; Cerebral peduncle; Fastigial nucleus; Folium VIIa; Substantia nigra (pars reticulata); Ventral tegmental area and substantia nigra (pars compacta); Reticular formation; Raphe nuclear complex; Superior cerebellar peduncle; VERMIS; Transpontine corticofugal tract (thick longitudinal bundles); Medial lemniscus; Fourth ventricle; Tuber VIIb; Pontocerebellar fibers (thin transverse bundles); Reticular formation; Medial longitudinal fasciculus; CENTRAL LOBE; Nodulus X; PONS; Uvula IX; Pyramis VIII; Middle cerebellar peduncle; INFERIOR LOBE; Pontine gray; POSTERIOR LOBE; Raphe nuclear complex (infiltrated by the Raphe migration); Choroid plexus; Arcuate nucleus (medulla, penetrated by Raphe migration); Prepositus nucleus; Area postrema?; Medial lemniscus; Medial accessory olive; Hypoglossal nucleus (XII); Inferior olive complex; Medial longitudinal fasciculus and tectospinal tract; Principal olivary nucleus; Gracile nucleus; MEDULLA; Gracile fasciculus; Pyramid (corticospinal tract); Pyramidal decussation; SPINAL CORD; Lateral corticospinal tract (medial edge); Ventral funiculus (spinal cord); Medial motor nucleus (spinal cord)

297

## PLATE 201A

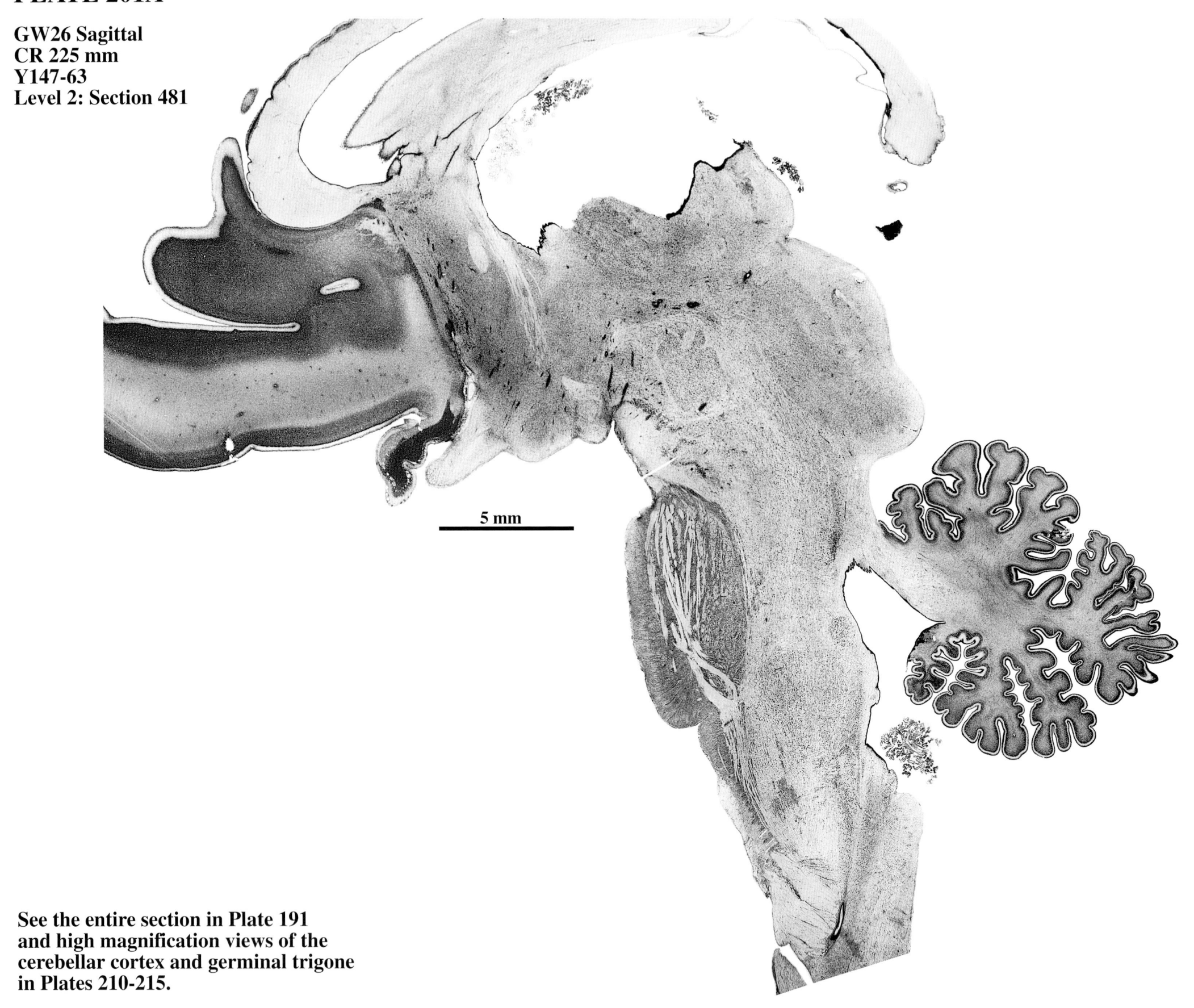

GW26 Sagittal
CR 225 mm
Y147-63
Level 2: Section 481

5 mm

See the entire section in Plate 191 and high magnification views of the cerebellar cortex and germinal trigone in Plates 210-215.

## PLATE 201B

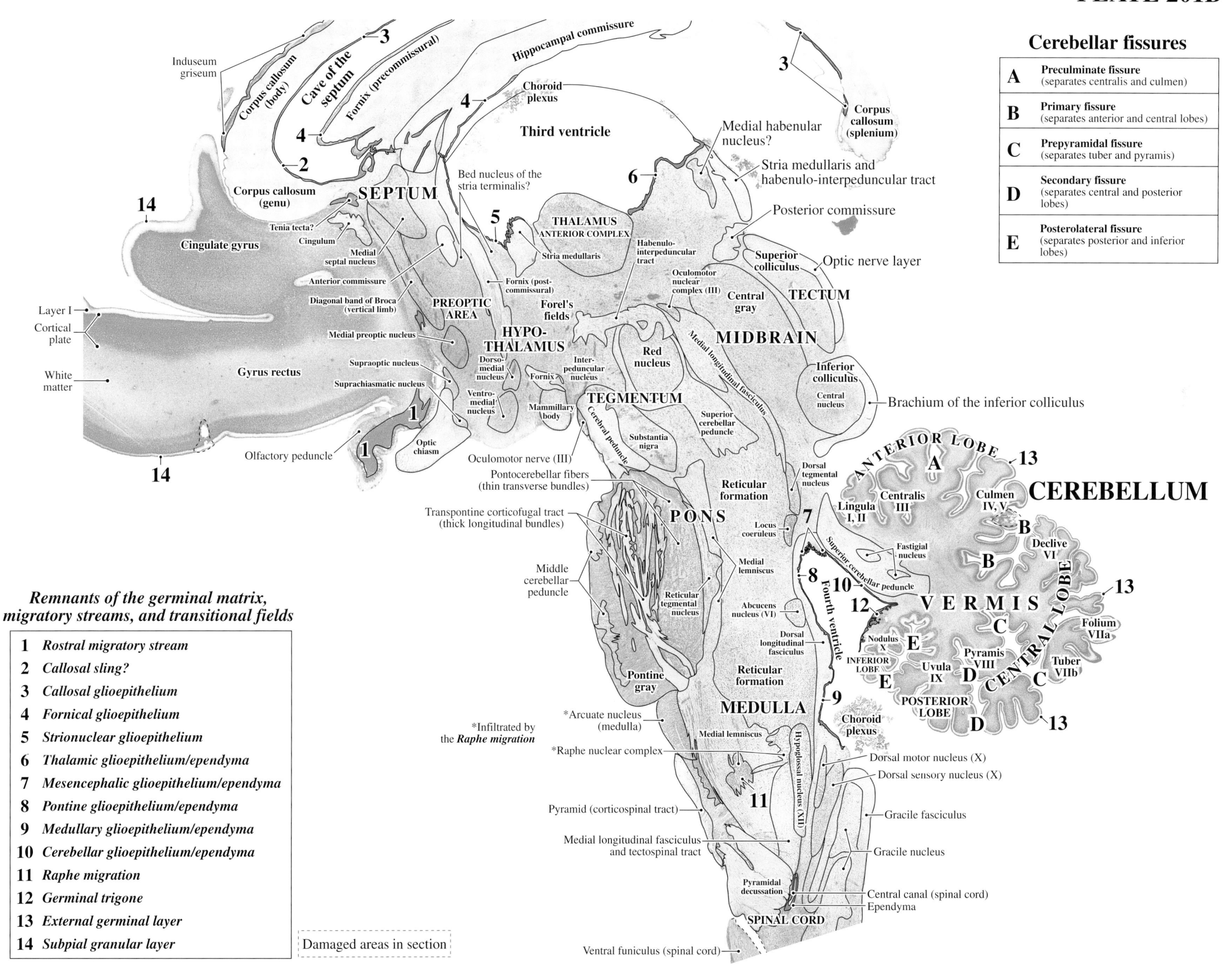

### Cerebellar fissures

| | | |
|---|---|---|
| **A** | Preculminate fissure | (separates centralis and culmen) |
| **B** | Primary fissure | (separates anterior and central lobes) |
| **C** | Prepyramidal fissure | (separates tuber and pyramis) |
| **D** | Secondary fissure | (separates central and posterior lobes) |
| **E** | Posterolateral fissure | (separates posterior and inferior lobes) |

*Remnants of the germinal matrix, migratory streams, and transitional fields*

1. *Rostral migratory stream*
2. *Callosal sling?*
3. *Callosal glioepithelium*
4. *Fornical glioepithelium*
5. *Strionuclear glioepithelium*
6. *Thalamic glioepithelium/ependyma*
7. *Mesencephalic glioepithelium/ependyma*
8. *Pontine glioepithelium/ependyma*
9. *Medullary glioepithelium/ependyma*
10. *Cerebellar glioepithelium/ependyma*
11. *Raphe migration*
12. *Germinal trigone*
13. *External germinal layer*
14. *Subpial granular layer*

Damaged areas in section

# PLATE 202A

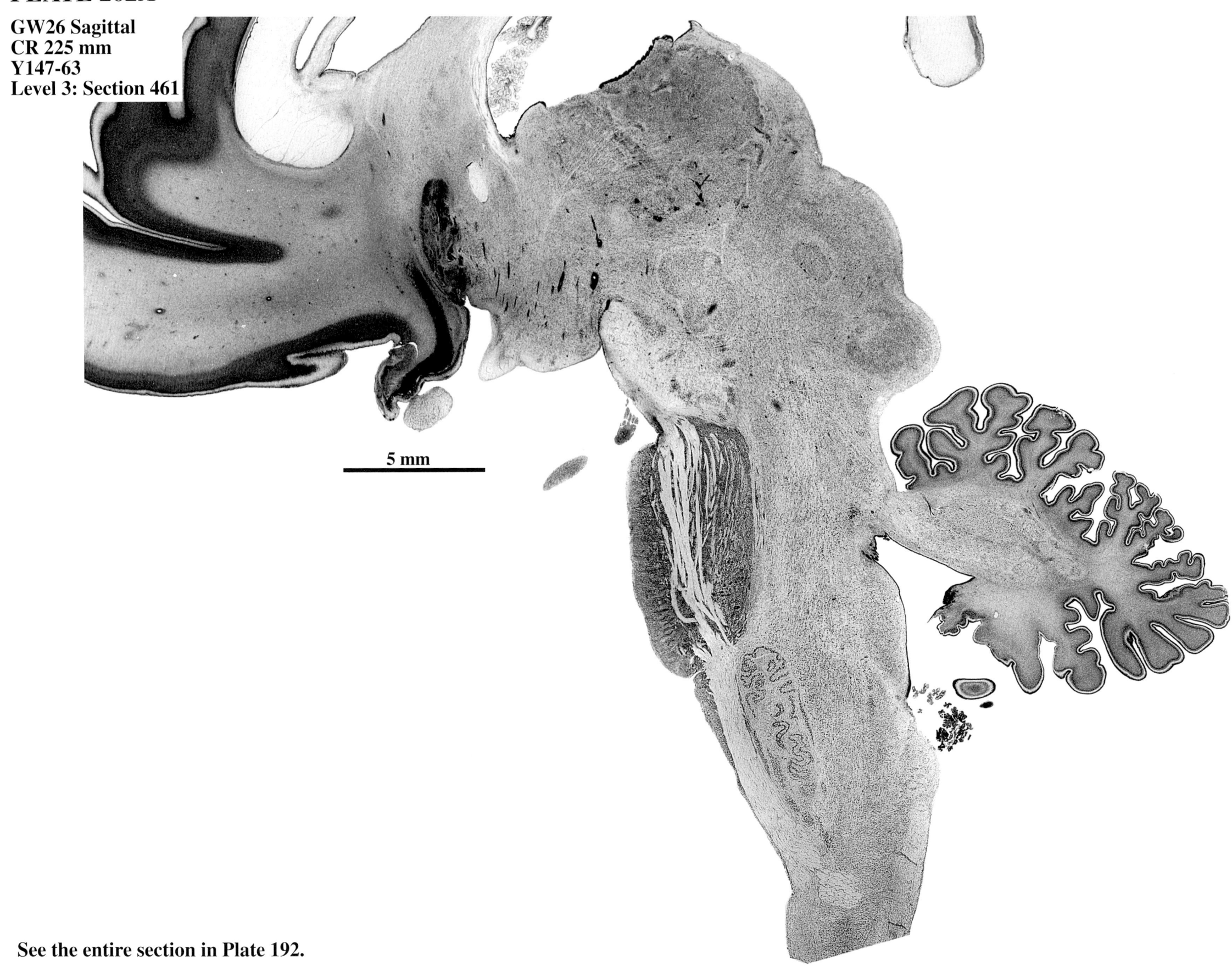

**GW26 Sagittal**
**CR 225 mm**
**Y147-63**
**Level 3: Section 461**

5 mm

See the entire section in Plate 192.

## PLATE 202B

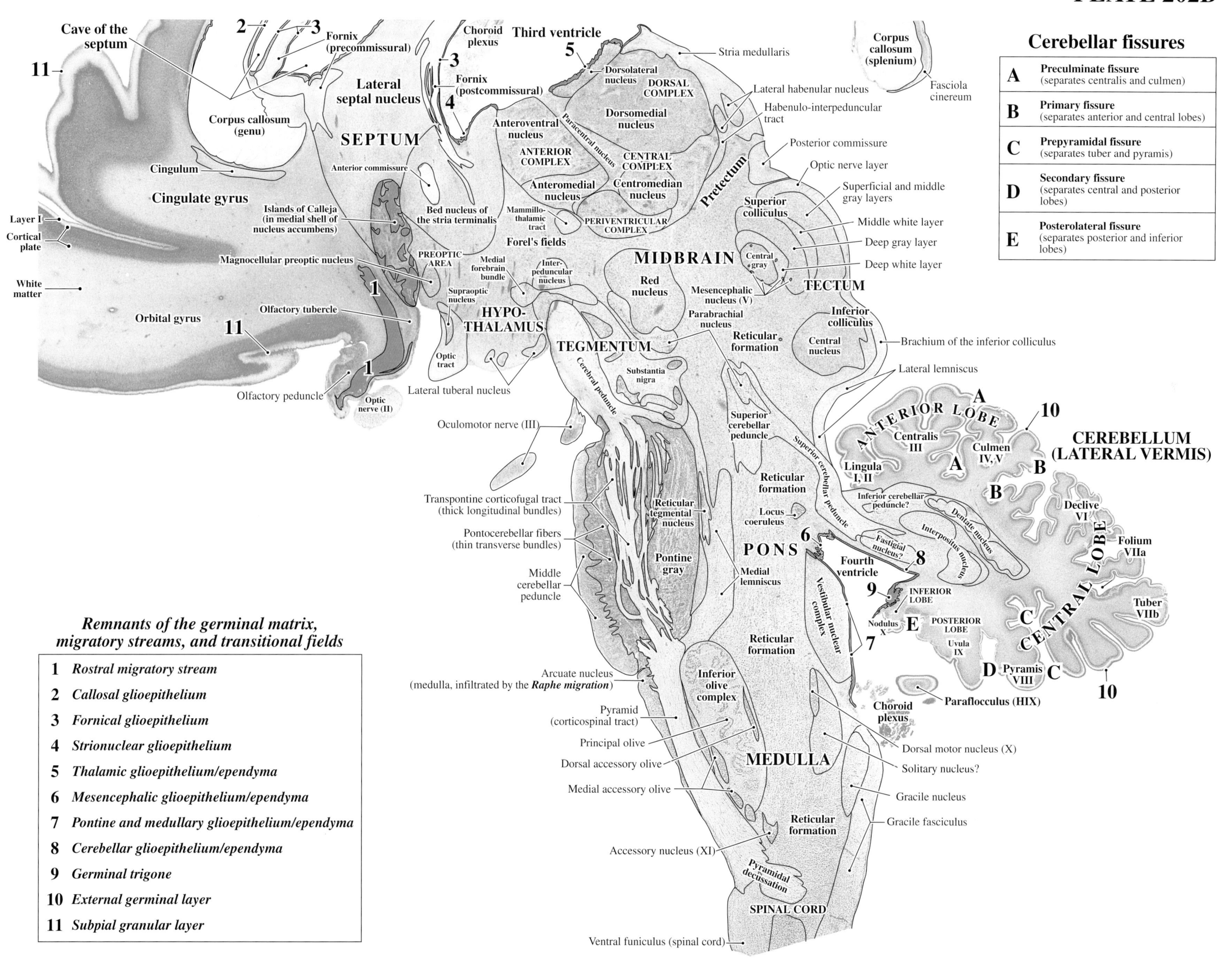

### Cerebellar fissures

| | | |
|---|---|---|
| A | Preculminate fissure | (separates centralis and culmen) |
| B | Primary fissure | (separates anterior and central lobes) |
| C | Prepyramidal fissure | (separates tuber and pyramis) |
| D | Secondary fissure | (separates central and posterior lobes) |
| E | Posterolateral fissure | (separates posterior and inferior lobes) |

### Remnants of the germinal matrix, migratory streams, and transitional fields

1. Rostral migratory stream
2. Callosal glioepithelium
3. Fornical glioepithelium
4. Strionuclear glioepithelium
5. Thalamic glioepithelium/ependyma
6. Mesencephalic glioepithelium/ependyma
7. Pontine and medullary glioepithelium/ependyma
8. Cerebellar glioepithelium/ependyma
9. Germinal trigone
10. External germinal layer
11. Subpial granular layer

**PLATE 203A**

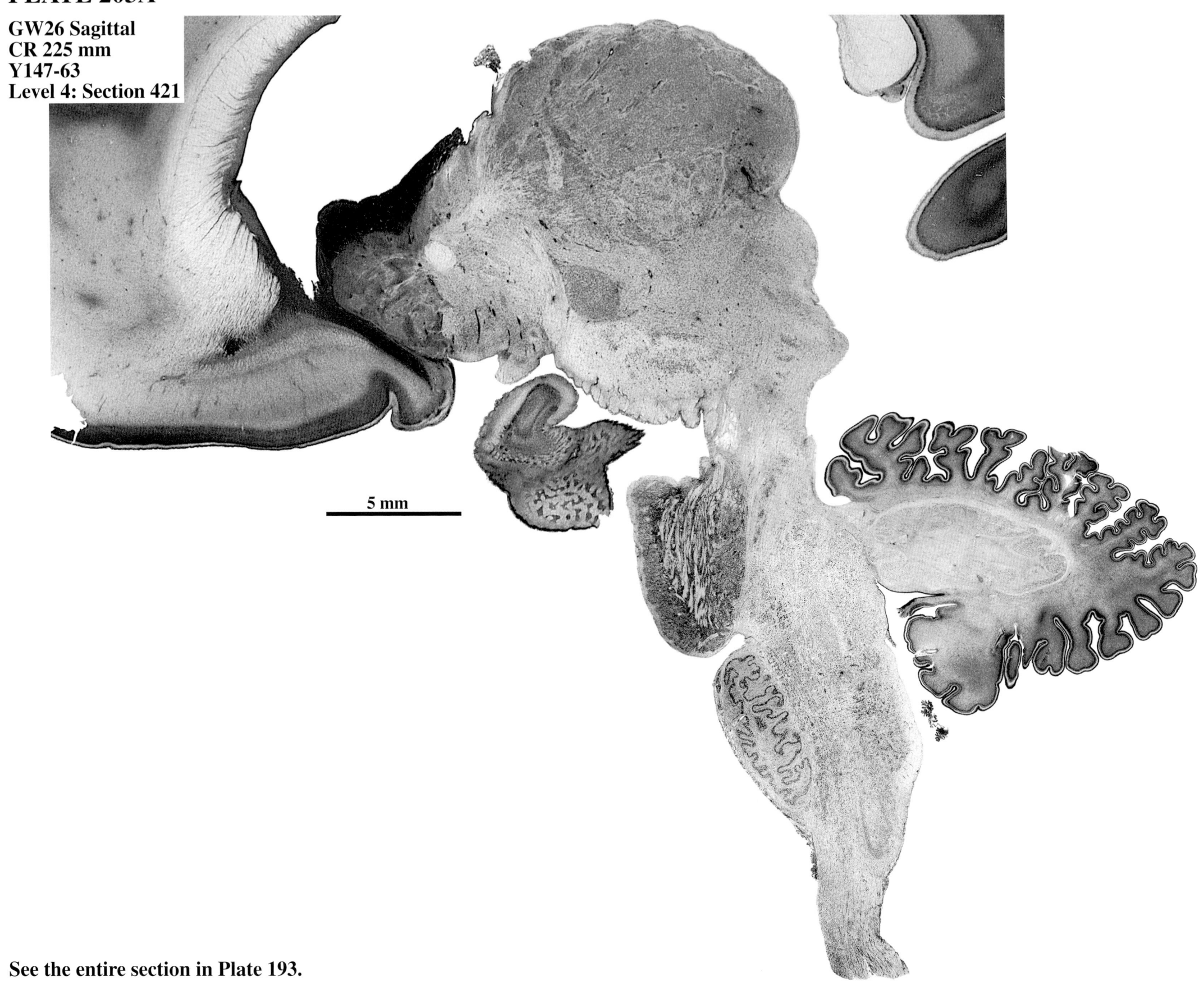

GW26 Sagittal
CR 225 mm
Y147-63
Level 4: Section 421

5 mm

See the entire section in Plate 193.

PLATE 203B

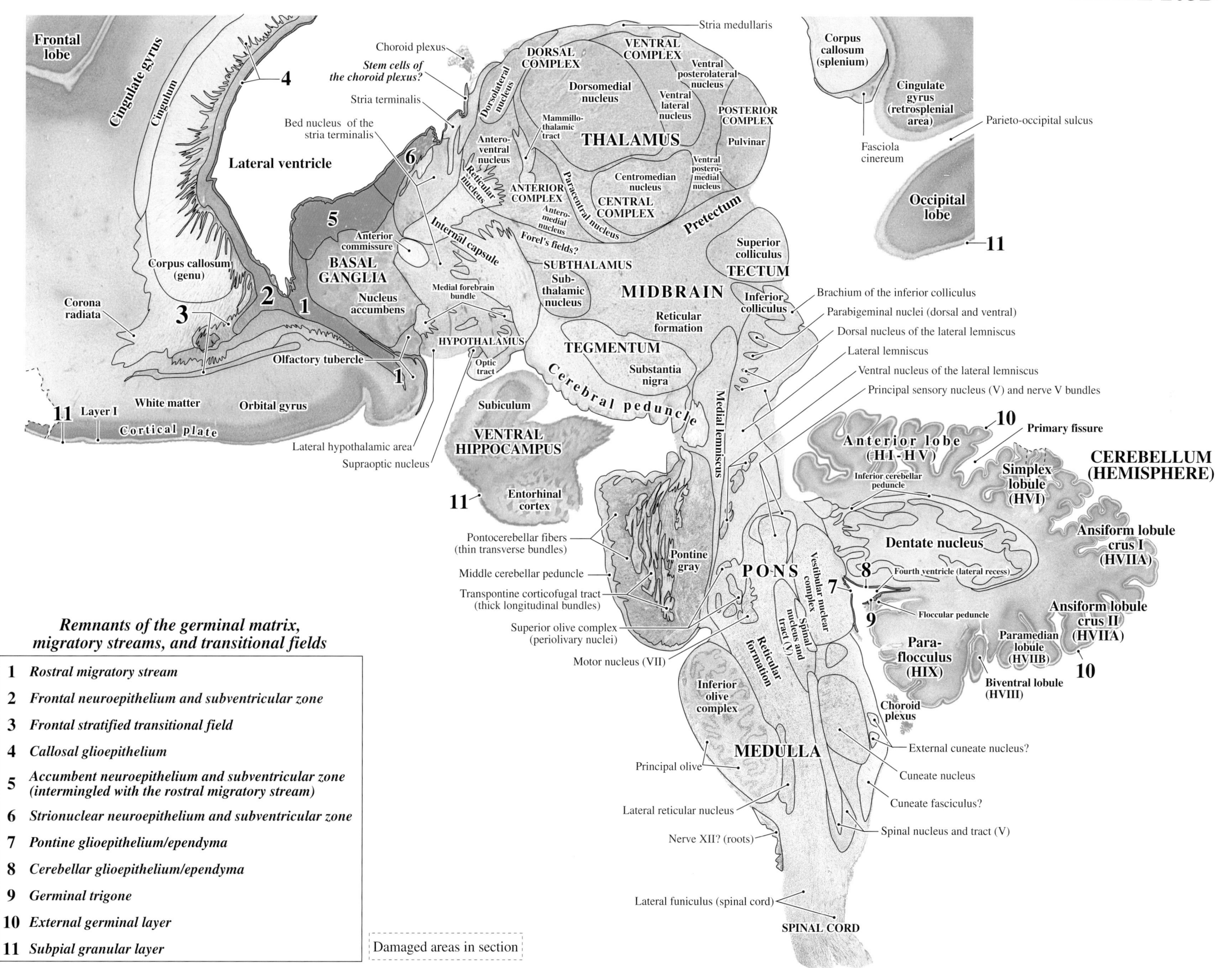

*Remnants of the germinal matrix, migratory streams, and transitional fields*

1. *Rostral migratory stream*
2. *Frontal neuroepithelium and subventricular zone*
3. *Frontal stratified transitional field*
4. *Callosal glioepithelium*
5. *Accumbent neuroepithelium and subventricular zone (intermingled with the rostral migratory stream)*
6. *Strionuclear neuroepithelium and subventricular zone*
7. *Pontine glioepithelium/ependyma*
8. *Cerebellar glioepithelium/ependyma*
9. *Germinal trigone*
10. *External germinal layer*
11. *Subpial granular layer*

Damaged areas in section

## PLATE 204A

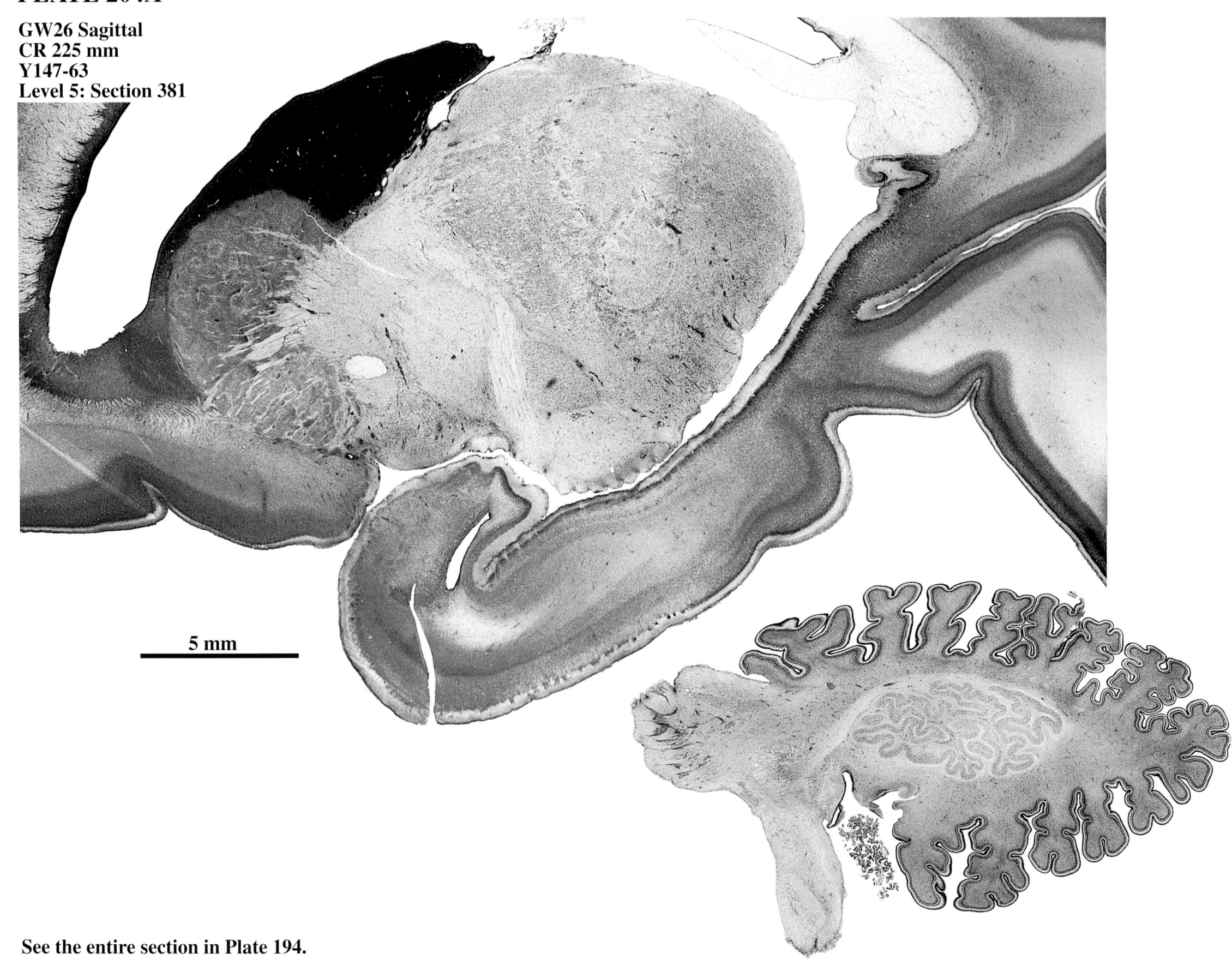

GW26 Sagittal
CR 225 mm
Y147-63
Level 5: Section 381

5 mm

See the entire section in Plate 194.

**PLATE 204B**

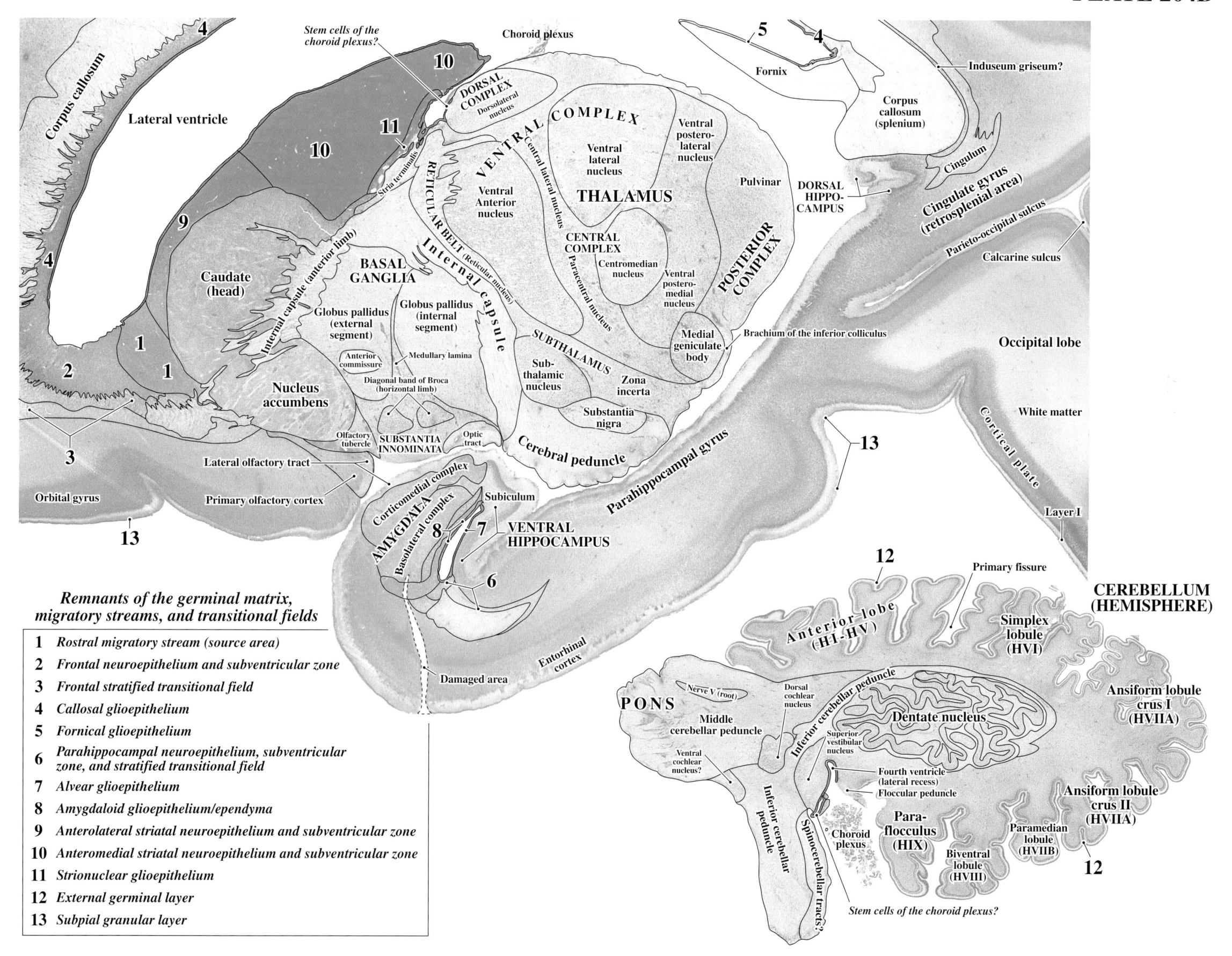

*Remnants of the germinal matrix, migratory streams, and transitional fields*

1. *Rostral migratory stream (source area)*
2. *Frontal neuroepithelium and subventricular zone*
3. *Frontal stratified transitional field*
4. *Callosal glioepithelium*
5. *Fornical glioepithelium*
6. *Parahippocampal neuroepithelium, subventricular zone, and stratified transitional field*
7. *Alvear glioepithelium*
8. *Amygdaloid glioepithelium/ependyma*
9. *Anterolateral striatal neuroepithelium and subventricular zone*
10. *Anteromedial striatal neuroepithelium and subventricular zone*
11. *Strionuclear glioepithelium*
12. *External germinal layer*
13. *Subpial granular layer*

## PLATE 205A

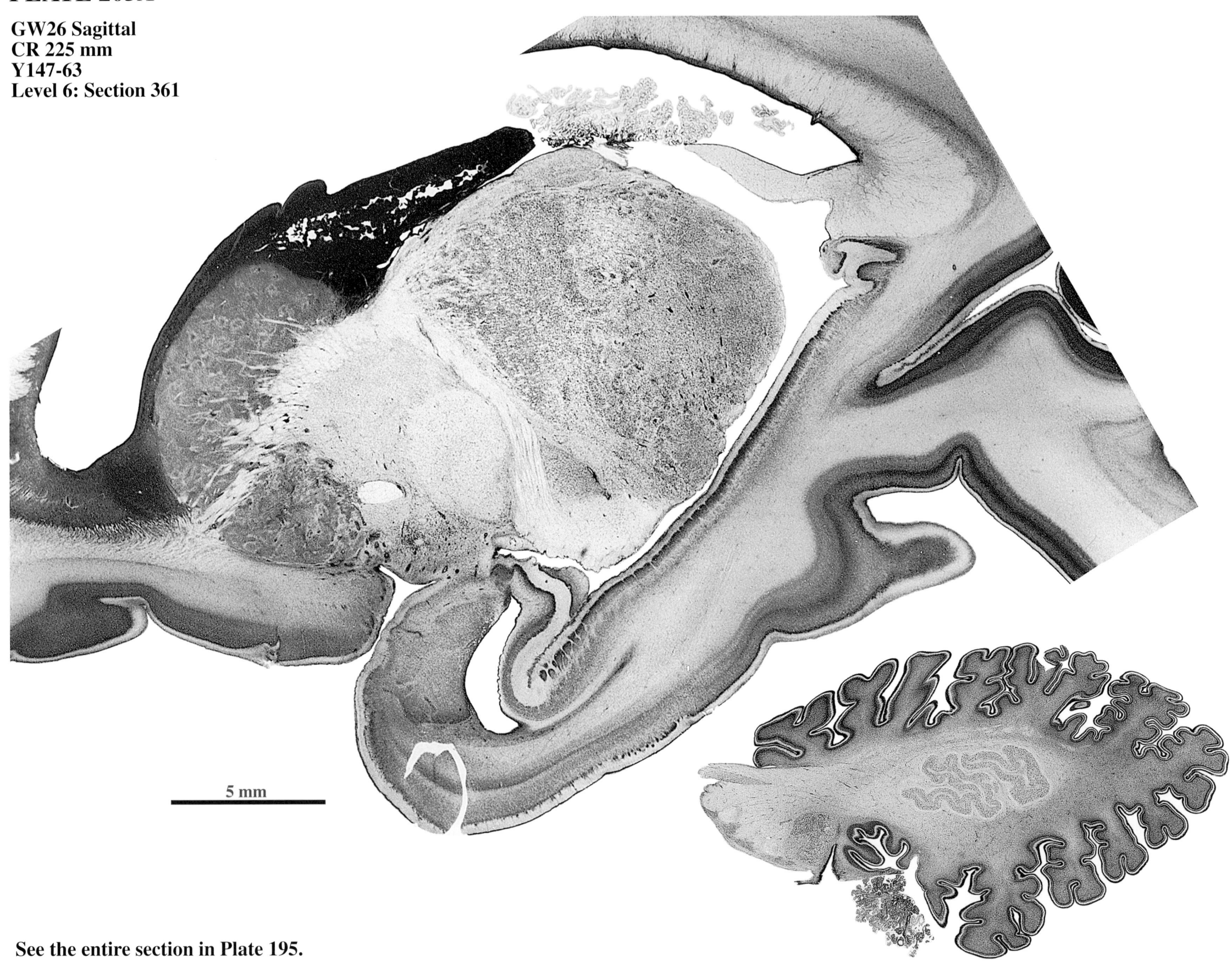

GW26 Sagittal
CR 225 mm
Y147-63
Level 6: Section 361

See the entire section in Plate 195.

**PLATE 205B**

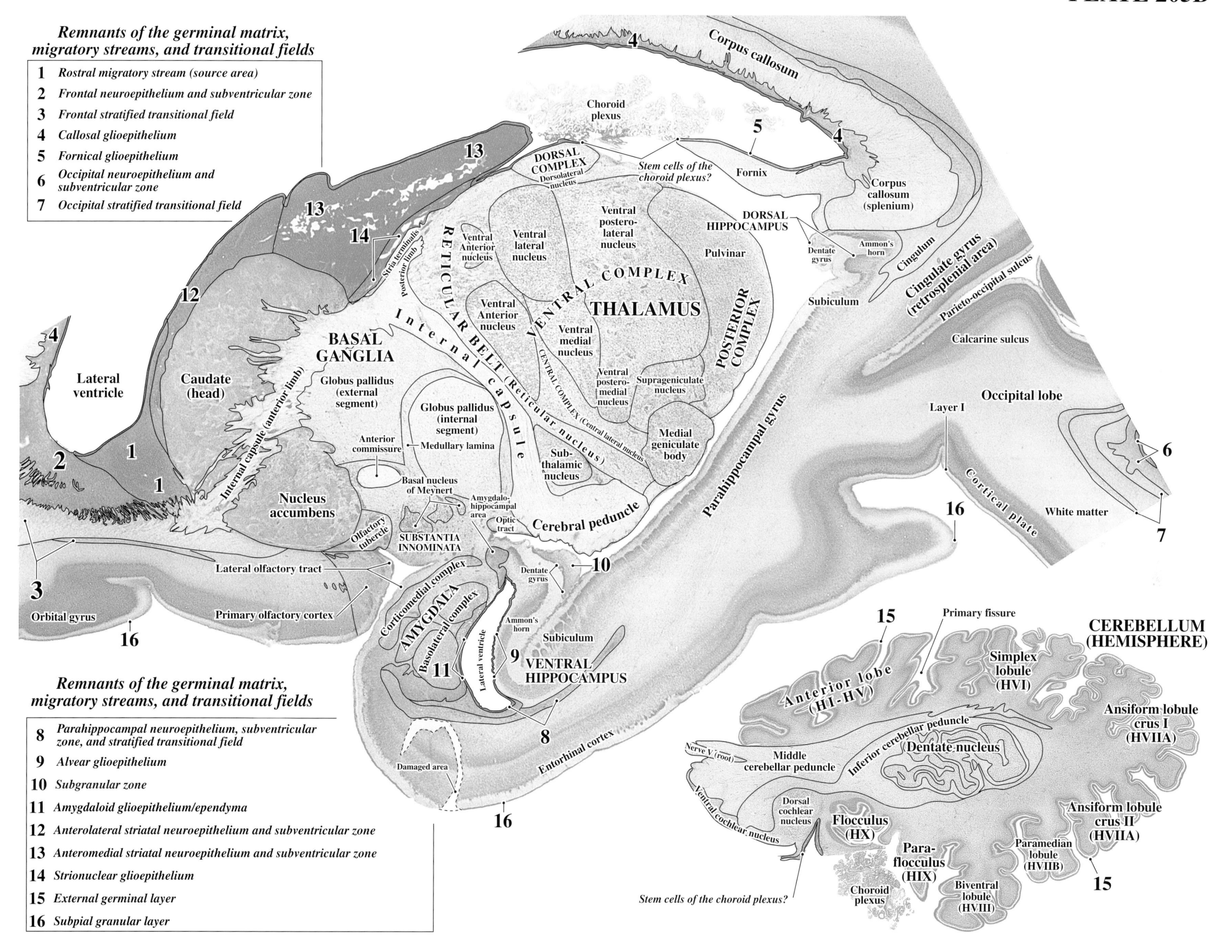

**Remnants of the germinal matrix, migratory streams, and transitional fields**

1. *Rostral migratory stream (source area)*
2. *Frontal neuroepithelium and subventricular zone*
3. *Frontal stratified transitional field*
4. *Callosal glioepithelium*
5. *Fornical glioepithelium*
6. *Occipital neuroepithelium and subventricular zone*
7. *Occipital stratified transitional field*

**Remnants of the germinal matrix, migratory streams, and transitional fields**

8. *Parahippocampal neuroepithelium, subventricular zone, and stratified transitional field*
9. *Alvear glioepithelium*
10. *Subgranular zone*
11. *Amygdaloid glioepithelium/ependyma*
12. *Anterolateral striatal neuroepithelium and subventricular zone*
13. *Anteromedial striatal neuroepithelium and subventricular zone*
14. *Strionuclear glioepithelium*
15. *External germinal layer*
16. *Subpial granular layer*

**PLATE 206A**

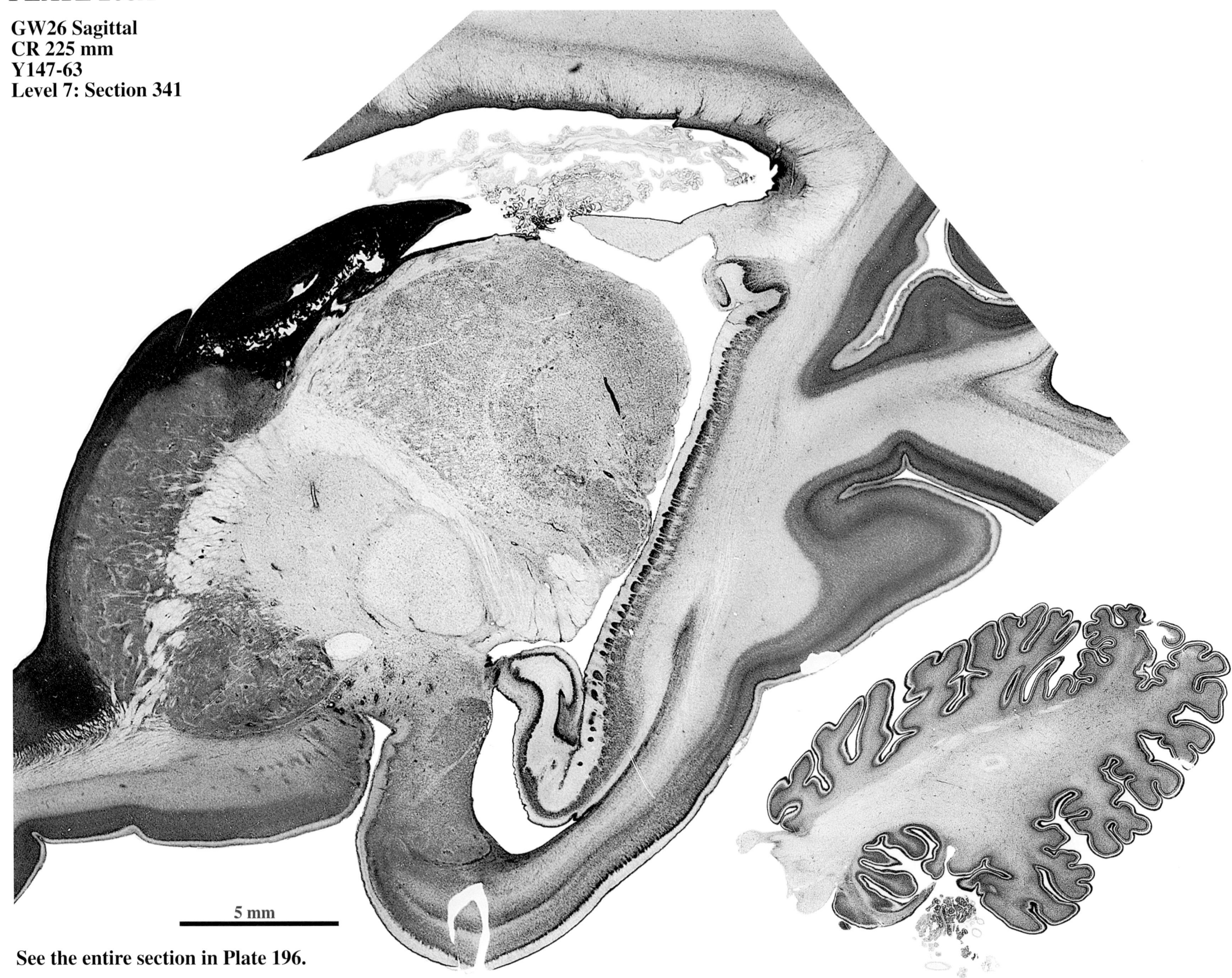

GW26 Sagittal
CR 225 mm
Y147-63
Level 7: Section 341

5 mm

See the entire section in Plate 196.

309

**PLATE 206B**

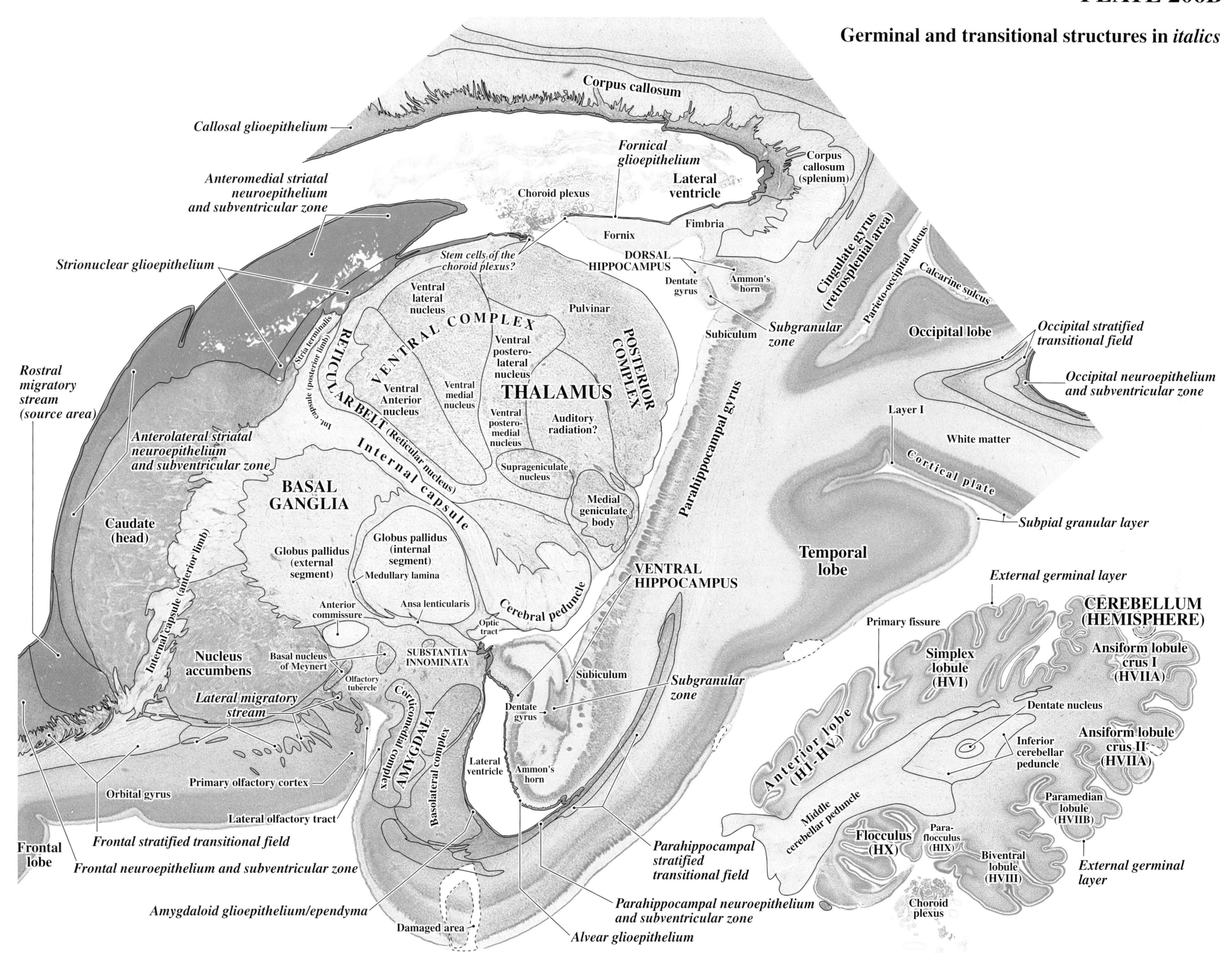

**Germinal and transitional structures in *italics***

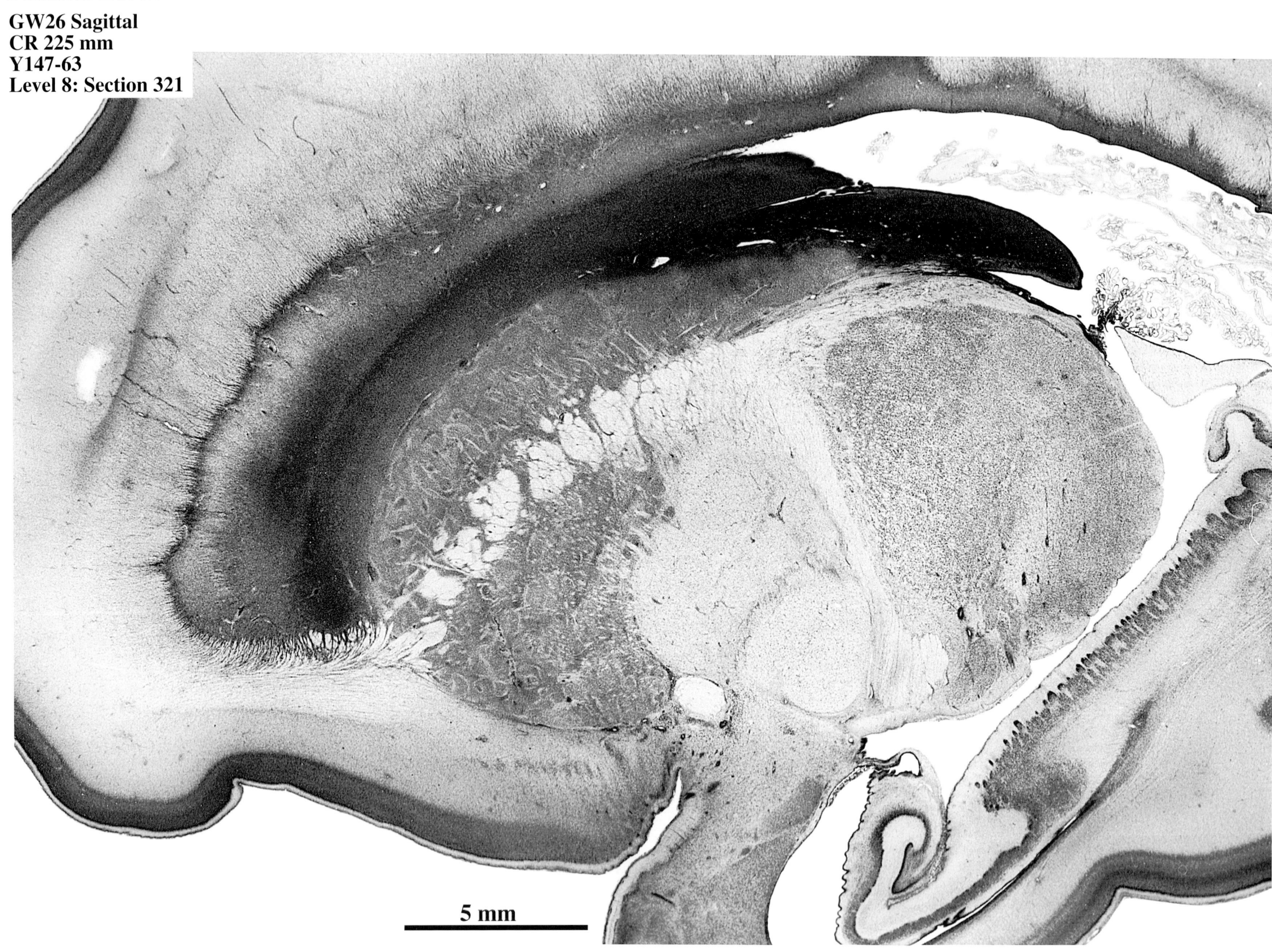

**PLATE 207A**
GW26 Sagittal
CR 225 mm
Y147-63
Level 8: Section 321

5 mm

See the entire section in Plate 197.

311

PLATE 207B

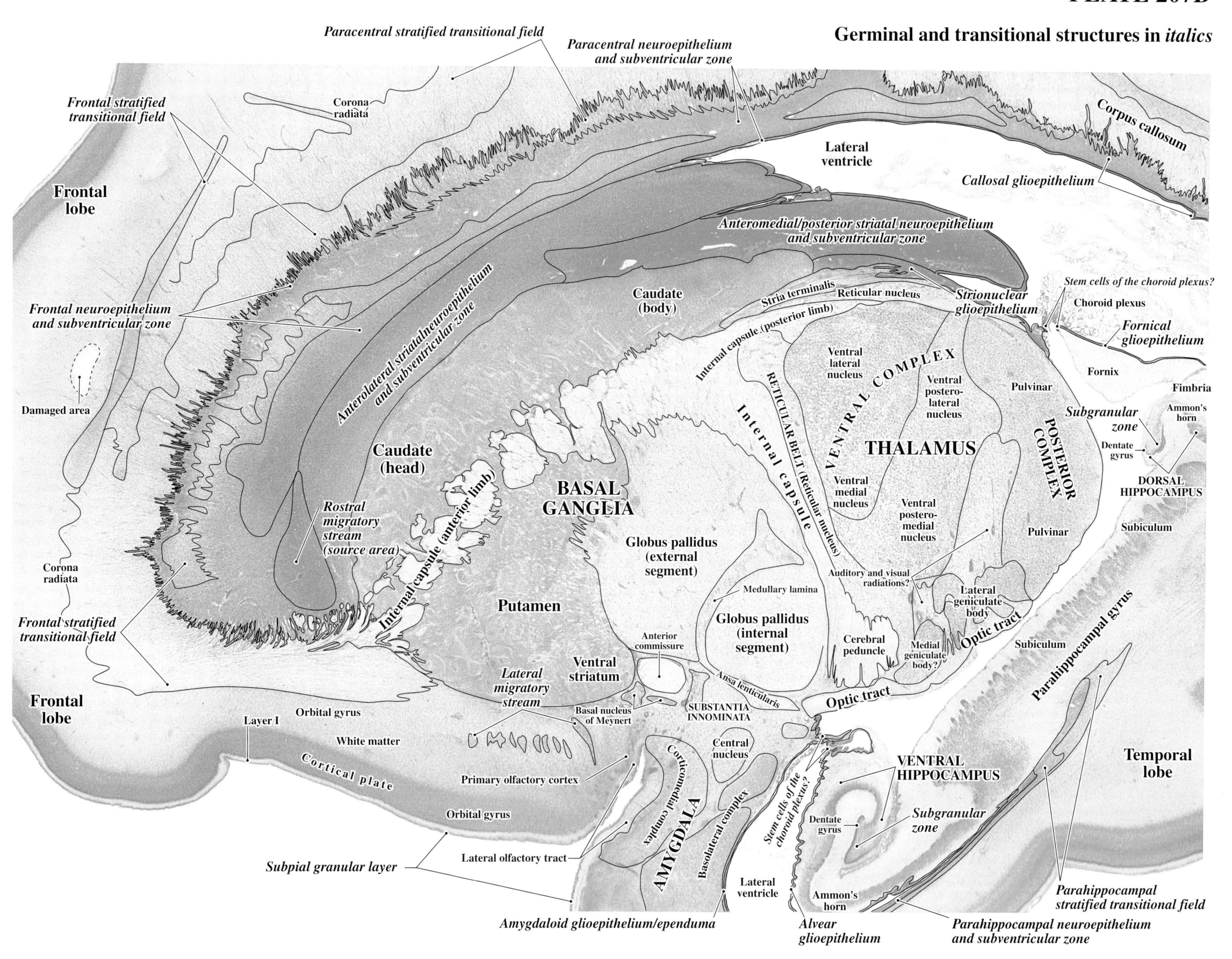

Germinal and transitional structures in *italics*

**PLATE 208A**
GW26 Sagittal
CR 225 mm
Y147-63
Level 9: Section 261

5 mm

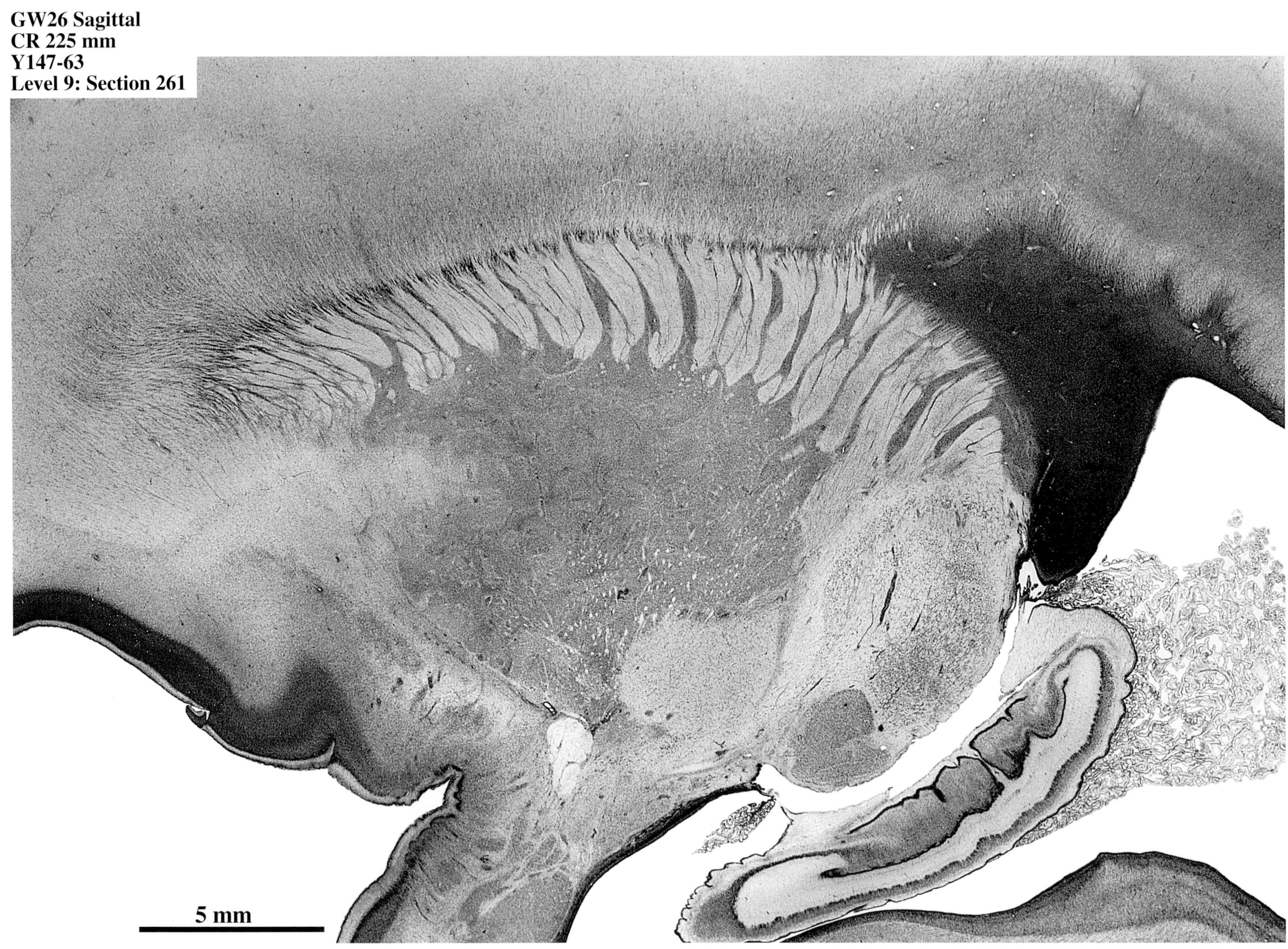

See the entire section in Plate 198.

### PLATE 208B

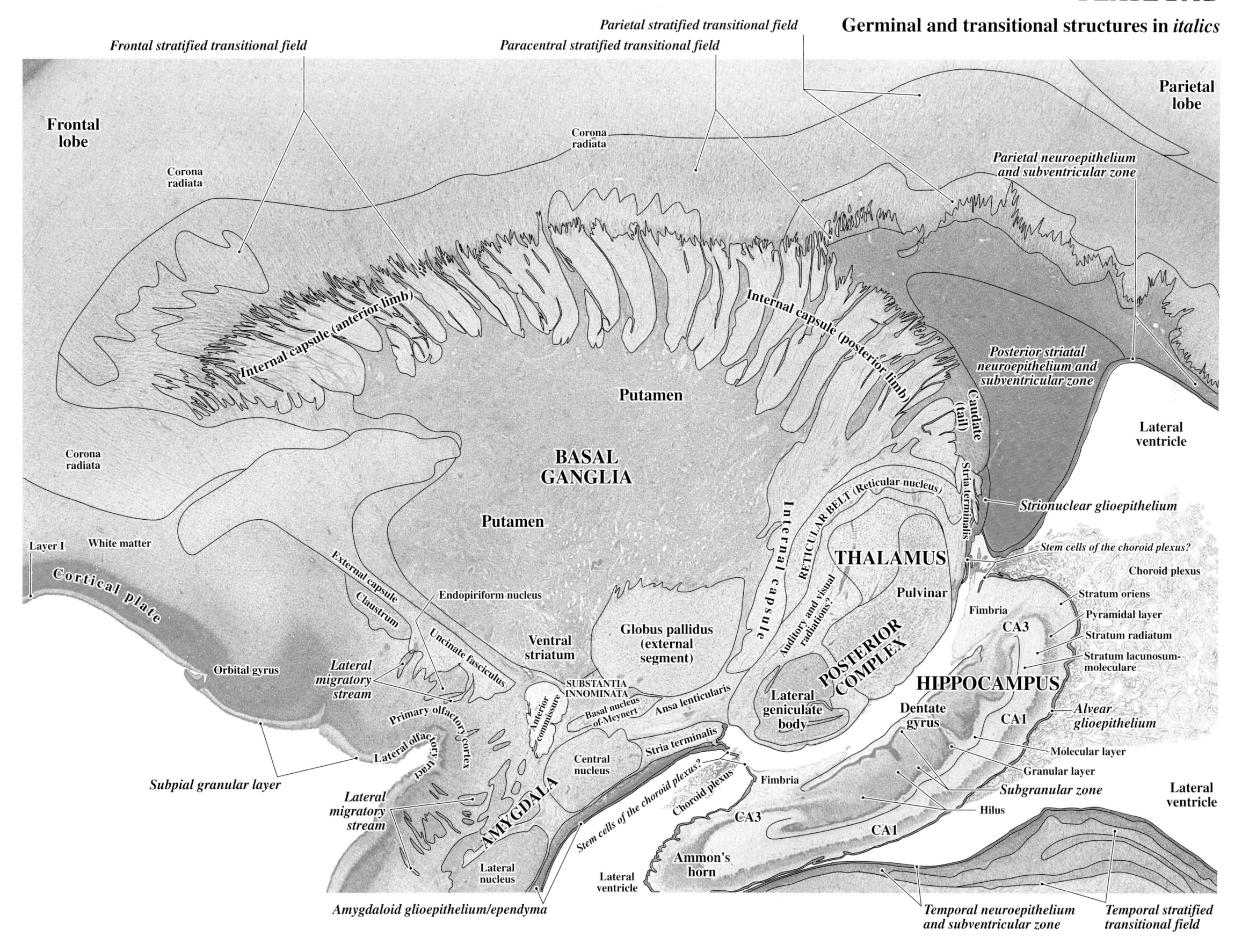

**Germinal and transitional structures in *italics***

## PLATE 209A

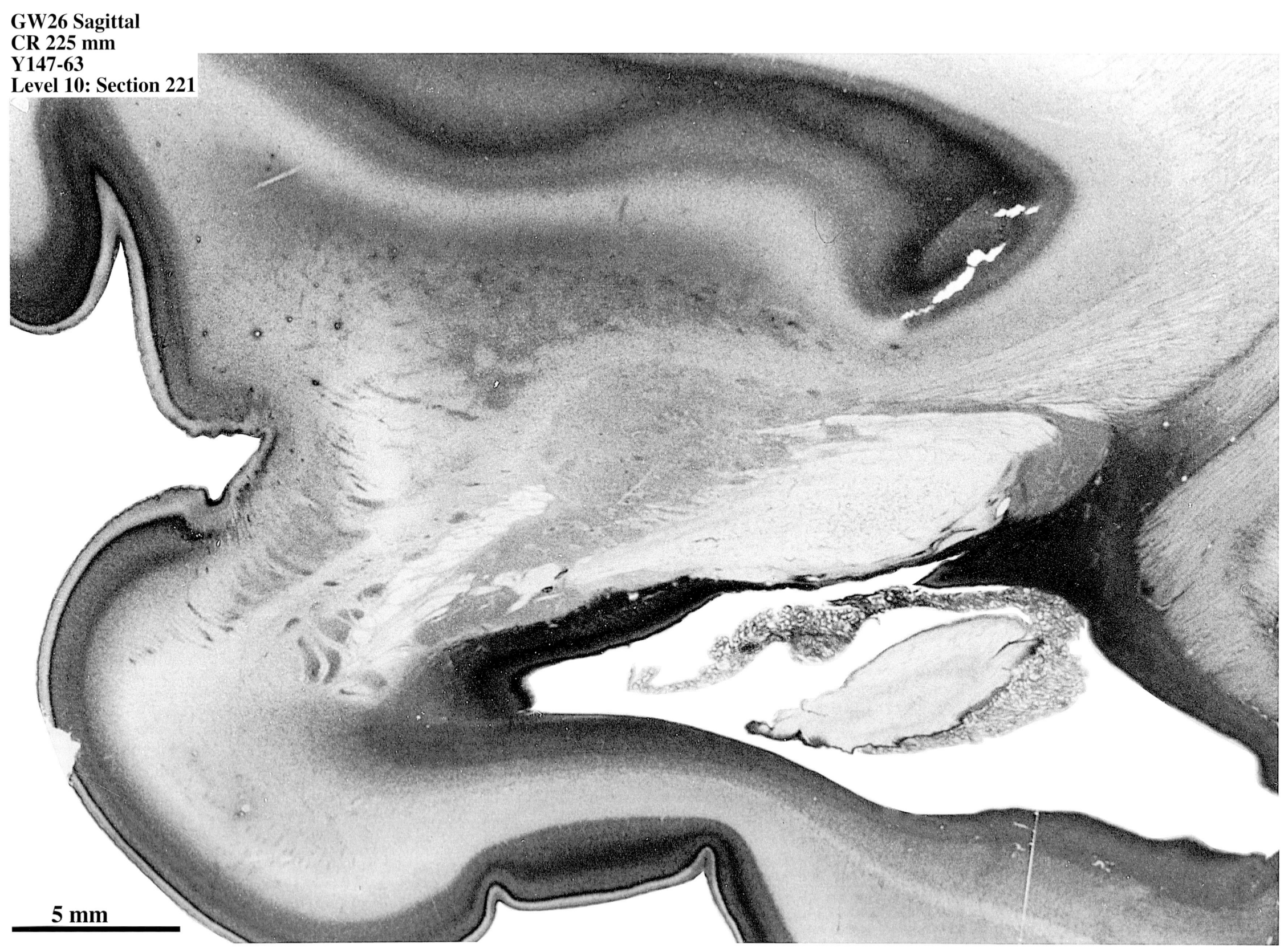

GW26 Sagittal
CR 225 mm
Y147-63
Level 10: Section 221

5 mm

See the entire section in Plate 199.

## PLATE 209B

**Germinal and transitional structures in *italics***

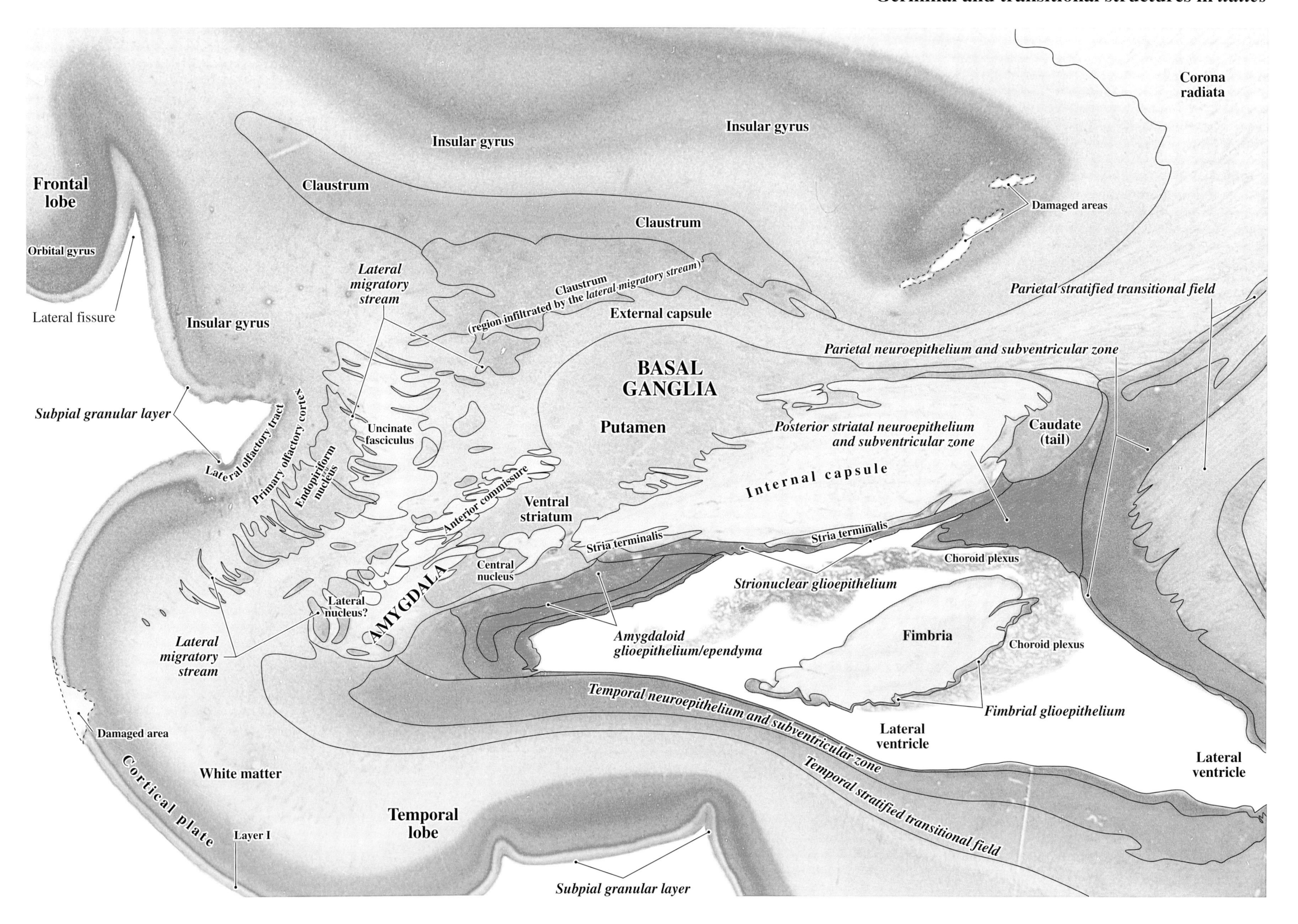

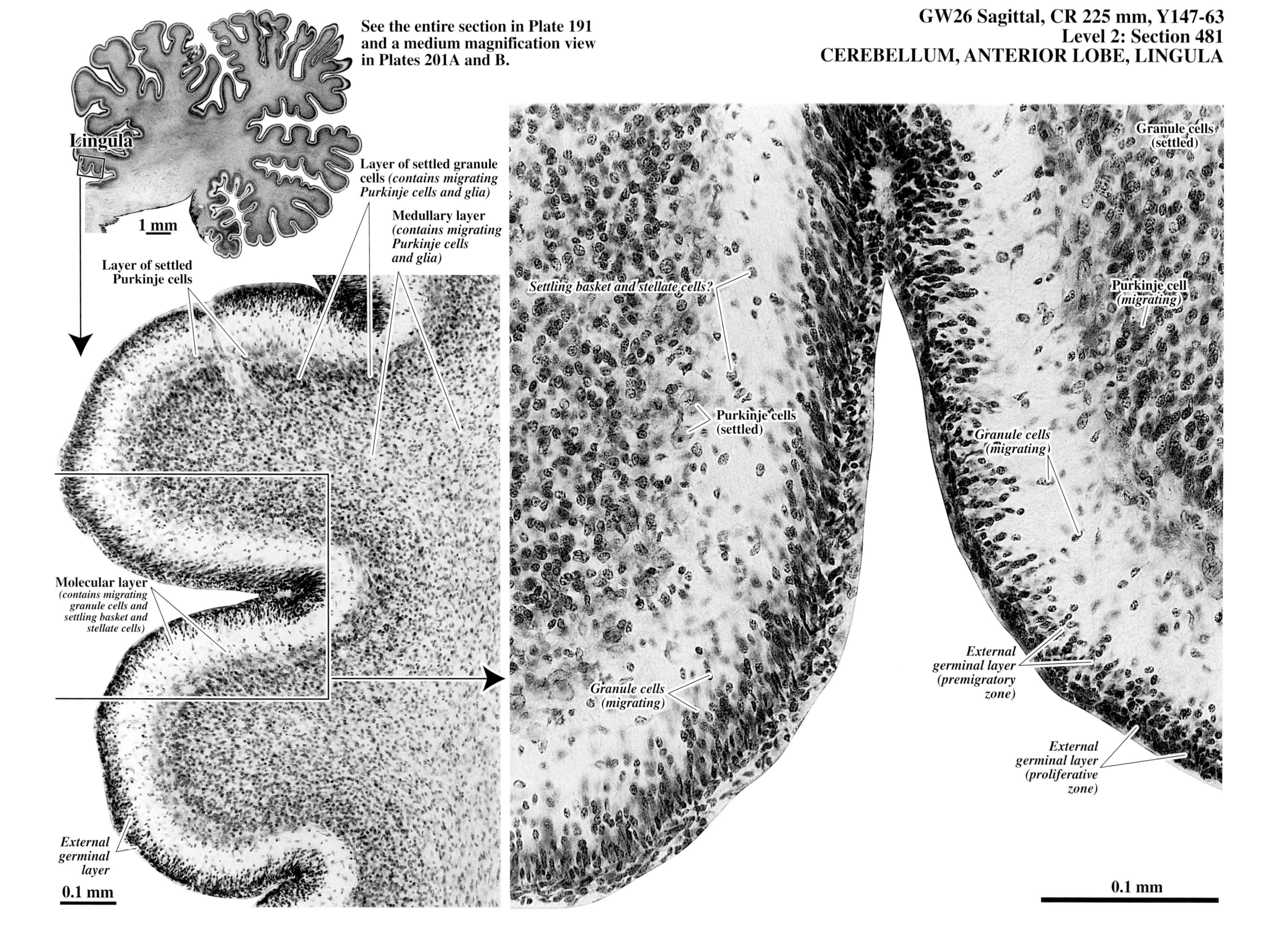

316

**PLATE 210**

See the entire section in Plate 191 and a medium magnification view in Plates 201A and B.

GW26 Sagittal, CR 225 mm, Y147-63
Level 2: Section 481
CEREBELLUM, ANTERIOR LOBE, LINGULA

**PLATE 211**

GW26 Sagittal, CR 225 mm, Y147-63
Level 2: Section 481
CEREBELLUM, ANTERIOR LOBE, CULMEN

See the entire section in Plate 191 and a medium magnification view in Plates 201A and B.

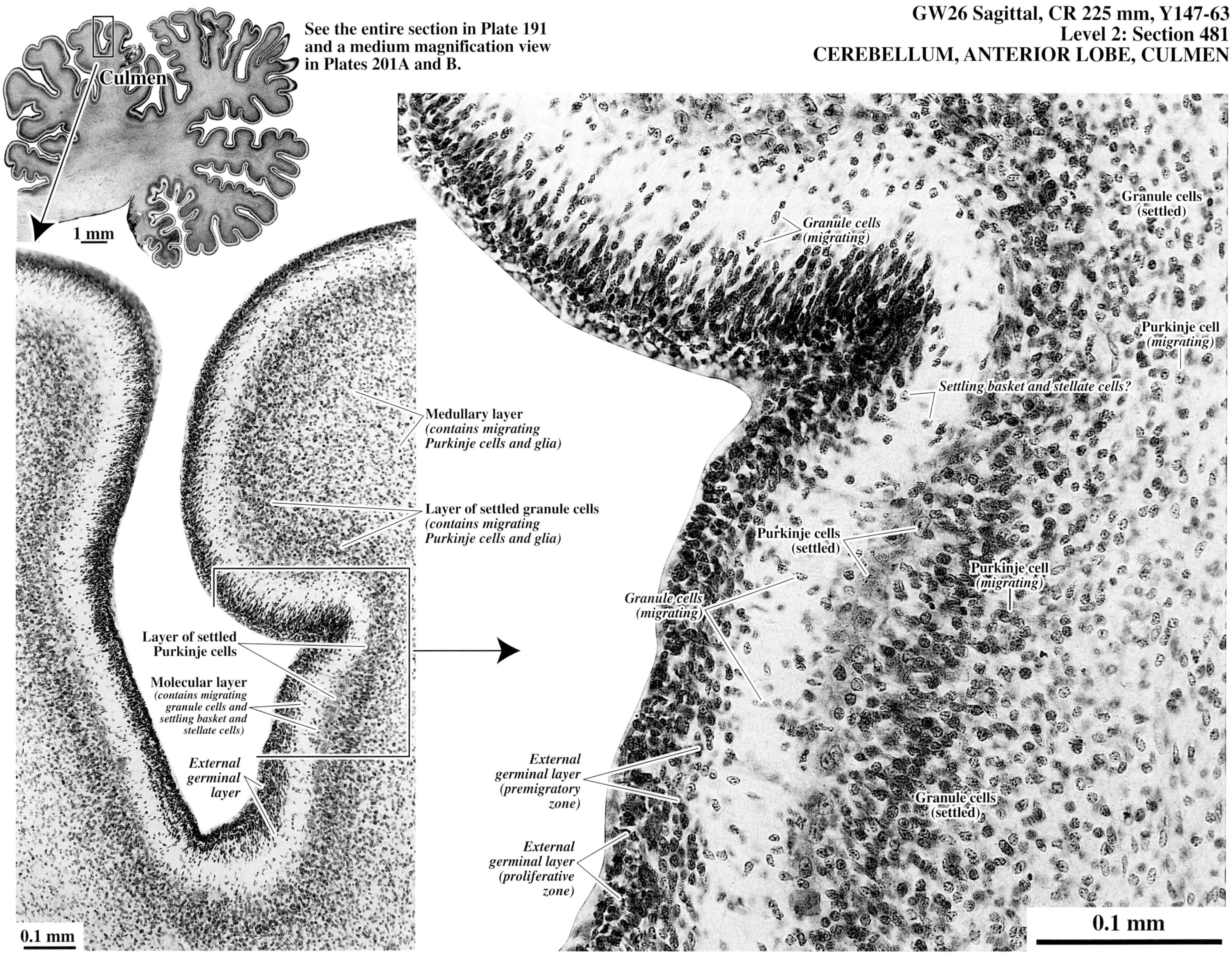

PLATE 212

GW26 Sagittal, CR 225 mm, Y147-63
Level 2: Section 481
CEREBELLUM, CENTRAL LOBE, DECLIVE

See the entire section in Plate 191 and a medium magnification view in Plates 201A and B.

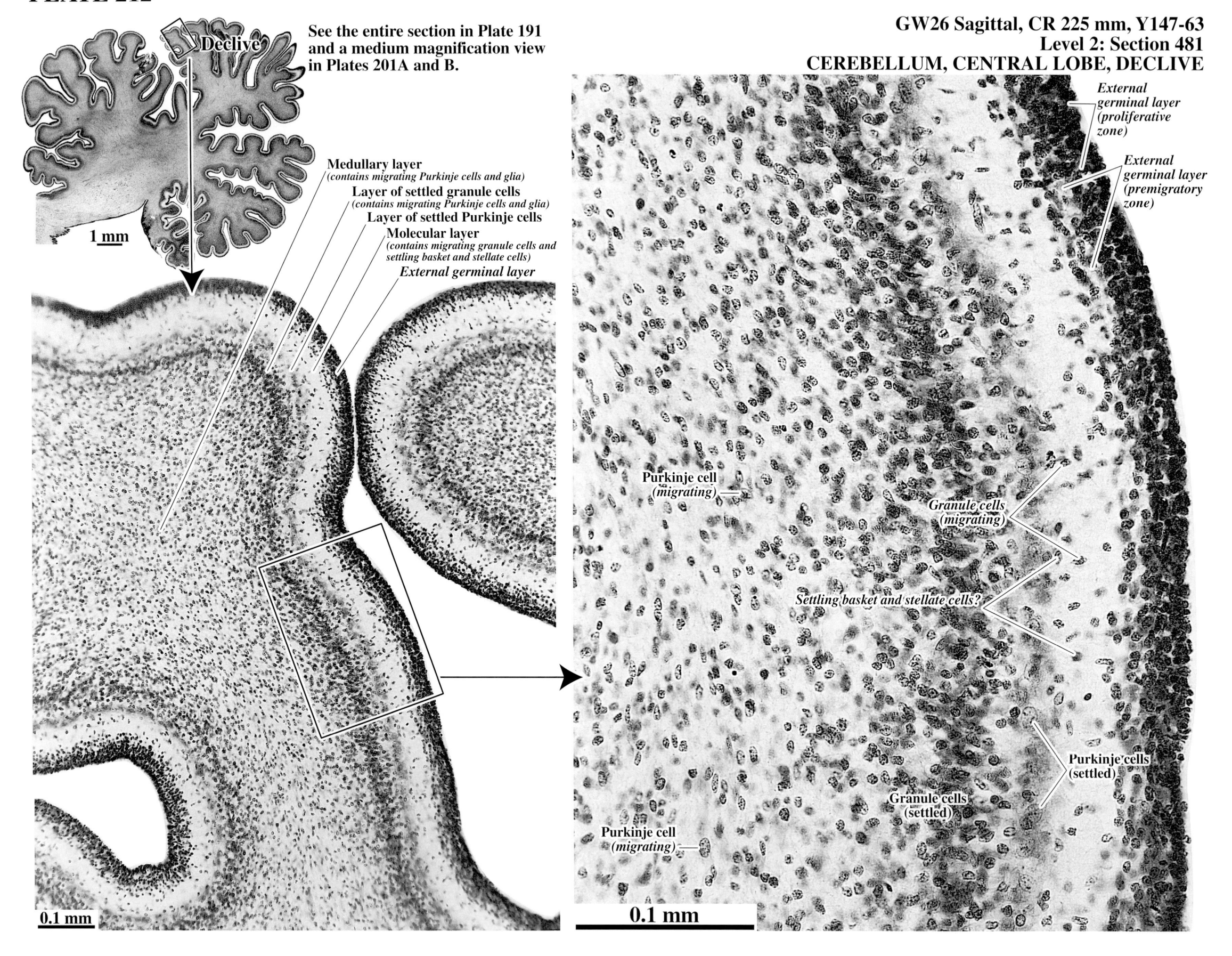

*Medullary layer*
(contains migrating Purkinje cells and glia)
*Layer of settled granule cells*
(contains migrating Purkinje cells and glia)
*Layer of settled Purkinje cells*
*Molecular layer*
(contains migrating granule cells and settling basket and stellate cells)
*External germinal layer*

*External germinal layer (proliferative zone)*
*External germinal layer (premigratory zone)*
*Granule cells (migrating)*
*Purkinje cell (migrating)*
*Settling basket and stellate cells?*
*Purkinje cells (settled)*
*Granule cells (settled)*
*Purkinje cell (migrating)*

# PLATE 213

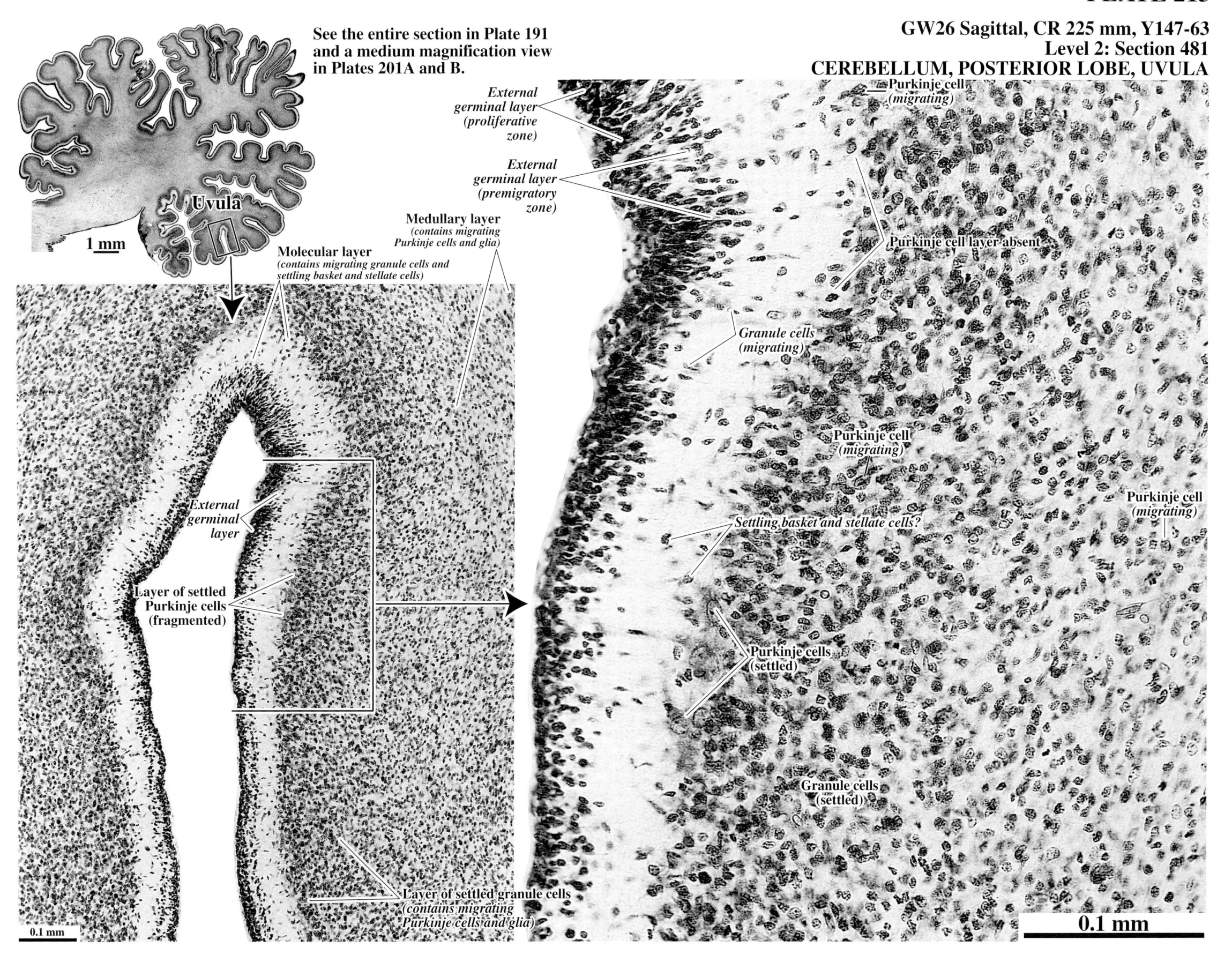

**GW26 Sagittal, CR 225 mm, Y147-63
Level 2: Section 481
CEREBELLUM, POSTERIOR LOBE, UVULA**

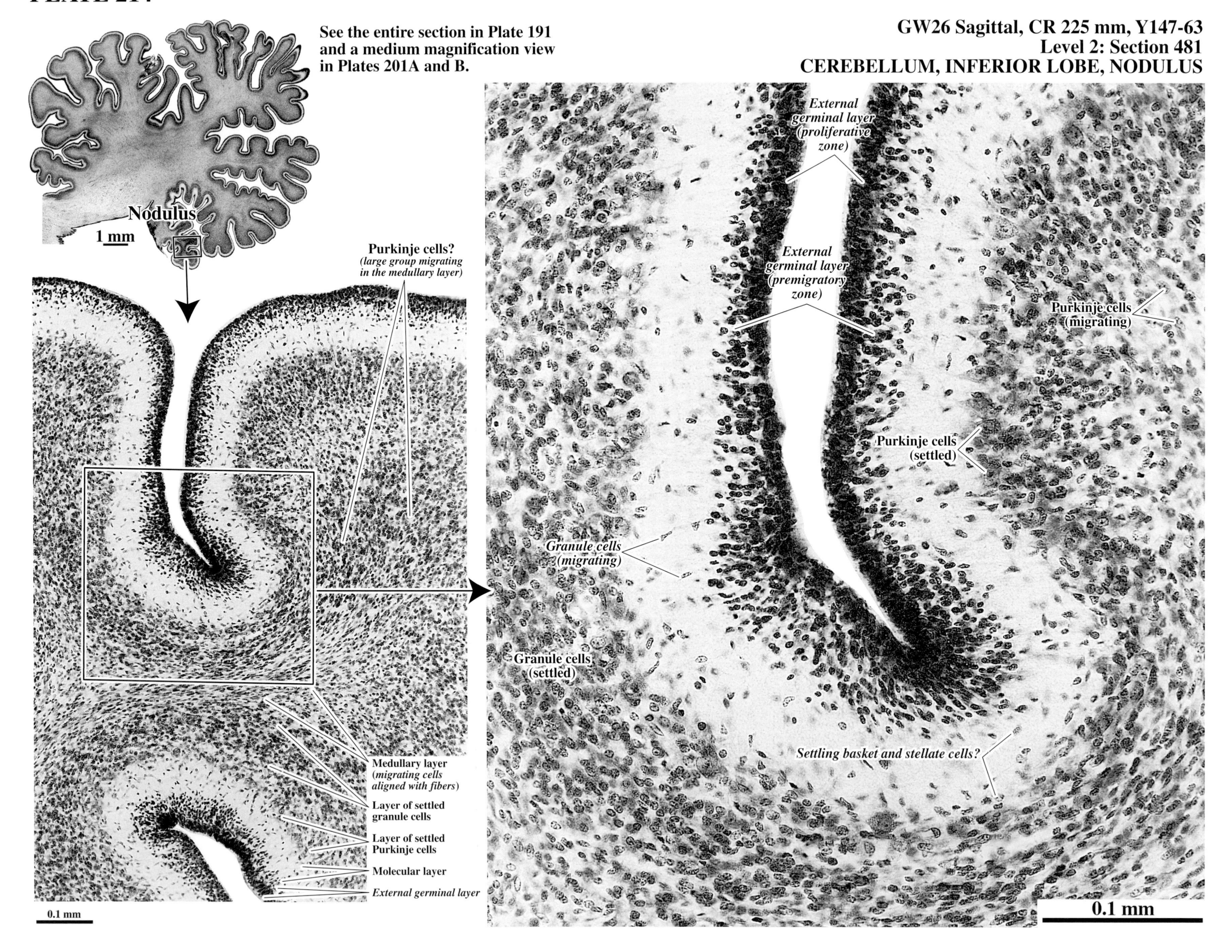

PLATE 214

See the entire section in Plate 191 and a medium magnification view in Plates 201A and B.

GW26 Sagittal, CR 225 mm, Y147-63
Level 2: Section 481
CEREBELLUM, INFERIOR LOBE, NODULUS

**PLATE 215**

GW26 Sagittal, CR 225 mm, Y147-63
Level 2: Section 481
**CEREBELLUM**
*Germinal Trigone*

See the entire section in Plate 191 and a medium magnification view in Plates 201A and B.

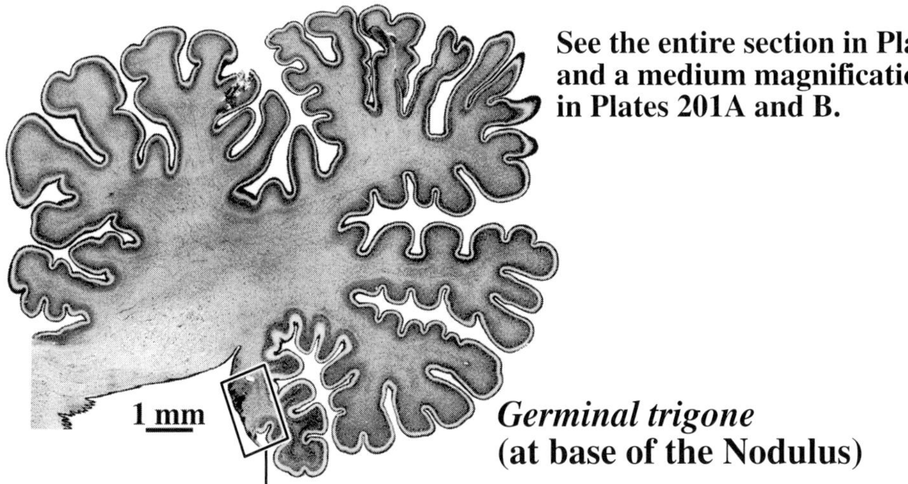

*Germinal trigone* (at base of the Nodulus)

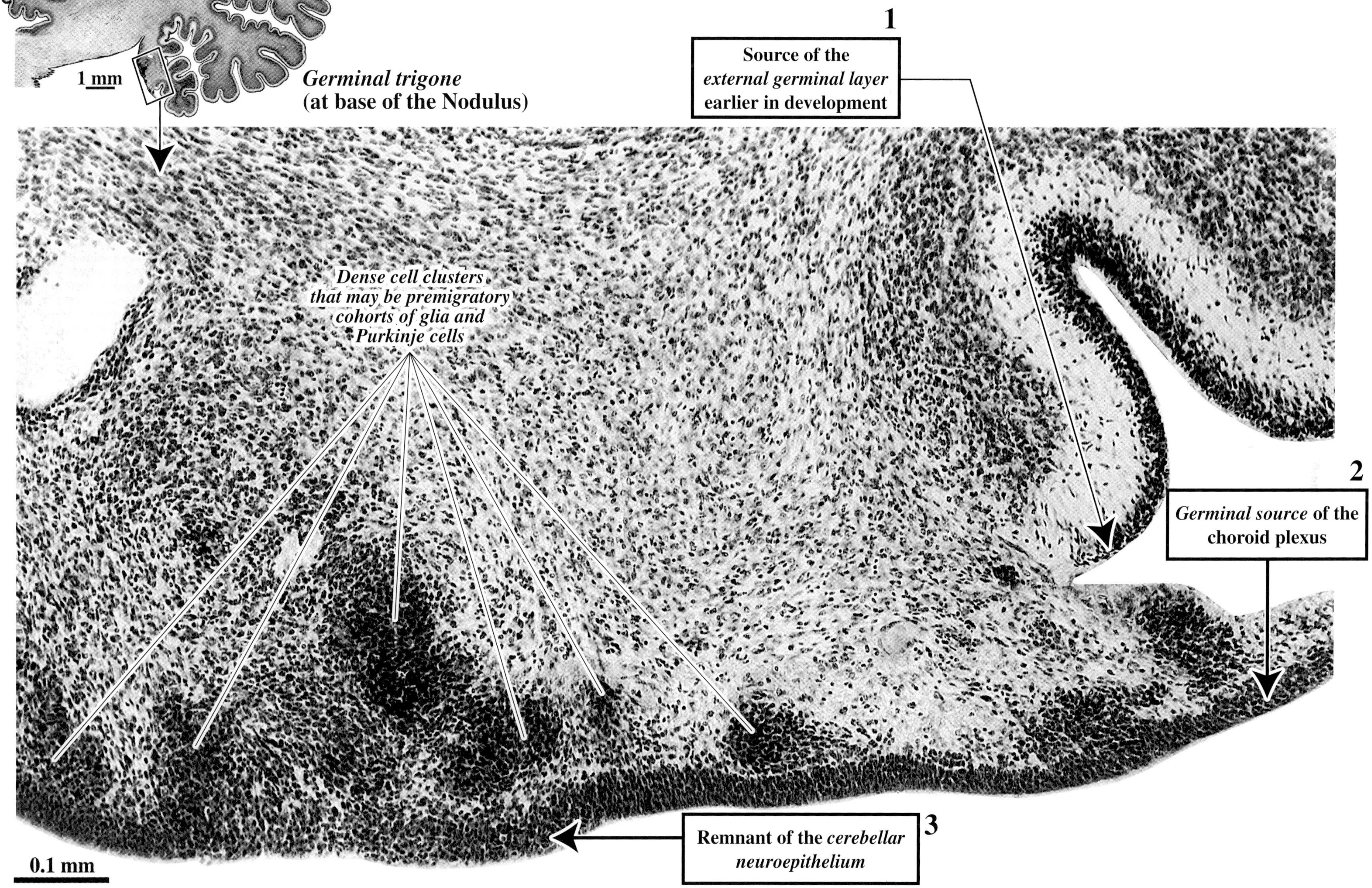

1 Source of the *external germinal layer* earlier in development

*Dense cell clusters that may be premigratory cohorts of glia and Purkinje cells*

2 *Germinal source* of the choroid plexus

3 Remnant of the *cerebellar neuroepithelium*

# PART IX: GW26 HORIZONTAL

This specimen is case number W-16-59 (Perinatal RPSL) in the Yakovlev Collection, a premature stillborn female infant. The brain is classified as a Normative Control in the Yakovlev Collection (Haleem, 1990). It was cut in the horizontal plane in 35-μm and 15-μm thick sections. Since there is no available photograph of this brain before it was embedded and cut, the photograph of the lateral view of another GW26 brain that Larroche published in 1967 (**Figure 12**) is similar to the features of the brain in Y16-59.

The approximate cutting plane of this brain is indicated in **Figure 13** (facing page) with lines superimposed on the GW26 brain from the Larroche (1967) series. The anterior part of each section (on the left) is dorsal to the posterior part (on the right). As in all other specimens, the sections chosen for illustration are more closely spaced to show small structures in the diencephalon, midbrain, pons, and medulla. Illustrated sections are spaced farther apart when they contain only large brain structures, such as the cerebral cortex, basal ganglia, and cerebellum. Photographs of 17 different Nissl-stained sections (**Levels 1-17**) are shown at low magnification in **Plates 216-229**. The core of the brain and the cerebellum are shown at high magnification in companion **Plates 230AB-244AB** for **Levels 3-17**.

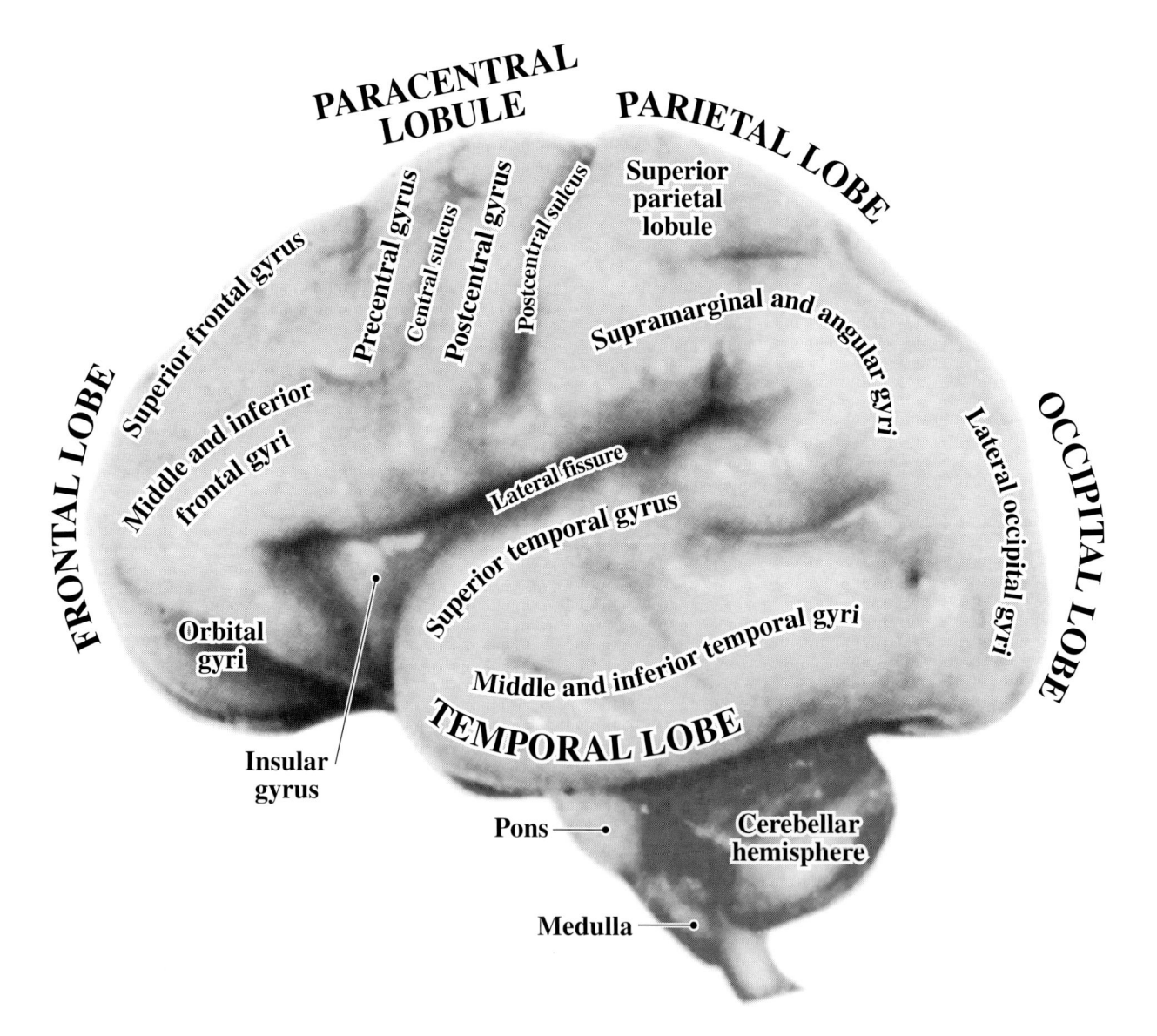

**Figure 12.** Lateral view of a GW26 brain with major structures in the cerebral hemispheres labeled. (From the photographic series of: J. C. Larroche (1967) Maturation morphologique du système nerveux central: ses rapports avec le développement pondéral du foetus et son age gestationnel. In: *Regional Development of the Brain in Early Life*, A. Minkowski (ed.), London: Blackwell, page 254.)

## GW26 HORIZONTAL SECTION PLANES

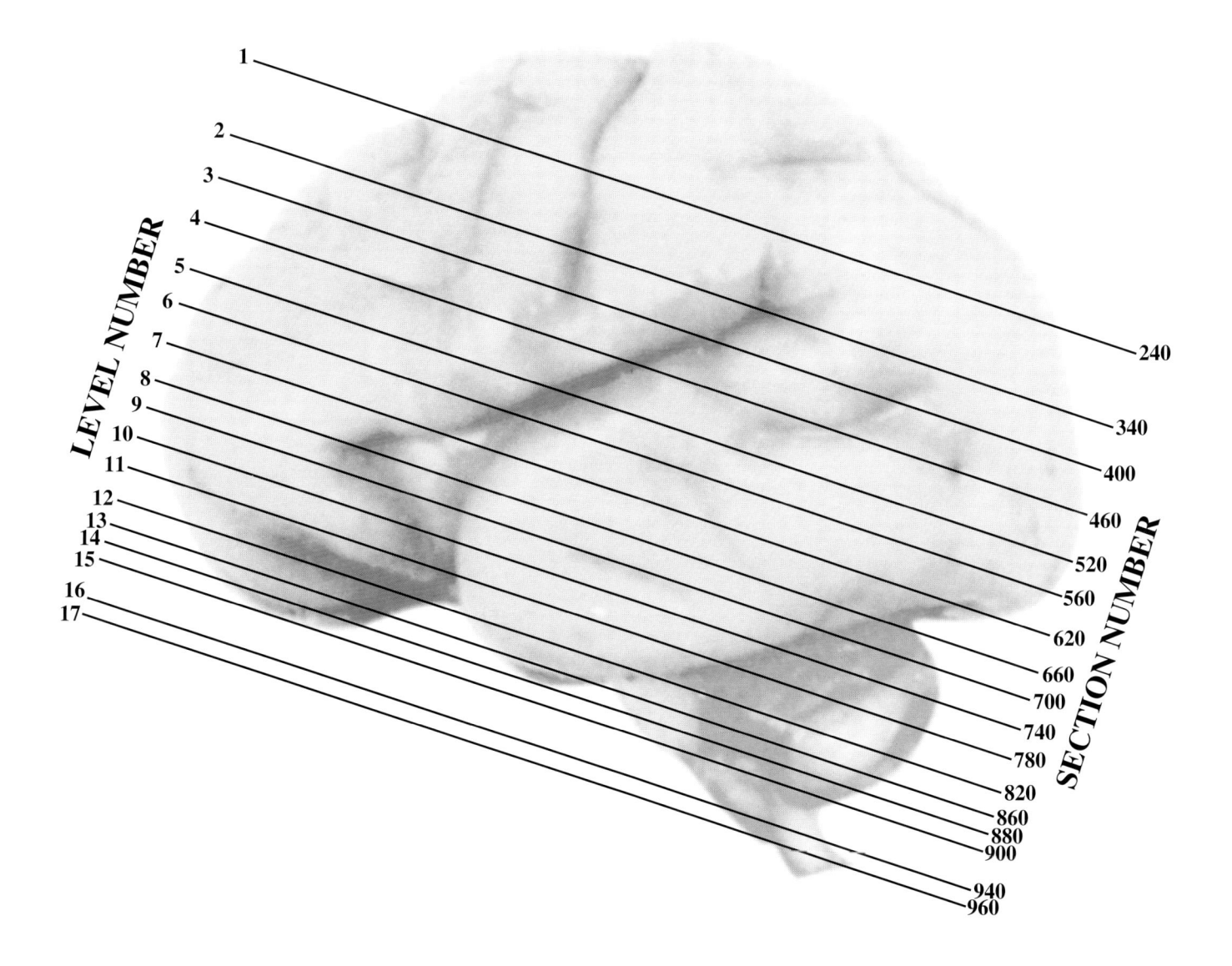

**Figure 13.** Lateral view of the same GW30 brain shown in **Figure 12** with the approximate locations and cutting angle of the sections of Y16-59. (From the photographic series of: J. C. Larroche (1967) Maturation morphologique du système nerveux central: ses rapports avec le développement pondéral du foetus et son age gestationnel. In: *Regional Development of the Brain in Early Life*, A. Minkowski (ed.), London: Blackwell, page 254.)

Y16-59 contains more prominent immature structures than in the older specimens. A densely staining and more thick *neuroepithelium/subventricular zone* (than the GW30 horizontal specimen) is generating neocortical interneurons and glia in all lobes of the cerebral cortex. The same thickness variations between the occipital and other lobes of the cerebral cortex are still there, however. Remnants of migrating and sojourning neurons and/or glia are visible in all lobes of the cerebral cortex as *stratified transitional fields*. Many neurons, glia, and their mitotic precursor cells are still migrating through the olfactory peduncle toward the olfactory bulb (*rostral migratory stream*) from a presumed source area in the germinal matrix at the junction between the cerebral cortex, striatum, and nucleus accumbens. Within the cerebral cortex, definite streams of neurons and glia are in the *lateral migratory stream* that percolates through the claustrum, endopiriform nucleus, external capsule, and uncinate fasciculus. These cells appear to be heading toward the insular cortex, primary olfactory cortex, temporal cortex, and basolateral parts of the amygdaloid complex. In the basal ganglia, there is a thick *neuroepithelium/subventricular zone* overlying the striatum and nucleus accumbens where neurons are being generated; at least three subdivisions (anteromedial, anterolateral, and posterior) can be distinguished in the striatal part. Another region of active neurogenesis in the telencephalon is the *subgranular zone* in the hilus of the dentate gyrus that is the source of granule cells. Other structures in the telencephalon, such as the septum, fornix, and Ammon's horn part of the hippocampus, have only a thin, darkly staining layer at the ventricle, and these are presumed to be generating glia, cells of the choroid plexus, and the ependymal lining of the ventricle.

Most of the structures in the diencephalon appear to be settled and are maturing, but the third ventricle is lined by a more densely staining *glioepithelium/ependyma* than in the older specimens. A convoluted *glioepithelium/ependyma* lines the cerebral aqueduct in the midbrain that continues into the anterior fourth ventricle. A smooth *glioepithelium/ependyma* lines the fourth ventricle through the remainder of the pons. A convoluted *glioepithelium/ependyma* lines the floor of the fourth ventricle through much of the medulla. The *external germinal layer* is prominent over the entire surface of the cerebellar cortex and is actively producing basket, stellate, and granule cells. The cerebellar cortex itself shows less definition between hemispheric lobules. The *germinal trigone* is at the base of the nodulus and along the floccular peduncle; choroid plexus cells and glia may still be originating here. In the lower medulla and spinal cord, this specimen is remarkable for showing dense myelination gliosis in the cuneate fasciculus.

# PLATE 216

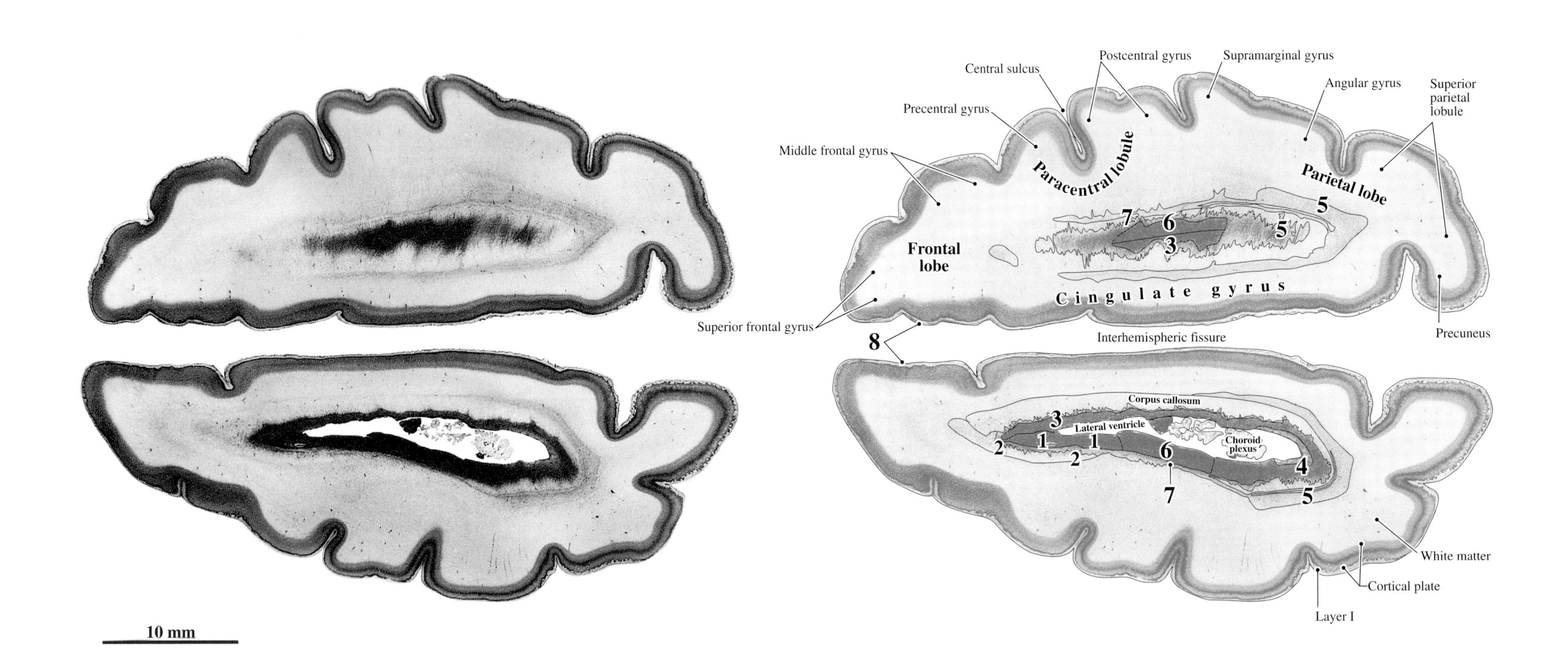

GW26 Horizontal
CR 235 mm
Y16-59
Level 1: Section 240

### Remnants of the germinal matrix, migratory streams, and transitional fields

| | | | |
|---|---|---|---|
| 1 | *Frontal neuroepithelium and subventricular zone* | 5 | *Parietal stratified transitional field* |
| 2 | *Frontal stratified transitional field* | 6 | *Paracentral neuroepithelium and subventricular zone* |
| 3 | *Callosal glioepithelium, neuroepithelium, subventricular zone, and stratified transitional field of the cingulate cortex* | 7 | *Paracentral stratified transitional field* |
| 4 | *Parietal neuroepithelium and subventricular zone* | 8 | *Subpial granular layer (cortical)* |

PLATE 217

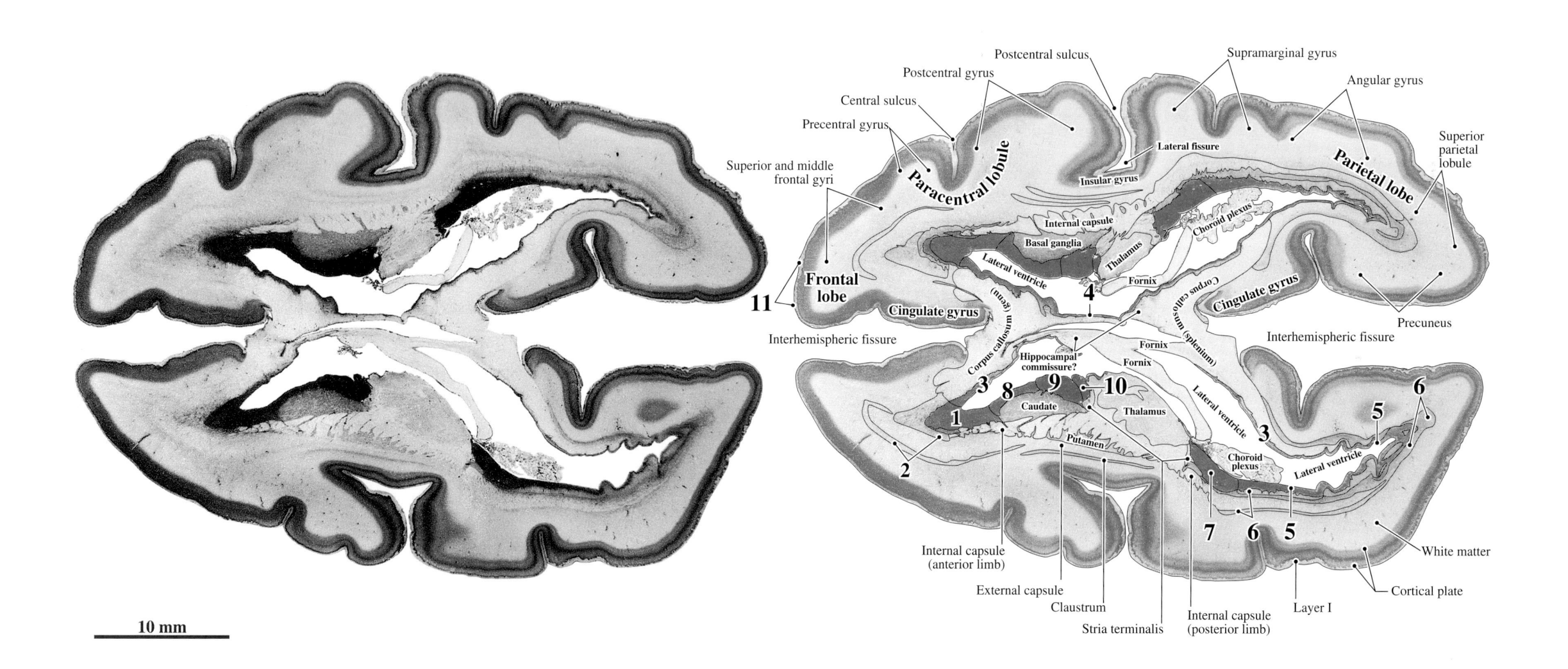

GW26 Horizontal
CR 235 mm
Y16-59
Level 2: Section 340

### Remnants of the germinal matrix, migratory streams, and transitional fields

| | | | |
|---|---|---|---|
| 1 | Frontal neuroepithelium and subventricular zone | 7 | Posterior striatal neuroepithelium and subventricular zone |
| 2 | Frontal stratified transitional field | 8 | Anterolateral striatal neuroepithelium and subventricular zone |
| 3 | Callosal glioepithelium | 9 | Anteromedial striatal neuroepithelium and subventricular zone |
| 4 | Fornical glioepithelium | 10 | Strionuclear glioepithelium |
| 5 | Parietal neuroepithelium and subventricular zone | 11 | Subpial granular layer (cortical) |
| 6 | Parietal stratified transitional field | | |

# PLATE 218

GW26 Horizontal
CR 235 mm
Y16-59
Level 3: Section 400

See detail of brain core
in Plates 230A and B.

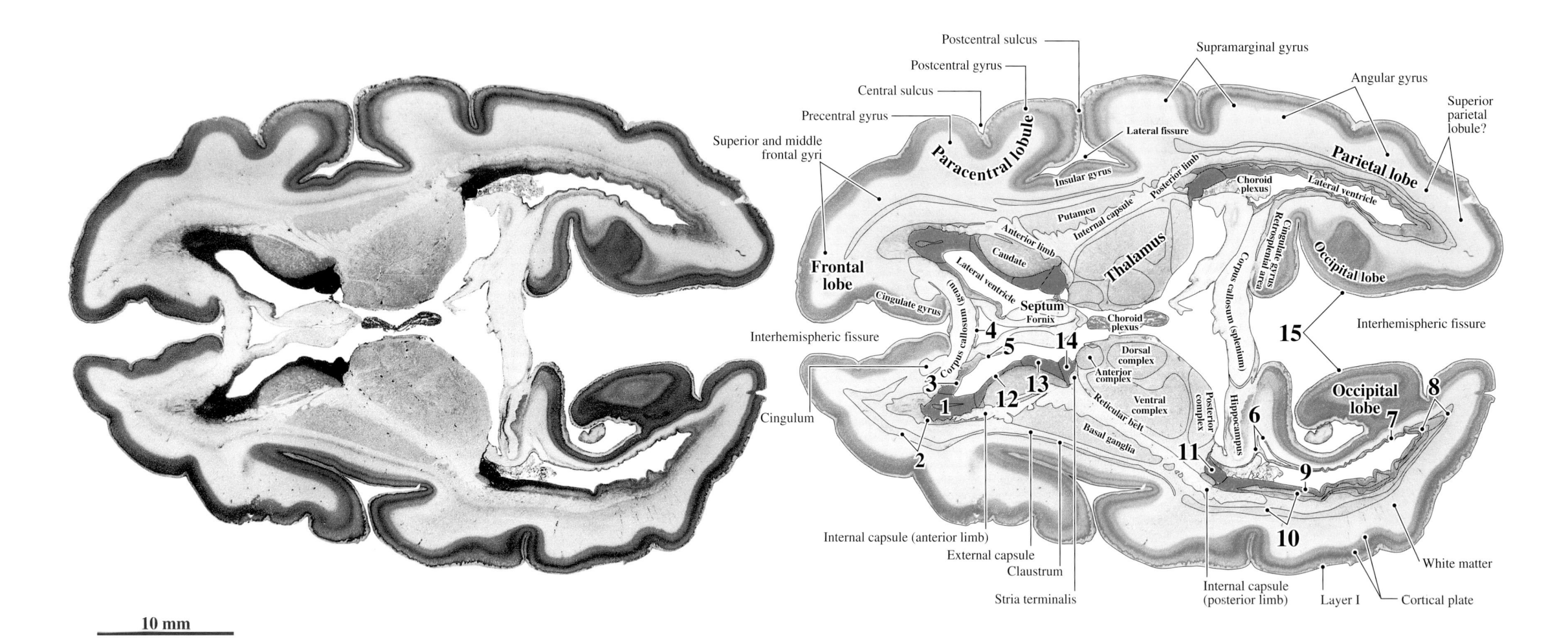

### Remnants of the germinal matrix, migratory streams, and transitional fields

| | | | |
|---|---|---|---|
| 1 | Frontal neuroepithelium and subventricular zone | 8 | Occipital stratified transitional field |
| 2 | Frontal stratified transitional field | 9 | Parietal neuroepithelium and subventricular zone |
| 3 | Callosal glioepithelium | 10 | Parietal stratified transitional field |
| 4 | Callosal sling | 11 | Posterior striatal neuroepithelium and subventricular zone |
| 5 | Fornical glioepithelium | 12 | Anterolateral striatal neuroepithelium and subventricular zone |
| 6 | Parahippocampal neuroepithelium, subventricular zone, and stratified transitional field | 13 | Anteromedial striatal neuroepithelium and subventricular zone |
| | | 14 | Strionuclear glioepithelium |
| 7 | Occipital neuroepithelium and subventricular zone | 15 | Subpial granular layer (cortical) |

## PLATE 219

**GW26 Horizontal**
**CR 235 mm**
**Y16-59**
**Level 4: Section 460**

See detail of brain core in Plates 231A and B.

### Remnants of the germinal matrix, migratory streams, and transitional fields

1. Frontal neuroepithelium and subventricular zone
2. Frontal stratified transitional field
3. Callosal glioepithelium
4. Fornical glioepithelium
5. Parahippocampal neuroepithelium, subventricular zone, and stratified transitional field
6. Occipital neuroepithelium and subventricular zone
7. Occipital stratified transitional field
8. Temporal neuroepithelium and subventricular zone
9. Temporal stratified transitional field
10. Alvear glioepithelium
11. Subgranular zone (dentate)
12. Posterior striatal neuroepithelium and subventricular zone
13. Anterolateral striatal neuroepithelium and subventricular zone
14. Anteromedial striatal neuroepithelium and subventricular zone
15. Strionuclear glioepithelium
16. Subpial granular layer (cortical)

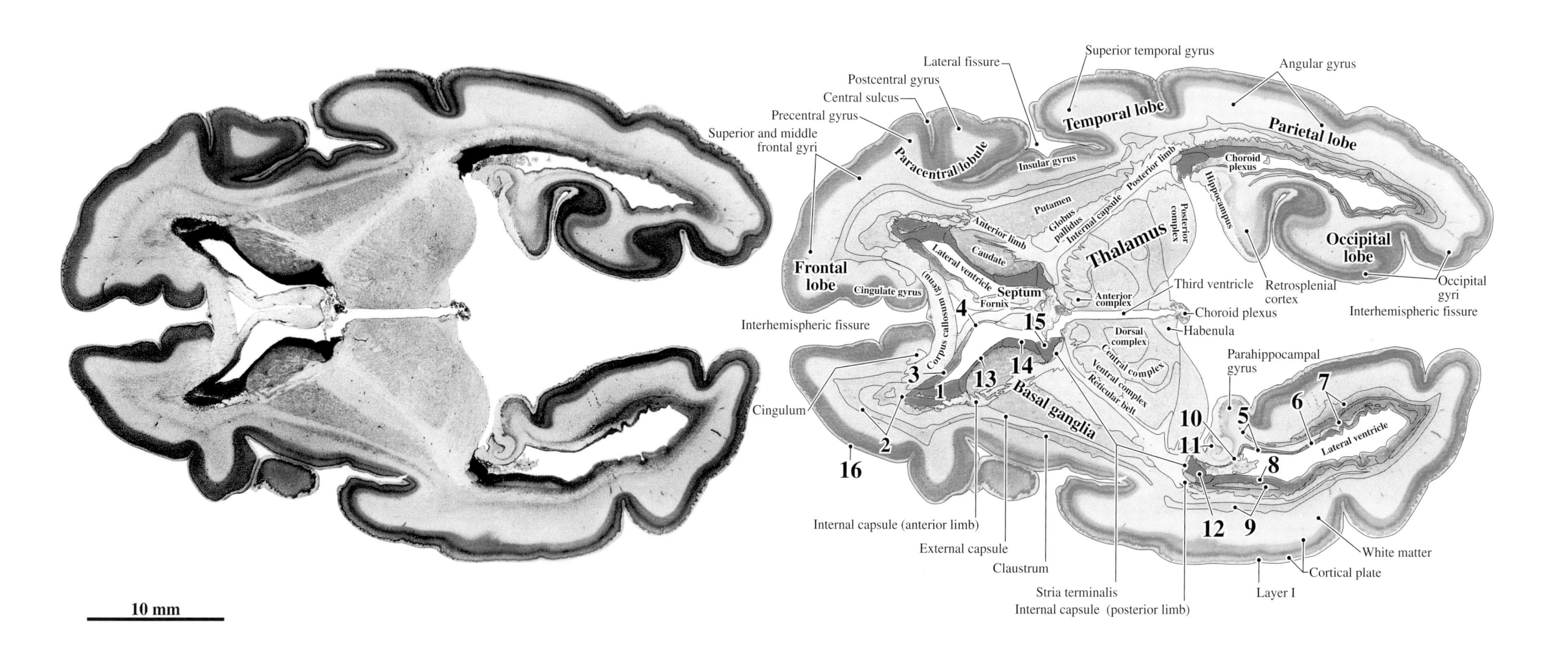

# PLATE 220

GW26 Horizontal
CR 235 mm
Y16-59
Level 5: Section 520

See detail of brain core
in Plates 232A and B.

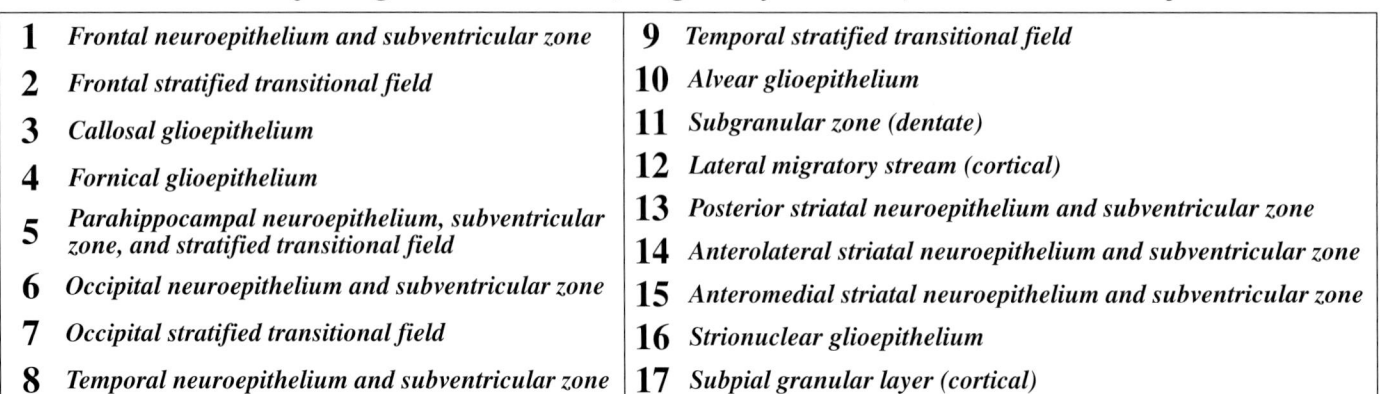

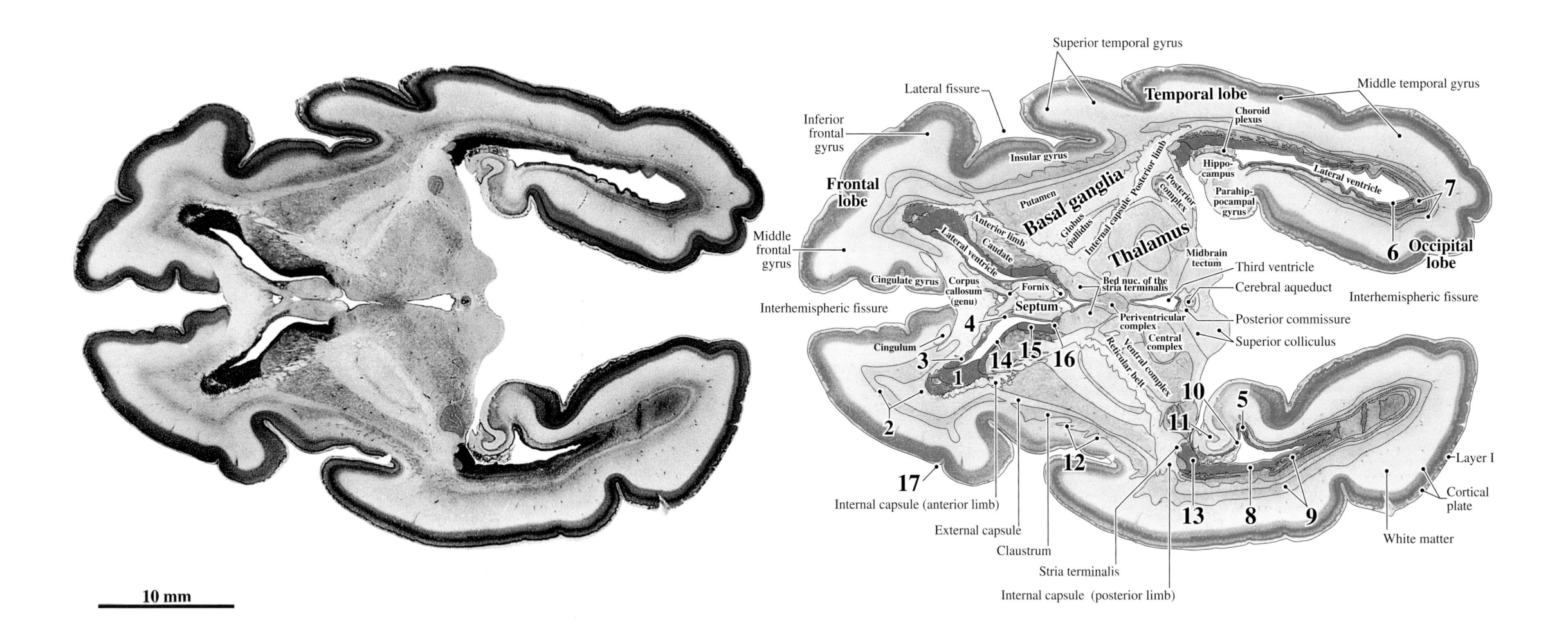

### Remnants of the germinal matrix, migratory streams, and transitional fields

| | | | |
|---|---|---|---|
| 1 | Frontal neuroepithelium and subventricular zone | 9 | Temporal stratified transitional field |
| 2 | Frontal stratified transitional field | 10 | Alvear glioepithelium |
| 3 | Callosal glioepithelium | 11 | Subgranular zone (dentate) |
| 4 | Fornical glioepithelium | 12 | Lateral migratory stream (cortical) |
| 5 | Parahippocampal neuroepithelium, subventricular zone, and stratified transitional field | 13 | Posterior striatal neuroepithelium and subventricular zone |
| | | 14 | Anterolateral striatal neuroepithelium and subventricular zone |
| 6 | Occipital neuroepithelium and subventricular zone | 15 | Anteromedial striatal neuroepithelium and subventricular zone |
| 7 | Occipital stratified transitional field | 16 | Strionuclear glioepithelium |
| 8 | Temporal neuroepithelium and subventricular zone | 17 | Subpial granular layer (cortical) |

10 mm

# PLATE 221

**GW26 Horizontal**
**CR 235 mm**
**Y16-59**
**Level 6: Section 560**

## Remnants of the germinal matrix, migratory streams, and transitional fields

| | | | |
|---|---|---|---|
| 1 | Rostral migratory stream (source area) | 9 | Alvear glioepithelium |
| 2 | Frontal neuroepithelium and subventricular zone | 10 | Subgranular zone (dentate) |
| 3 | Frontal stratified transitional field | 11 | Lateral migratory stream (cortical) |
| 4 | Callosal glioepithelium | 12 | Posterior striatal neuroepithelium and subventricular zone |
| 5 | Parahippocampal neuroepithelium, subventricular zone, and stratified transitional field | 13 | Anterolateral striatal neuroepithelium and subventricular zone |
| 6 | Occipital stratified transitional field | 14 | Anteromedial striatal neuroepithelium and subventricular zone |
| 7 | Temporal neuroepithelium and subventricular zone | 15 | Strionuclear glioepithelium |
| 8 | Temporal stratified transitional field | 16 | Subpial granular layer (cortical) |

See detail of brain core in Plates 233A and B.

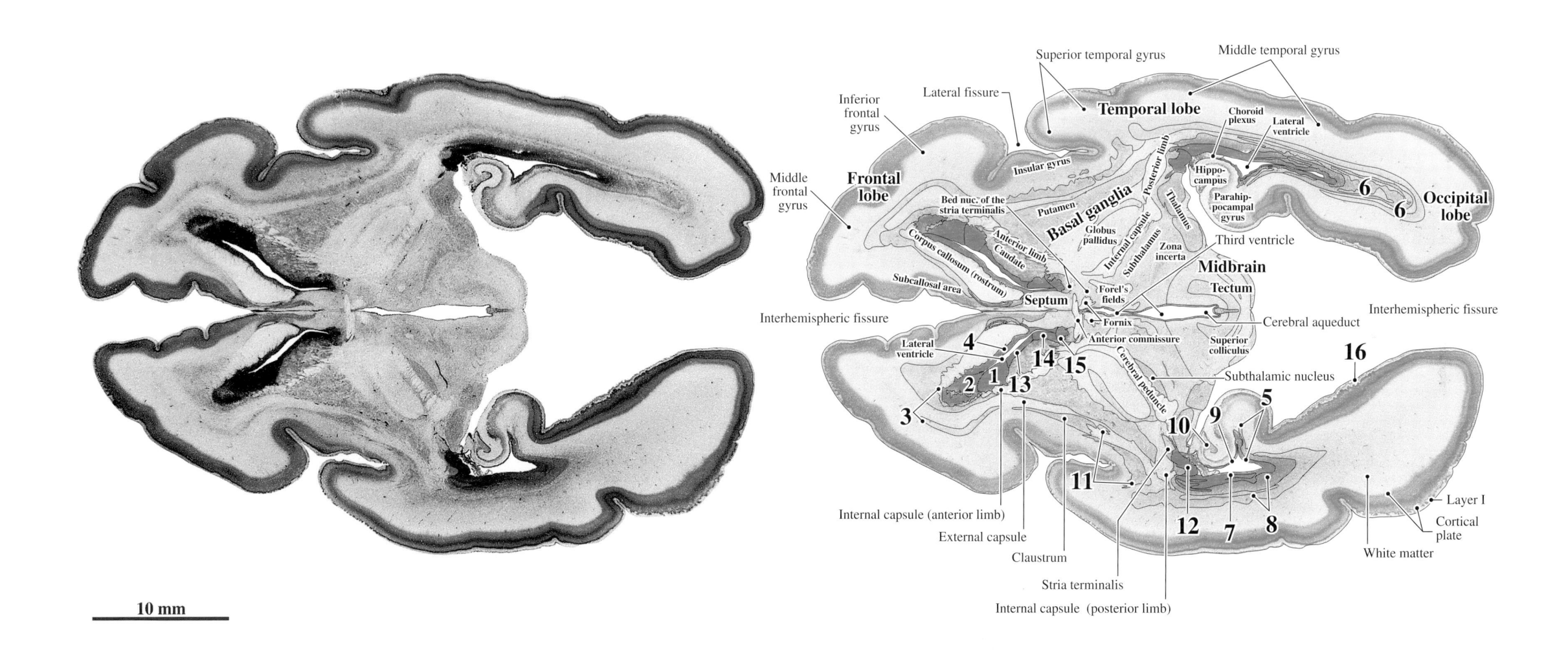

# PLATE 222

**GW26 Horizontal**
**CR 235 mm**
**Y16-59**
**Level 7: Section 620**

**See detail of brain core and cerebellum in Plates 234A and B.**

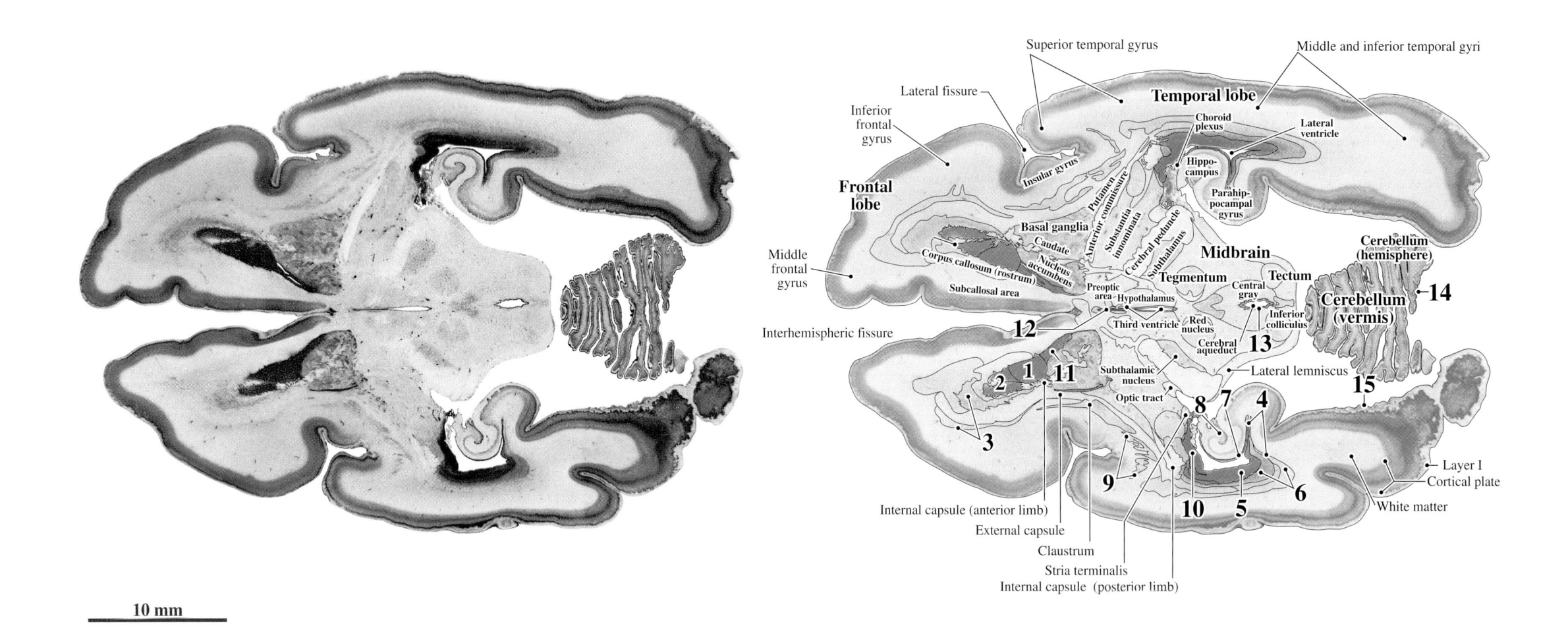

*Remnants of the germinal matrix, migratory streams, and transitional fields*

| | | | |
|---|---|---|---|
| 1 | *Rostral migratory stream (source area)* | 9 | *Lateral migratory stream (cortical)* |
| 2 | *Frontal neuroepithelium and subventricular zone* | 10 | *Posterior striatal neuroepithelium and subventricular zone* |
| 3 | *Frontal stratified transitional field* | 11 | *Accumbent neuroepithelium (intermingled with the rostral migratory stream)* |
| 4 | *Parahippocampal neuroepithelium, subventricular zone, and stratified transitional field* | 12 | *Diencephalic (hypothalamic) glioepithelium/ependyma* |
| 5 | *Temporal neuroepithelium and subventricular zone* | 13 | *Mesencephalic glioepithelium/ependyma* |
| 6 | *Temporal stratified transitional field* | 14 | *External germinal layer (cerebellar)* |
| 7 | *Alvear glioepithelium* | 15 | *Subpial granular layer (cortical)* |
| 8 | *Subgranular zone (dentate)* | | |

# PLATE 223

**GW26 Horizontal**
**CR 235 mm**
**Y16-59**
**Level 8: Section 660**

## Remnants of the germinal matrix, migratory streams, and transitional fields

| | | | |
|---|---|---|---|
| 1 | Rostral migratory stream | 8 | Subgranular zone (dentate) |
| 2 | Frontal neuroepithelium and subventricular zone | 9 | Lateral migratory stream (cortical) |
| 3 | Frontal stratified transitional field | 10 | Amygdaloid glioepithelium/ependyma |
| 4 | Parahippocampal neuroepithelium, subventricular zone, and stratified transitional field | 11 | Diencephalic (hypothalamic) glioepithelium/ependyma |
| 5 | Temporal neuroepithelium and subventricular zone | 12 | Mesencephalic glioepithelium/ependyma |
| 6 | Temporal stratified transitional field | 13 | External germinal layer (cerebellar) |
| 7 | Alvear glioepithelium | 14 | Subpial granular layer (cortical) |

**See detail of brain core and cerebellum in Plates 235A and B.**

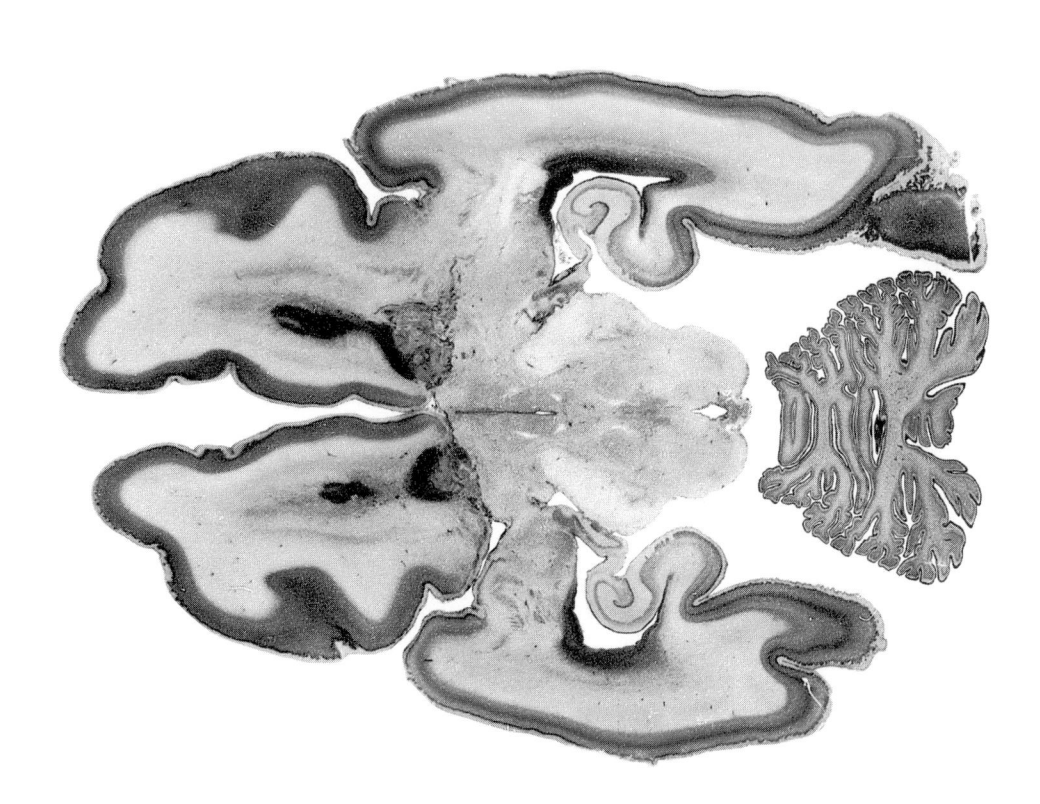

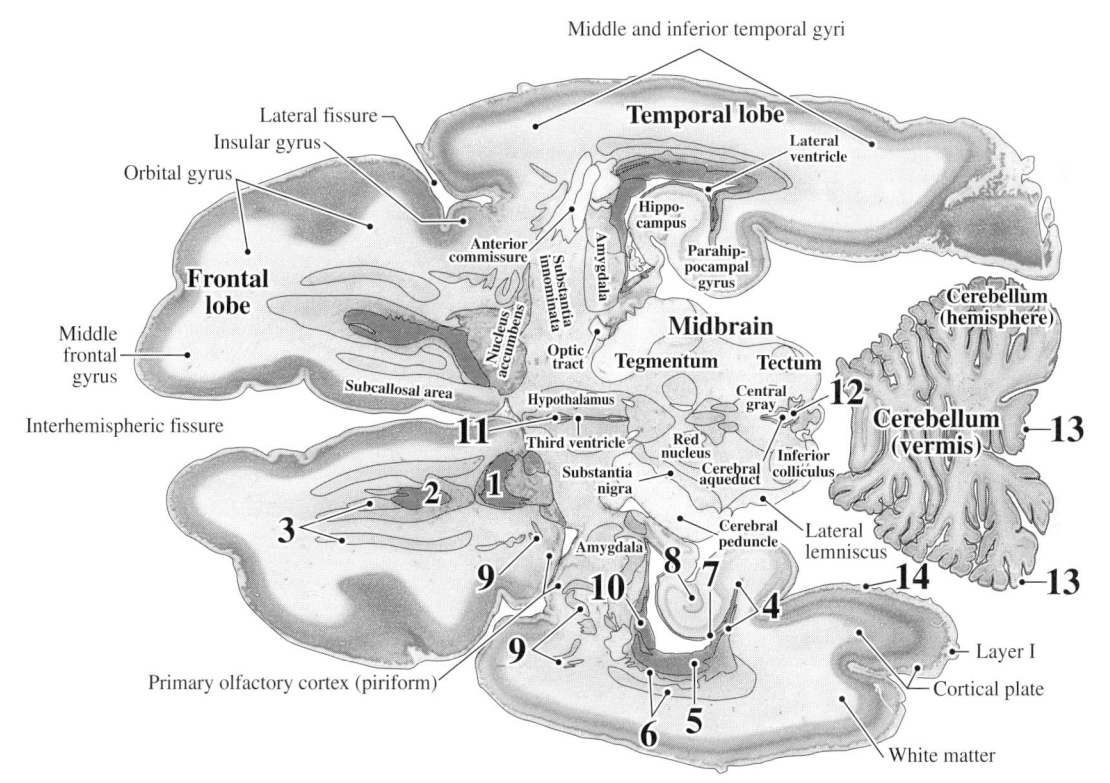

# PLATE 224

GW26 Horizontal
CR 235 mm
Y16-59
Level 9: Section 700

See detail of brain core and cerebellum in Plates 236A and B.

### Remnants of the germinal matrix, migratory streams, and transitional fields

| | | | |
|---|---|---|---|
| 1 | Rostral migratory stream | 8 | Subgranular zone (dentate) |
| 2 | Frontal neuroepithelium and subventricular zone | 9 | Lateral migratory stream (cortical) |
| 3 | Frontal stratified transitional field | 10 | Amygdaloid glioepithelium/ependyma |
| 4 | Parahippocampal neuroepithelium, subventricular zone, and stratified transitional field | 11 | Diencephalic (hypothalamic) glioepithelium/ependyma |
| 5 | Temporal neuroepithelium and subventricular zone | 12 | Mesencephalic glioepithelium/ependyma |
| 6 | Temporal stratified transitional field | 13 | External germinal layer (cerebellar) |
| 7 | Alvear glioepithelium | 14 | Subpial granular layer (cortical) |

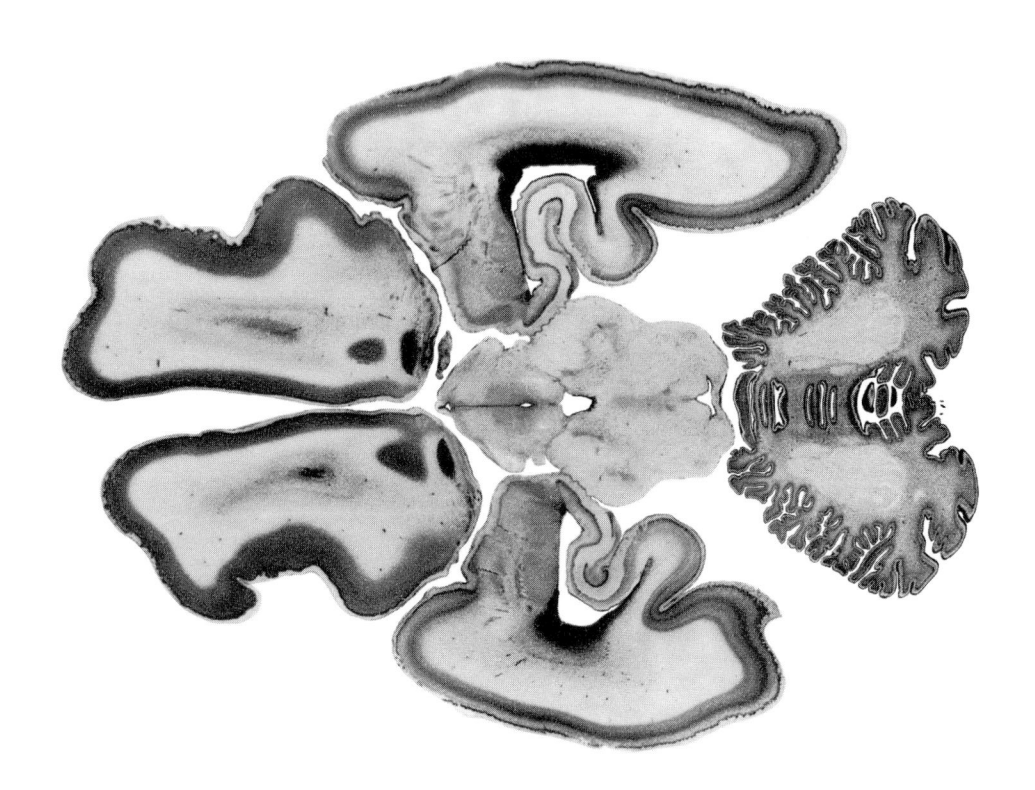

10 mm

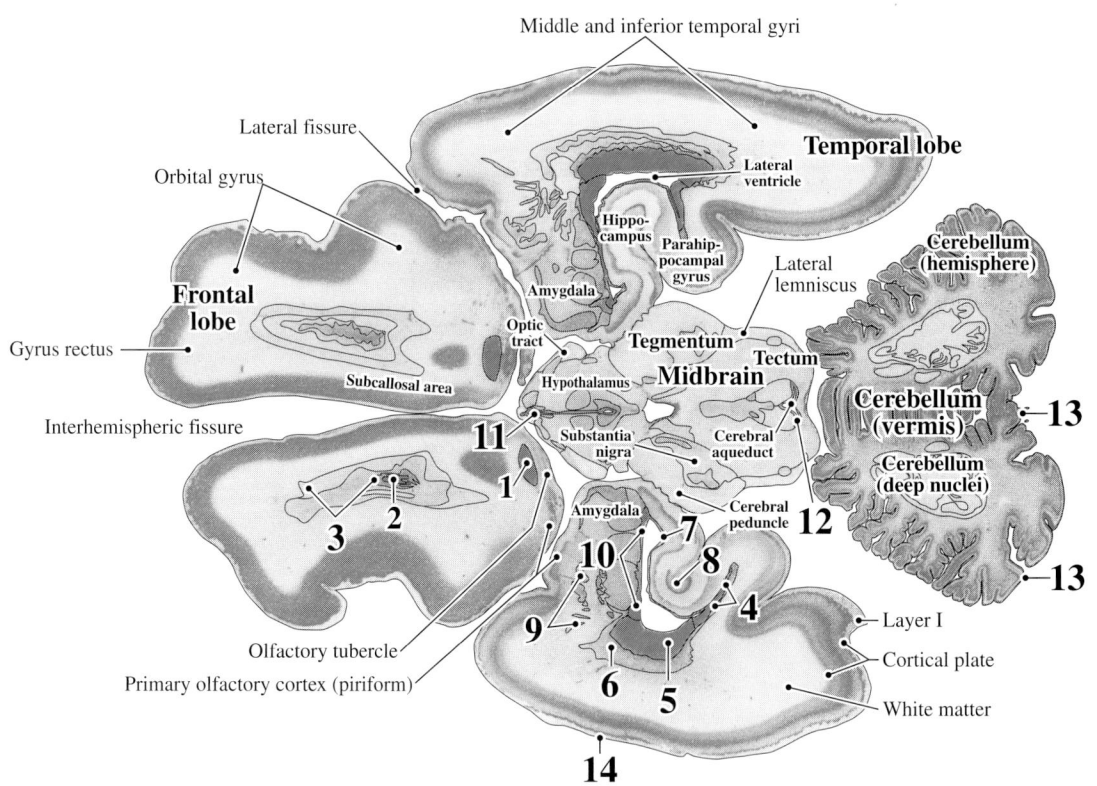

# PLATE 225

**GW26 Horizontal**
**CR 235 mm**
**Y16-59**
**Level 10: Section 740**

## *Remnants of the germinal matrix, migratory streams, and transitional fields*

| | | | |
|---|---|---|---|
| 1 | Rostral migratory stream | 7 | Subgranular zone (dentate) |
| 2 | Frontal stratified transitional field | 8 | Lateral migratory stream (cortical) |
| 3 | Parahippocampal neuroepithelium, subventricular zone, and stratified transitional field | 9 | Amygdaloid glioepithelium/ependyma |
| | | 10 | Diencephalic (hypothalamic) glioepithelium/ependyma |
| 4 | Temporal neuroepithelium and subventricular zone | 11 | Pontine glioepithelium/ependyma |
| 5 | Temporal stratified transitional field | 12 | External germinal layer (cerebellar) |
| 6 | Alvear glioepithelium | 13 | Subpial granular layer (cortical) |

**See detail of brain core and cerebellum in Plates 237A and B.**

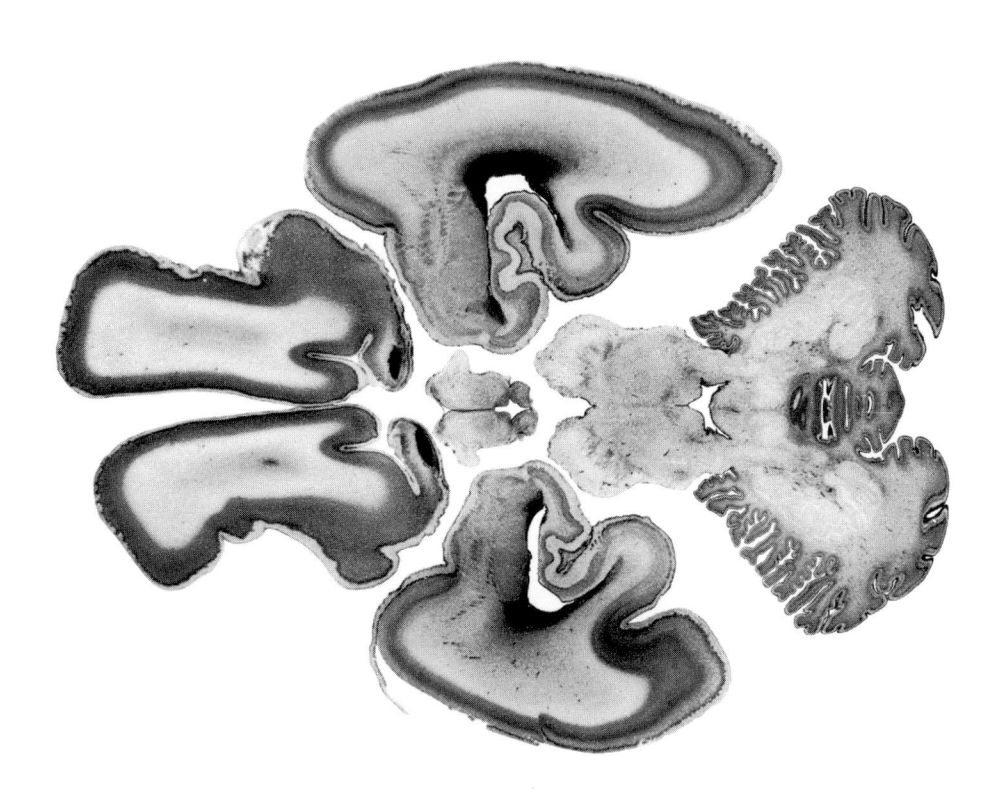

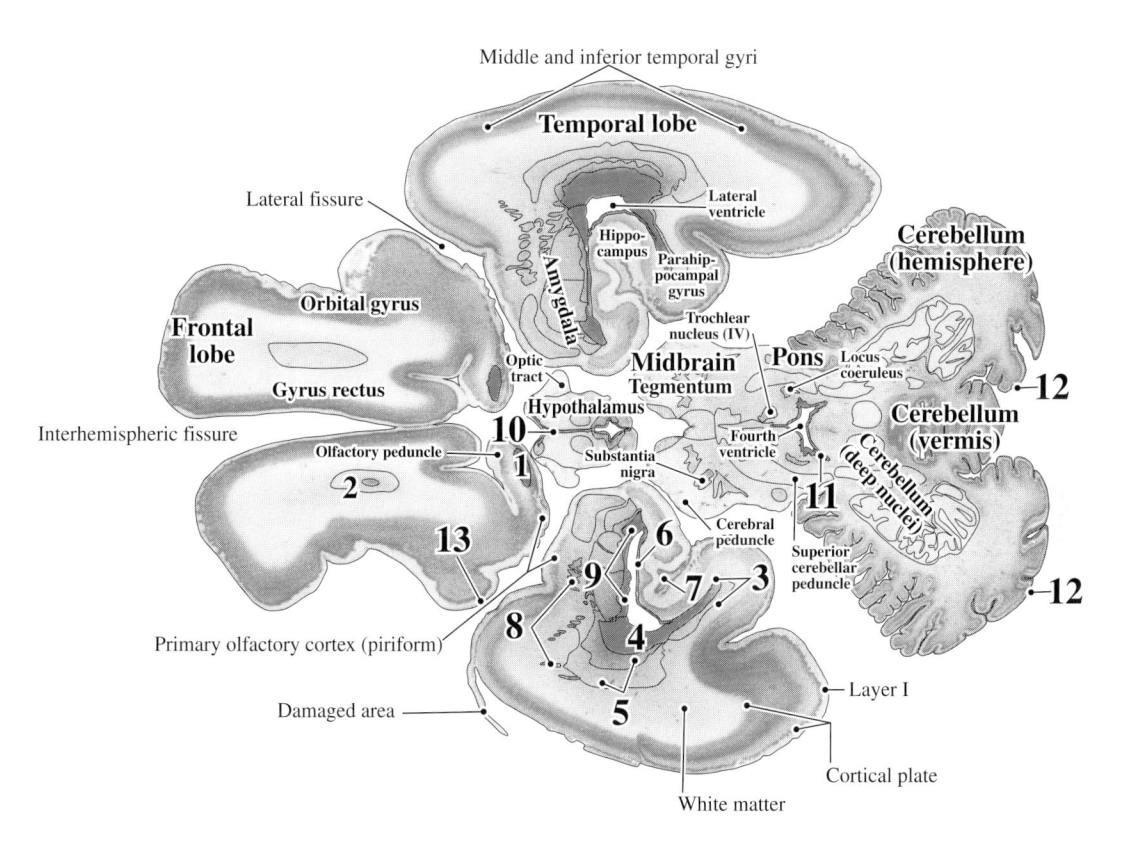

# PLATE 226

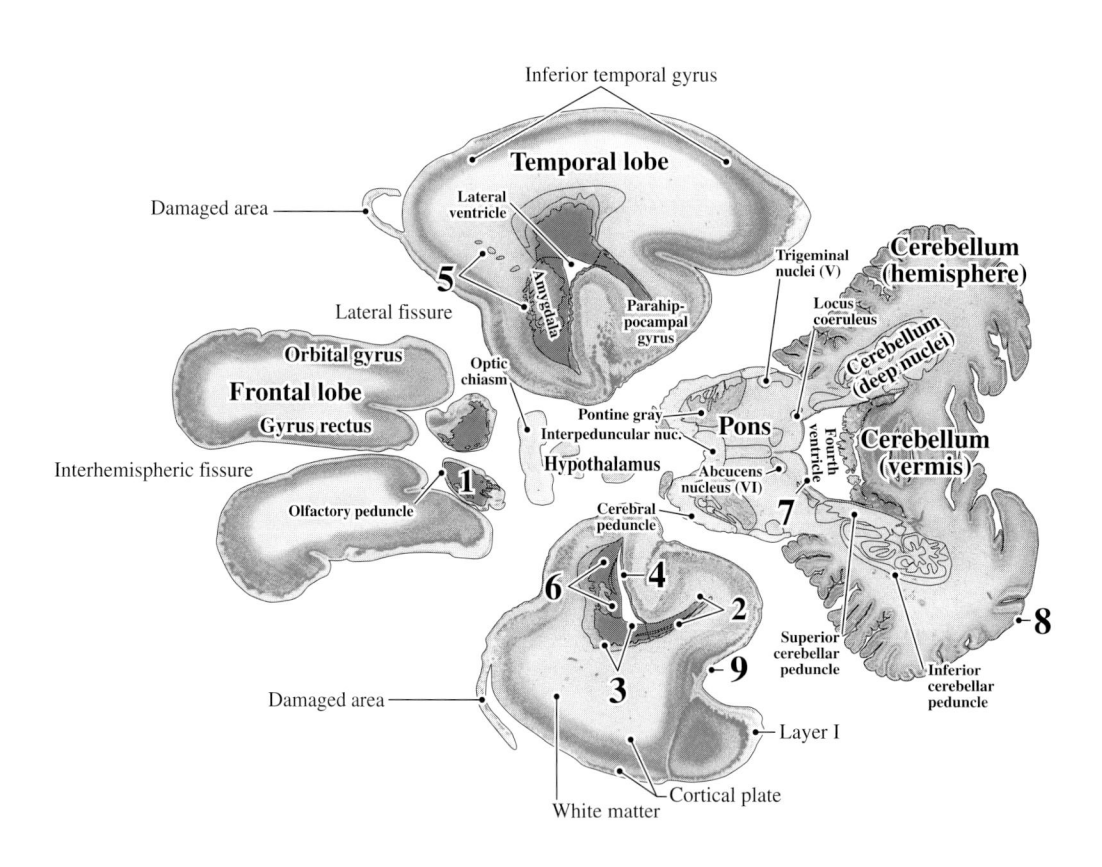

GW26 Horizontal
CR 235 mm
Y16-59
Level 11: Section 780

See detail of brain core and
cerebellum in Plates 238A and B.

*Remnants of the germinal matrix, migratory streams, and transitional fields*

| | | | |
|---|---|---|---|
| 1 | *Rostral migratory stream* | 5 | *Lateral migratory stream (cortical)* |
| 2 | *Parahippocampal neuroepithelium, subventricular zone, and stratified transitional field* | 6 | *Amygdaloid glioepithelium/ependyma* |
| 3 | *Temporal neuroepithelium, subventricular zone, and stratified transitional field* | 7 | *Pontine glioepithelium/ependyma* |
|   |   | 8 | *External germinal layer (cerebellar)* |
| 4 | *Alvear glioepithelium* | 9 | *Subpial granular layer (cortical)* |

10 mm

# PLATE 227

**GW26 Horizontal**
**CR 235 mm**
**Y16-59**
**Level 12: Section 820**

## Remnants of the germinal matrix, migratory streams, and transitional fields

| | | | |
|---|---|---|---|
| 1 | Rostral migratory stream | 4 | Amygdaloid glioepithelium/ependyma |
| 2 | Parahippocampal neuroepithelium, subventricular zone, and stratified transitional field | 5 | Medullary glioepithelium/ependyma |
| 3 | Temporal neuroepithelium, subventricular zone, and stratified transitional field | 6 | External germinal layer (cerebellar) |
| | | 7 | Subpial granular layer (cortical) |

**See detail of brain core and cerebellum in Plates 239A and B.**

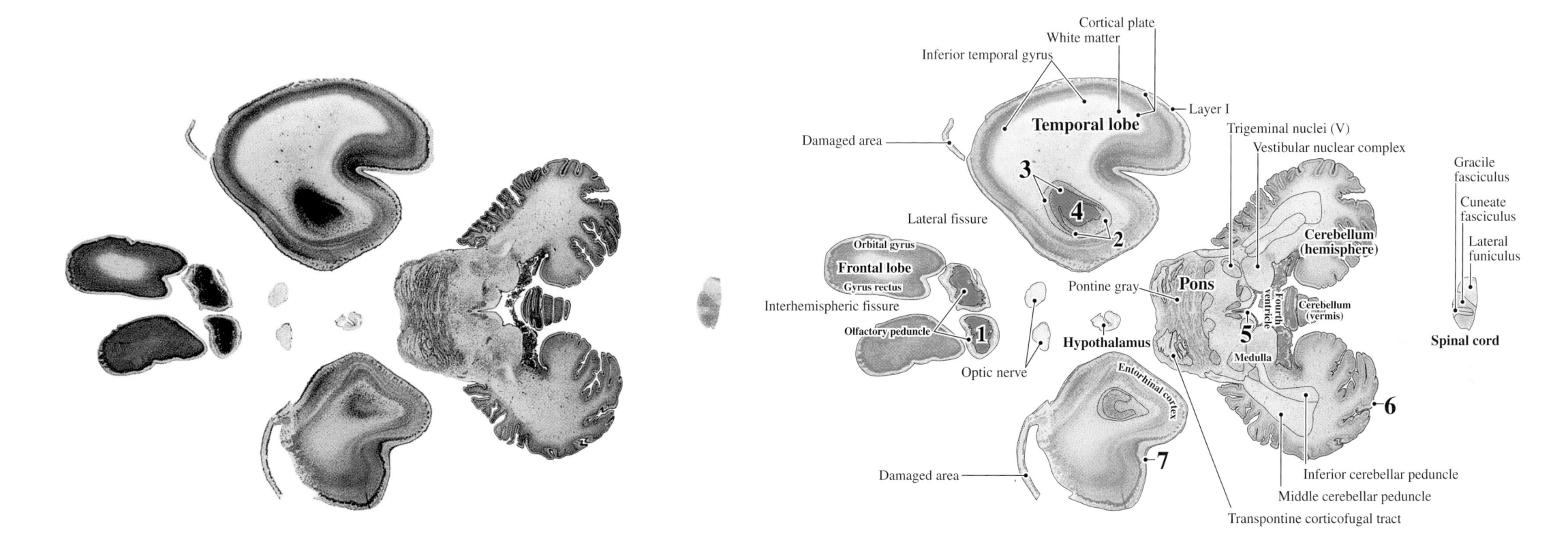

10 mm

# PLATE 228

GW26 Horizontal, CR 235 mm, Y16-59

**Remnants of the germinal matrix, migratory streams, and transitional fields**

| 1 | Rostral migratory stream | 3 | External germinal layer (cerebellar) | 5 | Dorsal funicular myelination gliosis |
|---|---|---|---|---|---|
| 2 | Medullary glioepithelium/ependyma | 4 | Subpial granular layer (cortical) | | |

## Level 13: Section 860

See detail of brain core and cerebellum in Plates 240A and B.

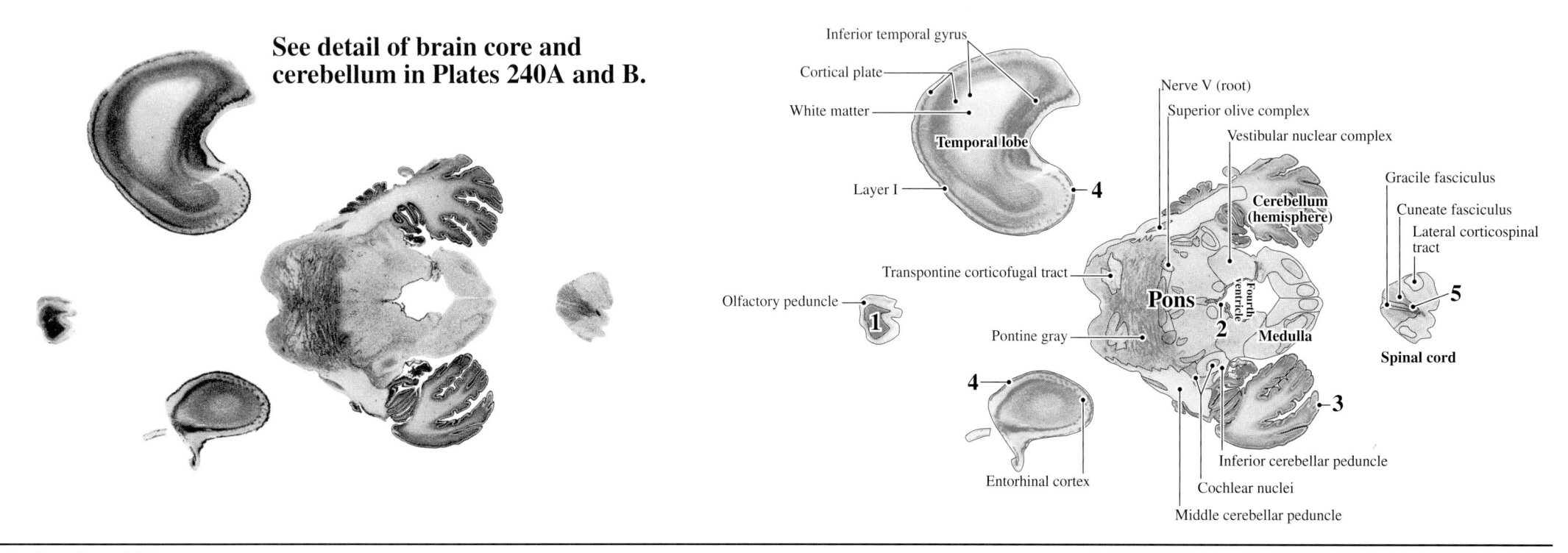

## Level 14: Section 880

See detail of brain core and cerebellum in Plates 241A and B.

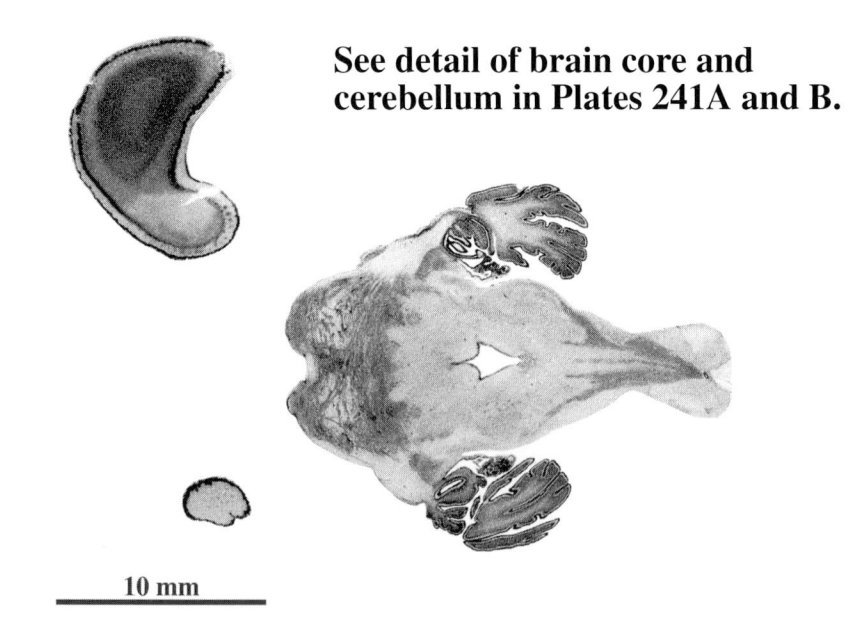

10 mm

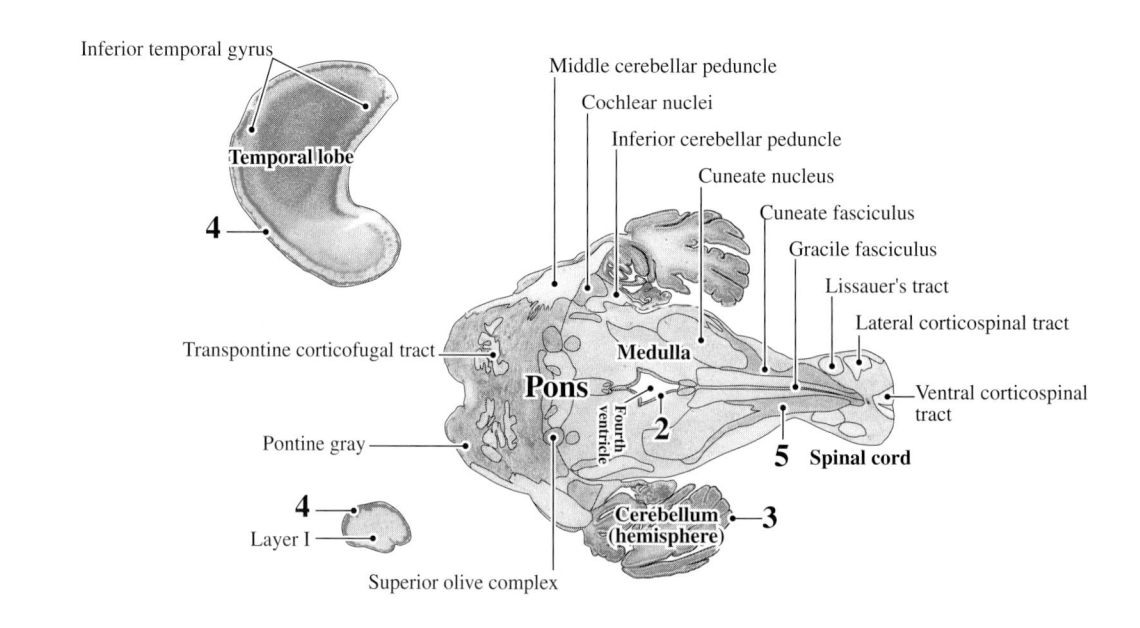

**PLATE 229**

*Remnants of the germinal matrix, migratory streams, and transitional fields*

GW26 Horizontal, CR 235 mm, Y16-59

| | | |
|---|---|---|
| **1** *Medullary glioepithelium/ependyma* | **3** *Subpial granular layer (cortical)* | **5** *Ventral funicular myelination gliosis* |
| **2** *External germinal layer (cerebellar)* | **4** *Dorsal funicular myelination gliosis* | |

**Level 15: Section 900**

See higher magnification view in Plates 242A and B.

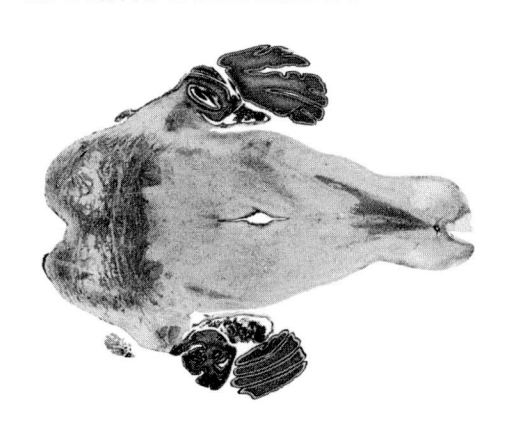

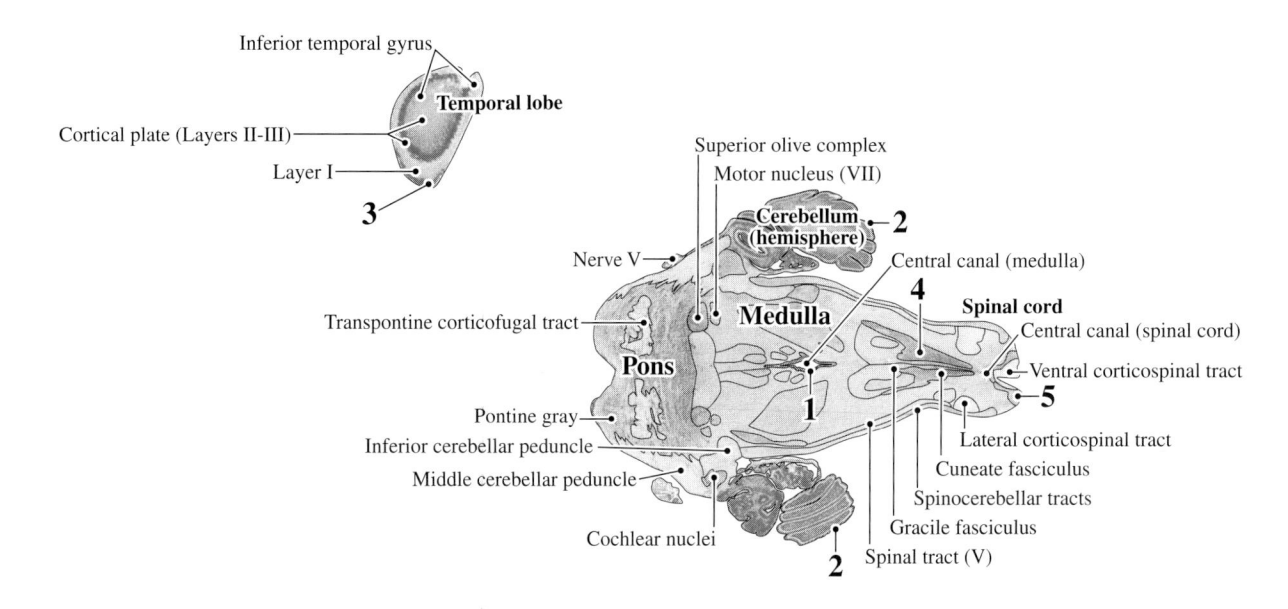

Labels: Inferior temporal gyrus, Temporal lobe, Cortical plate (Layers II-III), Superior olive complex, Layer I, Motor nucleus (VII), 3, Cerebellum (hemisphere), 2, Nerve V, Central canal (medulla), 4, Spinal cord, Transpontine corticofugal tract, Medulla, Central canal (spinal cord), Pons, 1, Ventral corticospinal tract, Pontine gray, 5, Inferior cerebellar peduncle, Lateral corticospinal tract, Middle cerebellar peduncle, Cuneate fasciculus, Spinocerebellar tracts, Cochlear nuclei, Gracile fasciculus, 2, Spinal tract (V)

**Level 16: Section 940**

See higher magnification view in Plates 243A and B.

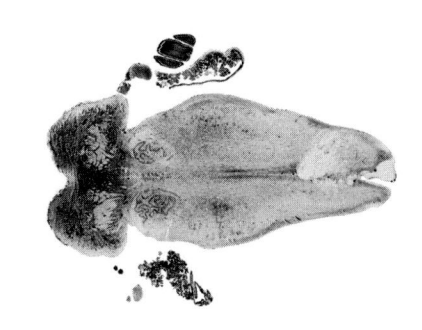

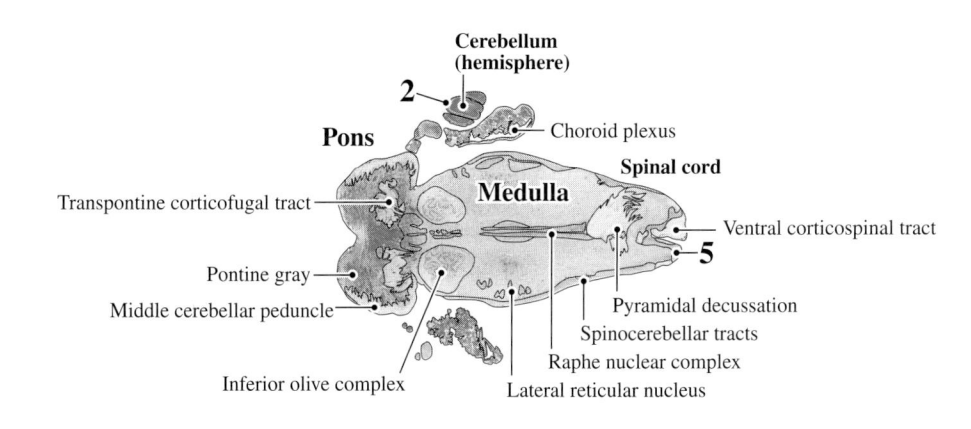

Labels: Cerebellum (hemisphere), 2, Pons, Choroid plexus, Spinal cord, Transpontine corticofugal tract, Medulla, Ventral corticospinal tract, 5, Pontine gray, Pyramidal decussation, Middle cerebellar peduncle, Spinocerebellar tracts, Raphe nuclear complex, Inferior olive complex, Lateral reticular nucleus

**Level 17: Section 960**

See higher magnification view in Plates 244A and B.

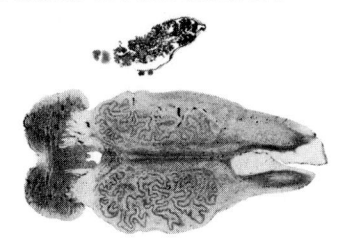

Labels: Pons, Medulla, Spinal cord, Pontine gray, Pyramidal decussation, Corticospinal tract, 5, Middle cerebellar peduncle, Raphe nuclear complex, Inferior olive complex

10 mm

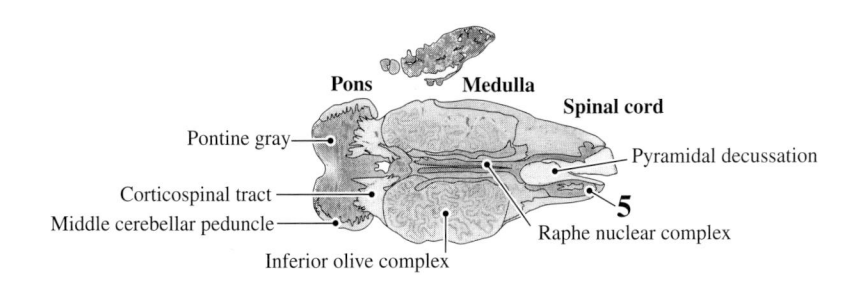

# PLATE 230A

**GW26 Horizontal, CR 235 mm, Y16-59**
**Level 3: Section 400**

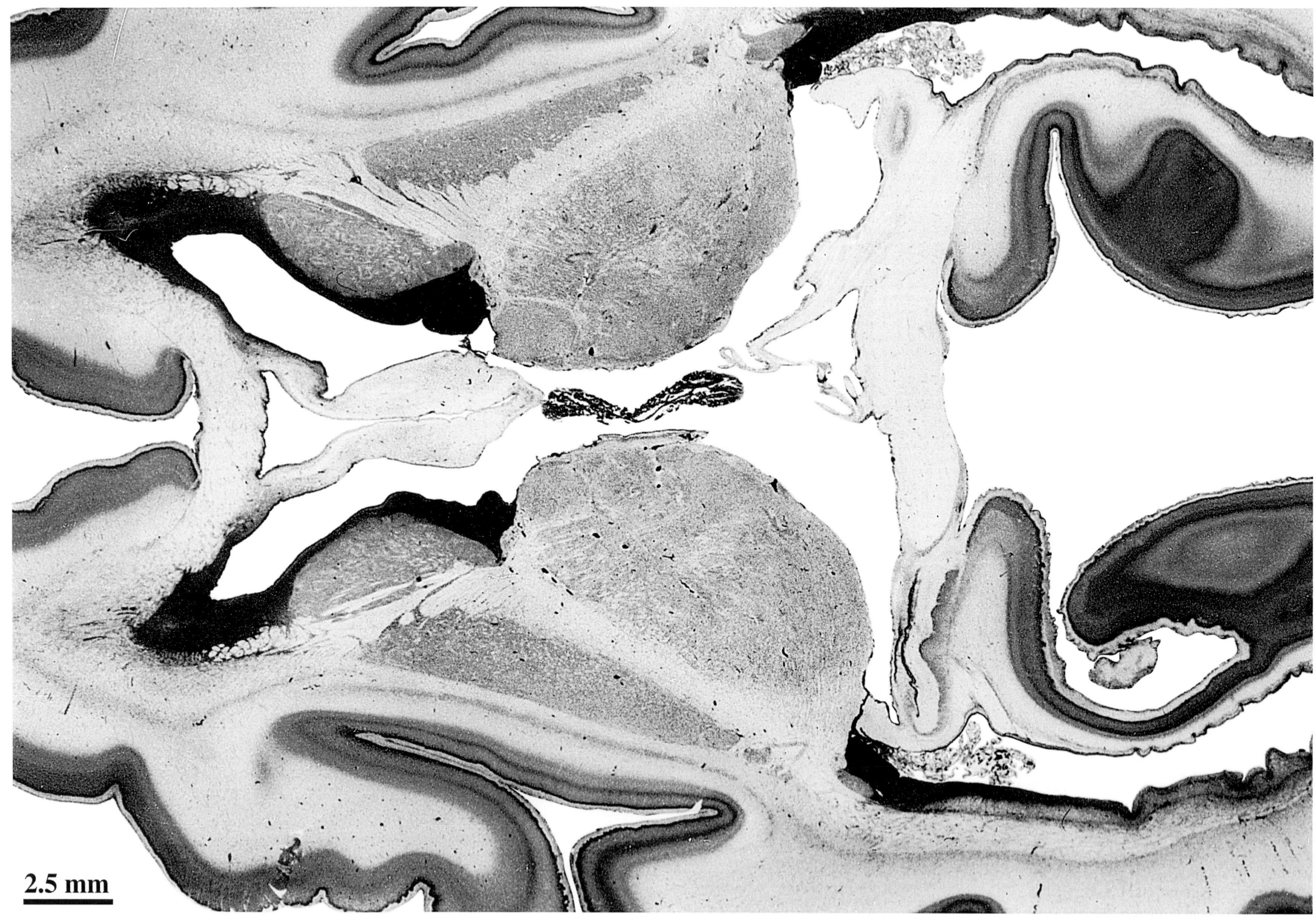

**2.5 mm**

See the entire section in Plate 218.

**PLATE 230B**

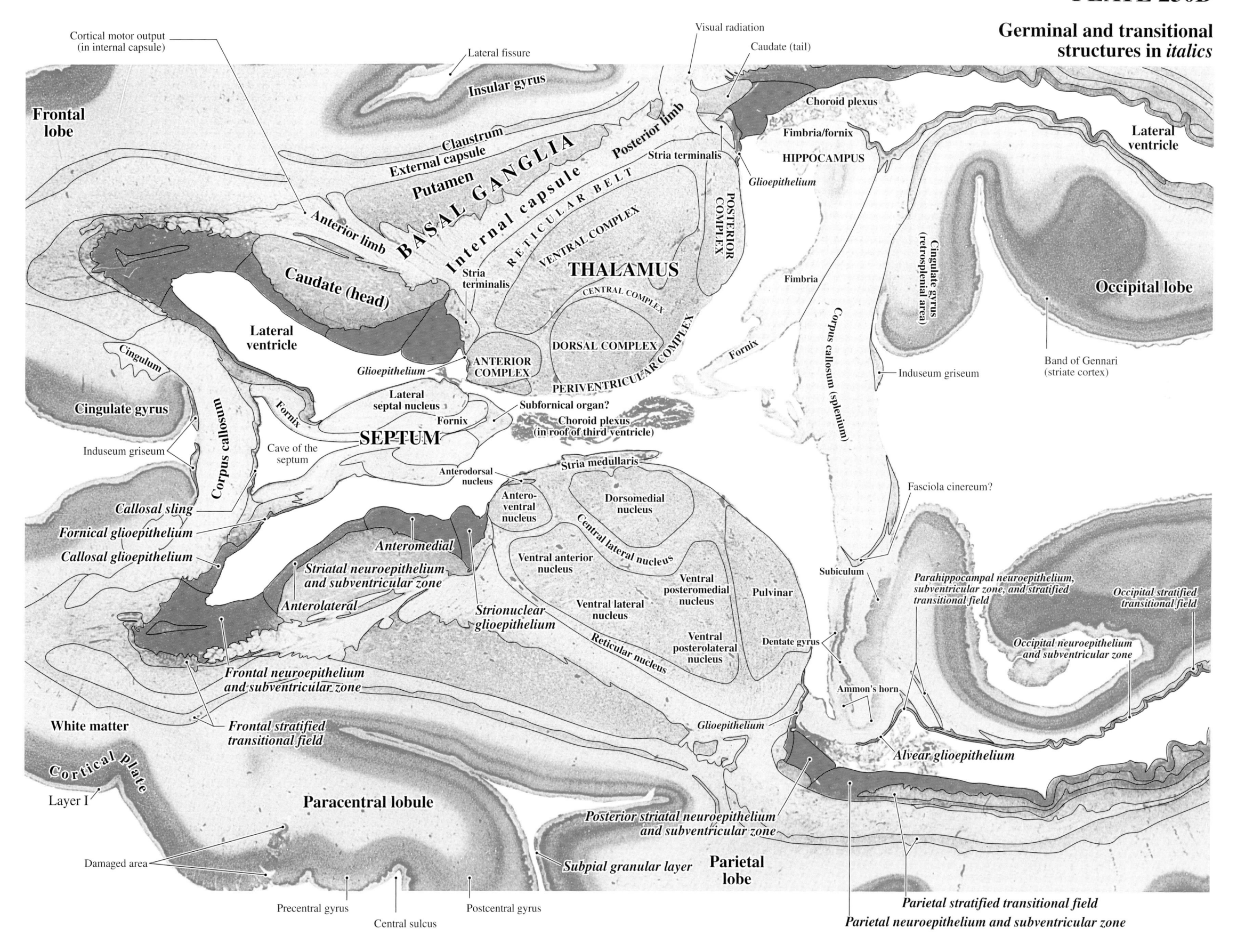

**Germinal and transitional structures in *italics***

## PLATE 231A

**GW26 Horizontal**
**CR 235 mm**
**Y16-59**
**Level 4: Section 460**

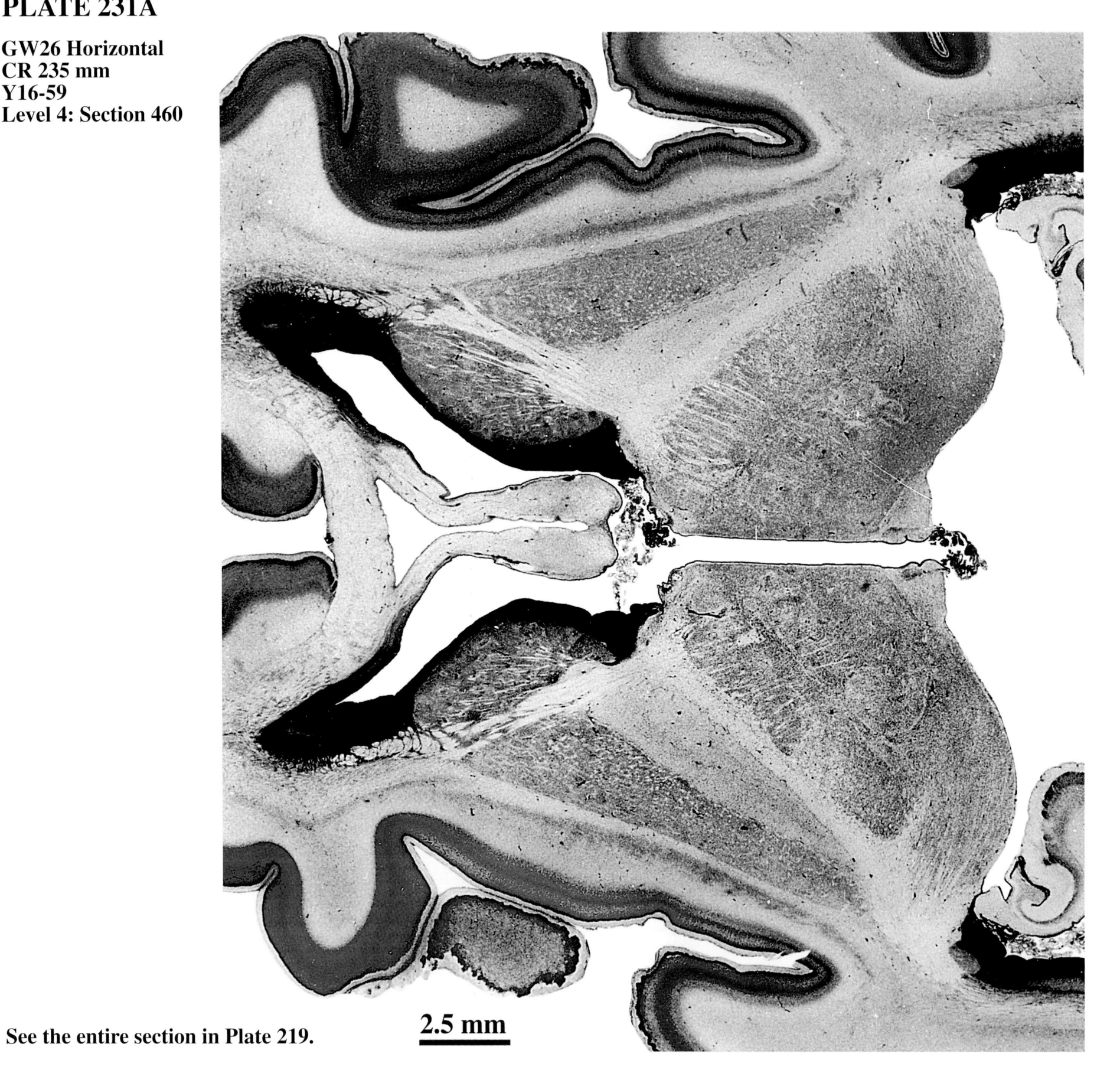

See the entire section in Plate 219.

2.5 mm

341

**PLATE 231B**

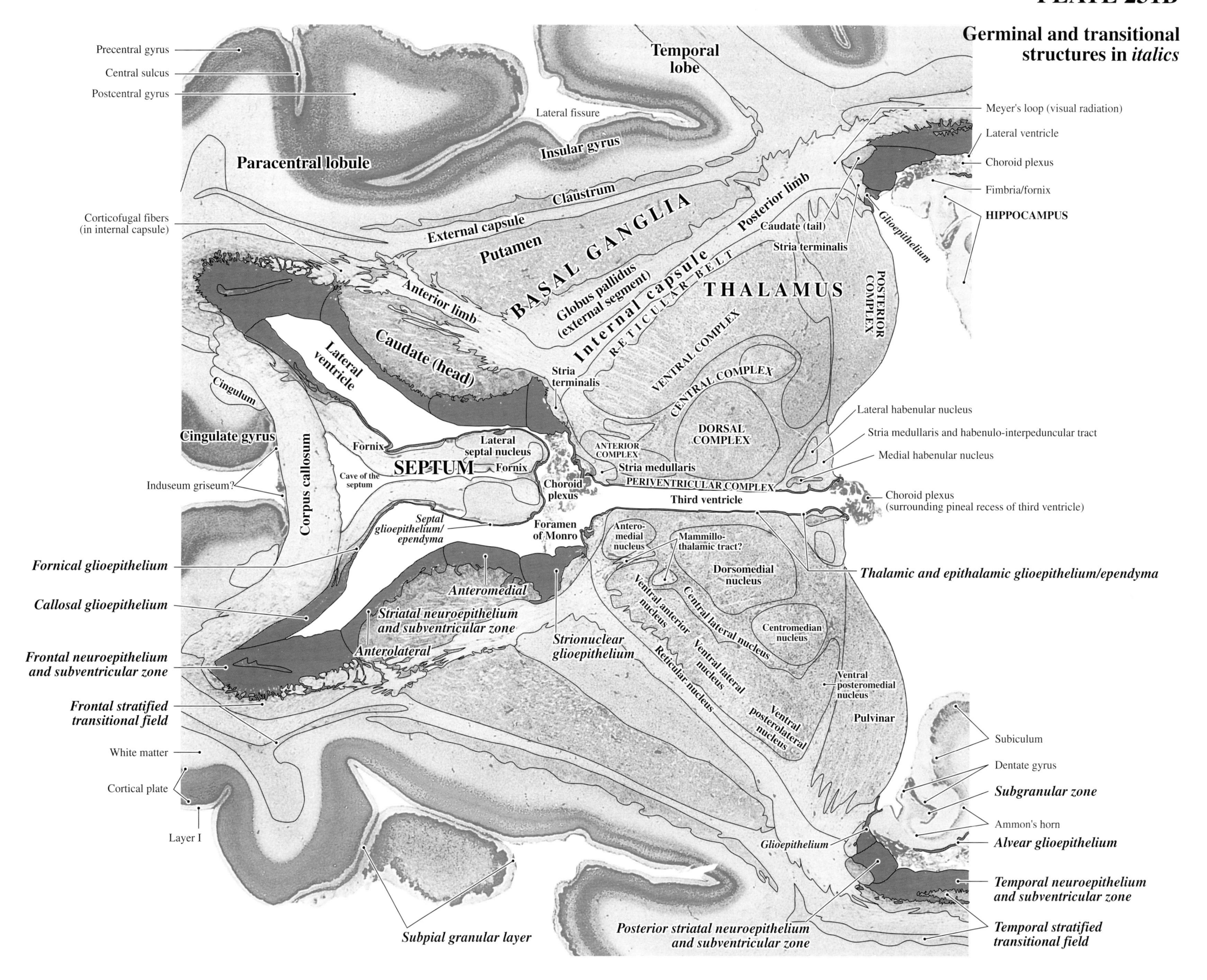

**Germinal and transitional structures in *italics***

**PLATE 232A**

GW26 Horizontal
CR 235 mm
Y16-59
Level 5: Section 520

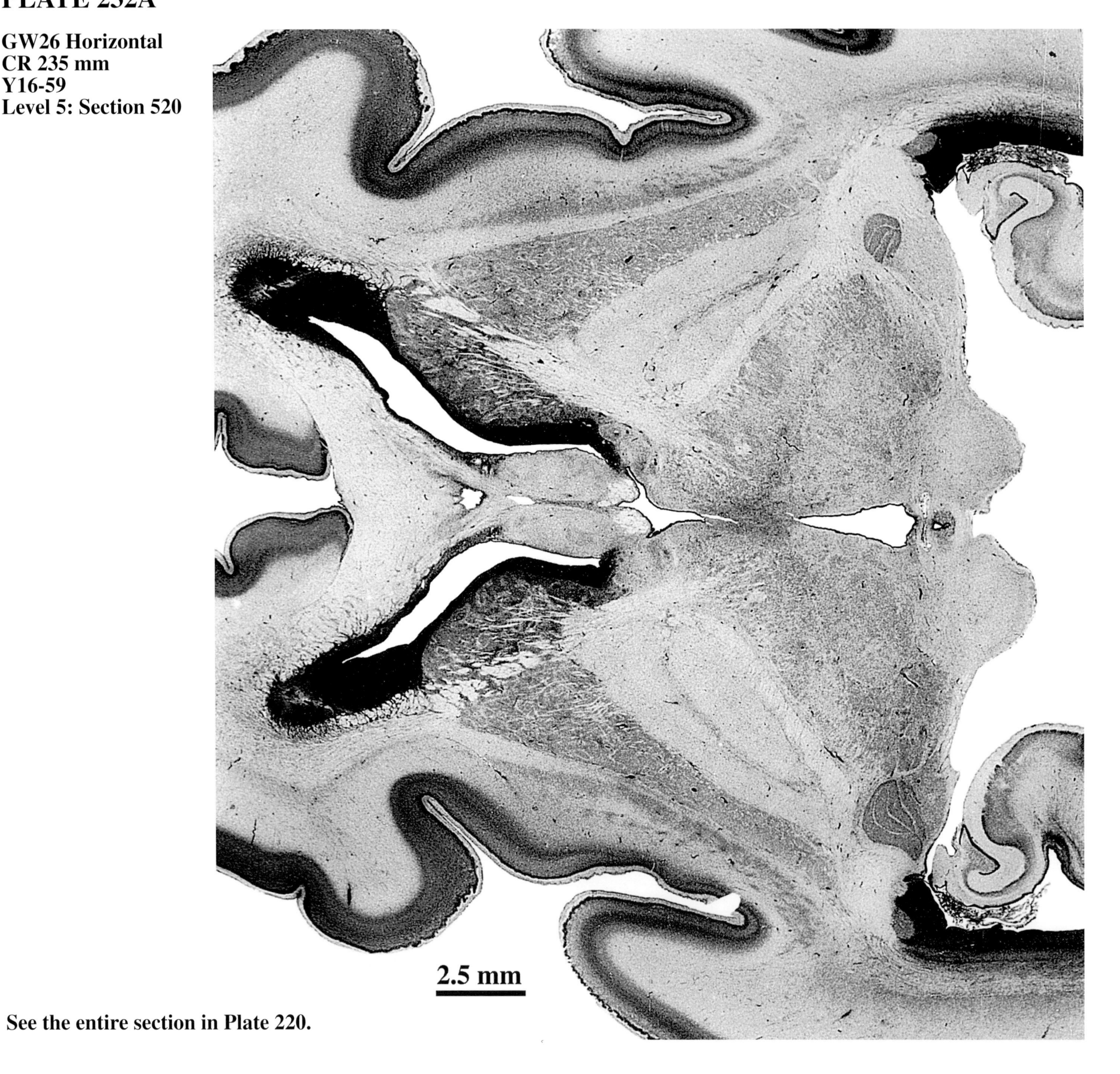

2.5 mm

See the entire section in Plate 220.

**PLATE 232B**

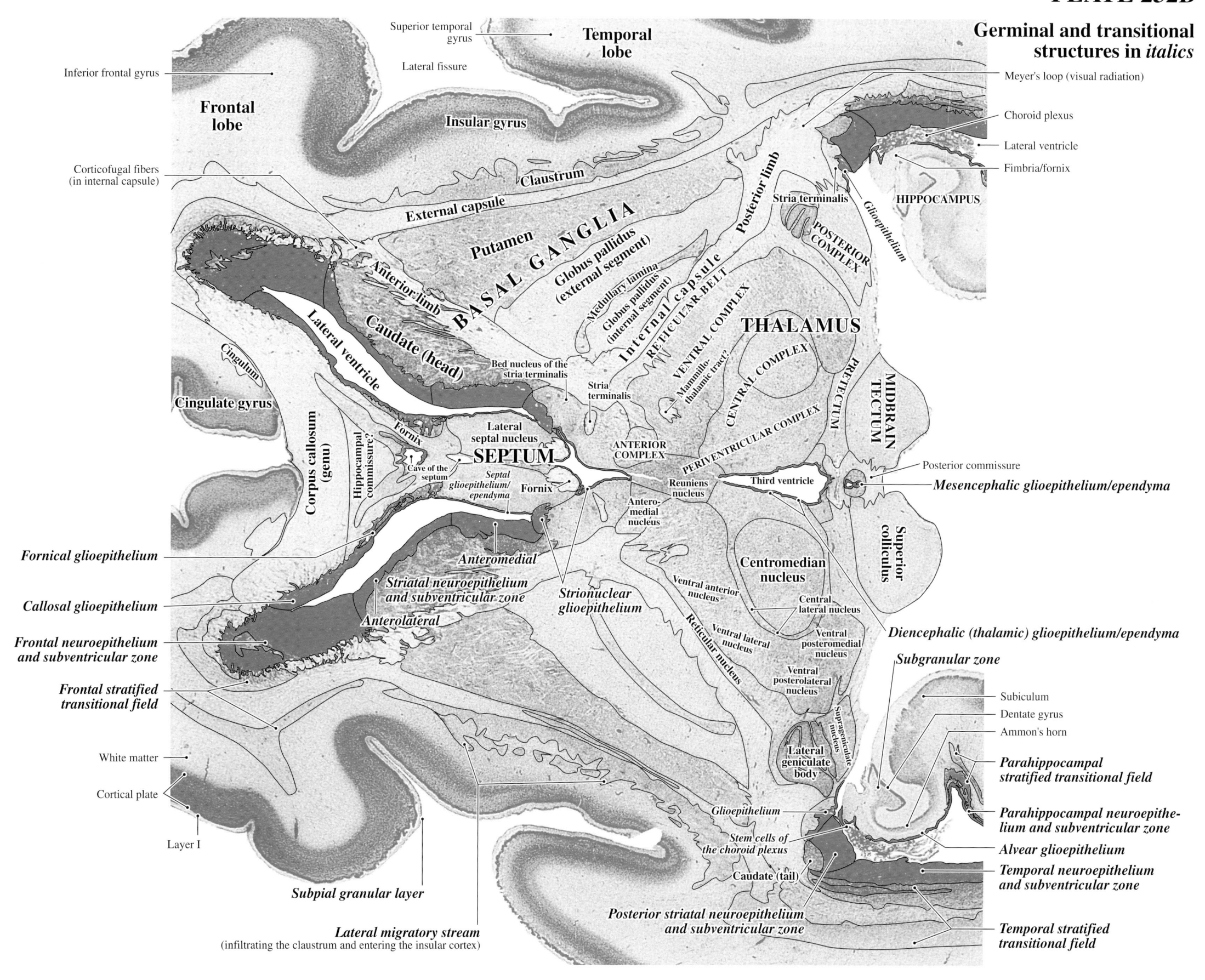

**Germinal and transitional structures in *italics***

## PLATE 233A

GW26 Horizontal
CR 235 mm
Y16-59
Level 6: Section 560

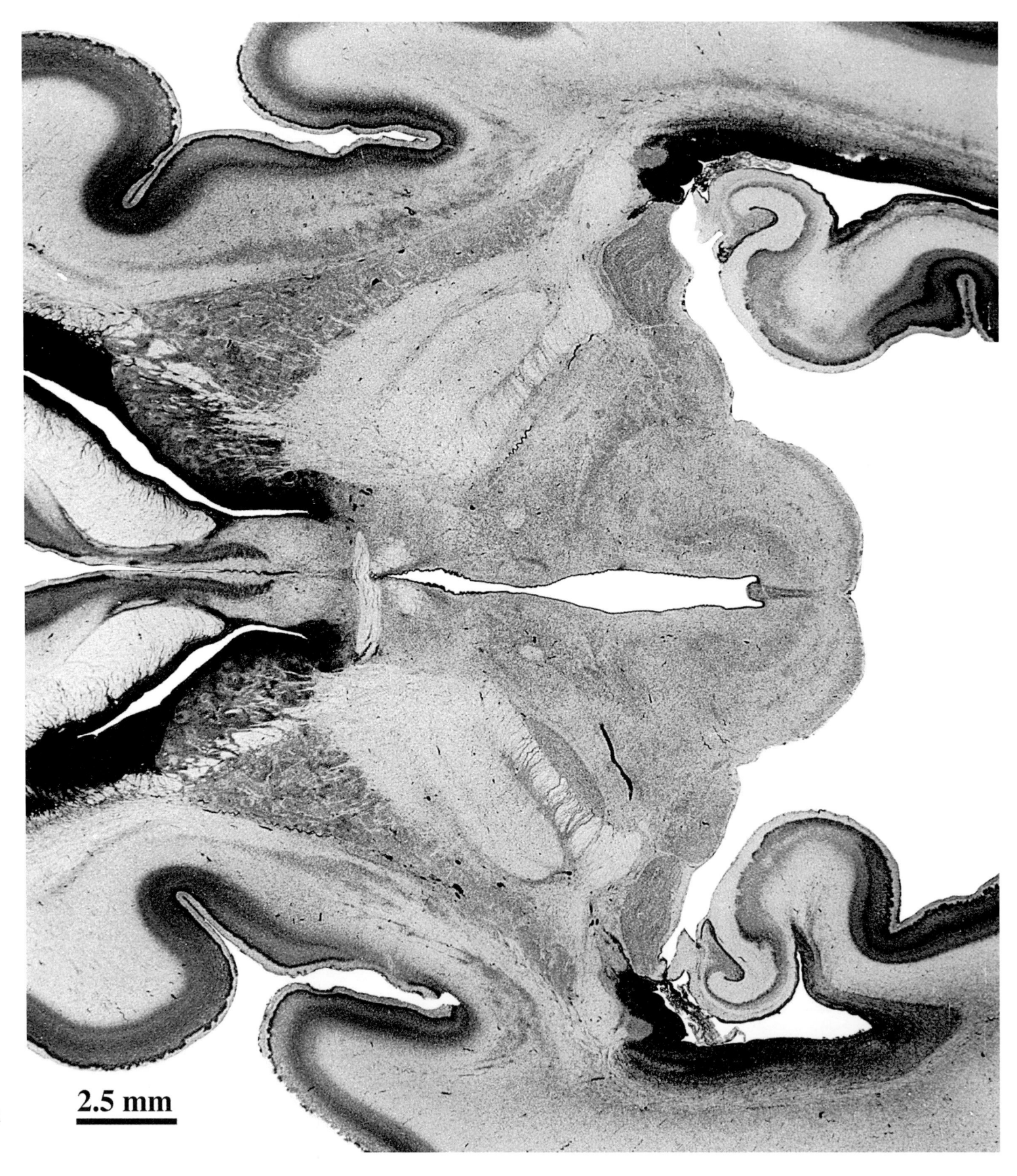

2.5 mm

See the entire section in Plate 221.

**PLATE 233B**

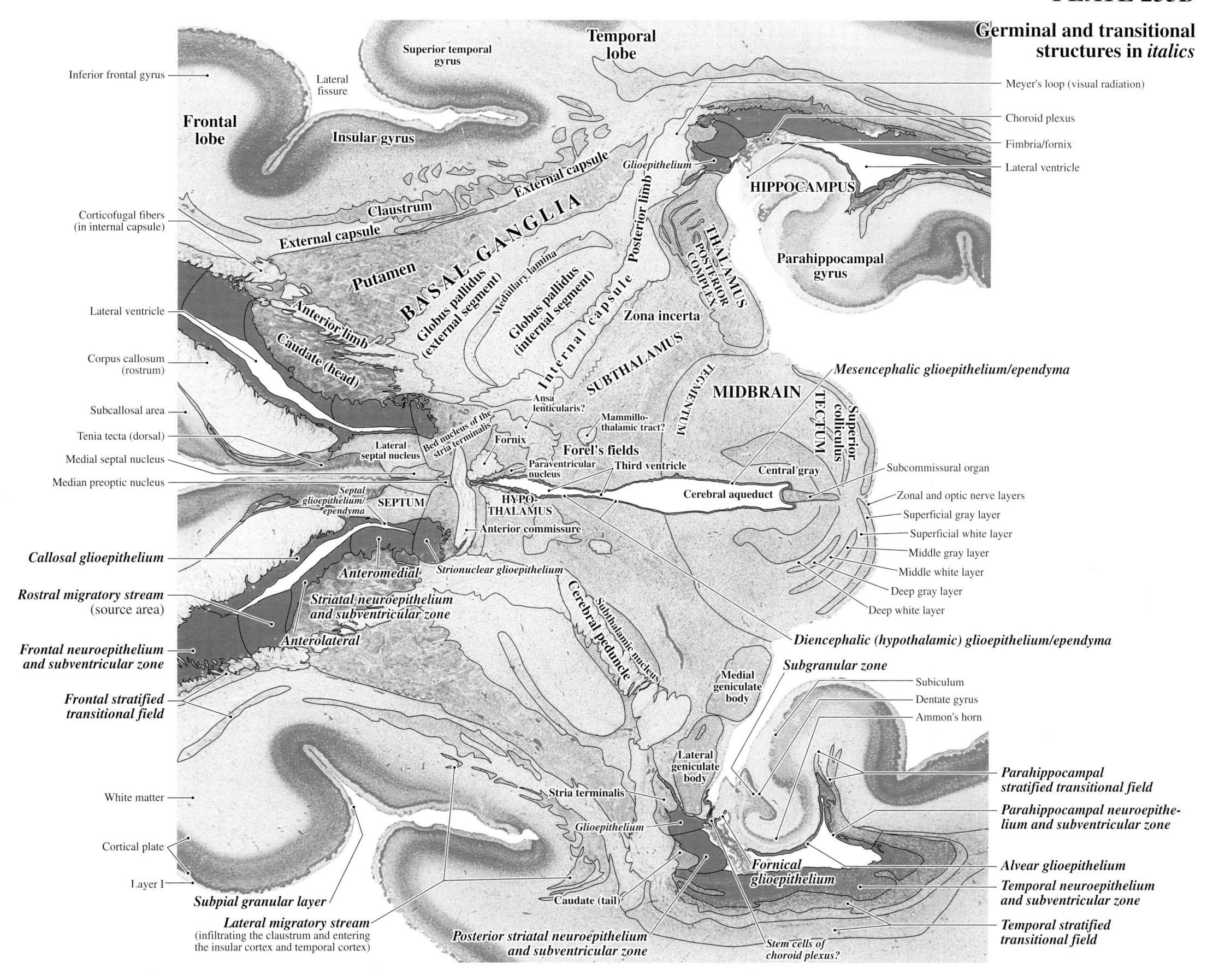

**Germinal and transitional structures in *italics***

## PLATE 234A

**GW26 Horizontal, CR 235 mm, Y16-59**
**Level 7: Section 620**

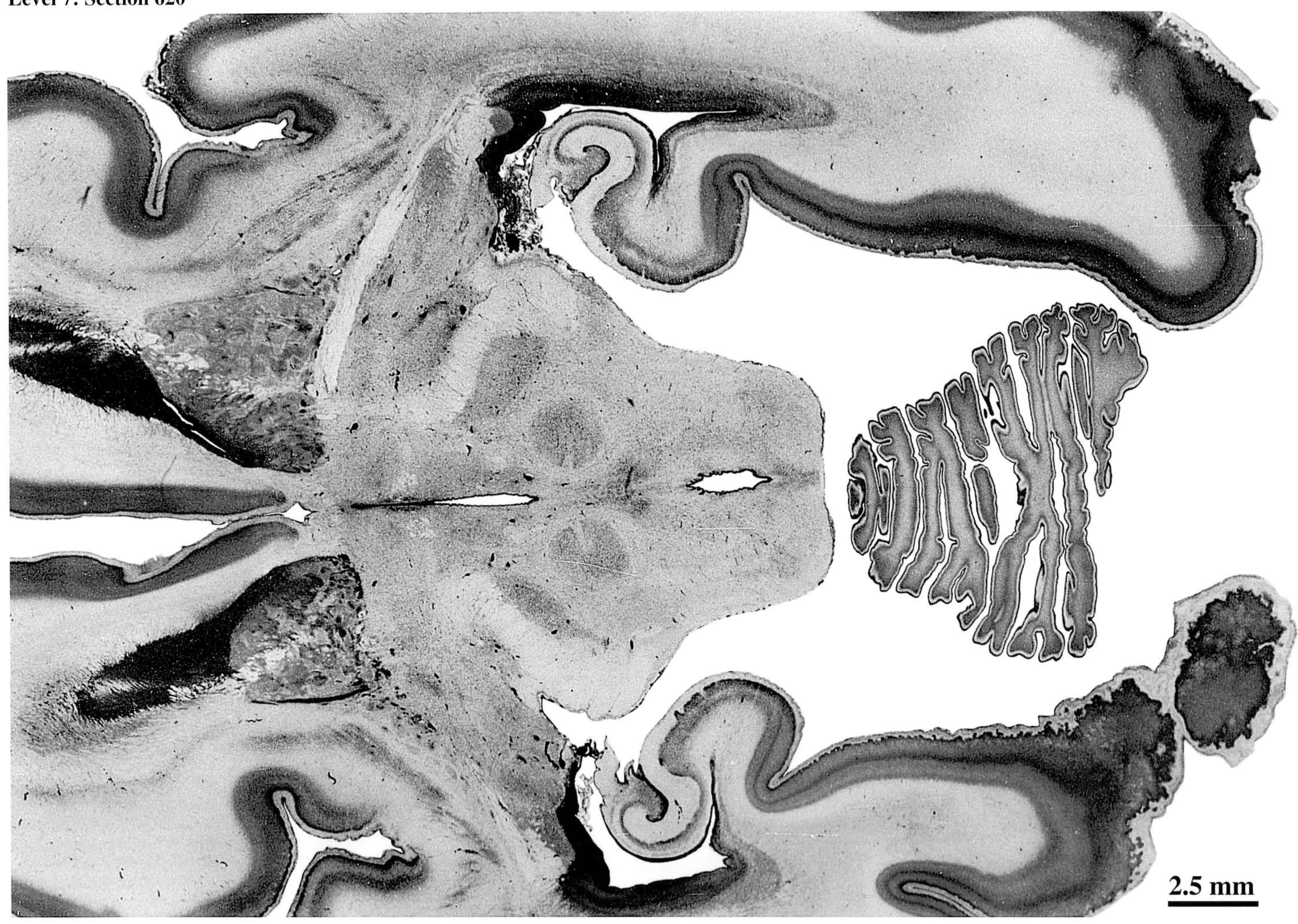

2.5 mm

See the entire section in Plate 222.

# PLATE 234B

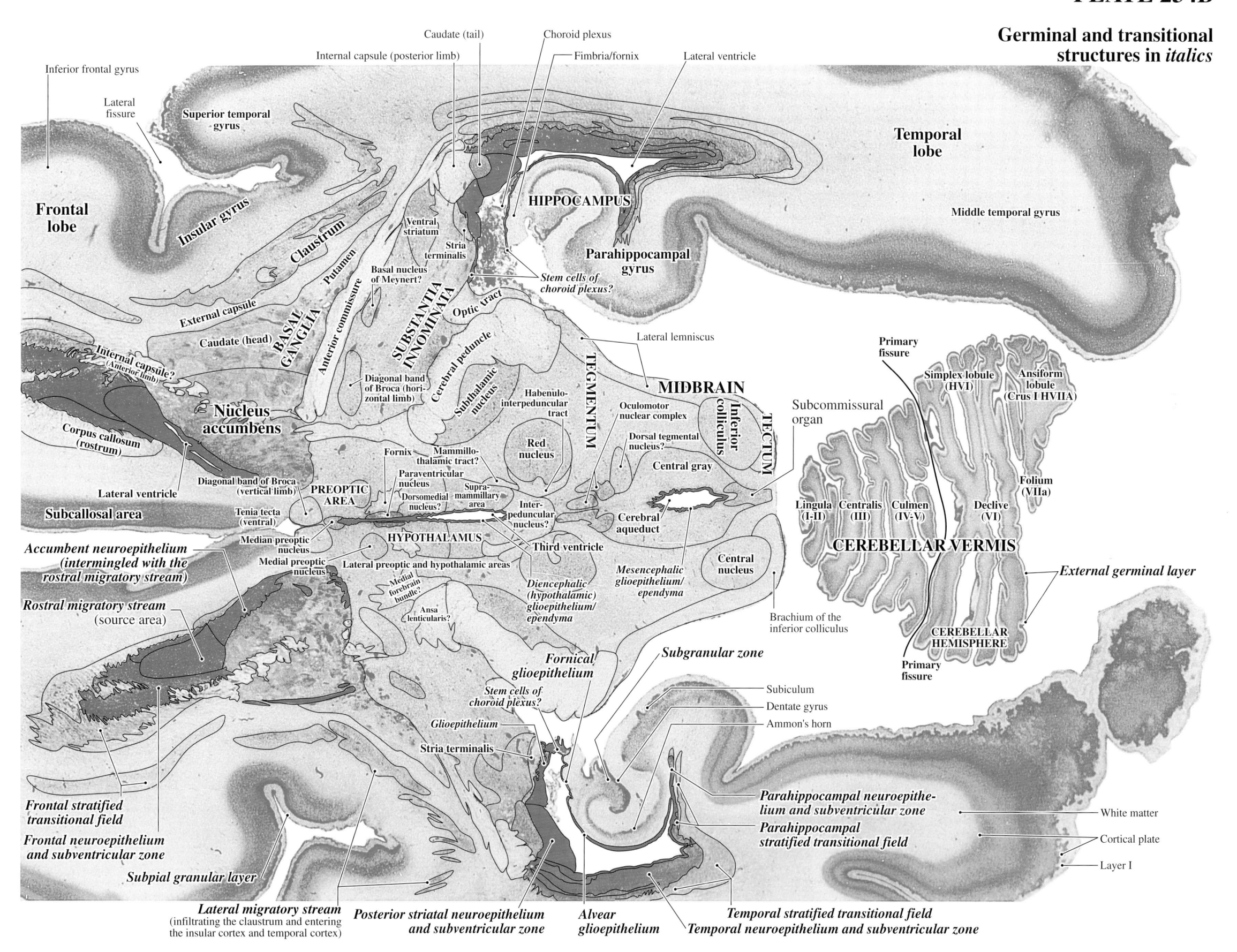

**Germinal and transitional structures in *italics***

## PLATE 235A

**GW26 Horizontal, CR 235 mm, Y16-59**
**Level 8: Section 660**

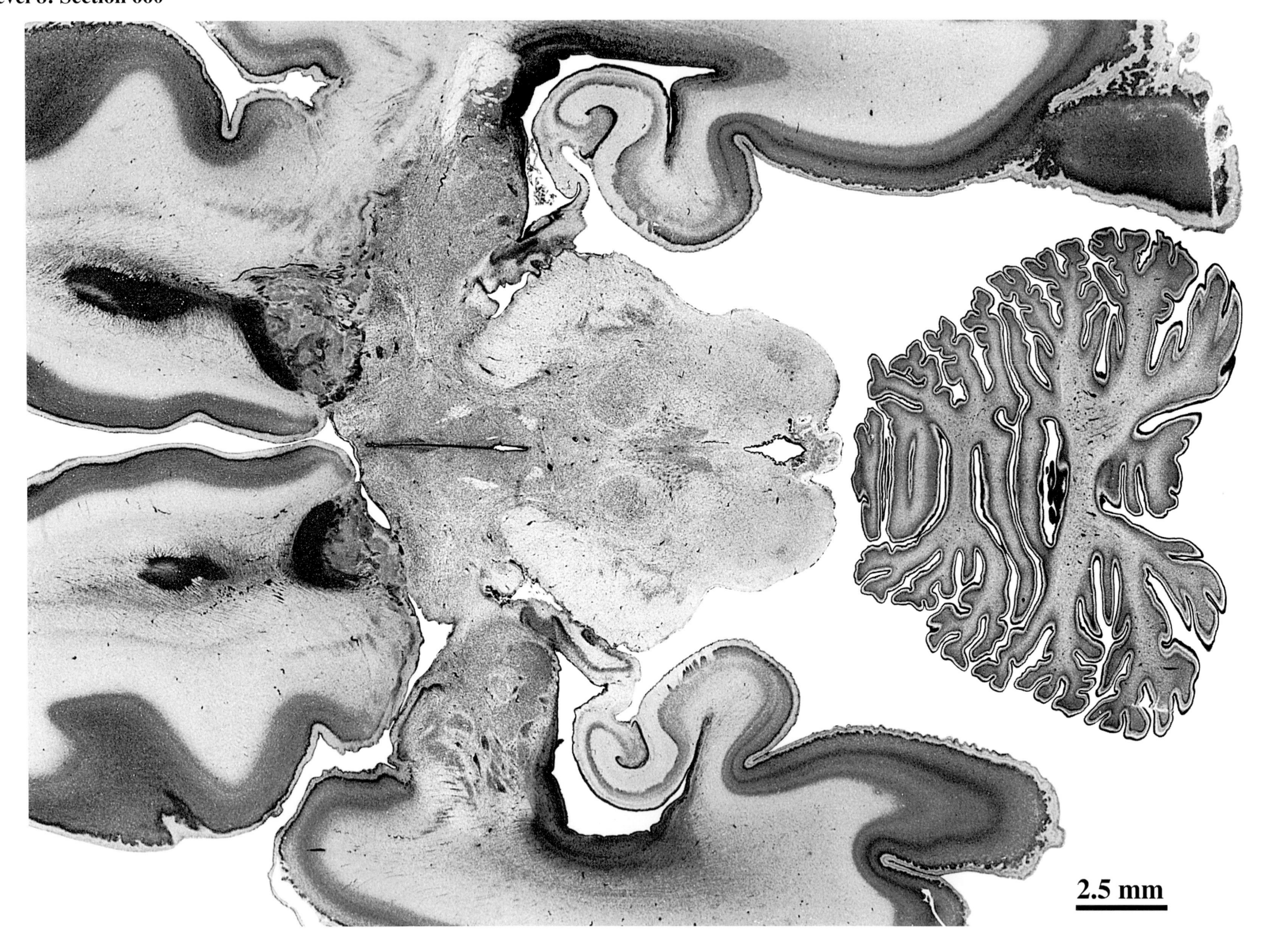

2.5 mm

See the entire section in Plate 223.

## PLATE 235B

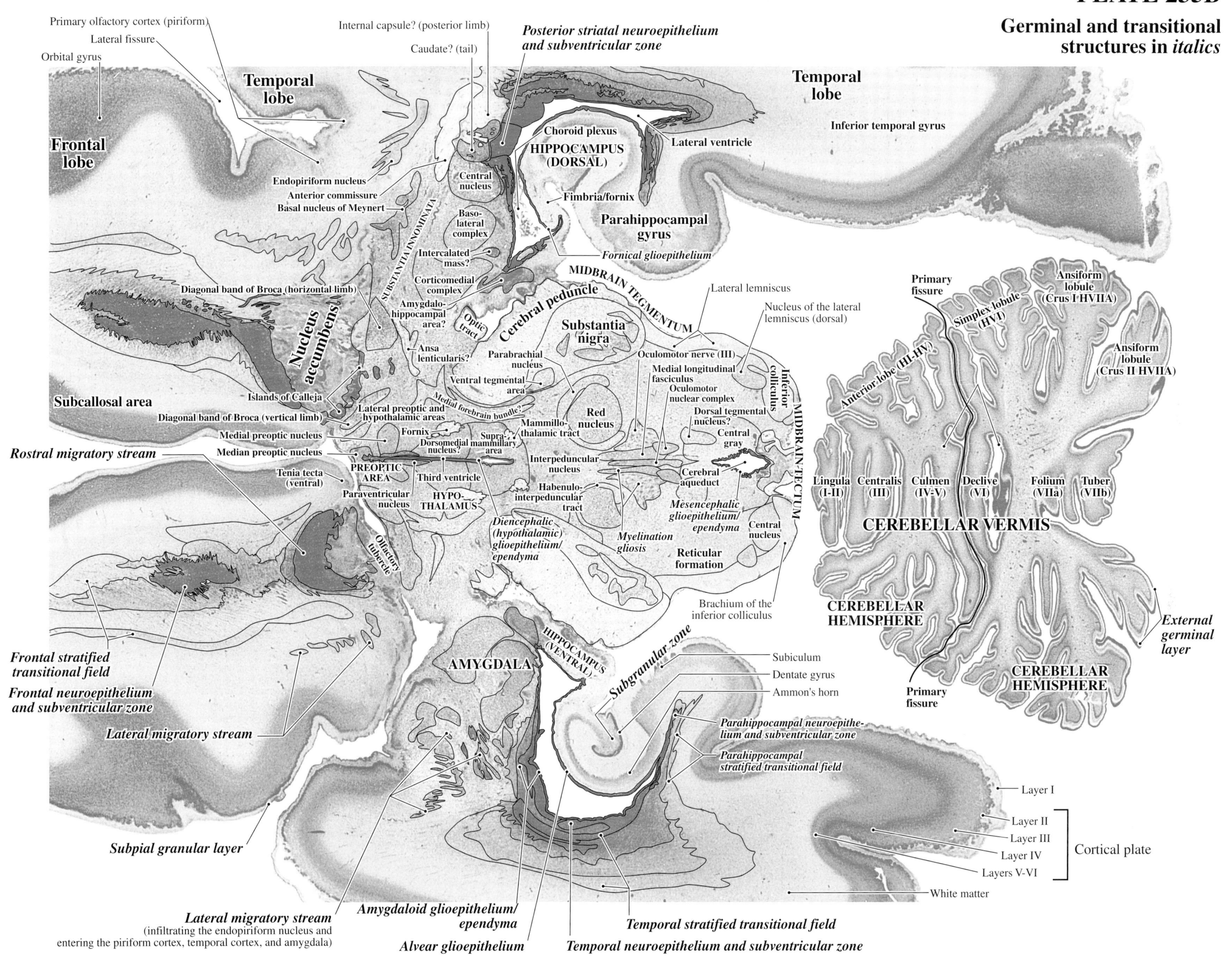

**Germinal and transitional structures in *italics***

## PLATE 236A

GW26 Horizontal, CR 235 mm, Y16-59
Level 9: Section 700

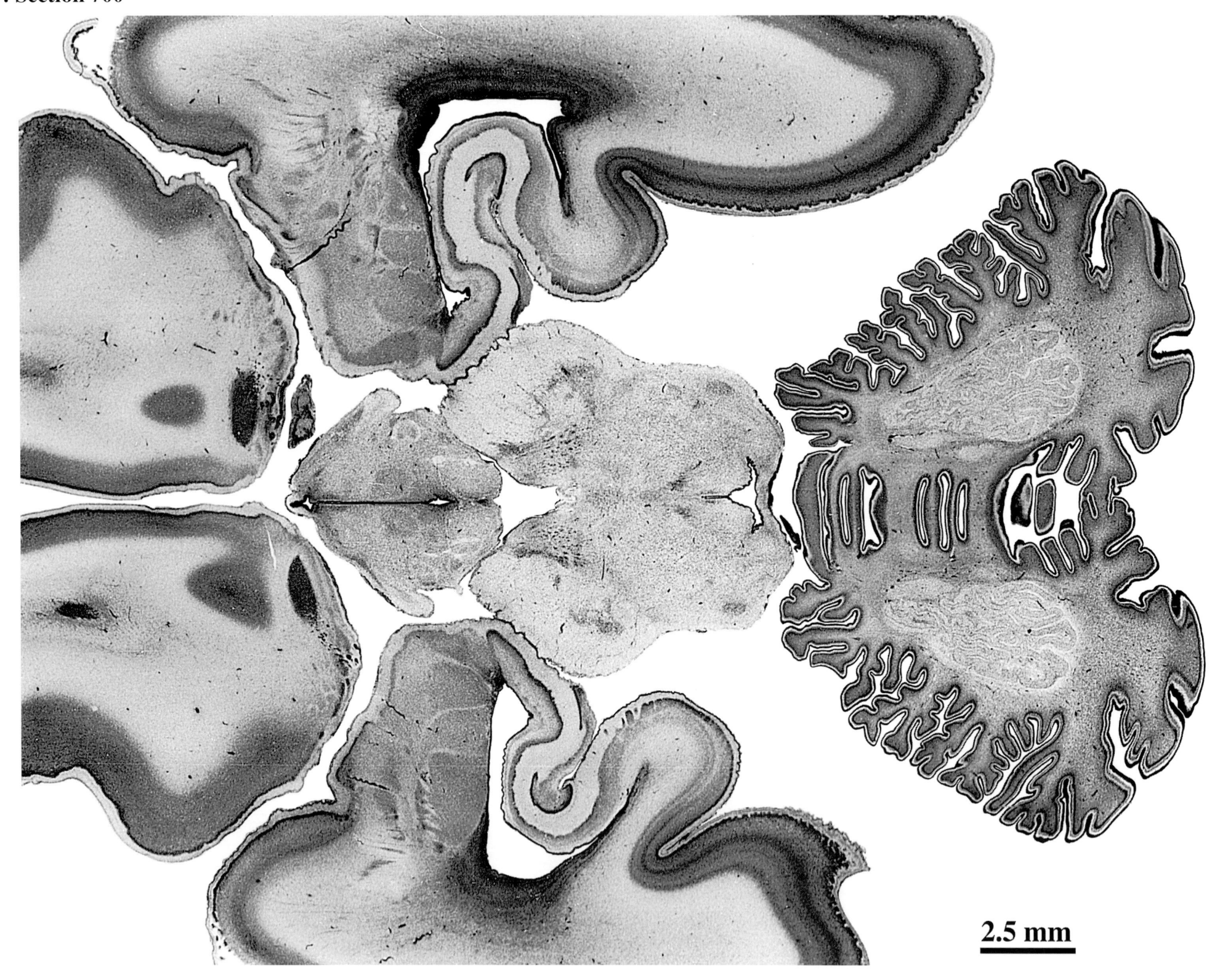

**2.5 mm**

See the entire section in Plate 224.

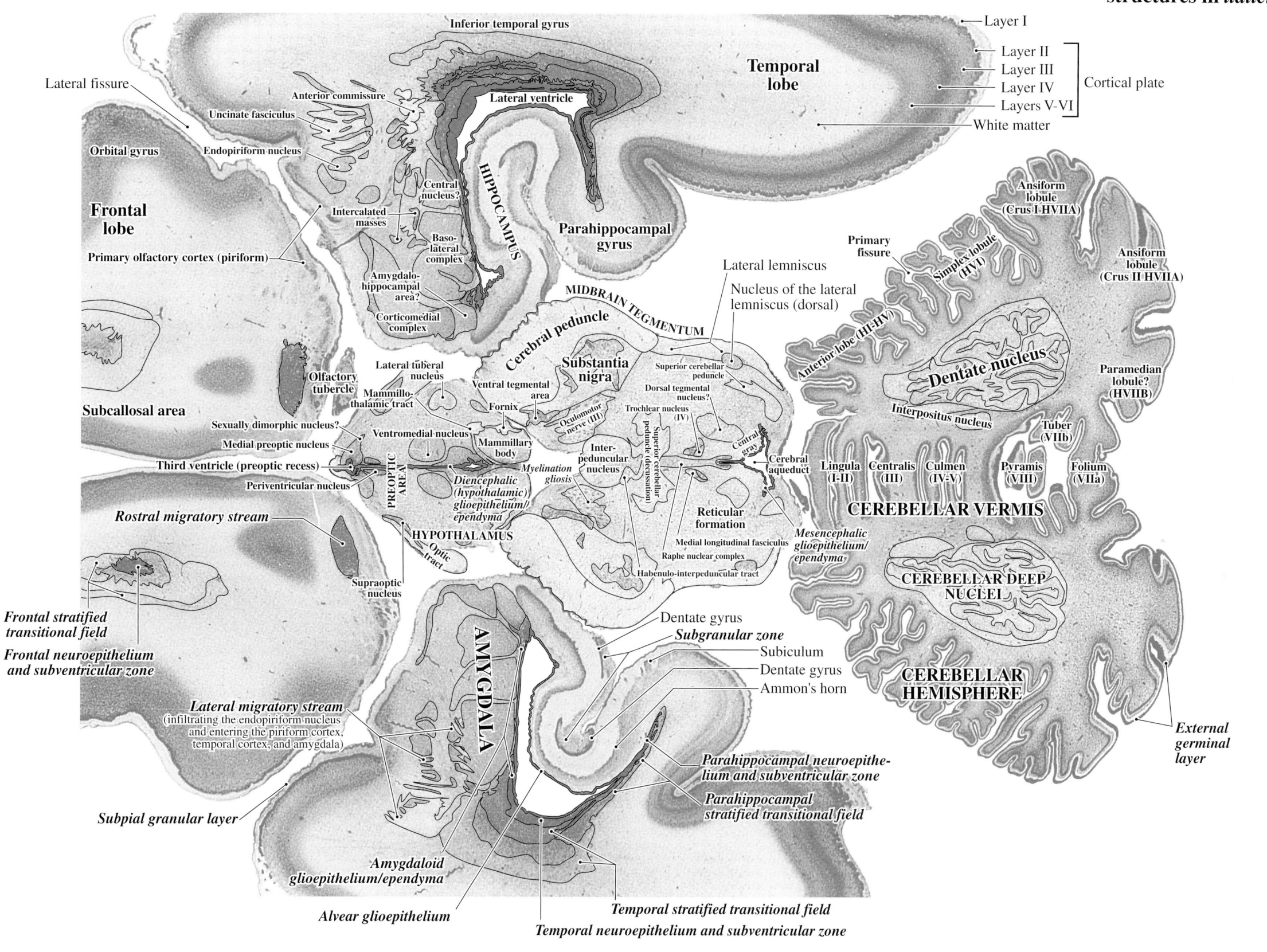

## PLATE 236B

**Germinal and transitional structures in *italics***

## PLATE 237A

**GW26 Horizontal, CR 235 mm, Y16-59**
**Level 10: Section 740**

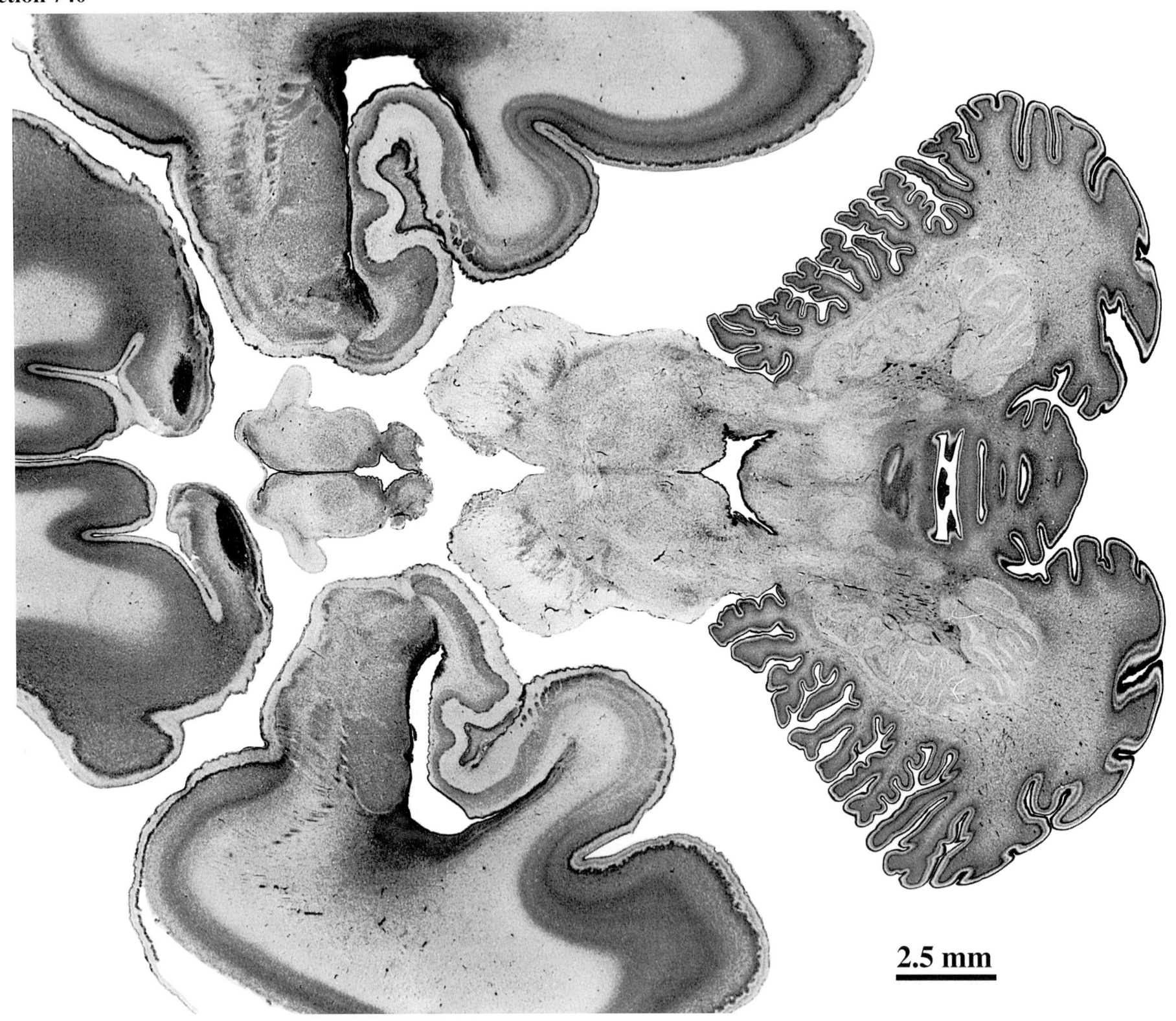

2.5 mm

See the entire section in Plate 225.

## PLATE 237B

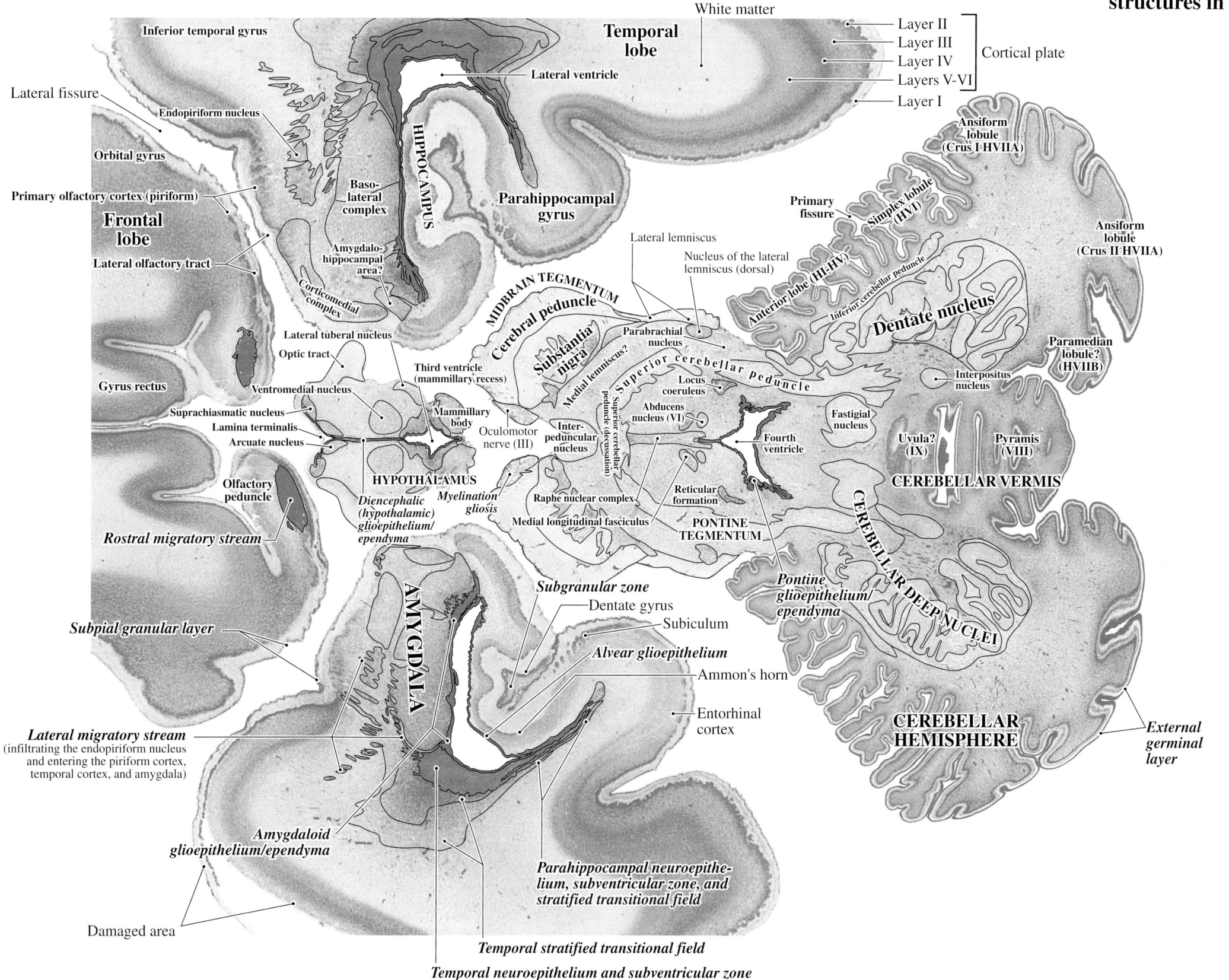

**Germinal and transitional structures in *italics***

354

# PLATE 238A

GW26 Horizontal, CR 235 mm, Y16-59
Level 11: Section 780

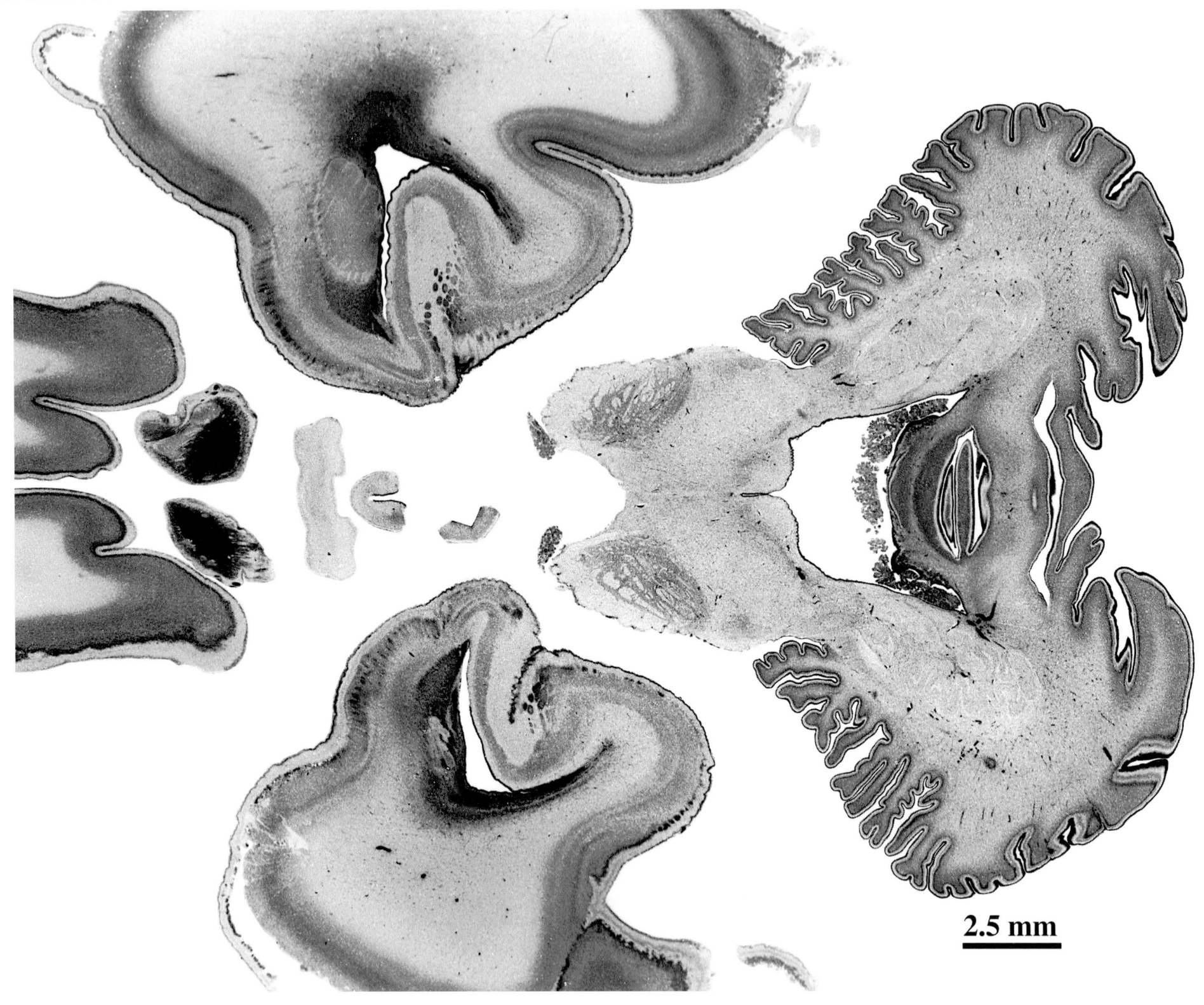

2.5 mm

See the entire section in Plate 226.

# PLATE 238B

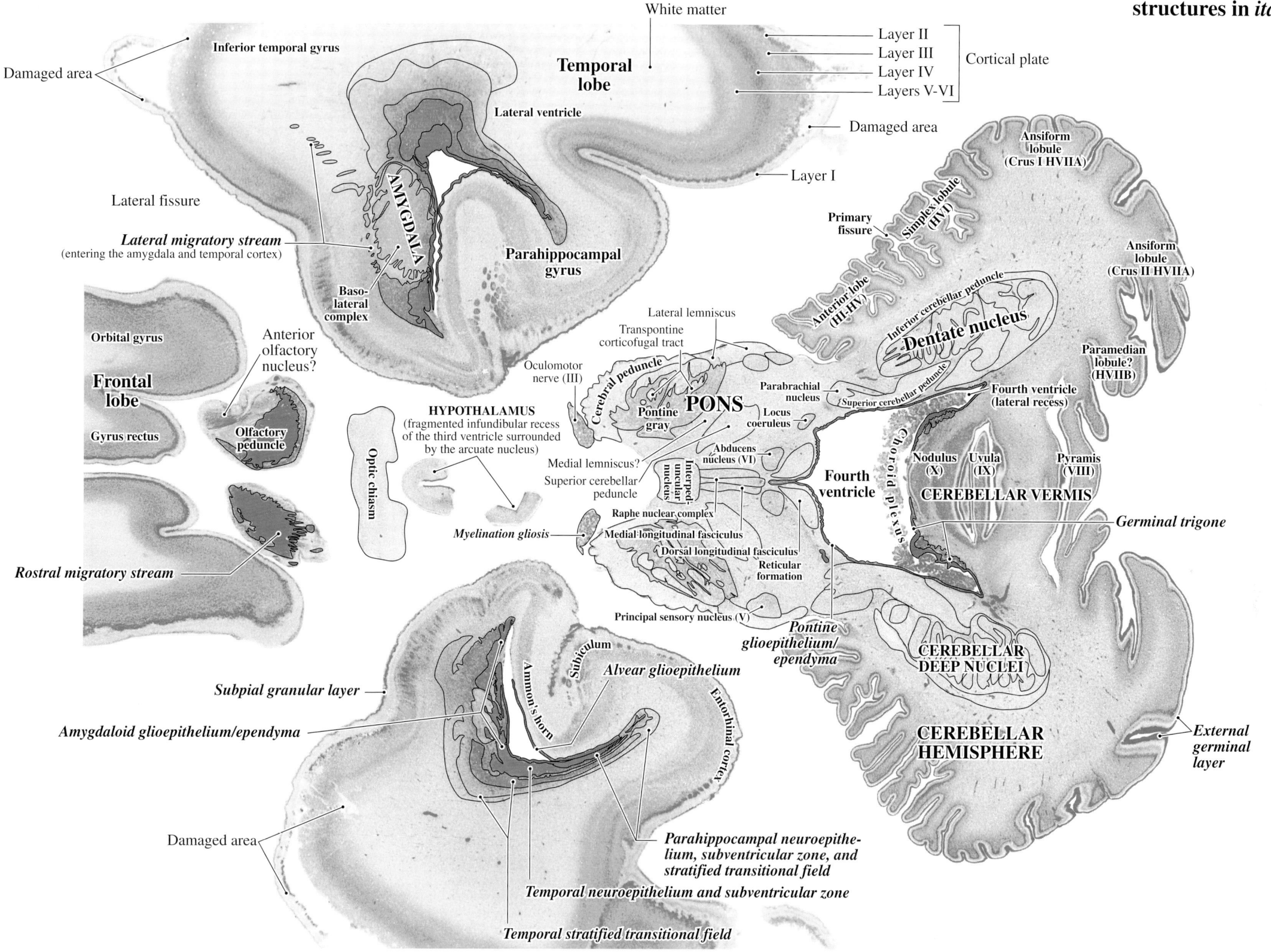

**Germinal and transitional structures in *italics***

## PLATE 239A

GW26 Horizontal
CR 235 mm
Y16-59
Level 12: Section 820

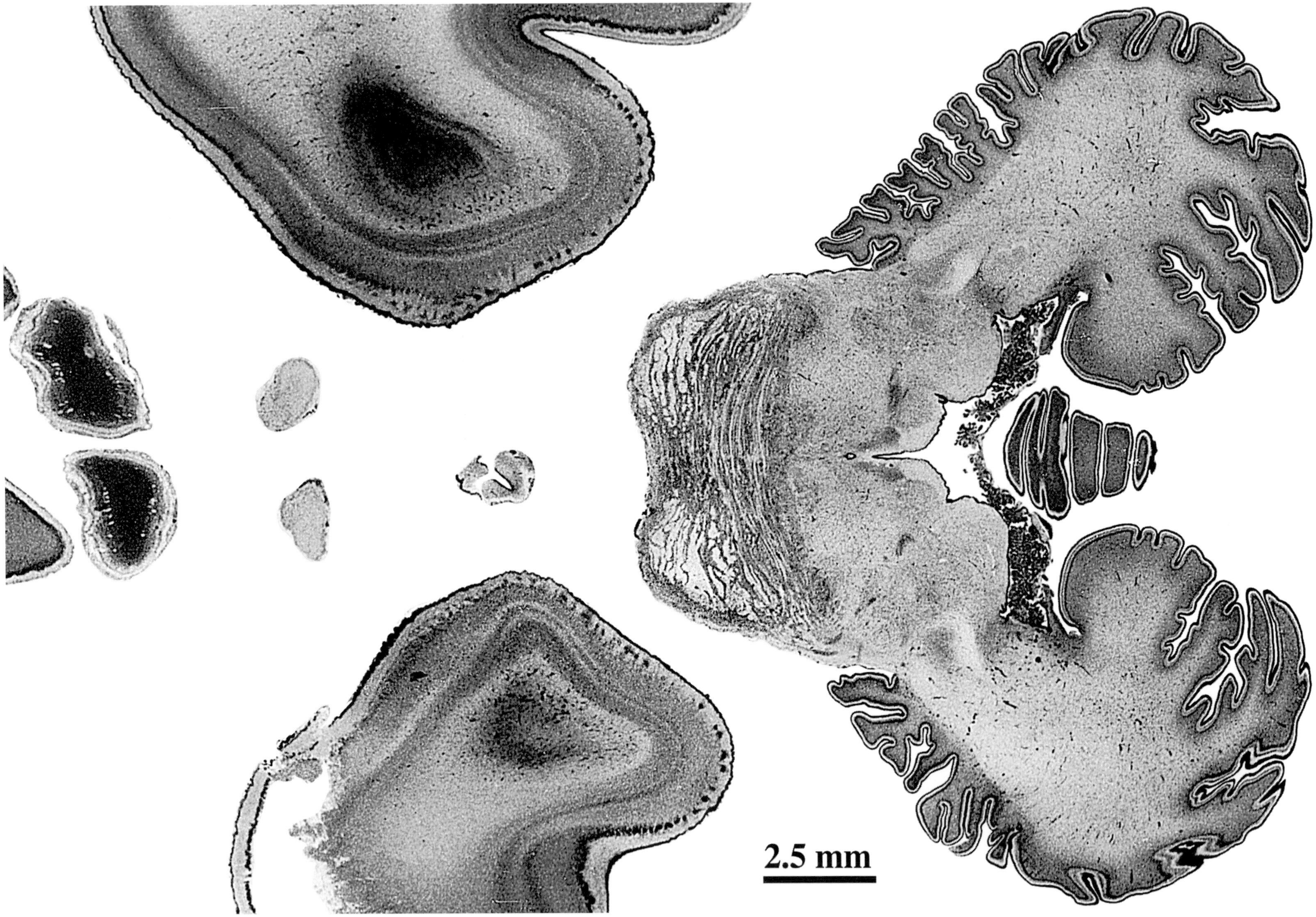

**2.5 mm**

See the entire section in Plate 227.

# PLATE 239B

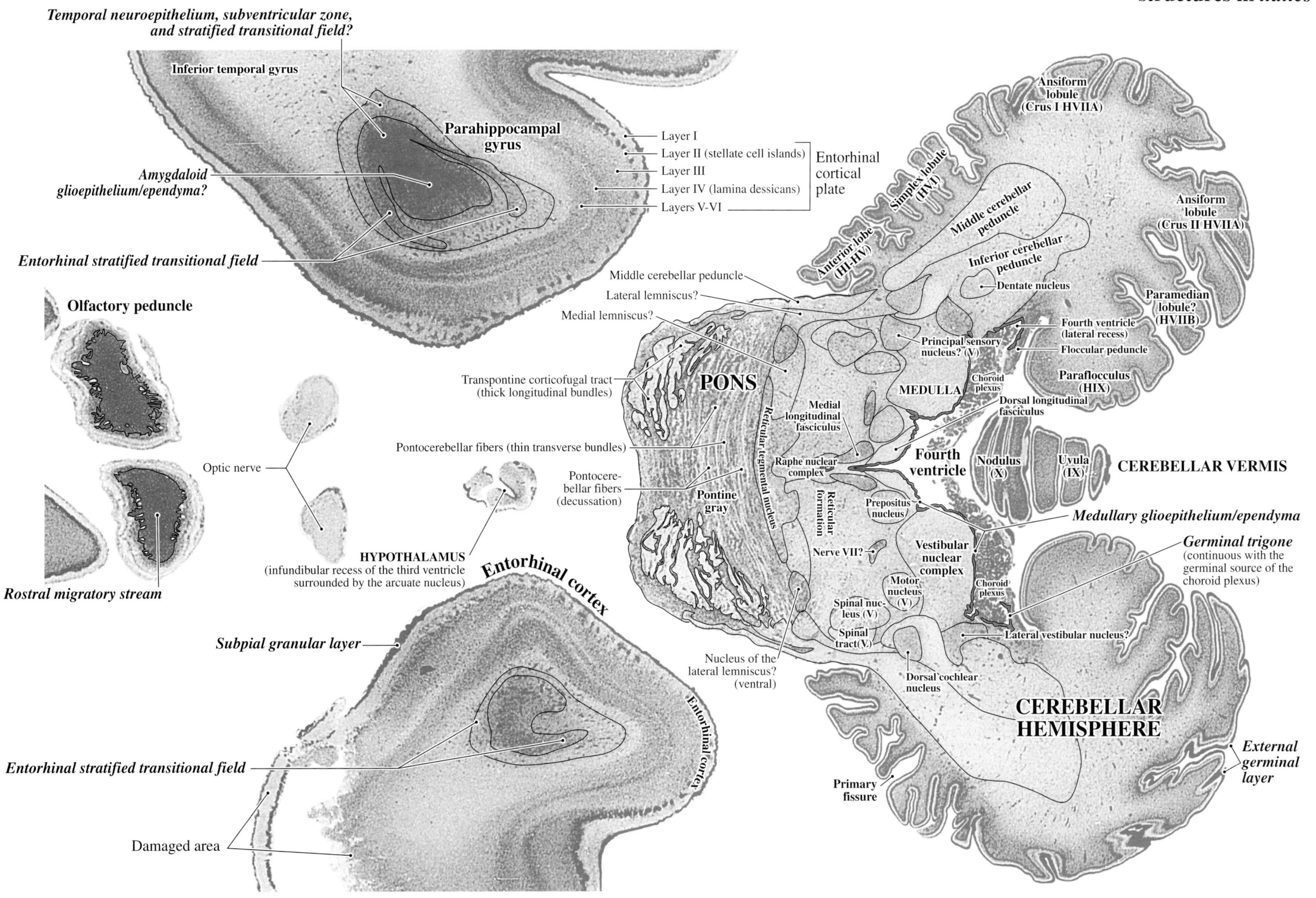

**Germinal and transitional structures in *italics***

*Temporal neuroepithelium, subventricular zone, and stratified transitional field?*

Inferior temporal gyrus

Parahippocampal gyrus

*Amygdaloid glioepithelium/ependyma?*

- Layer I
- Layer II (stellate cell islands)
- Layer III
- Layer IV (lamina dessicans)
- Layers V-VI

Entorhinal cortical plate

*Entorhinal stratified transitional field*

Olfactory peduncle

Optic nerve

*Rostral migratory stream*

HYPOTHALAMUS (infundibular recess of the third ventricle surrounded by the arcuate nucleus)

*Entorhinal cortex*

*Subpial granular layer*

*Entorhinal stratified transitional field*

Damaged area

Middle cerebellar peduncle
Lateral lemniscus?
Medial lemniscus?

Transpontine corticofugal tract (thick longitudinal bundles)

**PONS**

Pontocerebellar fibers (thin transverse bundles)

Pontocerebellar fibers (decussation)

Pontine gray

Raphe nuclear complex

Medial longitudinal fasciculus

Reticular tegmental nucleus

Reticular formation

Nerve VII?

Nucleus of the lateral lemniscus? (ventral)

Spinal nucleus (V)
Spinal tract (V)

Anterior lobe (HI-HV)
Simplex lobule (HVI)
Middle cerebellar peduncle
Inferior cerebellar peduncle
Dentate nucleus

Ansiform lobule (Crus I HVIIA)
Ansiform lobule (Crus II HVIIA)

Principal sensory nucleus? (V)

Fourth ventricle (lateral recess)
Floccular peduncle

Paramedian lobule? (HVIIB)

**MEDULLA**

Choroid plexus

Paraflocculus (HIX)

Dorsal longitudinal fasciculus

Prepositus nucleus

**Fourth ventricle**

Nodulus (X)
Uvula (IX)

**CEREBELLAR VERMIS**

*Medullary glioepithelium/ependyma*

Vestibular nuclear complex

Motor nucleus (V)

Choroid plexus

*Germinal trigone* (continuous with the germinal source of the choroid plexus)

Lateral vestibular nucleus?

Dorsal cochlear nucleus

**CEREBELLAR HEMISPHERE**

Primary fissure

*External germinal layer*

## PLATE 240A

**GW26 Horizontal**
**CR 235 mm**
**Y16-59**
**Level 13: Section 860**

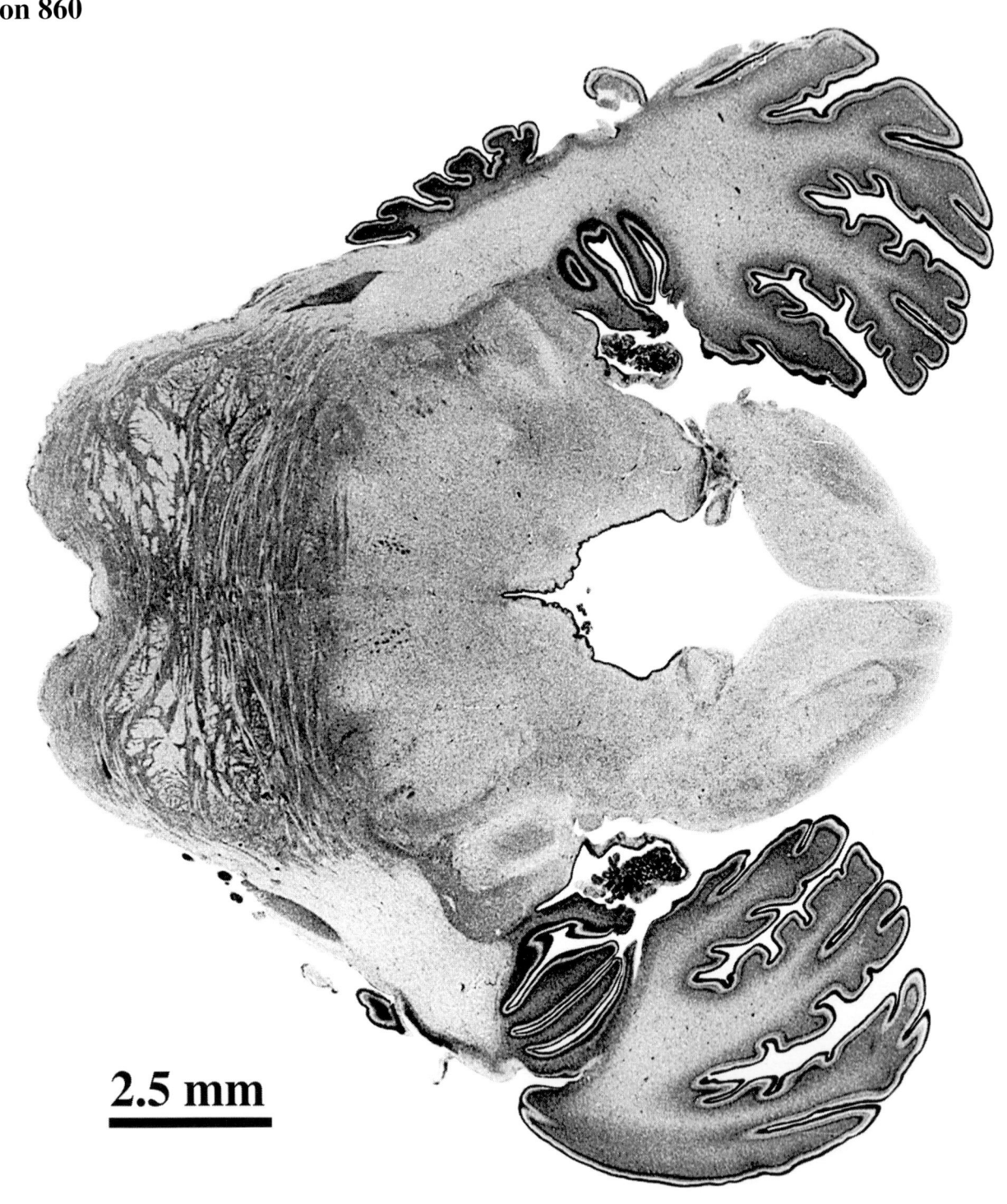

**2.5 mm**

See the entire section in Plate 228 (top).

## PLATE 240B

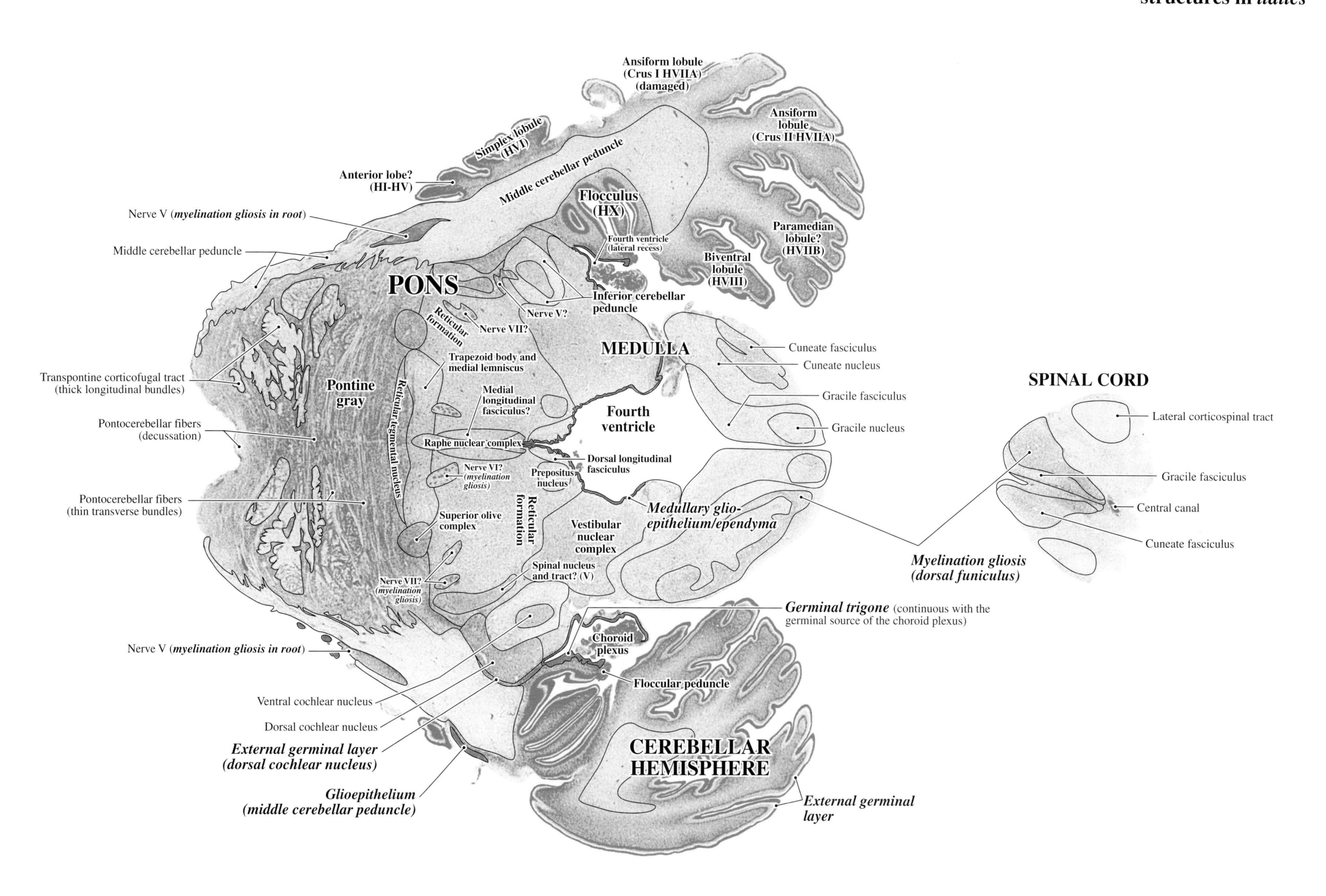

**Germinal and transitional structures in *italics***

**PLATE 241A**

GW26 Horizontal
CR 235 mm
Y16-59
Level 14: Section 880

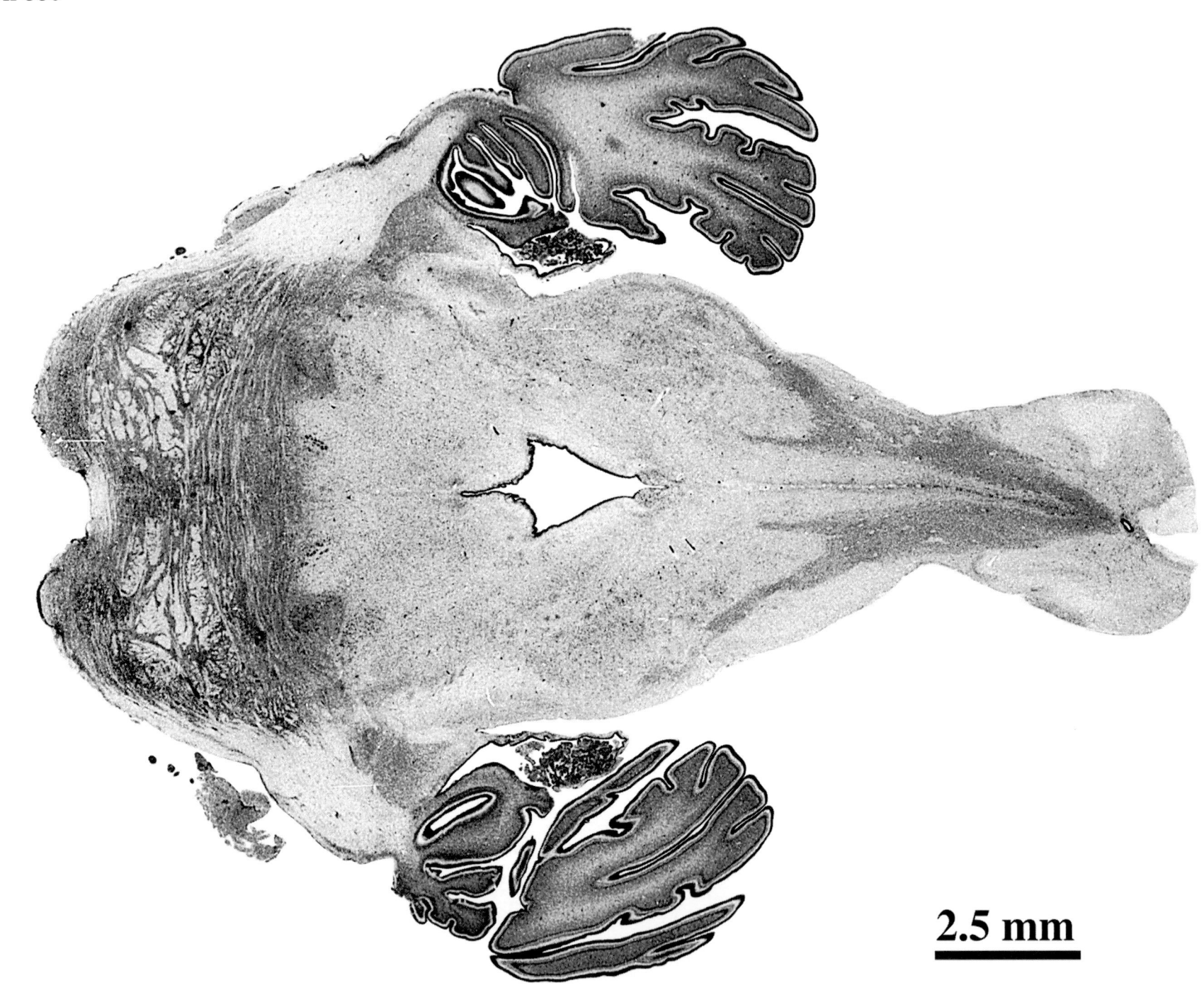

2.5 mm

See the entire section in Plate 228 (bottom).

## PLATE 241B

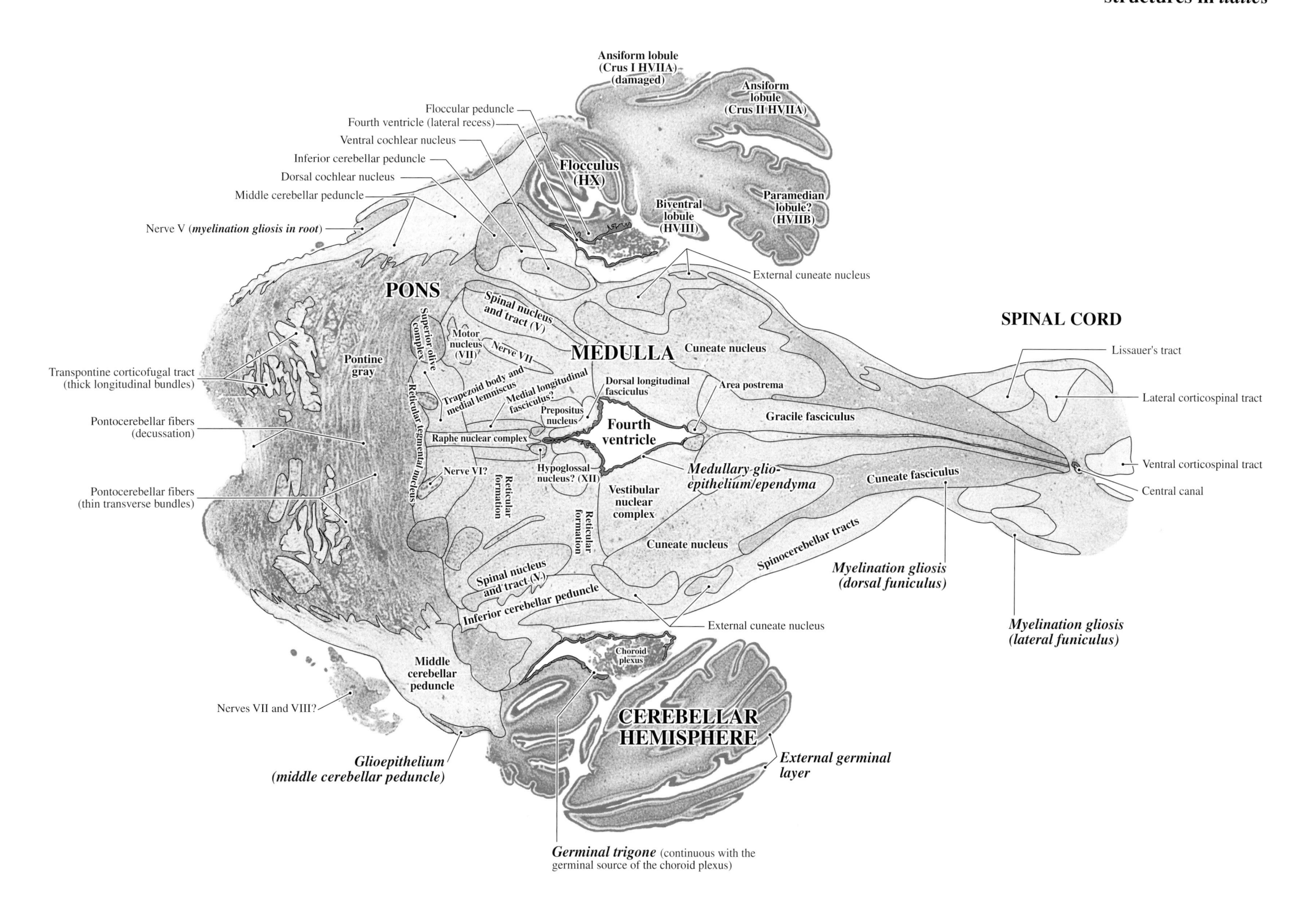

**Germinal and transitional structures in *italics***

**PLATE 242A**

GW26 Horizontal
CR 235 mm
Y16-59
Level 15: Section 900

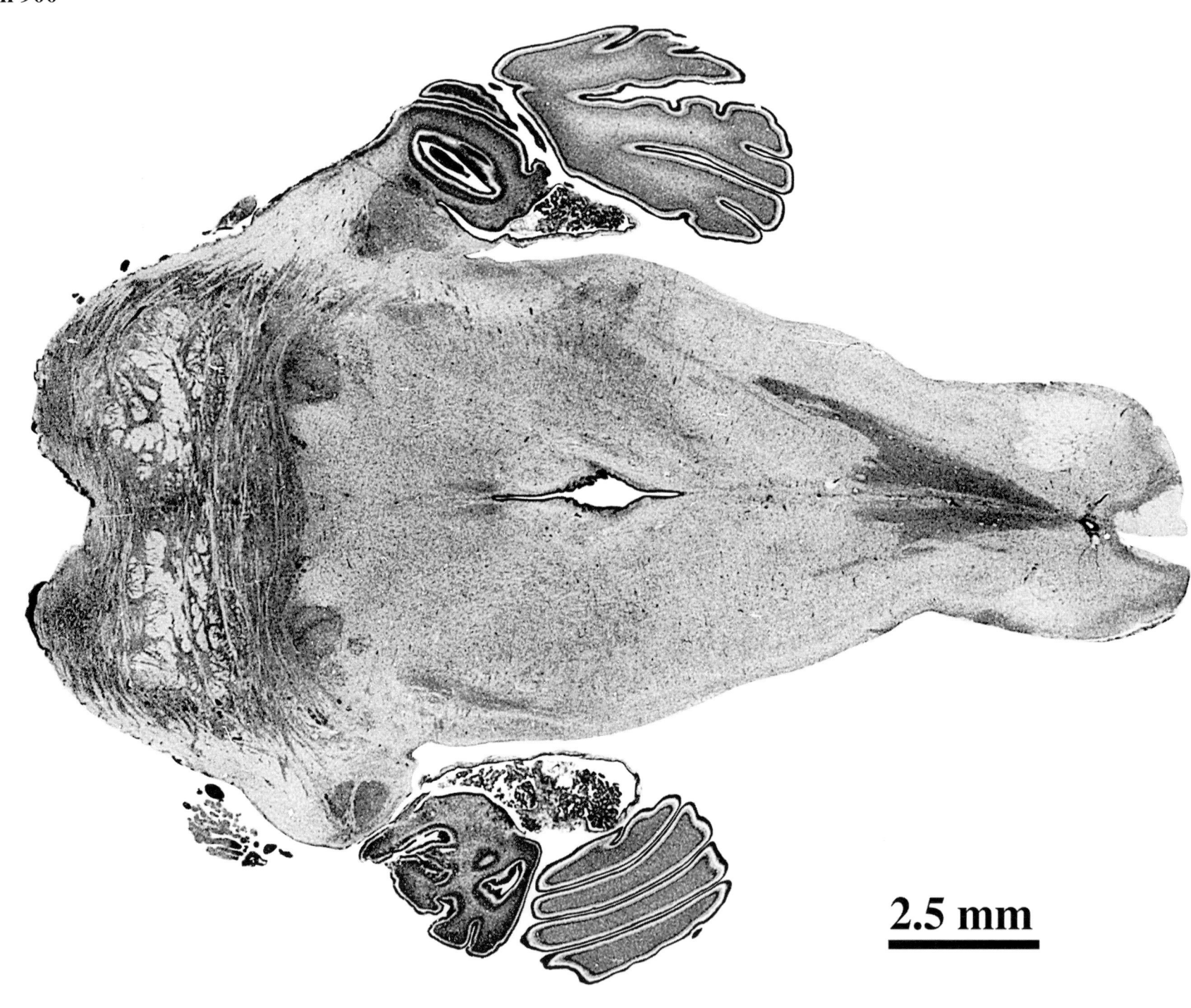

2.5 mm

See the entire section in Plate 229 (top).

### PLATE 242B

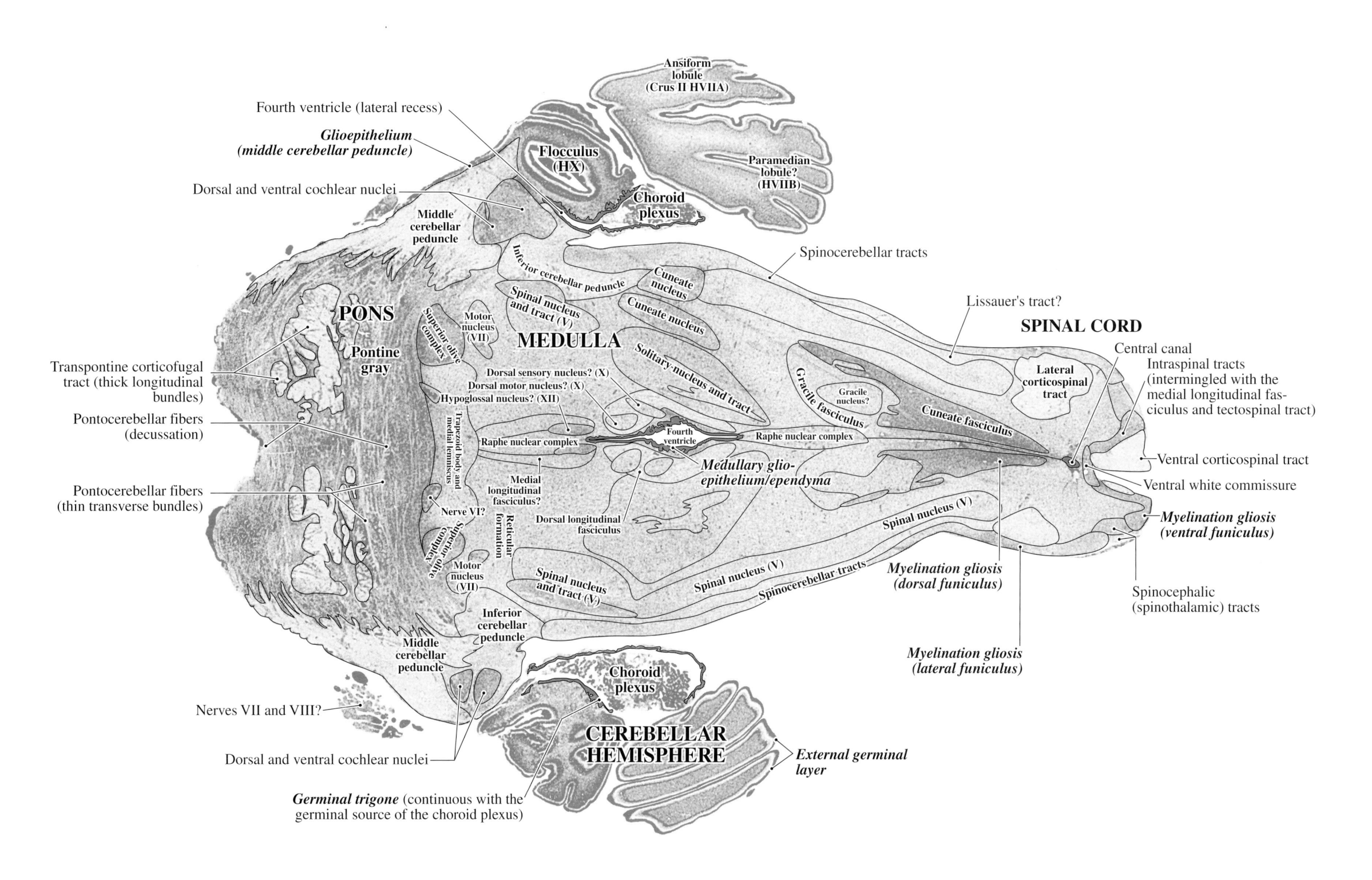

**Germinal and transitional structures in *italics***

## PLATE 243A

**GW26 Horizontal**
CR 235 mm
Y16-59
Level 16: Section 940

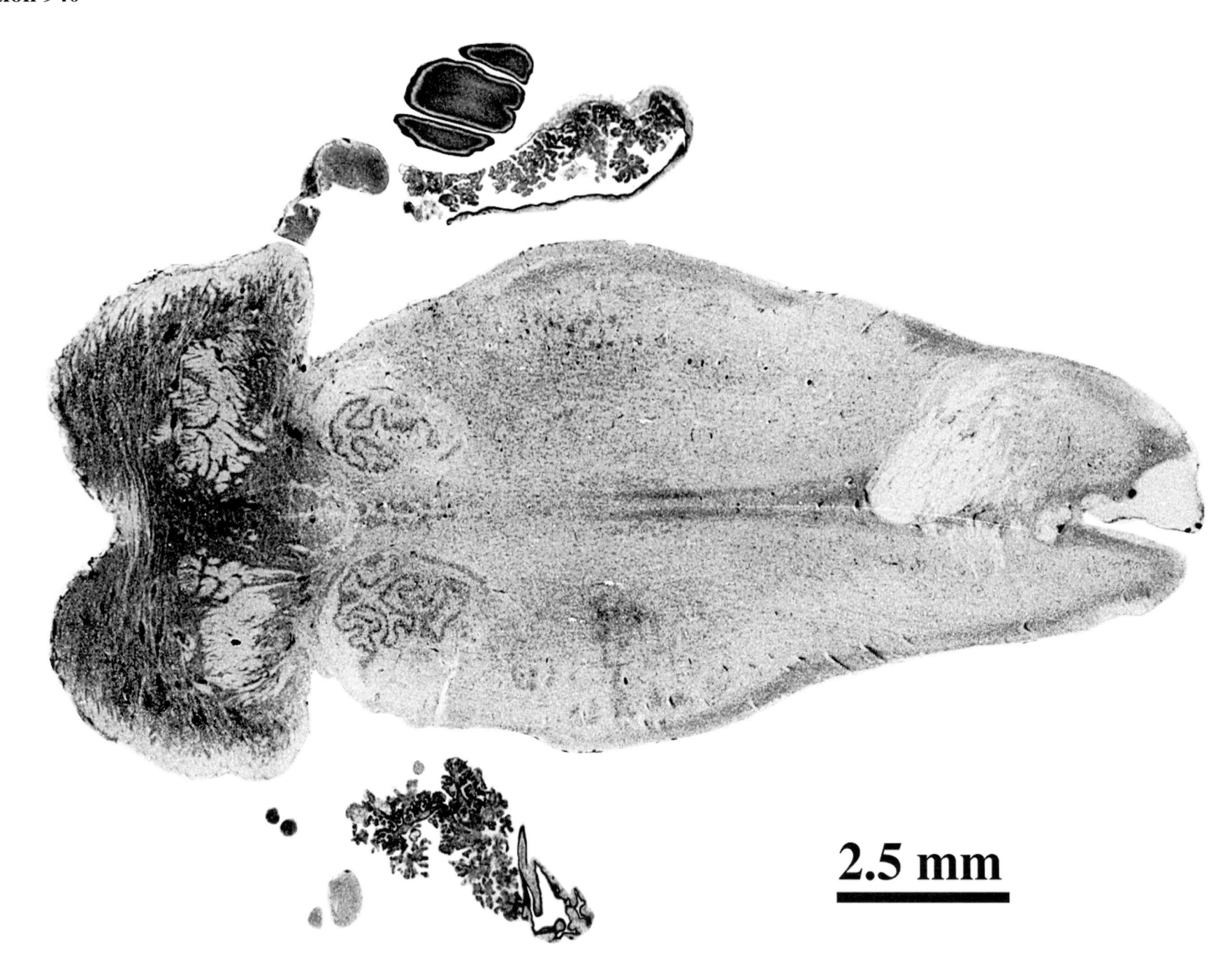

2.5 mm

See the entire section in Plate 229 (middle).

## PLATE 243B

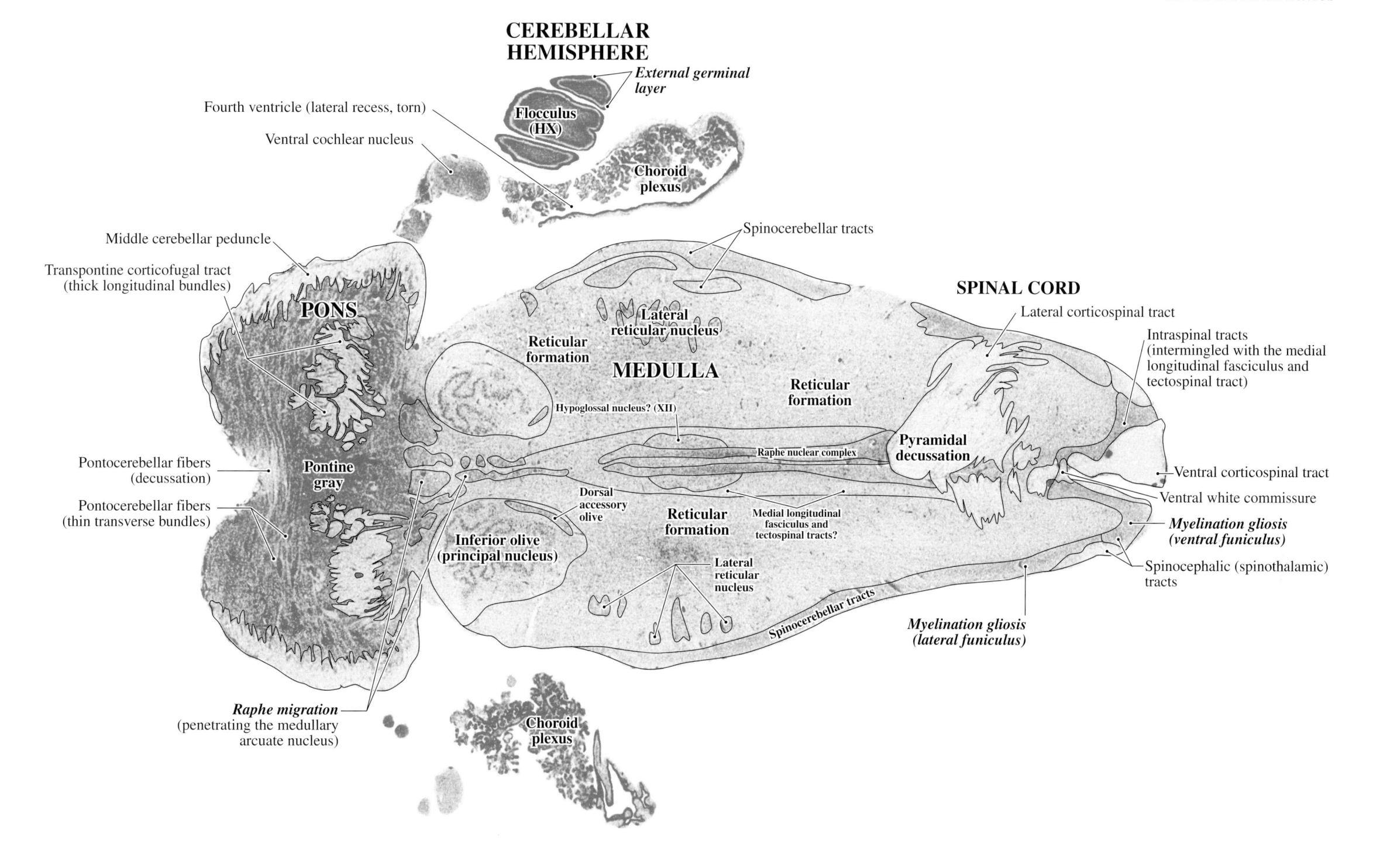

**Germinal and transitional structures in *italics***

*(Labels on the figure:)*

- CEREBELLAR HEMISPHERE
- *External germinal layer*
- Flocculus (HX)
- Fourth ventricle (lateral recess, torn)
- Ventral cochlear nucleus
- Choroid plexus
- Middle cerebellar peduncle
- Spinocerebellar tracts
- SPINAL CORD
- Transpontine corticofugal tract (thick longitudinal bundles)
- PONS
- Lateral reticular nucleus
- Lateral corticospinal tract
- Reticular formation
- MEDULLA
- Intraspinal tracts (intermingled with the medial longitudinal fasciculus and tectospinal tract)
- Hypoglossal nucleus? (XII)
- Reticular formation
- Pontocerebellar fibers (decussation)
- Pontine gray
- Raphe nuclear complex
- Pyramidal decussation
- Pontocerebellar fibers (thin transverse bundles)
- Dorsal accessory olive
- Ventral corticospinal tract
- Ventral white commissure
- Inferior olive (principal nucleus)
- Reticular formation
- Medial longitudinal fasciculus and tectospinal tracts?
- *Myelination gliosis (ventral funiculus)*
- Lateral reticular nucleus
- Spinocephalic (spinothalamic) tracts
- Spinocerebellar tracts
- *Myelination gliosis (lateral funiculus)*
- *Raphe migration* (penetrating the medullary arcuate nucleus)
- Choroid plexus

**PLATE 244A**

GW26 Horizontal
CR 235 mm
Y16-59
Level 17: Section 960

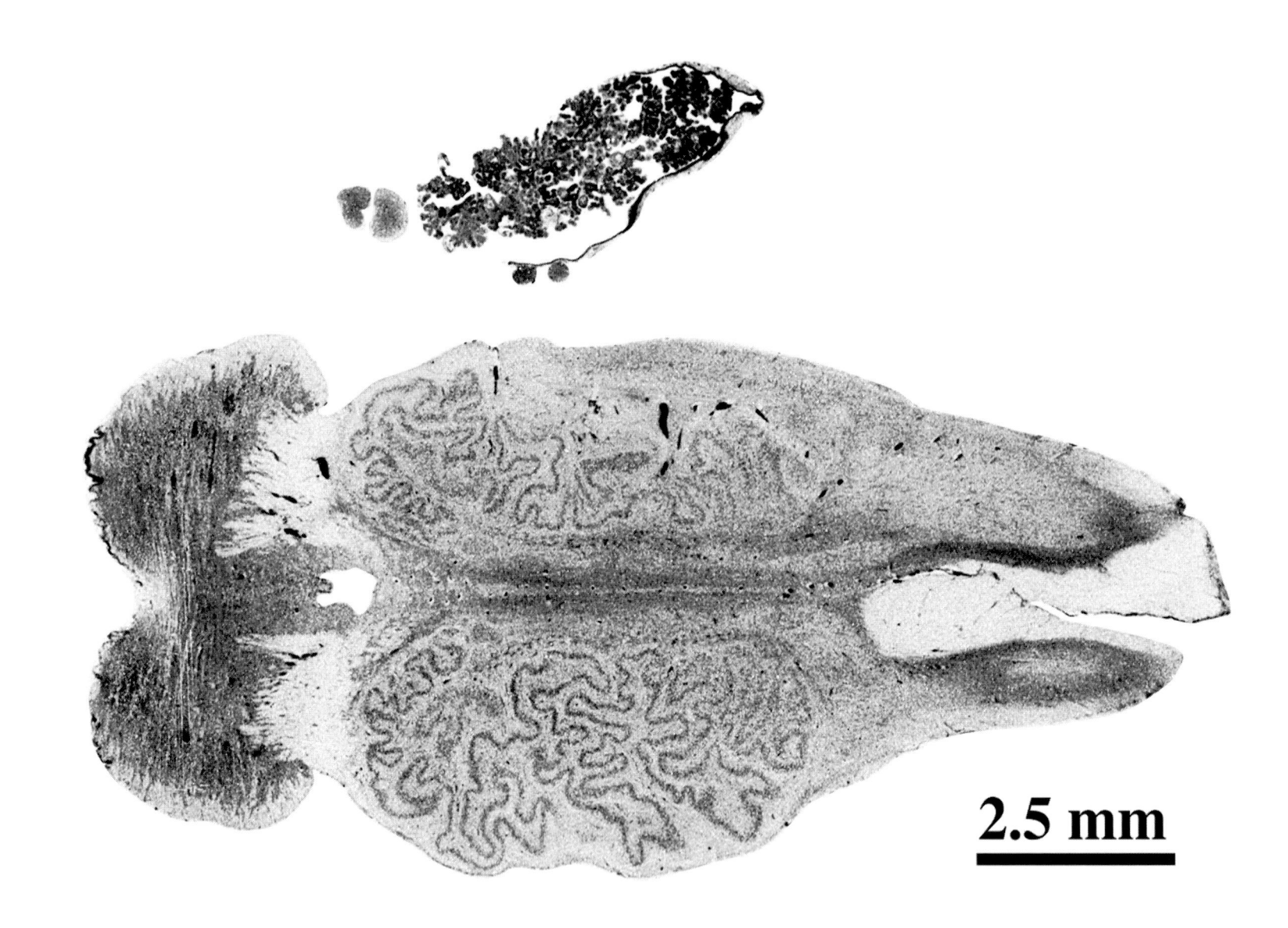

**2.5 mm**

See the entire section in Plate 229 (bottom).

## PLATE 244B

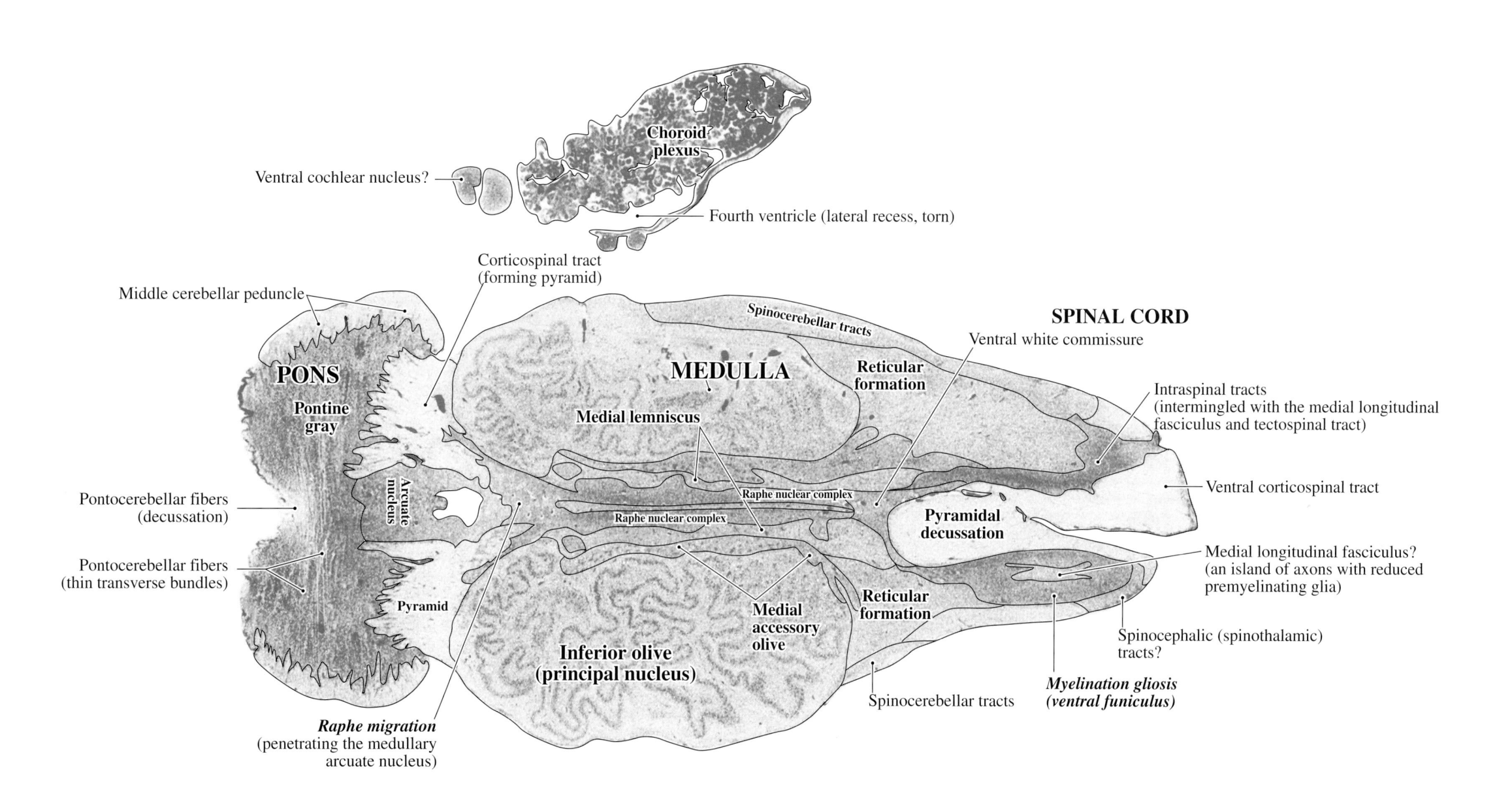

**Germinal and transitional structures in *italics***

# GLOSSARY

An asterisk in front of a term indicates that it has a separate entry in the Glossary with additional information. Terms referring to transient developmental structures are underlined.

## A

**Abducens nucleus (VI)** – An aggregate of cranial nerve motor neurons situated beneath the *fourth ventricle in the *pons. The nucleus receives input from the *vestibular nuclear complex and is the source of motor fibers of cranial *nerve VI that innervate the lateral rectus muscle of the eye.

**Accessory basal nucleus** – *See* **Basal accessory nucleus.**

**Accessory nucleus (XI)** – A column of motoneurons that extends from the region of the *nucleus ambiguus in the medulla to the 5th-6th segments of the cervical spinal cord. Its axons form *nerve XI that innervates the sternocleidomastoid and trapezius muscles.

<u>Accumbent neuroepithelium</u> – Putative germinal source of the neurons and neuroglia of the *nucleus accumbens. After cessation of neurogenesis, this germinal matrix is transformed into a *glioepithelium.

**Allocortex** – Cortical regions of the telencephalon with a "three-layered" cytoarchitectonic organization, such as the *hippocampus and *primary olfactory cortex. At most hippocampal sites, a central layer of densely packed neurons is sandwiched between an external and internal fibrous layer with scattered neurons. The allocortex is a phylogenetically older telencephalic system than the mammalian "six-layered" isocortex, or *neocortex.

<u>Alvear glioepithelium</u> – Putative source of the glia of the *alveus (alveolar path) of the *hippocampus.

**Alveus** – Deep layer of white matter that borders *Ammon's horn region of the hippocampus.

**Ammon's horn (hippocampus)** – Part of the *hippocampus that contains a prominent layer of large pyramidal cells that curves toward the *dentate gyrus. Layers distinguished in Ammon's horn are (from deep to superficial) the alveus, the stratum oriens, the pyramidal layer, the stratum radiatum, and the stratum lacunosum moleculare. Ammon's horn is subdivided into two large areas: CA1 adjacent to the *presubiculum, and CA3 adjacent to the *dentate gyrus. The two smaller regions are CA2 and CA4 (in the *hilus of the dentate gyrus).

**Amygdala** – A large structure in the *uncus of the temporal lobe that is a basal ganglionic component of the limbic system. The complex cell groupings in the amygdala have been put into two major subdivisions: the *corticomedial complex and the *basolateral complex. This structure is involved in linking emotions with olfactory, auditory, and visual sensory information and plays a major role in behavioral aggression.

**Amygdalohippocampal area** – Also called the cortical amygdaloid transition area, an elliptical mass of densely packed cells in the caudomedial amygdala between the ventral *subiculum and the *cortical nucleus. It is connected to the *bed nucleus of the stria terminalis and the *ventromedial hypothalamic nucleus.

**Ansa lenticularis** – Fiber tract that originates in the internal (medial) segment of the *globus pallidus, courses dorsal to the *zona incerta in *Forel's fields, and terminates in the *thalamus, in particular the *ventral complex and the *centromedian nucleus.

**Anterior amygdaloid area** – A region of small to medium-sized cells in the rostral half of the *amygdala that represents a transition zone between the *substantia innominata and the amygdaloid complex proper.

**Anterior commissure** – Large fiber bundle that crosses in the ventral telencephalon and interconnects several forebrain structures on the right and left sides, including the *olfactory bulb, the *primary olfactory cortex, the *entorhinal area, the *amygdala, and some components of the *temporal lobe.

**Anterior complex (thalamus)** – A group of anterior thalamic nuclei with related connections. Components of the anterior thalamic complex are the anterodorsal nucleus, the anteromedial nucleus, and the anteroventral nucleus. The afferents of the anterior complex come principally from the *hippocampal region and the *mammillary body. The ascending efferents terminate in the *cingulate gyrus while the descending efferents terminate in the *mammillary body. The anterior thalamic complex is considered a component of the "limbic system".

**Anterior cortical nucleus (amygdala)** – The anterior part of the large *cortical nucleus that has less dense cellular accumulation in layer II.

**Anterior corticospinal tract** – *See* **Ventral corticospinal tract**.

**Anterior lobe (cerebellum)** – Region of the vermis and hemispheres in front of the primary fissure. It contains the *lingula, *centralis and *culmen in the vermis, and their extensions in the hemispheres.

**Anterior olfactory nucleus** – A nucleus in the olfactory peduncle that has numerous anatomical connections to the *olfactory bulb and *primary olfactory cortex.

**Aqueduct** – Narrow region of the ventricular system situated between the *third ventricle rostrally and the *fourth ventricle caudally.

<u>Aqueduct (embryonic)</u> – During early development, the *neuroepithelium lining the hypertrophied aqueduct of the mesencephalon is the source of neurons and neuroglia of the *superior colliculus, the *inferior colliculus, the *central gray, the *tegmentum, and several components of the *isthmus.

**Arcuate nucleus (hypothalamus)** – A small-celled nucleus that surrounds the base of the third ventricle posteriorly. It contains releasing hormones and is involved in the central nervous regulation of the anterior pituitary gland.

**Arcuate nucleus (medulla)** – A small group of neurons in the ventral part of the *pyramids caudal to the *pontine gray. The chief afferents to this nucleus are from motor areas in the cerebral cortex. This nucleus provides output to the cerebellum. *See also* **Raphe migration**.

# GLOSSARY

**An asterisk in front of a term indicates that it has a separate entry in the Glossary with additional information. Terms referring to transient developmental structures are underlined.**

**Area postrema** – A circumventricular organ forming a dorsal eminence in the *fourth ventricle of the *medulla. It has connections with the hypothalamic *paraventricular nucleus, the *parabrachial nucleus of the pons and the *solitary nucleus of the medulla. It plays a role in the neuroendocrine regulation of feeding and drinking.

**Auditory radiation** – Thalamocortical fibers in the internal capsule and cortical white matter that originate in the *medial geniculate body and terminate in the auditory cortex (transverse gyrus of Heschl) in the *temporal lobe.

# B

**Basal accessory nucleus (amygdala)** – Also called the basomedial nucleus. This part of the *basolateral complex is located between the deep part of the *cortical nucleus and the large *basal nucleus. It contains small to medium-sized cells and some large cholinergic neurons. It is reciprocally connected to the *basal nucleus of Meynert, the rostral *temporal cortex, the prefrontal cortex, and the *orbital cortex; additional input comes from the *posterior complex of the thalamus. Its axons also terminate in the anterior *cingulate cortex, *insular cortex, *entorhinal cortex, *Ammon's horn of the hippocampus, *nucleus accumbens, *dorsomedial nucleus of the thalamus, and *ventromedial nucleus of the hypothalamus.

**Basal ganglia** – A broad term that includes three large ganglionic (subcortical) components of the telencephalon, the *caudate nucleus, the *putamen, and the *globus pallidus. The latter two are also referred to as the *striatum. The basal ganglia have been implicated in the coordination motor functions, and their diseases have been linked to Parkinsonism, Huntington's disease, and other motor abnormalities. The *substantia nigra and the *subthalamic nucleus are part of the basal ganglia circuitry.

**Basal nucleus (amygdala)** – The largest nucleus in the amygdala that forms a major part of the *basolateral complex. It is separated from the *lateral nucleus by a thin fibrous band. It contains small to medium-sized cells and many large cholinergic neurons. It is reciprocally connected to the *basal nucleus of Meynert, the cortex in the*temporal lobe, the *subiculum, and CA1 in *Ammon's horn. Additional input comes from the cortex in the *parahippocampal gyrus. Its axons also terminate in the cortex of the *occipital lobe, medial *frontal lobe, *insular gyrus, and anterior *cingulate gyrus. Additional projections are to the *bed nucleus of the stria terminalis, *nucleus accumbens, *striatum, *dorsomedial nucleus of the thalamus, and *lateral tuberal nucleus of the hypothalamus.

**Basal nucleus of Meynert** – Large-celled component of the *substantia innominata that provides cholinergic input to the *cerebral cortex.

**Basal telencephalon** – A general term sometimes used to refer to such *allocortical components of the telencephalon as the *primary olfactory cortex, and such noncortical regions as the *substantia innominata.

**Basolateral complex** – The largest and best differentiated part of the amygdala in man. It contains the *lateral nucleus, *basal nucleus, and *basal accessory nucleus.

**Bed nucleus of the stria terminalis** – A large subcortical telencephalic field with indistinct boundaries. It is situated medial to the *globus pallidus, lateral to the *septum, and is transected by the *anterior commissure; a thin portion extends back to the *amygdala adjacent to the *stria terminalis. It has its own germinal source, the *strionuclear neuroepithelium and glioepithelium. Only the glioepithelium is present during the third trimester. It receives major input from the *amygdala.

**Biventral lobule (cerebellum HVIII)** – A posterior and inferior hemispheric lobule, and a lateral extension of the vermal *pyramis. *See also* **Cerebellum (Hemisphere)**.

**Brachium of the inferior colliculus** – A fiber tract situated superficially in the fibrous covering of the *inferior colliculus. It is composed of ascending auditory fibers from the inferior colliculus and auditory nuclei in the *pons to the *medial geniculate body.

# C

**Calcarine sulcus** – Cortical fissure that extends from the *parieto-occipital sulcus anteriorly to the occipital pole posteriorly. Along its wall is located the primary visual projection area, the *striate cortex.

<u>**Callosal glioepithelium**</u> – Proliferative germinal matrix that lines the *corpus callosum in the fetal *cerebral cortex. It is the putative source of the myelinating oligodendroglia of the *corpus callosum.

**Callosal sling** – (formerly called glial sling) A thin layer of cells in the midline beneath the corpus callosum. During early development these cells precede the crossing of callosal axons across the midline and have been postulated to provide mechanical scaffolding. It has recently been discovered that these cells are neurons and not glia (Shu, T., Y. Li, A. Keller, and L. J. Richards, 2003, *Development*, 130:2929-2937).

**Caudate nucleus** – Elongated and arched component of the *basal ganglia beneath the *cerebral cortex. It abuts the lateral ventricle and extends from anterodorsal (its "head") to posteroventral (its "tail").

**Cave of the septum** – Midline, membrane-bound triangular space beneath the *corpus callosum and above the *septum. Unlike the cerebrospinal *ventricles, it is without an ependymal lining.

**Central autonomic area (spinal cord)** – Region of the spinal cord that surrounds the *central canal and is implicated in nociceptive and autonomic functions. It may be continuous rostrally with the periaqueductal *central gray.

**Central canal** – Portion of the ventricular system that extends caudally from the medullary *fourth ventricle to the sacral segments of the *spinal cord. During embryonic development, the proliferative *neuroepithelium lining this canal is the source of neurons and neuroglia of the spinal cord. After the cessation of neurogenesis and gliogenesis the shrunken central canal is lined by the *ependyma.

**Central complex (thalamus)** – A group of contiguous central thalamic nuclei, including the *centromedian, central lateral, and *paracentral nuclei.

**Central gray (periaqueductal)** – Oval shaped region in the core of the mesencephalon that surrounds the *aqueduct and is capped by the *superior colliculus and the *inferior colliculus.

**Central lobe (cerebellum VI-VIII)** – The vermal lobe that contains the latest maturing regions of the cerebellar cortex. It includes the *declive, *folium, *tuber, and *pyramis. It is separated from the *anterior lobe by the *primary fissure and from the *posterior lobe by the *posterolateral fissure. The lateral extension of the central lobe forms the main bulk of the hemisphere, including *crus I and *crus II of the ansiform lobule, *the biventral lobule, the *paramedian lobule, and the *simplex lobule.

**Central nucleus (amygdala)** – Part of the *corticomedial complex that is sometimes put in a class by itself. A large nucleus positioned lateral to the *medial nucleus that extends from

# GLOSSARY

**An asterisk in front of a term indicates that it has a separate entry in the Glossary with additional information. Terms referring to transient developmental structures are underlined.**

the *anterior amygdaloid area to the caudal pole of the amygdala where it blends with the *putamen and the tail of the *caudate nucleus. Its major inputs are from the hypothalamic *ventromedial nucleus, *lateral hypothalamic area, *substantia nigra, *ventral tegmental area, *parabrachial nucleus, and *posterior complex of the thalamus. It projects to the *lateral hypothalamic area, midline thalamus, and a wide array of brainstem targets, including the *central gray, *reticular formation (midbrain, pons, medulla), *substantia nigra, *solitary nuclear complex, and *dorsal motor nucleus (X).

**Central nucleus (inferior colliculus)** – Laminated core of the *inferior colliculus where auditory fibers of the *lateral lemniscus terminate in a tonotopic order.

**Central sulcus** – Large vertically oriented neocortical fissure between the *precentral gyrus and the *postcentral gyrus in the *paracentral lobule. The central sulcus divides the motor cortex (precentral gyrus) from the somatosensory cortex (postcentral gyrus).

**Central tegmental tract** – Large fiber bundle with heterogeneous composition that extends from the *red nucleus in the midbrain to the *inferior olive in the medulla.

**Centralis (cerebellum III)** – The middle lobule in the anterior lobe of the cerebellar vermis between the *lingula and the *culmen. Its posterodorsal border is defined by the *preculminate fissure. *See also* **Cerebellum (Vermis)**.

**Centromedian nucleus (thalamus)** – Large spherical structure surrounded by fibers of the internal medullary lamina, classified with the *central complex of the thalamus. It is a paleothalamic structure that has extensive connections with the *striatum and the midbrain *reticular formation.

**Cerebellar cortex** – The highly convoluted and laminated outer shell of the cerebellum. It is composed of five layers, the superficial *external germinal layer (embryonic), the *molecular layer, the *Purkinje cell layer, the *granular layer, and the deep fibrous *medullary layer. The Purkinje cells originate in the cerebellar *neuroepithelium of the fourth ventricle, whereas the basket and stellate cells of the molecular layer, and the granule cells of the granular layer originate in the external germinal layer. The medullary layer contains the afferents and efferents of the cerebellar cortex.

<u>Cerebellar glioepithelium</u> – Situated beneath the *superior cerebellar peduncle in the roof of the *fourth ventricle, this is a remnant of the cerebellar neuroepithelium that is present during earlier stages of brain development. It is possibly the source of glia of some of the cerebellar peduncles and the white matter of the *medullary layer.

**Cerebellum (deep nuclei)** – Three pairs of ganglionic structures in the depth of the cerebellum: the *fastigial nucleus (medial nucleus); the *interpositus nucleus (globose and emboliform, or intermediate nuclei); and the *dentate nucleus (lateral nucleus). The efferent fibers of cerebellar *Purkinje cells synapse with the neurons of the cerebellar deep nuclei which, in turn, are the source of cerebellofugal fibers that terminate in structures outside the cerebellum.

**Cerebellum (hemisphere)** – Portion of the cerebellar cortex situated on either side of the midline vermis. Its principal components are the *simplex lobule, *crus I and II of the ansiform lobule, the *flocculus, and the *paraflocculus.

**Cerebellum (vermis)** – Midline portion of the cerebellar cortex. It is divided by large fissures into 4 lobes with a total of 10 lobules: *anterior lobe (I-V), *central lobe (VI-VIII), *posterior lobe (IX), and *inferior lobe (X).

**Cerebral aqueduct** – *See* **Aqueduct**.

**Cerebral cortex** – The largest structure in the human brain that is composed of a cell sparse *layer I and cellular layers (II-VI) that vary in composition in specific regions of the *neocortex or *allocortex. The neocortex is subdivided into frontal, parietal, occipital, and temporal lobes, and the paracentral lobule. The primordium of cortical layers II-VI is the *cortical plate.

**Cerebral peduncle** – Fiber mass along the ventrolateral aspect of the diencephalon and mesencephalon. The term refers to fibers of the *corticofugal tract between the *internal capsule rostrally and the *transpontine corticofugal tract caudally.

**Choroid plexus** – Highly vascularized and arborized epithelial tissue of mesenchymal origin that secretes the cerebrospinal fluid circulating in the brain ventricles and subarachnoid spaces of the meninges. The *lateral, *third, and *fourth ventricles have their own choroid plexus arbors.

**Cingulate gyrus** – Medial *allocortical region that extends rostrocaudally above the corpus callosum. Its principal afferent connections are with nuclei of the thalamic *anterior complex, the *septum, the *amygdala, and the *frontal lobe. It is considered a component of the "limbic system."

<u>Cingulate neuroepithelium</u> – Putative source of the neurons and glia of the cingulate gyrus. It is flanked by the cingulate *subventricular zone and *stratified transitional field before the sojourning and migrating neurons settle in the cortical plate.

**Cingulum** – Large longitudinal fiber bundle situated above the *corpus callosum that follows the contours of the *cingulate gyrus. It contains efferents of the cingulate gyrus and long association fibers that interconnect it with the *hippocampal region and the *frontal lobe.

**Claustrum** – Subcortical gray matter adjacent to the *insula and separated from it and the *striatum by two fibrous layers, the extreme capsule and the *external capsule. There is some evidence that the claustrum is reciprocally connected with the *neocortex. During embryonic development it is in the path of the *lateral migratory stream.

**Cochlear nucleus (dorsal)** – This auditory nucleus occupies the external surface of the *inferior cerebellar peduncle and forms an eminence in the lateral part of the *fourth ventricle floor. Its neurons get input from primary auditory neurons in the cochlear spiral ganglion, which form the auditory component of cranial *nerve VIII. The dorsal cochlear nucleus sends ipsilateral and contralateral axons to the *lateral lemniscus that terminate in the *nuclei of the lateral lemniscus, the *inferior colliculus, and the *medial geniculate body.

**Cochlear nucleus (ventral)** – This nucleus surrounds the lateral and ventral parts of the *inferior cerebellar peduncle. Its neurons get input from primary auditory neurons in the spiral ganglion. Axons of this nucleus cross the midline in the *trapezoid body and terminate in the contralateral *nuclei of the lateral lemniscus, the *inferior colliculus, and the *medial geniculate body.

**Commissural nucleus (X)** – A small nucleus that is part of the sensory nuclei associated with the vagus nerve and the solitary complex. It lies caudal to the obex in the medulla where the *solitary nuclei on both sides of the brain are joined. *See also* **Nerve X**.

**Corona radiata** – Fan-shaped radiating mass of ascending (thalamocortical) and descending (corticofugal) fibers of the *internal capsule. It is continuous with the *white matter of the *cerebral cortex.

# GLOSSARY

**An asterisk in front of a term indicates that it has a separate entry in the Glossary with additional information. Terms referring to transient developmental structures are underlined.**

**Corpus callosum** – Large commissural fiber system that interconnects the two halves of the cerebral hemispheres. The frontal part of the corpus callosum, the genu, interconnects the anterior part of the frontal lobes. Its "body" interconnects the posterior *frontal lobe, the *paracentral lobule, and the *parietal lobe. The splenium contains the commissural fibers of the *temporal lobe and the *occipital lobe.

**Cortical nucleus (amygdala)** – Also called the periamygdaloid cortex, part of the *corticomedial complex in the superficial amygdala adjacent to the *medial nucleus. It is characterized by a cell-sparse layer I, a dense layer II, and a scattered layer III. It gets input from the *olfactory bulb, *primary olfactory cortex, CA1 in *Ammon's horn, and the *subiculum. It projects to the olfactory bulb, CA1, the *orbital gyrus, and the *dorsomedial nucleus of the thalamus.

<u>Cortical plate</u> – The densely packed cells of layers II-VI in the embryonic and fetal *cerebral cortex. It is situated between the preplate (the future molecular layer, or *layer I) and the subplate (the future layer VII).

**Corticomedial complex (amygdala)** – The part of the *amygdala that includes the *anterior amygdaloid complex, the *cortical nuclei, the *nucleus of the lateral olfactory tract, the *medial nucleus, and the *central nucleus.

**Corticofugal tract** – Collective term for the efferent fiber system that originates in the cerebral cortex and terminates in subcortical structures. It is known by different names as it passes from rostral to caudal: *internal capsule, *cerebral peduncle, *transpontine corticofugal tract, *pyramid, *pyramidal decussation, and *corticospinal tract.

**Corticospinal tract** – Component of the *corticofugal tract that targets the spinal cord. The bulk of the corticospinal fibers cross in the *pyramid of the medulla and descend from cervical to sacral levels in the *lateral corticospinal tract. A smaller component descends ipsilaterally in the *ventral corticospinal tract.

**Crus I, ansiform lobule (cerebellum HVIIA)** – The large anterior part of the ansiform lobule. It is a lateral extension of the *folium in the vermis.

**Crus II, ansiform lobule (cerebellum HVIIA)** – The smaller posterior part of the ansiform lobule that is a lateral extension of the *tuber in the vermis.

**Culmen (cerebellum IV-V)** – The largest lobule in the *anterior lobe of the cerebellar *vermis. It is separated from the *centralis by the *preculminate fissure and from the *declive in the *central lobe by the *primary fissure. The anterior lobe in the cerebellar hemispheres is mainly a lateral extension of the culmen.

**Cuneate fasciculus** – A large fiber tract in the dorsolateral spinal cord and caudal medulla. It is composed of primary sensory fibers of dorsal root ganglion cells that terminate topographically in the *cuneate nucleus. Nissl-stained sections of this fiber tract show prominent *myelination gliosis during the early third trimester.

**Cuneate nucleus** – Neurons that invade the *cuneate fasciculus from the ventral aspect to form a large mass in the posterolateral *medulla. Its chief input is from the *cuneate fasciculus and its output fibers cross the midline and enter the contralateral *medial lemniscus.

**Cuneus** – Wedge-shaped region of the medial *occipital lobe situated above the *calcarine sulcus and behind the *parietooccipital sulcus.

# D

**Declive (cerebellum VI)** – The most anterior lobule of the vermal *central lobe. Its anterior border is defined by the *primary fissure. The declive is continuous with the *simplex lobule in the hemisphere.

**Dentate granular layer** – *See* **Granular layer (dentate gyrus)**.

**Dentate gyrus** – Curved small-celled component of the *hippocampus, interlocked with the large-celled *Ammon's horn. It has an outer *molecular layer, a discrete *granular layer, and a *hilus that contains the *subgranular zone, a secondary germinal matrix.

**Dentate nucleus (cerebellum)** – Lobulated and largest of the *cerebellar deep nuclei, also known as the lateral cerebellar nucleus. Situated in the core of the cerebellar hemispheres, the dentate nucleus is the principal source of efferent fibers of the *superior cerebellar peduncle.

**Diagonal band of Broca** – Oblique nucleus situated ventral to the medial *septum. It is subdivided into a vertical limb dorsally and a horizontal limb ventrally.

<u>Diencephalic neuroepithelium</u> – The germinal matrix lining the embryonic *third ventricle. It is the source of neurons and glia of all components of the diencephalon. Its different subdivisions produce cells for the different regions or nuclei of the *thalamus, *hypothalamus, and *preoptic area.

**Dorsal accessory olive** – A dense band of neurons that arch over the principal nucleus in the *inferior olive of the *medulla. Its chief input is proprioceptive afferents from the *spinal cord, its efferents target the contralateral cerebellar *vermis, via the *inferior cerebellar peduncle.

**Dorsal complex (thalamus)** – Collective term for two structurally and functionally related dorsally situated thalamic regions, the *dorsomedial nucleus and the dorsolateral nucleus.

**Dorsal gray matter (spinal cord)** – Wing-shaped region of the spinal gray matter, the target of dorsal root afferents or their collaterals. Its principal component is the small-celled substantia gelatinosa. The neurons of the dorsal gray matter originate in the *neuroepithelium flanking the transient dorsal spinal canal.

**Dorsal longitudinal fasciculus** – A small fiber tract in the dorsomedial *medulla that extends to the ventromedial *central gray in the *midbrain. It relays visceral sensory and motor messages upstream and downstream between the medulla and the midbrain.

**Dorsal motor nucleus (X)** – A column of parasympathetic preganglionic motor neurons dorsolateral to the *hypoglossal nucleus. Their axons leave the brain in cranial *nerve X and terminate in the intramural parasympathetic ganglia supplying the viscera of the thoracic, pericardial, and abdominal cavities. *See also* **Nerve X**.

**Dorsal sensory nucleus (X)** – A medial nucleus in the *solitary nuclear complex of the *medulla that lies dorsolateral to the *dorsal motor nucleus (X) and is continuous with the *commissural nucleus (X). *See also* **Nerve X**.

**Dorsal tegmental nucleus** – Situated in the central gray dorsal to the *trochlear nucleus and extending caudally into the pons. It is targeted by fibers of the *mammillotegmental tract.

**Dorsal white matter (spinal cord)** – Medial fibrous component of the white matter situated between the wings of the *dorsal gray matter; also known as the dorsal column or the dorsal funiculus. It contains ascending somatosensory and proprioceptive fibers that terminate in the dorsal column nuclei of the medulla. In the upper spinal cord it has two distin-

# GLOSSARY

**An asterisk in front of a term indicates that it has a separate entry in the Glossary with additional information. Terms referring to transient developmental structures are underlined.**

guishable parts, the *gracile fasciculus medially, and the *cuneate fasciculus laterally.

**Dorsomedial nucleus (hypothalamus)** – Area situated above the more distinct *ventromedial nucleus of the hypothalamus. Its principal connections are with the *bed nucleus of the stria terminalis and the *septum.

**Dorsomedial nucleus (thalamus)** – Also known as the medial dorsal nucleus, this component of the *dorsal complex is situated between the internal medullary lamina and the periventricular gray, and has a large-celled and small-celled division. Its principal connections are with the *amygdala, the *hypothalamus, the *olfactory tubercle, and the *orbital gyrus. It is considered a paleothalamic component of the limbic system rather than as a neothalamic relay nucleus to the neocortex.

## E

**Endopiriform nucleus** – Small telencephalic nucleus deep to the *primary olfactory cortex and ventral to the *claustrum. The *lateral migratory stream percolates through this nucleus.

**Entorhinal cortex** – Multilayered cortical component of the *parahippocampal gyrus. It is bordered internally by the *subicular complex and is separated from the *neocortex by the rhinal sulcus. It is the source of the perforant pathway to the *hippocampus.

**Ependyma** – Layer of cuboidal cells that line the lumen of the permanent brain *ventricles and *central canal after dissolution of the proliferative *neuroepithelium.

**External capsule** – Slender fiber band situated between the *claustrum and the *putamen.

**External cuneate nucleus** – Situated lateral to the *cuneate nucleus in the *medulla, also known as the accessory cuneate nucleus, it is the source of the cuneocerebellar tract. The nucleus relays somesthetic and proprioceptive information from anterior regions of the body to the *cerebellum.

**External germinal layer (cerebellum)** – Subpial, secondary germinal matrix of the cerebellar cortex, the source of its granule, stellate, and basket cells. It has two components, an outer proliferative zone and an inner premigratory zone. It persists as a source of neurons over the surface of the human cerebellar cortex until the end of the second year of postnatal life.

## F

**Facial motor nucleus** – A large aggregate of somatic motor neurons in the ventrolateral pons dorsal to the *superior olive complex. It is the source of the motor fibers of *nerve VII that innervate the facial mimetic muscles. Subdivisions of this nucleus innervate different facial muscles.

**Fasciola cinereum** – Transitional region that wraps around the splenium of the *corpus callosum connecting a primordium of the hippocampal *dentate gyrus to the *induseum griseum.

**Fastigial nucleus (cerebellum)** – A deep nucleus of the *cerebellum, also known as the medial cerebellar nucleus. It is the target of Purkinje cell axons that originate in the *vermis. Its axons contribute to the large efferent system that leaves the cerebellum.

**Fimbria** – Large fiber tract of the *hippocampus that runs parallel with and then joins the *fornix.

**Floccular peduncle** – Stalk in the cerebellum that connects the *flocculus with the *nodulus. The *germinal trigone is often strung out along the base of this peduncle.

**Flocculus (cerebellum)** – Lateral extension of the vermal *nodulus in the *cerebellar hemisphere. The principal connections of the two, sometimes called the flocculonodular lobe, are with the vestibular system.

**Folium (cerebellum VIIa)** – Component of the *central lobe, anterior to the *tuber. The hemispheric extension of the folium is *crus I of the ansiform lobule. Together with the *tuber, this is a late-maturing region of the cerebellar cortex.

**Foramen of Monro** – Paired channels in the ventricular system that connect the *lateral ventricles in the telencephalon with the *third ventricle in the diencephalon.

**Forel's fields** – Subthalamic tegmental field ($H_1$ and $H_2$) with fiber bundles of the *ansa lenticularis, lenticular fasciculus, and subthalamic fasciculus.

**Fornical glioepithelium** – Germinal matrix lining the fornix that persists for some time after the dissolution of the hippocampal neuroepithelium. It is the putative germinal source of the oligodendrocytes of the fornix.

**Fornix** – The principal efferent fiber tract of the *hippocampus. The body of the fornix courses forward under the *corpus callosum and above the *thalamus, then turns downward and subdivides into a precommissural part that distributes fibers to the *septum, and a postcommissural part that curves behind the *anterior commissure. The postcommissural fibers traverse the diencephalon and most of them terminate in the nuclei of the thalamic *anterior complex and the *mammillary body.

**Fourth ventricle** – Region of the ventricular system in the *pons and *medulla between the *aqueduct rostrally and the *central canal caudally.

**Fourth ventricle (embryonic)** – The neuroepithelium surrounding the hypertrophied fourth ventricle and its several recesses is the source of neurons and glia of the *medulla, the *pons and the *cerebellum. Many of the neurons migrate over long distances to their settling sites.

**Frontal lobe** – Region of the cerebral cortex anterior to the *central sulcus. Some components of this region, e.g., *precentral gyrus and *orbital gyrus, are often excluded. Its major convolutions are the superior, middle and inferior frontal gyri. Minor convolutions are the gyrus rectus, the inferior rostral gyrus, and the superior rostral gyrus.

**Frontal neuroepithelium** – Putative source of the neurons and glia of the *frontal lobe. Before settling in the cortical plate, its cells form the frontal *subventricular zone and *stratified transitional field.

## G

**Germinal trigone (cerebellum)** – Proliferative germinal matrix in the fourth ventricle beneath the *nodulus that contains three prongs: the *cerebellar neuroepithelium/glioepithelium, the *choroid plexus, and the *external germinal layer.

**Glial sling (corpus callosum)** – *See* **Callosal sling**.

**Glioepithelium** – Fate-restricted germinal matrix in the developing brain, the presumed source of neuroglia (astrocytes and oligodendroglia). Glioepithelia are easiest to recognize without special glial markers at sites of considerable distance from neuronal aggregates or their migratory routes, such as along the *corpus callosum, the *fornix, and the *stria terminalis.

**Globus pallidus** – Component of the *striatum, situated medial to the *putamen, with an external (lateral) and internal (medial) segment. Pallidal fibers are the principal efferents

# GLOSSARY

**An asterisk in front of a term indicates that it has a separate entry in the Glossary with additional information. Terms referring to transient developmental structures are underlined.**

from the striatum to the *thalamus, the *subthalamus, and the *tegmentum.

**Gracile fasciculus** – A prominent fiber tract in the dorsomedial spinal cord and the caudal medulla. It is composed of axons from dorsal root ganglia that convey sensory input from lower parts of the body. The fibers terminate topographically in the *gracile nucleus. Nissl-stained sections of this fiber tract show prominent *myelination gliosis during the early third trimester.

**Gracile nucleus** – Gray mass in the core of the *gracile fasciculus that receives input from that fiber tract. Axons of these neurons cross the midline in the sensory decussation and enter the contralateral *medial lemniscus. The axons terminate topographically in the *ventral complex of the thalamus.

**Granular layer (cerebellar cortex)** – Situated beneath the monolayer of large Purkinje cells, it is composed principally of small, densely packed neurons, the granule cells. The embryonic source of cerebellar granule cells is the *external germinal layer. The axons of granule cells, the parallel fibers, are distributed superficially in the *molecular layer of the cerebellar cortex.

**Granular layer (dentate gyrus)** – Layer of the hippocampal *dentate gyrus with densely packed granule cells. The dendrites of granule cells are distributed in the *molecular layer, and their axons, the mossy fibers, synapse with the pyramidal cells of *Ammon's horn. Most of the granule cells originate late in development (late second trimester, third trimester, and postnatally) in the *subgranular zone of just beneath the granular layer.

**Gray matter** – General term for central nervous regions with a high concentration of neuronal cell bodies and unmyelinated nerve processes but few myelinated fibers. In fresh tissue, such regions appear gray. In histological preparations stained for neuronal cell bodies (perikarya) and their nuclei, the gray matter stands out against the *white matter by a high concentration of peikaryal profiles. In histological preparations that impregnate myelinated fibers, the gray matter appears pale against an opaque background.

## H

**Habenular nuclei** – Mediodorsally situated nuclei in the posterior *thalamus; sometimes distinguished from the thalamus proper as the epithalamus. There are two distinct habenular nuclei, the medial and the lateral. Habenular afferents come principally from the *septum and the *hypothalamus, and its efferents form the *habenulo-interpeduncular tract.

**Habenulo-interpeduncular tract** – A fiber bundle, also known as the fasciculus retroflexus, originating in the *habenular nuclei. It courses through the posterior *thalamus and terminates in the *interpeduncular nucleus of the midbrain.

**Hilus (dentate gyrus)** – Deep layer in the the *dentate gyrus that forms a V-shaped hook around the tip of field CA4 of *Ammon's horn; this layer contains large polymorph cells intermingled with small neurons. The *subgranular zone forms its superficial edge at the base of the *granular layer.

**Hippocampal neuroepithelium** – Putative source of the neurons and glia of the hippocampus that is present early in development. During the third trimester the hippocampal neuroepithelium is presumably transformed into the *alvear glioepithelium. It lines the internal wall of the hippocampo-amygdaloid fork of the *lateral ventricles opposite the *amygdaloid neuroepithelium, the *striatal neuroepithelium and the *temporal neuroepithelium. Its three distinctive components are the source of the large neurons of *Ammon's horn, the *subgranular zone that generates the smaller granule cells of the *dentate gyrus, and the *fornical glioepithelium.

**Hippocampal region** – An inclusive term (also called the hippocampal formation) that includes not only the *hippocampus proper but also the adjacent *subiculum, *presubiculum and *parasubiculum of the *parahippocampal gyrus.

**Hippocampus** – A distinctive allocortical (three-layered) region formed by the interlocking *dentate gyrus and *Ammon's horn. The hippocampus is continuous with the *subicular complex. The principal afferents of the hippocampus travel in the alveolar and perforant paths; its efferents leave by way of the *fimbria that join the *fornix.

**Hypoglossal nucleus (XII)** – A column of somatic motor neurons near the floor of the *fourth ventricle in the caudal medulla. Their axons form cranial *nerve XII that innervate the intrinsic and extrinsic muscles of the tongue.

**Hypothalamus** – Large diencephalic system that surrounds the ventral (or hypothalamic) *third ventricle. It is continuous anteriorly with the *preoptic area and merges caudally with the midbrain *tegmentum. The hypothalamus contains a large number of discrete nuclei, among them the *suprachiasmatic nucleus, the *supraoptic nucleus, the *paraventricular nucleus, the *arcuate nucleus, the *ventromedial nucleus, the *dorsomedial nucleus, the *lateral tuberal nucleus, and the *mammillary body. The supraoptic, paraventricular, and arcuate nuclei produce and release neurohormones. The hypothalamus is considered the head ganglion of the autonomic nervous system and, by way of its neural and hormonal links with the pituitary gland, of the neuroendocrine system.

## I

**Induseum griseum** – Small, medially situated extension of the *hippocampus that lies above the *corpus callosum throughout its entire length.

**Inferior cerebellar peduncle** – Large fiber tract (also known as the restiform body). It contains ascending afferents to the cerebellum from the *spinal cord (spinocerebellar tracts), the *external cuneate nucleus, the *inferior olive, and the *lateral reticular nucleus.

**Inferior colliculus** – Paired inferior hillocks of the midbrain *tectum that receive primary, secondary, and higher order auditory afferents. The output of the inferior colliculus is mainly to the *medial geniculate body in the thalamus, but some axons extend to the primary auditory cortex in the *temporal lobe.

**Inferior lobe (cerebellum X)** – A lobule in the *cerebellar vermis that is coextensive with the *nodulus. It is separated from the *posterior lobe by the posterolateral fissure. The *flocculus in the cerebellar hemisphere is a lateral extension of the nodulus, being connected to it by the floccular peduncle. The two are sometimes distinguished as the vestibulocerebellum.

**Inferior olive** – A distinctive region in the ventrolateral *medulla above the *pyramids and an important component of the precerebellar system. It contains three nuclei. The large principal nucleus is a convoluted structure with a laminar organization that receives input from the *spinal cord, the *red nucleus, motor areas of the *cerebral cortex, and the *central gray. Its axons cross the midline and enter the contralateral *inferior cerebellar peduncle and terminate in the *cerebellar cortex as climbing fibers. The *dorsal accessory olive and the *medial accessory olive are small dense bands of neurons above and medial to the principal nucleus.

# GLOSSARY

**An asterisk in front of a term indicates that it has a separate entry in the Glossary with additional information. Terms referring to transient developmental structures are underlined.**

**Inferior vestibular nucleus** – This nucleus begins caudally in the *medulla near the *external cuneate nucleus and extends rostrally along the medial border of the *inferior cerebellar peduncle. Its neurons receive afferents from the vestibular ganglion and their efferents target the *nodulus and the *flocculus. Some axons also join the *medial longitudinal fasciculus and terminate in cranial motor nuclei (III, IV, VI) that control extraocular muscles.

**Infundibular recess** – Small recess of the third ventricle that evaginates into the *infundibulum.

**Infundibulum** – Stalk extending from the mediobasal *hypothalamus that contains the median eminence and forms a link to the pituitary gland.

**Insula (insular cortex)** – Large buried neocortical region behind the lateral fissure. Its components (gyri breves, gyri longus) are continuous internally with the *frontal, *parietal and *temporal lobes.

**Intercalated cell groups (amygdala)** – Densely packed clusters of small cells that are typically embedded in the white matter between nuclei in the basolateral and corticomedial complexes in the *amygdala.

**Interhemispheric fissure** – Longitudinal cleft that separates the two cerebral hemispheres. It is traversed by the *corpus callosum.

**Internal capsule** – Massive fiber tract between the *thalamus and *striatum, composed of thalamocortical and corticofugal fibers. It is continuous with the *corona radiata rostrally, and the *cerebral peduncle caudally.

**Interpeduncular nucleus** – Midline mesencephalic structure above the interpeduncular fossa and between the *cerebral peduncles. It is the target of fibers of the *habenulo-interpeduncular tract.

**Interpositus nucleus (cerebellum)** – A deep cerebellar nucleus located between the *dentate nucleus and the *fastigial nucleus. It contains a lateral and a medial group of neurons, the emboliform near the *dentate nucleus, and the globosus near the *fastigial nucleus.

**Islands of Calleja** – Scattered clusters of small granule cells within the *olfactory tubercle and along the medial border of the *nucleus accumbens.

**Isthmal glioepithelium/ependyma** – The putative source of glia and the site of proliferating ependymal cells in the *isthmus that lines the posterior *aqueduct. This structure replaces the isthmal neuroepithelium.

**Isthmus** – Interconnecting bridge between the *midbrain and the *pons that is more prominent during earlier brain development than in the third trimester.

## L

**Lamina terminalis** – Anterior membranous border of the *third ventricle. It is the site where the anterior neuropore closes during early embryonic development.

**Lateral corticospinal tract** – A component of the larger *corticofugal tract that originates mainly from layer V in the motor area (including the large Betz pyramidal cells) and other areas of the cerebral cortex that controls skeletal muscles in the contralateral trunk and limbs. Axons carrying somatic motor output cross the midline in the *pyramidal decussation and descend through the entire length of the spinal cord in the dorsolateral part of the *lateral funiculus. Axons leave the tract in a topographically organized manner to terminate directly on spinal motoneurons in the ventral horn from cervical to sacral levels.

**Lateral fissure** – Deep, oblique sulcus (also known as the Sylvian fissure) that separates the *frontal lobe and the *paracentral lobule anterodorsally from the *temporal lobe ventrally.

**Lateral funiculus (spinal cord)** – Region of the spinal cord white matter that flanks the gray matter laterally. Major components of the lateral funiculus are the *lateral corticospinal tract, the *spinocephalic tract, and the *spinocerebellar tracts.

**Lateral geniculate body** – Distinct thalamic region composed of the laminated dorsal lateral geniculate nucleus and its fiber capsule. The principal afferents of this visual relay structure are ipsi- and contralateral fibers of the *optic tract; the principal target of its efferents is the ipsilateral *striate cortex. The connections of the lateral geniculate nucleus and the striate cortex are reciprocal.

**Lateral hypothalamic area** – An ill-defined fibrous region of the *hypothalamus with scattered neurons medial to the cerebral peduncle. It is traversed by many fiber tracts, including the *medial forebrain bundle. The region has been implicated in behavioral arousal and motivation.

**Lateral lemniscus** – The fiber tract on the lateral surface of the *pons that contains secondary auditory fibers from the dorsal and ventral *cochlear nuclei and higher-order auditory fibers from the *superior olivary complex. The dorsal and ventral *nuclei of the lateral lemniscus are embedded within the fiber tract.

**Lateral migratory stream (cortical)** – Tangentially-migrating neurons and glia in the developing *cerebral cortex that leave dorsal parts of the neocortical *neuroepithelium and migrate laterally and ventrally to insular and temporal cortical areas and other telencephalic structures that lack a nearby germinal matrix. The bulk of the lateral migratory stream follows a trajectory outlined by the receding *subventricular zone between the *basal ganglia and the lateral cortex.

**Lateral nucleus (amygdala)** – The most lateral nucleus in the *basolateral complex of the *amygdala. Cells from the *lateral migratory stream appear to enter the nucleus at the distinctive saw-toothed lateral edge. This nucleus is a major recipient of cortical input from the *temporal lobe, *occipital lobe, *insular gyrus, and *orbital gyrus. It also gets input from the *posterior complex of the thalamus. It projects to the *entorhinal cortex, *temporal cortex, *orbital cortex, *insular cortex, *nucleus accumbens, and *dorsomedial thalamic nucleus.

**Lateral preoptic area** – An anterior continuation of the lateral hypothalamic area in the *preoptic area.

**Lateral reticular nucleus** – A relatively discrete group of neurons in the caudal medulla, dorsolateral to the *inferior olive. The neurons of this precerebellar relay nucleus receive topographic exteroceptive and proprioceptive afferents from the *spinocephalic tract and project ipsilaterally to the *cerebellum via the *inferior cerebellar peduncle.

**Lateral septal nucleus** – An indistinct gray mass in the lateral *septum that is closely associated with the *fornix, which provides input to these neurons.

**Lateral tuberal nucleus (hypothalamus)** – Two or three distinct spherical masses near the inferior surface of the *lateral hypothalamic area. These neurons may play a role in thermoregulation and food intake.

**Lateral ventricle**s – Paired cavities of the ventricular system in the telencephalon. They are connected with the *third ventricle in the diencephalon by the *foramen of Monro.

# GLOSSARY

**An asterisk in front of a term indicates that it has a separate entry in the Glossary with additional information. Terms referring to transient developmental structures are underlined.**

<u>**Lateral ventricles (embryonic)**</u> – During embryonic development, the *neuroepithelium lining the hypertrophied lateral ventricles and the adjacent subventricular zone are the source of neurons and neuroglia in the telencephalon. In the mature brain the enduring shrunken regions of the lateral ventricles are lined by cells of the *ependyma.

**Lateral vestibular nucleus** – Also called Deiter's nucleus. A collection of large neurons lying dorsolaterally along the wall of the *fourth ventricle. It receives primary sensory input from the vestibular ganglion via cranial *nerve VIII and its large neurons are the source of the vestibulospinal tract. Smaller neurons send axons to vestibular areas of the cerebellum and to the *medial longitudinal fasciculus.

**Layer I (cerebral cortex)** – The cell-sparse layer beneath the pia in all parts of the *cerebral cortex. This is the first cortical layer to develop and contains the earliest generated cortical neurons, the Cajal-Retzius cells.

**Lingula (cerebellum I-II)** – The ventral lobule of the anterior lobe of the *cerebellar vermis with a tongue-like projection over the surface of the *superior cerebellar peduncle. This lobule does not extend to the hemispheres.

**Lingula (occipital lobe)** – Tongue-shaped gyrus beneath the *cuneus, ventral to the *calcarine fissure.

**Locus coeruleus** – Aggregate of large pigmented cells in the *pons. It is the major source of ascending and descending noradrenergic fibers that are widely distributed throughout the central nervous system.

## M

**Mammillary body** – Distinctive region in the posteroventral hypothalamus, composed of the medial and lateral mammillary nuclei. Its principal afferents are from the *septum and *subiculum that course in the *fornix; its efferents form the *mammillothalamic and *mammillotegmental tracts.

**Mammillotegmental tract** – Descending fiber bundle containing *mammillary body efferents to the brain stem, including the *dorsal tegmental nucleus.

**Mammillothalamic tract** – Ascending fiber bundle containing efferents of the *mammillary body that terminate in the thalamic *anterior complex.

**Massa intermedia (thalamus)** – Interthalamic bridge across the *third ventricle that contains several midline structures, including the *reuniens nucleus and the thalamic *periventricular complex.

**Medial accessory olive** – A small nucleus in the *inferior olive complex that contains densely packed neurons along the lateral border of the medial lemniscus. It receives proprioceptive input from the spinal cord and its efferents reach the contralateral cerebellum (mainly the vermis) by way of the *inferior cerebellar peduncle.

**Medial forebrain bundle** – A diffuse fiber tract that extends from the *olfactory tubercle, through the *lateral hypothalamic area, to the *substantia nigra in the midbrain *tegmentum. It is more conspicuous in the brains of lower animals than in the human brain.

**Medial geniculate body** – Principal thalamic relay station in the auditory pathway to the *cerebral cortex. Its afferents originate in the *trapezoid body, the *superior olivary complex, the *lateral lemniscus nuclei, and the *inferior colliculus. Its efferents form the *auditory radiation that terminates in the *temporal lobe. The connections with the cerebral cortex are reciprocal.

**Medial lemniscus** – Large fiber bundle conveying tactile and other somatosensory input to the thalamus. It originates in the *gracile and *cuneate nuclei in the *medulla, crosses to the opposite side (sensory decussation), ascends through the *pons and *midbrain, and terminates in the *ventral posterolateral and *ventral posteromedial nuclei of the thalamus.

**Medial lemniscus (decussation)** – Also known as the sensory decussation, a medial area in the lower *medulla, above the *pyramids, where axons of the medial lemniscus cross to the opposite side.

**Medial longitudinal fasciculus** – A dorsomedial tract in the *midbrain, *pons and *medulla that contains ascending and descending fibers coursing beneath the *oculomotor nuclear complex, *trochlear nucleus, *abducens nucleus, and *hypoglossal nucleus. It turns ventrally in the posterior medulla and extends into the ventral funiculus of the cervical spinal cord. Its ascending axons originate throughout the *vestibular nuclear complex. Its descending axons originate in the *medial vestibular nucleus and the *reticular formation.

**Medial nucleus (amygdala)** – Part of the *corticomedial complex that contains more loosely packed small-to-medium sized cells than the adjacent *cortical nucleus. It gets input from layer III of the *primary olfactory cortex, thalamic *periventricular complex, hypothalamic *ventromedial nucleus, and *lateral hypothalamic area. It projects to the *orbital cortex, thalamic *periventricular complex, and *ventromedial hypothalamic nucleus.

**Medial preoptic nucleus** – A rounded mass of neurons in the *medial preoptic area that plays a role in sexual functions.

**Medial septal nucleus** – An indistinct nucleus in the midline septum that is continuous with the vertical limb of the *diagonal band of Broca.

**Medial vestibular nucleus** – Component of the *vestibular nuclear complex situated underneath the *fourth ventricle medial to the other vestibular nuclei. Its neurons receive primary sensory input from the vestibular ganglion via cranial *nerve VIII and project to the *fastigial nucleus, the *flocculus *nodulus, and *uvula. Medial vestibular axons also extend through cervical levels of the spinal cord in the *medial longitudinal fasciculus.

**Median preoptic nucleus** – A small nucleus in the *preoptic area that forms a narrow cap around the *anterior commissure in the midline.

**Medulla** – Region of the neuraxis surrounding the posterior *fourth ventricle, also known as the medulla oblongata, bounded by the *pons rostrally and the *spinal cord caudally. An extremely heterogeneous region containing sensory, somatic motor, and visceral motor nuclei as well as large ascending, descending, and decussating fiber tracts.

<u>**Medullary neuroepithelium**</u> – Primary germinal matrix that lines the core and recesses of the posterior embryonic *fourth ventricle. Its several subdivisions are the source of neurons and glia of the different sensory, relay and motor nuclei of the medulla. During the third trimester, the medullary neuroepithelium has already been transformed into a glioepithelium/ependyma. *See also* <u>**Precerebellar neuroepithelium**</u>.

**Medullary layer (cerebellar cortex)** – The deep white matter in the *cerebellar cortex. It contains ascending climbing and mossy fibers and descending Purkinje cell axons.

<u>**Mesencephalic neuroepithelium**</u> – Primary germinal matrix that lines the embryonic *aqueduct. Its different subdivisions produce neurons and glia for the *superior colliculus, the *inferior colliculus, the *central gray, and the *tegmentum.

# GLOSSARY

**An asterisk in front of a term indicates that it has a separate entry in the Glossary with additional information. Terms referring to transient developmental structures are underlined.**

In the third trimester, this germinal matrix has already been transformed into a glioepithelium/ependyma.

**Mesencephalic nucleus (V)** – Large neurons scattered along the lateral border of the *central gray of the midbrain and pons. They are primary sensory neurons that enter the brain from the periphery early in development relaying proprioceptive information from the muscles of mastication. They may function in the reflex control of bite strength.

**Meyer's loop** – Part of the visual radiation that curves into the temporal lobe on its way to the *occipital lobe.

**Midbrain** – The most anterior part of the brainstem divided into the *tectum ("roof") and the *tegmentum ("floor").

**Midbrain tectum** – *See* **Tectum**.

**Midbrain tegmentum** – *See* **Tegmentum**.

**Middle cerebellar peduncle** – Massive tract of *pontocerebellar fibers that originate in the *pontine gray and enter the cerebellum posterolateral to the *inferior cerebellar peduncle.

**Molecular layer (cerebellar cortex)** – Superficial layer that contains the dendrites of Purkinje cells, the parallel fibers (axons) of granule cells, the climbing fibers of the *inferior olive, and the small basket and stellate cells and their processes.

**Molecular layer (dentate gyrus)** – Superficial fibrous layer of the hippocampal *dentate gyrus. It contains the dendrites of granule cells, some scattered neurons, and axon terminals from multiple sources.

**Motor nucleus, V** – *See* **Trigeminal, motor nucleus**.

**Motor nucleus, VII** *See* **Facial, motor nucleus**.

**Myelination gliosis** – Sometimes called premyelination gliosis. Transient increase in the concentration of proliferating glia in fiber tracts before the onset of myelination. It is evident during late fetal development in the *cuneate fasciculus and deep parts of the *gracile fasciculus.

# N

**Neocortex** – The "six-layered" cortex of the cerebral hemispheres, also known as the isocortex. Its principal divisions are the *frontal lobe, the *paracentral lobule, the *parietal lobe, the *occipital lobe, the *temporal lobe, and the *insula. The neocortex is considered to be a phylogenetically more recent laminated telencephalic structure than the "three-layered" *allocortex.

**Neocortical neuroepithelium** – Extensive primary germinal matrix that lines the dorsal and lateral walls of the hypertrophied embryonic *lateral ventricles. The nuclei of the dividing cells shuttle to the lumen of the ventricle to undergo mitotic division there. It is the ultimate source of neurons and glia in the neocortex and some other structures, such as the *primary olfactory cortex and the *basolateral complex of the amygdala.

**Neocortical stratified transitional field** – *See* **Stratified transitional field (cortical)**.

**Neocortical subventricular zone** – A *secondary germinal matrix situated either above the *neocortical neuroepithelium or, after dissolution of the *neuroepithelium, adjacent to the *lateral ventricle *ependyma. It is derived from the primary cortical neuroepithelium but is distinguished from it in that the nuclei of its proliferating cells do not shuttle to the lumen of the ventricle during mitosis.

**Nerve III (oculomotor)** – Cranial motor nerve originating in the *oculomotor nuclear complex. It innervates all the extraocular muscles – except the lateral rectus and superior oblique – and the skeletal muscles of the eyelid, the smooth sphincter muscles of the iris, and the ciliary muscles of the lens.

**Nerve IV (trochlear)** – Cranial motor nerve composed of axons of the *trochlear nucleus that innervates the superior oblique muscle of the eye. This nerve is unique because it exits from the dorsal surface of the *midbrain behind the *inferior colliculus.

**Nerve V (trigeminal)** – A mixed sensory and motor cranial nerve that has three peripheral branches, the ophthalmic, the maxillary, and the mandibular. All three branches contain peripheral sensory fibers from the trigeminal ganglion that terminate in the *trigeminal principal sensory nucleus, the *trigeminal spinal nucleus, and the substantia gelatinosa in upper cervical segments of the spinal cord. A bundle of fibers in the mandibular branch, originating in the *trigeminal motor nucleus, innervates the muscles of mastication.

**Nerve VI (abducens)** – A motor cranial nerve that originates in the *abducens nucleus and emerges near the midline at the caudal border of the *pons. The fibers innervate the lateral rectus muscle of the eye.

**Nerve VII (facial)** – A mixed sensory and motor nerve, the facial nerve has three components. Primary sensory gustatory fibers from the geniculate ganglion enter the solitary tract and terminate in the *solitary nucleus. Somatic motor fibers from the *facial motor nucleus innervate the muscles of facial expression. Visceral motor (parasympathetic) fibers from preganglionic neurons in the dorsolateral pontine *reticular formation (the indistinct salivatory nucleus) target the pterygopalatine and submandibular ganglia.

**Nerve VIII (cochlear, vestibular)** – A sensory cranial nerve that contains primary auditory afferents from the spiral ganglion in the cochlea and primary vestibular afferents from the vestibular (Scarpa's) ganglion. The auditory afferents terminate in the dorsal and ventral *cochlear nuclei. The vestibular afferents terminate in the nuclei of the *vestibular nuclear complex and some reach the cerebellar *nodulus and *flocculus.

**Nerve IX (glossopharyngeal)** – A mixed sensory and motor cranial nerve that exits the *medulla between the *inferior olive and the *inferior cerebellar peduncle. The sensory part of nerve IX relays gustatory input from the posterior third of the tongue and visceral sensory input from the tonsils, the Eustachian tube, and the carotid sinus via primary sensory neurons in the superior and inferior (nodose) ganglia. These fibers enter the solitary tract and terminate in the *solitary nucleus. The somatic motor part of nerve IX originates in the *nucleus ambiguus and innervates the pharyngeal and laryngeal muscles. The visceral motor fibers from parasympathetic preganglionic neurons in the salivatory nucleus terminate in the otic ganglion.

**Nerve X (vagus)** – A mixed sensory and motor nerve, with some somatic and many visceral afferents and efferents (parasympathetic) that are widely distributed throughout the body, including the pharynx, larynx, trachea, esophagus, and all the thoracic and abdominal viscera. Vagal afferents terminate in the *solitary nucleus and other medullary sites. Most of its preganglionic motor neurons are located in the *dorsal motor nucleus (X).

**Nerve XI (accessory)** – This nerve has a cranial and a spinal component. The cranial motor fibers innervate the intrinsic muscles of the larynx and the spinal motor fibers innervate the sternocleidomastoid and upper trapezius muscles. The spinal fibers originate in a column of motoneurons that

# GLOSSARY

**An asterisk in front of a term indicates that it has a separate entry in the Glossary with additional information. Terms referring to transient developmental structures are underlined.**

extends from the region of the *nucleus ambiguus to the 5th-6th segments of the cervical *spinal cord.

**Nerve XII (hypoglossal)** – A somatic motor cranial nerve that originates in the *hypoglossal nucleus and leaves the *medulla between the *pyramids and the *inferior olive. The fibers innervate the intrinsic and extrinsic muscles of the tongue.

<u>Neuroepithelium</u> – The primary source of the neurons and glia in the central nervous system. The neuroepithelial stem cells initially form the neural plate, which folds to form the neural groove. After closure of the neural tube caudally and of the cephalic vesicles rostrally, the *ventricles form and henceforth the nuclei of neuroepithelial stem cells shuttle to the ventricular lumen to undergo mitosis. The neuroepithelium is divisible into mosaic regions or patches that give rise to neurons and glia of distinct brain structures or cell types. The identified patches are listed seperately, e.g. *frontal neuroepithelium. The neuroepithelium is the source of several *secondary germinal matrices that also generate neurons. As neurogenesis ceases, the pluripotential neuroepithelium is transformed at many sites into a *glioepithelium, such as the *callosal glioepithelium or the *fornical glioepithelium, or into the *ependyma that lines the enduring ventricles.

**Nodulus (cerebellum X)** – Cerebellar lobule coextensive with the *inferior lobe of the cerebellar vermis. It is separated from the *posterior lobe by the *posterolateral fissure. It is continuous, via the floccular peduncle, with the *flocculus in the hemisphere. The nodulus gets extensive primary vestibular input from the vestibular ganglion and secondary input from the *vestibular nuclear complex.

**Nucleus accumbens** – Ganglionic component of the ventral telencephalon ventromedial to the *striatum. It is distinguished from the striatum by its cellular organization, neurochemical composition, and intimate connections with the *hypothalamus, *amygdala, and other parts of the limbic system.

**Nucleus ambiguus** – Aggregate of somatic motor neurons that form a thin column in the ventrolateral medulla. Its axons innervate the muscles of the larynx and pharynx via *nerve IX.

**Nucleus of the lateral lemniscus (dorsal)** – Interstitial neurons in the *lateral lemniscus ventral to the *inferior colliculus with input from the *cochlear nuclei and the *superior olivary complex. Its axons join the lateral lemniscus to terminate bilaterally in the *inferior colliculus and the *medial geniculate body.

**Nucleus of the lateral lemniscus (ventral)** – Interstitial neurons in the *lateral lemniscus lateral to the *superior olivary complex. Input and output are similar to the dorsal nucleus of the lateral lemniscus.

**Nucleus of the lateral olfactory tract (amygdala)** – A small mass of densely-stained neurons located deep to the lateral *olfactory tract and superficial to the medial part of the cortical nucleus. These neurons are reciprocally connected to the *olfactory bulb.

**Nucleus of Roller** – A large-celled nucleus ventral to the *hypoglossal nucleus. Sometimes classified as a component of the perihypoglossal nuclear complex.

## O

**Occipital lobe** – The most posterior region of the cerebral cortex, partially separated from the *parietal lobe by the *parietooccipital sulcus. It contains several visual areas with direct projections from the retina, by way of the *lateral geniculate body, or indirect projection by way of the *pulvinar. The distinctive primary visual projection area, the *striate cortex, lies along the wall of the *calcarine sulcus.

<u>Occipital neuroepithelium</u> – Putative source of the neurons and glia of the occipital lobe. It is flanked in the fetal neocortex by the occipital subventricular zone and the distinctive *stratified transitional field before all its neurons settle in the visual cortex.

**Oculomotor nerve (III)** – *See* **Nerve III**.

**Oculomotor nuclear complex** – Situated at the base of the *central gray beneath the *aqueduct, the cell columns of this complex extend from the anterior pole of the *superior colliculus rostrally to the *trochlear nucleus caudally. Its somatic motor nuclei innervate the medial rectus, inferior rectus, superior rectus and inferior oblique muscles of the eye, and are associated with the fibers of the *medial longitudinal fasciculus. Most prominent of its autonomic (preganglionic) components is the dorsally located Edinger-Westphal nucleus.

**Olfactory bulb** – Laminated brain structure where the first-order fibers of the olfactory nerve terminate and the second-order fibers of the lateral and medial *olfactory tracts originate. It is composed of three classes of neurons: large mitral cells, the intermediate tufted cells, and the small granule cells. The late-generated granule cells migrate to the olfactory bulb by way of the *rostral migratory stream.

**Olfactory peduncle** – A stalk that connects the olfactory bulb to the brain. It contains the *olfactory tract and the anterior olfactory nucleus. During the third trimester, the olfactory peduncle contains the *rostral migratory stream in its core.

**Olfactory tubercle** – Paleocortical area in the ventral telencephalon deep to the *diagonal band of Broca (vertical limb) and superficial to the *nucleus accumbens/ventral *striatum. It contains many dense cellular aggregates, the *islands of Calleja. Input comes mainly from the lateral *olfactory tract.

**Olfactory tract** – Large fiber bundle that originates in the *olfactory bulb and has two parts, the larger lateral olfactory stria and the smaller medial stria. The fibers of the lateral stria terminate in the *olfactory tubercle, the *primary olfactory cortex, and *corticomedial complex of the amygdala.

**Optic chiasm** – Site of the partial decussation of fibers of the *optic nerve. Fibers from the nasal half of each retina cross here to the opposite side while those from temporal half proceed uncrossed to form the *optic tract.

**Optic nerve** – The large second cranial nerve containing the axons of retinal ganglion cells. Beyond the *optic chiasm this nerve is called the *optic tract.

**Optic tract** – Large bundle of crossed and uncrossed retinal afferents. In the human brain the majority of the fibers terminate in the *dorsal lateral geniculate nucleus; others proceed to the *superior colliculus, the *suprachiasmatic nucleus, and the *pretectum.

**Orbital gyrus** – Ventromedial region of the *frontal lobe with afferents from the thalamic dorsomedial nucleus and efferents to the *preoptic area and the *hypothalamus. The region has been implicated in such visceromotor functions, as the regulation of blood pressure, respiratory rate, and gastric motility.

## P

**Parabigeminal nucleus** – This small nucleus, composed of a dorsal and ventral aggregate of cells in the lateral wall of the *midbrain tegmentum, receives second-order visual input from the *superior colliculus. It may be the homologue of the isthmo-optic nucleus of lower vertebrates.

**Parabrachial nucleus** – Dorsolateral pontine structure with indistinct boundaries that surrounds the *superior cerebellar

# GLOSSARY

**An asterisk in front of a term indicates that it has a separate entry in the Glossary with additional information. Terms referring to transient developmental structures are underlined.**

peduncle. Its principal input comes from the *solitary nucleus and its efferents target the *ventral posteromedial nucleus of the thalamus, the *amygdala, and the *insular cortex. This nucleus has been implicated in gustatory and related visceral functions.

**Paracentral lobule, cortical** – Inclusive term used for two neocortical convolutions surrounding the *central sulcus: the *precentral gyrus and the *postcentral gyrus. The term is used to distinguish these primary motor and sensory projection areas from the higher-order, integrative areas of the *frontal lobe anteriorly and the *parietal lobe posteriorly.

<u>Paracentral neuroepithelium</u> – Putative source of the neurons and glia of the *paracentral lobule in the developing neocortex. It is flanked by the paracentral *subventricular zone and the paracentral *stratified transitional field before all the neurons settle in the cortical plate.

**Paracentral nucleus, thalamic** – Component of the *periventricular complex of the thalamus.

**Parafascicular nucleus, thalamic** – Component of the *periventricular complex of the thalamus.

**Paraflocculus (cerebellum HIX)** – Also called the tonsil. Lobule of the cerebellar hemisphere that is connected to the *uvula in the vermis.

**Parahippocampal gyrus** – Transitional (allocortical-to-neocortical) area between *Ammon's horn of the hippocampus and the isocortex of the *temporal lobe. Its subdivisions are the *subicular complex and the *entorhinal cortex.

<u>Parahippocampal neuroepithelium</u> – Putative source of the neurons and glia of the parahippocampal gyrus. It is flanked by the parahippocampal *subventricular zone and the parahippocampal *stratified transitional field before all its neurons settle in the cortical plate.

**Parasubiculum** – *Allocortical component of the *parahippocampal gyrus between the *presubiculum and the *entorhinal cortex.

**Paratenial nucleus (thalamus)** – Component of the *periventricular complex of the thalamus.

**Paraventricular nucleus (hypothalamus)** – Prominent neuroendocrine structure abutting the third ventricle with a magnocellular and a parvocellular division. The large neurons of the paraventricular nucleus are the source of oxytocin and vasopressin that reach the posterior pituitary gland by axoplasmic flow. The small neurons of the nucleus are the source of releasing hormones conveyed to the portal vessels of the median eminence.

**Paraventricular nucleus (thalamus)** – Component of the *periventricular nuclear complex of the thalamus.

**Parietal lobe** – Region of the neocortex bounded anteriorly by the *postcentral sulcus and posteriorly by the *parieto-occipital sulcus. It subdivisions have been implicated in higher-level perceptual and cognitive functions, including language comprehension. Its major convolutions include the superior parietal lobule, the precuneus, the supramarginal gyrus, and the angular gyrus.

<u>Parietal neuroepithelium</u> – Putative source of the neurons and glia of the parietal lobe in the developing neocortex. It is flanked by the parietal *subventricular zone and the parietal *stratified transitional field before all the neurons settle in the cortical plate.

**Parieto-occipital sulcus** – Fissure that partially separates the *parietal lobe and the *occipital lobe.

**Parolfactory gyrus** – Anteromedial gyrus in the frontal lobe, situated beneath the genu of the corpus callosum and in front of the *anterior commissure.

**Peristriate cortex** – Area forming a belt around the *striate cortex and implicated in the higher-order processing of visual information.

**Periventricular complex (thalamus)** – Collective term for several thalamic nuclei lining the third ventricle, including the *paracentral, *parafascicular, *paratenial, *paraventricular, and *reuniens nuclei. Its principal connections are with limbic system structures and the *striatum; connections with the *neocortex are sparse.

**Pineal gland** – An endocrine gland connected by its stalk to the pineal recess of the dorsal *third ventricle. It secretes melatonin and other indoleamines. It is believed to receive indirect visual input from the retina.

**Piriform cortex** – *see* **Primary Olfactory Cortex**.

**Pons** – Region of the brain surrounding the anterior *fourth ventricle, situated between the *midbrain and the *medulla. It contains many ascending, descending and decussating fiber tracts, sensory and motor nuclei of the cranial nerves, the *reticular formation, the *raphe nuclear complex, and the *pontine gray.

**Pontine gray** – Large nucleus composed of a dense aggregate of small to medium-sized neurons in the ventral *pons. It contains the thick bundles of the longitudinally-oriented *corticofugal tract and the thinner bundles of the transversely-oriented *pontocerebellar fibers. Corticofugal axons that collateralize here are the principal afferents of the pontine gray neurons that are the source of the pontocerebellar fibers.

**Pontocerebellar fibers** – Axons of pontine gray neurons that terminate in the *granular layer of the *cerebellar cortex.

**Pontocerebellar fibers (decussation)** – Axons of pontine gray neurons that cross the midline and enter the contralateral *middle cerebellar peduncle.

**Postcentral gyrus** – Convolution of the *cerebral cortex between the *central sulcus and the postcentral sulcus. It is the primary somatosensory projection area of the cortex and is classified as a component of the *paracentral lobule.

**Posterior commissure** – Early developing decussating fiber tract in the dorsal *midbrain that interconnects several pretectal and tectal nuclei.

**Posterior complex (thalamus)** – Division of the thalamus that includes the *lateral geniculate body, the *medial geniculate body, and the *pulvinar. The nuclei of the thalamic posterior complex – together with the *ventral complex – constitute the dedicated sensory relay system that supplies the neocortex with high-resolution and fast visual, auditory and somesthetic information.

**Posterior lobe (cerebellum IX)** – The vermal lobe that is coextensive with the *uvula. It is separated from the *central lobe by the *secondary fissure and from the *inferior lobe by the *posterolateral fissure.

**Posterolateral fissure (cerebellum)** – The vermal fissure that separates the *inferior lobe and the *posterior lobe.

**Precentral gyrus** – Anterior convolution of the *paracentral lobule in the *cerebral cortex between the precentral sulcus and the *central sulcus. It constitutes the primary motor cortex with giant Betz pyramidal cells in layer V.

<u>Precerebellar neuroepithelium</u> – Dorsal medullary germinal matrix whose cells migrate rostrally to the pons and caudally to the

# GLOSSARY

**An asterisk in front of a term indicates that it has a separate entry in the Glossary with additional information. Terms referring to transient developmental structures are underlined.**

ventral medulla to settle in the precerebellar nuclei, including the *pontine gray, the *reticular tegmental nucleus, the *inferior olive, the *lateral reticular nucleus, and the *external cuneate nucleus. This part of the germinal matrix is a *glioepithelium/ependyma during the third trimester.

**Preculminate fissure (cerebellum)** – A fissure within the anterior lobe of the vermis that separates the *centralis from the *culmen.

**Premammillary area** – Region with ill-defined boundaries anterior to the *mammillary body in the *hypothalamus.

**Preoptic area** – Midline area surrounding the preoptic recess of the *third ventricle. It is contiguous anteriorly with the ventral telencephalon and blends posteriorly with the anterior *hypothalamus. It is implicated in the regulation of sexual behavior and reproductive functions.

**Prepositus nucleus** – Situated in the dorsomedial *medulla, it extends from the anterior part of the *hypoglossal nucleus to the posterior part of the *abducens nucleus.

**Prepyramidal fissure (cerebellum)** – Fissure in the vermal *central lobe that separates the *pyramis from the *tuber.

**Presubiculum** – *Allocortical component of the *parahippocampal gyrus between the *subiculum and the *parasubiculum.

**Pretectum** – Dorsal area between the posterior *thalamus and the *superior colliculus. Several of its nuclei are involved in visuomotor functions.

**Primary fissure (cerebellum)** – Dorsal cerebellar fissure that extends from the vermis to the hemispheres. In the vermis, it separates the *anterior lobe from the *central lobe. In the hemispheres, it separates HI-HV of the anterior lobe from the *simplex lobule (HVI).

**Primary olfactory cortex** – Paleocortical region, also called the piriform lobe or cortex, where fibers of the lateral olfactory tract terminate. It is situated rostral to the *entorhinal cortex of the *parahippocampal gyrus and includes the prepiriform area along the rhinal fissure and the periamygdaloid area.

**Primary visual cortex** – *See* **Striate cortex**.

**Principal sensory nucleus (V)** – *See* **Trigeminal, principal sensory nucleus**.

**Pulvinar (thalamus)** – Large nucleus of the thalamic *posterior complex. Its subdivisions send fibers to various regions of the *parietal lobe, *occipital lobe, *temporal lobe, and perhaps also to the frontal eye field.

**Purkinje cell layer (cerebellar cortex)** – The monolayer of cell bodies of Purkinje cells sandwiched between the superficial *molecular layer and the deep *granular layer. During development, Purkinje cells are piled on top of one another but gradually form a monolayer as the surface area of the cerebellar cortex expands.

**Putamen** – Lateral component of the *striatum. It lies between the *external capsule and the *globus pallidus. It is the major source of striatal efferents to the *thalamus, *subthalamic nucleus, *substantia nigra, and *tegmentum.

**Pyramid (corticospinal tract)** – Paired triangular-shaped masses of fibers on the ventral surface of the lower medulla that contain the *lateral and *ventral corticospinal tracts.

**Pyramidal decussation** – The region in the lower *medulla where the bulk of the fibers in the *pyramids cross the midline in a dorsolateral direction and descend in the spinal cord as the *lateral corticospinal tract.

**Pyramis (cerebellum VIII)** – Triangular vermal lobule in the posterior *central lobe that is continuous with the paramedian and *biventral lobules of the cerebellar hemispheres. It is bounded anteriorly by the *prepyramidal fissure and is separated from the *posterior lobe by the *secondary fissure.

# R

<u>Raphe migration</u> – Streams of cells that originate in the dorsal *medullary neuroepithelium and are distributed in the midline ventrally during late-fetal development. They appear to contribute cells to the medullary *arcuate nucleus.

**Raphe nuclear complex** – Several smaller and some larger cell aggregates that extend in and near the midline from the *midbrain rostrally to the *medulla caudally. The raphe cells are the principal source of serotonin-containing fibers distributed along the entire neuraxis from the forebrain to the *spinal cord. They are involved, as neuromodulators, in the regulation of sleep, wakefulness and emotional arousal.

**Red nucleus** – Spherical mesencephalic nucleus with a small-celled (parvocellular) and a large-celled (magnocellular) division. It is surrounded and traversed by fibers of the *superior cerebellar peduncle. Its afferents originate in the *cerebellar deep nuclei and the *cerebral cortex. Its descending efferents target various tegmental and medullary nuclei, and the *spinal cord (rubrospinal tract); its ascending fibers terminate in the *thalamus. The red nucleus is a component of the cerebellar feedback loop.

**Reticular belt (thalamus)** – Distinctive component of the thalamus; it is coextensive with the thalamic *reticular nucleus.

**Reticular formation** – A large collection of scattered neurons, enmeshed in a complex network of fibers, in the core of the *medulla, *pons, and *midbrain. The extensive axonal branches of neurons have a modulatory influence on the entire central nervous system.

**Reticular tegmental nucleus (pons)** – Situated dorsal to the *pontine gray, this precerebellar nucleus, also known as the nucleus reticularis tegmenti pontis, receives afferents from the *cerebellum and the *cerebral cortex. The principal target of its efferents is the cerebellum, where they terminate as mossy fibers.

**Reticular nucleus (thalamus)** – A thin belt of cells and fibers between the wall of the *thalamus and the *internal capsule. Virtually all fibers that interconnect the thalamus and the cerebral cortex traverse, and probably form connections with, the thalamic reticular nucleus.

**Retrofacial nucleus** – Small structure in the medulla caudal to the *facial motor nucleus and near the *nucleus ambiguus.

**Retrosplenial area (cingulate gyrus)** – Region of the cingulate gyrus that wraps around the splenium of the *corpus callosum.

**Reuniens nucleus (thalamus)** – Midline component of the *periventricular complex of the thalamus, situated in the *massa intermedia.

<u>Rostral migratory stream</u> – A large stream of mitotic and postmitotic cells in the forebrain extending from the rostral *lateral ventricle to the *olfactory bulb. It is a source of neurons – such as the late-generated granule cells of the olfactory bulb – and neuroglia, and persists after the *neuroepithelium has receded or disappeared at most brain sites.

# GLOSSARY

**An asterisk in front of a term indicates that it has a separate entry in the Glossary with additional information. Terms referring to transient developmental structures are underlined.**

## S

**Secondary fissure (cerebellum)** – Fissure in the midline vermis that separates the *posterior lobe from the *central lobe.

<u>**Secondary germinal matrix**</u> – Layer or field of proliferative precursors of neurons and glia outside the primary *neuroepithelium. These cells are progeny of the stem cells of the primary neuroepithelium that persist for varying periods after the neuroepithelium has disappeared. Examples of secondary germinal matrices are the *external germinal layer of the cerebellum, the *subventricular zone of the neocortex, the *subgranular zone of the dentate gyrus, the *rostral migratory stream, and the *striatal subventricular zone. Typically, the secondary germinal matrices are the source of late-generated small interneurons, or microneurons.

**Septum** – Midline telencephalic structure beneath the anterior *corpus callosum that is better developed in lower mammals than in man. It contains *medial and *lateral nuclei. Its principal connections are with the *hippocampus and the *hypothalamus by way of the *fornix. The septum is a focal component of the limbic system.

**Sexually dimorphic nucleus** – A small, dense cluster of neurons closely associated with the *medial preoptic nucleus in the *preoptic area.

**Simplex lobule (cerebellum HVI)** – Lobule in the cerebellar hemisphere that is continuous with the vermal *declive. It is delineated anteriorly by the *primary fissure and posteriorly by *crus I of the ansiform lobule.

**Solitary tract and nucleus** – This tract contains primary sensory fibers that reach the *medulla by way of cranial *nerves VII, IX, and X, conveying gustatory (VII and IX) and visceral sensory information (IX and X) to the solitary nucleus. The *dorsal sensory nucleus of X and the *commissural nucleus of X are part of this nuclear complex.

**Spinal cord** – Caudal tubular component of the central nervous system that surrounds the *central canal. Its continuous core of gray matter (the dorsal horn, intermediate gray and ventral horn) and surrounding white matter (the dorsal, *lateral and ventral funiculi) are divided into 31 segments by the discontinuous entry of dorsal root afferents and exit of ventral root efferents that form the mixed spinal nerves. The cervical enlargement and the lumbar enlargement of the spinal cord supply the nerves of the forelimbs and hind limbs, respectively.

**Spinal nucleus (V)** – *See* **Trigeminal, spinal nucleus**.

**Spinal tract (V)** – *See* **Trigeminal, spinal tract**.

**Spinocephalic tract** – Large ascending fiber tract in the *lateral funiculus of the spinal cord whose fibers or collaterals are widely distributed throughout the *medulla, *pons, *midbrain, *thalamus, and the limbic forebrain. The spinocephalic tract, which includes the spinoreticular, spinomesencephalic and spinothalamic tracts, conveys nociceptive and other affective stimuli to the brain. It is distinguished from the phylogenetically more recent ascending *cuneate and *gracile funiculi that convey primarily cognitive somesthetic information to the neocortex.

**Spinocerebellar tracts** – Several ascending pathways, including the dorsal and ventral spinocerebellar tracts in the *lateral funiculus that convey proprioceptive information from muscles and joints to the *cerebellum.

<u>**Stratified transitional field**</u> – Transient intermediate field in the fetal *cerebral cortex between the *neuroepithelium (or *subventricular zone) and the *cortical plate. It is a crisply laminated site with alternating fiber-rich and cell-rich layers that vary in their configuration in different lobes of the cerebral cortex; e.g., frontal stratified transitional field; occipital stratified transitional field. The different fibrous layers are continuous with the incoming thalamocortical afferents, the outgoing corticofugal efferents, and the decussating fibers of the *corpus callosum. The cellular layers are composed of sojourning and migrating neurons all of which eventually settle in the *cortical plate.

**Stria medullaris (thalamus)** – Mediodorsal fiber bundle in the diencephalon coursing in an anteroposterior direction and terminating in the *habenular nuclei.

<u>**Striatal neuroepithelium**</u> – Primary germinal source of neurons of the *caudate nucleus, *putamen, and *globus pallidus. It has a large anterolateral and anteromedial division, also known as the lateral and medial eminences, and a small posterior division that generates the neurons of the tail of the caudate nucleus.

<u>**Striatal subventricular zone**</u> – A large *secondary germinal matrix flanking the striatal neuroepithelium that generates the bulk of the neurons of *caudate nucleus, *putamen and *globus pallidus.

**Striate cortex** – Principal projection area of the *visual radiation, situated along the *calcarine sulcus. This distinctive thin cortex is rich in small granule cells and contains a white band (of Gennari) within layer IV.

**Stria terminalis** – Arched fiber bundle that originates in the corticomedial and basolateral *amygdala, courses along the medial surface of the *caudate nucleus, and terminates in the *bed nucleus of the stria terminalis, the anterior *hypothalamus, and the *preoptic area.

**Striatum** – Term used for two components of the *basal ganglia: the *caudate nucleus and the *putamen.

<u>**Strionuclear glioepithelium**</u> – Putative source of the glia of the stria terminalis, stria medullaris and possibly other nearby fiber tracts.

<u>**Strionuclear neuroepithelium**</u> – Putative germinal source of the neurons and glia of the *bed nucleus of the stria terminalis. It is situated beneath the *striatal neuroepithelium in a notch of the *lateral ventricle near the *foramen of Monro. Throughout the third trimester, this germinal site is most likely the source of glia, and is labeled as the *strionuclear glioepithelium.

**Subcommissural organ** – A highly vascularized circumventricular neuroendocrine organ located beneath the *posterior commissure in the roof of the posterior *third ventricle and anterior *aqueduct. It is devoid of neurons and may disappear from the human brain after birth.

**Subfornical organ** – A highly vascularized circumventricular neuroendocrine organ situated between the two columns of the fornix at the confluence of the *third ventricle and the *foramen of Monro. Among its afferents are fibers from the *preoptic area and the anterior *hypothalamus. Its efferents target the hypothalamic *paraventricular nucleus and the *supraoptic nucleus.

<u>**Subgranular zone (hippocampus)**</u> – A long-persisting secondary germinal matrix beneath the *granular layer of the dentate gyrus. It is a division of the *hippocampal neuroepithelium and is the source of late-generated dentate granule cells.

**Subicular complex** – Collective term for the *parasubiculum, the *presubiculum and the *subiculum of the *parahippocampal gyrus.

**Subiculum** – *Allocortical component of the *hippocampus, between CA1 and the *presubiculum.

# GLOSSARY

**An asterisk in front of a term indicates that it has a separate entry in the Glossary with additional information. Terms referring to transient developmental structures are underlined.**

<u>Subpial granular layer</u> – Transient cellular layer between *layer I and the pia in some regions of the developing *cerebral cortex. It may be a source of cortical glia.

**Substantia innominata** – Extensive telencephalic area with indistinct boundaries beneath the *globus pallidus. A prominent component of the substantia innominata is the *basal nucleus of Meynert.

**Substantia nigra** – Pigmented *midbrain tegmental structure abutting the base of the *internal capsule and the *cerebral peduncle. It has two components, the dopaminergic pars compacta, and the GABAergic pars reticulata. The substantia nigra receives afferents from the *striatum, *globus pallidus, the *bed nucleus of stria terminalis, the *central nucleus of the *amygdala, the *subthalamic nucleus, and the *raphe nuclear complex. Its dopaminergic efferents target the *striatum and some brainstem nuclei. Degeneration of the dopaminergic neurons of the substantia nigra have been implicated in Parkinson's disease.

**Subthalamic nucleus** – Biconvex diencephalic structure situated above the substantia nigra between the *zona incerta and the base of the *internal capsule. It has extensive reciprocal connections with the *globus pallidus, hence it is considered a component of the *basal ganglia circuitry. Subthalamic lesions produce persistent choroid movements (hemiballism) in the arms, legs and face.

<u>Subventricular zone</u> – Secondary germinal matrix, derived from the primary *neuroepithelium. The subventricular zone flanks the neuroepithelium during early development and flanks the ependyma when the neuroepithelium dissolves. The proliferative cells of the subventricular zone, unlike those of the neuroepithelium, do not shuttle to the lumen of the ventricle during mitosis. Two large subventricular zones are located in the *cerebral cortex and the *striatum.

**Superior cerebellar peduncle** – A large fiber tract of the *pons, also known as the brachium conjunctivum, that originates mainly in the *dentate nucleus and *interpositus nucleus. As the axons leave the cerebellum, they hug the dorsolateral wall of the pontine *fourth ventricle. As they approach the midbrain, they dip downward and cross the midline (decussation) posterior to the *red nucleus. Many of its axons terminate in the red nucleus, others continue rostrally and terminate in the *ventral anterior nucleus of the thalamus.

**Superior colliculus** – Anterior component of the *tectum, known in lower vertebrates as the optic lobe. A laminated structure, the superior colliculus is a direct target of optic nerve fibers. Its principal efferent outflow, by way of the tectobulbar and *tectospinal tracts, is to the medulla and spinal cord.

**Superior medullary velum** – Thin membrane that covers the roof of the anterior fourth ventricle near the posterior border of the *tectum. The root fibers of cranial *nerve IV decussate here and exit the brain.

**Superior olivary complex** – A group of neurons in the ventrolateral posterior *pons that receive auditory input from the dorsal and ventral *cochlear nuclei. Some of its axons cross the midline in the *trapezoid body. Ipsi- and contralateral fibers of the complex join the *lateral lemniscus and terminate in the *inferior colliculus and in the *medial geniculate body.

**Superior vestibular nucleus** – This nucleus lies dorsal and anterior to the *lateral vestibular nucleus. It receives input from the vestibular ganglion, and projects to the *cerebellum, mainly the *nodulus, the *flocculus, and the *uvula.

**Suprachiasmatic nucleus (hypothalamus)** – Small, paired midline structure above the *optic chiasm. It is implicated in the photic entrainment of the circadian rhythm.

**Suprageniculate nucleus (thalamus)** – Wedge-shaped structure between the *pulvinar and the *pretectum, and dorsal to the *medial geniculate body.

**Supramammillary area (hypothalamus)** – Hypothalamic region that caps the *mammillary body. Experimental studies in animals indicate that its cells project to the *dentate gyrus of the hippocampus.

**Supraoptic nucleus (hypothalamus)** – Located above the optic tract lateral to the optic chiasm. The large secretory neurons of this nucleus produce arginine vasopressin and oxytocin that are conveyed by axoplasmic flow to the posterior lobe of the pituitary gland.

# T

**Tectospinal tract** – Efferent fiber tract that originates in the deep layers of the *superior colliculus, courses in the ventral funiculus of the spinal cord and terminates in the ventral horn of the *spinal cord. The tract may trigger involuntary head rotation and body turning in response to visual stimulation.

**Tectum** – Inclusive term for the dorsal hillocks that cap the *central gray of the midbrain and is the target of visual, auditory and somatosensory afferents. Its two discrete components are the *superior colliculus and the *inferior colliculus.

**Tegmentum** – Ventral and ventrolateral region of the *midbrain and *pons with indistinct boundaries. It contains the *reticular formation, some discrete nuclei, and ascending, decussating, and descending fiber tracts. Some tegmental nuclei have been implicated in somatomotor and visceromotor functions.

**Temporal lobe** – Lateral portion of the cerebral hemispheres. It is separated by the *lateral fissure from the *frontal lobe anteriorly and the *paracentral lobule dorsally, but is continuous with the parietal lobe dorsally and the *occipital lobe posteriorly. Two horizontal fissures, the superior and middle temporal sulci, divide the temporal lobe into three convolutions, the superior, middle and inferior gyri. The primary auditory area lies buried in the floor of the lateral fissure in the anterior transverse gyrus (Heschl's gyrus).

<u>Temporal neuroepithelium</u> – Putative source of neurons and glia in the temporal lobe neocortex. It is flanked during fetal development by the temporal *subventricular zone and the temporal *stratified transitional field before all the neurons settle in the *cortical plate.

**Tenia tecta** – Components of the *cerebral cortex that extend into the dorsomedial *septum (dorsal tenia tecta) and medial *olfactory peduncle (ventral tenia tecta).

**Thalamocortical radiations** – Collective term for the large afferent tracts that proceed from the thalamic *posterior complex and *ventral complex, by way of the internal capsule, to the somatosensory and motor projection areas of the *paracentral lobule, the auditory projection area of the *temporal lobe, and the *visual projection area of the occipital lobe.

**Thalamus** – Massive dorsal diencephalic structure with several distinct and some indistinct nuclei. As a convenience, the thalamus is divided into the following nuclear regions: the *anterior complex, the *central complex, the *dorsal complex, the *periventricular complex, the *posterior complex, the *ventral complex, and the *reticular belt.

# GLOSSARY

**An asterisk in front of a term indicates that it has a separate entry in the Glossary with additional information. Terms referring to transient developmental structures are underlined.**

**Third ventricle** – Midline diencephalic component of the ventricular system. It is linked rostrally by the *foramen of Monro with the paired *lateral ventricles of the telencephalon, and caudally with the *aqueduct of the *midbrain. The dorsal region of the third ventricle is surrounded by the *thalamus, its ventral region and recesses by the *preoptic area and the *hypothalamus. During development, the neuroepithelium lining the third ventricle is the source of the neurons and glia of the diverse components the diencephalon.

**Trapezoid body** – A fiber tract extending from the ventral *cochlear nucleus to the contralateral *superior olivary complex. It contains second- and higher-order auditory fibers.

**Transpontine corticofugal tract** – Fibers of the corticofugal tract that traverse the *pontine gray nucleus. Collaterals of this tract synapse with neurons of the pontine gray that give rise to the *pontocerebellar fibers of the *middle cerebellar peduncle.

**Trigeminal, motor nucleus** – Aggregate of trigeminal somatic motor neurons situated medial to the *trigeminal principal sensory nucleus. It receives input from the *mesencephalic nucleus (V) and from motor areas of the *cerebral cortex via relays in the pontine *reticular formation. Its axons leave the brain in cranial *nerve V and join its mandibular branch that innervates the muscles of mastication.

**Trigeminal, principal sensory nucleus** – The second-order sensory neurons in the trigeminal system located dorsal and lateral to the incoming sensory root of cranial *nerve V. It receives topographic somatosensory input from the face and mouth, and its efferents cross the midline in the pons and proceed to the thalamic *ventral complex in close association with the *medial lemniscus.

**Trigeminal, spinal nucleus** – A continuation of the *trigeminal principal sensory nucleus that extends caudally through the *medulla to the second cervical level of the *spinal cord. It receives topographic input from the face via somatosensory neurons in the trigeminal ganglion. The axons of the trigeminal spinal nucleus cross the midline, join the *medial lemniscus and proceed to the *ventral complex of the thalamus.

**Trigeminal, spinal tract** – Primary sensory fibers of the trigeminal ganglion that convey touch and pressure information from the face. The axons enter the brain in the pons and proceed caudally, forming a lateral cap around the *trigeminal spinal nucleus where they terminate in a topographic order.

**Trochlear nucleus** – Aggregate of somatic motor neurons located posterior to the *oculomotor nuclear complex that innervate the superior oblique muscle of the eye by way of cranial *nerve IV.

**Tuber (cerebellum VIIb)** – The vermal lobule posterior to the *folium. The tuber is continuous with *crus II of the ansiform lobule (HVIIA) and the paramedian lobule (HVIIB) in the hemispheres. Together with the folium, the tuber is a late-maturing region of *cerebellar cortex.

## U

**Uncinate fasciculus** – A fiber tract in the temporal lobe that proceeds from the *uncus to the anterolateral *temporal lobe. It is infiltrated by cells in the *lateral migratory stream.

**Uncus** – The rostral cap of the *parahippocampal gyrus.

**Uvula (cerebellum IX)** – A large lobule in the *cerebellar vermis that is coextensive with the *posterior lobe, and is continuous with the *paraflocculus in the hemispheres. It is bounded anteriorly by the *secondary fissure, and posteriorly by the *posterolateral fissure. The uvula is part of the "vestibulo-cerebellum," receiving massive input from the vestibular nuclear complex as well as primary sensory input from the vestibular ganglion.

## V

**Ventral anterior nucleus (thalamus)** – The ventral anterior nucleus, to be distinguished from the *anteroventral nucleus (which is a component of the *anterior complex of the thalamus) is the most rostral component of the thalamic *ventral complex. Its afferents come mostly from the *globus pallidus and the *substantia nigra, and its efferents terminate in the *paracentral lobule of the neocortex. The ventral anterior nucleus may be the principal relay from the *basal ganglia to the *neocortex.

**Ventral complex (thalamus)** – A group of structurally and functionally related ventrolateral and ventral nuclei of the thalamus, including the *ventral anterior, *ventral lateral, *ventral posterolateral, and *ventral posteromedial nuclei. The ventral thalamic complex is the principal topographically organized relay system of direct (lemniscal) and indirect (cerebellar and striatal) somatosensory and proprioceptive input to the sensory and motor areas of the *neocortex.

**Ventral corticospinal tract** – A small complement of ipsilateral corticospinal fibers that do not cross in the *pyramidal decussation and descend in the ventral funiculus of the *spinal cord.

**Ventral lateral nucleus (thalamus)** – Situated caudal to the *ventral anterior nucleus, this component of the thalamic *ventral complex is the target of input from the *superior cerebellar peduncle and the *red nucleus. Its somatotopically organized efferents terminate in the motor cortex and adjacent areas. The ventral lateral nucleus may be the principal relay from the *cerebellum to the *neocortex.

**Ventral posterolateral nucleus (thalamus)** – Situated caudal to the *ventral lateral nucleus, this region of the thalamic *ventral complex is the target of fibers of the *medial lemniscus that originate in the *cuneate nucleus and the *gracile nucleus and convey somatosensory information from the trunk and the extremities. Its efferents form the somesthetic radiation that terminates in a precise topographic order in the medial part of the *postcentral gyrus. The ventral posterolateral nucleus is the principal thalamic relay of somesthetic input from the trunk and limbs to the *neocortex.

**Ventral posteromedial nucleus (thalamus)** – Situated between the *ventral posterolateral nucleus and the *centromedian nucleus, this nucleus receives afferents from the *trigeminal sensory nuclei and the *parabrachial nucleus that convey sensory information from the face, the tongue, the oral cavity, and the neck. The efferents of this nucleus terminate in a precise topographic order in the lateral part of the *postcentral gyrus. The ventral posteromedial nucleus is the principal thalamic relay of somatosensory and gustatory input from the neck, head, and mouth to the *neocortex.

**Ventral tegmental area** – Medial area flanking the *substantia nigra and containing a high concentration of dopaminergic neurons, much like the substantia nigra, pars compacta. It is generally considered as a component of the substantia nigra complex.

**Ventricles** – The cerebrospinal fluid-filled system in the core of the brain that is divided into the paired *lateral ventricles in the telencephalon, the single *third ventricle in the diencephalon, the *aqueduct in the mesencephalon, and the *fourth ventricle in the pons and medulla. The fourth ventricle is

# GLOSSARY

**An asterisk in front of a term indicates that it has a separate entry in the Glossary with additional information. Terms referring to transient developmental structures are underlined.**

continuous caudally with the *central canal of the spinal cord. The lateral, third and fourth ventricles have a prominent *choroid plexus.

<u>Ventricles, embryonic</u> – The ventricular system is proportionally larger during embryonic and fetal development than in the mature central nervous system. Initially it is lined throughout by the stem cells of the *neuroepithelium. As neurogenesis declines, the ventricles shrink considerably, and altogether disappear in many locations. At some sites, the proliferative *subventricular zone replaces the neuroepithelium, while at others the enduring ventricles are lined by the specialized cells of the *ependyma.

**Ventromedial nucleus (hypothalamus)** – Large spherical nucleus that flanks the third ventricle and is surrounded by a fibrous shell. It has reciprocal connections with the *amygdala, the *bed nucleus of the stria terminalis, the *septum, and the *subiculum. It has been implicated in motivational functions related to feeding and sexual behavior.

**Vermis** – *See* **Cerebellum (vermis)**.

**Vestibular nuclear complex** – A large area in the dorsal medulla, composed of the *medial, the *lateral, the *superior, and the *inferior vestibular nuclei. These nuclei get primary sensory input from the vestibular ganglion; their efferents join the *medial longitudinal fasciculus and form the vestibulospinal tract.

**Visual radiation** – Thalamocortical fibers beneath the cerebral *white matter that originate in the *lateral geniculate body and terminate in the *striate cortex of the occipital lobe.

## W

**White matter** – General term for extensive regions in the brain and spinal cord composed of myelinated fiber tracts but few or no neuronal cell bodies. In histological preparations with myelin stains, the white matter appears black. In laminated brain regions, as in the *cerebral cortex and the *cerebellar cortex, the white matter is called the medullary layer.

## Z

**Zona incerta** – Sheet of *gray matter at the base of the *thalamus. It is generally considered to be a rostral extension of the *tegmentum. *Forel's fields are contiguous with the zona incerta.